W9-BVY-486

Student's Solutions Manual

Intermediate Algebra

Concepts and Applications

FOURTH EDITION

Student's Solutions Manual

Intermediate Algebra

Concepts and Applications

FOURTH EDITION

BITTINGER/KEEDY/ELLENBOGEN

Judith A. Penna

ADDISON-WESLEY PUBLISHING COMPANY
Reading, Massachusetts • Menlo Park, California • New York
Don Mills, Ontario • Wokingham, England • Amsterdam • Bonn
Sydney • Singapore • Tokyo • Madrid • San Juan • Milan • Paris

Reproduced by Addison-Wesley from camera-ready copy supplied by the authors.

ISBN: 0-201-53788-5

7 8 9 10-CRS-9695

TABLE OF CONTENTS

Chapter 1 1

Chapter 2 29

Chapter 3 65

Chapter 4 109

Chapter 5 141

Chapter 6 173

Chapter 7 209

Chapter 8 241

Chapter 9 295

Chapter 10 329

Chapter 11 355

Special thanks are extended to Patsy Hammond for her excellent typing and to Pam Smith for her careful proofreading. Their patience, efficiency, and good humor made the author's work much easier.

Student's Solutions Manual

Intermediate Algebra

Concepts and Applications

FOURTH EDITION

Exercise Set 1.1

1. Four less than some number
   Let n represent the number. Then we have
   $n - 4$.

2. $n + 6$, or $6 + n$

3. Twice a number
   Let x represent the number. Then we have
   $2x$.

4. $5n$

5. Thirty-two percent of some number
   Let n represent the number. Then we have
   $32\%n$, or $0.32n$.

6. $0.47n$

7. Seven more than half of a number
   Let x represent the number. Then we have
   $\frac{1}{2}x + 7$, or $7 + \frac{1}{2}x$, or $\frac{x}{2} + 7$, or $7 + \frac{x}{2}$.

8. $2x - 8$

9. Four less than nineteen percent of some number
   Let t represent the number. Then we have
   $0.19t - 4$.

10. $0.82x + 3$

11. Five more than the difference of two numbers
    Let x represent the first number and y the second. Then we have
    $x - y + 5$, or $5 + x - y$.

12. $x + y - 6$

13. Four less than the product of two numbers
    Let x and y represent the numbers. Then we have
    $xy - 4$, or $yx - 4$.

14. $xy + 10$

15. One more than thirty-five percent of some number
    Let t represent the number. Then we have
    $0.35t + 1$, or $1 + 0.35t$.

16. $0.08x - 2$

17. Substitute and carry out the multiplication and addition:
    $$\begin{aligned} 2x + y &= 2\cdot 3 + 7 \\ &= 6 + 7 \\ &= 13 \end{aligned}$$

18. 16

19. Substitute and multiply:
    $$\begin{aligned} 7abc &= 7\cdot 2\cdot 1\cdot 3 \\ &= 42 \end{aligned}$$

20. 20

21. Substitute and carry out the operations indicated:
    $$\begin{aligned} 8mn - p &= 8\cdot 1\cdot 2 - 9 \\ &= 16 - 9 \\ &= 7 \end{aligned}$$

22. 47

23. Substitute and carry out the operations indicated:
    $$\begin{aligned} 5ab \div c &= 5\cdot 4\cdot 2 \div 10 \\ &= 40 \div 10 \\ &= 4 \end{aligned}$$

24. 6

25. Substitute and carry out the operations indicated:
    $$\begin{aligned} 2ab - a &= 2\cdot 5\cdot 3 - 5 \\ &= 30 - 5 \\ &= 25 \end{aligned}$$

26. 21

27. Substitute and carry out the operations indicated:
    $$\begin{aligned} pqr \div q &= 2\cdot 3\cdot 2 \div 3 \\ &= 12 \div 3 \\ &= 4 \end{aligned}$$

28. 3

29. We substitute 3 for x and see whether we get a true sentence.
    $$\begin{array}{c|l} x - 2 & 1 \\ \hline 3 - 2 \ ? & 1 \\ 1 & \end{array}$$
    We get a true sentence, so 3 is a solution.

30. Yes

31. We substitute 7 for a and see whether we get a true sentence.
    $$\begin{array}{c|l} 4 + a & 13 \\ \hline 4 + 7 \ ? & 13 \\ 11 & \end{array}$$
    We do not get a true sentence, so 7 is not a solution.

32. No

33. We substitute 6 for y and see whether we get a true sentence.

$$\begin{array}{c|c} 13 - y & 7 \\ \hline 13 - 6 & ?\ 7 \\ 7 & \end{array}$$

We get a true sentence, so 6 is a solution.

34. No

35. We substitute 5 for x and see whether we get a true sentence.

$$\begin{array}{c|c} 2x + 3 & 13 \\ \hline 2\cdot5 + 3 & ?\ 13 \\ 10 + 3 & \\ 13 & \end{array}$$

We get a true sentence, so 5 is a solution.

36. Yes

37. We substitute 0.4 for n and see whether we get a true sentence.

$$\begin{array}{c|c} 8n - 1.7 & 2.5 \\ \hline 8(0.4) - 1.7 & ?\ 2.5 \\ 3.2 - 1.7 & \\ 1.5 & \end{array}$$

We do not get a true sentence, so 0.4 is not a solution.

38. Yes

39. We substitute $\frac{17}{3}$ for x and see whether we get a true sentence.

$$\begin{array}{c|c} 3x - 4 & 13 \\ \hline 3\left(\frac{17}{3}\right) - 4 & ?\ 13 \\ 17 - 4 & \\ 13 & \end{array}$$

We get a true sentence, so $\frac{17}{3}$ is a solution.

40. Yes

41. List the letters in the set: {a, e, i, o, u}, or {a, e, i, o, u, y}

42. {Sunday, Monday, Tuesday, Wednesday, Thursday, Friday, Saturday}

43. List the numbers in the set: {2, 4, 6, 8, . . .}

44. {1, 3, 5, 7, . . .}

45. List the numbers in the set: {5, 10, 15, 20, . . .}

46. {3, 6, 9, 12, . . .}

47. Specify the conditions under which a number is in the set: {x|x is an odd number between 10 and 30}

48. {x|x is a multiple of 4 between 22 and 45}

49. Specify the conditions under which a number is in the set: {x|x is a whole number less than 5}

50. {x|x is an integer less than 3 and greater than -4}

51. Specify the conditions under which a number is in the set: {n|n is a multiple of 5 between 7 and 79}

52. {x|x is an even number between 9 and 99}

53. Since 7 is an element of the set of natural numbers, the statement is true.

54. False

55. Since 5.1 is not an element of the set of integers, the statement is false.

56. True

57. Since $-7.\overline{4}$ is an element of the set of rational numbers, the statement is true.

58. True

59. Since $\sqrt{7}$ is an element of the set of real numbers, the statement is true.

60. True

61. Since 7.1 is not an element of the set of whole numbers, the statement is true.

62. True

63. Since all the elements of the set of natural numbers are also elements of the set of integers, the statement is true.

64. True

65. Since all the elements of the set of natural numbers are also elements of the set of whole numbers, the statement is true.

66. False

67. Since the set of whole numbers has an element, namely 0, that is not an element of the set of natural numbers, the statement is false.

68. True

69. Since all the elements of the set of rational numbers are also elements of the set of real numbers, the statement is true.

70. True

71.

72. 

73. Three times the sum of the two numbers

Let x and y represent the numbers. The sum of the two numbers is x + y. Then three times their sum is 3(x + y).

74. $\frac{1}{2}(x - y)$

75. The quotient of the difference of two numbers and their sum

Let n and m represent the numbers. Their difference is n - m, and their sum is n + m. The quotient desired is then $\frac{n - m}{n + m}$.

76. (x + y)(x - y)

77. The only whole number that is not also a natural number is 0. Using roster notation to name the set, we have {0}.

78. {-1, -2, -3, . . .}

79. Observe that 2(x + 3) is equivalent to 2x + 6, so the equation is an identity (is true for all real numbers). Using set-builder notation to name the set, we have $\{x | x \text{ is a real number}\}$.

80. $\{x | x \text{ is a real number}\}$

81. Recall from geometry that when a right triangle has legs of length 2 and 3, the length of the hypotenuse is $\sqrt{2^2 + 3^2} = \sqrt{4 + 9} = \sqrt{13}$. We draw such a triangle:

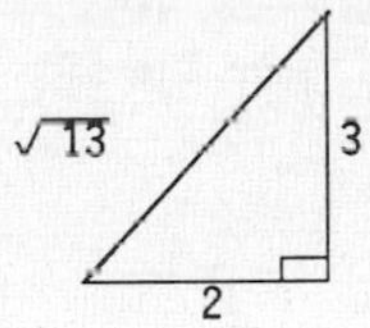

Exercise Set 1.2

1. $|-7| = 7$ -7 is 7 units from 0.

2. 9

3. $|9| = 9$ 9 is 9 units from 0.

4. 12

5. $|-6.2| = 6.2$ -6.2 is 6.2 units from 0.

6. 7.9

7. $|0| = 0$ 0 is 0 units from itself.

8. $3\frac{3}{4}$

9. $\left|1\frac{7}{8}\right| = 1\frac{7}{8}$ $1\frac{7}{8}$ is $1\frac{7}{8}$ units from 0.

10. 0.91

11. $|-4.21| = 4.21$ -4.21 is 4.21 units from 0.

12. 5.309

13. $-9 \leq -1$ -9 is less than or equal to -1, a true statement since -9 is to the left of -1 on the number line.

14. -1 is less than or equal to -5; false

15. $-7 > 1$ -7 is greater than 1, a false statement since -7 is to the left of 1 on the number line.

16. 7 is greater than or equal to -2; true

17. $3 \geq -5$ 3 is greater than or equal to -5, a true statement since -5 is to the left of 3.

18. 9 is less than or equal to 9; true

19. $-9 < -4$ -9 is less than -4, a true statement since -9 is to the left of -4

20. 7 is greater than or equal to -8; true

21. $-4 \geq -4$ -4 is greater than or equal to -4. Since -4 = -4 is true, $-4 \geq -4$ is true.

22. 2 is less than 2; false

23. $-5 < -5$ -5 is less than -5, a false statement since -5 does not lie to the left of itself.

24. -2 is greater than -12; true

25. 5 + 12

Two positive numbers: Add the numbers, getting 17. The answer is positive, 17.

26. 16

27. -4 + (-7)

Two negative numbers: Add the absolute values, getting 11. The answer is negative, -11.

28. -11

29. -5.9 + 2.7

A negative and a positive number: The absolute values are 5.9 and 2.7. Subtract 2.7 from 5.9 to get 3.2. The larger absolute value came from the negative number, so the answer is negative, -3.2.

30. 5.4

31. $\frac{2}{7} + \left(-\frac{3}{5}\right) = \frac{10}{35} + \left(-\frac{21}{35}\right)$

A positive and a negative number. The absolute values are $\frac{10}{35}$ and $\frac{21}{35}$. Subtract $\frac{10}{35}$ from $\frac{21}{35}$ to get $\frac{11}{35}$. The larger absolute value came from the negative number, so the answer is negative, $-\frac{11}{35}$.

32. $-\frac{1}{40}$

33. -4.9 + (-3.6)

Two negative numbers: Add the absolute values, getting 8.5. The answer is negative, -8.5.

34. -9.6

35. $-\frac{1}{9} + \frac{2}{3} = -\frac{1}{9} + \frac{6}{9}$

A negative and a positive number. The absolute values are $\frac{1}{9}$ and $\frac{6}{9}$. Subtract $\frac{1}{9}$ from $\frac{6}{9}$ to get $\frac{5}{9}$. The larger absolute value came from the positive number, so the answer is positive, $\frac{5}{9}$.

36. $\frac{3}{10}$

37. 0 + (-4.5)

One number is zero: The sum is the other number, -4.5.

38. -3.19

39. -7.24 + 7.24

A negative and a positive number: The numbers have the same absolute value, 7.24, so the answer is 0.

40. 0

41. 15.9 + (-22.3)

A positive and a negative number: The absolute values are 15.9 and 22.3. Subtract 15.9 from 22.3 to get 6.4. The larger absolute value came from the negative number, so the answer is negative, -6.4.

42. -6.6

43. The opposite of 7.29 is -7.29, because -7.29 + 7.29 = 0.

44. -5.43

45. The opposite of -4.8 is 4.8, because -4.8 + 4.8 = 0.

46. 8.1

47. The opposite of 0 is 0, because 0 + 0 = 0.

48. $2\frac{3}{4}$

49. The opposite of $-6\frac{1}{3}$ is $6\frac{1}{3}$, because $-6\frac{1}{3} + 6\frac{1}{3} = 0$.

50. $-4\frac{1}{5}$

51. If x = 7, then -x = -7. (The opposite of 7 is -7.)

52. -3

53. If x = -2.7, then -x = -(-2.7) = 2.7. (The opposite of -2.7 is 2.7.)

54. 1.9

55. If x = 1.79, then -x = -1.79. (The opposite of 1.79 is -1.79.)

56. -3.14

57. If x = 0, then -x = 0. (The opposite of 0 is 0.)

58. 1

59. If x = -0.03, then -x = -(-0.03) = 0.03. (The opposite of -0.03 is 0.03.)

60. 1.09

61. 9 - 7 = 9 + (-7) Change the sign and add.
= 2

62. 5

63. 4 - 9 = 4 + (-9) Change the sign and add.
= -5

64. -7

65. -6 - (-10) = -6 + 10 Change the sign and add.
= 4

66. 6

67. -4 - 13 = -4 + (-13) = -17

68. -15

69. 2.7 - 5.8 = 2.7 + (-5.8) = -3.1

70. -0.5

71. $-\frac{3}{5} - \frac{1}{2} = -\frac{3}{5} + \left(-\frac{1}{2}\right)$

$= -\frac{6}{10} + \left(-\frac{5}{10}\right)$ Finding a common denominator

$= -\frac{11}{10}$

72. $-\frac{13}{15}$

73. $-3.9 - (-6.8) = -3.9 + 6.8 = 2.9$

74. -1.1

75. $0 - (-7.9) = 0 + 7.9 = 7.9$

76. -5.3

77. $(-4)7$

Two numbers with unlike signs: Multiply their absolute values, getting 28. The answer is negative, -28.

78. -45

79. $(-3)(-8)$

Two numbers with the same sign: Multiply their absolute values, getting 24. The answer is positive, 24.

80. 56

81. $(4.2)(-5)$

Two numbers with unlike signs: Multiply their absolute values, getting 21. The answer is negative, -21.

82. -28

83. $(-7.2)(1)$

Two numbers with unlike signs: Multiply their absolute values, getting 7.2. The answer is negative, -7.2.

84. 5.9

85. $15.2 \times 0 = 0$

86. 0

87. $(-3.2) \times (-1.7)$

Two numbers with the same sign: Multiply their absolute values, getting 5.44. The answer is positive, 5.44.

88. 8.17

89. $\frac{-10}{-2}$

Two numbers with the same sign: Divide their absolute values, getting 5. The answer is positive, 5.

90. 5

91. $\frac{-100}{20}$

Two numbers with unlike signs: Divide their absolute values, getting 5. The answer is negative, -5.

92. -10

93. $\frac{73}{-1}$

Two numbers with unlike signs: Divide their absolute values, getting 73. The answer is negative, -73.

94. -62

95. $\frac{0}{-7} = 0$

96. 0

97. $\frac{-42}{-6}$

Two numbers with the same sign: Divide their absolute values, getting 7. The answer is positive, 7.

98. 8

99. The reciprocal of 5 is $\frac{1}{5}$, because $5 \cdot \frac{1}{5} = 1$.

100. $\frac{1}{3}$

101. The reciprocal of -9 is $\frac{1}{-9}$, or $-\frac{1}{9}$, because $-9\left(-\frac{1}{9}\right) = 1$.

102. $-\frac{1}{7}$

103. The reciprocal of $\frac{2}{3}$ is $\frac{3}{2}$, because $\frac{2}{3} \cdot \frac{3}{2} = 1$.

104. $\frac{7}{4}$

105. The reciprocal of $-\frac{3}{11}$ is $-\frac{11}{3}$, because $-\frac{3}{11}\left(-\frac{11}{3}\right) = 1$.

106. $-\frac{3}{7}$

107. $\frac{2}{3} \div \frac{4}{5} = \frac{2}{3} \cdot \frac{5}{4}$ Multiplying by the reciprocal of 4/5

$= \frac{10}{12}$, or $\frac{5}{6}$

108. $\frac{5}{21}$

109. $\left(-\frac{3}{5}\right) \div \frac{1}{2} = -\frac{3}{5} \cdot \frac{2}{1}$ Multiplying by the reciprocal of 1/2

$= -\frac{6}{5}$

110. $-\frac{12}{7}$

111. $-\frac{2}{9} \div \left(-\frac{3}{4}\right) = -\frac{2}{9} \cdot \left(-\frac{4}{3}\right) = \frac{8}{27}$

112. $-\frac{7}{22}$

113. $-\frac{3}{8} \div 1 = -\frac{3}{8} \cdot \frac{1}{1} = -\frac{3}{8}$

114. $\frac{2}{7}$

115. $\frac{7}{3} \div (-1) = \frac{7}{3}\left(-\frac{1}{1}\right) = -\frac{7}{3}$

116. $-\frac{5}{4}$

117. Since 35 = 7·5 we use $\frac{5}{5}$ as a name for 1:

$$\frac{3x}{7} = \frac{3x}{7} \cdot \frac{5}{5} = \frac{3x \cdot 5}{35} = \frac{15x}{35}$$

118. $\frac{16a}{40}$

119. Since 32 = 8·4, we use $\frac{4}{4}$ as a name for 1:

$$-\frac{3a}{8} = -\frac{3a}{8} \cdot \frac{4}{4} = -\frac{3a \cdot 4}{32} = -\frac{12a}{32}$$

120. $-\frac{44x}{33}$

121. $\frac{50x}{5} = \frac{10x \cdot 5}{1 \cdot 5}$ Factor numerator and denominator after identifying a common factor of 5

$= \frac{10x}{1} \cdot \frac{5}{5}$ Factor the fractional expression, with one factor of 1

$= 10x$ Leave off the factor of 1

122. 25x

123. $\frac{56a}{7} = \frac{8a \cdot 7}{1 \cdot 7} = \frac{8a}{1} \cdot \frac{7}{7} = 8a$

124. 8a

125. $5y + 2x$

$= 2x + 5y$ Commutative law of addition

126. x + 4y

127. $2x - 3y = 2x + (-3y)$

$= -3y + 2x$ Commutative law of addition

128. -4x + 5y

129. $\frac{x}{2} \cdot \frac{y}{3}$

$= \frac{y}{3} \cdot \frac{x}{2}$ Commutative law of multiplication

130. $\frac{y}{4} \cdot 3$

131. $2 \cdot (8x)$

$= (2 \cdot 8)x$ Associative law of multiplication

132. (x·3)y

133. $x + (2y + 5)$

$= (x + 2y) + 5$ Associative law of addition

134. 3y + (4 + 10)

135. $3(a + 1)$

$= 3 \cdot a + 3 \cdot 1$ Using the distributive law

$= 3a + 3$

136. 8x + 8

137. $4(x - y)$

$= 4 \cdot x - 4 \cdot y$ Using the distributive law

$= 4x - 4y$

138. 9a - 9b

139. $-5(2a + 3b)$

$= -5 \cdot 2a + (-5) \cdot 3b$

$= -10a - 15b$

140. -6c - 10d

141. $2a(b - c + d)$

$= 2a \cdot b - 2a \cdot c + 2a \cdot d$

$= 2ab - 2ac + 2ad$

142. 5xy - 5xz + 5xw

143. $8x + 8y$

$= 8 \cdot x + 8 \cdot y$

$= 8(x + y)$

144. 7(a + b)

145. $9p - 9$

$= 9 \cdot p - 9 \cdot 1$

$= 9(p - 1)$

146. 12(x - 1)

147. $7x - 21$

$= 7 \cdot x - 7 \cdot 3$

$= 7(x - 3)$

148. 6(y - 6)

149. $2x - 2y + 2z$

$= 2 \cdot x - 2 \cdot y + 2 \cdot z$

$= 2(x - y + z)$

150. 3(x + y - z)

151. Five less than seventy percent of a number

Let x represent the number. Then we have 0.7x - 5.

152. $\frac{1}{2}x + 2$

153. Substitute and carry out the operations indicated:

$$7xy - z = 7\cdot2\cdot1 - 20 = 14 - 20 = -6$$

154. 23

155. 

156. 

157. $\frac{2x + 10}{3x + 15} = \frac{2(x + 5)}{3(x + 5)} = \frac{2}{3} \cdot \frac{x + 5}{x + 5} = \frac{2}{3}$

158. $\frac{2}{3}$

159. $\frac{6x - 12}{3x + 6} = \frac{3\cdot2(x - 2)}{3(x + 2)} = \frac{3}{3} \cdot \frac{2(x - 2)}{x + 2} = \frac{2(x - 2)}{x + 2}$

160. $\frac{x - 2}{2x + 3}$

Exercise Set 1.3

1. $3x = 12$ and $2x = 8$

   When x is replaced by 4, both equations are true, and for any other replacement, both equations are false. Thus the equations are equivalent.

2. Yes

3. $2x - 1 = -7$ and $x = -3$

   When x is replaced by -3, both equations are true, and for any other replacement, both equations are false. Thus the equations are equivalent.

4. Yes

5. $x + 5 = 11$ and $3x = 18$

   When x is replaced by 6, both equations are true, and for any other replacement, both equations are false. Thus the equations are equivalent.

6. No

7. $13 - x = 4$ and $2x = 20$

   When x is replaced by 9, the first equation is true, but the second equation is false. Thus the equations are not equivalent.

8. Yes

9. $5x = 2x$ and $\frac{4}{x} = 3$

   When x is replaced by 0, the first equation is true, but the second equation is not defined. Thus the equations are not equivalent.

10. No

11.
$$x - 5.2 = 9.4$$
$$x - 5.2 + 5.2 = 9.4 + 5.2 \quad \text{Addition principle; adding 5.2}$$
$$x + 0 = 9.4 + 5.2 \quad \text{Law of opposites}$$
$$x = 14.6$$

Check:

$x - 5.2 = 9.4$	
$14.6 - 5.2 \;?\; 9.4$	
9.4	TRUE

The solution is 14.6.

12. 6.9

13.
$$9y = 72$$
$$\frac{1}{9} \cdot 9y = \frac{1}{9} \cdot 72 \quad \text{Multiplication principle; multiplying by } \tfrac{1}{9}\text{, the reciprocal of 9}$$
$$1\cdot y = 8$$
$$y = 8$$

Check:

$9y = 72$	
$9\cdot8 \;?\; 72$	
72	TRUE

The solution is 8.

14. 9

15.
$$4x - 12 = 60$$
$$4x - 12 + 12 = 60 + 12$$
$$4x = 72$$
$$\frac{1}{4} \cdot 4x = \frac{1}{4} \cdot 72$$
$$1\cdot x = \frac{72}{4}$$
$$x = 18$$

The solution is 18.

Check:

$4x - 12 = 60$	
$4\cdot18 - 12 \;?\; 60$	
$72 - 12$	
60	TRUE

16. 19

17.
$$5y + 3 = 28$$
$$5y + 3 + (-3) = 28 + (-3)$$
$$5y = 25$$
$$\frac{1}{5} \cdot 5y = \frac{1}{5} \cdot 25$$
$$1\cdot y = \frac{25}{5}$$
$$y = 5$$

The solution is 5.

Check:

$5y + 3 = 28$	
$5\cdot5 + 3 \;?\; 28$	
$25 + 3$	
28	TRUE

18. 9

19. $2y - 11 = 37$
$2y - 11 + 11 = 37 + 11$
$2y = 48$
$\frac{1}{2} \cdot 2y = \frac{1}{2} \cdot 48$
$1 \cdot y = \frac{48}{2}$
$y = 24$
The solution is 24.

Check:
$2y - 11 = 37$
$2 \cdot 24 - 11 \ ? \ 37$
$48 - 11$
37 | TRUE

20. 14

21. $-4x - 7 = -35$
$-4x - 7 + 7 = -35 + 7$
$-4x = -28$
$-\frac{1}{4} \cdot (-4x) = -\frac{1}{4} \cdot (-28)$
$1 \cdot x = \frac{28}{4}$
$x = 7$
The solution is 7.

Check:
$-4x - 7 = -35$
$-4 \cdot 7 - 7 \ ? \ -35$
$-28 - 7$
-35 | TRUE

22. 11

23. $4a + 5a$
$= (4 + 5)a$
$= 9a$

24. 12x

25. $8b - 11b$
$= (8 - 11)b$
$= -3b$

26. -3c

27. $14y + y$
$= 14y + 1y$
$= (14 + 1)y$
$= 15y$

28. 14x

29. $12a - a$
$= 12a - 1a$
$= (12 - 1)a$
$= 11a$

30. 14x

31. $t - 9t$
$= 1t - 9t$
$= (1 - 9)t$
$= -8t$

32. -5x

33. $5x - 3x + 8x$
$= (5 - 3 + 8)x$
$= 10x$

34. -6x

35. $5x - 8y + 3x$
$= (5 + 3)x - 8y$
$= 8x - 8y$

36. 13a - 10b

37. $7c + 8d - 5c + 2d$
$= (7 - 5)c + (8 + 2)d$
$= 2c + 10d$

38. 7a + 9b

39. $4x - 7 + 18x + 25$
$= (4 + 18)x + (-7 + 25)$
$= 22x + 18$

40. 9p + 12

41. $13x + 14y - 11x - 47y$
$= (13 - 11)x + (14 - 47)y$
$= 2x - 33y$

42. 5a - 21b

43. $5x + 2x = 56$
$7x = 56$
$\frac{1}{7} \cdot 7x = \frac{1}{7} \cdot 56$
$x = 8$
The solution is 8.

Check:
$5x + 2x = 56$
$5 \cdot 8 + 2 \cdot 8 \ ? \ 56$
$40 + 16$
56 | TRUE

44. 12

45. $9y - 7y = 42$
$2y = 42$
$\frac{1}{2} \cdot 2y = \frac{1}{2} \cdot 42$
$y = 21$
The solution is 21.

Check:
$9y - 7y = 42$
$9 \cdot 21 - 7 \cdot 21 \ ? \ 42$
$189 - 147$
42 | TRUE

46. 13

47. $-6y - 10y = -32$
$-16y = -32$
$-\frac{1}{16} \cdot (-16y) = -\frac{1}{16} \cdot (-32)$
$y = 2$

Check:

$-6y - 10y$	$= -32$
$-6\cdot 2 - 10\cdot 2$	? -32
$-12 - 20$	
-32	TRUE

The solution is 2.

48. -2

49.
$$7y - 1 = 23 - 5y$$
$$7y - 1 + 5y = 23 - 5y + 5y$$
$$12y - 1 = 23$$
$$12y - 1 + 1 = 23 + 1$$
$$12y = 24$$
$$\frac{1}{12}\cdot 12y = \frac{1}{12}\cdot 24$$
$$y = 2$$

Check:

$7y - 1$	$= 23 - 5y$
$7\cdot 2 - 1$	? $23 - 5\cdot 2$
$14 - 1$	$23 - 10$
13	13 TRUE

The solution is 2.

50. -6

51.
$$5 - 4a = a - 13$$
$$5 - 4a + 4a = a - 13 + 4a$$
$$5 = 5a - 13$$
$$5 + 13 = 5a - 13 + 13$$
$$18 = 5a$$
$$\frac{1}{5}\cdot 18 = \frac{1}{5}\cdot 5a$$
$$\frac{18}{5} = a$$

Check:

$5 - 4a$	$= a - 13$
$5 - 4\cdot\frac{18}{5}$	? $\frac{18}{5} - 13$
$\frac{25}{5} - \frac{72}{5}$	$\frac{18}{5} - \frac{65}{5}$
$-\frac{47}{5}$	$-\frac{47}{5}$ TRUE

The solution is $\frac{18}{5}$.

52. 4

53.
$$3m - 7 = -7 - 4m - m$$
$$3m - 7 = -7 - 5m$$
$$3m - 7 + 5m = -7 - 5m + 5m$$
$$8m - 7 = -7$$
$$8m - 7 + 7 = -7 + 7$$
$$8m = 0$$
$$\frac{1}{8}\cdot 8m = \frac{1}{8}\cdot 0$$
$$m = 0$$

Check:

$3m - 7$	$= -7 - 4m - m$
$3\cdot 0 - 7$	? $-7 - 4\cdot 0 - 0$
$0 - 7$	$-7 - 0 - 0$
-7	-7 TRUE

The solution is 0.

54. 0

55.
$$5r - 2 + 3r = 2r + 6 - 4r$$
$$8r - 2 = 6 - 2r$$
$$8r - 2 + 2r = 6 - 2r + 2r$$
$$10r - 2 = 6$$
$$10r - 2 + 2 = 6 + 2$$
$$10r = 8$$
$$\frac{1}{10}\cdot 10r = \frac{1}{10}\cdot 8$$
$$r = \frac{8}{10}$$
$$r = \frac{4}{5}$$

Check:

$5r - 2 + 3r$	$= 2r + 6 - 4r$
$5\cdot\frac{4}{5} - 2 + 3\cdot\frac{4}{5}$	? $2\cdot\frac{4}{5} + 6 - 4\cdot\frac{4}{5}$
$\frac{20}{5} - \frac{10}{5} + \frac{12}{5}$	$\frac{8}{5} + \frac{30}{5} - \frac{16}{5}$
$\frac{22}{5}$	$\frac{22}{5}$ TRUE

The solution is $\frac{4}{5}$.

56. -8

57.
$$\frac{1}{4} + \frac{3}{8}y = \frac{3}{4}$$
$$\frac{1}{4} + \frac{3}{8}y - \frac{1}{4} = \frac{3}{4} - \frac{1}{4}$$
$$\frac{3}{8}y = \frac{1}{2}$$
$$\frac{8}{3}\cdot\frac{3}{8}y = \frac{8}{3}\cdot\frac{1}{2}$$
$$y = \frac{4}{3}$$

Check:

$\frac{1}{4} + \frac{3}{8}y$	$= \frac{3}{4}$
$\frac{1}{4} + \frac{3}{8}\cdot\frac{4}{3}$	? $\frac{3}{4}$
$\frac{1}{4} + \frac{1}{2}$	
$\frac{3}{4}$	TRUE

The solution is $\frac{4}{3}$.

58. 2

59. $$-\frac{5}{2}x + \frac{1}{2} = -18$$
$$-\frac{5}{2}x + \frac{1}{2} - \frac{1}{2} = -18 - \frac{1}{2}$$
$$-\frac{5}{2}x = -\frac{37}{2}$$
$$-\frac{2}{5}\left(-\frac{5}{2}x\right) = -\frac{2}{5}\left(-\frac{37}{2}\right)$$
$$x = \frac{37}{5}$$

Check:

$$\begin{array}{r|l} \multicolumn{2}{c}{-\frac{5}{2}x + \frac{1}{2} = -18} \\ \hline -\frac{5}{2}\cdot\frac{37}{5} + \frac{1}{2} & ?\ -18 \\ -\frac{37}{2} + \frac{1}{2} & \\ -\frac{36}{2} & \\ -18 & \quad \text{TRUE} \end{array}$$

The solution is $\frac{37}{5}$.

50. $\frac{49}{9}$

61. $$0.8t - 0.3t = 6.5$$
$$0.5t = 6.5$$
$$\frac{1}{0.5}(0.5t) = \frac{1}{0.5}(6.5)$$
$$t = \frac{6.5}{0.5}$$
$$t = 13$$

The solution is 13.

Check:

$$\begin{array}{r|l} \multicolumn{2}{c}{0.8t - 0.3t = 6.5} \\ \hline 0.8(13) - 0.3(13) & ?\ 6.5 \\ 10.4 - 3.9 & \\ 6.5 & \quad \text{TRUE} \end{array}$$

62. -5.02

63. $a - (2a + 5)$
$= a - 2a - 5$
$= -a - 5$

64. $-4x - 9$

65. $4m - (3m - 1)$
$= 4m - 3m + 1$
$= m + 1$

66. $a + 3$

67. $3d - 7 - (5 - 2d)$
$= 3d - 7 - 5 + 2d$
$= 5d - 12$

68. $13x - 16$

69. $-2(x + 3) - 5(x - 4)$
$= -2x - 6 - 5x + 20$
$= -7x + 14$

70. $-15y - 45$

71. $5x - 7(2x - 3)$
$= 5x - 14x + 21$
$= -9x + 21$

72. $-12y + 24$

73. $9a - [7 - 5(7a - 3)]$
$= 9a - [7 - 35a + 15]$
$= 9a - [22 - 35a]$
$= 9a - 22 + 35a$
$= 44a - 22$

74. $47b - 51$

75. $5\{-2 + 3[4 - 2(3 + 5)]\}$
$= 5\{-2 + 3[4 - 2\cdot 8]\}$
$= 5\{-2 + 3[4 - 16]\}$
$= 5\{-2 + 3[-12]\}$
$= 5\{-2 - 36\}$
$= 5\{-38\}$
$= -190$

76. -1449

77. $2y + \{7[3(2y - 5) - (8y + 7)] + 9\}$
$= 2y + \{7[6y - 15 - 8y - 7] + 9\}$
$= 2y + \{7[-2y - 22] + 9\}$
$= 2y + \{-14y - 154 + 9\}$
$= 2y + \{-14y - 145\}$
$= 2y - 14y - 145$
$= -12y - 145$

78. $-11b + 217$

79. $$2(x + 6) = 8x$$
$$2x + 12 = 8x$$
$$12 = 6x$$
$$2 = x$$

Check:

$$\begin{array}{r|l} \multicolumn{2}{c}{2(x + 6) = 8x} \\ \hline 2(2 + 6) & ?\ 8\cdot 2 \\ 2\cdot 8 & 16 \\ 16 & \quad \text{TRUE} \end{array}$$

The solution is 2.

80. 3

81. $80 = 10(3t + 2)$
$80 = 30t + 20$
$60 = 30t$
$2 = t$

Check:
$80 = 10(3t + 2)$
$80 \ ?\ 10(3\cdot 2 + 2)$
$10(6 + 2)$
$10\cdot 8$
80 TRUE

The solution is 2.

82. 1

83. $180(n - 2) = 900$
$180n - 360 = 900$
$180n = 1260$
$n = 7$

Check:
$180(n - 2) = 900$
$180(7 - 2) \ ?\ 900$
$180\cdot 5$
900 TRUE

The solution is 7.

84. 7

85. $5y - (2y - 10) = 25$
$5y - 2y + 10 = 25$
$3y + 10 = 25$
$3y = 15$
$y = 5$

Check:
$5y - (2y - 10) = 25$
$5\cdot 5 - (2\cdot 5 - 10) \ ?\ 25$
$25 - (10 - 10)$
$25 - 0$
25 TRUE

The solution is 5.

86. 7

87. $0.7(3x + 6) = 1.1 - (x + 2)$
$2.1x + 4.2 = 1.1 - x - 2$
$2.1x + 4.2 = -x - 0.9$
$3.1x + 4.2 = -0.9$
$3.1x = -5.1$
$x = -\frac{5.1}{3.1}$
$x = -\frac{51}{31}$

The number $-\frac{51}{31}$ checks and is the solution.

88. $\frac{39}{14}$

89. $\frac{1}{8}(16y + 8) - 17 = -\frac{1}{4}(8y - 16)$
$2y + 1 - 17 = -2y + 4$
$2y - 16 = -2y + 4$
$4y - 16 = 4$
$4y = 20$
$y = 5$

The number 5 checks and is the solution.

90. 6

91. $a + (a - 3) = (a + 2) - (a + 1)$
$a + a - 3 = a + 2 - a - 1$
$2a - 3 = 1$
$2a = 4$
$a = 2$

Check:
$a + (a - 3) = (a + 2) - (a + 1)$
$2 + (2 - 3) \ ?\ (2 + 2) - (2 + 1)$
$2 - 1 \mid 4 - 3$
$1 \mid 1$ TRUE

The solution is 2.

92. -7.4

93. $5[2 + 3(x - 1)] = 4$
$5[2 + 3x - 3] = 4$
$5[-1 + 3x] = 4$
$-5 + 15x = 4$
$15x = 9$
$\frac{1}{15}\cdot 15x = \frac{1}{15}\cdot 9$
$x = \frac{3}{5}$

Check:
$5[2 + 3(x - 1)] = 4$
$5\left[2 + 3\left(\frac{3}{5} - 1\right)\right] \ ?\ 4$
$5\left[2 + 3\left(-\frac{2}{5}\right)\right]$
$5\left[2 - \frac{6}{5}\right]$
$5\left[\frac{4}{5}\right]$
4 TRUE

The solution is $\frac{3}{5}$.

94. -9

95. $5 + 2(x - 3) = 2[5 - 4(x + 2)]$
$5 + 2x - 6 = 2[5 - 4x - 8]$
$2x - 1 = 2[-4x - 3]$
$2x - 1 = -8x - 6$
$2x = -8x - 5$
$10x = -5$
$\frac{1}{10}\cdot 10x = \frac{1}{10}(-5)$
$x = -\frac{1}{2}$

Check:
$5 + 2(x - 3) = 2[5 - 4(x + 2)]$
$5 + 2\left(-\frac{1}{2} - 3\right) \ ?\ 2\left[5 - 4\left(-\frac{1}{2} + 2\right)\right]$
$5 + 2\left(-\frac{7}{2}\right) \mid 2\left[5 - 4\left(\frac{3}{2}\right)\right]$
$5 - 7 \mid 2[5 - 6]$
$-2 \mid 2[-1]$
$\mid -2$ TRUE

The solution is $-\frac{1}{2}$.

96. $\frac{23}{8}$

97.
$$\begin{aligned}
2\{9 - 3[2x - 4(x + 1)]\} &= 5(2x + 8)\\
2\{9 - 3[2x - 4x - 4]\} &= 10x + 40\\
2\{9 - 3[-2x - 4]\} &= 10x + 40\\
2\{9 + 6x + 12\} &= 10x + 40\\
2\{21 + 6x\} &= 10x + 40\\
42 + 12x &= 10x + 40\\
12x &= 10x - 2\\
2x &= -2\\
x &= -1
\end{aligned}$$

Check:

$2\{9 - 3[2x - 4(x + 1)]\}$	$= 5(2x + 8)$	
$2\{9 - 3[2(-1) - 4(-1 + 1)]\}$	? $5(2\cdot -1 + 8)$	
$2\{9 - 3[-2 - 4\cdot 0]\}$	$5(-2 + 8)$	
$2\{9 - 3[-2 + 0]\}$	$5(6)$	
$2\{9 - 3[-2]\}$	30	
$2\{9 + 6\}$		
$2\{15\}$		
30		TRUE

The solution is -1.

98. $\frac{1}{7}$

99.
$$\begin{aligned}
4x - 2x - 2 &= 2x\\
2x - 2 &= 2x\\
-2x + 2x - 2 &= -2x + 2x\\
-2 &= 0
\end{aligned}$$

Since the original equation is equivalent to the false equation $-2 = 0$, there is no solution. The solution set is Ø.

100. The set of all real numbers

101.
$$\begin{aligned}
2 + 9x &= 3(3x + 1) - 1\\
2 + 9x &= 9x + 3 - 1\\
2 + 9x &= 9x + 2\\
2 + 9x - 9x &= 9x + 2 - 9x\\
2 &= 2
\end{aligned}$$

The original equation is equivalent to the equation $2 = 2$ which is true for all real numbers. Thus the solution set is the set of all real numbers.

102. Ø

103.
$$\begin{aligned}
-8x + 5 &= 14 - 8x\\
-8x + 5 + 8x &= 14 - 8x + 8x\\
5 &= 14
\end{aligned}$$

Since the original equation is equivalent to the false equation $5 = 14$, there is no solution. The solution set is Ø.

104. The set of all real numbers

105.
$$\begin{aligned}
2\{9 - 3[-2x - 4]\} &= 12x + 42\\
2\{9 + 6x + 12\} &= 12x + 42\\
2\{21 + 6x\} &= 12x + 42\\
42 + 12x &= 12x + 42\\
42 + 12x - 12x &= 12x + 42 - 12x\\
42 &= 42
\end{aligned}$$

The original equation is equivalent to the equation $42 = 42$, which is true for all real numbers. Thus the solution set is the set of all real numbers.

106. The set of all real numbers

107. Roster notation: List the numbers in the set.

{1, 2, 3, 4, 5, 6, 7, 8, 9}

Set-builder notation: Specify the conditions under which a number is in the set.

{x|x is a positive integer less than 10}

108. {-8, -7, -6, -5, -4, -3, -2, -1};

{x|x is a negative integer greater than -9}

109. $-7 - (-5.3) = -7 + 5.3 = -1.7$

(Change the sign of the number being subtracted and add.)

110. -5.3

111. $(-9)(-6) = 54$

(The product of two numbers with the same sign is positive.)

112. 36

113. $(-12) \div (-3) = 4$

(The quotient of two numbers with the same sign is positive.)

114. 3

115.

116.

117.
$$\begin{aligned}
0.0008x &= 0.00000564\\
x &= \frac{0.00000564}{0.0008}\\
x &= 0.00705
\end{aligned}$$

118. 0.0279282

119.
$$\begin{aligned}
4.23x - 17.898 &= -1.65x - 42.454\\
5.88x - 17.898 &= -42.454\\
5.88x &= -24.556\\
x &= -\frac{24.556}{5.88}\\
x &= -4.1762
\end{aligned}$$

120. 0.2140224

121. $x - \{3x - [2x - (5x - (7x - 1))]\} = x + 7$

$x - \{3x - [2x - (5x - 7x + 1)]\} = x + 7$

$x - \{3x - [2x - (-2x + 1)]\} = x + 7$

$x - \{3x - [2x + 2x - 1]\} = x + 7$

$x - \{3x - [4x - 1]\} = x + 7$

$x - \{3x - 4x + 1\} = x + 7$

$x - \{-x + 1\} = x + 7$

$x + x - 1 = x + 7$

$2x - 1 = x + 7$

$x - 1 = 7$

$x = 8$

The check is left to the student. The solution is 8.

122. 4

123. $17 - 3\{5 + 2[x - 2]\} + 4\{x - 3(x + 7)\} = 9\{x + 3[2 + 3(4 - x)]\}$

$17 - 3\{5 + 2x - 4\} + 4\{x - 3x - 21\} = 9\{x + 3[2 + 12 - 3x]\}$

$17 - 3\{1 + 2x\} + 4\{-2x - 21\} = 9\{x + 3[14 - 3x]\}$

$17 - 3 - 6x - 8x - 84 = 9\{x + 42 - 9x\}$

$-14x - 70 = 9\{-8x + 42\}$

$-14x - 70 = -72x + 378$

$58x - 70 = 378$

$58x = 448$

$x = \frac{448}{58}$, or $\frac{224}{29}$

The check is left to the student. The solution is $\frac{224}{29}$.

124. $\frac{19}{46}$

Exercise Set 1.4

1. Familiarize: There are two numbers involved, and we want to find both of them. We can let x represent the first number and note that the second number is 9 more than the first. Also, the sum of the numbers is 81.

   Translate: The second number can be named x + 9. We translate to an equation:

   First Number plus Second Number is 81.

   $x + x + 9 = 81$

2. Let x and x + 11 be the numbers; $x + (x + 11) = 95$

3. Familiarize: Let t represent the time required, and list the relevant information in a table.

Deirdre's speed in still water	5 km/h
Speed of current	3.2 km/h
Deirdre's speed up river	(5 - 3.2) km/h
Distance to be traveled	2.7 km
Time required	t

   Translate: Speed × Time = Distance

   $(5 - 3.2) \times t = 2.7$

4. Let t be Francis' time; $(4 - 1.5)t = 3.75$

5. Familiarize: Let t represent the time required, and list the relevant information in a table.

Boat's speed in still water	12 km/h
Speed of current	3 km/h
Boat's speed down river	(12 + 3) km/h
Distance to be traveled	35 km
Time required	t

   Translate: Speed × Time = Distance

   $(12 + 3) \times t = 35$

6. Let t be the boat's time; $(14 + 7)t = 56$

7. Familiarize: There are three angle measures involved, and we want to find all three. We can let x represent the smallest angle measure and note that the second is one more than x and the third is one more than the second, or two more than x. We also note that the sum of the three angle measures must be 180°.

   Translate: The three angles measures are x, x + 1, and x + 2. We translate to an equation:

   First plus second plus third is 180°.

   $x + (x + 1) + (x + 2) = 180$

8. Let x, x + 2, and x + 4 be the angle measures; $x + (x + 2) + (x + 4) = 180$

9. Familiarize: Let t represent the time required. Note that the plane must climb 29,000 - 8000, or 21,000 ft.

   Translate: Speed × Time = Distance

   $3500 \times t = 21{,}000$

10. Let x represent the longer length; $x + \frac{2}{3}x = 10$

11. Familiarize: Let x represent the measure of the second angle. Then the first angle is three times x, and the third is 12° less than twice x. The sum of the three angle measures is 180°.

Translate: The first angle is 3x, the second is x, and the third is 2x - 12. Translate to an equation:

First	plus	second	plus	third	is	180°.
3x	+	x	+	(2x - 12)	=	180

12. Let x represent the measure of the second angle;

$4x + x + (2x + 5) = 180$

13. Familiarize: Note that each odd integer is two more than the one preceding it. If we let n represent the first odd integer, then the second is 2 more than the first and the third is 2 more than the second, or 4 more than the first. We are told that the sum of the first, twice the second, and three times the third is 70.

Translate: The three odd integers are n, n + 2, and n + 4. Translate to an equation.

First	plus	two times second	plus	three times third	is	70.
n	+	2(n + 2)	+	3(n + 4)	=	70

14. Let x represent the first number;

$2x + 3(x + 2) = 76$

15. Familiarize: Recall that the perimeter of each square is four times the length of a side. If s represents the length of a side of the smaller square, then (s + 2) represents the length of a side of the larger square. The sum of the two perimeters must be 100 cm.

Translate:

Perimeter of smaller square	plus	perimeter of larger square	is	100 cm.
4s	+	4(s + 2)	=	100

16. Let x represent the length of one piece;

$\left(\frac{x}{4}\right)^2 = \left(\frac{100 - x}{4}\right)^2 + 144$

17. Familiarize: If we let x represent the first number, then the second is six less than 3 times x and the third is two more than $\frac{2}{3}$ of the second. The sum of the three numbers is 172.

Translate:

First	plus	second	plus	third	is	172.
x	+	3x - 6	+	$\left[\frac{2}{3}(3x - 6) + 2\right]$	=	172

18. Let x represent the price of the least expensive set;

$(x + 20) + (x + 6\cdot 20) = x + 12\cdot 20$

19. Familiarize: After the next test there will be six test scores. The average of the six scores is their sum divided by 6. We let x represent the next test score.

Translate:

Average of the six scores is 88.

$$\frac{93 + 89 + 72 + 80 + 96 + x}{6} = 88$$

20. Let x represent the percent of total change;

$1.2(1.3)(0.8) - 1 = x$

21. Familiarize and translate: Let x represent the unknown number.

Three times	some number	is 12.3.
3 ·	x	= 12.3

Carry out: $3x = 12.3$

$x = 4.1$

Check: The product of 3 and 4.1 is 12.3, so the answer checks.

State: The other number is 4.1.

22. 37.5

23. Familiarize and translate: Let x represent the number.

Some number	less 12.1	is 38.2.
x	- 12.1	= 38.2

Carry out: $x - 12.1 = 38.2$

$x = 50.3$

Check: 50.3 less 12.1 is 38.2, so the answer checks.

State: The number is 50.3.

24. 156.7

25. Familiarize and translate: Let y represent the unknown number.

128 is 0.4 of what number?

$128 = 0.4\cdot y$

Carry out: $128 = 0.4y$

$320 = y$

Check: 0.4 of 320 is 128, so the answer checks.

State: The number is 320.

26. 1368

27. Familiarize: If we let x represent the smaller number, then the larger is x + 12. The sum of the numbers is 114. We want to find the larger number.

Translate:

Smaller number	plus	larger number	is 114.
x	+	(x + 12)	= 114

<u>Carry out</u>: $x + x + 12 = 114$

$$2x + 12 = 114$$
$$2x = 102$$
$$x = 51$$

If $x = 51$, then $x + 12 = 51 + 12$, or 63.

<u>Check</u>: The larger number, 63, is 12 more than the smaller number, 51. Also, $51 + 63 = 114$. The numbers check.

<u>State</u>: The larger number is 63.

28. 13.5

29. <u>Familiarize</u>: If we let x represent the first number, then the second number is 2x. The sum of the numbers is 495. We want to find both numbers.

<u>Translate</u>:

First plus second is 495.

$$x + 2x = 495$$

<u>Carry out</u>: $x + 2x = 495$

$$3x = 495$$
$$x = 165$$

If $x = 165$, then $2x = 2\cdot165$, or 330.

<u>Check</u>: 330 is two times 165. Also, $165 + 330 = 495$. The numbers check.

<u>State</u>: The numbers are 165 and 330.

30. $78\frac{2}{3}$ and $393\frac{1}{3}$

31. The Familiarize and Translate steps were completed in Exercise 4.

<u>Carry out</u>: $(4 - 1.5)t = 3.75$

$$2.5t = 3.75$$
$$t = 1.5$$

<u>Check</u>: Deirdre's speed up river is $(4 - 1.5)$, or 2.5 km/h. In 1.5 hr she will travel 2.5(1.5), or 3.75 km. The answer checks.

<u>State</u>: It will take Deirdre 1.5 hr to swim 3.75 km up river.

32. $1\frac{1}{2}$ hr

33. The Familiarize and Translate steps were completed in Exercise 14.

<u>Carry out</u>: $2x + 3(x + 2) = 76$

$$2x + 3x + 6 = 76$$
$$5x + 6 = 76$$
$$5x = 70$$
$$x = 14$$

If $x = 14$, the $x + 2 = 14 + 2$, or 16.

<u>Check</u>: 14 and 16 are consecutive even integers. Two times 14 plus three times 16 is $2\cdot14 + 3\cdot16$, or $28 + 48$, or 76. The numbers check.

<u>State</u>: The numbers are 14 and 16.

34. 9, 11, and 13

35. The Familiarize and Translate steps were completed in Exercise 10.

<u>Carry out</u>: $x + \frac{2}{3}x = 10$

$$\frac{5}{3}x = 10$$
$$x = 6$$

If $x = 6$, $\frac{2}{3}x = \frac{2}{3}\cdot 6$, or 4.

<u>Check</u>: 4 is $\frac{2}{3}$ of 6. Also, $4 + 6 = 10$. The numbers check.

<u>State</u>: The wire should be cut into a 6 m piece and a 4 m piece.

36. 6 min

37. The Familiarize and Translate steps were completed in Exercise 11.

<u>Carry out</u>: $3x + x + 2x - 12 = 180$

$$6x - 12 = 180$$
$$6x = 192$$
$$x = 32$$

If $x = 32$, $3x = 3\cdot32$, or 96, and $2x - 12 = 2\cdot32 - 12$, or 52.

<u>Check</u>: The first angle, 96°, is three times the second angle, 32°. The third angle, 52°, is 12° less than twice the second angle, 32°. Also, $96° + 32° + 52° = 180°$. The numbers check.

<u>State</u>: The angle measures are 96°, 32°, and 52°.

38. 46 cm, 54 cm

39. $3[2x - (5 + 4x)] = 5 - 7x$

$$3[2x - 5 - 4x] = 5 - 7x$$
$$3[-2x - 5] = 5 - 7x$$
$$-6x - 15 = 5 - 7x$$
$$x - 15 = 5$$
$$x = 20$$

The solution is 20.

40. -19

41. $4 + 2(a - 7) = 2(a - 5)$

$$4 + 2a - 14 = 2a - 10$$
$$-10 + 2a = 2a - 10$$
$$-10 = -10$$

The solution is all real numbers.

42. No solution

43.

44.

45. Familiarize: If we let x represent the height, then the three sides are x + 1, x + 2, and x + 3, with x + 3 representing the base. The perimeter is 42 in.

Translate: The perimeter is 42 in.

$$(x + 1) + (x + 2) + (x + 3) = 42$$

Carry out:
$$x + 1 + x + 2 + x + 3 = 42$$
$$3x + 6 = 42$$
$$3x = 36$$
$$x = 12$$

If x = 12, then x + 1 = 12 + 1, or 13; x + 2 = 12 + 2, or 14; and x + 3 = 12 + 3, or 15.

Check: 12, 13, 14, and 15 are four consecutive integers. Also, 13 + 14 + 15 = 42. The numbers check. The height is 12 in., and the base is 15 in.

Now we find the area of the triangle.

$$\text{Area} = \frac{1}{2}\cdot \text{base}\cdot \text{height}$$
$$= \frac{1}{2}\cdot 15 \text{ in.}\cdot 12 \text{ in.}$$
$$= 90 \text{ in}^2$$

State: The area of the triangle is 90 in^2.

46. $\dfrac{100n}{100 - n}$

47. Familiarize: Let p represent the population at the start of 1986. The population at the end of 1986 is 108% of p, or 1.08p. At the end of 1987 the population is 110% of the population at the end of 1986, or 1.1(1.08p). At the end of 1988 the population is 111% of the population at the end of 1987, or 1.11(1.1)(1.08p).

Translate:

Population at the end of 1988 is 1,582,416.

$$1.11(1.1)(1.08p) = 1{,}582{,}416$$

Carry out:
$$1.11(1.1)(1.08p) = 1{,}582{,}416$$
$$1.31868p = 1{,}582{,}416$$
$$p = 1{,}200{,}000$$

Check: If the population was 1,200,000 at the beginning of 1986, then at the end of 1986 the population would have been 1.08(1,200,000), or 1,296,000. At the end of 1987 the population would have been 1.1(1,296,000), or 1,425,600, and at the end of 1988 it would have been 1.11(1,425,600), or 1,582,416. The answer checks.

State: The population at the start of 1986 was 1,200,000.

48. 10 points

Exercise Set 1.5

1. $3^2\cdot 3^5 = 3^{2+5} = 3^7$

2. 2^{11}

3. $5^6\cdot 5^3 = 5^{6+3} = 5^9$

4. 6^8

5. $a^3\cdot a^0 = a^{3+0} = a^3$

6. x^5

7. $5x^4\cdot 3x^2 = 5\cdot 3\cdot x^4\cdot x^2 = 15x^{4+2} = 15x^6$

8. $8a^{10}$

9. $(-3m^4)(-7m^9) = (-3)(-7)m^4\cdot m^9 = 21m^{4+9} = 21m^{13}$

10. $-14a^9$

11. $(x^3y^4)(x^7y^6z^0) = (x^3x^7)(y^4y^6)(z^0) = x^{3+7}y^{4+6}\cdot 1 = x^{10}y^{10}$

12. $m^{10}n^{12}$

13. $\dfrac{a^9}{a^3} = a^{9-3} = a^6$

14. x^9

15. $\dfrac{8x^7}{4x^4} = \dfrac{8}{4}\cdot x^{7-4} = 2x^3$

16. $4a^{16}$

17. $\dfrac{m^7n^9}{m^2n^5} = m^{7-2}\cdot n^{9-5} = m^5n^4$

18. m^8n^3

19. $\dfrac{35x^8y^5}{7x^2y} = \dfrac{35}{7}\cdot x^{8-2}\cdot y^{5-1} = 5x^6y^4$

20. $9x^6y^6$

21. $\dfrac{-49a^5b^{12}}{7a^2b^2} = \dfrac{-49}{7}\cdot a^{5-2}\cdot b^{12-2} = -7a^3b^{10}$

22. $-6ab^3$

23. $\dfrac{18x^6y^7z^9}{-6x^2yz^3} = \dfrac{18}{-6}\cdot x^{6-2}\cdot y^{7-1}\cdot z^{9-3} = -3x^4y^6z^6$

24. $-4x^4y^8z^{11}$

25. $6^{-3} = \dfrac{1}{6^3}$

26. $\dfrac{1}{8^4}$

27. $9^{-5} = \dfrac{1}{9^5}$

28. $\dfrac{1}{16^2}$

29. $(-11)^{-1} = \dfrac{1}{(-11)^1}$, or $\dfrac{1}{-11}$

30. $\dfrac{1}{(-4)^3}$

31. $(5x)^{-3} = \frac{1}{(5x)^3}$

32. $\frac{1}{(4xy)^5}$

33. $x^2y^{-3} = x^2 \cdot \frac{1}{y^3} = \frac{x^2}{y^3}$

34. $\frac{2a^2}{b^5}$

35. $x^2y^{-2} = x^2 \cdot \frac{1}{y^2} = \frac{x^2}{y^2}$

36. $\frac{a^2c^4}{b^3d^5}$

37. $\frac{x^3}{y^{-2}} = x^3 \cdot \frac{1}{y^{-2}} = x^3y^2$

38. xy^4z^3

39. $\frac{y^{-5}}{x^2} = \frac{1}{x^2} \cdot y^{-5} = \frac{1}{x^2} \cdot \frac{1}{y^5} = \frac{1}{x^2y^5}$

40. $\frac{1}{3x^5z^4}$

41. $\frac{y^{-5}}{x^{-3}} = \frac{1}{x^{-3}} \cdot y^{-5} = x^3 \cdot \frac{1}{y^5} = \frac{x^3}{y^5}$

42. $\frac{y^7}{(7x)^4}$

43. $\frac{x^{-2}y^7}{z^{-4}} = x^{-2} \cdot y^7 \cdot \frac{1}{z^{-4}} = \frac{1}{x^2} \cdot y^7 \cdot z^4 = \frac{y^7z^4}{x^2}$

44. $\frac{x^2y^4}{z^3}$

45. $\frac{1}{3^4} = 3^{-4}$

46. 9^{-2}

47. $\frac{1}{(-16)^2} = (-16)^{-2}$

48. $(-8)^{-6}$

49. $6^4 = \frac{1}{6^{-4}}$

50. $\frac{1}{8^{-5}}$

51. $6x^2 = 6 \cdot \frac{1}{x^{-2}} = \frac{6}{x^{-2}}$

52. $\frac{-4}{y^{-5}}$

53. $\frac{1}{(5y)^3} = (5y)^{-3}$

54. $(5x)^{-5}$

55. $\frac{1}{3y^4} = \frac{1}{3} \cdot \frac{1}{y^4} = \frac{1}{3} \cdot y^{-4} = \frac{y^{-4}}{3}$

56. $\frac{b^{-3}}{4}$

57. $8^{-6} \cdot 8^2 = 8^{-6+2} = 8^{-4}$, or $\frac{1}{8^4}$

58. 9^{-2}, or $\frac{1}{9^2}$

59. $8^{-2} \cdot 8^{-4} = 8^{-2+(-4)} = 8^{-6}$, or $\frac{1}{8^6}$

60. 9^{-7}, or $\frac{1}{9^7}$

61. $b^2 \cdot b^{-5} = b^{2+(-5)} = b^{-3}$, or $\frac{1}{b^3}$

62. a

63. $a^{-3} \cdot a^4 \cdot a^2 = a^{-3+4+2} = a^3$

64. 1

65. $(14m^2n^3)(-2m^3n^2) = 14 \cdot (-2) \cdot m^2 \cdot m^3 \cdot n^3 \cdot n^2$
$= -28m^{2+3}n^{3+2}$
$= -28m^5n^5$

66. $-18x^7y$

67. $(-2x^{-3})(7x^{-8}) = -2 \cdot 7 \cdot x^{-3} \cdot x^{-8} = -14x^{-3+(-8)}$
$= -14x^{-11}$, or $\frac{-14}{x^{11}}$

68. $-24x^{-12}y$, or $-\frac{24y}{x^{12}}$

69. $5^{a+1} \cdot 5^{2a-1} = 5^{a+1+2a-1} = 5^{3a}$

70. 7^{2a+b+1}

71. $\frac{4^3}{4^{-2}} = 4^{3-(-2)} = 4^{3+2} = 4^5$

72. 5^{11}

73. $\frac{10^{-3}}{10^6} = 10^{-3-6} = 10^{-9}$, or $\frac{1}{10^9}$

74. 12^{-12}, or $\frac{1}{12^{12}}$

75. $\frac{9^{-4}}{9^{-6}} = 9^{-4-(-6)} = 9^{-4+6} = 9^2$, or 81

76. 2^{-2}, or $\frac{1}{2^2}$

77. $\frac{a^3}{a^{-2}} = a^{3-(-2)} = a^{3+2} = a^5$

78. y^9

79. $\frac{9a^2}{3ab^3} = \frac{9}{3} \cdot a^{2-1} \cdot b^{0-3} = 3ab^{-3}$, or $\frac{3a}{b^3}$

80. $-3ab^2$

81. $\frac{-24x^6y^7}{18x^{-3}y^9} = \frac{-24}{18}x^{6-(-3)}y^{7-9} = -\frac{4}{3}x^9y^{-2}$, or $-\frac{4x^9}{3y^2}$

82. $-\frac{7}{4}a^{-4}b^2$, or $-\frac{7b^2}{4a^4}$

83. $\frac{-18x^{-2}y^3}{-12x^{-5}y^5} = \frac{-18}{-12}x^{-2-(-5)}y^{3-5} = \frac{3}{2}x^3y^{-2}$, or $\frac{3x^3}{2y^2}$

84. $\frac{7}{9}a^{16}b^5$

85. $\frac{10^{2a}}{10^a} = 10^{2a-a} = 10^a$

86. 11^{-2b} (The sign of -2b depends on the sign of b.)

87. $(4^3)^2 = 4^{3\cdot 2} = 4^6$

88. 5^{20}

89. $(8^4)^{-3} = 8^{4(-3)} = 8^{-12}$, or $\frac{1}{8^{12}}$

90. 9^{-12}, or $\frac{1}{9^{12}}$

91. $(6^{-4})^{-3} = 6^{-4(-3)} = 6^{12}$

92. 7^{40}

93. $(3x^2y^2)^3 = 3^3(x^2)^3(y^2)^3 = 27x^6y^6$

94. $32a^{15}b^{20}$

95. $(-2x^3y^{-4})^{-2} = (-2)^{-2}(x^3)^{-2}(y^{-4})^{-2} = (-2)^{-2}x^{-6}y^8$,

or $\frac{1}{4}x^{-6}y^8$, or $\frac{y^8}{4x^6}$

96. $-\frac{1}{27}a^{-6}b^{15}$, or $-\frac{b^{15}}{27a^6}$

97. $(-6a^{-2}b^3c)^{-2} = (-6)^{-2}(a^{-2})^{-2}(b^3)^{-2}c^{-2}$

$= (-6)^{-2}a^4b^{-6}c^{-2}$, or $\frac{1}{36}a^4b^{-6}c^{-2}$, or $\frac{a^4}{36b^6c^2}$

98. $\frac{1}{(-8)^4}x^{16}y^{-20}z^{-8}$, or $\frac{x^{16}}{(-8)^4y^{20}z^8}$

99. $\frac{(5a^3b)^2}{10a^2b} = \frac{5^2(a^3)^2b^2}{10a^2b} = \frac{25a^6b^2}{10a^2b} = \frac{25}{10}\cdot a^{6-2}\cdot b^{2-1}$

$= \frac{5}{2}a^4b$, or $\frac{5a^4b}{2}$

100. $\frac{9}{2}x^8y^9$

101. $\left[\frac{2x^3y^{-2}}{3y^{-3}}\right]^3 = \frac{(2x^3y^{-2})^3}{(3y^{-3})^3} = \frac{2^3(x^3)^3(y^{-2})^3}{3^3(y^{-3})^3} = \frac{8x^9y^{-6}}{27y^{-9}} =$

$\frac{8x^9y^3}{27}$

102. $\frac{625x^{-20}y^{24}}{256}$, or $\frac{625y^{24}}{256x^{20}}$

103. $\left[\frac{30x^5y^{-7}}{6x^{-2}y^{-6}}\right]^0 = 1$ (Any nonzero real number raised to the zero power is 1.)

104. $9a^{-4}b^{-10}$, or $\frac{9}{a^4b^{10}}$

105. $\left[\frac{5x^0y^{-7}}{2x^{-2}y^4}\right]^{-2} = \frac{(5x^0y^{-7})^{-2}}{(2x^{-2}y^4)^{-2}} = \frac{5^{-2}x^0y^{14}}{2^{-2}x^4y^{-8}} =$

$\frac{2^2y^{22}}{5^2x^4} = \frac{4}{25}x^{-4}y^{22}$, or $\frac{4y^{22}}{25x^4}$

106. 1

107. $5 + 2\cdot 3^2 = 5 + 2\cdot 9$ Evaluating the exponential expression

$= 5 + 18$ Multiplying

$= 23$ Adding

108. -3

109. $12 - (9 - 3\cdot 2^3) = 12 - (9 - 3\cdot 8)$ Working within the parentheses first

$= 12 - (9 - 24)$

$= 12 - (-15)$

$= 12 + 15$

$= 27$

110. -3

111. $\frac{5\cdot 2 - 4^2}{27 - 2^4} = \frac{5\cdot 2 - 16}{27 - 16} = \frac{10 - 16}{11} = \frac{-6}{11}$, or $-\frac{6}{11}$

112. $-\frac{4}{17}$

113. $\frac{3^4 - (5 - 3)^4}{1 - 2^3} = \frac{3^4 - 2^4}{1 - 8} = \frac{81 - 16}{-7} = \frac{65}{-7}$, or $-\frac{65}{7}$

114. $\frac{55}{2}$

115. $5^3 - [2(4^2 - 3^2 - 6)]^3 = 5^3 - [2(16 - 9 - 6)]^3 =$

$5^3 - [2\cdot 1]^3 = 5^3 - 2^3 = 125 - 8 = 117$

116. 13

117. $|2^2 - 7|^3 + 1 = |4 - 7|^3 + 1 = |-3|^3 + 1 =$

$3^3 + 1 = 27 + 1 = 28$

118. 79

119. $30 - (-5)^2 + 15 \div (-3)\cdot 2$

$= 30 - 25 + 15 \div (-3)\cdot 2$ Evaluating the exponential expression

$= 30 - 25 - 5\cdot 2$ Dividing

$= 30 - 25 - 10$ Multiplying

$= -5$ Subtracting

120. 0

121. $12 - (7 - 5) + 4 \div 3 \cdot 2^3$

$= 12 - 2 + 4 \div 3 \cdot 8$ Working within the parentheses and evaluating the exponential expression

$= 12 - 2 + \frac{4}{3} \cdot 8$ Dividing

$= 12 - 2 + \frac{32}{3}$ Multiplying

$= \frac{62}{3}$, or $20\frac{2}{3}$ Subtracting and adding

122. $24\frac{1}{2}$

123. $\frac{9 + 3 \cdot 4 + 1}{2}$

$\boxed{9}\ \boxed{+}\ \boxed{3}\ \boxed{\times}\ \boxed{4}\ \boxed{+}\ \boxed{1}\ \boxed{=}\ \boxed{\div}\ \boxed{2}\ \boxed{=}$

124. $\boxed{7}\ \boxed{-}\ \boxed{2}\ \boxed{\times}\ \boxed{3}\ \boxed{+}\ \boxed{5}\ \boxed{=}\ \boxed{\div}\ \boxed{4}\ \boxed{=}$

125. $8 - 2 \cdot 3 + \frac{4}{15}$

$\boxed{8}\ \boxed{-}\ \boxed{2}\ \boxed{\times}\ \boxed{3}\ \boxed{+}\ \boxed{4}\ \boxed{\div}\ \boxed{1}\ \boxed{5}\ \boxed{=}$

126. $\boxed{1}\ \boxed{2}\ \boxed{+}\ \boxed{3}\ \boxed{\times}\ \boxed{2}\ \boxed{-}\ \boxed{7}\ \boxed{\div}\ \boxed{1}\ \boxed{2}\ \boxed{=}$

127. $\frac{9 - 2^8 + 4}{3}$

$\boxed{9}\ \boxed{-}\ \boxed{2}\ \boxed{x^y}\ \boxed{8}\ \boxed{+}\ \boxed{4}\ \boxed{=}\ \boxed{\div}\ \boxed{3}\ \boxed{=}$

128. $\boxed{1}\ \boxed{5}\ \boxed{-}\ \boxed{3}\ \boxed{x^y}\ \boxed{7}\ \boxed{+}\ \boxed{8}\ \boxed{=}\ \boxed{\div}$

$\boxed{1}\ \boxed{4}\ \boxed{=}$

129. $\left(7 - \frac{2}{3}\right)^4 - \frac{5}{8}$

$\boxed{7}\ \boxed{-}\ \boxed{2}\ \boxed{\div}\ \boxed{3}\ \boxed{=}\ \boxed{x^y}\ \boxed{4}\ \boxed{-}\ \boxed{5}\ \boxed{\div}\ \boxed{8}\ \boxed{=}$

130. $\boxed{9}\ \boxed{-}\ \boxed{4}\ \boxed{\div}\ \boxed{7}\ \boxed{=}\ \boxed{x^y}\ \boxed{5}\ \boxed{+}\ \boxed{3}\ \boxed{\div}$

$\boxed{1}\ \boxed{1}\ \boxed{=}$

131. $x + (2xy - z)^2 = 3 + (2 \cdot 3(-4) - (-20))^2 =$
$3 + (-24 + 20)^2 = 3 + (-4)^2 = 3 + 16 = 19$

132. -8

133. Familiarize and Translate: Let x represent the first integer, $x + 2$ the second, and $x + 4$ the third.

First plus second plus third is 183.

$x + (x + 2) + (x + 4) = 183$

Carry out: $x + x + 2 + x + 4 = 183$

$3x + 6 = 183$

$3x = 177$

$x = 59$

If $x = 59$, then $x + 2 = 59 + 2$, or 61, and $x + 4 = 59 + 4$, or 63.

Check: The check is left to the student.

State: The integers are 59, 61, and 63.

134. 36, 38, and 40

135.

136.

137. $\frac{9a^{x-2}}{3a^{2x+2}} = \frac{9}{3} \cdot a^{x-2-(2x+2)} = 3a^{x-2-2x-2} = 3a^{-x-4}$

138. $-3x^{2a-1}$

139. $[7y(7 - 8)^{-2} - 8y(8 - 7)^{-2}]^{-(-2)^2}$

$= [7y(7 - 8)^{-2} - 8y(8 - 7)^{-2}]^{-4}$

$= [7y(-1)^{-2} - 8y(1)^{-2}]^{-4}$

$= \left[\frac{7y}{(-1)^2} - \frac{8y}{1^2}\right]^{-4}$

$= \left[\frac{7y}{1} - \frac{8y}{1}\right]^{-4}$

$= [7y - 8y]^{-4}$

$= [-y]^{-4}$

$= \frac{1}{(-y)^4}$

$= \frac{1}{y^4}$

140. 8^{-2abc}

141. $(3^{a+2})^a = 3^{(a+2)(a)} = 3^{a^2+2a}$

142. 12^{6b-2ab}

143. $\frac{4x^{2a+3}y^{2b-1}}{2x^{a+1}y^{b+1}} = \frac{4}{2} \cdot x^{2a+3-(a+1)}y^{2b-1-(b+1)} =$
$2x^{2a+3-a-1}y^{2b-1-b-1} = 2x^{a+2}y^{b-2}$

144. $-5x^{2b}y^{-2a}$

145. $\frac{(2^{-2})^a \cdot (2^b)^{-a}}{(2^{-2})^{-b}(2^b)^{-2a}} = \frac{2^{-2a} \cdot 2^{-ab}}{2^{2b} \cdot 2^{-2ab}} = \frac{2^{-2a-ab}}{2^{2b-2ab}} =$
$2^{-2a-ab-(2b-2ab)} = 2^{-2a-ab-2b+2ab} = 2^{-2a-2b+ab}$

146. $-4x^{10}y^8$

147. $\frac{3^{q+3} - 3^2(3^q)}{3(3^{q+4})} = \frac{3^{q+3} - 3^{q+2}}{3^{q+5}} = \frac{3^{q+2}(3 - 1)}{3^{q+2}(3^3)} =$
$\frac{2}{3^3} = \frac{2}{27}$

148. $\frac{a^{-14ac}}{b^{27ac}}$

Exercise Set 1.6

1. $A = \ell w$

$\frac{1}{\ell} \cdot A = \frac{1}{\ell} \cdot \ell w$ Multiplying by $\frac{1}{\ell}$

$\frac{A}{\ell} = w$ Simplifying

2. $a = \frac{F}{m}$

3. $W = EI$

$\frac{1}{E} \cdot W = \frac{1}{E} \cdot EI$ Multiplying by $\frac{1}{E}$

$\frac{W}{E} = I$ Simplifying

4. $E = \frac{W}{I}$

5. $d = rt$

$d \cdot \frac{1}{t} = rt \cdot \frac{1}{t}$

$\frac{d}{t} = r$

6. $t = \frac{d}{r}$

7. $V = \ell wh$

$V \cdot \frac{1}{wh} = \ell wh \cdot \frac{1}{wh}$

$\frac{V}{wh} = \ell$

8. $r = \frac{I}{Pt}$

9. $E = mc^2$

$E \cdot \frac{1}{c^2} = mc^2 \cdot \frac{1}{c^2}$

$\frac{E}{c^2} = m$

10. $c^2 = \frac{E}{m}$

11. $P = 2\ell + 2w$

$P - 2w = 2\ell + 2w - 2w$

$P - 2w = 2\ell$

$\frac{1}{2}(P - 2w) = \frac{1}{2} \cdot 2\ell$

$\frac{P - 2w}{2} = \ell$, or $\frac{P}{2} - w = \ell$

12. $w = \frac{P - 2\ell}{2}$

13. $c^2 = a^2 + b^2$

$c^2 - b^2 = a^2 + b^2 - b^2$

$c^2 - b^2 = a^2$

14. $b^2 = c^2 - a^2$

15. $A = \pi r^2$

$\frac{1}{\pi} \cdot A = \frac{1}{\pi} \cdot \pi r^2$

$\frac{A}{\pi} = r^2$

16. $\pi = \frac{A}{r^2}$

17. $W = \frac{11}{2}(h - 40)$

$\frac{2}{11} \cdot W = \frac{2}{11} \cdot \frac{11}{2}(h - 40)$

$\frac{2}{11}W = h - 40$

$\frac{2}{11}W + 40 = h - 40 + 40$

$\frac{2}{11}W + 40 = h$

18. $F = \frac{9}{5}C + 32$

19. $V = \frac{4}{3}\pi r^3$

$V = \frac{4\pi}{3}r^3$

$\frac{3}{4\pi} \cdot V = \frac{3}{4\pi} \cdot \frac{4\pi}{3}r^3$

$\frac{3V}{4\pi} = r^3$

20. $\pi = \frac{3V}{4r^3}$

21. $A = \frac{h}{2}(b_1 + b_2)$

$2 \cdot A = 2 \cdot \frac{h}{2}(b_1 + b_2)$

$2A = h(b_1 + b_2)$

$2A \cdot \frac{1}{b_1 + b_2} = h(b_1 + b_2) \cdot \frac{1}{b_1 + b_2}$

$\frac{2A}{b_1 + b_2} = h$

22. $b_2 = \frac{2A}{h} - b_1$

23. $F = \frac{mv^2}{r}$

$F = m \cdot \frac{v^2}{r}$

$F \cdot \frac{r}{v^2} = m \cdot \frac{v^2}{r} \cdot \frac{r}{v^2}$

$\frac{Fr}{v^2} = m$

24. $v^2 = \frac{rF}{m}$

25. $A = \frac{q_1 + q_2 + q_3}{n}$

$n \cdot A = n \cdot \frac{q_1 + q_2 + q_3}{n}$ Clearing the fraction

$nA = q_1 + q_2 + q_3$

$nA \cdot \frac{1}{A} = (q_1 + q_2 + q_3) \cdot \frac{1}{A}$

$n = \frac{q_1 + q_2 + q_3}{A}$

26. $d = \frac{s + t}{r}$

27. $v = \frac{d_2 - d_1}{t}$

$t \cdot v = t \cdot \frac{d_2 - d_1}{t}$

$tv = d_2 - d_1$

$tv \cdot \frac{1}{v} = (d_2 - d_1) \cdot \frac{1}{v}$

$t = \frac{d_2 - d_1}{v}$

28. $m = \frac{s_2 - s_1}{v}$

29. $v = \frac{d_2 - d_1}{t}$

$t \cdot v = t \cdot \frac{d_2 - d_1}{t}$

$tv = d_2 - d_1$

$tv - d_2 = d_2 - d_1 - d_2$

$tv - d_2 = -d_1$

$-1 \cdot (tv - d_2) = -1 \cdot (-d_1)$

$-tv + d_2 = d_1$,

or $d_2 - tv = d_1$

30. $s_1 = s_2 - vm$

31. $r = m + mnp$

$r = m(1 + np)$ Factoring

$r \cdot \frac{1}{1 + np} = m(1 + np) \cdot \frac{1}{1 + np}$

$\frac{r}{1 + np} = m$

32. $x = \frac{p}{1 - yz}$

33. $y = ab - ac^2$

$y = a(b - c^2)$

$y \cdot \frac{1}{b - c^2} = a(b - c^2) \cdot \frac{1}{b - c^2}$

$\frac{y}{b - c^2} = a$

34. $m = \frac{d}{n - p^3}$

35. Translate. The formula for the area of a parallelogram with base b and height h is

$A = bh.$

Since we want to find the length of the base, we solve the formula for b.

$A = bh$

$A \cdot \frac{1}{h} = bh \cdot \frac{1}{h}$

$\frac{A}{h} = b$

We now have a formula that says to find b, we divide A by h.

Carry out. Substitute 72 for A and 6 for h in the formula and calculate:

$\frac{72}{6} = b$

$12 = b$

We leave the check to the student. The base of the parallelogram is 12 cm.

36. 6 cm

37. Translate. Thurnau's formula is

$P = 9.337da - 299.$

Since we want to find the diameter of the fetus' head, we solve for d.

$P = 9.337da - 299$

$P + 299 = 9.337da$

$\frac{P + 299}{9.337a} = d$

Carry out. Substitute 1614 for P and 24.1 for a in the formula and calculate:

$\frac{1614 + 299}{9.337(24.1)} = d$

$8.5 \approx d$

The check is left to the student. The diameter of the fetus' head at 29 weeks is about 8.5 cm.

38. 27.6 cm

39. Translate. The simple interest formula is $I = Prt$, where I is the amount of interest, P is the principal, r is the interest rate, and t is the time in years. What we want to know is the amount to invest, so we solve the formula for P.

$I = Prt$

$\frac{I}{rt} = P$

We now have a formula that gives P.

Carry out. We substitute \$110 for I, 7% (or 0.07) for r and 1 for t.

$P = \frac{I}{rt}$

$P = \frac{110}{0.07 \cdot 1}$

$P = 1571.43$ Rounding to the nearest cent

The check is left to the student. You will have to invest \$1571.43.

40. 6.4%

41. Translate. The formula for the area of a trapezoid is $A = \frac{1}{2}h(b_1 + b_2)$, where A is the area, h is the height, and b_1 and b_2 are the bases. The unknown dimension in this problem is h, so we solve the formula for h.

$$A = \frac{1}{2}h(b_1 + b_2)$$

$$2A = h(b_1 + b_2)$$

$$\frac{2A}{b_1 + b_2} = h$$

We now have a formula that gives h.

Carry out. We substitute 90 for A, 8 for b_1, and 12 for b_2.

$$h = \frac{2A}{b_1 + b_2}$$

$$h = \frac{2 \cdot 90}{8 + 12}$$

$$h = \frac{180}{20}$$

$$h = 9$$

The check is left to the student. The unknown dimension is 9 ft.

42. 25 ft

43. Translate. The formula $A = P + Prt$ tells us how much a principal P, in dollars, will be worth when invested at simple interest at a rate r for t years. We want to find the length of the investment period, so we solve the formula for t.

$$A = P + Prt$$

$$A - P = Prt$$

$$\frac{A - P}{Pr} = t$$

Carry out. Substitute 2608 for A, 1600 for P and 9%(or 0.09) for r.

$$\frac{A - P}{Pr} = t$$

$$\frac{2608 - 1600}{1600(0.09)} = t$$

$$\frac{1008}{144} = t$$

$$7 = t$$

The check is left to the student. It will take 7 years for the investment to be worth $2608.

44. 6 years

45. Translate. According to Goiten's model,

$$I = 1.08\,\frac{T}{N}.$$

Since we want to find the number of scheduled appointments, we solve for N.

$$I = 1.08\,\frac{T}{N}$$

$$NI = 1.08T$$

$$N = \frac{1.08T}{I}$$

Carry out. Substitute 480 for T (8 hr = 8·60 min = 480 min) and 15 for I and calculate:

$$N = \frac{1.08(480)}{15}$$

$$N = 34.56$$

The check is left to the student. Since fractional parts of appointments make no sense, we round down. The doctor should schedule 34 appointments in one day.

46. 7.7 hr

47. Translate. We solve the formula $D = \frac{m}{V}$ for m. (See Example 5.)

$$D = \frac{m}{V}$$

$$DV = m$$

Carry out. Substitute 7.5 for D and 1230 for V.

$$DV = m$$

$$7.5(1230) = m$$

$$9225 = m$$

The check is left to the student. The anchor's mass is 9225 g.

48. 358.85205 g

49. Familiarize and Translate. Recall that "%" means "× 0.01." Let n represent the percent.

What percent of 5800 is 4176?

$$n \quad \times 0.01 \quad \times 5800 \quad = 4176$$

Carry out. $n \times 0.01 \times 5800 = 4176$

$$58n = 4176$$

$$n = \frac{4176}{58}$$

$$n = 72$$

The check is left to the student.

State. 4176 is 72% of 5800.

50. $-8a + 13b$

51. $-72.5 - (-14.06) = -72.5 + 14.06 = -58.44$

52. 3

53.

54.

55. We will use the formulas for density and for the volume of a right circular cylinder.

Translate. Solving the formula $D = \frac{m}{V}$ for V, we get $V = \frac{m}{D}$. Also, the volume of a right circular cylinder with radius r and height h is given by $V = \pi r^2 h$, so we have $\pi r^2 h = \frac{m}{D}$. Solve for h:

$$h = \frac{m}{\pi r^2 D}$$

Note also that the radius of a penny is $\frac{1.85}{2}$, or 0.925.

Carry out. Substitute 8.93 for D, 177.6 for m, 0.925 for r, and 3.14 for π:

$$h = \frac{m}{\pi r^2 D}$$

$$h = \frac{177.6}{3.14(0.925)^2(8.93)}$$

$$h \approx 7.4$$

The check is left to the student. The roll of pennies is about 7.4 cm tall.

56. 610.55 cm

57. From Example 2 we know that $t = \frac{I}{Pr}$.

Carry out. Substitute 6 for I, 200 for P, and 12% (or 0.12) for r.

$$t = \frac{I}{Pr}$$

$$t = \frac{6}{200(0.12)}$$

$$t = \frac{6}{24}$$

$$t = \frac{1}{4}$$

The check is left to the student. It would take $\frac{1}{4}$ year, or 3 months $\left[\frac{1}{4} \times 12 \text{ months}\right]$.

58. 11%

59.
$$s = v_1t + \frac{1}{2}at^2$$
$$s - v_1t = \frac{1}{2}at^2$$
$$2(s - v_1t) = at^2$$
$$\frac{2(s - v_1t)}{t^2} = a,$$
or $$\frac{2s - 2v_1t}{t^2} = a$$

60. $\ell = \frac{A - w^2}{4w}$

61.
$$\frac{P_1V_1}{T_1} = \frac{P_2V_2}{T_2}$$
$$\frac{T_1}{P_1} \cdot \frac{P_1V_1}{T_1} = \frac{T_1}{P_1} \cdot \frac{P_2V_2}{T_2}$$
$$V_1 = \frac{T_1P_2V_2}{P_1T_2}$$

62. $T_2 = \frac{T_1P_2V_2}{P_1V_1}$

63.
$$x = \frac{a}{b + c}$$
$$x(b + c) = a$$
$$xb + xc = a$$
$$xc = a - xb$$
$$c = \frac{a - xb}{x}, \text{ or } \frac{a}{x} - b$$

64. $d = \frac{me^2}{f}$

Exercise Set 1.7

1. $47{,}000{,}000{,}000 = 47{,}000{,}000{,}000 \times (10^{-10} \times 10^{10})$
$= (47{,}000{,}000{,}000 \times 10^{-10}) \times 10^{10}$
$= 4.7 \times 10^{10}$

2. 2.6×10^{12}

3. $863{,}000{,}000{,}000{,}000{,}000$
$= 863{,}000{,}000{,}000{,}000{,}000 \times (10^{-17} \times 10^{17})$
$= (863{,}000{,}000{,}000{,}000{,}000 \times 10^{-17}) \times 10^{17}$
$= 8.63 \times 10^{17}$

4. 9.57×10^{17}

5. $0.000000016 = 0.000000016 \times (10^8 \times 10^{-8})$
$= (0.000000016 \times 10^8) \times 10^{-8}$
$= 1.6 \times 10^{-8}$

6. 2.63×10^{-7}

7. $0.00000000007 = 0.00000000007 \times (10^{11} \times 10^{-11})$
$= (0.00000000007 \times 10^{11}) \times 10^{-11}$
$= 7 \times 10^{-11}$

8. 9×10^{-11}

9. $407{,}000{,}000{,}000 = 407{,}000{,}000{,}000 \times (10^{-11} \times 10^{11})$
$= (407{,}000{,}000{,}000 \times 10^{-11}) \times 10^{11}$
$= 4.07 \times 10^{11}$

10. 3.09×10^{12}

11. $0.000000603 = 0.000000603 \times (10^7 \times 10^{-7})$
$= (0.000000603 \times 10^7) \times 10^{-7}$
$= 6.03 \times 10^{-7}$

12. 8.02×10^{-9}

13. $492{,}700{,}000{,}000 = 492{,}700{,}000{,}000 \times (10^{-11} \times 10^{11})$
$= (492{,}700{,}000{,}000 \times 10^{-11}) \times 10^{11}$
$= 4.927 \times 10^{11}$

14. 9.534×10^{11}

15. $4 \times 10^{-4} = 0.0004$ Moving the decimal point 4 places to the left

16. 0.00005

17. $6.73 \times 10^{8} = 673{,}000{,}000$ Moving the decimal point 8 places to the right

18. 92,400,000

19. $8.923 \times 10^{-10} = 0.0000000008923$ Moving the decimal point 10 places to the left

20. 0.07034

21. $9.03 \times 10^{10} = 90{,}300{,}000{,}000$ Moving the decimal point 10 places to the right

22. 1,010,000,000,000

23. $4.037 \times 10^{-8} = 0.00000004037$ Moving the decimal point 8 places to the left

24. 0.000000003007

25. $8.007 \times 10^{12} = 8{,}007{,}000{,}000{,}000$ Moving the decimal point 12 places to the right

26. 90,010,000,000

27. $(2.3 \times 10^{6})(4.2 \times 10^{-11})$
$= (2.3 \times 4.2)(10^{6} \times 10^{-11})$
$= 9.66 \times 10^{-5}$
$= 9.7 \times 10^{-5}$ Rounding to 2 significant digits

28. 3.4×10^{-4}

29. $(2.34 \times 10^{-8})(5.7 \times 10^{-4})$
$= (2.34 \times 5.7)(10^{-8} \times 10^{-4})$
$= 13.338 \times 10^{-12}$
$= (1.3338 \times 10^{1}) \times 10^{-12}$
$= 1.3338 \times (10^{1} \times 10^{-12})$
$= 1.3338 \times 10^{-11}$
$= 1.3 \times 10^{-11}$ Rounding to 2 significant digits

30. 2.7×10^{-11}

31. $(3.2 \times 10^{6})(2.6 \times 10^{4}) = (3.2 \times 2.6)(10^{6} \times 10^{4})$
$= 8.32 \times 10^{10}$
$= 8.3 \times 10^{10}$
(2 significant digits)

32. 3.14×10^{16}

33. $(3.01 \times 10^{-5})(6.5 \times 10^{7})$
$= (3.01 \times 6.5)(10^{-5} \times 10^{7})$
$= 19.565 \times 10^{2}$
$= (1.9565 \times 10^{1}) \times 10^{2}$
$= 1.9565 \times (10^{1} \times 10^{2})$
$= 1.9565 \times 10^{3}$
$= 2.0 \times 10^{3}$ Rounding to 2 significant digits

34. 3.1×10^{-4}

35. $(5.01 \times 10^{-7})(3.02 \times 10^{-6})$
$= (5.01 \times 3.02)(10^{-7} \times 10^{-6})$
$= 15.1302 \times 10^{-13}$
$= (1.51302 \times 10^{1}) \times 10^{-13}$
$= 1.51302 \times (10^{1} \times 10^{-13})$
$= 1.51302 \times 10^{-12}$
$= 1.51 \times 10^{-12}$ Rounding to 3 significant digits

36. 6.34×10^{-15}

37. $\dfrac{8.5 \times 10^{8}}{3.4 \times 10^{5}} = \dfrac{8.5}{3.4} \times \dfrac{10^{8}}{10^{5}}$
$= 2.5 \times 10^{3}$

38. 1.5×10^{3}

39. $\dfrac{4.0 \times 10^{-6}}{8.0 \times 10^{-3}} = \dfrac{4.0}{8.0} \times \dfrac{10^{-6}}{10^{-3}}$
$= 0.5 \times 10^{-3} = (5 \times 10^{-1}) \times 10^{-3}$
$= 5.0 \times (10^{-1} \times 10^{-3}) = 5.0 \times 10^{-4}$
We write 5.0 to indicate 2 significant digits.

40. 3.0×10^{-5}

41. $\dfrac{12.6 \times 10^{8}}{4.2 \times 10^{-3}} = \dfrac{12.6}{4.2} \times \dfrac{10^{8}}{10^{-3}}$
$= 3.0 \times 10^{11}$ We write 3.0 to indicate 2 significant digits.

42. 4.0×10^{-16}

43. $\dfrac{2.42 \times 10^{5}}{1.21 \times 10^{-5}} = \dfrac{2.42}{1.21} \times \dfrac{10^{5}}{10^{-5}}$
$= 2.00 \times 10^{10}$ We write 2.00 to indicate 3 significant digits.

44. 3.00×10^{-22}

45. $\dfrac{4.7 \times 10^{-9}}{2.35 \times 10^{7}} = \dfrac{4.7}{2.35} \times \dfrac{10^{-9}}{10^{7}}$
$= 2.0 \times 10^{-16}$ We write 2.0 to indicate 2 significant digits.

46. 2.00×10^{26}

47. $\dfrac{1.05 \times 10^{-6}}{4.2 \times 10^{-7}} = \dfrac{1.05}{4.2} \times \dfrac{10^{-6}}{10^{-7}}$

$= 0.25 \times 10^{1}$

$= (2.5 \times 10^{-1}) \times 10^{1}$

$= 2.5 \times (10^{-1} \times 10^{1})$

$= 2.5 \times 10^{0}$

$= 2.5 \times 1$

$= 2.5$

48. 1.3×10^{-4}

49. $\dfrac{(6.1 \times 10^{4})(7.2 \times 10^{-6})}{9.8 \times 10^{-4}} = \dfrac{6.1 \times 7.2}{9.8} \times \dfrac{10^{4} \times 10^{-6}}{10^{-4}}$

$= 4.48 \times 10^{2}$

$= 4.5 \times 10^{2}$ Rounding to 2 significant digits

50. 1.5×10^{-17}

51. $\dfrac{780{,}000{,}000 \times 0.00071}{0.000005} = \dfrac{(7.8 \times 10^{8}) \times (7.1 \times 10^{-4})}{5 \times 10^{-6}}$

$= \dfrac{7.8 \times 7.1}{5} \times \dfrac{10^{8} \times 10^{-4}}{10^{-6}}$

$= 11.076 \times 10^{10}$

$= 1.1076 \times 10^{1} \times 10^{10}$

$= 1.1076 \times (10^{1} \times 10^{10})$

$= 1 \times 10^{11}$ (1 significant digit)

52. 3.2×10

53. $\dfrac{43{,}000{,}000 \times 0.095}{63{,}000} = \dfrac{(4.3 \times 10^{7}) \times (9.5 \times 10^{-2})}{6.3 \times 10^{4}}$

$= \dfrac{4.3 \times 9.5}{6.3} \times \dfrac{10^{7} \times 10^{-2}}{10^{4}}$

$= 6.5 \times 10$ Rounding to 2 significant digits

54. $1. \times 10^{3}$

55. Familiarize. We will let y represent the number of light years from one end of the galaxy to the other (that is, the number of light years in the diameter of the galaxy). Recall from Example 7 that 1 light year = 5.88×10^{12} mi.

Translate. Note that the diameter of the galaxy is $(5.88 \times 10^{12})y$ mi. We are also told that this distance is 5.88×10^{17} mi. Since these quantities represent the same number, we write the equation

$$(5.88 \times 10^{12})y = 5.88 \times 10^{17}.$$

Carry out. Solve the equation:

$$(5.88 \times 10^{12})y = 5.88 \times 10^{17}$$

$$\frac{1}{5.88 \times 10^{12}}(5.88 \times 10^{12})y = \frac{1}{5.88 \times 10^{12}} \times 5.88 \times 10^{17}$$

$$y = \frac{5.88 \times 10^{17}}{5.88 \times 10^{12}}$$

$$y = \frac{5.88}{5.88} \times \frac{10^{17}}{10^{12}}$$

$y = 1.00 \times 10^{5}$ There are 3 significant digits.

The check is left to the student.

State. It is 1.00×10^{5}, or 100,000, light years from one end of the Milky Way galaxy to the other.

56. 8.00 light years

57. Familiarize. We are told that 1 Angstrom = 10^{-10} m, one parsec ≈ 3.26 light years, and 1 light year = 9.46×10^{15} m. Let a represent the number of Angstroms in one parsec.

Translate. The length of one parsec is $a \times 10^{-10}$ m. It can also be expressed as 3.26 light years, or $3.26 \times 9.46 \times 10^{15}$ m. Since these quantities represent the same number, we can write the equation

$$a \times 10^{-10} = 3.26 \times 9.46 \times 10^{15}.$$

Carry out. Solve the equation:

$$a \times 10^{-10} = 3.26 \times 9.46 \times 10^{15}$$

$$a \times 10^{-10} \times \frac{1}{10^{-10}} = 3.26 \times 9.46 \times 10^{15} \times \frac{1}{10^{-10}}$$

$$a = \frac{3.26 \times 9.46 \times 10^{15}}{10^{-10}}$$

$= (3.26 \times 9.46) \times \dfrac{10^{15}}{10^{-10}}$

$= 30.8396 \times 10^{25}$

$= (3.08396 \times 10) \times 10^{25}$

$= 3.08396 \times (10 \times 10^{25})$

$\approx 3.08 \times 10^{26}$ Rounding to 3 significant digits

The check is left to the student.

State. There are about 3.08×10^{26} Angstroms in one parsec.

58. 3.08×10^{13} km

59. Familiarize. We have a very long cylinder. Its length is the average distance from the earth to the sun, 1.5×10^{11} m, and the diameter of its base is 3Å. We will use the formula for the volume of a cylinder, $V = \pi r^2 h$. (See Example 8.)

Translate. We will express all distances in Angstroms.

Height (length): $1.5 \times 10^{11}\text{ m} = \frac{1.5 \times 10^{11}}{10^{-10}}$ Å, or

1.5×10^{21} Å

Diameter: 3Å

The radius is half the diameter:

Radius: $\frac{1}{2} \times 3\text{Å} = 1.5$ Å

Now substitute into the formula (using 3.14 for π):

$V = \pi r^2 h$

$V = 3.14 \times 1.5^2 \times 1.5 \times 10^{21}$

Carry out. Do the calculations.

$V = 3.14 \times 1.5^2 \times 1.5 \times 10^{21}$

$= 10.5975 \times 10^{21}$

$= 1.05975 \times 10^{22}$

$= 1 \times 10^{22}$ Rounding to 1 significant digit

Check. In this case, about all we can do is recheck the translation and the calculations.

State. The volume of the sunbeam is about 1×10^{22} cu Å.

60. 3×10^{22} cu Å

61. Familiarize. When the roll of plastic is unrolled, its length, width, and height (thickness) are 30 m, 1 m, and 0.8 mm, respectively. We will use the formula for the volume of a rectangular solid, $V = \ell wh$, where ℓ is length, w is width, and h is height.

Translate. We will express all dimensions in meters.

Length: 30 m

Width: 1 m

Height: $0.8\text{ mm} = 0.8 \times 10^{-3}$ m, or 8×10^{-4} m

Substitute into the formula:

$V = \ell wh$

$= 30 \times 1 \times 8 \times 10^{-4}$

Carry out. Do the calculations.

$V = 30 \times 1 \times 8 \times 10^{-4}$

$= 240 \times 10^{-4}$

$= 2.4 \times 10^{-2}$

$= 2. \times 10^{-2}$ Rounding to 1 significant digit; the decimal point emphasizes that we didn't round to the nearest tenth.

Check. Recheck the translation and the calculations.

State. The volume of plastic in a roll is $2. \times 10^{-2}\text{ m}^3$, or 0.02 m^3.

62. 1.5×10^{12} mi

63. Familiarize. We can use a circle whose radius is the average distance of the earth from the sun to approximate the earth's orbit about the sun. The distance the earth travels in a yearly orbit about the sun is given by the circumference of that circle. Recall that the formula for the circumference of a circle is $C = 2\pi r$, where r is the radius.

Translate. Substitute 3.14 for π and 9.3×10^7 for r in the formula.

$C = 2\pi r$

$= 2 \times 3.14 \times 9.3 \times 10^7$

Carry out. Do the calculations.

$C = 2 \times 3.14 \times 9.3 \times 10^7$

$= 58.404 \times 10^7$

$= 5.8404 \times 10^8$

$= 5.8 \times 10^8$ Rounding to 2 significant digits

Check. Recheck the translation and the calculations.

State. The earth travels approximately 5.8×10^8 mi in a yearly orbit about the sun.

64. \$67,000

65. $-\frac{5}{6} - \left(-\frac{3}{4}\right) = -\frac{5}{6} + \frac{3}{4} = -\frac{10}{12} + \frac{9}{12} = -\frac{1}{12}$

66. 30.96

67. $-2(x - 3) - 3(4 - x) = -2x + 6 - 12 + 3x$

$= (-2x + 3x) + (6 - 12)$

$= x - 6$

68. $\frac{7}{4}$

69.

70.

71. The larger number is the one in which the power of ten has the larger exponent. Since -90 is larger than -91, $8 \cdot 10^{-90}$ is larger than $9 \cdot 10^{-91}$.

$8 \cdot 10^{-90} - 9 \cdot 10^{-91} = 10^{-90}(8 - 9 \cdot 10^{-1})$

$= 10^{-90}(8 - 0.9)$

$= 7.1 \times 10^{-90}$

72. 1.25×10^{22}

73. $(4096)^{0.05}(4096)^{0.2} = 4096^{0.25}$

$= (2^{12})^{0.25}$

$= 2^3$

$= 8$

74. 1

75. Familiarize. Observe that there are 2^{n-1} grains of sand on the nth square of the chessboard. Let g represent this quantity. Recall that a chessboard has 64 squares. Note also that $2^{10} \approx 10^3$.

Translate. We write the equation

$$g = 2^{n-1}.$$

To find the number of grains of sand on the last (or 64th) square, substitute 64 for n: $g = 2^{64-1}$

Carry out. Do the calculations, expressing the result in scientific notation.

$$g = 2^{64-1} = 2^{63} = 2^3(2^{10})^6$$

$$\approx 2^3(10^3)^6 \approx 8 \times 10^{18}$$

Check. Recheck the translation and the calculations.

State. Approximately 8×10^{18} grains of sand are required for the last square.

Exercise Set 2.1

1.

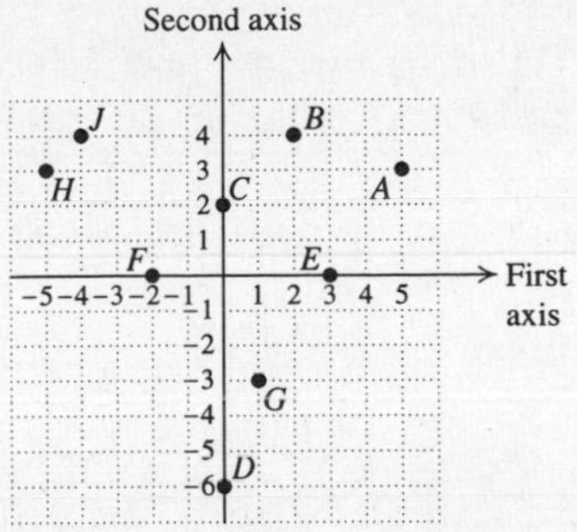

A(5,3) is 5 units right and 3 units up.
B(2,4) is 2 units right and 4 units up.
C(0,2) is 0 units left or right and 2 units up.
D(0,-6) is 0 units left or right and 6 units down.
E(3,0) is 3 units right and 0 units up or down.
F(-2,0) is 2 units left and 0 units up or down.
G(1,-3) is 1 unit right and 3 units down.
H(-5,3) is 5 units left and 3 units up.
J(-4,4) is 4 units left and 4 units up.

2.

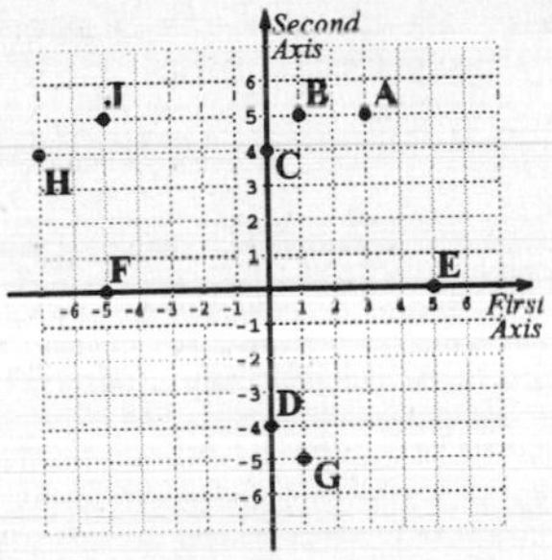

3.

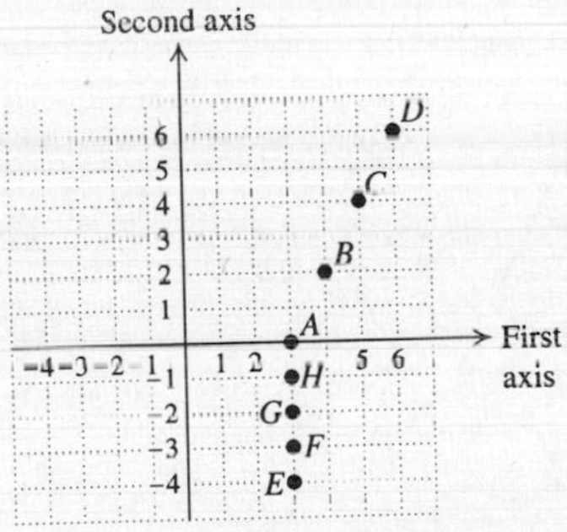

A(3,0) is 3 units right and 0 units up or down.
B(4,2) is 4 units right and 2 units up.
C(5,4) is 5 units right and 4 units up.
D(6,6) is 6 units right and 6 units up.
E(3,-4) is 3 units right and 4 units down.
F(3,-3) is 3 units right and 3 units down.
G(3,-2) is 3 units right and 2 units down.
H(3,-1) is 3 units right and 1 unit down.

4.

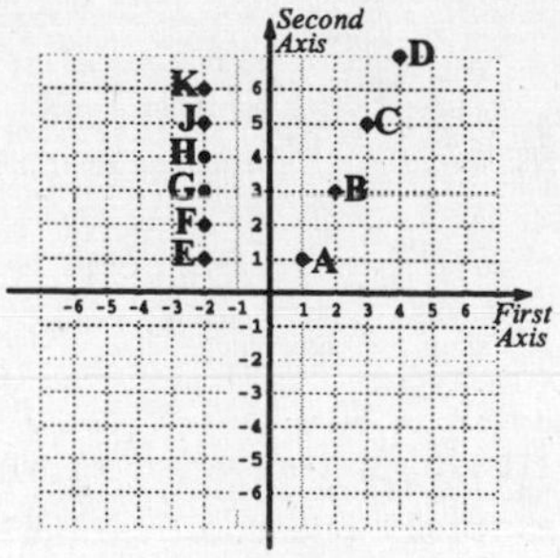

5.

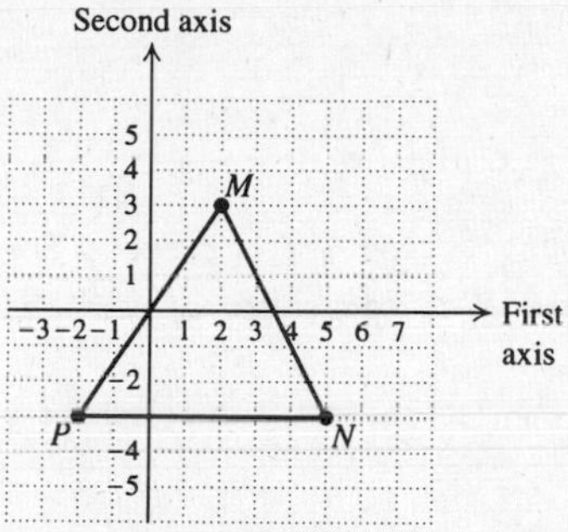

A triangle is formed. The area of a triangle is found by using the formula $A = \frac{1}{2}bh$. In this triangle the base and height are 7 units and 6 units, respectively.

$A = \frac{1}{2}bh = \frac{1}{2}\cdot 7\cdot 6 = \frac{42}{2} = 21$ square units

6.

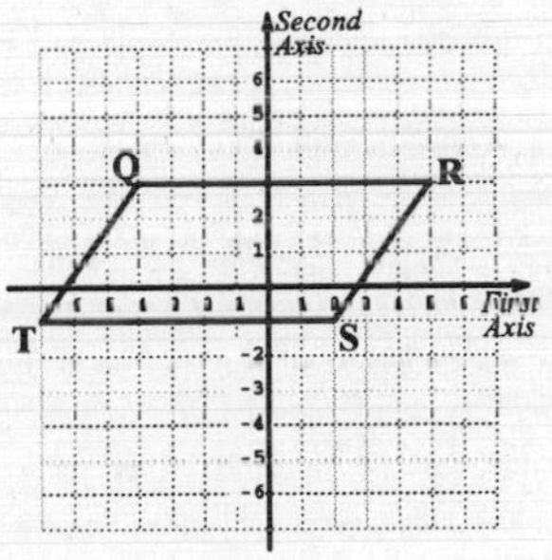

A parallelogram is formed.

A = 36 square units

7. Both coordinates are negative, so the point (-3,-5) is in quadrant III.

8. I

9. The first coordinate is negative and the second positive, so the point (-6,1) is in quadrant II.

10. IV

11. Both coordinates are positive, so the point $\left(3,\frac{1}{2}\right)$ is in quadrant I.

12. III

13. The first coordinate is positive and the second negative, so the point (7,-0.2) is in quadrant IV.

14. II

15. $y = 2x - 3$

$-1 \; ? \; 2\cdot1 - 3$ Substituting 1 for x and -1 for y

$\quad\; | \; -1$ (alphabetical order of variables)

Since $-1 = -1$ is true, (1,-1) is a solution of $y = 2x - 3$.

16. Yes

17. $3s + t = 4$

$3\cdot3 + 4 \; ? \; 4$ Substituting 3 for s and 4 for t

$9 + 4$ (alphabetical order of variables)

13

Since $13 = 4$ is false, (3,4) is not a solution of $3s + t = 4$.

18. No

19. $4x - y = 7$

$4\cdot3 - 5 \; ? \; 7$ Substituting 3 for x and 5 for y

$12 - 5$ (alphabetical order of variables)

7

Since $7 = 7$ is true, (3,5) is a solution of $4x - y = 7$.

20. Yes

21. $2a + 5b = 3$

$2\cdot0 + 5\cdot\frac{3}{5} \; ? \; 3$ Substituting 0 for a and $\frac{3}{5}$ for b

$0 + 3$ (alphabetical order of variables)

3

Since $3 = 3$ is true, $\left(0,\frac{3}{5}\right)$ is a solution of $2a + 5b = 3$.

22. Yes

23. $4r + 3s = 5$

$4\cdot2 + 3\cdot(-1) \; ? \; 5$ Substituting 2 for r and -1 for s (alphabetical order of variables)

$8 - 3$

5

Since $5 = 5$ is true, (2,-1) is a solution of $4r + 3s = 5$.

24. Yes

25. $3x - 2y = -4$

$3\cdot3 - 2\cdot2 \; ? \; -4$ Substituting 3 for x and 2 for y

$9 - 4$ (alphabetical order of variables)

5

Since $5 = -4$ is false, (3,2) is not a solution of $3x - 2y = -4$.

26. No

27. $y = 3x^2$

$3 \; ? \; 3(-1)^2$ Substituting -1 for x and 3 for y

$3\cdot1$ (alphabetical order of variables)

3

Since $3 = 3$ is true, (-1,3) is a solution of $y = 3x^2$.

28. No

29. $5s^2 - t = 7$

$5(2)^2 - 3 \; ? \; 7$ Substituting 2 for s and 3 for t

$5\cdot4 - 3$ (alphabetical order of variables)

$20 - 3$

17

Since $17 = 7$ is false, (2,3) is not a solution of $5s^2 - t = 7$.

30. Yes

31. $y = -2x$

To find an ordered pair, we choose any number for x and then determine y by substitution.

When $x = 0$, $y = -2\cdot0 = 0$.

When $x = 3$, $y = -2\cdot3 = -6$.

When $x = -2$, $y = -2\cdot(-2) = 4$.

x	y	(x,y)
0	0	(0,0)
3	-6	(3,-6)
-2	4	(-2,4)

Plot these points, draw the line they determine, and label the graph $y = -2x$.

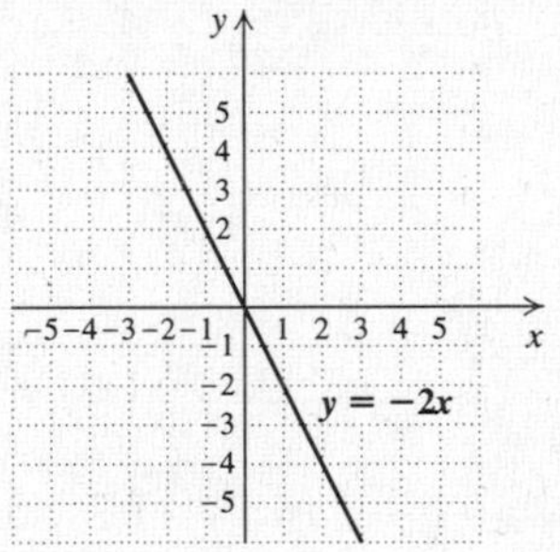

32.

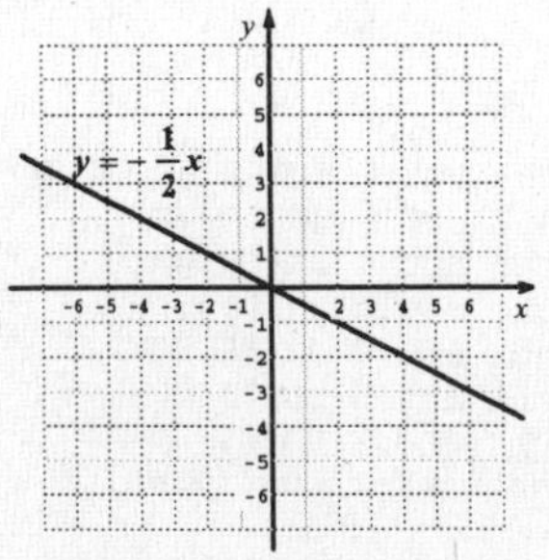

33. y = x + 3

To find an ordered pair we choose any number for x and then determine y. For example, if we choose 1 for x, then y = 1 + 3, or 4. We find several ordered pairs, plot them, and draw the line.

x	y	(x,y)
1	4	(1,4)
2	5	(2,5)
-1	2	(-1,2)
-3	0	(-3,0)

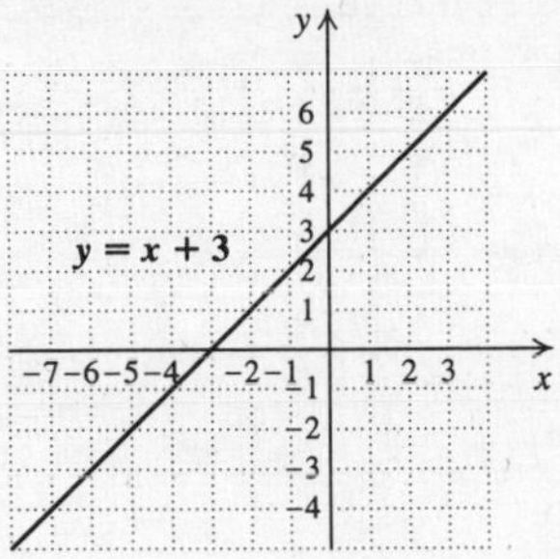

34.

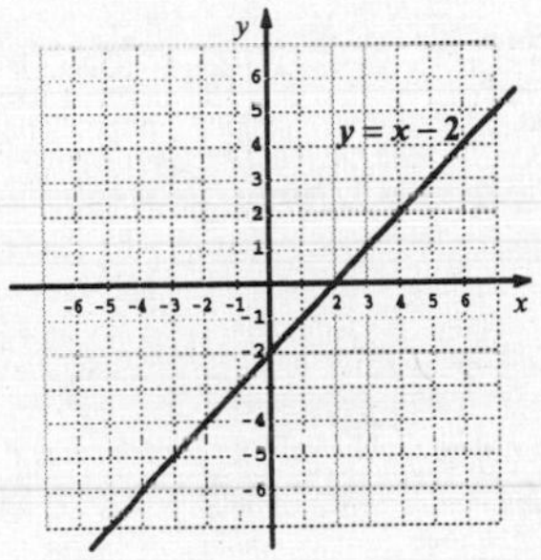

35. y = 3x - 2

To find an ordered pair, we choose any number for x and then determine y. For example, if x = 2, then y = 3·2 - 2 = 6 - 2 = 4. We find several ordered pairs, plot them, and draw the line.

x	y	(x,y)
2	4	(2,4)
0	-2	(0,-2)
-1	-5	(-1,-5)
1	1	(1,1)

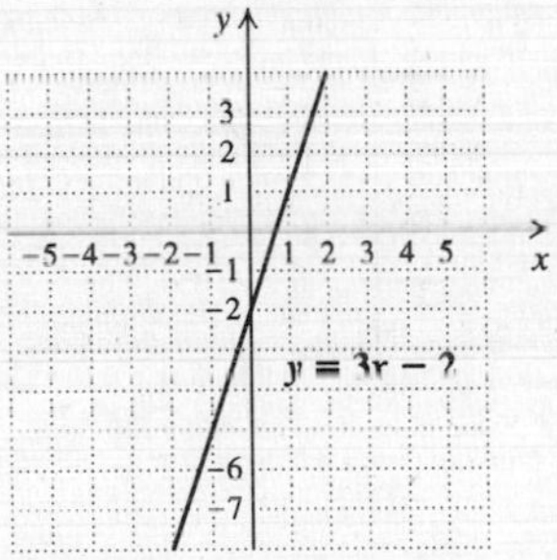

36.

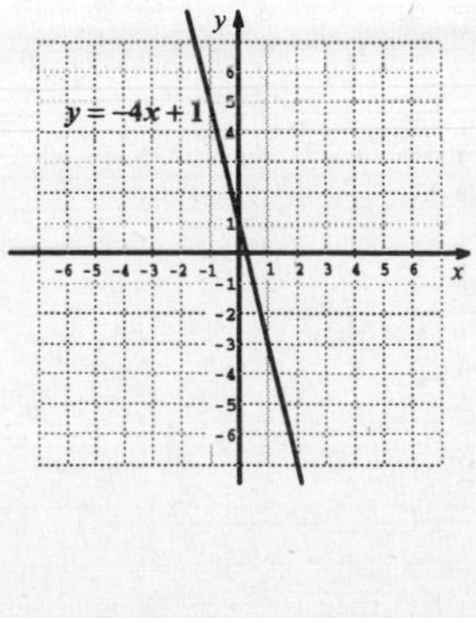

37. y = -2x + 3

To find an ordered pair, we choose any number for x and then determine y. For example, if x = 1, then y = -2·1 + 3 = -2 + 3 = 1. We find several ordered pairs, plot them, and draw the line.

x	y
1	1
3	-3
-1	5
0	3

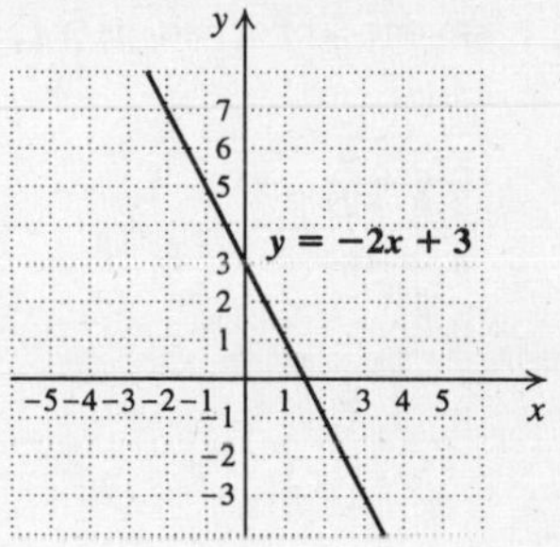

38.

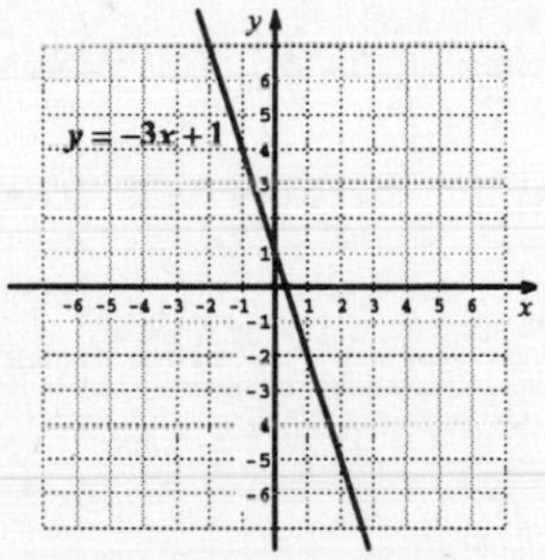

39. $y = \frac{2}{3}x + 1$

To find an ordered pair, we choose any number for x and then determine y. For example, if x = 3, then $y = \frac{2}{3} \cdot 3 + 1 = 2 + 1 = 3$. We find several ordered pairs, plot them, and draw the line.

x	y
3	3
0	1
-3	-1

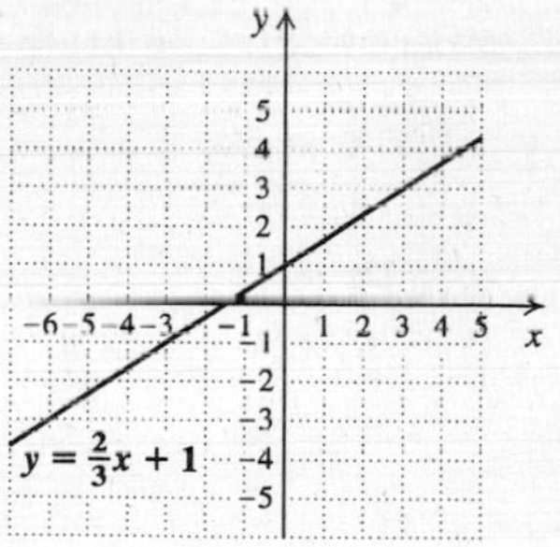

40.

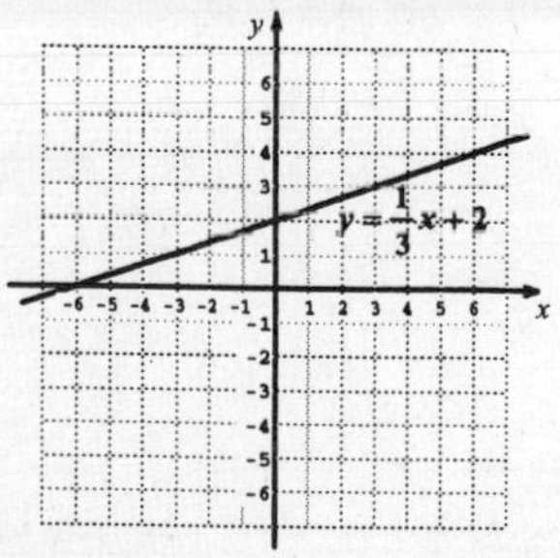

41. $y = -\frac{3}{2}x + 1$

To find an ordered pair, we choose any number for x and then determine y. For example, if x = 2, then $y = -\frac{3}{2} \cdot 2 + 1 = -3 + 1 = -2$. We find several ordered pairs, plot them, and draw the line.

x	y
2	-2
4	-5
0	1
-2	4

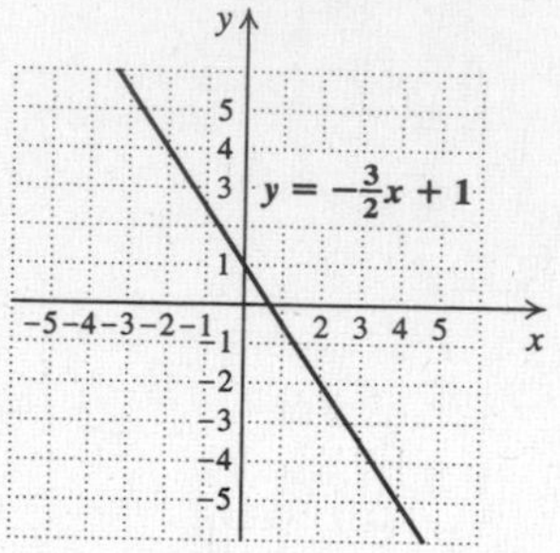

42.

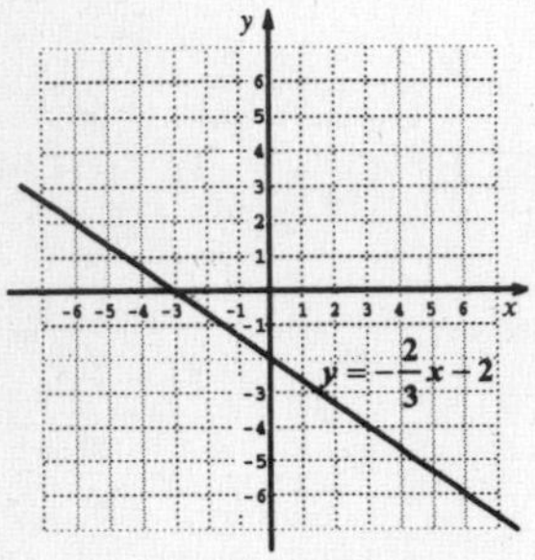

43. $y = \frac{3}{4}x + 1$

To find an ordered pair, we choose any number for x and then determine y. For example, if x = 4, $y = \frac{3}{4} \cdot 4 + 1 = 3 + 1 = 4$. We find several ordered pairs, plot them and draw the line.

x	y
4	4
0	1
-4	-2

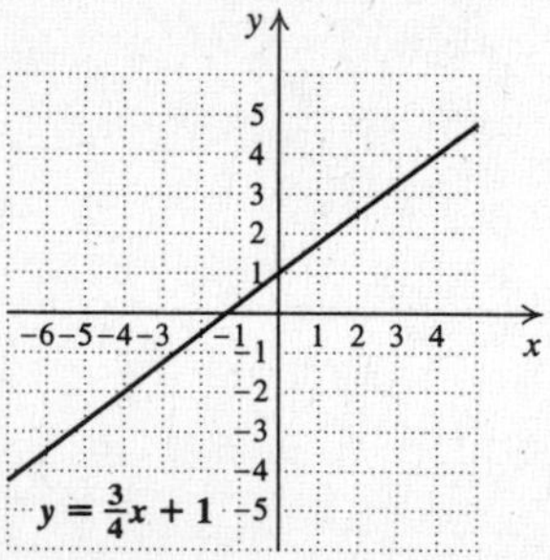

44.

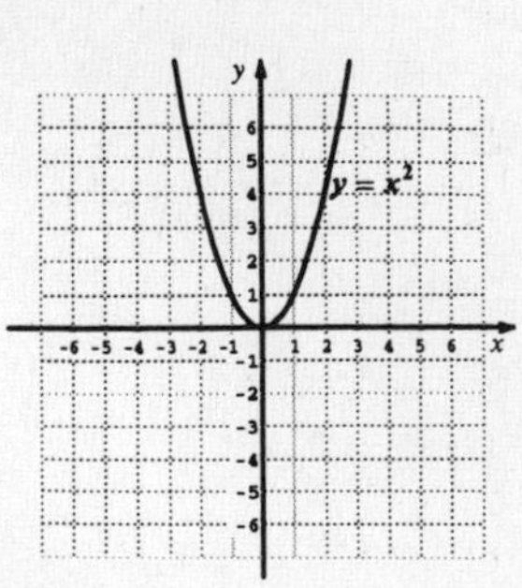

45. $y = -x^2$

To find an ordered pair, we choose any number for x and then determine y. For example, if x = 2, then $y = -(2)^2 = -4$. We find several ordered pairs, plot them, and connect them with a smooth curve.

x	y
2	-4
1	-1
0	0
-1	-1
-2	-4

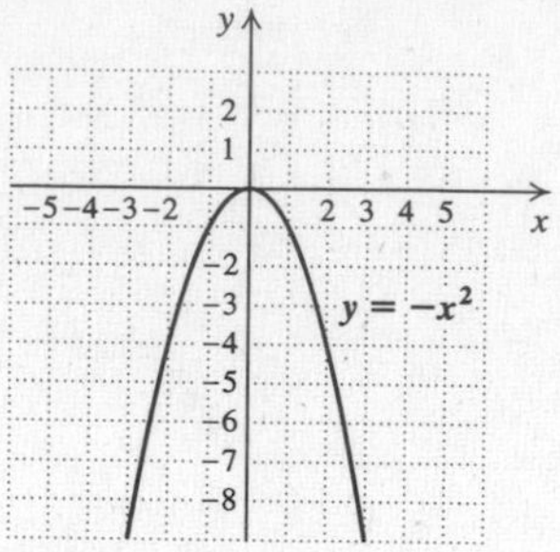

46.

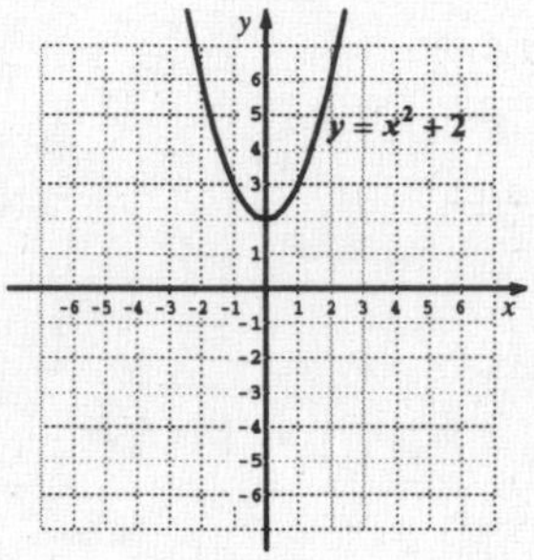

47. $y = x^2 - 2$

To find an ordered pair, we choose any number for x and then determine y. For example, if x = 2, $y = 2^2 - 2 = 4 - 2 = 2$. We find several ordered pairs, plot them, and connect them with a smooth curve.

x	y
2	2
1	-1
0	-2
-1	-1
-2	2

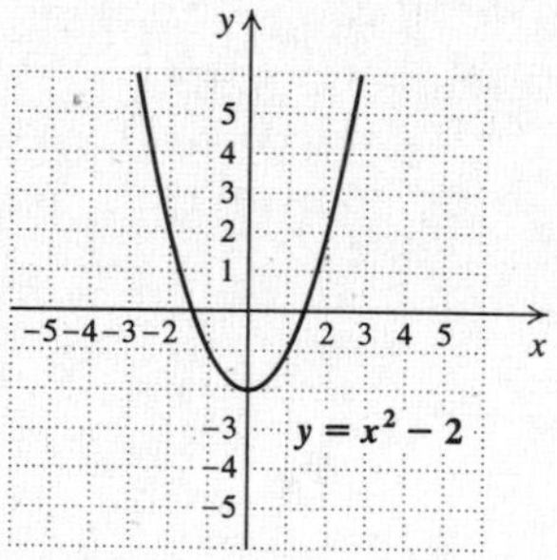

48.

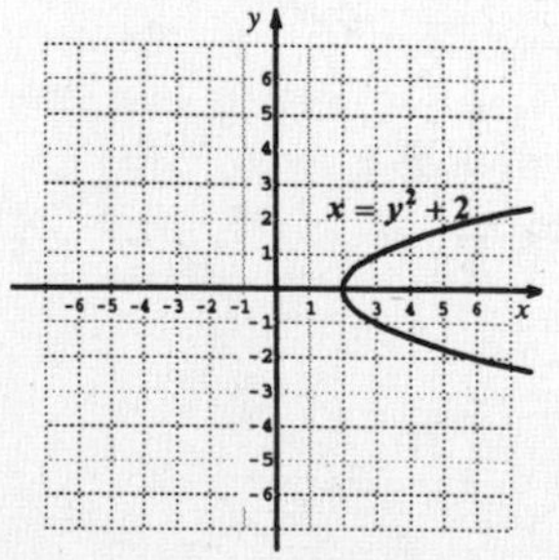

49. $y = |x| + 2$

We select x-values and find the corresponding y-values. The table lists some ordered pairs.

x	y
3	5
1	3
0	2
-1	3
-3	5

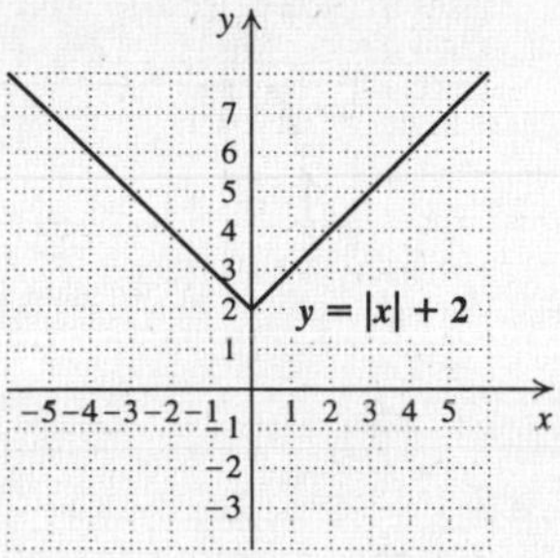

We plot these points. Note that the graph is V-shaped, centered at (0,2).

50.

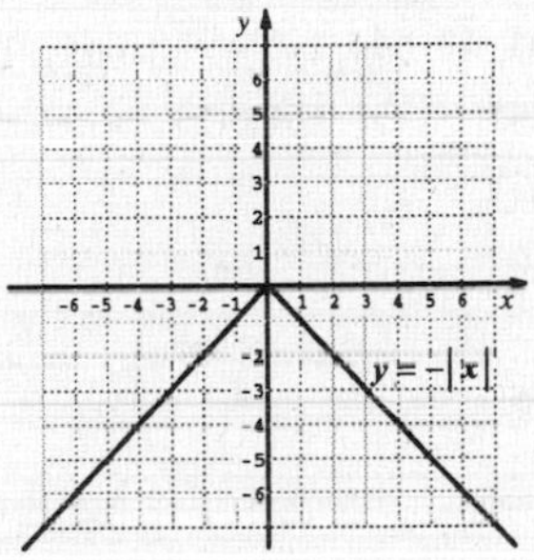

51. $y = 3 - x^2$

To find an ordered pair, we choose any number for x and then determine y. For example, if x = 2, $y = 3 - 2^2 = 3 - 4 = -1$. We find several ordered pairs, plot them, and connect them with a smooth curve.

x	y
2	-1
1	2
0	3
-1	2
-2	-1

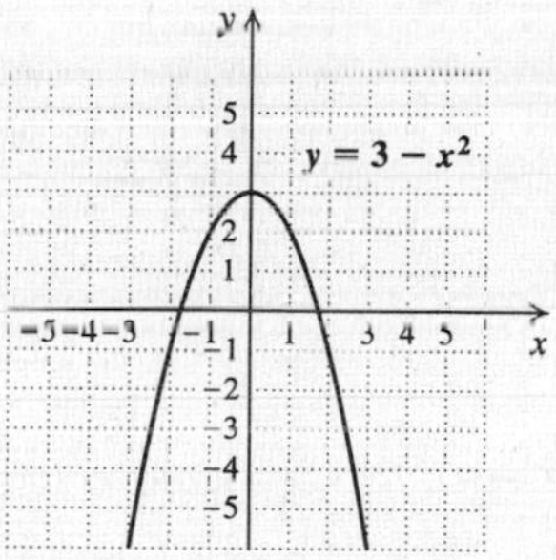

52.

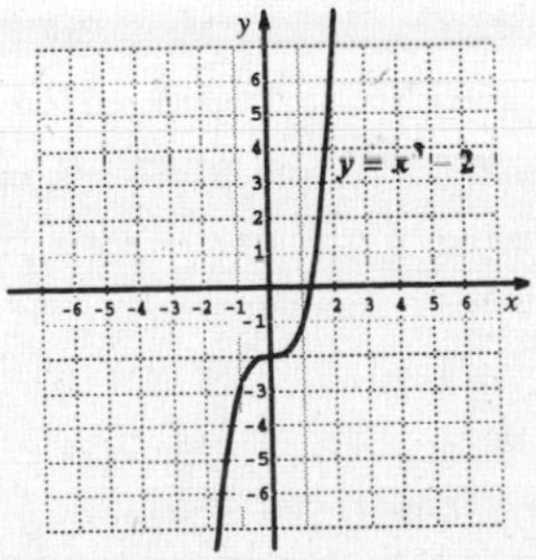

53. $y = -\frac{1}{x}$

We select x-values and find the corresponding y-values. The table lists some ordered pairs.

x	y
4	$-\frac{1}{4}$
2	$-\frac{1}{2}$
1	-1
$\frac{1}{2}$	-2
$-\frac{1}{2}$	2
-1	1
-2	$\frac{1}{2}$
-4	$\frac{1}{4}$

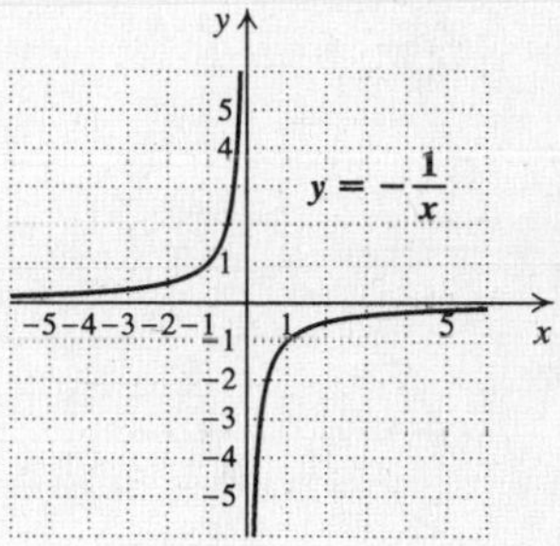

We plot these points. Note that we cannot use 0 as a first-coordinate, since -1/0 is undefined. Thus the graph has two branches, one on each side of the y-axis.

54.

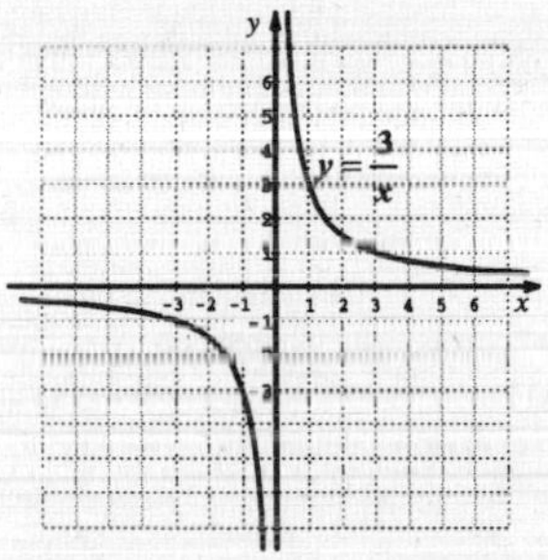

55. Familiarize. The formula for the area of a triangle with base b and height h is $A = \frac{1}{2}bh$.

Translate. Substitute 156 for A and 12 for b in the formula.

$$A = \frac{1}{2}bh$$

$$156 = \frac{1}{2} \cdot 12 \cdot h$$

Carry out. Solve the equation.

$$156 = \frac{1}{2} \cdot 12 \cdot h$$

$$156 = 6h$$

$$26 = h$$

Check. Left to the student.

State. The triangle should be 26 ft tall.

56. 11%

57. $-3.9 - (-2.5) = -3.9 + 2.5 = -1.4$

58. 5

59\.

60\.

61\.

62\. (-1,-2)

63\. Substitute $-\frac{1}{3}$ for x and $\frac{1}{4}$ for y in each equation.

a)
$$-\frac{3}{2}x - 3y = -\frac{1}{4}$$
$$-\frac{3}{2}\left(-\frac{1}{3}\right) - 3\left(\frac{1}{4}\right) \ ? \ -\frac{1}{4}$$
$$\frac{1}{2} - \frac{3}{4}$$
$$-\frac{1}{4}$$

Since $-\frac{1}{4} = -\frac{1}{4}$ is true, $\left(-\frac{1}{3}, \frac{1}{4}\right)$ is a solution.

b)
$$8y - 15x = \frac{7}{2}$$
$$8\left(\frac{1}{4}\right) - 15\left(-\frac{1}{3}\right) \ ? \ \frac{7}{2}$$
$$2 + 5$$
$$7$$

Since $7 = \frac{7}{2}$ is false, $\left(-\frac{1}{3}, \frac{1}{4}\right)$ is not a solution.

c)
$$0.16y = -0.09x + 0.1$$
$$0.16\left(\frac{1}{4}\right) \ ? \ -0.09\left(-\frac{1}{3}\right) + 0.1$$
$$0.04 \ \big| \ 0.03 + 0.1$$
$$0.13$$

Since 0.04 = 0.13 is false, $\left(-\frac{1}{3}, \frac{1}{4}\right)$ is not a solution.

d)
$$2(-y + 2) - \frac{1}{4}(3x - 1) = 4$$
$$2\left(-\frac{1}{4} + 2\right) - \frac{1}{4}\left[3\left(-\frac{1}{3}\right) - 1\right] \ ? \ 4$$
$$2\left(\frac{7}{4}\right) - \frac{1}{4}(-2)$$
$$\frac{14}{4} + \frac{2}{4}$$
$$\frac{16}{4}$$
$$4$$

Since 4 = 4 is true, $\left(-\frac{1}{3}, \frac{1}{4}\right)$ is a solution.

64\.

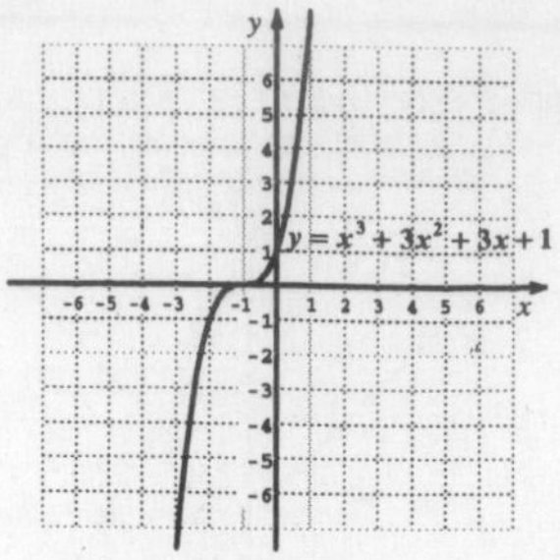

65\. $y = \frac{1}{x - 2}$

Choose x-values from -1 to 5, and use a calculator to find the corresponding y-values. (Note that we cannot choose 2 as a first coordinate since $\frac{1}{2 - 2}$, or $\frac{1}{0}$, is not defined.) Plot the points, and draw the graph. Note that it has two branches, one on each side of a vertical line through (2,0).

x	y
-1	$-0.\overline{3}$
-0.5	-0.4
0	-0.5
1	-1
1.9	-10
2.1	10
2.5	2
3	1
4	0.5
5	$0.\overline{3}$

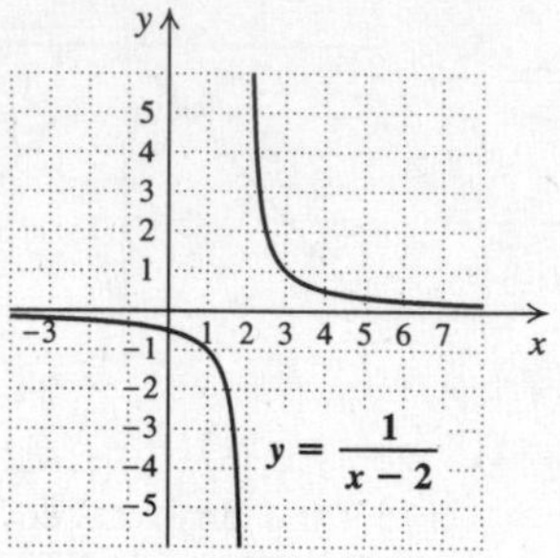

66\.

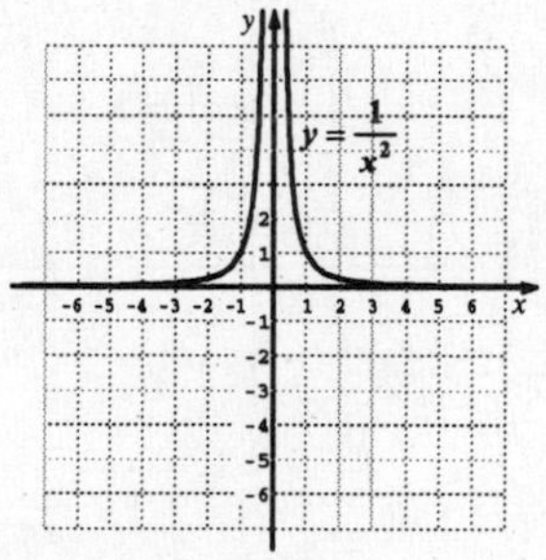

67\. $y = -\frac{1}{x^2}$

Choose x-values from -3 to 3, and use a calculator to find the corresponding y-values. [Note that we cannot choose 0 as a first coordinate, since $-\frac{1}{0^2}$, or $-\frac{1}{0}$, is not defined.] Plot the points, and draw the graph. Note that it has two branches, one on each side of the y-axis.

x	y
-3	$-\frac{1}{9}$
-2	$-\frac{1}{4}$
-1	-1
-0.5	-4
-0.25	-16
0.25	-16
0.5	-4
1	-1
2	$-\frac{1}{4}$
3	$-\frac{1}{9}$

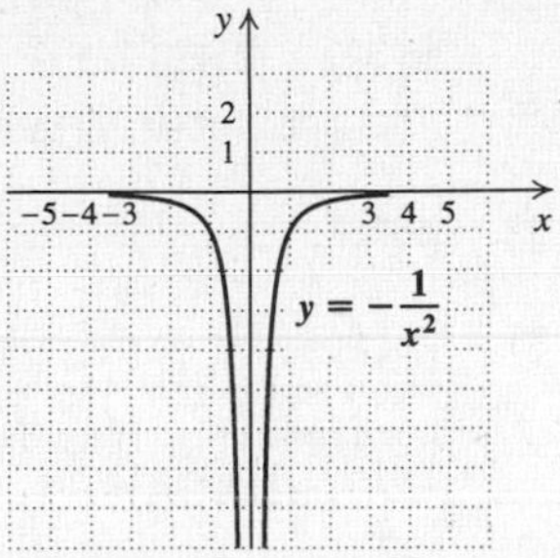

68. (-1,-4), (4,1)

69.

70. a) $y = -12.4x + 7.8$

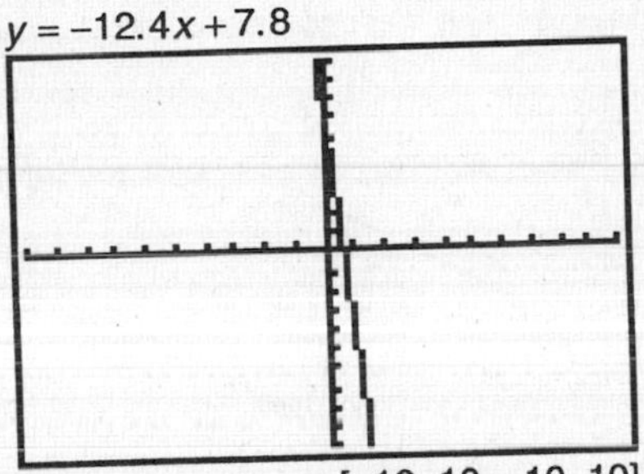

[-10, 10, -10, 10]

b) $y = -3.5x^2 + 6x - 8$

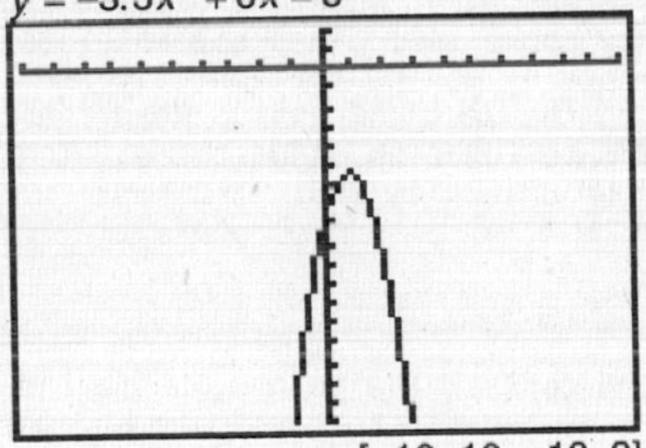

[-10, 10, -18, 2]

c) $y = (x - 3.4)^3 + 5.6$

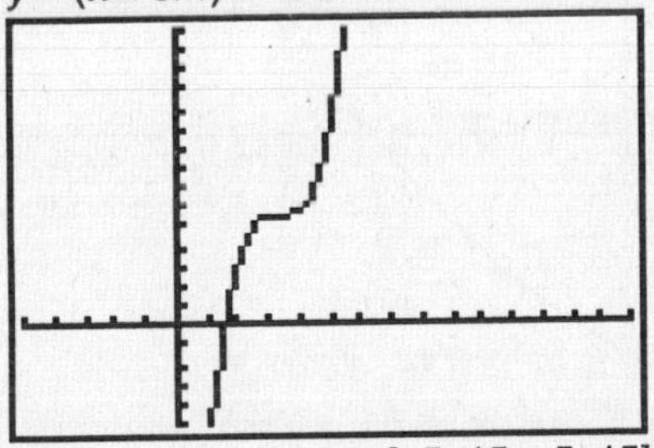

[-5, 15, -5, 15]

71. a), b), c) See the answer section in the text.

Exercise Set 2.2

1. The correspondence is not a function, because a member of the domain (a) corresponds to more than one member of the range.

2. Yes

3. The correspondence is a function, because each member of the domain corresponds to just one member of the range.

4. Yes

5. The correspondence is a function, because each member of the domain corresponds to just one member of the range.

6. No

7. The correspondence is not a function, because a member of the domain (Viola) corresponds to more than one member of the range.

8. No

9. The correspondence is a function, because each member of the domain corresponds to just one member of the range.

10. Yes

11. This correspondence is a function, because each class member has only one seat number.

12. Function

13. This correspondence is a function, because each shape has only one number for its area.

14. Function

15. The correspondence is not a function, since it is reasonable to assume that at least one person in the town has more than one aunt.

    The correspondence is a relation, since it is reasonable to assume that each person in the town has at least one aunt.

16. A relation but not a function

17. $g(x) = x + 1$
    a) $g(0) = 0 + 1 = 1$
    b) $g(-4) = -4 + 1 = -3$
    c) $g(-7) = -7 + 1 = -6$
    d) $g(8) = 8 + 1 = 9$
    e) $g(a + 2) = a + 2 + 1 = a + 3$

18. a) 0, b) 4, c) -7, d) -8, e) a - 5

19. $f(n) = 5n^2 + 4$

a) $f(0) = 5(0)^2 + 4 = 0 + 4 = 4$

b) $f(-1) = 5(-1)^2 + 4 = 5 + 4 = 9$

c) $f(3) = 5(3)^2 + 4 = 45 + 4 = 49$

d) $f(t) = 5(t)^2 + 4 = 5t^2 + 4$

e) $f(2a) = 5(2a)^2 + 4 = 5\cdot 4a^2 + 4 = 20a^2 + 4$

20. a) -2, b) 1, c) 25, d) $3t^2 - 2$, e) $12a^2 - 2$

21. $g(r) = 3r^2 + 2r - 1$

a) $g(2) = 3(2)^2 + 2(2) - 1 = 12 + 4 - 1 = 15$

b) $g(3) = 3(3)^2 + 2(3) - 1 = 27 + 6 - 1 = 32$

c) $g(-3) = 3(-3)^2 + 2(-3) - 1 = 27 - 6 - 1 = 20$

d) $g(1) = 3(1)^2 + 2(1) - 1 = 3 + 2 - 1 = 4$

e) $g(3r) = 3(3r)^2 + 2(3r) - 1 = 3\cdot 9r^2 + 6r - 1 = 27r^2 + 6r - 1$

22. a) 35, b) 2, c) 7, d) 20, e) $36r^2 - 3r + 2$

23. $f(x) = \dfrac{x - 3}{2x - 5}$

a) $f(0) = \dfrac{0 - 3}{2\cdot 0 - 5} = \dfrac{-3}{0 - 5} = \dfrac{-3}{-5} = \dfrac{3}{5}$

b) $f(4) = \dfrac{4 - 3}{2\cdot 4 - 5} = \dfrac{1}{8 - 5} = \dfrac{1}{3}$

c) $f(-1) = \dfrac{-1 - 3}{2(-1) - 5} = \dfrac{-4}{-2 - 5} = \dfrac{-4}{-7} = \dfrac{4}{7}$

d) $f(3) = \dfrac{3 - 3}{2\cdot 3 - 5} = \dfrac{0}{6 - 5} = \dfrac{0}{1} = 0$

e) $f(x + 2) = \dfrac{x + 2 - 3}{2(x + 2) - 5} = \dfrac{x - 1}{2x + 4 - 5} = \dfrac{x - 1}{2x - 1}$

24. a) $\frac{26}{25}$, b) $\frac{2}{9}$, c) undefined, d) $-\frac{7}{3}$, e) $\dfrac{3x + 5}{2x + 11}$

25. $A(s) = s^2 \dfrac{\sqrt{3}}{4}$

$A(4) = 4^2 \dfrac{\sqrt{3}}{4} = 4\sqrt{3}$

The area is $4\sqrt{3}$ cm².

26. $9\sqrt{3}$ in²

27. $V(r) = 4\pi r^2$

$V(3) = 4\pi(3)^2 = 36\pi$

The area is 36π in² $\approx$ 113.04 in².

28. 314 cm²

29. $F(C) = \frac{9}{5}C + 32$

$F(-10) = \frac{9}{5}(-10) + 32 = -18 + 32 = 14$

The equivalent temperature is 14° F.

30. 41° F

31. $H(x) = 2.75x + 71.48$

$H(32) = 2.75(32) + 71.48 = 159.48$

The predicted height is 159.48 cm.

32. 167.73 cm

33. Locate the point that is directly above 225. Then estimate its second coordinate by moving horizontally from the point to the vertical axis. The rate is about 75 per 10,000 men.

34. 125 per 10,000 men

35. Locate the point that is directly above 1989. Then estimate its second coordinate by moving horizontally from the point to the vertical axis. The number of wood bats sold in 1989 was about 1.4 million.

36. 0.7 million

37. Plot and connect the points, using body weight as the first coordinate and the corresponding number of drinks as the second coordinate.

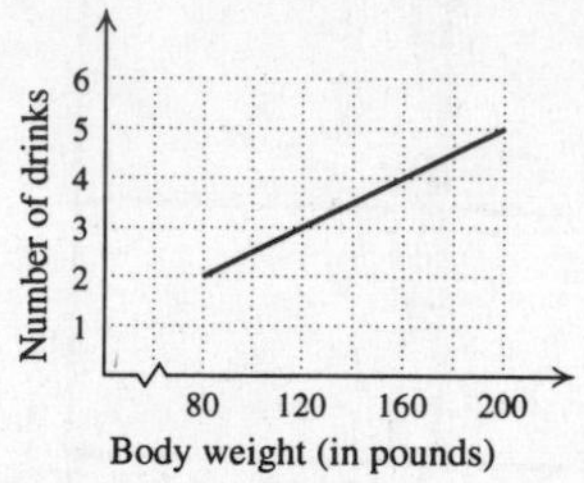

To estimate the number of drinks that a 140-lb person would have to drink to be considered intoxicated, first locate the point that is directly above 140. Then estimate its second coordinate by moving horizontally from the point to the vertical axis. Read the approximate function value there. The estimated number of drinks is 3.5.

38. 3 drinks

39. Plot and connect the points, using the counter reading as the first coordinate and the time of tape as the second coordinate.

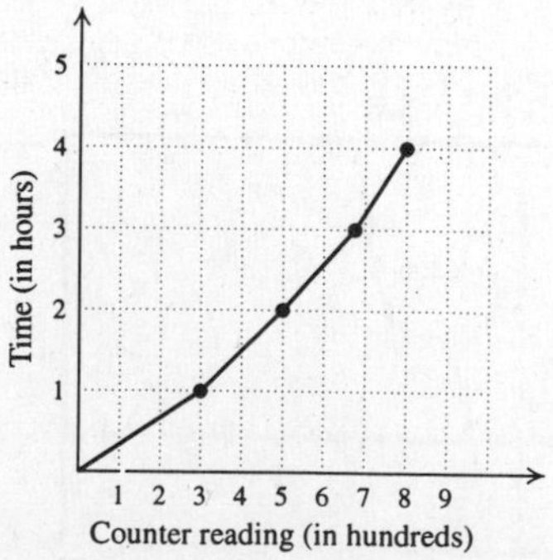

To estimate the time elapsed when the counter has reached 600, first locate the point that is directly above 600. Then estimate its second coordinate by moving horizontally from the point to the vertical axis. Read the approximate function value there. The time elapsed is about 2.5 hr.

40. About 0.5 hr

41. Plot and connect the points, using the year as the first coordinate and the population as the second.

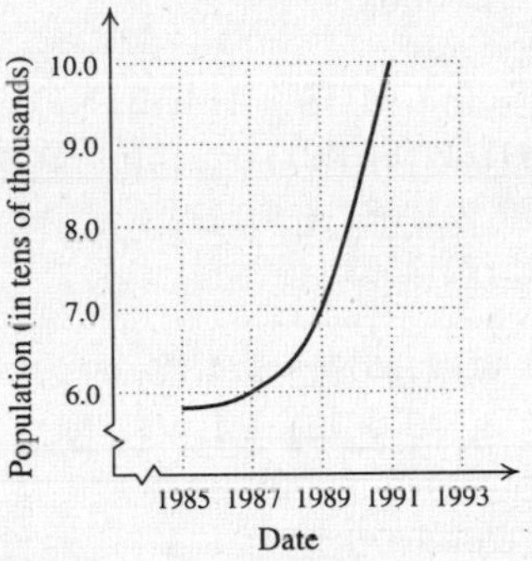

To estimate what the population was in 1988, first locate the point that is directly above 1988. Then estimate its second coordinate by moving horizontally from the point to the vertical axis. Read the approximate function value there. The population was about 64,000.

42. About 150,000

43. Plot and connect the points, using the year as the first coordinate and the sales total as the second coordinate.

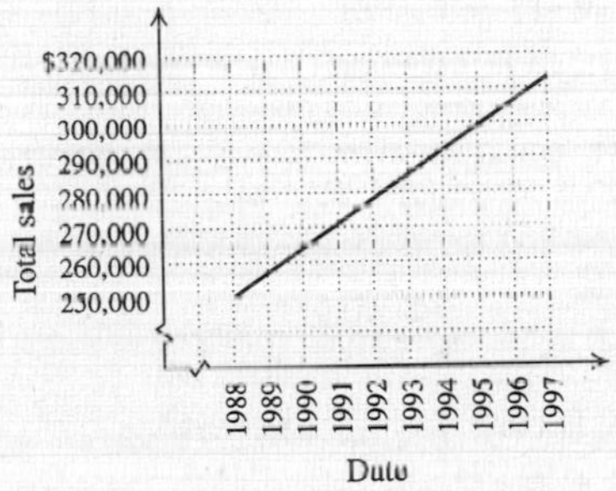

To predict the total sales for 1997, first locate the point directly above 1997. Then estimate its second coordinate by moving horizontally to the vertical axis. Read the approximate function value there. The predicted 1997 sales total is about $310,000.

44. About $270,000

45. We can use the vertical line test:

If it is possible for a vertical line to intersect a graph more than once, the graph is not the graph of a function.

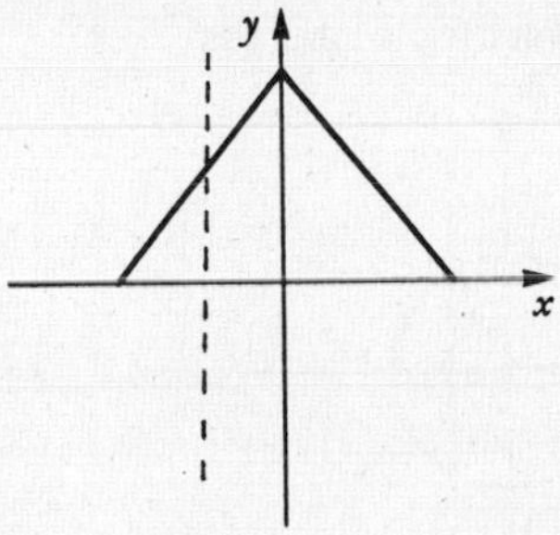

Visualize moving this vertical line across the graph. Ask yourself the question:

Will this line ever intersect the graph more than once?

If the answer is yes, the graph is not a graph of a function. If the answer is no, the graph is a graph of a function.

In this problem the vertical line will not intersect the graph more than once. Thus, the graph is a graph of a function.

46. No

47. We can use the vertical line test:

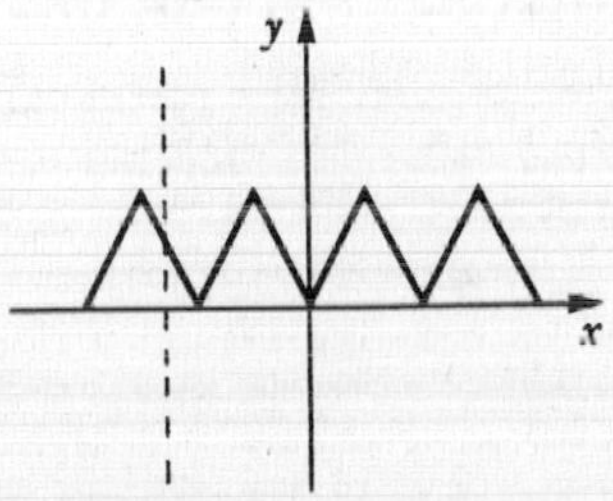

Visualize moving this vertical line across the graph. The vertical line will not intersect the graph more than once. Thus, the graph is a graph of a function.

48. No

49. We can use the vertical line test:

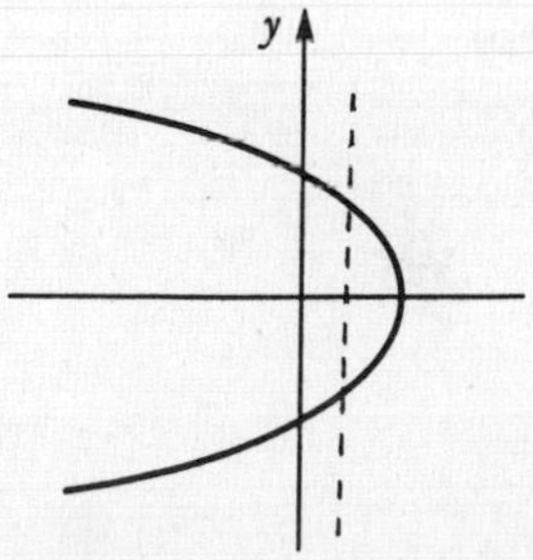

It is possible for a vertical line to intersect

the graph more than once. Thus this is not the graph of a function.

50. Yes

51. We can use the vertical line test.

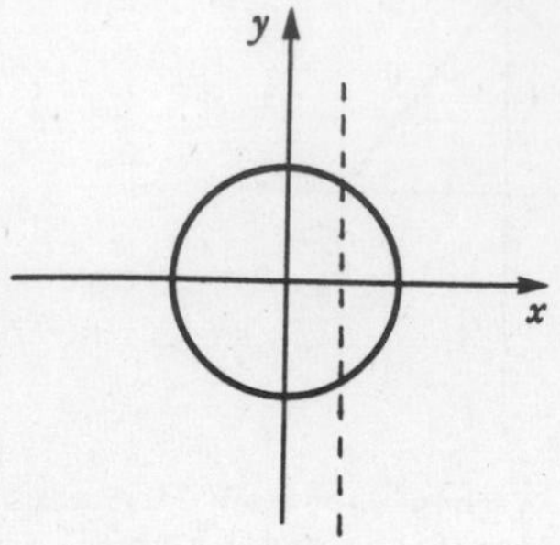

It is possible for a vertical line to intersect the graph more than once. Thus this is not a graph of a function.

52. Yes

53. Familiarize. If x represents the first integer, then x + 2 represents the second integer, and x + 4 represents the third.

Translate. We write an equation.

First integer	plus	two times the second	plus	three times the third	is	124.
x	+	2(x + 2)	+	3(x + 4)	=	124

Carry out. Solve the equation.

$$x + 2(x + 2) + 3(x + 4) = 124$$
$$x + 2x + 4 + 3x + 12 = 124$$
$$6x + 16 = 124$$
$$6x = 108$$
$$x = 18$$

If x = 18, then x + 2 = 18 + 2, or 20, and x + 4 = 18 + 4, or 22.

Check. 18, 20, and 22 are consecutive even integers. Also, $18 + 2\cdot 20 + 3\cdot 22 = 18 + 40 + 66 = 124$. The numbers check.

State. The integers are 18, 20, and 22.

54. 20.175% increase

55.
$$S = 2\ell h + 2\ell w + 2wh$$
$$S - 2wh = 2\ell h + 2\ell w$$
$$S - 2wh = \ell(2h + 2w)$$
$$\frac{S - 2wh}{2h + 2w} = \ell$$

56. $\frac{5}{3}$

57.

58.

59.
$$f(x) = 4.3x^2 - 1.4x$$

a) $f(1.034) = 4.3(1.034)^2 - 1.4(1.034)$
$= 4.3(1.069156) - 1.4(1.034)$
$= 4.5973708 - 1.4476$
$= 3.1497708$

b) $f(-3.441) = 4.3(-3.441)^2 - 1.4(-3.441)$
$= 4.3(11.840481) - 1.4(-3.441)$
$= 50.9140683 + 4.8174$
$= 55.7314683$

c) $f(27.35) = 4.3(27.35)^2 - 1.4(27.35)$
$= 4.3(748.0225) - 1.4(27.35)$
$= 3216.49675 - 38.29$
$= 3178.20675$

d) $f(-16.31) = 4.3(-16.31)^2 - 1.4(-16.31)$
$= 4.3(266.0161) - 1.4(-16.31)$
$= 1143.86923 + 22.834$
$= 1166.70323$

60. a) 11,394.477, b) -582,136.93,
c) 3.5018862, d) -554.4995

61. We know that (-1,-7) and (3,8) are both solutions of g(x) = mx + b. Substituting, we have

$-7 = m(-1) + b$, or $-7 = -m + b$,
and $8 = m(3) + b$, or $8 = 3m + b$.

Solve the first equation for b and substitute that expression into the second equation.

$-7 = -m + b$ First equation
$m - 7 = b$ Solving for b
$8 = 3m + b$ Second equation
$8 = 3m + (m - 7)$ Substituting
$8 = 3m + m - 7$
$8 = 4m - 7$
$15 = 4m$
$\frac{15}{4} = m$

We know that m - 7 = b, so $\frac{15}{4} - 7 = b$, or $-\frac{13}{4} = b$.

Then we have $m = \frac{15}{4}$, $b = -\frac{13}{4}$. We can express g(x) as $g(x) = \frac{15}{4}x - \frac{13}{4}$.

62. 35

63.

64. 22 mm

65. Locate the highest point on the graph. For this graph, there are two equally high points that are higher than all the other points of the graph. Then move down a vertical line through each point to the horizontal axis and read the corresponding times.

The times are 2 min, 40 sec and 5 min, 40 sec.

66.

67. From Exercise 65 we know that the first largest contraction occurred at 2 min, 40 sec, and the second occurred at 5 min, 40 sec. Find the time between the contractions:

5 min	40 sec
- 2 min	40 sec
3 min	

Then the frequency of the largest contractions is 1 contraction every 3 min.

68.

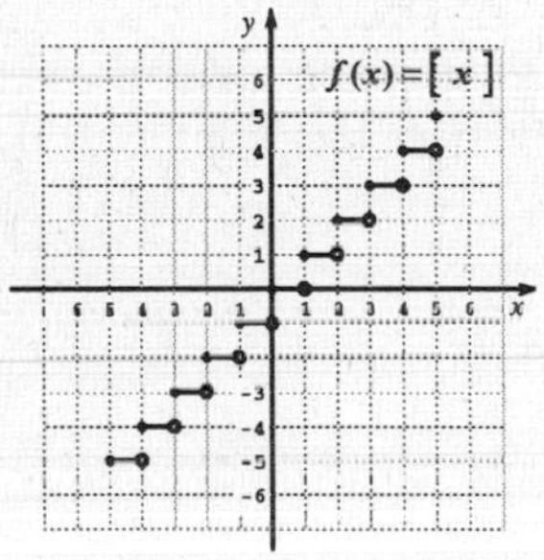

Exercise Set 2.3

1. $y = 4x + 5$

↑ ↑

$y = mx + b$

The slope is 4, and the y-intercept is (0,5), or 5.

2. Slope is 5; y-intercept is 3.

3. $f(x) = -2x - 6$

↑ ↑

$f(x) = mx + b$

The slope is -2, and the y-intercept is (0,-6), or -6.

4. Slope is -5; y-intercept is 7.

5. $y = -\frac{3}{8}x - 0.2$

↑ ↑

$y = mx + b$

The slope is $-\frac{3}{8}$, and the y-intercept is (0,-0.2), or -0.2.

6. Slope is $\frac{15}{7}$; y-intercept is 2.2.

7. $g(x) = 0.5x - 9$

↑ ↑

$g(x) = m x + b$

The slope is 0.5, and the y-intercept is (0,-9), or -9.

8. Slope is -3.1; y-intercept is 5.

9. $y = 7$

Think of this as $y = 0 \cdot x + 7$.

↑ ↑

$y = m x + b$

The slope is 0, and the y-intercept is (0,7), or 7.

10. Slope is 0; y-intercept is -2.

11. Use the slope-intercept equation, $y = mx + b$, with $m = \frac{2}{3}$ and $b = -7$.

$y = mx + b$

$y = \frac{2}{3}x + (-7)$

$y = \frac{2}{3}x - 7$

12. $y = -\frac{3}{4}x + 5$

13. Use the slope-intercept equation, $y = mx + b$, with $m = -4$ and $b = 2$.

$y = mx + b$

$y = -4x + 2$

14. $y = 2x - 1$

15. Use the slope-intercept equation, $y = mx + b$, with $m = -\frac{7}{9}$ and $b = 3$.

$y = mx + b$

$y = -\frac{7}{9}x + 3$

16. $y = -\frac{4}{11}x + 9$

17. Use the slope-intercept equation, $y = mx + b$, with $m = 5$ and $b = \frac{1}{2}$.

$y = mx + b$

$y = 5x + \frac{1}{2}$

18. $y = 6x + \frac{2}{3}$

19. $y = \frac{5}{2}x + 1$

Slope is $\frac{5}{2}$; y-intercept (0,1).

From the y-intercept, we go up 5 units and to the right 2 units. This gives us the point (2,6). We can now draw the graph.

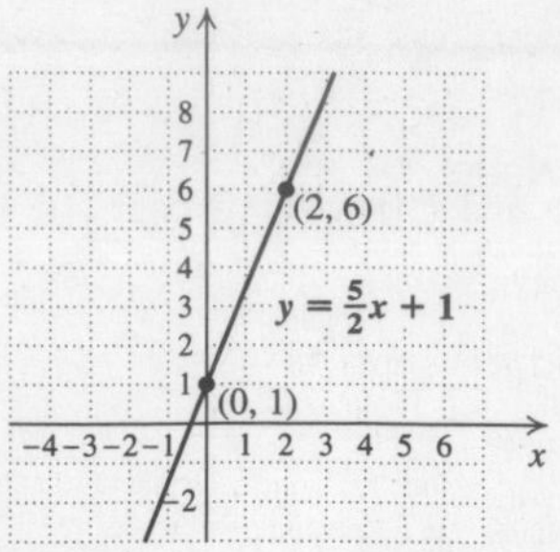

As a check, we can rename the slope and find another point.

$$\frac{5}{2} = \frac{5}{2} \cdot \frac{-1}{-1} = \frac{-5}{-2}$$

From the y-intercept, we go <u>down</u> 5 units and to the <u>left</u> 2 units. This gives us the point (-2,-4). Plot this point to see if it is on the line.

<u>20</u>. Slope is $\frac{2}{5}$; y-intercept is (0,4).

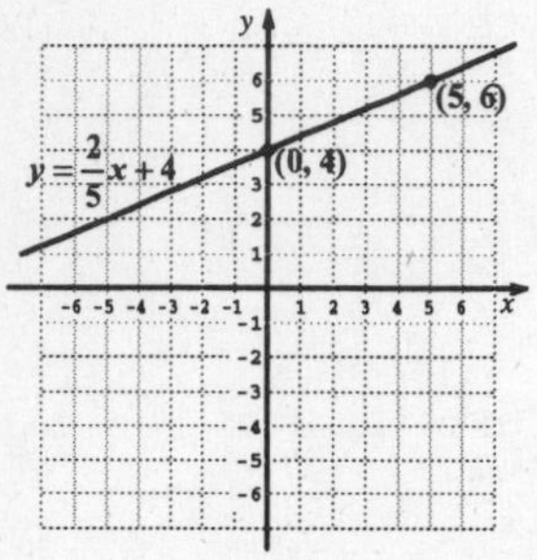

<u>21</u>. $y = -\frac{5}{2}x + 4$

Slope is $-\frac{5}{2}$, or $\frac{-5}{2}$; y-intercept is (0,4).

From the y-intercept, we go <u>down</u> 5 units and to the <u>right</u> 2 units. This gives us the point (2,-1). We can now draw the graph.

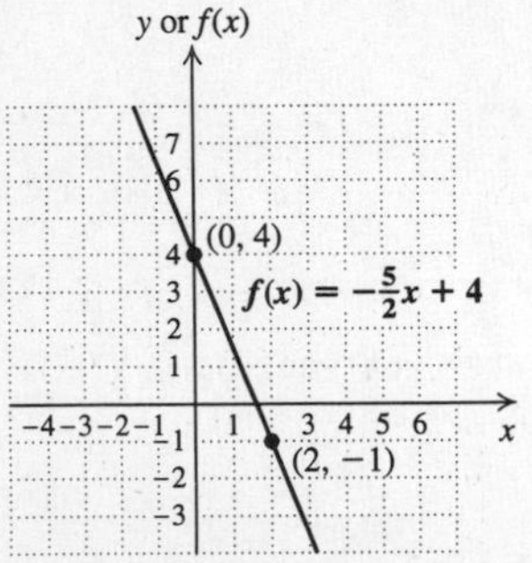

As a check, we can rename the slope and find another point.

$$\frac{-5}{2} = \frac{-5}{2} \cdot \frac{2}{2} = \frac{-10}{4}$$

From the y-intercept, we go <u>down</u> 10 units and to the <u>right</u> 4 units. This gives us the point (4,-6). Plot this point to see if it is on the line.

<u>22</u>. Slope is $-\frac{2}{5}$; y-intercept is (0,3).

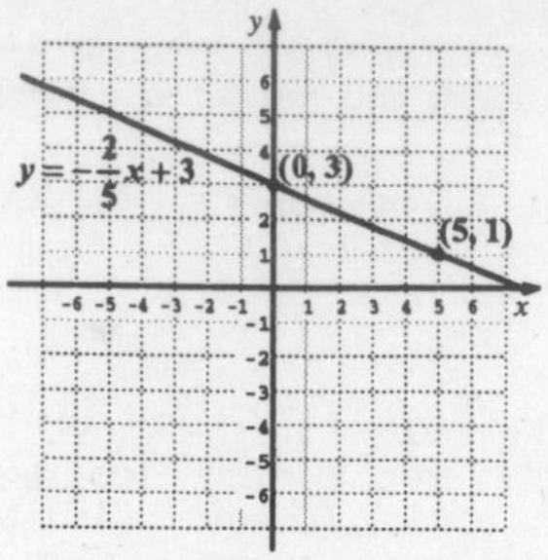

<u>23</u>. Convert to a slope-intercept equation:

$$2x - y = 5$$
$$-y = -2x + 5$$
$$y = 2x - 5$$

Slope is 2, or $\frac{2}{1}$; y-intercept is (0,-5).

From the y-intercept, we go <u>up</u> 2 units and to the <u>right</u> 1 unit. This gives us the point (1,-3). We can now draw the graph.

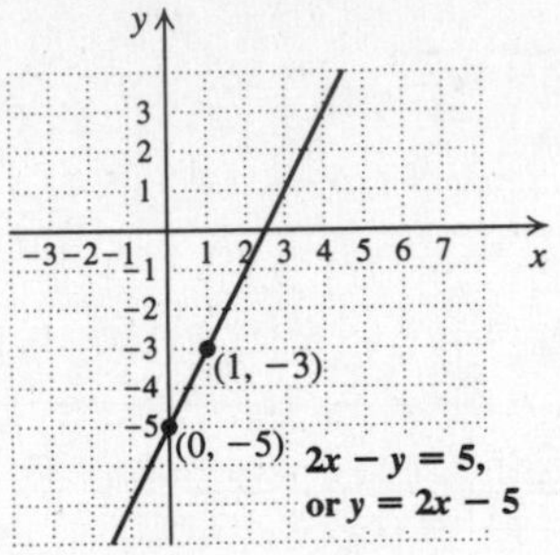

As a check, we can rename the slope and find another point.

$$2 = \frac{2}{1} \cdot \frac{3}{3} = \frac{6}{3}$$

From the y-intercept, we go <u>up</u> 6 units and to the <u>right</u> 3 units. This gives us the point (3,1). Plot this point to see if it is on the line.

<u>24</u>. Slope is -2; y-intercept is (0,4).

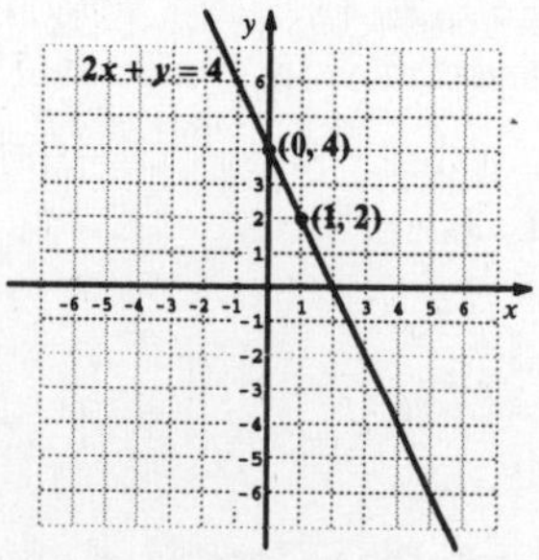

<u>25</u>. $y = \frac{1}{3}x + 6$

Slope is $\frac{1}{3}$; y-intercept is (0,6).

From the y-intercept, we go <u>up</u> 1 unit and to the <u>right</u> 3 units. This gives us the point (3,7). We can now draw the graph.

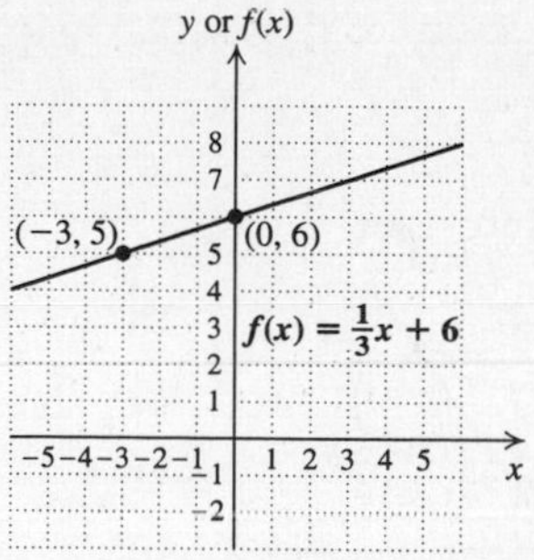

As a check, we can rename the slope and find another point.

$$\frac{1}{3} = \frac{1}{3} \cdot \frac{-1}{-1} = \frac{-1}{-3}$$

From the y-intercept, we go <u>down</u> 1 unit and to the <u>left</u> 3 units. This gives us the point (-3,5). Plot this point to see if it is on the line.

<u>26</u>. Slope is -3; y-intercept is (0,6).

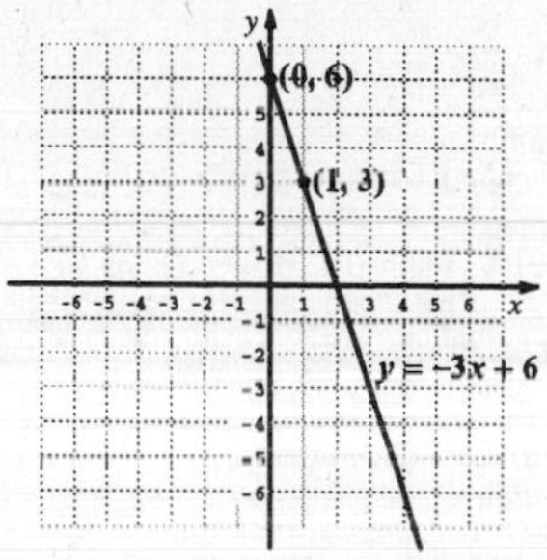

<u>27</u>. Convert to a slope intercept equation:

$$7y + 2x = 7$$
$$7y = -2x + 7$$
$$y = \frac{1}{7}(-2x + 7)$$
$$y = -\frac{2}{7}x + 1$$

Slope is $-\frac{2}{7}$ or $\frac{-2}{7}$; y-intercept is (0, 1).

From the y-intercept we go <u>down</u> 2 units and to the <u>right</u> 7 units. This gives us the point (7, -1). We can now draw the graph.

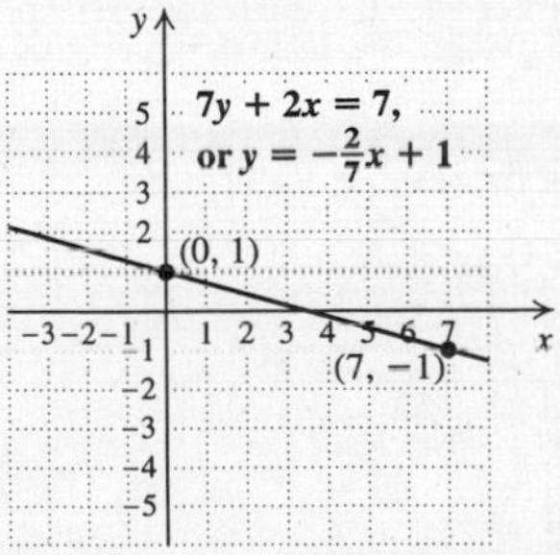

As a check we can rename the slope and find another point.

$$\frac{-2}{7} = \frac{-2}{7} \cdot \frac{-1}{-1} = \frac{2}{-7}$$

From the y-intercept, we go <u>up</u> 2 units and to the <u>left</u> 7 units. This gives us the point (-7, 3). Plot this point to see if it is on the line.

<u>28</u>. Slope is $\frac{1}{4}$; y-intercept is (0, -5).

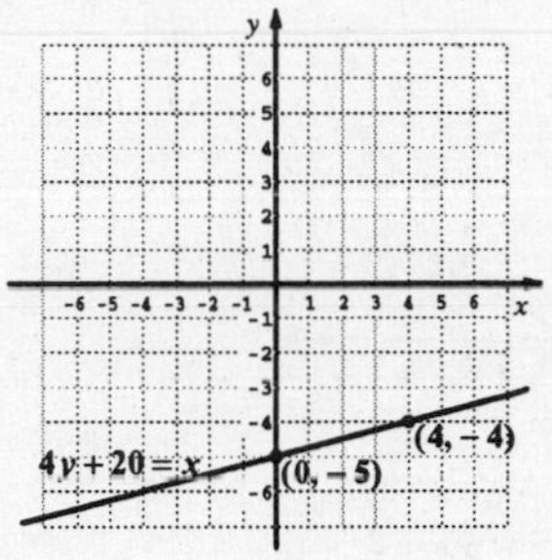

<u>29</u>. y = -0.25x + 2

Slope is -0.25, or $\frac{-1}{4}$; y-intercept is (0,2).

From the y-intercept, we go <u>down</u> 1 unit and to the <u>right</u> 4 units. This gives us the point (4,1). We can now draw the graph.

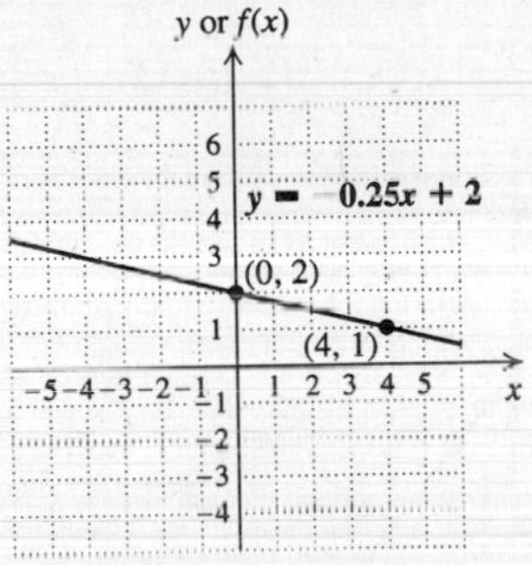

As a check, we can rename the slope and find another point.

$$\frac{-1}{4} = \frac{-1}{4} \cdot \frac{-1}{-1} = \frac{1}{-4}$$

From the y-intercept, we go <u>up</u> 1 unit and to the <u>left</u> 4 units. This gives us the point (-4,3). Plot this point to see if it is on the line.

<u>30</u>. Slope is 1.5, or $\frac{3}{2}$; y-intercept is (0,-3).

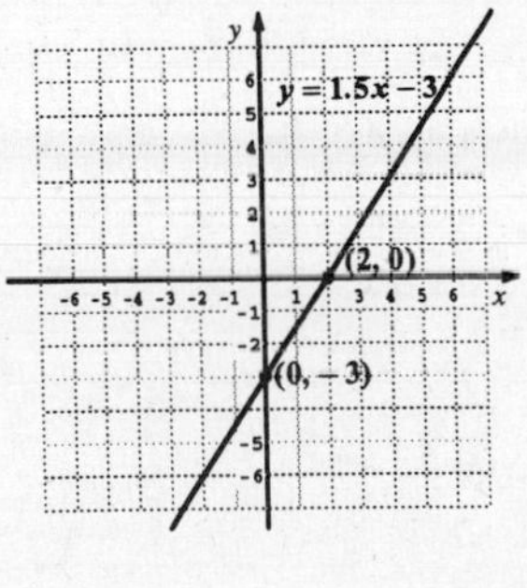

31. $y = \frac{4}{5}x - 2$

Slope is $\frac{4}{5}$; y-intercept is (0,-2).

From the y-intercept, we go up 4 units and to the right 5 units. This gives us the point (5,2). We can now draw the graph.

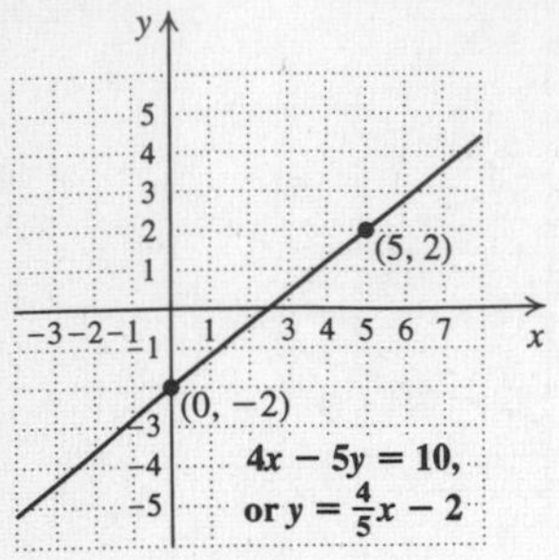

As a check, we choose some other value for x, say -5, and determine y:

$$y = \frac{4}{5}(-5) - 2 = -4 - 2 = -6$$

We plot the point (-5,-6) and see if it is on the line.

32. Slope is $-\frac{5}{4}$; y-intercept is (0,1).

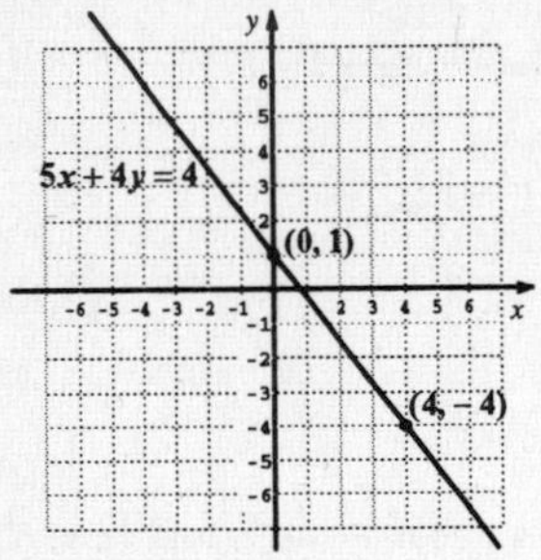

33. $f(x) = \frac{5}{4}x - 2$

Slope is $\frac{5}{4}$; y-intercept is (0,-2).

From the y-intercept, we go up 5 units and to the right 4 units. This gives us the point (4,3). We can now draw the graph.

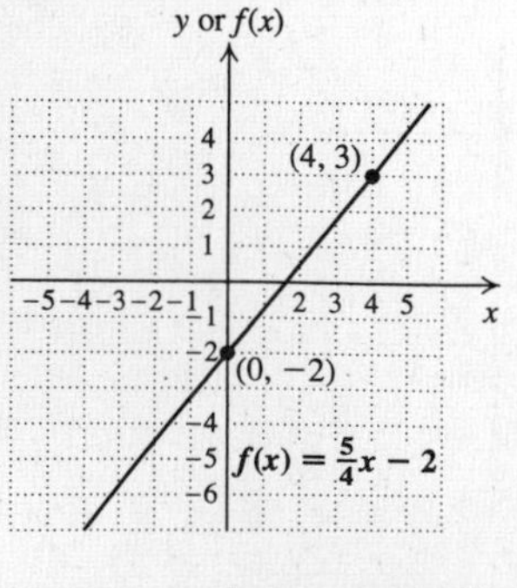

As a check, we choose some other value for x, say -2, and determine f(x):

$$f(x) = \frac{5}{4}(-2) - 2 = -\frac{5}{2} - 2 = -\frac{9}{2}$$

We plot the point $\left(-2, -\frac{9}{2}\right)$ and see if it is on the line.

34. Slope is $\frac{4}{3}$; y-intercept is (0,2).

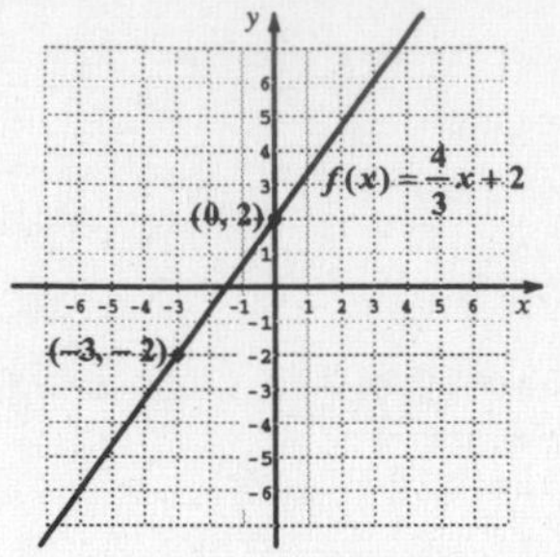

35. Convert to a slope-intercept equation:

$$12 - 4f(x) = 3x$$
$$-4f(x) = 3x - 12$$
$$f(x) = -\frac{1}{4}(3x - 12)$$
$$f(x) = -\frac{3}{4}x + 3$$

Slope is $-\frac{3}{4}$, or $\frac{-3}{4}$; y-intercept is (0, 3).

From the y-intercept, we go down 3 units and to the right 4 units. This gives us the point (4, 0). We can now draw the graph.

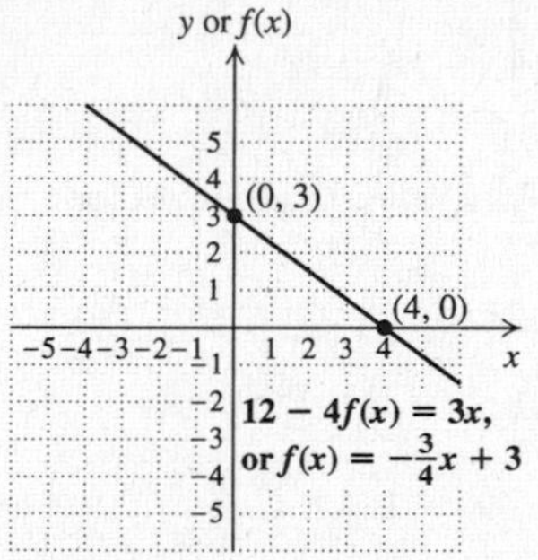

As a check, we choose some other value for x, say -4, and determine f(x):

$$f(-4) = -\frac{3}{4}(-4) + 3 = 3 + 3 = 6$$

We plot the point (-4, 6) and see if it is on the line.

36. Slope is $-\frac{2}{5}$; y-intercept is (0, -3).

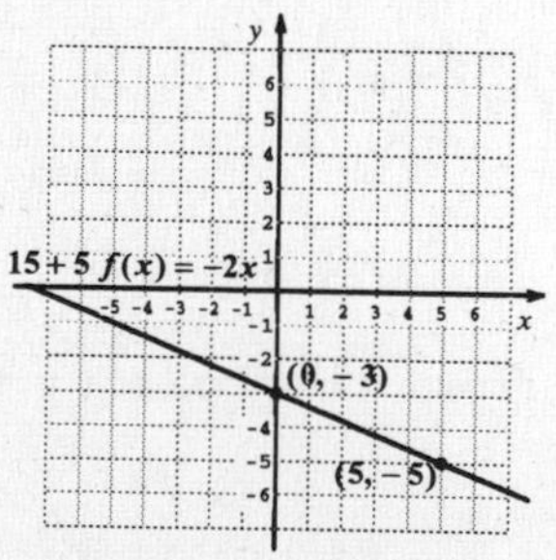

37. f(x) = 4

Think of this as f(x) = 0·x + 4.

Slope is 0; y-intercept is (0,4).

From the y-intercept we do not go up or down. We can go right or left any non-zero number of units, since 0/a = 0, a ≠ 0. Let's choose to go right 3 units. This gives us the point (3,4). We can now draw the graph.

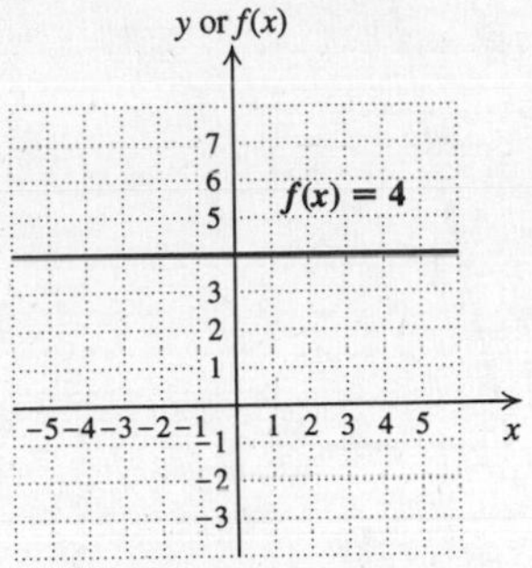

As a check, we choose some other value for x, say -4, and determine f(x):

f(x) = 0(-4) + 4 = 0 + 4 = 4.

We plot the point (-4,4) and see if it is on the line.

38. Slope is 0; y-intercept is (0,-1).

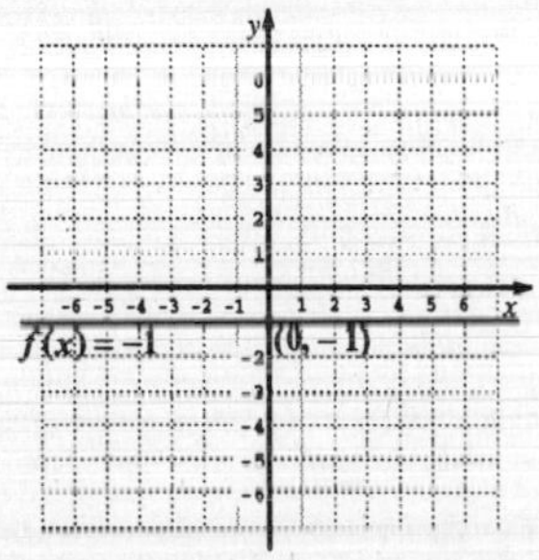

39. a) f(t) = 2.6t + 17.8

Slope is 2.6, or $\frac{26}{10}$, or $\frac{13}{5}$; y-intercept is (0, 17.8). We locate the y-intercept (0, 17.8) and from there count up 13 units and to the right 5 units. This gives us the point (5, 30.8). We can now draw the graph. As a check, we note that the pair (10, 43.8) also satisfies the equation. (We do not plot points for t < 0, since it does not make sense to talk about a negative number of years after 1975.)

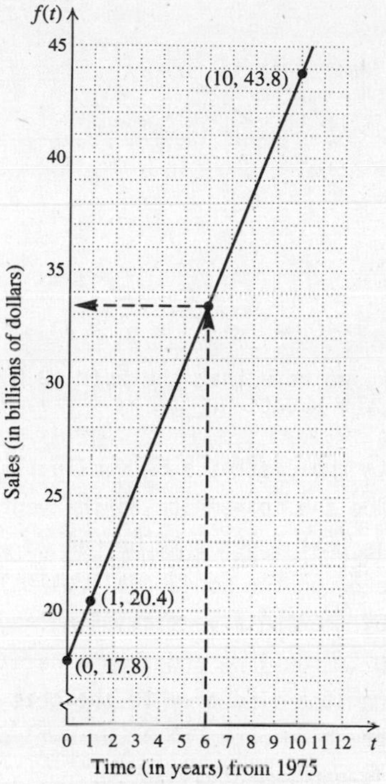

b) Since 1981 is 6 years after 1975, we determine the coordinate on the vertical axis that appears to be paired with the coordinate 6 on the horizontal axis. We do so by drawing a vertical line segment from 6 up to the line and then a horizontal segment from the line to the vertical axis. (See the graph in part (a).) We estimate that the total sales of cotton goods in 1981 were approximately $33 billion. An exact calculation is made by substituting into the function formula:

f(6) = 2.6(6) + 17.8 = $33.4 billion

40. a)

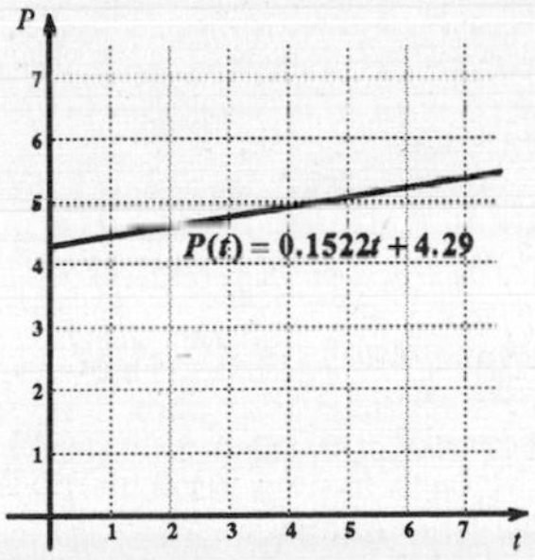

b) Approximately $6

41. a) C(m) = 0.75m + 2

Slope is 0.75, or $\frac{3}{4}$; y-intercept is (0,2).

From the y-intercept we go up 3 units and to the right 4 units. This gives us the point (4,5). We can now draw the graph. (We do not plot points for m < 0, since it does not make sense to talk about a taxi ride of a negative number of miles.)

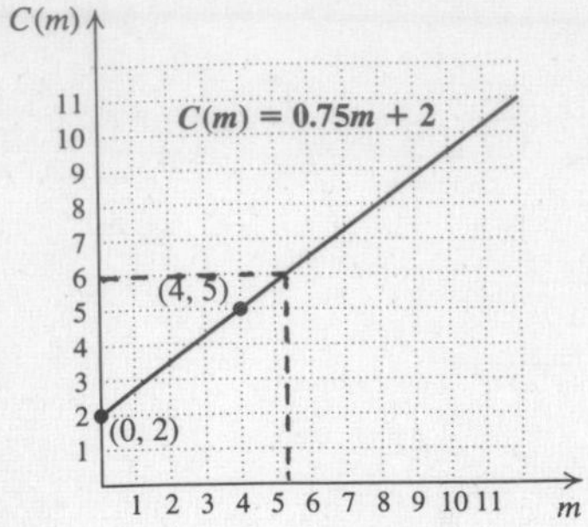

As a check, we note that the pair (8,8) also satisfies the equation.

b) To estimate the cost of a $5\frac{1}{2}$ - mile taxi ride, we determine the coordinate on the vertical (cost) axis that appears to be paired with the coordinate $5\frac{1}{2}$ on the horizontal (miles) axis. We do so by drawing a vertical line segment from $5\frac{1}{2}$ up to the line and then a horizontal segment from the line over to the cost axis. (See the graph in part (a).) We estimate that the cost is approximately \$6.00.

42. a)

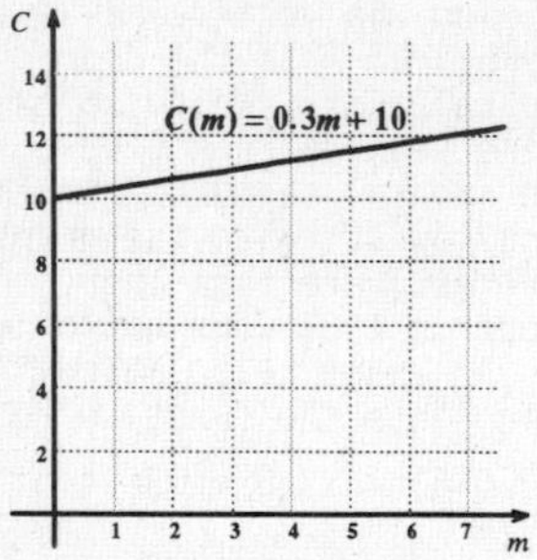

b) \$22

43. a) $D(t) = \frac{1}{5}t + 20$

Slope is $\frac{1}{5}$; y-intercept is (0,20).

From the y-intercept go up 1 unit and to the right 5 units. This gives us the point (5,21). We can now draw the graph.

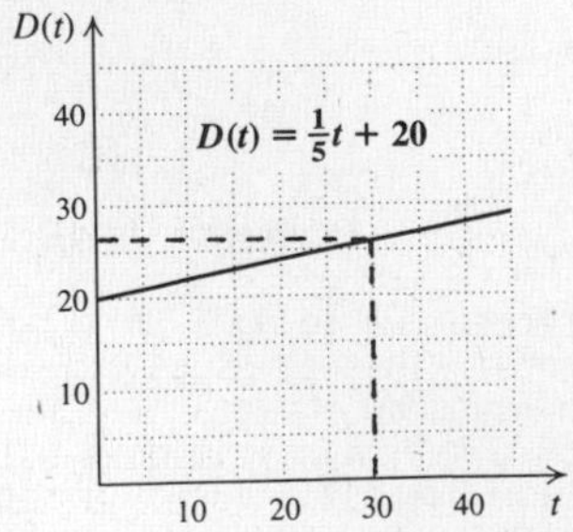

As a check, we note that the point (25,25) also satisfies the equation.

b) To predict the demand for gas in 1995 (35 years after 1960), draw a vertical line segment from 35 on the horizontal (time) axis up to the line and then a horizontal segment from the line over to the vertical (demand) axis. (See the graph in part (a).) We estimate that the demand will be about 27 quadrillion joules.

44. a)

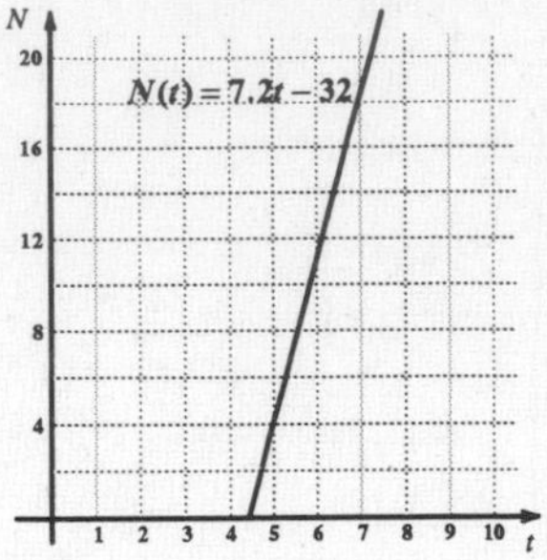

b) 4 chirps per minute

45. a) C(m) = 0.80m + 1

Slope is 0.80, or $\frac{4}{5}$; y-intercept is (0,1).

From the y-intercept go up 4 units and to the right 5 units. This gives us the point (5,5). We can now draw the graph.

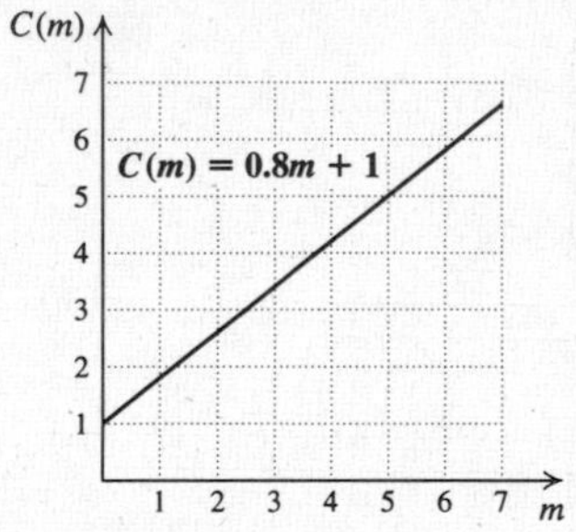

As a check, we note that the point (10,9) also satisfies the equation.

b) Draw a vertical line segment from 4 on the horizontal (minutes) axis up to the line and then a horizontal line over to the vertical (cost) axis. (See the graph in part (a).) We predict that the cost of a 4-min call is about \$4.25.

c) Draw a horizontal line segment from \$7.40 (or 7.4) on the vertical (cost) axis over to the line and then down to the horizontal (minutes) axis. (See the graph in part (a).) We determine that a phone call about 8 min long can be made for \$7.40.

46. a)

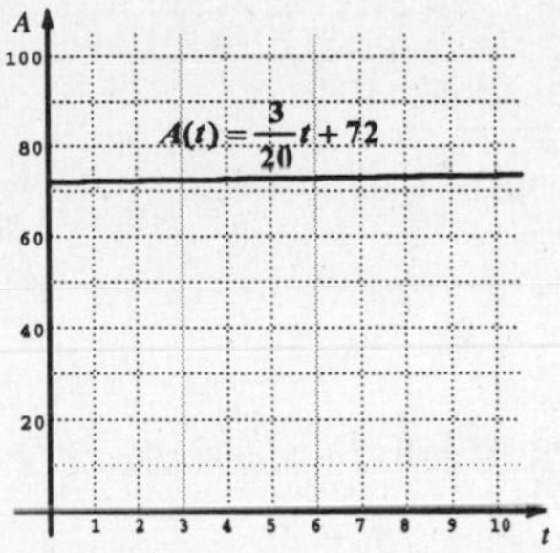

b) About 80 yr

c) 1990

47. We first solve for y.

$3y = 9$

$y = 3$

Since x is missing, any number for x will do. Thus all ordered pairs (x,3) are solutions. A few solutions are listed below. Plot these and draw the line.

x	y
-2	3
0	3
3	3

(y must be 3; x can be any number.)

The graph is a line parallel to the x-axis with y-intercept (0,3).

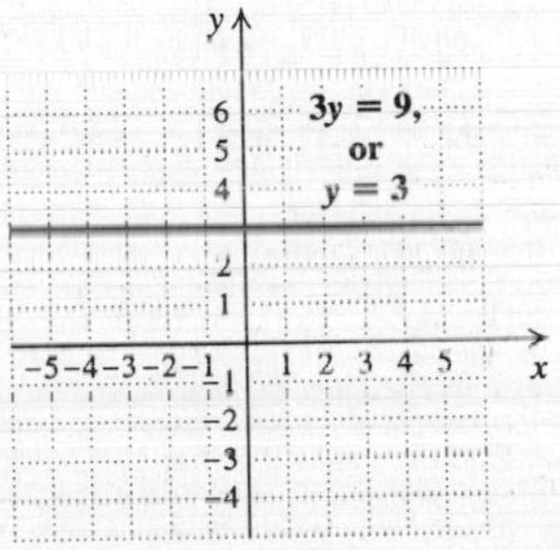

48.

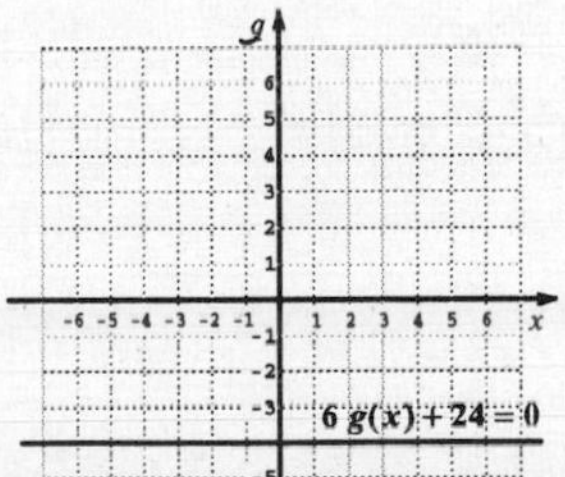

49. We first solve for x.

$3x = -15$

$x = -5$

Since y is missing, any number for y will do. Thus all ordered pairs (-5,y) are solutions. A few solutions are listed below. Plot these and draw the line.

x	y
-5	-3
-5	1
-5	4

(x must be -5; y can be any number.)

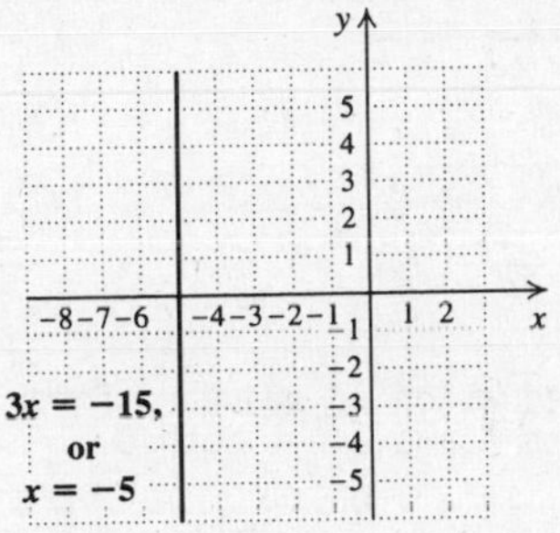

The graph is a line parallel to the y-axis with x-intercept (-5,0).

50.

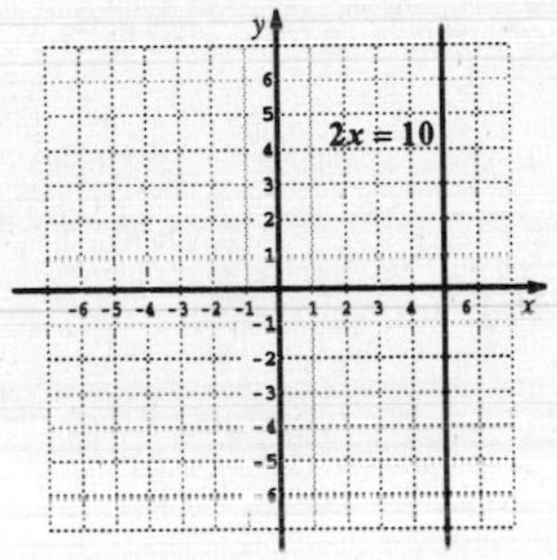

51. We first solve for g(x).

$4g(x) + 3x = 12 + 3x$

$4g(x) = 12$

$g(x) = 3$

Since x is missing, any number for x will do. Thus all ordered pairs (x,3) are solutions. A few solutions are listed below. Plot these and draw the line.

x	g(x)
-4	3
-1	3
0	3

(g(x) must be 3; x can be any number.)

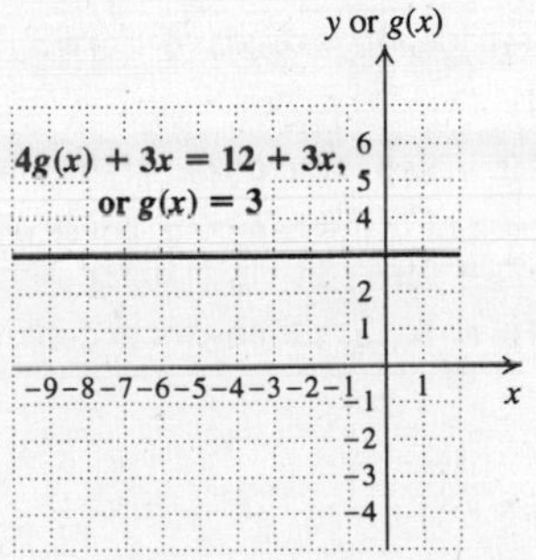

The graph is a line parallel to the x-axis with y-intercept (0,3).

52. 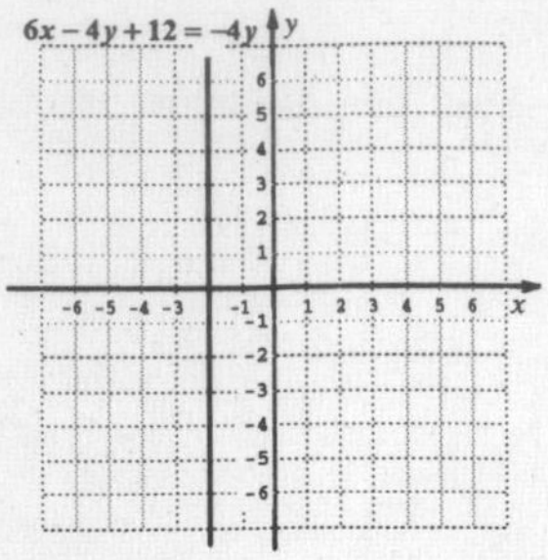

53. Try to put the equation in standard form, $Ax + By = C$, with A and B not both 0.

$$3x + 5f(x) + 15 = 0$$
$$3x + 5y + 15 = 0 \quad \text{Replace } f(x) \text{ with } y$$
$$3x + 5y = -15$$

The equation can be put in standard form with $A = 3$, $B = 5$, and $C = -15$. It is linear.

Now convert standard form to a slope-intercept equation, $y = mx + b$.

$$3x + 5y = -15$$
$$5y = -3x - 15$$
$$\frac{1}{5} \cdot 5y = \frac{1}{5}(-3x - 15)$$
$$y = -\frac{3}{5}x - 3$$

The slope is $-\frac{3}{5}$.

54. Linear; $\frac{5}{3}$

55. Try to put the equation in standard form, $Ax + By = C$, with A and B not both 0.

$$3x - 12 = 0$$
$$3x = 12$$
$$x = 4$$

The equation can be put in standard form with $A = 1$, $B = 0$, and $C = 4$. It is linear. (It is the equation of a vertical line.)

56. Linear; 0

57. Try to put the equation in standard form.

$$2x + 4g(x) = 19$$
$$2x + 4y = 19 \quad \text{Replace } g(x) \text{ with } y$$

The equation can be put in standard form with $A = 2$, $B = 4$, and $C = 19$. It is linear.

Convert standard form to a slope-intercept equation.

$$2x + 4y = 19$$
$$4y = -2x + 19$$
$$\frac{1}{4} \cdot 4y = \frac{1}{4}(-2x + 19)$$
$$y = -\frac{1}{2}x + \frac{19}{4}$$

The slope is $-\frac{1}{2}$.

58. Not linear

59. $5x - 4xy = 12$

The equation is not linear, because it has an xy-term.

60. Not linear

61. Try to put the equation in standard form.

$$\frac{3y}{4x} = 5y + 2$$
$$4x \cdot \frac{3y}{4x} = 4x(5y + 2)$$
$$3y = 20xy + 8x$$
$$-8x + 3y - 20xy = 0$$

The equation is not linear, because it has an xy-term.

62. Not linear

63. Try to put the equation in standard form.

$$f(x) = x^3$$
$$y = x^3 \quad \text{Replace } f(x) \text{ with } y$$
$$-x^3 + y = 0$$

The equation is not linear, because it has an x^3-term.

64. Not linear

65.
$$9\{2x - 3[5x + 2(-3x + y^0 - 2)]\}$$
$$= 9\{2x - 3[5x + 2(-3x + 1 - 2)]\} \quad (y^0 = 1)$$
$$= 9\{2x - 3[5x + 2(-3x - 1)]\}$$
$$= 9\{2x - 3[5x - 6x - 2]\}$$
$$= 9\{2x - 3[-x - 2]\}$$
$$= 9\{2x + 3x + 6\}$$
$$= 9\{5x + 6\}$$
$$= 45x + 54$$

66. $-\frac{27}{14}$

67.

68.

69. We first solve for y.

$$ry = -5x + p$$
$$y = -\frac{5}{r}x + \frac{p}{r}$$

The slope is $-\frac{5}{r}$, and the y-intercept is $\left(0, \frac{p}{r}\right)$.

70. Slope is $\frac{p}{4}$; y-intercept is $\left(0, \frac{9r + 2}{4}\right)$.

71. We first solve for y.

$rx + py = s$

$py = -rx + s$

$y = -\frac{r}{p}x + \frac{s}{p}$

The slope is $-\frac{r}{p}$, and the y-intercept is $\left(0, \frac{s}{p}\right)$.

72. Slope is $-\frac{r}{r + p}$; y-intercept is $\left(0, \frac{s}{r + p}\right)$.

73. $rx + 3y = p - s$

The equation is in standard form with $A = r$, $B = 3$, and $C = p - s$. It is linear.

74. Linear

75. Try to put the equation in standard form.

$r^2x = py + 5$

$r^2x - py = 5$

The equation is in standard form with $A = r^2$, $B = -p$, and $C = 5$. It is linear.

76. Linear

77. We know that the y-intercept is (0,3.1). We also know that to get another point on the graph, (2,7.8), we go up 4.7 units (7.8 - 3.1 = 4.7) and to the right 2 units (2 - 0 = 2). Then the slope of the line is $\frac{4.7}{2}$, or $\frac{47}{20}$.

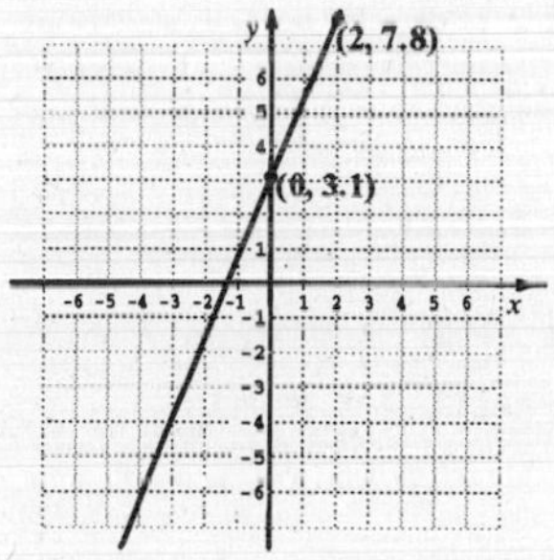

To write an equation for the function, use the slope-intercept equation with $m = \frac{47}{20}$ and $b = 3.1$:

$y = mx + b$

$y = \frac{47}{20}x + 3.1$

78. $y = -\frac{5}{3}x + \frac{14}{3}$

79. See the answer section in the text.

Exercise Set 2.4

1. Graph x - 2 = y.

To find the y-intercept, let x = 0.

$x - 2 = y$

$0 - 2 = y$, or $-2 = y$

The y-intercept is (0,-2).

To find the x-intercept, let y = 0.

$x - 2 = y$

$x - 2 = 0$, or $x = 2$

The x-intercept is (2,0).

Plot these points and draw the line. A third point could be used as a check.

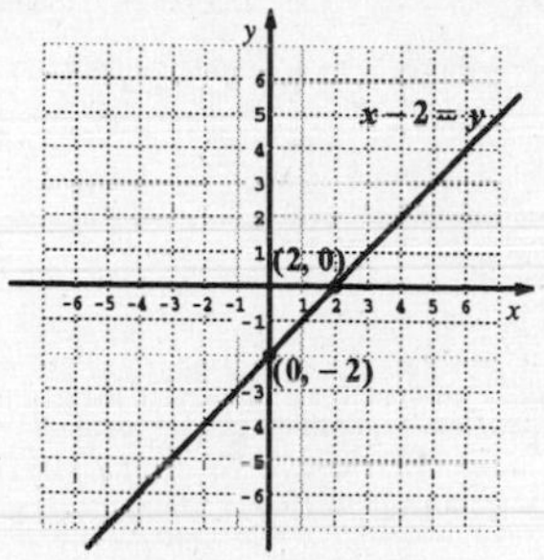

2.

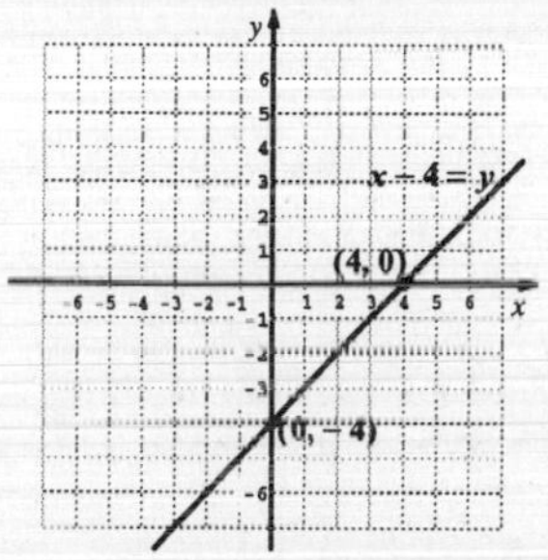

3. Graph 3x - 1 = y.

To find the y-intercept, let x = 0.

$3x - 1 = y$

$3\cdot 0 - 1 = y$, or $-1 = y$

The y-intercept is (0,-1).

To find the x-intercept, let y = 0.

$3x - 1 = y$

$3x - 1 = 0$

$3x = 1$

$x = \frac{1}{3}$

The x-intercept is $\left(\frac{1}{3}, 0\right)$.

Plot these points and draw the line. A third point could be used as a check.

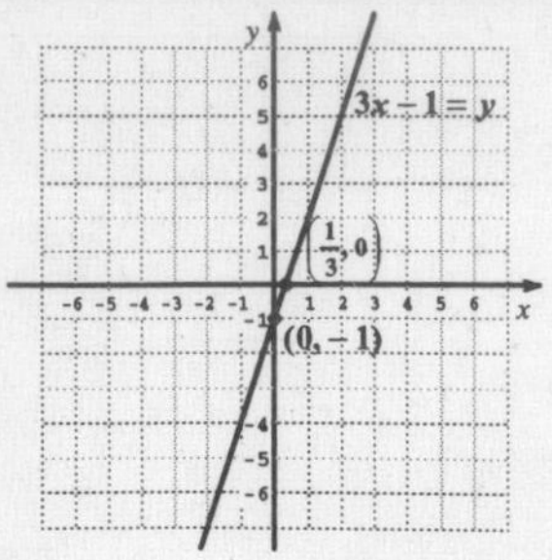

4.

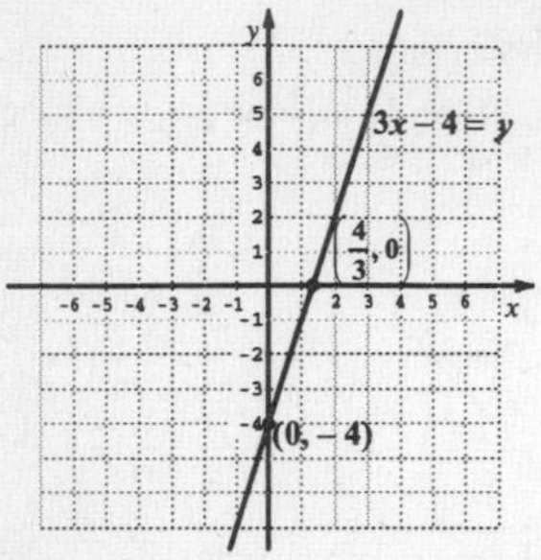

5. Graph 5x - 4y = 20.

To find the y-intercept, let x = 0.

$$5x - 4y = 20$$
$$5 \cdot 0 - 4y = 20$$
$$-4y = 20$$
$$y = -5$$

The y-intercept is (0,-5).

To find the x-intercept, let y = 0.

$$5x - 4y = 20$$
$$5x - 4 \cdot 0 = 20$$
$$5x = 20$$
$$x = 4$$

The x-intercept is (4,0).

Plot these points and draw the line. A third point could be used as a check.

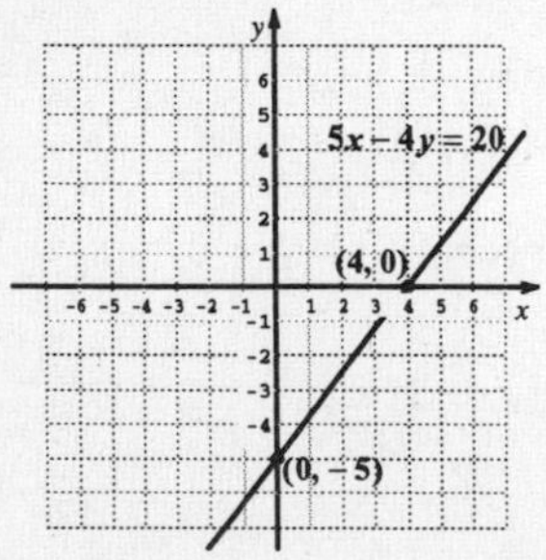

6.

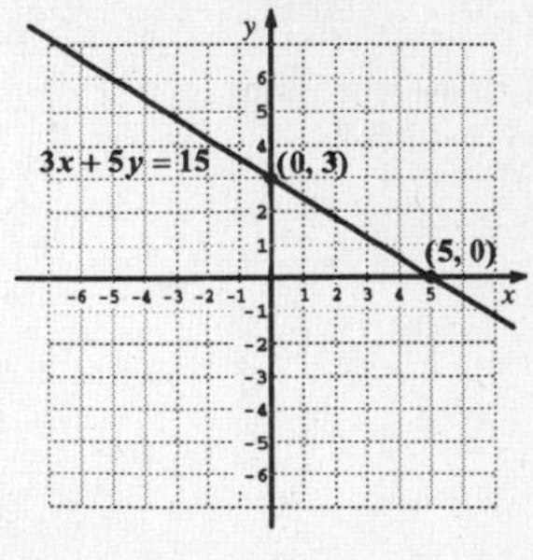

7. Graph y = -5 - 5x.

To find the y-intercept, let x = 0.

$$y = -5 - 5x$$
$$y = -5 - 5 \cdot 0$$
$$y = -5$$

(0,-5) is the y-intercept.

To find the x-intercept, let y = 0.

$$y = -5 - 5x$$
$$0 = -5 - 5x$$
$$5x = -5$$
$$x = -1$$

(-1,0) is the x-intercept.

Plot these points and draw the line. A third point could be used as a check.

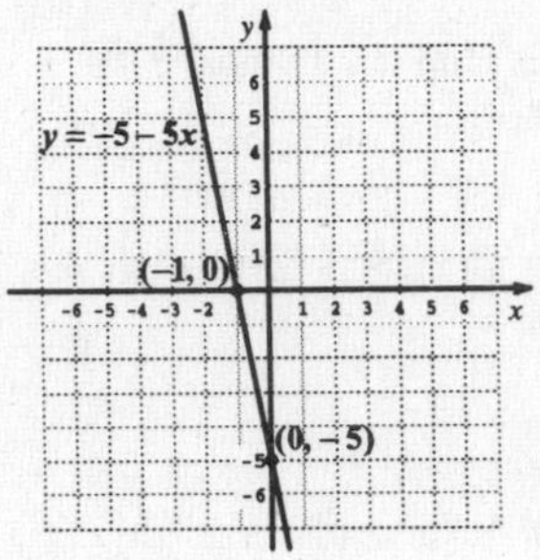

8.

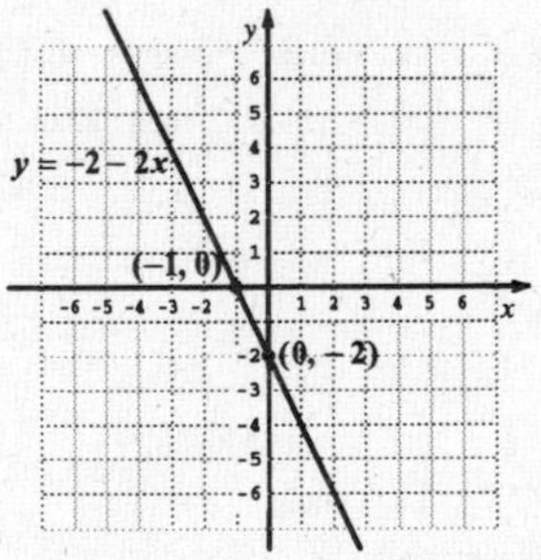

9. Graph 5y = -15 + 3x.

To find the y-intercept, let x = 0.

$$5y = -15 + 3x$$
$$5y = -15 + 3 \cdot 0$$
$$5y = -15$$
$$y = -3$$

(0,-3) is the y-intercept.

To find the x-intercept, let y = 0.

$$5y = -15 + 3x$$
$$5 \cdot 0 = -15 + 3x$$
$$15 = 3x$$
$$5 = x$$

(5,0) is the x-intercept.

Plot these points and draw the line. A third point could be used as a check.

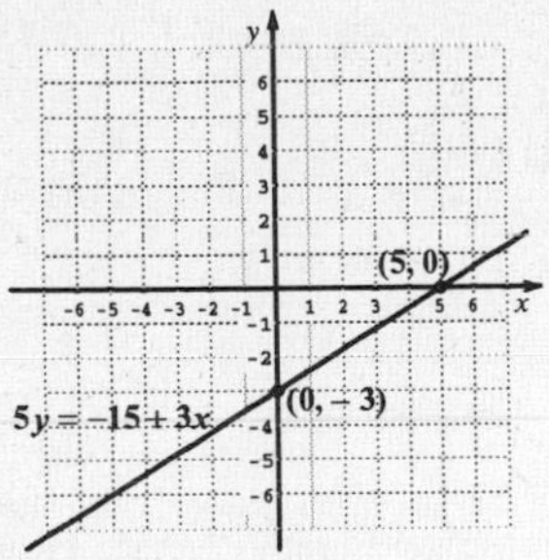

12.

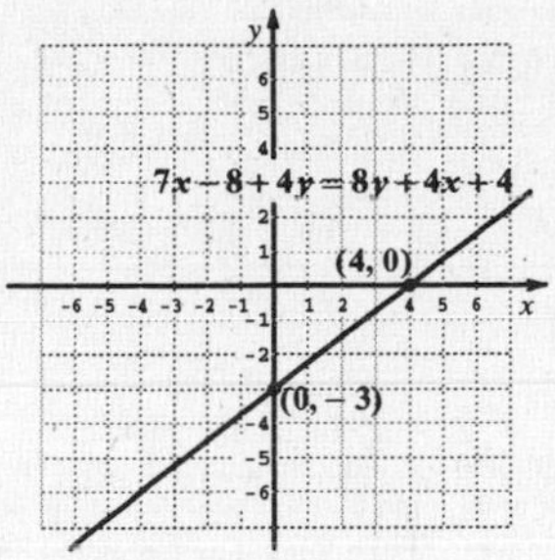

10.

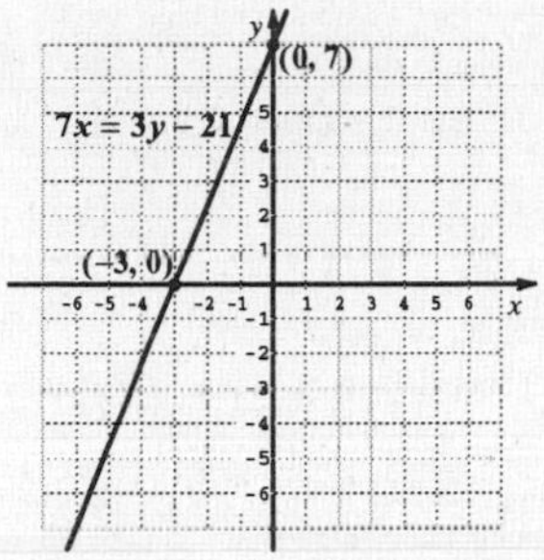

11. $6x - 7 + 3y = 9x - 2y + 8$

$-3x + 5y = 15$

Graph $-3x + 5y = 15$

To find the y-intercept, let $x = 0$.

$-3x + 5y = 15$

$-3\cdot 0 + 5y = 15$

$5y = 15$

$y = 3$

(0,3) is the y-intercept.

To find the x-intercept, let $y = 0$.

$-3x + 5y = 15$

$-3x + 5\cdot 0 = 15$

$3x = 15$

$x = -5$

(-5,0) is the x-intercept.

Plot these points and draw the line. A third point could be used as a check.

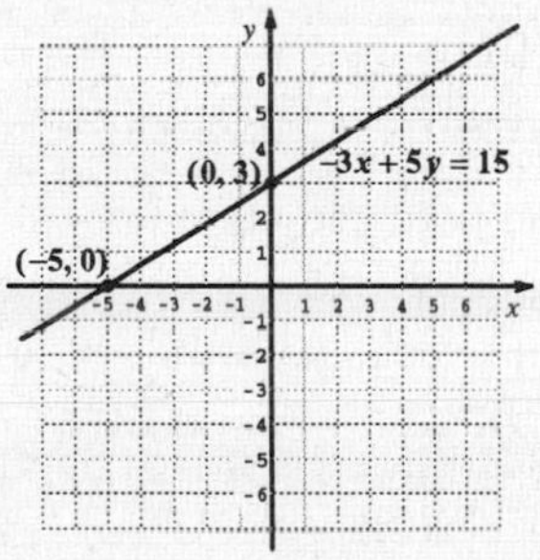

13. $1.4y - 3.5x = -9.8$

$14y - 35x = -98$ Multiplying by 10

$2y - 5x = -14$ Multiplying by $\frac{1}{7}$

Graph $2y - 5x = -14$.

To find the y-intercept, let $x = 0$.

$2y - 5x = -14$

$2y - 5\cdot 0 = -14$

$2y = -14$

$y = -7$

(0,-7) is the y-intercept.

To find the x-intercept, let $y = 0$.

$2y - 5x = -14$

$2\cdot 0 - 5x = -14$

$-5x = -14$

$x = 2.8$

(2.8, 0) is the x-intercept.

Plot these points and draw the line. A third point could be used as a check.

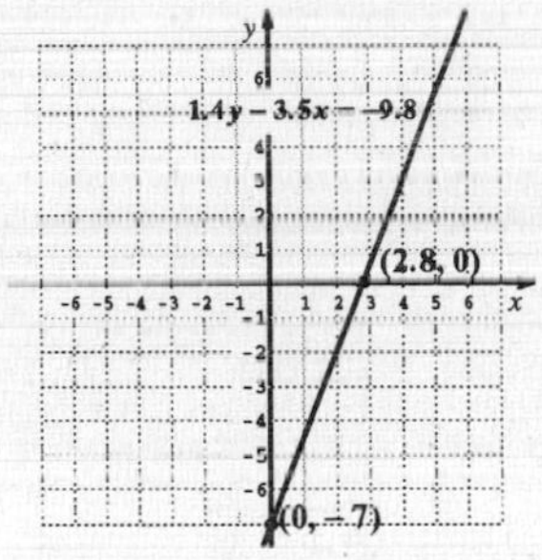

14.

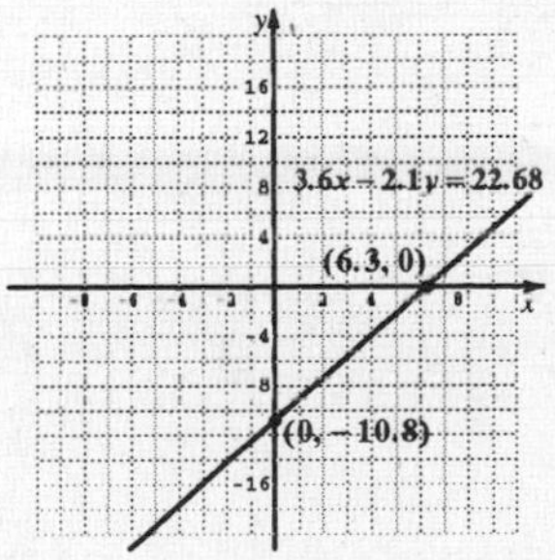

15. Graph $5x + 2y = 7$

To find the y-intercept, let $x = 0$.

$$5x + 2y = 7$$
$$5 \cdot 0 + 2y = 7$$
$$2y = 7$$
$$y = \frac{7}{2}$$

$\left(0, \frac{7}{2}\right)$ is the y-intercept.

To find the x-intercept, let $y = 0$.

$$5x + 2y = 7$$
$$5x + 2 \cdot 0 = 7$$
$$5x = 7$$
$$x = \frac{7}{5}$$

$\left(\frac{7}{5}, 0\right)$ is the x-intercept.

Plot these points and draw the line. A third point could be used as a check.

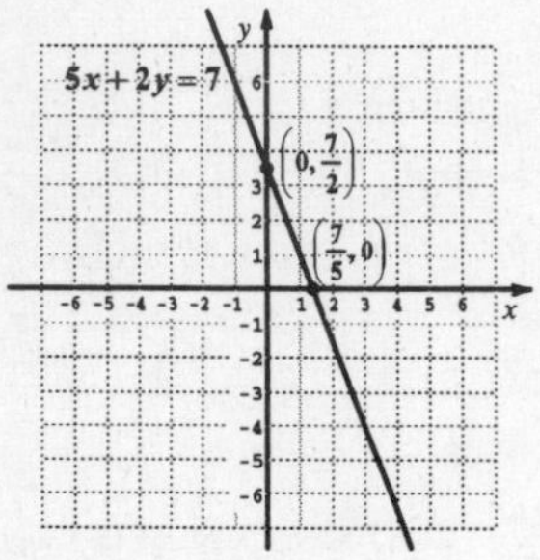

16.

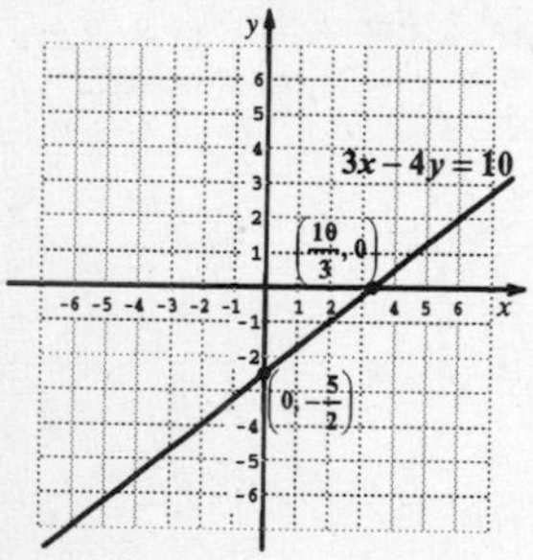

17. Let $(6,9) = (x_1,y_1)$ and $(4,5) = (x_2,y_2)$.

Slope $= \frac{y_2 - y_1}{x_2 - x_1} = \frac{5 - 9}{4 - 6} = \frac{-4}{-2} = 2$

18. $\frac{4}{3}$

19. Let $(3,8) = (x_1,y_1)$ and $(9,-4) = (x_2,y_2)$.

Slope $= \frac{y_2 - y_1}{x_2 - x_1} = \frac{-4 - 8}{9 - 3} = \frac{-12}{6} = -2$

20. $\frac{3}{26}$

21. Let $(-8,-7) = (x_1,y_1)$ and $(-9,-12) = (x_2,y_2)$.

Slope $= \frac{y_2 - y_1}{x_2 - x_1} = \frac{-12 - (-7)}{-9 - (-8)} = \frac{-5}{-1} = 5$

22. $-\frac{3}{4}$

23. Let $(-16.3,12.4) = (x_1,y_1)$ and $(-5.2,8.7) = (x_2,y_2)$.

Slope $= \frac{y_2 - y_1}{x_2 - x_1} = \frac{8.7 - 12.4}{-5.2 - (-16.3)} = \frac{-3.7}{11.1} = -\frac{37}{111} = -\frac{1}{3}$

24. $\frac{98}{269}$

25. Let $(3.2,-12.8) = (x_1,y_1)$ and $(3.2,2.4) = (x_2,y_2)$.

Slope $= \frac{y_2 - y_1}{x_2 - x_1} = \frac{2.4 - (-12.8)}{3.2 - 3.2} = \frac{15.2}{0}$

Since we cannot divide by 0, the slope is undefined.

26. Undefined

27. Let $(7,3.4) = (x_1,y_1)$ and $(-1,3.4) = (x_2,y_2)$.

Slope $= \frac{y_2 - y_1}{x_2 - x_1} = \frac{3.4 - 3.4}{-1 - 7} = \frac{0}{-8} = 0$

28. 0

29. Let $(0,9.1) = (x_1,y_1)$ and $(9.1,0) = (x_2,y_2)$.

Slope $= \frac{y_2 - y_1}{x_2 - x_1} = \frac{0 - 9.1}{9.1 - 0} = \frac{-9.1}{9.1} = -1$

30. 1

31. Familiarize. We might familiarize ourselves with this problem in one of two ways. First, we might use the formula $d = r \cdot t$. (See Example 4.) Second, we might use a graph to give a "picture" of the problem.

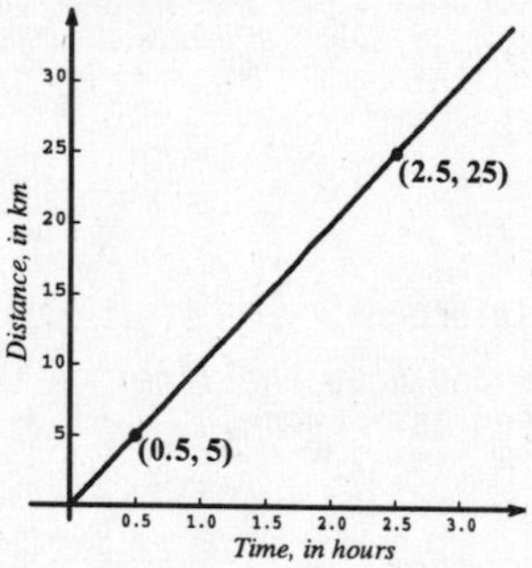

Translate. Using the first approach, we could substitute either 0.5 for t and 5 for d or 2.5 for t and 25 for d. We have the equation

$$5 = r(0.5) \quad \text{or} \quad 25 = r(2.5).$$

Using the second approach, we could write the equation

rate = change in distance/change in time or

$$r = (25 - 5)/(2.5 - 0.5).$$

Carry out. Using the first approach, we solve

$$5 = r(0.5) \quad \text{or} \quad 25 = r(2.5)$$

to obtain $r = 10$ km/hr.

Using the second approach, we solve

$$r = \frac{25 - 5}{2.5 - 0.5} = \frac{20}{2} = 10 \text{ km/hr}.$$

Check. If the rate is 10 km/hr, in 0.5 hr the marathoner will have run 0.5(10) = 5 km, and in 2.5 hr, 2.5(10) = 25 km. The answer checks.

State. The marathoner's speed is 10 km/hr.

32. 12 km/hr

33. Familiarize. We can use a graph to give a "picture" of the problem.

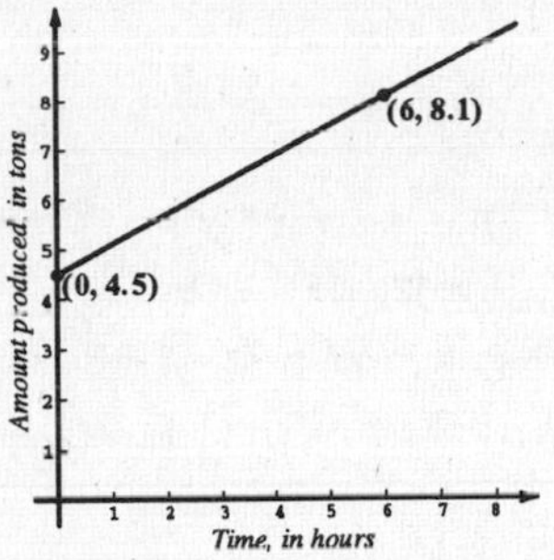

Translate. We can write an equation:

rate of production = change in amount/change in time

or $r = (8.1 - 4.5)/(6 - 0)$

Carry out. Solve the equation.

$$r = \frac{8.1 - 4.5}{6 - 0} = \frac{3.6}{6} = 0.6 \text{ ton/hr}$$

Check. If the rate of production is 0.6 ton/hr, in 6 hr then 0.6(6) = 3.6 tons are produced. If 4.5 tons had already been produced at the beginning of the run, then the total amount after 6 hours is 4.5 + 3.6 = 8.1 tons. The answer checks.

State. The rate of production is 0.6 ton/hr.

34. $\frac{5}{96}$ of the house per hour

35. Familiarize. We can use a graph to give a "picture" of the problem. Express both times in minutes.

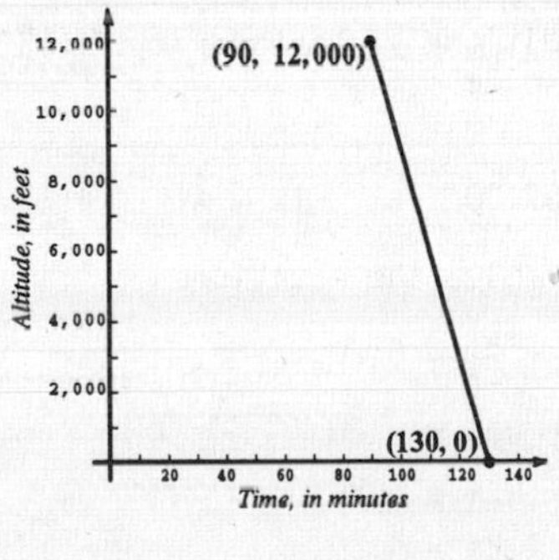

Translate. We can write an equation:

rate = change in altitude/change in time

or $r = (0 - 12{,}000)/(130 - 90)$

Carry out. Solve the equation.

$$r = \frac{0 - 12{,}000}{130 - 90} = \frac{-12{,}000}{40} = -300 \text{ ft/min}$$

The negative result indicates that the plane is descending. We can say the rate is -300 ft/min or the rate of descent is 300 ft/min.

Check. If the rate of descent is 300 ft/min, then in 40 min $\left[2 \text{ hr } 10 \text{ min} - 1\frac{1}{2} \text{ hr}\right]$ the plane descends 300(40) = 12,000 ft. The answer checks.

State. The rate of descent is 300 ft/min.

36. $1166\frac{2}{3}$ ft/hr

37. Familiarize. We can use a graph to give a "picture" of the problem.

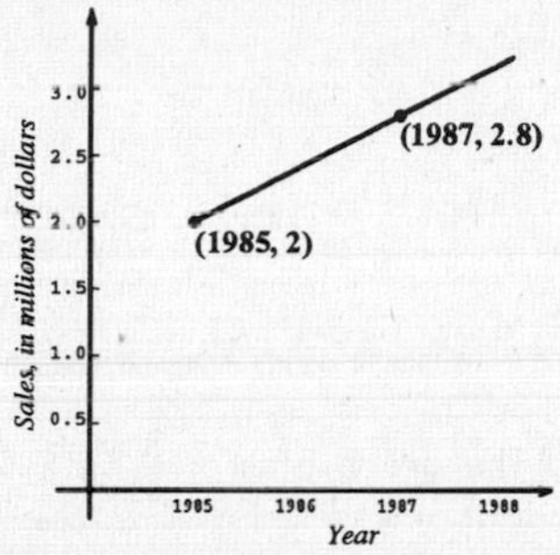

Translate. We can write an equation:

rate = change in sales/change in time

or $r = (2.8 - 2.0)/(1987 - 1985)$

Carry out. Solve the equation.

$$r = \frac{2.8 - 2.0}{1987 - 1985} = \frac{0.8}{2} = \$0.4 \text{ million/yr}$$

Check. If the rate at which sales are increasing is \$0.4 million/yr, then in 2 yr the sales increase 0.4(2) = \$0.8 million. The answer checks.

State. Sales are increasing at the rate of \$0.4 million/yr.

38. 1.367 million/yr

39. $5x - 6 = 15$

$5x = 21$

$x = \frac{21}{5}$ The graph of $x = \frac{21}{5}$ is a vertical line.

Since $5x - 6 = 15$ is equivalent to $x = \frac{21}{5}$, the slope of the line $5x - 6 = 15$ is undefined.

40. $\frac{3}{5}$

41. $3x = 12 + y$

$3x - 12 = y$

$y = 3x - 12$ $(y = mx + b)$

The slope is 3.

42. Undefined

43. $5y = 6$

$y = \frac{6}{5}$ The graph of $y = \frac{6}{5}$ is a horizontal line.

Since $5y = 6$ is equivalent to $y = \frac{6}{5}$, the slope of the line $5y = 6$ is 0.

44. 0

45. $5x - 7y = 30$

$-7y = -5x + 30$

$y = \frac{5}{7}x - \frac{30}{7}$ $(y = mx + b)$

The slope is $\frac{5}{7}$.

46. $\frac{2}{3}$

47. $12 - 4x = 9 + x$

$3 = 5x$

$\frac{3}{5} = x$ The graph of $x = \frac{3}{5}$ is a vertical line.

Since $12 - 4x = 9 + x$ is equivalent to $x = \frac{3}{5}$, the slope of the line $12 - 4x = 9 + x$ is undefined.

48. Undefined

49. $2y - 4 = 35 + x$

$2y = x + 39$

$y = \frac{1}{2}x + \frac{39}{2}$ $(y = mx + b)$

The slope is $\frac{1}{2}$.

50. -2

51. $3y + x = 3y + 2$

$x = 2$ The graph of $x = 2$ is a vertical line.

Since $3y + x = 3y + 2$ is equivalent to $x = 2$, the slope of the line $3y + x = 3y + 2$ is undefined.

52. Undefined

53. $4y + 8x = 6$

$4y = -8x + 6$

$y = -2x + \frac{3}{2}$ $(y = mx + b)$

The slope is -2.

54. $-\frac{6}{5}$

55. $y - 6 = 14$

$y = 20$ The graph of $y = 20$ is a horizontal line.

Since $y - 6 = 14$ is equivalent to $y = 20$, the slope of the line $y - 6 = 14$ is 0.

56. 0

57. $3y - 2x = 5 + 9y - 2x$

$3y = 5 + 9y$

$0 = 5 + 6y$

$-5 = 6y$

$-\frac{5}{6} = y$ The graph of $y = -\frac{5}{6}$ is a horizontal line.

Since $3y - 2x = 5 + 9y - 2x$ is equivalent to $y = -\frac{5}{6}$, the slope of the line $3y - 2x = 5 + 9y - 2x$ is 0.

58. 0

59. $7x - 3y = -2x + 1$

$-3y = -9x + 1$

$y = 3x - \frac{1}{3}$ $(y = mx + b)$

The slope is 3.

60. $\frac{3}{2}$

61. Familiarize. The cost of the taxi ride is $1 the first $\frac{1}{2}$ mile and $1.20 per mile [30¢ per $\frac{1}{4}$ mile = $1.20 per mile] for the mileage beyond the first $\frac{1}{2}$ mile. We let x represent the number of miles from Johnson Street to Elm Street. Then $x - \frac{1}{2}$ represents the number of miles beyond the first $\frac{1}{2}$ mile.

Translate. Write an equation.

Cost for first $\frac{1}{2}$ mile	+	Cost per additional mile	·	Mileage beyond first $\frac{1}{2}$ mile	=	Total cost
$1	+	$1.20	·	$\left(x - \frac{1}{2}\right)$	=	$5.20

Carry out. We solve the equation.

$$1 + 1.20\left(x - \frac{1}{2}\right) = 5.20$$
$$1 + 1.2x - 0.6 = 5.20$$
$$1.2x + 0.4 = 5.20$$
$$1.2x = 4.80$$
$$x = \frac{4.80}{1.2}$$
$$x = 4$$

Check. The first $\frac{1}{2}$ mile costs $1. The remaining $3\frac{1}{2}$ miles $\left(4 - \frac{1}{2} = 3\frac{1}{2}\right)$ cost $1.20 per mile which is $1.20 \times 3\frac{1}{2}$, or $4.20. The total cost is

\$1 + \$4.20 or \$5.20. The value checks.

State. It is 4 miles from Johnson Street to Elm Street.

62. $F = \dfrac{f(c - v_s)}{c - v_0}$

63. 

64. 

65. Observe that for each point (x,y), $y = -\frac{1}{25}x$. To find other points of the line, choose any number for x (other than -100 and 0, which are already given) and find the corresponding y-value. When x = 50, $y = -\frac{1}{25}(50)$, or -2, giving the point (50,-2). When x = 25, $y = -\frac{1}{25}(25)$, or -1, giving the point (25,-1). When x = -25, $y = -\frac{1}{25}(-25)$, or 1, giving the point (-25,1). When x = -50, $y = -\frac{1}{25}(-50)$, or 2, giving the point (-50,2). There are many more correct answers.

66. $y = \frac{4}{5}x - 4$

67. $y = mx + b \quad (m \neq 0)$

To find the x-intercept, let y = 0.

$$y = mx + b$$
$$0 = mx + b$$
$$-b = mx$$
$$-\frac{b}{m} = x$$

Thus, the x-intercept is $\left(-\frac{b}{m},0\right)$

68. $\frac{5}{8}$

69. a) Let $(5b,-6c) = (x_1,y_1)$ and $(b,-c) = (x_2,y_2)$.

$$\text{Slope} = \frac{y_2 - y_1}{x_2 - x_1} = \frac{-c - (-6c)}{b - 5b} = \frac{5c}{-4b} = -\frac{5c}{4b}$$

b) Let $(b,d) = (x_1,y_1)$ and $(b,d + e) = (x_2,y_2)$.

$$\text{Slope} = \frac{y_2 - y_1}{x_2 - x_1} = \frac{(d + e) - d}{b - b} = \frac{e}{0}$$

Since we cannot divide by 0, the slope is undefined.

c) Let $(c + f,a + d) = (x_1,y_1)$ and $(c - f,-a - d) = (x_2,y_2)$

$$\text{Slope} = \frac{y_2 - y_1}{x_2 - x_1} = \frac{(-a - d) - (a + d)}{(c - f) - (c + f)}$$
$$= \frac{-a - d - a - d}{c - f - c - f}$$
$$= \frac{-2a - 2d}{-2f}$$
$$= \frac{-2(a + d)}{-2f}$$
$$= \frac{a + d}{f}$$

70. The slope of equation B is $\frac{1}{2}$ the slope of equation A.

Exercise Set 2.5

1. $y - y_1 = m(x - x_1)$ Point-slope equation

$y - 2 = 4(x - 3)$ Substituting 4 for m, 3 for x_1, and 2 for y_1

We graph the equation by plotting (3,2), counting off a slope of $\frac{4}{1}$, and drawing a line.

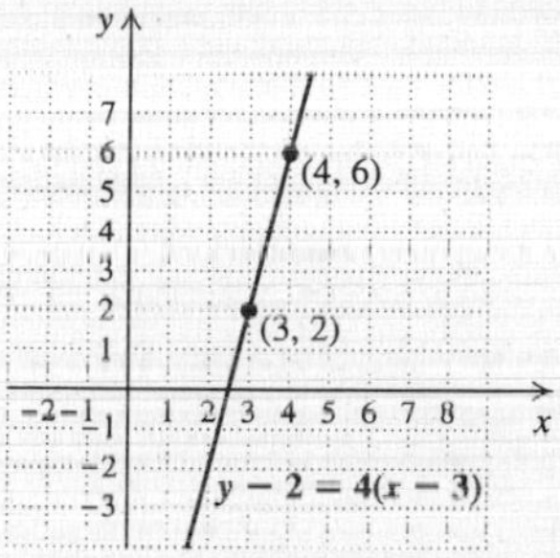

2. $y - 4 = 5(x - 5)$

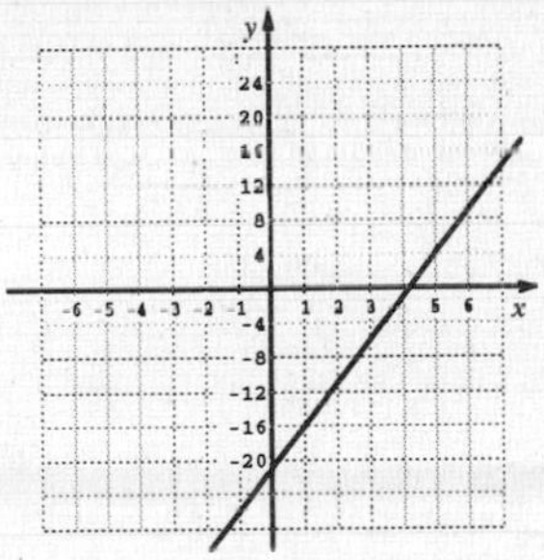

3. $y - y_1 = m(x - x_1)$ Point-slope equation

$y - 7 = -2(x - 4)$ Substituting -2 for m, 4 for x_1, and 7 for y_1

We graph the equation by plotting (4,7), counting off a slope of $\frac{-2}{1}$, and drawing a line.

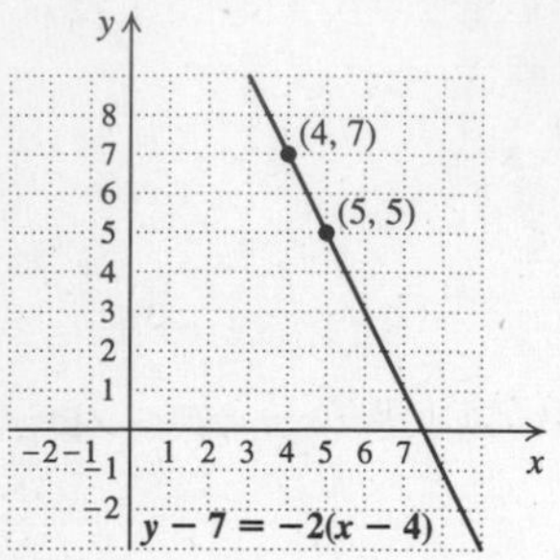

4. $y - 3 = -3(x - 7)$

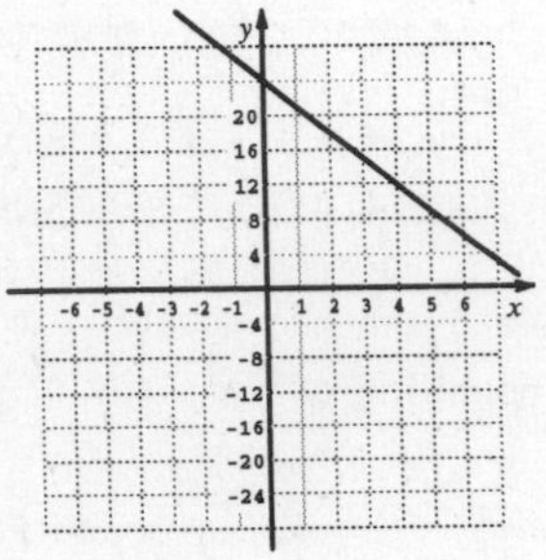

5. $y - y_1 = m(x - x_1)$ Point-slope equation

$y - (-4) = 3[x - (-2)]$ Substituting 3 for m, -2 for x_1, and -4 for y_1

We graph the equation by plotting (-2,-4), counting off a slope of $\frac{3}{1}$, and drawing a line.

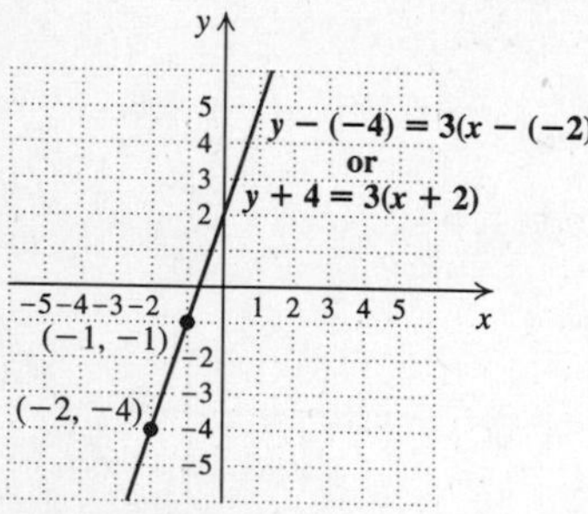

6. $y - (-7) = x - (-5)$

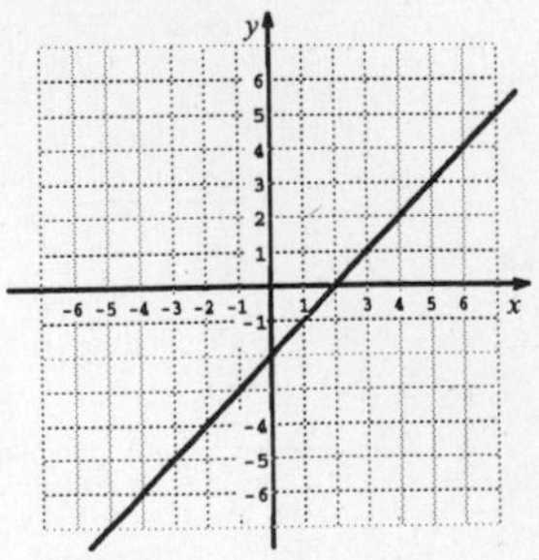

7. $y - y_1 = m(x - x_1)$ Point-slope equation

$y - 0 = -2(x - 8)$ Substituting -2 for m, 8 for x_1, and 0 for y_1

We graph the equation by plotting (8,0), counting off a slope of $\frac{2}{-1}$, and drawing a line.

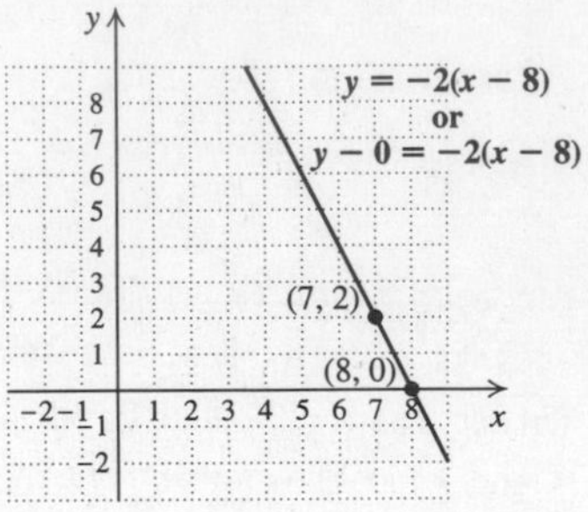

8. $y - 0 = -3[x - (-2)]$

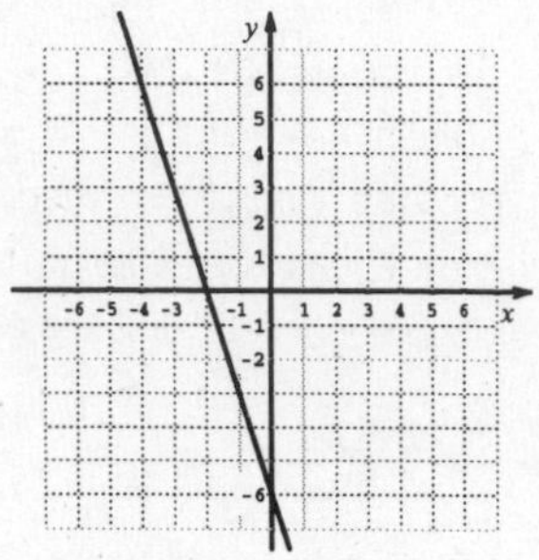

9. $y - y_1 = m(x - x_1)$ Point-slope equation

$y - (-7) = 0(x - 0)$ Substituting 0 for m, 0 for x_1, and -7 for y_1

Since the slope is 0, we graph the equation by drawing a horizontal line through (0,-7).

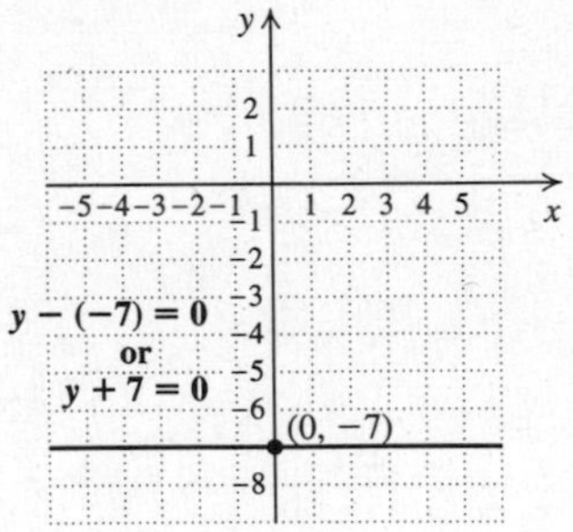

10. $y - 4 = 0(x - 0)$

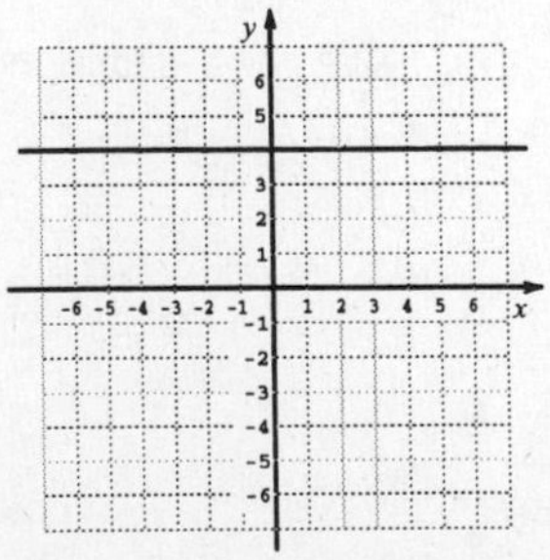

11. $y - y_1 = m(x - x_1)$ Point-slope equation

$y - (-1) = \frac{3}{4}(x - 5)$ Substituting $\frac{3}{4}$ for m, 5 for x_1, and -1 for y_1

We graph the equation by plotting (5,-1), counting off a slope of $\frac{3}{4}$, and drawing a line.

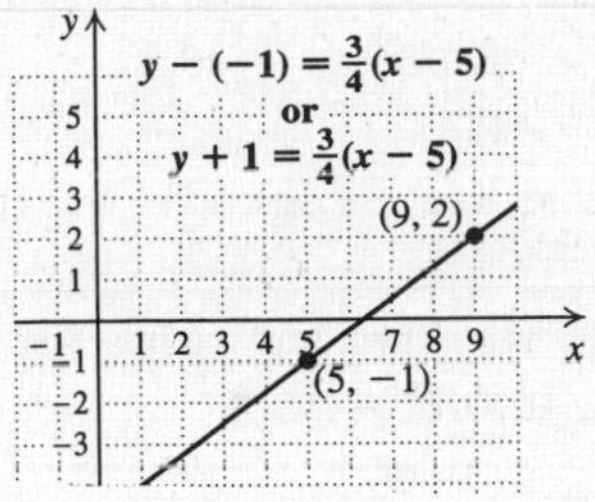

12. $y - 7 = \frac{2}{5}[x - (-3)]$

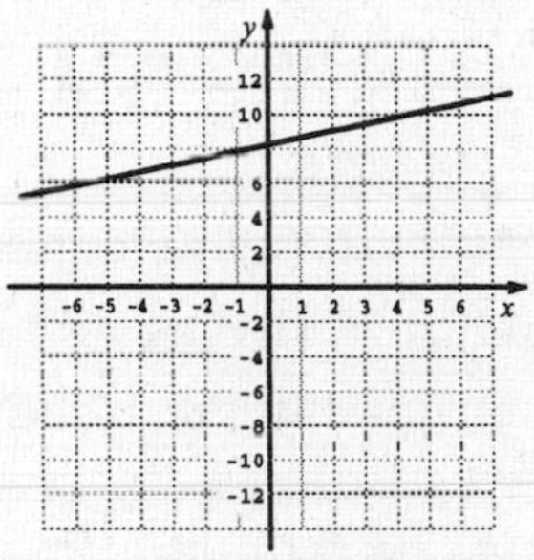

13. $y - y_1 = m(x - x_1)$ Point-slope equation

$y - (-3) = 5(x - 2)$ Substituting 5 for m, 2 for x_1, and -3 for y_1

$y + 3 = 5x - 10$ Simplifying

$y = 5x - 13$ Subtracting 3 on both sides

14. $y = -4x + 1$

15. $y - y_1 = m(x - x_1)$ Point-slope equation

$y - (-7) = -\frac{2}{3}(x - 4)$ Substituting $-\frac{2}{3}$ for m, 4 for x_1, and -7 for y_1

$y + 7 = -\frac{2}{3}x + \frac{8}{3}$ Simplifying

$y = -\frac{2}{3}x - \frac{13}{3}$ Subtracting 7 on both sides

16. $y = \frac{3}{7}x + \frac{13}{7}$

17. $y - y_1 = m(x - x_1)$ Point-slope equation

$y - (-4) = -0.6[x - (-3)]$ Substituting -0.6 for m, -3 for x_1, and -4 for y_1

$y + 4 = -0.6(x + 3)$

$y + 4 = -0.6x - 1.8$

$y = -0.6x - 5.8$

18. $y = -3.1x + 13.5$

19. First find the slope of the line:

$$m = \frac{6 - 4}{5 - 1} = \frac{2}{4} = \frac{1}{2}$$

Use the point-slope equation with $m = \frac{1}{2}$ and $(1,4) = (x_1,y_1)$. (We could let $(5,6) = (x_1,y_1)$ instead and obtain an equivalent equation.)

$y - 4 = \frac{1}{2}(x - 1)$

$y - 4 = \frac{1}{2}x - \frac{1}{2}$

$y = \frac{1}{2}x + \frac{7}{2}$

$f(x) = \frac{1}{2}x + \frac{7}{2}$ Using function notation

20. $f(x) = -\frac{5}{2}x + 11$

21. First find the slope of the line:

$$m = \frac{2 - (-1)}{2 - (-1)} = \frac{2 + 1}{2 + 1} = \frac{3}{3} = 1$$

Use the point slope equation with m = 1 and $(-1,-1) = (x_1,y_1)$.

$y - (-1) = 1[(x - (-1)]$

$y + 1 = x + 1$

$y = x$

$f(x) = x$ Using function notation

22. $f(x) = x$

23. First find the slope of the line:

$$m = \frac{5 - 0}{0 - (-2)} = \frac{5}{2}$$

Use the point slope equation with $m = \frac{5}{2}$ and $(-2,0) = (x_1,y_1)$.

$y - 0 = \frac{5}{2}[x - (-2)]$

$y = \frac{5}{2}(x + 2)$

$y = \frac{5}{2}x + 5$

$f(x) = \frac{5}{2}x + 5$

24. $f(x) = \frac{1}{2}x - 3$

25. First find the slope of the line:

$$m = \frac{3 - 5}{-5 - 3} = \frac{-2}{-8} = \frac{1}{4}$$

Use the point-slope equation with $m = \frac{1}{4}$ and $(3,5) = (x_1,y_1)$.

$$y - 5 = \frac{1}{4}(x - 3)$$
$$y - 5 = \frac{1}{4}x - \frac{3}{4}$$
$$y = \frac{1}{4}x + \frac{17}{4}$$
$$f(x) = \frac{1}{4}x + \frac{17}{4}$$

26. $f(x) = \frac{1}{5}x + \frac{26}{5}$

27. First find the slope of the line:

$$m = \frac{2 - 0}{5 - 0} = \frac{2}{5}$$

Use the point-slope equation with $m = \frac{2}{5}$ and $(0,0) = (x_1,y_1)$.

$$y - 0 = \frac{2}{5}(x - 0)$$
$$y = \frac{2}{5}x$$
$$f(x) = \frac{2}{5}x$$

28. $f(x) = \frac{3}{7}x$

29. First find the slope of the line:

$$m = \frac{-1 - (-7)}{-2 - (-4)} = \frac{-1 + 7}{-2 + 4} = \frac{6}{2} = 3$$

Use the point-slope equation with $m = 3$ and $(-4,-7) = (x_1,y_1)$.

$$y - (-7) = 3[x - (-4)]$$
$$y + 7 = 3(x + 4)$$
$$y + 7 = 3x + 12$$
$$y = 3x + 5$$
$$f(x) = 3x + 5$$

30. $f(x) = \frac{3}{2}x$

31. a) We form pairs of the type (t,R) where t is the number of years since 1930 and R is the record. We have two pairs, (0,46.8) and (40,43.8). These are two points on the graph of the linear function we are seeking. We use the point-slope form to write an equation relating R and t:

$$m = \frac{43.8 - 46.8}{40 - 0} = \frac{-3}{40} = -0.075$$

$$R - 46.8 = -0.075(t - 0)$$
$$R - 46.8 = -0.075t$$
$$R = -0.075t + 46.8$$
$$R(t) = -0.075t + 46.8 \quad \text{Using function notation}$$

b) To predict the record in 1995, we find R(65):

$$R(65) = -0.075(65) + 46.8$$
$$= 41.9 \quad \text{(Using 3 significant digits)}$$

We predict that the record will be 41.9 seconds in 1995.

To predict the record in 1998, we find R(68):

$$R(68) = -0.075(68) + 46.8$$
$$= 41.7$$

We predict that the record will be 41.7 seconds in 1998.

c) Substitute 40 for R(t) and solve for t:

$$40 = -0.075t + 46.8$$
$$-6.8 = -0.075t$$
$$91 \approx t$$

The record will be 40 seconds about 91 years after 1930, or in 2021.

32. a) $R(t) = -0.0075t + 3.85$

b) 3.34 minutes; 3.31 minutes

c) One-third of the way through 2003

33. a) We form the pairs (0, 132.7) and (4, 150). We use the point-slope form to write an equation relating A and t:

$$m = \frac{150 - 132.7}{4 - 0} = \frac{17.3}{4} = 4.325$$

$$A - 132.7 = 4.325(t - 0)$$
$$A - 132.7 = 4.325t$$
$$A = 4.325t + 132.7$$
$$A(t) = 4.325t + 132.7 \quad \text{Using function notation}$$

b) To predict the amount of PAC contributions in 1998, we find A(12):

$$A(12) = 4.325(12) + 132.7$$
$$= 184.6$$

We predict that the amount of PAC contributions in 1998 will be \$184.6 million.

34. a) $A(p) = -2.5p + 21.5$

b) 11.5 million lb

35. a) We form the pairs (0, 14.5) and (8, 23.5). We use the point-slope form to write an equation relating N and t:

$$m = \frac{23.5 - 14.5}{8 - 0} = \frac{9}{8} = 1.125$$

$$N - 14.5 = 1.125(t - 0)$$
$$N - 14.5 = 1.125t$$
$$N = 1.125t + 14.5$$
$$N(t) = 1.125t + 14.5 \quad \text{Using function notation}$$

b) To predict the amount recycled in 1997, we find N(17):

$$N(17) = 1.125(17) + 14.5$$
$$= 33.625$$

We predict that 33.625 million tons of garbage will be recycled in 1997.

36. a) $A(p) = 2p - 7$

b) 1 million lb

37. Familiarize. We form the pairs (1000, 101,000) and (1250, 126,000) and plot these data points, choosing suitable scales on the two axes. We will let a represent the amount spent on radio advertising and I represent the resulting sales increase.

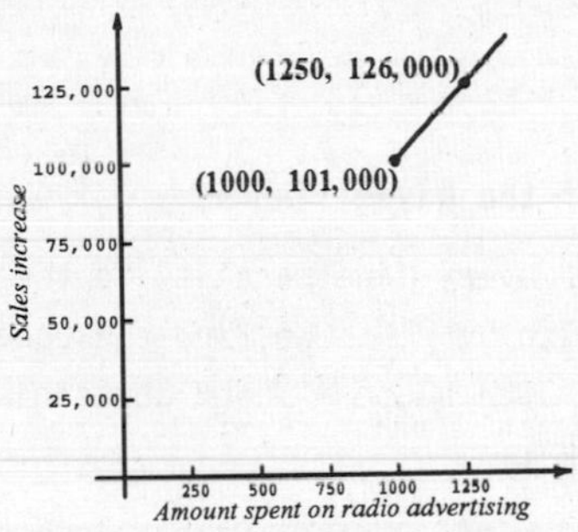

Translate. We seek an equation relating I and a. We find the slope and use the point-slope form:

$$m = \frac{126,000 - 101,000}{1250 - 1000} = \frac{25,000}{250} = 100$$

$$I - 101,000 = 100(a - 1000)$$
$$I - 101,000 = 100a - 100,000$$
$$I = 100a + 1000$$

Carry out. Using function notation, we have

$$I(a) = 100a + 1000.$$

To predict the increase in sales when \$1500 is spent on radio advertising, we find

$$I(1500) = 100 \cdot 1500 + 1000$$
$$= 151,000.$$

To predict the increase in sales when \$2000 is spent on radio advertising, we find

$$I(2000) = 100 \cdot 2000 + 1000$$
$$= 201,000$$

Check. We observe that the results seem reasonable since (1500, 151,000) and (2000, 201,000) appear to lie on the graph.

State. When \$1500 is spent on radio advertising, sales increase by \$151,000. When \$2000 is spent on radio advertising, sales increase by \$201,000.

38. 31,205 items

39. a) We form the pairs (0, 65) and (20, 68). We use the point-slope form to write an equation relating E and t:

$$m = \frac{68 - 65}{20 - 0} = \frac{3}{20} = 0.15$$

$$E - 65 = 0.15(t - 0)$$
$$E - 65 = 0.15t$$
$$E = 0.15t + 65$$
$$E(t) = 0.15t + 65 \quad \text{Using function notation}$$

b) To predict the life expectancy of males in the year 2000, we find E(50):

$$E(50) = 0.15(50) + 65$$
$$= 72.5$$

We predict that the life expectancy of males in the year 2000 will be 72.5 years.

40. a) $P(d) = 0.03d + 1$

b) 21.7 atm

41. We first solve for y and determine the slope of each line.

$$x + 6 = y$$
$$y = x + 6 \quad \text{Reversing the order}$$

The slope of $y = x + 6$ is 1.

$$y - x = -2$$
$$y = x - 2$$

The slope of $y = x - 2$ is 1.

The slopes are the same; the lines are parallel.

42. Yes

43. We first solve for y and determine the slope of each line.

$$y + 3 = 5x$$
$$y = 5x - 3$$

The slope of $y = 5x - 3$ is 5.

$$3x - y = -2$$
$$3x + 2 = y$$
$$y = 3x + 2 \quad \text{Reversing the order}$$

The slope of $y = 3x + 2$ is 3.

The slopes are not the same; the lines are not parallel.

44. No

45. We determine the slope of each line.

The slope of $y = 3x + 9$ is 3.

$$2y = 6x - 2$$
$$y = 3x - 1$$

The slope of $y = 3x - 1$ is 3.

The slopes are the same; the lines are parallel.

46. Yes

47. First solve the equation for y and determine the slope of the given line.

$x + 2y = 6$ Given line

$2y = -x + 6$

$y = -\frac{1}{2}x + 3$

The slope of the given line is $-\frac{1}{2}$.

The slope of every line parallel to the given line must also be $-\frac{1}{2}$. We find the equation of the line with slope $-\frac{1}{2}$ and containing the point (3,7).

$y - y_1 = m(x - x_1)$ Point-slope equation

$y - 7 = -\frac{1}{2}(x - 3)$ Substituting

$y - 7 = -\frac{1}{2}x + \frac{3}{2}$

$y = -\frac{1}{2}x + \frac{17}{2}$

48. $y = 3x + 3$

49. First solve the equation for y and determine the slope of the given line.

$5x - 7y = 8$ Given line

$5x - 8 = 7y$

$\frac{5}{7}x - \frac{8}{7} = y$

$y = \frac{5}{7}x - \frac{8}{7}$

The slope of the given line is $\frac{5}{7}$.

The slope of every line parallel to the given line must also be $\frac{5}{7}$. We find the equation of the line with slope $\frac{5}{7}$ and containing the point (2,-1).

$y - y_1 = m(x - x_1)$ Point-slope equation

$y - (-1) = \frac{5}{7}(x - 2)$ Substituting

$y + 1 = \frac{5}{7}x - \frac{10}{7}$

$y = \frac{5}{7}x - \frac{17}{7}$

50. $y = -2x - 13$

51. First solve the equation for y and determine the slope of the given line.

$3x - 9y = 2$ Given line

$3x - 2 = 9y$

$\frac{1}{3}x - \frac{2}{9} = y$

The slope of the given line is $\frac{1}{3}$.

The slope of every line parallel to the given line must also be $\frac{1}{3}$. We find the equation of the line with slope $\frac{1}{3}$ and containing the point (-6,2).

$y - y_1 = m(x - x_1)$ Point-slope equation

$y - 2 = \frac{1}{3}[x - (-6)]$ Substituting

$y - 2 = \frac{1}{3}(x + 6)$

$y - 2 = \frac{1}{3}x + 2$

$y = \frac{1}{3}x + 4$

52. $y = -\frac{5}{2}x - \frac{35}{2}$

53. First solve the equation for y and determine the slope of the given line.

$3x + 2y = -7$ Given line

$2y = -3x - 7$

$y = -\frac{3}{2}x - \frac{7}{2}$

The slope of the given line is $-\frac{3}{2}$.

The slope of every line parallel to the given line must also be $-\frac{3}{2}$. We find the equation of the line with slope $-\frac{3}{2}$ and containing the point (-3,-2).

$y - y_1 = m(x - x_1)$ Point-slope equation

$y - (-2) = -\frac{3}{2}[x - (-3)]$ Substituting

$y + 2 = -\frac{3}{2}(x + 3)$

$y + 2 = -\frac{3}{2}x - \frac{9}{2}$

$y = -\frac{3}{2}x - \frac{13}{2}$

54. $y = \frac{6}{5}x + \frac{39}{5}$

55. We determine the slope of each line.

The slope of $y = 4x - 5$ is 4.

$4y = 8 - x$

$4y = -x + 8$

$y = -\frac{1}{4}x + 2$

The slope of $4y = 8 - x$ is $-\frac{1}{4}$.

The product of their slopes is $4\left(-\frac{1}{4}\right)$, or -1; the lines are perpendicular.

56. No

57. We determine the slope of each line.

$x + 2y = 5$

$2y = -x + 5$

$y = -\frac{1}{2}x + \frac{5}{2}$

The slope of $x + 2y = 5$ is $-\frac{1}{2}$.

$$2x + 4y = 8$$
$$4y = -2x + 8$$
$$y = -\frac{1}{2}x + 2.$$

The slope of $2x + 4y = 8$ is $-\frac{1}{2}$.

The product of their slopes is $\left(-\frac{1}{2}\right)\left(-\frac{1}{2}\right)$, or $\frac{1}{4}$; the lines are not perpendicular. For the lines to be perpendicular, the product must be -1.

58. Yes

59. First solve the equation for y and determine the slope of the given line.

$2x + y = -3$ Given line

$y = -2x - 3$

The slope of the given line is -2.

The slope of a perpendicular line is given by the opposite of the reciprocal of -2, $\frac{1}{2}$.

We find the equation of the line with slope $\frac{1}{2}$ containing the point (2,5).

$y - y_1 = m(x - x_1)$ Point-slope equation

$y - 5 = \frac{1}{2}(x - 2)$ Substituting

$y - 5 = \frac{1}{2}x - 1$

$y = \frac{1}{2}x + 4$

60. $y = -3x + 12$

61. First solve the equation for y and determine the slope of the given line.

$3x + 4y = 5$ Given line

$4y = -3x + 5$

$y = -\frac{3}{4}x + \frac{5}{4}$

The slope of the given line is $-\frac{3}{4}$.

The slope of a perpendicular line is given by the opposite of the reciprocal of $-\frac{3}{4}$, $\frac{4}{3}$.

We find the equation of the line with slope $\frac{4}{3}$ and containing the point (3,-2).

$y - y_1 = m(x - x_1)$ Point-slope equation

$y - (-2) = \frac{4}{3}(x - 3)$ Substituting

$y + 2 = \frac{4}{3}x - 4$

$y = \frac{4}{3}x - 6$

62. $y = -\frac{2}{5}x - \frac{31}{5}$

63. First solve the equation for y and determine the slope of the given line.

$2x + 5y = 7$ Given line

$5y = -2x + 7$

$y = -\frac{2}{5}x + \frac{7}{5}$

The slope of the given line is $-\frac{2}{5}$.

The slope of a perpendicular line is given by the opposite of the reciprocal of $-\frac{2}{5}$, $\frac{5}{2}$.

We find the equation of the line with slope $\frac{5}{2}$ and containing the point (0,9).

$y - y_1 = m(x - x_1)$ Point-slope equation

$y - 9 = \frac{5}{2}(x - 0)$ Substituting

$y - 9 = \frac{5}{2}x$

$y = \frac{5}{2}x + 9$

64. $y = -2x - 10$

65. First solve the equation for y and find the slope of the given line.

$3x - 5y = 6$

$-5y = -3x + 6$

$y = \frac{3}{5}x - \frac{6}{5}$

The slope of the given line is $\frac{3}{5}$. The slope of a perpendicular line is given by the opposite of the reciprocal of $\frac{3}{5}$, $-\frac{5}{3}$.

We find the equation of the line with slope $-\frac{5}{3}$ and containing the point (-4,-7).

$y - y_1 = m(x - x_1)$ Point-slope equation

$y - (-7) = -\frac{5}{3}[x - (-4)]$

$y + 7 = -\frac{5}{3}(x + 4)$

$y + 7 = -\frac{5}{3}x - \frac{20}{3}$

$y = -\frac{5}{3}x - \frac{41}{3}$

66. $y = -\frac{2}{7}x + \frac{27}{7}$

67. Familiarize. We find the sales tax by multiplying the listed price by the tax rate. We let x represent the price and 5%x represent the sales tax.

Translate. Listed price + Sales tax = Total price

x + 5%x = \$36.75

Carry out. We solve the equation.

$x + 0.05x = 36.75$

$1.05x = 36.75$

$x = \frac{36.75}{1.05}$

$x = 35$

Check. If the listed price is \$35, then the sales tax is 5%·\$35, or \$1.75. The total price is 35 + 1.75, or \$36.75. The value checks.

State. The price of the radio before the tax is \$35.

68. 69 points

69. Familiarize. Let n represent the number.

Translate.

15% of what number is 12.4?

15% × n = 12.4

Carry out. Solve the equation.

$15\%n = 12.4$

$0.15n = 12.4$

$15n = 1240$ Multiplying by 100 to clear decimals

$n = \frac{1240}{15}$

$n = \frac{248}{3}$, or $82\frac{2}{3}$

Check. Left to the student.

State. 15% of $82\frac{2}{3}$ is 12.4.

70. $-\frac{13}{12}$

71. ◈

72. ◈

73. Familiarize. We form the pairs (32, 0) and (212, 100) and plot these data points, choosing suitable scales on the two axes. We will let F represent the Fahrenheit temperature and C represent the Celsius temperature.

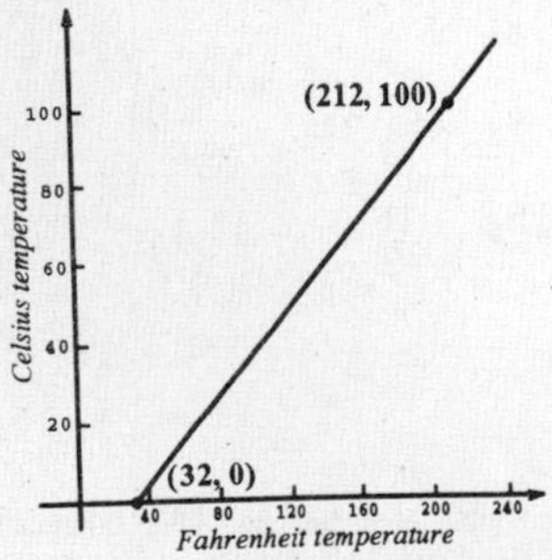

Translate. We seek an equation relating C and F. We find the slope and use the point-slope form:

$$m = \frac{100 - 0}{212 - 32} = \frac{100}{180} = \frac{5}{9}$$

$C - 0 = \frac{5}{9}(F - 32)$

$C = \frac{5}{9}(F - 32)$

$C = \frac{5}{9}F - \frac{160}{9}$

Carry out. Using function notation, we have

$C(F) = \frac{5}{9}F - \frac{160}{9}$.

To find the Celsius temperature that corresponds to 70° F, we find

$C(70) = \frac{5}{9}(70) - \frac{160}{9}$

≈ 21.1

Check. We observe that the result seems reasonable since (70, 21.1) appears to lie on the graph.

State. A temperature of about 21.1° C corresponds to 70° F.

74. \$1300

75. a) We have two pairs, (-1,3) and (2,4). Use the point-slope form:

$$m = \frac{4 - 3}{2 - (-1)} = \frac{1}{2 + 1} = \frac{1}{3}$$

$y - 3 = \frac{1}{3}[x - (-1)]$

$y - 3 = \frac{1}{3}(x + 1)$

$y - 3 = \frac{1}{3}x + \frac{1}{3}$

$y = \frac{1}{3}x + \frac{10}{3}$

$f(x) = \frac{1}{3}x + \frac{10}{3}$ Using function notation

b) $f(x) = \frac{1}{3}(3) + \frac{10}{3} = 1 + \frac{10}{3} = \frac{13}{3}$

c) $f(a) = \frac{1}{3}(a) + \frac{10}{3} = \frac{a}{3} + \frac{10}{3}$

If f(a) = 100, we have

$100 = \frac{a}{3} + \frac{10}{3}$

$300 = a + 10$ Multiplying by 3

$290 = a$

76. a) $g(x) = x - 8$

b) -10

c) 83

77. Find the slope of $5y - kx = 7$:

$5y - kx = 7$

$5y = kx + 7$

$y = \frac{k}{5}x + \frac{7}{5}$

The slope is $\frac{k}{5}$.

Find the slope of the line containing the points (7,-3) and (-2,5):

$$m = \frac{5 - (-3)}{-2 - 7} = \frac{5 + 3}{-9} = -\frac{8}{9}$$

If the lines are parallel, their slopes must be equal:

$$\frac{k}{5} = -\frac{8}{9}$$

$$k = -\frac{40}{9}$$

78. 7

Exercise Set 2.6

1. Since f(1) = -3·1 + 1 = -2 and g(1) = 1^2 + 2 = 3, we have f(1) + g(1) = -2 + 3 = 1.

2. 1

3. Since f(-1) = -3(-1) + 1 = 4 and g(-1) = $(-1)^2$ + 2 = 3 we have f(-1) + g(-1) = 4 + 3 = 7.

4. 13

5. Since f(-7) = -3(-7) + 1 = 22 and g(-7) = $(-7)^2$ + 2 = 51, we have f(-7) - g(-7) = 22 - 51 = -29.

6. -11

7. Since f(5) = -3(5) + 1 = -14 and g(5) = 5^2 + 2 = 27, we have f(5) - g(5) = -14 - 27 = -41.

8. -29

9. Since f(2) = -3·2 + 1 = -5 and g(2) = 2^2 + 2 = 6, we have f(2)·g(2) = -5·6 = -30.

10. -88

11. Since f(-3) = -3(-3) + 1 = 10 and g(-3) = $(-3)^2$ + 2 = 11, we have f(-3)·g(-3) = 10·11 = 110.

12. 234

13. Since f(0) = -3·0 + 1 = 1 and g(0) = 0^2 + 2 = 2, we have f(0)/g(0) = 1/2.

14. -2/3

15. Using our work in Exercise 11, we have

f(-3)/g(-3) = 10/11.

16. 13/18

17.
$$(F + G)(x) = F(x) + G(x)$$
$$= x^2 - 3 + 4 - x$$
$$= x^2 - x + 1 \quad \text{Combining like terms}$$
$$(F + G)(-3) = (-3)^2 - (-3) + 1 = 9 + 3 + 1 = 13$$

18. 7

19. Using our work in Exercise 17, we have

(F + G)(x) = $x^2 - x + 1$.

20. $a^2 - a + 1$

21.
$$(F - G)(x) = F(x) - G(x)$$
$$= x^2 - 3 - (4 - x)$$
$$= x^2 - 3 - 4 + x$$
$$= x^2 + x - 7$$
$$(F - G)(-4) = (-4)^2 + (-4) - 7 = 16 - 4 - 7 = 5$$

22. 13

23.
$$(F\cdot G)(x) = F(x)\cdot G(x)$$
$$= (x^2 - 3)(4 - x)$$
$$= 4x^2 - x^3 - 12 + 3x$$
$$(F\cdot G)(2) = 4\cdot 2^2 - 2^3 - 12 + 3\cdot 2 =$$
$$16 - 8 - 12 + 6 = 2$$

24. 6

25. Using our work in Exercise 23, we have:
$$(F\cdot G)(-3) = 4(-3)^2 - (-3)^3 - 12 + 3(-3)$$
$$= 4\cdot 9 - (-27) - 12 - 9$$
$$= 36 + 27 - 12 - 9$$
$$= 42$$

26. 104

27.
$$(F/G)(x) = F(x)/G(x)$$
$$= \frac{x^2 - 3}{4 - x}$$
$$(F/G)(0) = \frac{0^2 - 3}{4 - 0} = -\frac{3}{4}$$

28. $-\frac{2}{3}$

29. Using our work in Exercise 27, we have
$$(F/G)(-2) = \frac{(-2)^2 - 3}{4 - (-2)} = \frac{4 - 3}{4 + 2} = \frac{1}{6}.$$

30. $-\frac{2}{5}$

31. The domain of f and of g is all real numbers. Thus, Domain of f + g = Domain of f - g = Domain of f·g = {x|x is a real number}.

32. {x|x is a real number}

33. Because division by 0 is undefined, we have
Domain of f = {x|x is a real number and $x \neq 2$},
and
Domain of g = {x|x is a real number}.
Thus, Domain of f + g = Domain of f - g = Domain of f·g = {x|x is a real number and $x \neq 2$}.

34. {x|x is a real number and $x \neq 4$}

35. Because division by 0 is undefined, we have
Domain of f = {x|x is a real number and $x \neq 0$},
and
Domain of g = {x|x is a real number}.
Thus, Domain of f + g = Domain of f - g = Domain of f·g = {x|x is a real number and $x \neq 0$}.

36. {x|x is a real number and $x \neq 0$}

37. Because division by 0 is undefined, we have
Domain of f = {x|x is a real number and $x \neq 1$},
and
Domain of g = {x|x is a real number}.
Thus, Domain of f + g = Domain of f - g = Domain of f·g = {x|x is a real number and $x \neq 1$}.

38. {x|x is a real number and $x \neq 5$}

39. Because division by 0 is undefined, we have
Domain of f = {x|x is a real number and $x \neq 2$},
and
Domain of g = {x|x is a real number and $x \neq 4$}.
Thus, Domain of f + g = Domain of f - g = Domain of f·g = {x|x is a real number and $x \neq 2$ and $x \neq 4$}.

40. {x|x is a real number and $x \neq 3$ and $x \neq 2$}

41. Because division by 0 is undefined, we have
Domain of f = {x|x is a real number and $x \neq -2$},
and
Domain of g = {x|x is a real number and $x \neq 4$}.
Thus, Domain of f + g = Domain of f - g = Domain of f·g = {x|x is a real number and $x \neq -2$ and $x \neq 4$}.

42. {x|x is a real number and $x \neq 3$ and $x \neq 5$}

43. Domain of f = Domain of g = {x|x is a real number}.
Since g(x) = 0 when x - 3 = 0, we have g(x) = 0 when x = 3. We conclude that Domain of f/g = {x|x is a real number and $x \neq 3$}.

44. {x|x is a real number and $x \neq 5$}

45. Domain of f = Domain of g = {x|x is a real number}.
Since g(x) = 0 when 2x - 8 = 0, we have g(x) = 0 when x = 4. We conclude that Domain of f/g = {x|x is a real number and $x \neq 4$}.

46. {x|x is a real number and $x \neq 3$}

47. Domain of f = {x|x is a real number and $x \neq 4$}.
Domain of g = {x|x is a real number}.
Since g(x) = 0 when 5 - x = 0, we have g(x) = 0 when x = 5. We conclude that Domain of f/g = {x|x is a real number and $x \neq 4$ and $x \neq 5$}.

48. {x|x is a real number and $x \neq 2$ and $x \neq 7$}

49. Domain of f = {x|x is a real number and $x \neq -1$}.
Domain of g = {x|x is a real number}.
Since g(x) = 0 when 2x + 5 = 0, we have g(x) = 0 when $x = -\frac{5}{2}$. We conclude that Domain of f/g = $\left\{x \middle| x \text{ is a real number and } x \neq -1 \text{ and } x \neq -\frac{5}{2}\right\}$.

50. $\left\{x \middle| x \text{ is a real number and } x \neq 2 \text{ and } x \neq -\frac{7}{3}\right\}$

51. Domain of f = {x|x is a real number and $x \neq 4$}
Domain of g = {x|x is a real number}. Since g(x) = 0 when $3x^2 = 0$, we have g(x) = 0 when x = 0. We conclude that Domain of f/g = {x|x is a real number and $x \neq 4$ and $x \neq 0$}.

52. {x|x is a real number and $x \neq 0$ and $x \neq -3$}

53. Domain of f = {x|x is a real number}.
Domain of g = {x|x is a real number and $x \neq 4$}.
Since g(x) = 0 when $\frac{x - 1}{x - 4} = 0$, we have g(x) = 0 when x = 1. We conclude that Domain of f/g = {x|x is a real number and $x \neq 4$ and $x \neq 1$}.

54. {x|x is a real number and $x \neq -2$ and $x \neq -3$}

55. Domain of f = {x|x is a real number and $x \neq 2$}.
Domain of g = {x|x is a real number and $x \neq 4$}.
Since g(x) = 0 when $\frac{x - 3}{x - 4} = 0$, we have g(x) = 0 when x = 3. We conclude that Domain of f/g = {x|x is a real number and $x \neq 2$, $x \neq 4$, and $x \neq 3$}.

56. {x|x is a real number and $x \neq -3$, $x \neq -2$, and $x \neq -1$}

57. From the graph we see that Domain of F = {x|$0 \leqslant x \leqslant 9$} and Domain of G = {x|$3 \leqslant x \leqslant 10$}. Then Domain of F + G = {x|$3 \leqslant x \leqslant 9$}. Since G(x) is never 0, Domain of F/G = {x|$3 \leqslant x \leqslant 9$}.

58. {x|$3 \leqslant x \leqslant 9$ and $x \neq 6$ and $x \neq 8$}

59. We use $(F + G)(x) = F(x) + G(x)$.

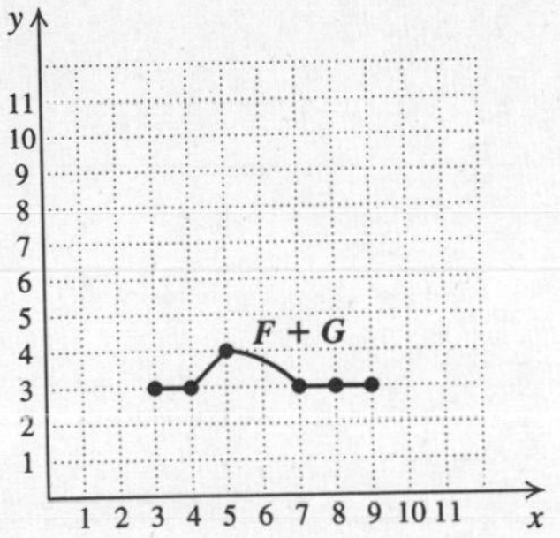

60.

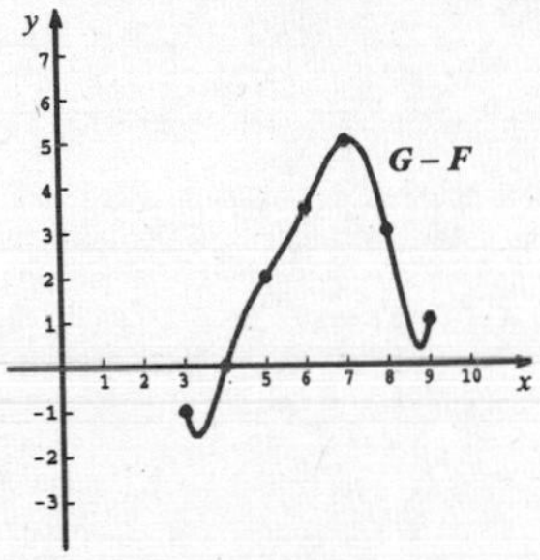

61. $5x - 7 = 0$

$5x = 7$ Adding 7 on both sides

$x = \frac{7}{5}$ Multiplying by $\frac{7}{5}$ on both sides

62. -37

63. Familiarize. The average score on n tests is the sum of the n test scores divided by n. If the average on 4 tests is 78.5, then the sum of the 4 scores is 4(78.5), or 314. Let s represent the score required on the fifth test to raise the average to 80.

Translate.

Sum of 5 scores divided by 5 is 80.

$(314 + s) \div 5 = 80$

Carry out. Solve the equation.

$\frac{314 + s}{5} = 80$

$314 + s = 400$

$s = 86$

Check. If the sum of the 5 scores is 314 + 86, or 400, then the average is 400/5, or 80. The value checks.

State. A score of 86 is needed on the fifth test to raise the average to 80.

64. 6.31×10^{-6}

65.

66.

67. Domain of p = $\{x|x$ is a real number and $x \neq 1\}$ (since p is not defined for $x = 1$).

Domain of q = $\{x|x$ is a real number and $x \neq 2\}$. Since $q(x) = 0$ when $\frac{x - 3}{x - 2} = 0$, we have $q(x) = 0$ when $x = 3$. We conclude that Domain of p/q = $\{x|x$ is a real number and $x \neq 1$, $x \neq 2$, and $x \neq 3\}$.

68. $\left\{x \middle| x \text{ is a real number and } x \neq \frac{3}{2} \text{ and } -1 < x < 5\right\}$

69. The domain of each function is the set of first coordinates for that function.

Domain of f = $\{-2,-1,0,1,2\}$, and

Domain of g = $\{-4,-3,-2,-1,0,1\}$.

Domain of f + g = Domain of f - g = Domain of f·g = $\{-2,-1,0,1\}$.

Since $g(-1) = 0$, we conclude that Domain of f/g = $\{-2,0,1\}$.

70. 5; 15; 2/3

71. Domain of F = $\{x|x$ is a real number and $x \neq 4\}$.

Domain of G = $\{x|x$ is a real number and $x \neq 3\}$.

Since $G(x) = 0$ when $\frac{x^2 - 4}{x - 3} = 0$, we have $G(x) = 0$ when $x^2 - 4 = 0$, or when $x = 2$ or $x = -2$. We conclude that Domain of F/G = $\{x|x$ is a real number and $x \neq 4$, $x \neq 3$, $x \neq 2$, and $x \neq -2\}$.

72. $\left\{x \middle| x \text{ is a real number and } x \neq -\frac{5}{2}, x \neq -3, x \neq 1, \text{ and } x \neq -1\right\}$.

73. Answers may vary. $f(x) = \frac{1}{x + 2}$, $g(x) = \frac{1}{x - 5}$

74. Answers may vary.

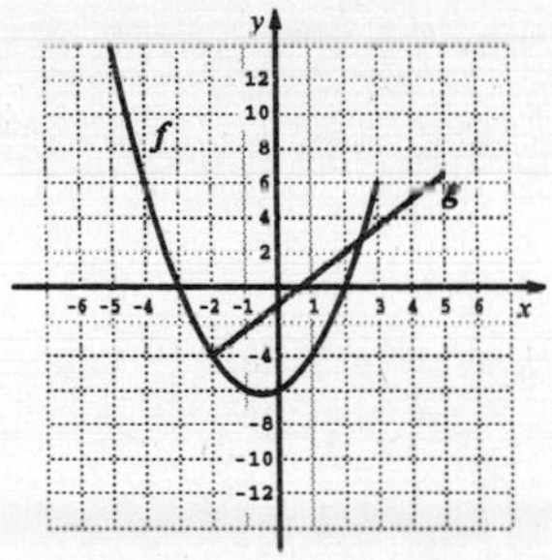

75. a) $(k + \ell + n)(1983) = k(1983) + \ell(1983) + n(1983) \approx 67$ million

b) $(\ell + n)(1986) = \ell(1986) + n(1986) \approx 50$ million

c) $(\ell + k)(1986) = (k + \ell + n)(1986) - n(1986) \approx 80 - 30 \approx 50$ million

Exercise Set 3.1

1. Familiarize. Let x = the first number and y = the second number.

Translate.

The sum of two numbers is -42.

Rewording: The first number	plus	the second number	is	-42.
x	$+$	y	$=$	-42
The first number	minus	the second number	is	52.
x	$-$	y	$=$	52

We have a system of equations:

$x + y = -42,$

$x - y = 52$

2. $x - y = 11,$

$3x + 2y = 123$

3. Familiarize. Let x = the number of white scarves sold and y = the number of printed scarves sold.

Translate: Organize the information in a table.

Kind of scarf	White	Printed	Total
Number sold	x	y	40
Price	$4.95	$7.95	
Amount taken in	4.95x	7.95y	282

The "Number sold" row of the table gives us one equation:

$x + y = 40$

The "Amount taken in" row gives us a second equation:

$4.95x + 7.95y = 282$

We have a system of equations:

$x + y = 40,$

$4.95x + 7.95y = 282$

We can multiply the second equation on both sides by 100 to clear the decimals:

$x + y = 40,$

$495x + 795y = 28,200$

4. $x + y = 45,$

$8.50x + 9.75y = 398.75$

5. Familiarize. Let x = the measure of the first angle and y = the measure of the second angle.

Translate.

Two angles are complementary.

Rewording: The sum of the measures is 90°.

$x + y = 90$

The sum of the measures of the first angle angle and half the second angle is 64°.

Rewording: The measure of the first angle	plus	half the measure of second angle	is 64°.	
x	$+$	$\frac{1}{2}y$	$=$	64

We have a system of equations:

$x + y = 90,$

$x + \frac{1}{2}y = 64$

6. $x + y = 180,$

$x = 2y - 3$

7. Familiarize. Let x = the number of children's plates and y = the number of adult's plates served.

Translate. We organize the information in a table.

Kind of plate	Children's	Adult's	Total
Number sold	x	y	250
Price	$3.50	$7.00	
Amount collected	3.50x	7.00y	1347.50

The "Number sold" row of the table gives us one equation:

$x + y = 250$

The "Amount collected" row gives us a second equation:

$3.50x + 7.00y = 1347.50$

We have a system of equations:

$x + y = 250,$

$3.50x + 7.00y = 1347.50$

We can multiply both sides of the second equation by 10 to clear the decimals:

$x + y = 250,$

$35x + 70y = 13,475$

8. $x + y = 18,$

$2x + y = 30$

9. Familiarize. The tennis court is a rectangle with perimeter 228 ft. Let ℓ = length and w = width. Recall that for a rectangle with length ℓ and width w, the perimeter P is given by $P = 2\ell + 2w$.

Translate. The formula for perimeter gives us one equation:

$2\ell + 2w = 228$

The statement relating width and length gives us a second equation:

Width is 42 ft less than length.

$w = \ell - 42$

We have a system of equations:

$2w + 2\ell = 228,$

$w = \ell - 42$

10. $2\ell + 2w = 288,$

$\ell = 44 + w$

11. Familiarize. Let x = the number of 2-pointers made and y = the number of 3-pointers made.

Translate. We organize the information in a table.

Type of score	2-pointer	3-pointer	Total
Number scored	x	y	40
Points scored	$2x$	$3y$	89

The "Number scored" row of the table gives us one equation:

$$x + y = 40$$

The "Points scored" row gives us a second equation:

$$2x + 3y = 89$$

We have a system of equations:

$$x + y = 40,$$
$$2x + 3y = 89$$

12. $x + y = 77,$

$3.00 + 1.50y = 213$

13. Familiarize. Let x = the number of units of lumber produced and y = the number of units of plywood produced in a day.

Translate. We organize the information in a table.

Product	Lumber	Plywood	Total
Number of units produced	x	y	400
Profit per unit	\$20	\$30	
Amount of profit	$20x$	$30y$	11,000

The "Number of units produced" row of the table gives us one equation:

$$x + y = 400$$

The "Amount of profit" row gives us a second equation:

$$20x + 30y = 11{,}000$$

We have a system of equations:

$$x + y = 400,$$
$$20x + 30y = 11{,}000$$

14. $2x + y = 60,$

$x = 9 + y$

15. Familiarize. Let x = number of 30-sec commercials and y = number of 60-sec commercials. The total time used by x 30-sec commercials is $30x$; the total time used by y 60-sec commercials is $60y$. Also note that 10 min = 10×60, or 600 sec.

Translate.

Total number of commercials is 12.

$$x + y = 12$$

Total commercial time is 10 min, or 600 sec.

$$30x + 60y = 600$$

We have a system of equations:

$$x + y = 12,$$
$$30x + 60y = 600$$

16. $x + y = 152,$

$x = 5 + 6y$

17. We use alphabetical order for the variables. We replace x by 1 and y by 2.

$4x - y = 2$			$10x - 3y = 4$		
$4\cdot1 - 2$	?	2	$10\cdot1 - 3\cdot2$	?	4
$4 - 2$			$10 - 6$		
2		TRUE	4		TRUE

The pair (1,2) makes both equations true, so it is a solution of the system.

18. Yes

19. We use alphabetical order for the variables. We replace x by 2 and y by 5.

$y = 3x - 1$			$2x + y = 4$		
5	?	$3\cdot2 - 1$	$2\cdot2 + 5$	?	4
		$6 - 1$	$4 + 5$		
		5 TRUE	9		FALSE

The pair (2,5) is not a solution of $2x + y = 4$. Therefore it is not a solution of the system of equations.

20. No

21. We replace x by 1 and y by 5.

$x + y = 6$			$y = 2x + 3$		
$1 + 5$	?	6	5	?	$2\cdot1 + 3$
6		TRUE			$2 + 3$
					5 TRUE

The pair (1,5) makes both equations true, so it is a solution of the system.

22. Yes

23. We replace a by 2 and b by -7.

$3a + b = -1$		$2a - 3b = -8$	
$3\cdot 2 + (-7)$? -1		$2\cdot 2 - 3\cdot(-7)$? -8	
$6 - 7$		$4 + 21$	
-1	TRUE	25	FALSE

The pair (2,-7) is not a solution of 2a - 3b = -8. Therefore it is not a solution of the system of equations.

24. No

25. We replace x by 3 and y by 1.

$3x + 4y = 13$		$5x - 4y = 11$	
$3\cdot 3 + 4\cdot 1$? 13		$5\cdot 3 - 4\cdot 1$? 11	
$9 + 4$		$15 - 4$	
13	TRUE	11	TRUE

The pair (3,1) makes both equations true, so it is a solution of the system.

26. No

27. Graph both lines on the same set of axes.

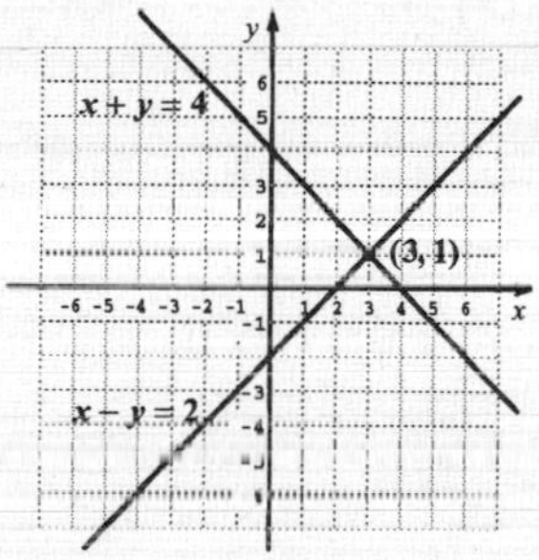

The solution (point of intersection) seems to be the point (3,1).

Check:

$x + y = 4$		$x - y = 2$	
$3 + 1$? 4		$3 - 1$? 2	
4	TRUE	2	TRUE

The solution is (3,1).

28. (4,1)

29. Graph both lines on the same set of axes.

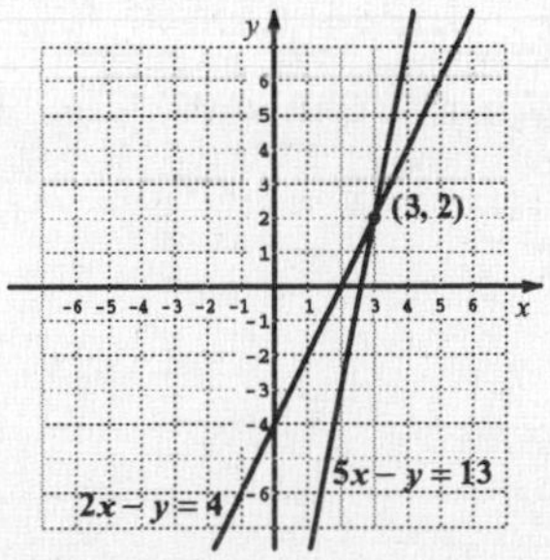

The solution (point of intersection) seems to be the point (3,2).

Check:

$2x - y = 4$		$5x - y = 13$	
$2\cdot 3 - 2$? 4		$5\cdot 3 - 2$? 13	
$6 - 2$		$15 - 2$	
4	TRUE	13	TRUE

The solution is (3,2).

30. (2,-1)

31. Graph both lines on the same set of axes.

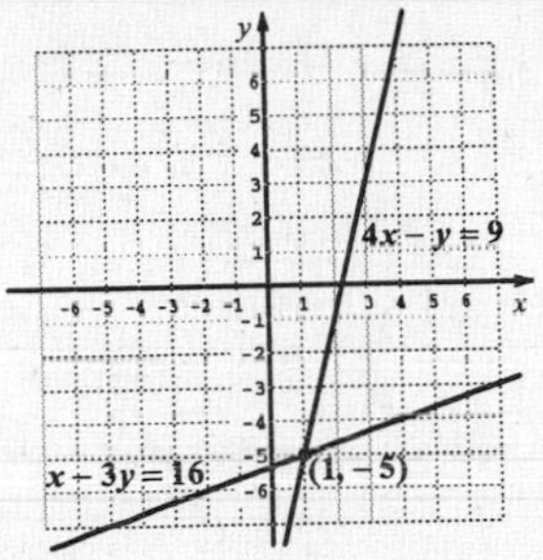

The solution seems to be the point (1,-5).

Check:

$4x - y = 9$		$x - 3y = 16$	
$4\cdot 1 - (-5)$? 9		$1 - 3(-5)$? 16	
$4 + 5$		$1 + 15$	
9	TRUE	16	TRUE

The solution is (1,-5).

32. (4,3)

33. Graph both lines on the same set of axes.

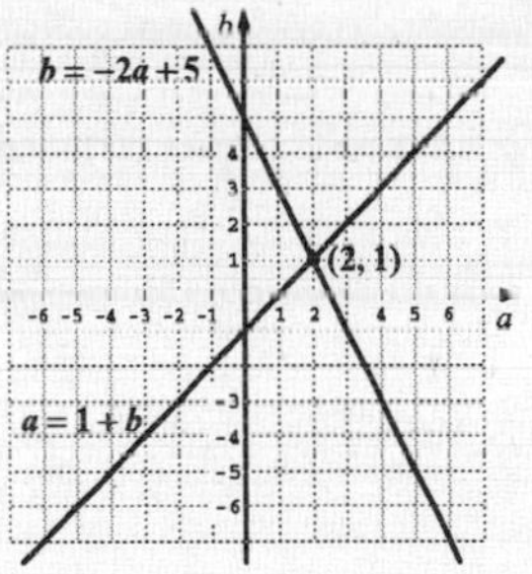

The solution seems to be the point (2,1).

Check:

$a = 1 + b$		$b = -2a + 5$	
2 ? $1 + 1$		1 ? $-2\cdot 2 + 5$	
2	TRUE	$-4 + 5$	
		1	TRUE

The solution is (2,1).

34. (-3,-2)

35. Graph both lines on the same set of axes.

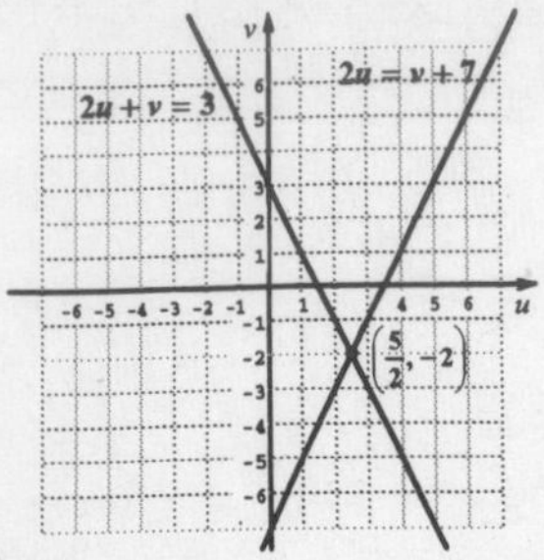

The solution seems to be $\left(\frac{5}{2}, -2\right)$.

Check:

$$\begin{array}{c|c} 2u + v & 3 \\ \hline 2 \cdot \frac{5}{2} + (-2) & 3 \\ 5 - 2 & \\ 3 & \end{array} \quad \text{TRUE}$$

$$\begin{array}{c|c} 2u & v + 7 \\ \hline 2 \cdot \frac{5}{2} & -2 + 7 \\ 5 & 5 \end{array} \quad \text{TRUE}$$

The solution is $\left(\frac{5}{2}, -2\right)$.

36. (7,2)

37. Graph both lines on the same set of axes.

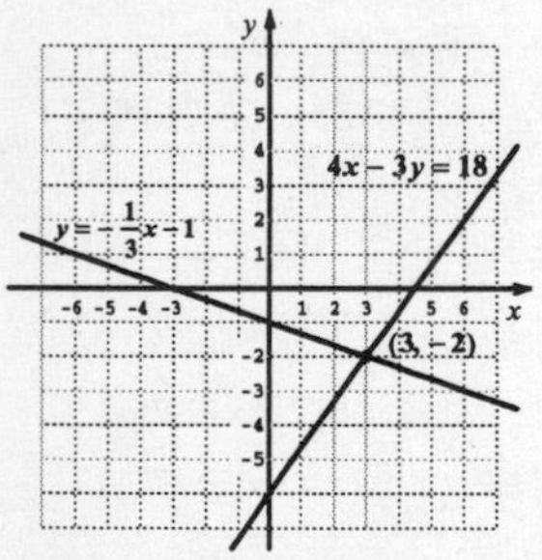

The ordered pair (3,-2) checks in both equations. It is the solution.

38. (4,0)

39. Graph both lines on the same set of axes.

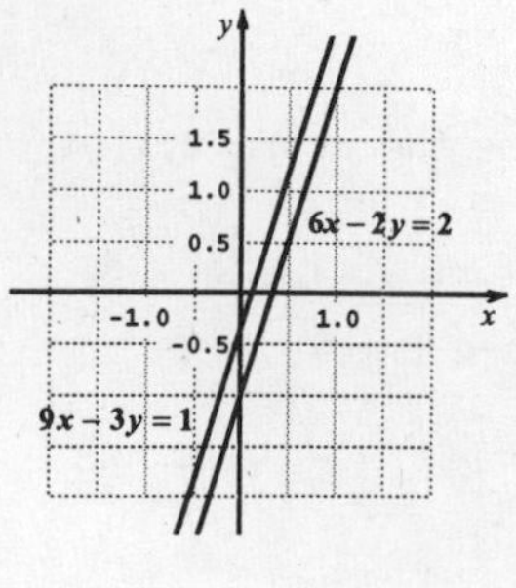

The lines are parallel. There solution set is $\emptyset$.

40. $\emptyset$

41. Graph both lines on the same set of axes.

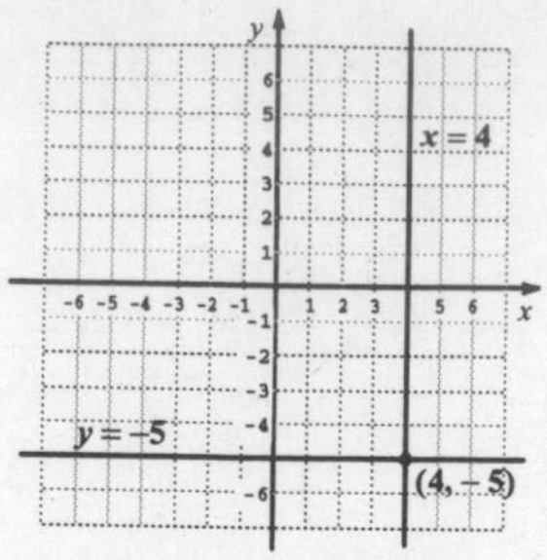

The ordered pair (4,-5) checks in both equations. It is the solution.

42. (-3,2)

43. Graph both lines on the same set of axes.

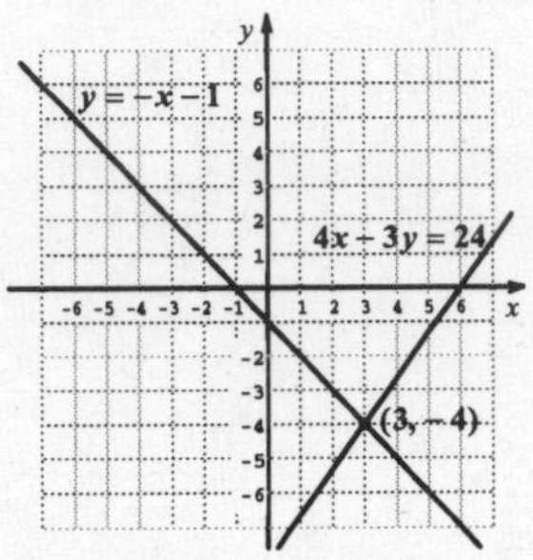

The ordered pair (3,-4) checks in both equations. It is the solution.

44. $\left(\frac{1}{2}, 3\right)$

45. Graph both lines on the same set of axes.

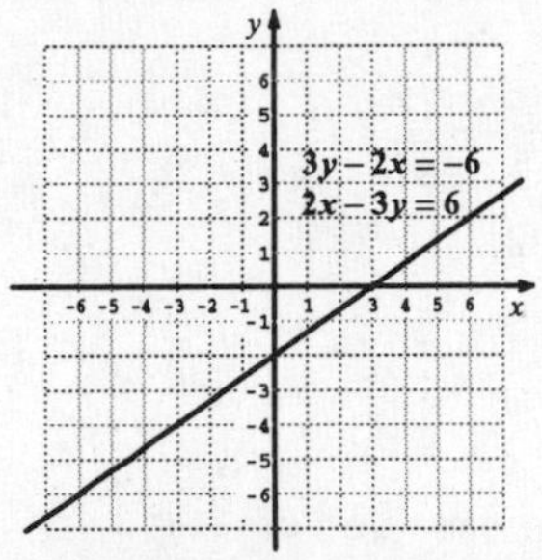

The graphs are the same. Any solution of one equation is a solution of the other. Each equation has infinitely many solutions. The solution set is the set of all pairs (x,y) for which 2x - 3y = 6, or $\{(x,y) \mid 2x - 3y = 6\}$. (In place of 2x - 3y = 6 we could have used 3y - 2x = -6 since the two equations are equivalent.)

46. $\{(x,y) \mid y = 3 - x\}$

47. A system of equations is consistent if it has at least one solution. Of the systems under consideration, only the one in Exercise 39 has no solution. Therefore, all except the system in Exercise 39 are consistent.

48. All except 40

49. A system of two equations in two variables is dependent if it has infinitely many solutions. Only the system in Exercise 45 is dependent.

50. 46

51. $3x + 4 = x - 2$

$2x + 4 = -2$ Adding $-x$ on both sides

$2x = -6$ Adding -4 on both sides

$x = -3$ Multiplying by $\frac{1}{2}$ on both sides

52. -35

53. $4x - 5x = 8x - 9 + 11x$

$-x = 19x - 9$ Collecting like terms

$-20x = -9$ Adding $-19x$ on both sides

$x = \frac{9}{20}$ Multiplying by $-\frac{1}{20}$ on both sides

54. $b = a - 4Q$

55.

56.

57. 1983 and 1990

58. 1989

59. 1987

60. $A = -\frac{17}{4}$, $B = -\frac{12}{5}$

61. a) There are many correct answers. One can be found by expressing the sum and difference of the two numbers:

$x + y = 6,$

$x - y = 4$

b) There are many correct answers. For example, write an equation in two variables. Then write a second equation by multiplying the left side of the first equation by one constant and multiplying the right side by another constant.

$x + y = 1,$

$2x + 2y = 3$

c) There are many correct answers. One can be found by writing an equation in two variables and then writing a constant multiple of that equation:

$x + y = 1,$

$2x + 2y = 2$

62. a) (4,-5); answers may vary

b) Infinitely many

63. Familiarize. Let b = Burt's age and s = his son's age. Ten years ago, Burt's age was $b - 10$ and his son's age was $s - 10$.

Translate.

Burt is twice as old as his son.

$b = 2s$

Ten years ago Burt was three times as old as his son.

$b - 10 = 3(s - 10)$

We have a system of equations:

$b = 2s,$

$b - 10 = 3(s - 10)$

64. $x + y = 46,$

$x - 2 = 2.5(y - 2)$

65. Familiarize. Let ℓ = the length and w = the width. If you cut 6 in. off the width, the new width is $w - 6$.

Translate.

The perimeter is 156 in.

$2\ell + 2w = 156$

If you cut 6 in. off the width the length becomes four times the width.

$\ell = 4(w - 6)$

We have a system of equations:

$2\ell + 2w = 156,$

$\ell = 4(w - 6)$

66. $w + \ell = 104,$

$w - 8 = 3\ell$

67. Graph both equations on the same set of axes.

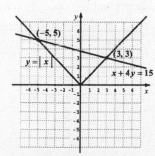

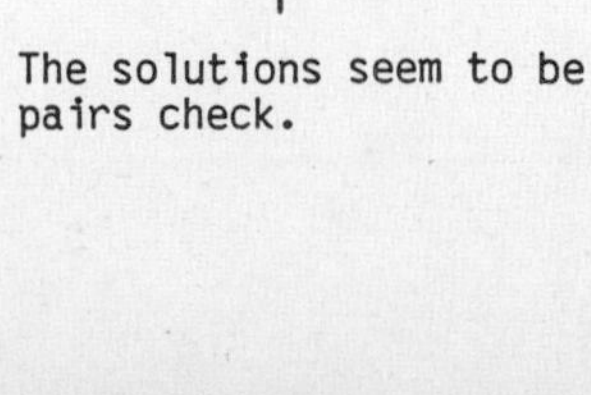

The solutions seem to be (-5,5) and (3,3). Both pairs check.

68. (0,0), (1,1)

69. (-0.39, -1.10)

70. (0.07, -7.95)

71. (-0.13, 0.67)

72. (0.02, 1.25)

Exercise Set 3.2

1. $3x + 5y = 3,$ (1)
   $x = 8 - 4y$ (2)

   We substitute 8 - 4y for x in the first equation and solve for y.

   $3x + 5y = 3$ (1)
   $3(8 - 4y) + 5y = 3$ Substituting
   $24 - 12y + 5y = 3$
   $24 - 7y = 3$
   $-7y = -21$
   $y = 3$

   Next we substitute 3 for y in either equation of the original system and solve for x.

   $x = 8 - 4y$ (2)
   $x = 8 - 4\cdot 3$ Substituting
   $x = 8 - 12$
   $x = -4$

   We check the ordered pair (-4,3).

$3x + 5y = 3$		$x = 8 - 4y$	
$3(-4) + 5\cdot 3 \ ? \ 3$		$-4 \ ? \ 8 - 4\cdot 3$	
$-12 + 15$			$8 - 12$
3	TRUE	-4	TRUE

   Since (-4,3) checks, it is the solution.

2. (2,-3)

3. $9x - 2y = 3,$ (1)
   $3x - 6 = y$ (2)

   We substitute 3x - 6 for y in the first equation and solve for x.

   $9x - 2y = 3$ (1)
   $9x - 2(3x - 6) = 3$ Substituting
   $9x - 6x + 12 = 3$
   $3x + 12 = 3$
   $3x = -9$
   $x = -3$

   Next we substitute -3 for x in either equation of the original system and solve for y.

   $3x - 6 = y$ (2)
   $3(-3) - 6 = y$ Substituting
   $-9 - 6 = y$
   $-15 = y$

   We check the ordered pair (-3,-15).

$9x - 2y = 3$		$3x - 6 = y$	
$9(-3) - 2(-15) \ ? \ 3$		$3(-3) - 6 \ ? \ -15$	
$-27 + 30$		$-9 - 6$	
3	TRUE	-15	TRUE

   Since (-3,-15) checks, it is the solution.

4. $\left(\frac{21}{5}, \frac{12}{5}\right)$

5. $5m + n = 8,$ (1)
   $3m - 4n = 14$ (2)

   We solve the first equation for n.

   $5m + n = 8$ (1)
   $n = 8 - 5m$ (3)

   We substitute 8 - 5m for n in the second equation and solve for m.

   $3m - 4n = 14$ (2)
   $3m - 4(8 - 5m) = 14$ Substituting
   $3m - 32 + 20m = 14$
   $23m - 32 = 14$
   $23m = 46$
   $m = 2$

   Now we substitute 2 for m in Equation (1), (2), or (3). It is easiest to use Equation (3) since it is already solved for n.

   $n = 8 - 5m = 8 - 5(2) = 8 - 10 = -2$

   We check the ordered pair (2,-2).

$5m + n = 8$		$3m - 4n = 14$	
$5\cdot 2 + (-2) \ ? \ 8$		$3\cdot 2 - 4(-2) \ ? \ 14$	
$10 - 2$		$6 + 8$	
8	TRUE	14	TRUE

   Since (2,-2) checks, it is the solution.

6. (2,-7)

7. $4x + 12y = 4,$ (1)
   $-5x + y = 11$ (2)

   We solve the second equation for y.

   $-5x + y = 11$ (2)
   $y = 5x + 11$ (3)

   We substitute 5x + 11 for y in the first equation and solve for x.

   $4x + 12y = 4$ (1)
   $4x + 12(5x + 11) = 4$ Substituting
   $4x + 60x + 132 = 4$
   $64x + 132 = 4$
   $64x = -128$
   $x = -2$

   Now substitute -2 for x in Equation (3).

   $y = 5x + 11 = 5(-2) + 11 = -10 + 11 = 1$

We check the ordered pair (-2,1).

$4x + 12y = 4$		$-5x + y = 11$	
$4(-2) + 12\cdot 1 \ ?\ 4$		$-5(-2) + 1 \ ?\ 11$	
$-8 + 12$		$10 + 1$	
4	TRUE	11	TRUE

Since (-2,1) checks, it is the solution.

8. (4,-1)

9. $3x - y = 1,$ (1)
$2x + 2y = 2$ (2)

We solve the first equation for y.

$3x - y = 1$ (1)

$3x - 1 = y$ (3)

We substitute 3x - 1 for y in the second equation and solve for x.

$$\begin{aligned} 2x + 2y &= 2 && (2)\\ 2x + 2(3x - 1) &= 2 && \text{Substituting}\\ 2x + 6x - 2 &= 2\\ 8x - 2 &= 2\\ 8x &= 4\\ x &= \frac{1}{2} \end{aligned}$$

Next we substitute $\frac{1}{2}$ for x in Equation (3).

$$y = 3x - 1 = 3\cdot\frac{1}{2} - 1 = \frac{3}{2} - 1 = \frac{1}{2}$$

We check the ordered pair $\left(\frac{1}{2},\frac{1}{2}\right)$.

$3x - y = 1$		$2x + 2y = 2$	
$3\cdot\frac{1}{2} - \frac{1}{2} \ ?\ 1$		$2\cdot\frac{1}{2} + 2\cdot\frac{1}{2} \ ?\ 2$	
$\frac{3}{2} - \frac{1}{2}$		$1 + 1$	
1	TRUE	2	TRUE

Since $\left(\frac{1}{2},\frac{1}{2}\right)$ checks, it is the solution.

10. (3,-2)

11. $3x - y = 7,$ (1)
$2x + 2y = 5$ (2)

We solve the first equation for y.

$3x - y = 7$ (1)

$3x - 7 = y$ (3)

We substitute 3x - 7 for y in the second equation and solve for x.

$$\begin{aligned} 2x + 2y &= 5 && (2)\\ 2x + 2(3x - 7) &= 5 && \text{Substituting}\\ 2x + 6x - 14 &= 5\\ 8x - 14 &= 5\\ 8x &= 19\\ x &= \frac{19}{8} \end{aligned}$$

Now we substitute $\frac{19}{8}$ for x in Equation (3).

$$y = 3x - 7 = 3\cdot\frac{19}{8} - 7 = \frac{57}{8} - 7 = \frac{1}{8}$$

The ordered pair $\left(\frac{19}{8},\frac{1}{8}\right)$ checks in both equations. It is the solution.

12. $\left(\frac{25}{23},-\frac{11}{23}\right)$

13. $x + 2y = 6,$ (1)
$x = 4 - 2y$ (2)

We substitute 4 - 2y for x in the first equation and solve for y.

$$\begin{aligned} x + 2y &= 6 && (1)\\ (4 - 2y) + 2y &= 6 && \text{Substituting}\\ 4 - 2y + 2y &= 6\\ 4 &= 6 && \text{Collecting like terms} \end{aligned}$$

We have a false equation. Therefore, there is no solution. The solution set is Ø.

14. Ø

15. $x - 3 = y,$ (1)
$2x - 2y = 6$ (2)

Substitute x - 3 for y in the second equation and solve for x.

$$\begin{aligned} 2x - 2y &= 6 && (2)\\ 2x - 2(x - 3) &= 6 && \text{Substituting}\\ 2x - 2x + 6 &= 6\\ 6 &= 6 && \text{Collecting like terms} \end{aligned}$$

We have an equation that is true for all numbers x and y. The system is dependent and has an infinite number of solutions. Using (1), we can express the solution set as $\{(x,y) | x - 3 = y\}$.

16. $\{(x,y) | 3y = x - 2\}$

17.
$$\begin{aligned} x + 3y &= 7\\ -x + 4y &= 7\\ \hline 0 + 7y &= 14 && \text{Adding}\\ 7y &= 14\\ y &= 2 \end{aligned}$$

Substitute 2 for y in one of the original equations and solve for x.

$$\begin{aligned} x + 3y &= 7\\ x + 3\cdot 2 &= 7 && \text{Substituting}\\ x + 6 &= 7\\ x &= 1 \end{aligned}$$

Check: For (1,2)

$x + 3y = 7$		$-x + 4y = 7$	
$1 + 3\cdot 2 \ ?\ 7$		$-1 + 4\cdot 2 \ ?\ 7$	
$1 + 6$		$-1 + 8$	
7	TRUE	7	TRUE

Since (1,2) checks, it is the solution.

18. (2,7)

19. $2x + y = 6$
$\underline{x - y = 3}$
$3x + 0 = 9$ Adding
$3x = 9$
$x = 3$

Substitute 3 for x in one of the original equations and solve for y.

$2x + y = 6$
$2 \cdot 3 + y = 6$ Substituting
$6 + y = 6$
$y = 0$

We obtain (3,0). This checks, so it is the solution.

20. (10,2)

21. $9x + 3y = -3$
$\underline{2x - 3y = -8}$
$11x + 0 = -11$ Adding
$11x = -11$
$x = -1$

Substitute -1 for x in one of the original equations and solve for y.

$9x + 3y = -3$
$9(-1) + 3y = -3$ Substituting
$-9 + 3y = -3$
$3y = 6$
$y = 2$

We obtain (-1,2). This checks, so it is the solution.

22. $\left(\frac{1}{2},-5\right)$

23. $5x + 3y = 19,$ (1)
$2x - 5y = 11$ (2)

We multiply twice to make two terms become opposites.

From (1): $25x + 15y = 95$ Multiplying by 5
From (2): $\underline{6x - 15y = 33}$ Multiplying by 3
$31x = 128$ Adding
$x = \frac{128}{31}$

Substitute $\frac{128}{31}$ for x in one of the original equations and solve for y.

$5x + 3y = 19$
$5 \cdot \frac{128}{31} + 3y = 19$ Substituting
$\frac{640}{31} + 3y = \frac{589}{31}$
$3y = -\frac{51}{31}$
$\frac{1}{3} \cdot 3y = \frac{1}{3} \cdot \left(-\frac{51}{31}\right)$
$y = -\frac{17}{31}$

We obtain $\left(\frac{128}{31},-\frac{17}{31}\right)$. This checks, so it is the solution.

24. $\left(\frac{10}{21},\frac{11}{14}\right)$

25. $5r - 3s = 24,$ (1)
$3r + 5s = 28$ (2)

We multiply twice to make two terms become opposites.

From (1): $25r - 15s = 120$ Multiplying by 5
From (2): $\underline{9r + 15s = 84}$ Multiplying by 3
$34r = 204$ Adding
$r = 6$

Substitute 6 for r in one of the original equations and solve for s.

$3r + 5s = 28$
$3 \cdot 6 + 5s = 28$ Substituting
$18 + 5s = 28$
$5s = 10$
$s = 2$

We obtain (6,2). This checks, so it is the solution.

26. (1,3)

27. $0.3x - 0.2y = 4,$
$0.2x + 0.3y = 1$

We first multiply each equation by 10 to clear decimals.

$3x - 2y = 40$ (1)
$2x + 3y = 10$ (2)

We use the multiplication principle with both equations of the resulting system.

From (1): $9x - 6y = 120$ Multiplying by 3
From (2): $\underline{4x + 6y = 20}$ Multiplying by 2
$13x = 140$ Adding
$x = \frac{140}{13}$

Substitute $\frac{140}{13}$ for x in one of the equations in which the decimals were cleared and solve for y.

$2x + 3y = 10$

$2 \cdot \frac{140}{13} + 3y = 10$ Substituting

$\frac{280}{13} + 3y = \frac{130}{13}$

$3y = -\frac{150}{13}$

$y = -\frac{50}{13}$

We obtain $\left(\frac{140}{13}, -\frac{50}{13}\right)$. This checks, so it is the solution.

28. (2,3)

29. $\frac{1}{2}x + \frac{1}{3}y = 4,$ (1)

$\frac{1}{4}x + \frac{1}{3}y = 3$ (2)

We first multiply each equation by the LCM of the denominators to clear fractions.

From (1): $3x + 2y = 24$ Multiplying by 6

From (2): $3x + 4y = 36$ Multiplying by 12

We multiply by -1 on both sides of the first equation and then add.

$-3x - 2y = -24$ Multiplying by -1

$3x + 4y = 36$

$2y = 12$ Adding

$y = 6$

Substitute 6 for y in one of the equations in which the fractions were cleared and solve for x.

$3x + 2y = 24$

$3x + 2\cdot 6 = 24$ Substituting

$3x + 12 = 24$

$3x = 12$

$x = 4$

We obtain (4,6). This checks, so it is the solution.

30. (-21,21)

31. $\frac{2}{5}x + \frac{1}{2}y = 2,$ (1)

$\frac{1}{2}x - \frac{1}{6}y = 3$ (2)

We first multiply each equation by the LCM of the denominators to clear fractions.

From (1): $4x + 5y = 20$ Multiplying by 10

From (2): $3x - y = 18$ Multiplying by 6

We multiply by 5 on both sides of the second equation and then add.

$4x + 5y = 20$

$15x - 5y = 90$ Multiplying by 5

$19x = 110$ Adding

$x = \frac{110}{19}$

Substitute $\frac{110}{19}$ for x in one of the equations in which the fractions were cleared and solve for y.

$3x - y = 18$

$3\left(\frac{110}{19}\right) - y = 18$ Substituting

$\frac{330}{19} - y = \frac{342}{19}$

$-y = \frac{12}{19}$

$y = -\frac{12}{19}$

We obtain $\left(\frac{110}{19}, -\frac{12}{19}\right)$. This checks, so it is the solution.

32. (12,15)

33. $2x + 3y = 1,$

$4x + 6y = 2$

Multiply the first equation by -2 and then add.

$-4x - 6y = -2$

$4x + 6y = 2$

$0 = 0$ Adding

We have an equation that is true for all numbers x and y. The system is dependent and has an infinite number of solutions. The solution set can be expressed using either equation. Using the first equation, we have $\{(x,y) | 2x + 3y = 1\}$.

34. $\{(x,y) | 3x - 2y = 1\}$

35. $2x - 4y = 5,$

$2x - 4y = 6$

Multiply the first equation by -1 and then add.

$-2x + 4y = -5$

$2x - 4y = 6$

$0 = 1$

We have a false equation. The system has no solution. The solution set is $\emptyset$.

36. $\emptyset$

37. $5x - 9y = 7,$

$7y - 3x = -5$

We first write the second equation in the form $Ax + By = C$.

$5x - 9y = 7$ (1)

$-3x + 7y = -5$ (2)

We use the multiplication principle with both equations and then add.

From (1): $15x - 27y = 21$ Multiplying by 3

From (2): $-15x + 35y = -25$ Multiplying by 5

$8y = -4$ Adding

$y = -\frac{1}{2}$

Substitute $-\frac{1}{2}$ for y in one of the original equations and solve for x.

$$5x - 9y = 7 \quad (1)$$

$$5x - 9\left(-\frac{1}{2}\right) = 7 \quad \text{Substituting}$$

$$5x + \frac{9}{2} = \frac{14}{2}$$

$$5x = \frac{5}{2}$$

$$x = \frac{1}{2}$$

We obtain $\left(\frac{1}{2},-\frac{1}{2}\right)$. This checks, so it is the solution.

38. (-2,-9)

39. $3(a - b) = 15,$

$4a = b + 1$

We first write each equation in the form $Ax + By = C$.

$$3a - 3b = 15$$

$$4a - b = 1$$

We multiply by -3 on both sides of the second equation and then add.

$$\begin{aligned} 3a - 3b &= 15 \\ -12a + 3b &= -3 \quad \text{Multiplying by -3} \\ \hline -9a \quad &= 12 \end{aligned}$$

$$a = -\frac{12}{9}$$

$$a = -\frac{4}{3}$$

Substitute $-\frac{4}{3}$ for a in one of the equations and solve for b.

$$4a - b = 1$$

$$4\left(-\frac{4}{3}\right) - b = 1$$

$$-\frac{16}{3} - b = \frac{3}{3}$$

$$-b = \frac{19}{3}$$

$$b = -\frac{19}{3}$$

We obtain $\left(-\frac{4}{3},-\frac{19}{3}\right)$. This checks, so it is the solution.

40. (30,6)

41. $x - \frac{1}{10}y = 100,$

$y - \frac{1}{10}x = -100$

We first write the second equation in the form $Ax + By = C$.

$$x - \frac{1}{10}y = 100$$

$$-\frac{1}{10}x + y = -100$$

Next we multiply each equation by 10 to clear fractions.

$$10x - y = 1000$$

$$-x + 10y = -1000$$

We multiply by 10 on both sides of the first equation and then add.

$$\begin{aligned} 100x - 10y &= 10{,}000 \quad \text{Multiplying by 10} \\ -x + 10y &= -1000 \\ \hline 99x \quad &= 9000 \end{aligned}$$

$$x = \frac{9000}{99}$$

$$x = \frac{1000}{11}$$

Substitute $\frac{1000}{11}$ for x in one of the equations in which the fractions were cleared and solve for y.

$$10x - y = 1000$$

$$10\left(\frac{1000}{11}\right) - y = 1000 \quad \text{Substituting}$$

$$\frac{10{,}000}{11} - y = \frac{11{,}000}{11}$$

$$-y = \frac{1000}{11}$$

$$y = -\frac{1000}{11}$$

We obtain $\left(\frac{1000}{11},-\frac{1000}{11}\right)$. This checks, so it is the solution.

42. (-20,20)

43. $0.05x + 0.25y = 22,$

$0.15x + 0.05y = 24$

We first multiply each equation by 100 to clear decimals.

$$5x + 25y = 2200$$

$$15x + 5y = 2400$$

We multiply by -5 on both sides of the second equation and add.

$$\begin{aligned} 5x + 25y &= 2200 \\ -75x - 25y &= -12{,}000 \quad \text{Multiplying by -5} \\ \hline -70x \quad &= -9800 \quad \text{Adding} \end{aligned}$$

$$x = \frac{-9800}{-70}$$

$$x = 140$$

Substitute 140 for x in one of the equations in which the decimals were cleared and solve for y.

$$5x + 25y = 2200$$

$$5\cdot 140 + 25y = 2200 \quad \text{Substituting}$$

$$700 + 25y = 2200$$

$$25y = 1500$$

$$y = 60$$

We obtain (140,60). This checks, so it is the solution.

44. (10,5)

45. Familiarize. Recall the formula for simple interest, $I = Prt$, where I = interest, P = principal, r = interest rate, and t = time (in years).

Translate. Substitute \$17.60 for I, \$320 for P, and $\frac{1}{2}$ for t.

$$I = Prt$$

$$17.60 = 320r\left(\frac{1}{2}\right)$$

Carry out. Solve the equation.

$$17.60 = 320r\left(\frac{1}{2}\right)$$

$$17.60 = 160r$$

$$\frac{17.60}{160} = r$$

$$0.11 = r$$

Check. $320(0.11)\left(\frac{1}{2}\right) = 17.60$. The answer checks.

State. The rate of interest is 0.11, or 11%.

46. $-15y - 39$

47.

48.

49. $3.5x - 2.1y = 106.2,$
$4.1x + 16.7y = -106.28$

Since this is a calculator exercise, you may choose not to clear the decimals. We will do so here, however.

$35x - 21y = 1062$ Multiplying by 10
$410x + 1670y = -10{,}628$ Multiplying by 100

Multiply twice to make two terms become opposites.

$58{,}450x - 35{,}070y = 1{,}773{,}540$ Multiplying by 1670
$8610x + 35{,}070y = -223{,}188$ Multiplying by 21
$67{,}060x = 1{,}550{,}352$ Adding
$x \approx 23.118879$

Substitute 23.118879 for x in one of the equations in which the decimals were cleared and solve for y.

$35x - 21y = 1062$
$35(23.118879) - 21y = 1062$ Substituting
$809.160765 - 21y = 1062$
$-21y = 252.839235$
$y \approx -12.039964$

The numbers check, so the solution is (23.118879,-12.039964).

50. $\left(-\frac{32}{17},\frac{38}{17}\right)$

51. $5x + 2y = a,$
$x - y = b$

We multiply by 2 on both sides of the second equation and then add.

$5x + 2y = a$
$2x - 2y = 2b$ Multiplying by 2
$7x = a + 2b$ Adding
$x = \frac{a + 2b}{7}$

Next we multiply by -5 on both sides of the second equation and then add.

$5x + 2y = a$
$-5x + 5y = -5b$ Multiplying by -5
$7y = a - 5b$
$y = \frac{a - 5b}{7}$

We obtain $\left(\frac{a + 2b}{7},\frac{a - 5b}{7}\right)$. This checks, so it is the solution.

52. $a = 5$, $b = 2$

53. $(0,-3)$ and $\left(-\frac{3}{2},6\right)$ are two solutions of $px - qy = -1$.

Substitute 0 for x and -3 for y.

$p\cdot 0 - q\cdot(-3) = -1$
$3q = -1$
$q = -\frac{1}{3}$

Substitute $-\frac{3}{2}$ for x and 6 for y.

$p\cdot\left(-\frac{3}{2}\right) - q\cdot 6 = -1$
$-\frac{3}{2}p - 6q = -1$

Substitute $-\frac{1}{3}$ for q and solve for p.

$-\frac{3}{2}x - 6\cdot\left(-\frac{1}{3}\right) = -1$
$-\frac{3}{2}p + 2 = -1$
$-\frac{3}{2}p = -3$
$-\frac{2}{3}\cdot\left(-\frac{3}{2}p\right) = -\frac{2}{3}\cdot(-3)$
$p = 2$

Thus, $p = 2$ and $q = -\frac{1}{3}$.

54. $m = -\frac{1}{2}$, $b = \frac{5}{2}$

55. $\frac{1}{x} - \frac{3}{y} = 2,$ or $\frac{1}{x} - 3 \cdot \frac{1}{y} = 2,$

$\frac{6}{x} + \frac{5}{y} = -34$ $\quad 6 \cdot \frac{1}{x} + 5 \cdot \frac{1}{y} = -34$

Substitute u for $\frac{1}{x}$ and v for $\frac{1}{y}$.

$u - 3v = 2,$ (1)

$6u + 5v = -34$ (2)

$-6u + 18v = -12$ Multiplying (1) by -6

$6u + 5v = -34$ (2)

$23v = -46$ Adding

$v = -2$

Substitute -2 for v in (1).

$u - 3v = 2$

$u - 3(-2) = 2$

$u = -4$

If $u = -4$, then $\frac{1}{x} = -4$, so $x = -\frac{1}{4}$.

If $v = -2$, then $\frac{1}{y} = -2$, so $y = -\frac{1}{2}$.

The solution is $\left(-\frac{1}{4}, -\frac{1}{2}\right)$.

56. $\left(-\frac{1}{5}, \frac{1}{10}\right)$

Exercise Set 3.3

1. The Familiarize and Translate steps were done in Exercise 1 of Exercise Set 3.1.

Carry out. We solve the system of equations

$x + y = -42,$

$x - y = 52$

where x = the first number and y = the second number. We use the elimination method.

$x + y = -42$

$x - y = 52$

$2x = 10$ Adding

$x = 5$

Substitute 5 for x in one of the equations and solve for y.

$x + y = -42$

$5 + y = -42$

$y = -47$

Check. The sum of the numbers is 5 + (-47), or -42. The difference is 5 - (-47), or 52. The numbers check.

State. The numbers are 5 and -47.

2. 29 and 18

3. The Familiarize and Translate steps were done in Exercise 3 of Exercise Set 3.1.

Carry out. We solve the system of equations

$x + y = 40,$

$495x + 795y = 28{,}200$

where x = the number of white scarves sold and y = the number of printed scarves sold. We use the elimination method. We eliminate x by multiplying the first equation by -495 and then adding it to the second.

$-495x - 495y = -19{,}800$ Multiplying by -495

$495x + 795y = 28{,}200$

$300y = 8400$ Adding

$y = 28$

Substitute 28 for y in one of the original equations and solve for x.

$x + 28 = 40$

$x = 12$

Check. The number of scarves sold is 12 + 28, or 40.

Money from white scarves = $4.95 × 12 = $ 59.40

Money from printed scarves = $7.95 × 28 = $222.60

Total = $282.00

The numbers check.

State. 12 white scarves and 28 printed scarves were sold.

4. 32 of the $8.50 pens, 13 of the $9.75 pens

5. The Familiarize and Translate steps were done in Exercise 5 of Exercise Set 3.1.

Carry out. We solve the system of equations

$x + y = 90,$ (1)

$x + \frac{1}{2}y = 64$ (2)

where x = the measure of the first angle and y = the measure of the second angle. We use elimination.

$x + y = 90$ (1)

$-x - \frac{1}{2}y = -64$ Multiplying (2) by -1

$\frac{1}{2}y = 26$

$y = 52$

Substitute 52 for y in (1) and solve for x.

$x + 52 = 90$

$x = 38$

Check. The sum of the angle measures is $38° + 52°$, or $90°$, so the angles are complementary. The sum of the measures of the first angle and half the second angle is $38° + \frac{1}{2} \cdot 52°$, or $38° + 26°$, or $64°$. The numbers check.

State. The measures of the angles are 38° and 52°.

6. 119°, 61°

7. The Familiarize and Translate steps were done in Exercise 7 of Exercise Set 3.1.

Carry out. We solve the system of equations

$x + y = 250,$ (1)

$35x + 70y = 13,475$ (2)

where x = the number of children's plates and y = the number of adult's plates served. We use elimination.

$-35x - 35y = -8750$ Multiplying (1) by -35

$35x + 70y = 13,475$

$35y = 4725$ Adding

$y = 135$

Substitute 135 for y in (1) and solve for x.

$x + 135 = 250$

$x = 115$

Check. Number of dinners: 115 + 135 = 250.

Money from children's plates: \$3.50(115) = \$402.50

Money from adult's plates: \$7.00(135) = \$945.00

Total money: \$402.50 + \$945.00 = \$1347.50

The numbers check.

State. 115 children's plates and 135 adult's plates were served.

8. 12 field goals, 6 free throws

9. The Familiarize and Translate steps were done in Exercise 9 of Exercise Set 3.1.

Carry out. We solve the system of equations

$2w + 2\ell = 228,$ (1)

$w = \ell - 42$ (2)

where w = the width and ℓ = the length. We use substitution.

$2(\ell - 42) + 2\ell = 228$ Substituting ℓ - 42 for w in (1)

$2\ell - 84 + 2\ell = 228$

$4\ell - 84 = 228$

$4\ell = 312$

$\ell = 78$

$w = 78 - 42$ Substituting 78 for ℓ in (2)

$w = 36$

Check. The perimeter is 2·36 + 2·78, or 72 + 156, or 228. Since 78 - 42 = 36, the width is 42 ft less than the length. The numbers check.

State. The width is 36 ft, and the length is 78 ft.

10. Length: 94 ft, width: 50 ft

11. The Familiarize and Translate steps were done in Exercise 11 of Exercise Set 3.1.

Carry out. We solve the system of equations

$x + y = 40,$ (1)

$2x + 3y = 89$ (2)

where x = the number of 2-pointers and y = the number of 3-pointers. We use elimination.

$-2x - 2y = -80$ Multiplying (1) by -2

$2x + 3y = 89$

$y = 9$ Adding

Substitute 9 for y in (1) and solve for x.

$x + 9 = 40$

$x = 31$

Check. Total field goals: 31 + 9 = 40

Points from 2-pointers: 2·31 = 62

Points from 3-pointers: 3·9 = 27

Total points: 62 + 27 = 89

The numbers check.

State. The Cougars made 31 2-pointers and 9 3-pointers.

12. 65 general interest films, 12 children's films

13. The Familiarize and Translate steps were done in Exercise 13 of Exercise Set 3.1.

Carry out. We solve the system of equations

$x + y = 400,$ (1)

$20x + 30y = 11,000$ (2)

where x = the number of units of lumber produced and y = the number of units of plywood produced. We use elimination.

$-20x - 20y = -8000$ Multiplying (1) by -20

$20x + 30y = 11,000$

$10y = 3000$ Adding

$y = 300$

Substitute 300 for y in (1) and solve for x.

$x + 300 = 400$

$x = 100$

Check. Total output: 100 + 300 = 400 units.

Profit from lumber: \$20·100 = \$2000

Profit from plywood: \$30·300 = \$9000

Total profit: \$2000 + \$9000 = \$11,000

State. The lumber company must produce 100 units of lumber and 300 units of plywood.

14. 23 wins, 14 ties

15. The Familiarize and Translate steps were done in Exercise 15 of Exercise Set 3.1.

Carry out. We solve the system of equations

$x + y = 12,$ (1)

$30x + 60y = 600$ (2)

where x = the number of 30-sec commercials and y = the number of 60-sec commercials. We use elimination.

$-30x - 30y = -360$ Multiplying (1) by -30

$30x + 60y = 600$

$30y = 240$ Adding

$y = 8$

Substitute 8 for y in (1) and solve for x.

$x + 8 = 12$

$x = 4$

Check. Total number of commercials: $4 + 8 = 12$

Commercial time for 30-sec commercials:

$30 \cdot 4 = 120$ sec

Commercial time for 60-sec commercials:

$60 \cdot 8 = 480$ sec

Total commercial time: 120 sec + 480 sec = 600 sec, or 10 min.

The numbers check.

State. There were 4 30-sec commercials and 8 60-sec commercials.

16. 131 coach-class seats, 21 first-class seats

17. Familiarize. Let x = the number of pounds of soybean meal and y = the number of pounds of corn meal in the mixture.

Translate. We organize the information in a table.

	Soybean	Corn	Mixture
Pounds of meal	x	y	350
Percent of protein	16%	9%	12%
Amount of protein	0.16x	0.09y	0.12 × 350 or 42

The "Pounds of meal" row gives us one equation: $x + y = 350$

The last row gives us a second equation: $0.16x + 0.09y = 42$

After clearing decimals, we have this system:

$x + y = 350$ (1)

$16x + 9y = 4200$ (2)

Carry out. We use the elimination method to solve the system of equations.

$-9x - 9y = -3150$ Multiplying (1) by -9

$16x + 9y = 4200$

$7x = 1050$

$x = 150$

$150 + y = 350$ Substituting 150 for x in (1)

$y = 200$

Check. The total number of pounds is 150 + 200, or 350. Also, 16% of 150 is 24, and 9% of 200 is 18. Their total is 42. The numbers check.

State. 150 lb of soybean meal and 200 lb of corn meal should be mixed.

18. 4 L of 25%, 6 L of 50%

19. Familiarize. Let x = the number of liters of Arctic Antifreeze and y = the number of liters of Frost-No-More in the mixture.

Translate. We organize the information in a table.

	Arctic Antifreeze	Frost-No-more	Mixture
Number of liters	x	y	20
Percent of alcohol	18%	10%	15%
Amount of alcohol	0.18x	0.10y	0.15 × 20, or 3

The "Number of liters" row gives us one equation: $x + y = 20$

The last row gives us a second equation: $0.18x + 0.10y = 3$

After clearing decimals, we have this system:

$x + y = 20,$ (1)

$18x + 10y = 300$ (2)

Carry out. We use the elimination method to solve the system of equations.

$-10x - 10y = -200$ Multiplying (1) by -10

$18x + 10y = 300$

$8x = 100$

$x = 12\frac{1}{2}$

$12\frac{1}{2} + y = 20$ Substituting $12\frac{1}{2}$ for x in (1)

$y = 7\frac{1}{2}$

Check. Total liters: $12\frac{1}{2} + 7\frac{1}{2} = 20$

Total amount of alcohol: $18\% \times 12\frac{1}{2} + 10\% \times 7\frac{1}{2} = 2.25 + 0.75 = 3$

Percentage of alcohol in mixture:

$$\frac{\text{Total amount of alcohol}}{\text{Total liters in mixture}} = \frac{3}{20} = 0.15 = 15\%$$

The numbers check.

State. $12\frac{1}{2}$ L of Arctic Antifreeze and $7\frac{1}{2}$ L of Frost-No-More should be used.

20. 5 liters of each

21. Familiarize. Let x = the amount invested at 9% and y = the amount invested at 10%. Recall the formula for simple interest:

Interest = Principal · Rate · Time

Translate. We organize the information in a table.

	First Investment	Second Investment	Total
Principal	x	y	$15,000
Interest Rate	9%	10%	
Time	1 year	1 year	
Interest	0.09x	0.10y	$1432

The "Principal" row gives us one equation: $x + y = 15{,}000$

The last row gives us a second equation: $0.09x + 0.10y = 1432$

After clearing the decimals we have this system:

$x + y = 15{,}000$ (1)

$9x + 10y = 143{,}200$ (2)

Carry out. We use the elimination method to solve the system of equations.

$-9x - 9y = -135{,}000$ Multiplying (1) by -9

$9x + 10y = 143{,}200$

$y = 8200$

$x + 8200 = 15{,}000$ Substituting 8200 for y in (1)

$x = 6800$

Check. Total investment: $6800 + $8200 = $15,000

Total interest: 9% × $6800 + 10% × $8200 = $612 + $820 = $1432

The numbers check.,

State. $6800 is invested at 9%, and $8200 is invested at 10%.

22. $4100 at 14%, $4700 at 16%

23. Familiarize. Let x = the amount invested at 10% and y = the amount invested at 12%. Recall the formula for simple interest.

Interest = Principal · Rate · Time

Translate. We organize the information in a table.

	First Investment	Second Investment	Total
Principal	x	y	$27,000
Interest Rate	10%	12%	
Time	1 year	1 year	
Interest	0.10x	0.12y	$2990

The "Principal" row gives us one equation: $x + y = 27{,}000$

The last row gives us a second equation: $0.10x + 0.12y = 2990$

After clearing the decimals we have this system:

$x + y = 27{,}000$ (1)

$10x + 12y = 299{,}000$ (2)

Carry out. We use the elimination method to solve the system of equations.

$-10x - 10y = -270{,}000$ Multiplying (1) by -10

$10x + 12y = 299{,}000$

$2y = 29{,}000$

$y = 14{,}500$

$x + 14{,}500 = 27{,}000$ Substituting 14,500 for y in (1)

$x = 12{,}500$

Check. Total investment: $12,500 + $14,500 = $27,000

Total interest: 10% × $12,500 + 12% × $14,500 = $1250 + $1740 = $2990

The numbers check.

State. $12,500 is invested at 10%, and $14,500 is invested at 12%.

24. $725 at 12%, $425 at 11%

25. Familiarize. Let a = the number of adult's tickets sold and c = the number of children's tickets sold.

Translate. We organize the information in a table.

	Adult's tickets	Children's tickets	Total
Number sold	a	c	117
Price	$1.25	$0.75	
Amount taken in	1.25a	0.75c	129.75

The "Number sold" row gives us one equation: $a + c = 117$

The last row gives us a second equation: $1.25a + 0.75c = 129.75$

After clearing the decimals we have this system:

$a + c = 117,$ (1)

$125a + 75c = 12{,}975$ (2)

Carry out. We use the elimination method to solve the system of equations:

$-75a - 75c = -8775$ Multiplying (1) by -75

$125a + 75c = 12{,}975$

$50a = 4200$

$a = 84$

$84 + c = 117$ Substituting 84 for a in (1)

$c = 33$

Check. Total tickets sold: $84 + 33 = 117$

Total taken in: $\$1.25 \times 84 + \$0.75 \times 33 =$

$\$105 + \$24.75 = \$129.75$

The numbers check.

State. 84 adult's tickets and 33 children's tickets were sold.

26. 8 white, 22 yellow

27. Familiarize. Let x = Paula's age now and y = Bob's age now. Four years from now, Paula's and Bob's ages will be x + 4 and y + 4, respectively.

Translate.

Paula's age now	is	12 years more than	Bob's age now
x	$=$	12	$+ \quad y$

Four years from now

Bob's age	will be	$\frac{2}{3}$ of	Paula's age.
$y + 4$	$=$	$\frac{2}{3} \cdot$	$(x + 4)$

We have a system of equations:

$x = 12 + y,$ (1)

$y + 4 = \frac{2}{3}(x + 4)$ (2)

Carry out. We use the substitution method.

$y + 4 = \frac{2}{3}(12 + y + 4)$ Substituting 12 + y for x in (2)

$y + 4 = \frac{2}{3}(16 + y)$

$3(y + 4) = 2(16 + y)$ Clearing the fraction

$3y + 12 = 32 + 2y$

$y = 20$

$x = 12 + 20$ Substituting 20 for y in (1)

$x = 32$

Check. Bob's age plus 12 years is 20 + 12, or 32, which is Paula's age. Four years from now Bob will be 24 and Paula will be 36, and 24 is $\frac{2}{3}$ of 36. The numbers check.

State. Now Paula is 32 years old and Bob is 20 years old.

28. Carlos: 28, Maria: 20

29. Familiarize. We first make a drawing. We let ℓ = length and w = width.

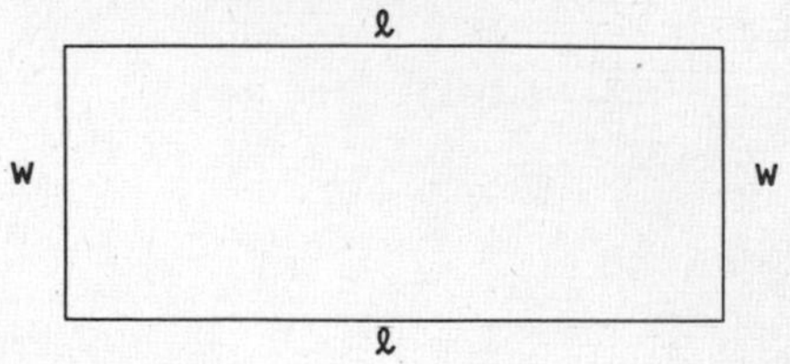

The formula for perimeter is $P = 2\ell + 2w$.

Translate. The perimeter is 190 m.

$2\ell + 2w = 194$

The width is one fourth of the length.

$w \quad = \quad \frac{1}{4} \quad \cdot \quad \ell$

We have a system of equations:

$2\ell + 2w = 190,$ (1)

$w = \frac{1}{4}\ell$ (2)

Carry out. We use the substitution method.

$2\ell + 2 \cdot \frac{1}{4}\ell = 190$ Substituting $\frac{1}{4}\ell$ for w in (1)

$2\ell + \frac{1}{2}\ell = 190$

$\frac{5}{2}\ell = 190$

$\ell = 76$

$w = \frac{1}{4} \cdot 76$ Substituting 76 for ℓ in (2)

$w = 19$

Check. The perimeter of a rectangle with width 19 m and length 76 m is $2 \cdot 76 + 2 \cdot 19 = 152 + 38 = 190$ m. Also, 19 is one fourth of 76. The numbers check.

State. The width is 19 m and the length is 76 m.

30. Length: 78 yd, width: 19 yd

31. Familiarize. Let x = the number of \$5 bills and y = the number of \$1 bills. The total value of the \$5 bills is 5x, and the total value of the \$1 bills is $1 \cdot y$, or y.

Translate.

The total number of bills is 22.

$x + y \quad = \quad 22$

The total value of the bills is \$50.

$5x + y \quad = \quad 50$

We have a system of equations:

$x + y = 22,$ (1)

$5x + y = 50$ (2)

Carry out. We use the elimination method.

$-x - y = -22$ Multiplying (1) by -1

$\underline{5x + y = \ 50}$

$4x \quad = 28$

$x = 7$

$7 + y = 22$ Substituting 7 for x in (1)

$y = 15$

Check. Total number of bills: $7 + 15 = 22$

Total value of bills: $\$5 \cdot 7 + \$1 \cdot 15 = \$35 + \$15 = \$50$.

The numbers check.

State. There are 7 \$5 bills and 15 \$1 bills.

32. 17 quarters, 13 fifty-cent pieces

33. Familiarize. We first make a drawing.

Slow train
d kilometers 75 km/h (t + 2) hours

Fast train
d kilometers 125 km/h t hours

From the drawing we see that the distances are the same. Now complete the chart.

d = r · t

	Distance	Rate	Time	
Slow train	d	75	t + 2	→ d = 75(t + 2)
Fast train	d	125	t	→ d = 125t

Translate. Using d = rt in each row of the table, we get a system of equations:

$d = 75(t + 2),$

$d = 125t$

Carry out. We solve the system of equations.

$125t = 75(t + 2)$ Using substitution

$125t = 75t + 150$

$50t = 150$

$t = 3$

The time for the fast train should be 3 hr, and the time for the slow train 3 + 2, or 5 hr.

Then $d = 125t = 125 \cdot 3 = 375$

Check. At 125 km/h, in 3 hr the fast train will travel $125 \cdot 3 = 375$ km. At 75 km/h, in 5 hr the slow train will travel $75 \cdot 5 = 375$ km. The numbers check.

State. The trains will meet 375 km from the station.

34. 3 hr

35. Familiarize. We first make a drawing.

Chicago Indianapolis
110 km/h t hours t hours 90 km/h

350 km

The sum of the distances is 350 km. The times are the same. We let t = the time, d = the distance from Chicago, and 350 - d = the distance from Indianapolis. We organize the information in a table.

d = r · t

Motorcycle	Distance	Rate	Time	
From Chicago	d	110	t	→ d = 110t
From Indpls.	350 - d	90	t	→ 350 - d = 90t

Translate. Using d = rt in each row of the table, we get a system of equations:

$d = 110t,$ (1)

$350 - d = 90t$ (2)

Carry out. We use the substitution method.

$350 - 110t = 90t$ Substituting 110t for d in (2)

$350 = 200t$

$\frac{350}{200} = t$

$\frac{7}{4} = t$

Check. The motorcycle from Chicago will travel $110 \cdot \frac{7}{4}$, or 192.5 km. The motorcycle from Indianapolis will travel $90 \cdot \frac{7}{4}$, or 157.5 km. The sum of the two distances is 192.5 + 157.5, or 350 km. The value checks.

State. In $1\frac{3}{4}$ hours the motorcycles will meet.

36. 2 hr

37. Familiarize. We first make a drawing. Let d = the distance and r = the speed of the boat in still water. Then when the boat travels downstream its speed is r + 6, and its speed upstream is r - 6. From the drawing we see that the distances are the same.

Downstream, 6 mph current
d mi, r + 6, 3 hr

Upstream, 6 mph current
d mi, r - 6, 5 hr

Organize the information in a table.

d = r · t

	Distance	Rate	Time	
Downstream	d	r + 6	3	→ d = (r + 6)3
Upstream	d	r - 6	5	→ d = (r - 6)5

Translate. Using d = rt in each row of the table, we get a system of equations:

$d = 3r + 18,$

$d = 5r - 30$

Carry out. Solve the system of equations.

$3r + 18 = 5r - 30$ Using substitution

$18 = 2r - 30$

$48 = 2r$

$24 = r$

Check. When r = 24, r + 6 = 30, and the distance traveled in 3 hr is $30 \cdot 3 = 90$ mi. Also, r - 6 = 18, and the distance traveled in 5 hr is $18 \cdot 5 = 90$ mi. The value checks.

State. The speed of the boat in still water is 24 mph.

38. 14 km/h

39. $-3(x - 7) - 2[x - (4 + 3x)]$

$= -3(x - 7) - 2[x - 4 - 3x]$	Removing the innermost parentheses
$= -3(x - 7) - 2(-2x - 4)$	Simplifying
$= -3x + 21 + 4x + 8$	Removing parentheses
$= x + 29$	Simplifying

40. $\frac{27}{4}$

41. We use the point-slope equation.

$$y - y_1 = m(x - x_1)$$

$$y - (-5) = -\frac{3}{4}(x - 2) \quad \text{Substituting}$$

$$y + 5 = -\frac{3}{4}x + \frac{3}{2}$$

$$y = -\frac{3}{4}x - \frac{7}{2}$$

42.

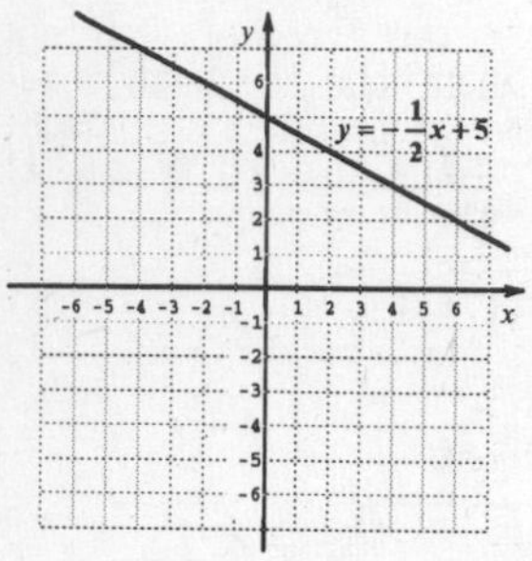

43. The Familiarize and Translate steps were done in Exercise 63 of Exercise Set 3.1.

Carry out. We solve the system of equations

$$x = 2y, \quad (1)$$

$$x + 20 = 3y \quad (2)$$

where x = Burl's age now and y = his son's age now.

$$2y + 20 = 3y \quad \text{Substituting } 2y \text{ for } x \text{ in (2)}$$

$$20 = y$$

$$x = 2\cdot 20 \quad \text{Substituting 20 for } y \text{ in (1)}$$

$$x = 40$$

Check. Burl's age now, 40, is twice his son's age now, 20. Ten years ago Burl was 30 and his son was 10, and $30 = 3\cdot 10$. The numbers check.

State. Now Burl is 40 and his son is 20.

44. Lou: 32 years, Juanita: 14 years

45. The Familiarize and Translate steps were done in Exercise 65 of Exercise Set 3.1.

Carry out. We solve the system of equations

$$2\ell + 2w = 156, \quad (1)$$

$$\ell = 4(w - 6) \quad (2)$$

where ℓ = length and w = width.

$$2\cdot 4(w - 6) + 2w = 156 \quad \text{Substituting } 4(w - 6) \text{ for } \ell \text{ in (1)}$$

$$8w - 48 + 2w = 156$$

$$10w - 48 = 156$$

$$10w = 204$$

$$w = \frac{204}{10}, \text{ or } \frac{102}{5}$$

$$\ell = 4\left[\frac{102}{5} - 6\right] \quad \text{Substituting } \frac{102}{5} \text{ for } w \text{ in (2)}$$

$$\ell = 4\left[\frac{102}{5} - \frac{30}{5}\right]$$

$$\ell = 4\left[\frac{72}{5}\right]$$

$$\ell = \frac{288}{5}$$

Check. The perimeter of a rectangle with width $\frac{102}{5}$ in. and length $\frac{288}{5}$ in. is $2\left[\frac{288}{5}\right] + 2\left[\frac{102}{5}\right] = \frac{576}{5} + \frac{204}{5} = \frac{780}{5} = 156$ in. If 6 in. is cut off the width, the new width is $\frac{102}{5} - 6 = \frac{102}{5} - \frac{30}{5} = \frac{72}{5}$. The length, $\frac{288}{5}$, is $4\left[\frac{72}{5}\right]$. The numbers check.

State. The original piece of posterboard has width $\frac{102}{5}$ in. and length $\frac{288}{5}$ in.

46. 80 wins, 24 losses

47. Familiarize. Let x = the amount of the original solution that remains after some of the original solution is drained and replaced with pure antifreeze. Let y = the amount of the original solution that is drained and replaced with pure antifreeze.

Translate. We organize the information in a table. Keep in mind that the table contains information regarding the solution after some of the original solution is drained and replaced with pure antifreeze.

	Original Solution	Pure Antifreeze	New Mixture
Amount of solution	x	y	16 L
Percent of antifreeze	30%	100%	50%
Amount of antifreeze in solution	0.3x	$1\cdot y$, or y	0.5(16), or 8

The "Amount of solution" row gives us one equation:
$x + y = 16$

The last row gives us a second equation:
$0.3x + y = 8$

After clearing the decimal we have the following system of equations:

$$x + y = 16, \quad (1)$$

$$3x + 10y = 80 \quad (2)$$

Carry out. We use the elimination method.

$-3x - 3y = -48$ Multiplying (1) by -3

$3x + 10y = 80$

$7y = 32$

$y = \frac{32}{7}$, or $4\frac{4}{7}$

Although the problem only asks for the amount of pure antifreeze added, we will also find x in order to check.

$x + 4\frac{4}{7} = 16$ Substituting $4\frac{4}{7}$ for y in (1)

$x = 11\frac{3}{7}$

Check. Total amount of new mixture: $11\frac{3}{7} + 4\frac{4}{7} =$

16 L

Amount of antifreeze in new mixture:

$0.3\left(11\frac{3}{7}\right) + 4\frac{4}{7} = \frac{3}{10} \cdot \frac{80}{7} + \frac{32}{7} = \frac{56}{7} = 8$ L

The numbers check.

State. Michelle should drain $4\frac{4}{7}$ L of the original solution and replace it with pure antifreeze.

48. 4 km

49. Familiarize. Let x = the ten's digit and y = the unit's digit. Then the number is 10x + y. If the digits are interchanged, the new number is 10y + x.

Translate.

Ten's digit is 2 more than 3 times unit's digit.

$x \quad = \quad 2 \quad + \quad 3 \quad \cdot \quad y$

If the digits are interchanged,

new number is half of given number minus 13.

$10y + x \quad = \quad \frac{1}{2} \quad \cdot \quad (10x + y) \quad - \quad 13$

The system of equations is

$x = 2 + 3y,$ (1)

$10y + x = \frac{1}{2}(10x + y) - 13$ (2)

Carry out. We use the substitution method. Substitute 2 + 3y for x in (2).

$10y + (2 + 3y) = \frac{1}{2}[10(2 + 3y) + y] - 13$

$13y + 2 = \frac{1}{2}[20 + 30y + y] - 13$

$13y + 2 = \frac{1}{2}[20 + 31y] - 13$

$13y + 2 = 10 + \frac{31}{2}y - 13$

$13y + 2 = \frac{31}{2}y - 3$

$5 = \frac{5}{2}y$

$2 = y$

$x = 2 + 3 \cdot 2$ Substituting 2 for y in (1)

$x = 2 + 6$

$x = 8$

Check. If x = 8 and y = 2, the given number is 82 and the new number is 28. In the given number the ten's digit, 8, is two more than three times the unit's digit, 2. The new number is 13 less than one-half the given number: $28 = \frac{1}{2}(82) - 13$. The values check.

State. The given integer is 82.

50. 180

51. Familiarize. We first make a drawing. Let r_1 = speed of the first train and r_2 = the speed of the second train. If the first train leaves at 9 A.M. and the second at 10 A.M., we have:

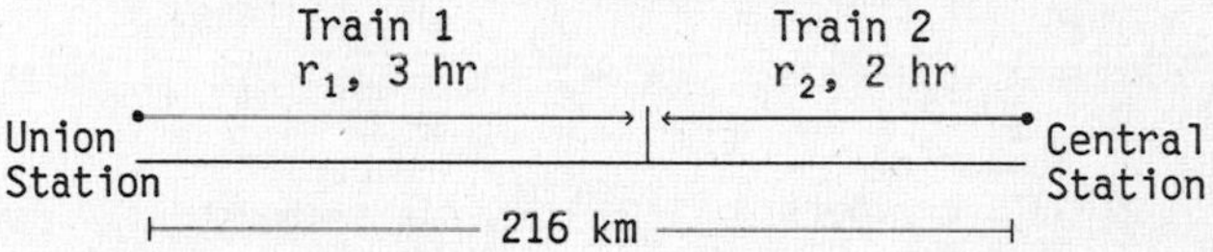

If the second train leaves at 9 A.M. and the first at 10:30 A.M. we have:

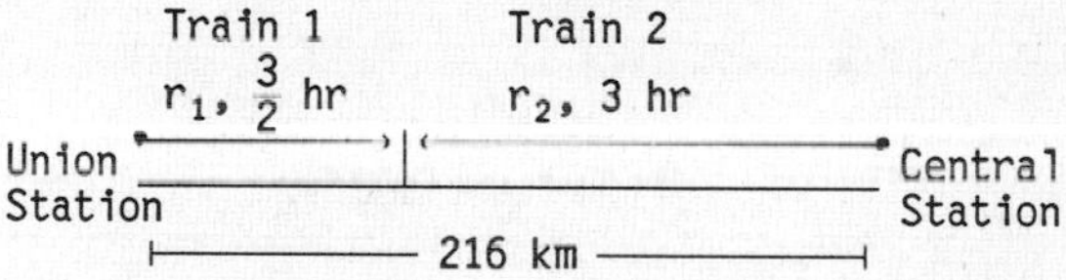

The total distance traveled in each case is 216 km and is equal to the sum of the distances traveled by each train.

Translate. We will use the formula d = rt. For each situation we have:

Total distance	is	Train 1's distance	plus	Train 2's distance.
216	=	$3r_1$	+	$2r_2$
and 216	=	$\frac{3}{2}r_1$	+	$3r_2$

Clearing the fraction, we have this system:

$216 = 3r_1 + 2r_2,$ (1)

$432 = 3r_1 + 6r_2$ (2)

Carry out. Solve the system of equations.

$-216 = -3r_1 - 2r_2$ Multiplying (1) by -1

$432 = 3r_1 + 6r_2$

$216 = 4r_2$

$54 = r_2$

$216 = 3r_1 + 2(54)$ Substituting 54 for r_2 in (1)

$216 = 3r_1 + 108$

$108 = 3r_1$

$36 = r_1$

Check. If Train 1 travels for 3 hr at 36 km/h and Train 2 travels for 2 hr at 54 km/h, the total distance traveled is $3 \cdot 36 + 2 \cdot 54 = 108 + 108 = 216$ km. If Train 1 travels for $\frac{3}{2}$ hr at 36 km/h and Train 2 travels for 3 hr at 54 km/h, the total distance traveled is $\frac{3}{2} \cdot 36 + 3 \cdot 54 = 54 + 162 = 216$ km. The numbers check.

State. The speed of the first train is 36 km/h, and the speed of the second train is 54 km/h.

52. City: 261 mi, highway: 204 mi

53. Familiarize. Let g = the number of girls and b = the number of boys. Then Phyllis has b brothers and g - 1 sisters, and Phil has b - 1 brothers and g sisters.

Translate.

Number of Phyllis' brothers	is twice	number of her sisters.
b	$= 2 \cdot$	$(g - 1)$

Number of Phil's brothers	is same as	number of his sisters.
$b - 1$	$=$	g

We have a system of equations:

$b = 2(g - 1)$, (1)

$b - 1 = g$ (2)

Carry out. We use the substitution method.

$b = 2[(b - 1) - 1]$ Substituting b - 1 for g in (1)

$b = 2(b - 2)$

$b = 2b - 4$

$4 = b$

$4 - 1 = g$ Substituting 4 for b in (2)

$3 = g$

Check. If there are 3 girls and 4 boys in the family, then Phyllis has 4 brothers and 2 sisters. She has twice as many brothers as sisters. Also, Phil has 3 brothers and 3 sisters, the same number of each. The numbers check.

State. There are 3 girls and 4 boys in the family.

54.

Exercise Set 3.4

1. Substitute (1,-2,3) into the three equations, using alphabetical order.

$$\begin{array}{c|l} x + y + z = 2 & \\ \hline 1 + (-2) + 3 \; ? \; 2 & \\ 2 & \text{TRUE} \end{array}$$

$$\begin{array}{c|l} x - 2y - z = 2 & \\ \hline 1 - 2(-2) - 3 \; ? \; 2 & \\ 1 + 4 - 3 & \\ 2 & \text{TRUE} \end{array}$$

$$\begin{array}{c|l} 3x + 2y + z = 2 & \\ \hline 3 \cdot 1 + 2(-2) + 3 \; ? \; 2 & \\ 3 - 4 + 3 & \\ 2 & \text{TRUE} \end{array}$$

The triple (1,-2,3) makes all three equations true, so it is a solution.

2. No

3. $x + y + z = 6$, (1)
$2x - y + 3z = 9$, (2)
$-x + 2y + 2z = 9$ (3)

1., 2. The equations are already in standard form with no fractions or decimals.

3. Add equations (1) and (2) to eliminate y:

$x + y + z = 6$ (1)
$2x - y + 3z = 9$ (2)
$3x + 4z = 15$ (4) Adding

4. Use a different pair of equations and eliminate y:

$4x - 2y + 6z = 18$ Multiplying (2) by 2
$-x + 2y + 2z = 9$ (3)
$3x + 8z = 27$ (5)

5. Now solve the system of equations (4) and (5).

$3x + 4z = 15$ (4)
$3x + 8z = 27$ (5)

$-3x - 4z = -15$ Multiplying (4) by -1
$3x + 8z = 27$
$4z = 12$
$z = 3$

$3x + 4 \cdot 3 = 15$ Substituting 3 for z in (4)
$3x + 12 = 15$
$3x = 3$
$x = 1$

6. Substitute in one of the original equations to find y.

$1 + y + 3 = 6$ Substituting 1 for x and 3 for z in (1)
$y + 4 = 6$
$y = 2$

We obtain (1,2,3). This checks, so it is the solution.

4. (4,0,2)

5. $2x - y - 3z = -1$, (1)
$2x - y + z = -9$, (2)
$x + 2y - 4z = 17$ (3)

1., 2. The equation is already in standard form with no fractions or decimals.

3., 4. We eliminate z from two different pairs of equations.

$2x - y - 3z = -1$ (1)

$6x - 3y + 3z = -27$ Multiplying (2) by 3

$8x - 4y = -28$ (4) Adding

$8x - 4y + 4z = -36$ Multiplying (2) by 4

$x + 2y - 4z = 17$

$9x - 2y = -19$ (5) Adding

5. Now solve the system of equations (4) and (5).

$8x - 4y = -28$ (4)

$9x - 2y = -19$ (5)

$8x - 4y = -28$ (4)

$-18x + 4y = 38$ Multiplying (5) by -2

$-10x = 10$ Adding

$x = -1$

$8(-1) - 4y = -28$ Substituting -1 for x in (4)

$-8 - 4y = -28$

$-4y = -20$

$y = 5$

6. Substitute in one of the original equations to find z.

$2(-1) - 5 + z = -9$ Substituting -1 for x and 5 for y in (2)

$-2 - 5 + z = -9$

$-7 + z = -9$

$z = -2$

We obtain (-1,5,-2). This checks, so it is the solution.

6. (2,-2,2)

7. $2x - 3y + z = 5,$ (1)

$x + 3y + 8z = 22,$ (2)

$3x - y + 2z = 12$ (3)

1., 2. The equation is already in standard form with no fractions or decimals.

3., 4. We eliminate y from two different pairs of equations.

$2x - 3y + z = 5$ (1)

$x + 3y + 8z = 22$ (2)

$3x + 9z = 27$ (4) Adding

$x + 3y + 8z = 22$ (2)

$9x - 3y + 6z = 36$ Multiplying (3) by 3

$10x + 14z = 58$ (5) Adding

5. Solve the system of equations (4) and (5).

$3x + 9z = 27$ (4)

$10x + 14z = 58$ (5)

$30x + 90z = 270$ Multiplying (4) by 10

$-30x - 42z = -174$ Multiplying (5) by -3

$48z = 96$ Adding

$z = 2$

$3x + 9 \cdot 2 = 27$ Substituting 2 for z in (4)

$3x + 18 = 27$

$3x = 9$

$x = 3$

6. Substitute in one of the original equations to find y.

$2 \cdot 3 - 3y + 2 = 5$ Substituting 3 for x and 2 for z in (1)

$-3y + 8 = 5$

$-3y = -3$

$y = 1$

We obtain (3,1,2). This checks, so it is the solution.

8. (3,-2,1)

9. $3a - 2b + 7c = 13,$ (1)

$a + 8b - 6c = -47,$ (2)

$7a - 9b - 9c = -3$ (3)

1., 2. The equations are already in standard form with no fractions or decimals.

3., 4. We eliminate a from two different pairs of equations.

$3a - 2b + 7c = 13$ (1)

$-3a - 24b + 18c = 141$ Multiplying (2) by -3

$-26b + 25c = 154$ (4) Adding

$-7a - 56b + 42c = 329$ Multiplying (2) by -7

$7a - 9b - 9c = -3$ (3)

$-65b + 33c = 326$ (5) Adding

5. Now solve the system of equations (4) and (5).

$-26b + 25c = 154$ (4)

$-65b + 33c = 326$ (5)

$-130b + 125c = 770$ Multiplying (4) by 5

$130b - 66c = -652$ Multiplying (5) by -2

$59c = 118$

$c = 2$

$-26b + 25 \cdot 2 = 154$ Substituting 2 for c in (4)

$-26b + 50 = 154$

$-26b = 104$

$b = -4$

6. Substitute in one of the original equations to find a.

$a + 8(-4) - 6(2) = -47$ Substituting -4 for b and 2 for c in (2)

$a - 32 - 12 = -47$

$a - 44 = -47$

$a = -3$

We obtain (-3,-4,2). This checks, so it is the solution.

10. (7,-3,-4)

11. $2x + 3y + z = 17,$ (1)

$x - 3y + 2z = -8,$ (2)

$5x - 2y + 3z = 5$ (3)

1., 2. The equation is already in standard form with no fractions or decimals.

3., 4. We eliminate y from two different pairs of equations.

$2x + 3y + z = 17$ (1)

$x - 3y + 2z = -8$ (2)

$3x + 3z = 9$ (4) Adding

$4x + 6y + 2z = 34$ Multiplying (1) by 2

$15x - 6y + 9z = 15$ Multiplying (3) by 3

$19x + 11z = 49$ (5) Adding

5. Now solve the system of equations (4) and (5).

$3x + 3z = 9$ (4)

$19x + 11z = 49$ (5)

$33x + 33z = 99$ Multiplying (4) by 11

$-57x - 33z = -147$ Multiplying (5) by -3

$-24x = -48$

$x = 2$

$3\cdot 2 + 3z = 9$ Substituting 2 for x in (4)

$6 + 3z = 9$

$3z = 3$

$z = 1$

6. Substitute in one of the original equations to find y.

$2\cdot 2 + 3y + 1 = 17$ Substituting 2 for x and 1 for z in (1)

$3y + 5 = 17$

$3y = 12$

$y = 4$

We obtain (2,4,1). This checks, so it is the solution.

12. (2,1,3)

13. $2x + y + z = -2,$ (1)

$2x - y + 3z = 6,$ (2)

$3x - 5y + 4z = 7$ (3)

1., 2. The equations are already in standard form with no fractions or decimals.

3., 4. We eliminate y from two different pairs of equations.

$2x + y + z = -2$ (1)

$2x - y + 3z = 6$ (2)

$4x + 4z = 4$ (4) Adding

$10x + 5y + 5z = -10$ Multiplying (1) by 5

$3x - 5y + 4z = 7$ (3)

$13x + 9z = -3$ (5) Adding

5. Now solve the system of equations (4) and (5).

$4x + 4z = 4$ (4)

$13x + 9z = -3$ (5)

$36x + 36z = 36$ Multiplying (4) by 9

$-52x - 36z = 12$ Multiplying (5) by -4

$-16x = 48$ Adding

$x = -3$

$4(-3) + 4z = 4$ Substituting -3 for x in (4)

$-12 + 4z = 4$

$4z = 16$

$z = 4$

6. Substitute in one of the original equations to find y.

$2(-3) + y + 4 = -2$ Substituting -3 for x and 4 for z in (1)

$y - 2 = -2$

$y = 0$

We obtain (-3,0,4). This checks, so it is the solution.

14. (2,-5,6)

15. $x - y + z = 4,$ (1)

$5x + 2y - 3z = 2,$ (2)

$4x + 3y - 4z = -2$ (3)

1., 2. The equations are already in standard form with no fractions or decimals.

3., 4. We eliminate z from two different pairs of equations.

$3x - 3y + 3z = 12$ Multiplying (1) by 3

$5x + 2y - 3z = 2$ (2)

$8x - y = 14$ (4) Adding

$4x - 4y + 4z = 16$ Multiplying (1) by 4

$4x + 3y - 4z = -2$ (3)

$8x - y = 14$ (5) Adding

5. Now solve the system of equations (4) and (5).

$8x - y = 14$ (4)

$8x - y = 14$ (5)

$8x - y = 14$ (4)

$-8x + y = -14$ Multiplying (5) by -1

$0 = 0$ (6)

Equation (6) indicates that equations (1), (2), and (3) are dependent. (Note that if equation (1) is subtracted from equation (2), the result is equation (3).)

16. The system is dependent.

17. $4x - y - z = 4,$ (1)

$2x + y + z = -1,$ (2)

$6x - 3y - 2z = 3$ (3)

1., 2. The equations are already in standard form with no fractions or decimals.

3. Add equations (1) and (2) to eliminate y.

$4x - y - z = 4$ (1)

$2x + y + z = -1$ (2)

$6x = 3$ (4) Adding

4. At this point we can either continue by eliminating y from a second pair of equations or we can solve (4) for x and substitute that value in a different pair of the original equations to obtain a system of two equations in two variables. We take the second option.

$6x = 3$ (4)

$x = \frac{1}{2}$

Substitute $\frac{1}{2}$ for x in (1):

$4\left(\frac{1}{2}\right) - y - z = 4$

$2 - y - z = 4$

$-y - z = 2$ (5)

Substitute $\frac{1}{2}$ for x in (3):

$6\left(\frac{1}{2}\right) - 3y - 2z = 3$

$3 - 3y - 2z = 3$

$-3y - 2z = 0$ (6)

5. Solve the system of equations (5) and (6).

$2y + 2z = -4$ Multiplying (5) by -2

$-3y - 2z = 0$ (6)

$-y = -4$

$y = 4$

6. Substitute to find z.

$-4 - z = 2$ Substituting 4 for y in (5)

$-z = 6$

$z = -6$

We obtain $\left(\frac{1}{2},4,-6\right)$. This checks, so it is the solution.

18. (3,-5,8)

19. $2r + 3s + 12t = 4,$ (1)

$4r - 6s + 6t = 1,$ (2)

$r + s + t = 1$ (3)

1., 2. The equations are already in standard form with no fractions or decimals.

3., 4. We eliminate s from two different pairs of equations.

$4r + 6s + 24t = 8$ Multiplying (1) by 2

$4r - 6s + 6t = 1$ (2)

$8r + 30t = 9$ (4) Adding

$4r - 6s + 6t = 1$ (2)

$6r + 6s + 6t = 6$ Multiplying (3) by 6

$10r + 12t = 7$ (5) Adding

5. Solve the system of equations (4) and (5).

$40r + 150t = 45$ Multiplying (4) by 5

$-40r - 48t = -28$ Multiplying (5) by -4

$102t = 17$

$t = \frac{17}{102}$

$t = \frac{1}{6}$

$8r + 30\left(\frac{1}{6}\right) = 9$ Substituting $\frac{1}{6}$ for t in (4)

$8r + 5 = 9$

$8r = 4$

$r = \frac{1}{2}$

6. Substitute in one of the original equations to find s.

$\frac{1}{2} + s + \frac{1}{6} = 1$ Substituting $\frac{1}{2}$ for r and $\frac{1}{6}$ for t in (3)

$s + \frac{2}{3} = 1$

$s = \frac{1}{3}$

We obtain $\left(\frac{1}{2},\frac{1}{3},\frac{1}{6}\right)$. This checks, so it is the solution.

20. $\left(\frac{3}{5},\frac{2}{3},-3\right)$

21. $4a + 9b = 8,$ (1)

$8a + 6c = -1,$ (2)

$6b + 6c = -1$ (3)

1., 2. The equations are already in standard form with no fractions or decimals.

3., 4. Note that there is no c in equation (1). We will use equations (2) and (3) to obtain another equation with no c term.

$8a + 6c = -1$ (2)

$- 6b - 6c = 1$ Multiplying (3) by -1

$8a - 6b = 0$ (4) Adding

5. Now solve the system of equations (1) and (4).

$$\begin{aligned} -8a - 18b &= -16 \quad \text{Multiplying (1) by -2} \\ \underline{8a - 6b} &= \underline{0} \\ -24b &= -16 \\ b &= \frac{2}{3} \end{aligned}$$

$$\begin{aligned} 8a - 6\left(\frac{2}{3}\right) &= 0 \quad \text{Substituting } \tfrac{2}{3} \text{ for b in (4)} \\ 8a - 4 &= 0 \\ 8a &= 4 \\ a &= \frac{1}{2} \end{aligned}$$

6. Substitute in equation (2) or (3) to find c.

$$\begin{aligned} 8\left(\frac{1}{2}\right) + 6c &= -1 \quad \text{Substituting } \tfrac{1}{2} \text{ for a in (2)} \\ 4 + 6c &= -1 \\ 6c &= -5 \\ c &= -\frac{5}{6} \end{aligned}$$

We obtain $\left(\frac{1}{2},\frac{2}{3},-\frac{5}{6}\right)$. This checks, so it is the solution.

22. $\left(4,\frac{1}{2},-\frac{1}{2}\right)$

23.
$$\begin{aligned} x + y + z &= 57, \quad (1) \\ -2x + y &= 3, \quad (2) \\ x - z &= 6 \quad (3) \end{aligned}$$

1., 2. The equations are already in standard form with no fractions or decimals.

3., 4. Note that there is no z in equation (2). We will use equations (1) and (3) to obtain another equation with no z term.

$$\begin{aligned} x + y + z &= 57 \quad (1) \\ \underline{x - z} &= \underline{6} \quad (3) \\ 2x + y &= 63 \quad (4) \end{aligned}$$

5. Now solve the system of equations (2) and (4).

$$\begin{aligned} -2x + y &= 3 \quad (2) \\ \underline{2x + y} &= \underline{63} \quad (4) \\ 2y &= 66 \\ y &= 33 \end{aligned}$$

$$\begin{aligned} 2x + 33 &= 63 \quad \text{Substituting 33 for y in (4)} \\ 2x &= 30 \\ x &= 15 \end{aligned}$$

6. Substitute in equation (1) or (3) to find z.

$$\begin{aligned} 15 - z &= 6 \quad \text{Substituting 15 for x in (3)} \\ 9 &= z \end{aligned}$$

We obtain (15,33,9). This checks, so it is the solution.

24. (17,9,79)

25.
$$\begin{aligned} 2a - 3b &= 2, \quad (1) \\ 7a + 4c &= \frac{3}{4}, \quad (2) \\ -3b + 2c &= 1 \quad (3) \end{aligned}$$

1. The equations are already in standard form.

2. Multiply equation (2) by 4 to clear the fraction. The resulting system is

$$\begin{aligned} 2a - 3b &= 2, \quad (1) \\ 28a + 16c &= 3, \quad (4) \\ -3b + 2c &= 1 \quad (3) \end{aligned}$$

3. Note that there is no b in equation (2). We will use equations (1) and (3) to obtain another equation with no b term.

$$\begin{aligned} 2a - 3b &= 2 \quad (1) \\ \underline{3b - 2c} &= \underline{-1} \quad \text{Multiplying (3) by -1} \\ 2a - 2c &= 1 \quad (5) \end{aligned}$$

5. Now solve the system of equations (4) and (5).

$$\begin{aligned} 28a + 16c &= 3 \quad (4) \\ \underline{16a - 16c} &= \underline{8} \quad \text{Multiplying (5) by 8} \\ 44a &= 11 \\ a &= \frac{1}{4} \end{aligned}$$

$$\begin{aligned} 2 \cdot \frac{1}{4} - 2c &= 1 \quad \text{Substituting } \tfrac{1}{4} \text{ for a in (5)} \\ \frac{1}{2} - 2c &= 1 \\ -2c &= \frac{1}{2} \\ c &= -\frac{1}{4} \end{aligned}$$

6. Substitute in equation (1) or (3) to find b.

$$\begin{aligned} 2\left(\frac{1}{4}\right) - 3b &= 2 \quad \text{Substituting } \tfrac{1}{4} \text{ for a in (1)} \\ \frac{1}{2} - 3b &= 2 \\ -3b &= \frac{3}{2} \\ b &= -\frac{1}{2} \end{aligned}$$

We obtain $\left(\frac{1}{4},-\frac{1}{2},-\frac{1}{4}\right)$. This checks, so it is the solution.

26. (3,4,-1)

27.
$$\begin{aligned} x + y + z &= 180, \quad (1) \\ y &= 2 + 3x, \quad (2) \\ z &= 80 + x \quad (3) \end{aligned}$$

1. Only equation (1) is in standard form. Rewrite the system with all equations in standard form.

$$\begin{aligned} x + y + z &= 180, \quad (1) \\ -3x + y &= 2, \quad (4) \\ -x + z &= 80 \quad (5) \end{aligned}$$

2. There are no fractions or decimals.

3., 4. Note that there is no z in equation (4). We will use equations (1) and (5) to obtain another equation with no z term.

$$\begin{array}{ll} x + y + z = 180 & (1) \\ \underline{x \qquad - z = -80} & \text{Multiplying (5) by -1} \\ 2x + y \qquad = 100 & (6) \end{array}$$

5. Now solve the system of equations (4) and (6).

$$\begin{array}{ll} -3x + y = 2 & (4) \\ \underline{-2x - y = -100} & \text{Multiplying (6) by -1} \\ -5x \qquad = -98 & \\ x = \frac{98}{5} & \\ -3 \cdot \frac{98}{5} + y = 2 & \text{Substituting } \frac{98}{5} \text{ for x in (4)} \\ -\frac{294}{5} + y = 2 & \\ y = \frac{304}{5} & \end{array}$$

6. Substitute in equation (1) or (5) to find z.

$$\begin{array}{ll} -\frac{98}{5} + z = 80 & \text{Substituting } -\frac{98}{5} \text{ for x in (5)} \\ z = \frac{498}{5} & \end{array}$$

We obtain $\left(\frac{98}{5}, \frac{304}{5}, \frac{498}{5}\right)$. This checks, so it is the solution.

28. (2,5, 3)

29.
$$\begin{array}{ll} x \qquad + z = 0, & (1) \\ x + y + 2z = 3, & (2) \\ y + z = 2 & (3) \end{array}$$

1., 2. The equations are already in standard form with no fractions or decimals.

3., 4. The variable y is missing in equation (1). We use equations (2) and (3) to obtain another equation with no y term.

$$\begin{array}{ll} x + y + 2z = 3 & (2) \\ \underline{-y - z = -2} & \text{Multiplying (3) by -1} \\ x \qquad + z = 1 & \text{(4) Adding} \end{array}$$

5. Now solve the system of equations (1) and (4).

$$\begin{array}{ll} x + z = 0 & (1) \\ \underline{-x - z = -1} & \text{Multiplying (4) by -1} \\ 0 = -1 & \text{Adding} \end{array}$$

We get a false equation. There is no solution. The solution set is $\emptyset$.

30. $\emptyset$

31.
$$\begin{array}{ll} x + y + z = 1, & (1) \\ -x + 2y + z = 2, & (2) \\ 2x - y \qquad = -1 & (3) \end{array}$$

1., 2. The equations are already in standard form with no fractions or decimals.

3. Use equations (1) and (2) to eliminate z:

$$\begin{array}{ll} x + y + z = 1 & (1) \\ \underline{x - 2y - z = -2} & \text{Multiplying (2) by -1} \\ 2x - y \qquad = -1 & \text{(4) Adding} \end{array}$$

Equations (3) and (4) are identical, so the system is dependent. (We have seen that if equation (2) is multiplied by -1 and added to equation (1), the result is equation (3).)

32. The system is dependent.

33.
$$\begin{aligned} F &= \tfrac{1}{2}t(c - d) \\ 2F &= t(c - d) \\ 2F &= tc - td \\ 2F + td &= tc \\ \frac{2F + td}{t} &= c, \text{ or} \\ \frac{2F}{t} + d &= c \end{aligned}$$

34. $d = \dfrac{2F - tc}{-t}$, or $-\dfrac{2F}{t} + c$

35.

36.

37.
$$\frac{x + 2}{3} - \frac{y + 4}{2} + \frac{z + 1}{6} = 0,$$
$$\frac{x - 4}{3} + \frac{y + 1}{4} - \frac{z - 2}{2} = -1,$$
$$\frac{x + 1}{2} + \frac{y}{2} + \frac{z - 1}{4} = \frac{3}{4}$$

1., 2. We clear fractions and write each equation in standard form.

To clear fractions, we multiply each equation by the LCM of its denominators. The LCM's are 6, 12, and 4, respectively.

$$\begin{aligned} 2(x + 2) - 3(y + 4) + (z + 1) &= 0 \\ 2x + 4 - 3y - 12 + z + 1 &= 0 \\ 2x - 3y + z &= 7 \end{aligned}$$

$$\begin{aligned} 4(x - 4) + 3(y + 1) - 6(z - 2) &= -12 \\ 4x - 16 + 3y + 3 - 6z + 12 &= -12 \\ 4x + 3y - 6z &= -11 \end{aligned}$$

$$\begin{aligned} 2(x + 1) + 2(y) + (z - 1) &= 3 \\ 2x + 2 + 2y + z - 1 &= 3 \\ 2x + 2y + z &= 2 \end{aligned}$$

The resulting system is

$$\begin{array}{ll} 2x - 3y + z = 7, & (1) \\ 4x + 3y - 6z = -11, & (2) \\ 2x + 2y + z = 2 & (3) \end{array}$$

3., 4. We eliminate z from two different pairs of equations.

$$\begin{array}{rll} 12x - 18y + 6z = 42 & \text{Multiplying (1) by 6} \\ \underline{4x + 3y - 6z = -11} & (2) \\ 16x - 15y \quad = 31 & (4) & \text{Adding} \end{array}$$

$$\begin{array}{rll} 2x - 3y + z = 7 & (1) \\ \underline{-2x - 2y - z = -2} & \text{Multiplying (3) by -1} \\ -5y \quad = 5 & (5) & \text{Adding} \end{array}$$

5. Solve (5) for y: $-5y = 5$

$y = -1$

Substitute -1 for y in (4):

$16x - 15(-1) = 31$

$16x + 15 = 31$

$16x = 16$

$x = 1$

6. Substitute 1 for x and -1 for y in (1):

$2\cdot 1 - 3(-1) + z = 7$

$5 + z = 7$

$z = 2$

We obtain (1,-1,2). This checks, so it is the solution.

38. (1,1,1)

39. $w + x + y + z = 2,$ (1)

$w + 2x + 2y + 4z = 1,$ (2)

$w - x + y + z = 6,$ (3)

$w - 3x - y + z = 2,$ (4)

The equations are already in standard form with no fractions or decimals.

Start by eliminating w from three different pairs of equations.

$$\begin{array}{rll} w + x + y + z = 2 & (1) \\ \underline{-w - 2x - 2y - 4z = -1} & \text{Multiplying (2) by -1} \\ -x - y - 3z = 1 & (5) & \text{Adding} \end{array}$$

$$\begin{array}{rll} w + x + y + z = 2 & (1) \\ \underline{-w + x - y - z = -6} & \text{Multiplying (3) by -1} \\ 2x \quad = -4 & (6) & \text{Adding} \end{array}$$

$$\begin{array}{rll} w + x + y + z = 2 & (1) \\ \underline{-w + 3x + y - z = -2} & \text{Multiplying (4) by -1} \\ 4x + 2y \quad = 0 & (7) & \text{Adding} \end{array}$$

We can solve (6) for x:

$2x = -4$

$x = -2$

Substitute -2 for x in (7):

$4(-2) + 2y = 0$

$-8 + 2y = 0$

$2y = 8$

$y = 4$

Substitute -2 for x and 4 for y in (5):

$-(-2) - 4 - 3z = 1$

$-2 - 3z = 1$

$-3z = 3$

$z = -1$

Substitute -2 for x, 4 for y, and -1 for z in (1):

$w - 2 + 4 - 1 = 2$

$w + 1 = 2$

$w = 1$

We obtain (1,-2,4,-1). This checks, so it is the solution.

40. (-3,-1,0,4)

41. $\frac{2}{x} - \frac{1}{y} - \frac{3}{z} = -1,$

$\frac{2}{x} - \frac{1}{y} + \frac{1}{z} = -9,$

$\frac{1}{x} + \frac{2}{y} - \frac{4}{z} = 17$

Let u represent $\frac{1}{x}$, v represent $\frac{1}{y}$, and w represent $\frac{1}{z}$. Substituting, we have

$2u - v - 3w = -1,$ (1)

$2u - v + w = -9,$ (2)

$u + 2v - 4w = 17.$ (3)

1., 2. The equations in u, v, and w are in standard form with no fractions or decimals.

3., 4. We eliminate v from two different pairs of equations.

$$\begin{array}{rll} 2u - v - 3w = -1 & (1) \\ \underline{-2u + v - w = 9} & \text{Multiplying (2) by -1} \\ -4w = 8 & (4) & \text{Adding} \end{array}$$

$$\begin{array}{rll} 4u - 2v - 6w = -2 & \text{Multiplying (1) by 2} \\ \underline{u + 2v - 4w = 17} & (3) \\ 5u \quad - 10w = 15 & (5) & \text{Adding} \end{array}$$

5. We can solve (4) for w:

$-4w = 8$

$w = -2$

Substitute -2 for w in (5):

$5u - 10(-2) = 15$

$5u + 20 = 15$

$5u = -5$

$u = -1$

6. Substitute -1 for u and -2 for w in (1):

$2(-1) - v - 3(-2) = -1$

$-v + 4 = -1$

$-v = -5$

$v = 5$

Solve for x, y, and z. We substitute -1 for u, 5 for v and -2 for w.

$$u = \frac{1}{x} \qquad v = \frac{1}{y} \qquad w = \frac{1}{z}$$

$$-1 = \frac{1}{x} \qquad 5 = \frac{1}{y} \qquad -2 = \frac{1}{z}$$

$$x = -1 \qquad y = \frac{1}{5} \qquad z = -\frac{1}{2}$$

We obtain $\left(-1,\frac{1}{5},-\frac{1}{2}\right)$. This checks, so it is the solution.

42. $\left(-\frac{1}{2},-1,-\frac{1}{3}\right)$

43. $x - 3y + 2z = 1,$ (1)

$2x + y - z = 3,$ (2)

$9x - 6y + 3z = k$ (3)

Eliminate z from two different pairs of equations.

$x - 3y + 2z = 1$ (1)

$4x + 2y - 2z = 6$ Multiplying (2) by 2

$5x - y = 7$ (4)

$6x + 3y - 3z = 9$ Multiplying (2) by 3

$9x - 6y + 3z = k$ (3)

$15x - 3y = 9 + k$ (5)

Solve the system of equations (4) and (5).

$-15x + 3y = -21$ Multiplying (4) by -3

$15x - 3y = 9 + k$

$0 = -12 + k$ (6)

The system is dependent for the value of k that makes equation (6) true. We solve for k:

$0 = -12 + k$

$12 = k$

44. 14

45. $Ax + By + Cz = 12$

Three solutions are $\left(1,\frac{3}{4},3\right)$, $\left(\frac{4}{3},1,2\right)$, and (2,1,1). We substitute for x, y, and z and solve for A, B, and C.

$A + \frac{3}{4}B + 3C = 12,$

$\frac{4}{3}A + B + 2C = 12,$

$2A + B + C = 12$

1., 2. The equations are in standard form. We clear fractions.

$4A + 3B + 12C = 48,$ (1)

$4A + 3B + 6C = 36,$ (2)

$2A + B + C = 12$ (3)

3. Use equations (1) and (2) to eliminate A.

$4A + 3B + 12C = 48$ (1)

$-4A - 3B - 6C = -36$ Multiplying (2) by -1

$6C = 12$

$C = 2$

4. At this point we can use another point of equations to eliminate A, or we can substitute 2 for C in either equation (1) or (2) and in equation (3) to get a system of equations in B and C. We choose the latter option.

$4A + 3B = 24$ (4) Substituting in (1)

$2A + B = 10$ (5) Substituting in (3)

5., 6. Solve the system of equations (4) and (5).

$4A + 3B = 24$ (4)

$-4A - 2B = -20$ Multiplying (5) by -2

$B = 4$

$2A + 4 = 10$ Substituting 4 for B in (5)

$2A = 6$

$A = 3$

The solution is (3,4,2), so the equation is $3x + 4y + 2z = 12$.

46. $z = 8 - 2x - 4y$

Exercise Set 3.5

1. Familiarize. Let x = the first number, y = the second number, and z = the third number.

Translate.

The sum of three numbers is 105.

$x + y + z = 105$

The third number	is	ten times the second number	less	eleven.
z	=	10y	-	11

Twice the first number	is	seven more than	three times the second number.
2x	=	7 +	3y

We now have a system of equations.

$x + y + z = 105$ or $x + y + z = 105$

$z = 10y - 11$ $\quad -10y + z = -11$

$2x = 7 + 3y$ $\quad 2x - 3y = 7$

Carry out. Solving the system we get (17,9,79).

Check. The sum of the three numbers is 105. Ten times the second number is 90, and 11 less than 90 is 79, the third number. Three times the second number is 27, and 7 more than 27 is 34, which is twice the first number. The numbers check.

State. The numbers are 17, 9, and 79.

2. 16, 19, 22

3. Familiarize. Let x = the first number, y = the second number, and z = the third number.

Translate.

The sum of the three numbers is 5.

$x + y + z = 5$

The first number minus the second plus the third is 1.

$x - y + z = 1$

The first number minus the third is 3 more than the second.

$x - z = y + 3$

We now have a system of equations.

$x + y + z = 5$ or $x + y + z = 5$

$x - y + z = 1$ $\quad$ $x - y + z = 1$

$x - z = y + 3$ $\quad$ $x - y - z = 3$

Carry out. Solving the system we get (4,2,-1).

Check. The sum of the numbers is 5. The first minus the second plus the third is 4 - 2 + (-1), or 1. The first minus the third is 5, which is three more than the second. The numbers check.

State. The numbers are 4, 2, and -1.

4. 8, 21, -3

5. Familiarize. We first make a drawing.

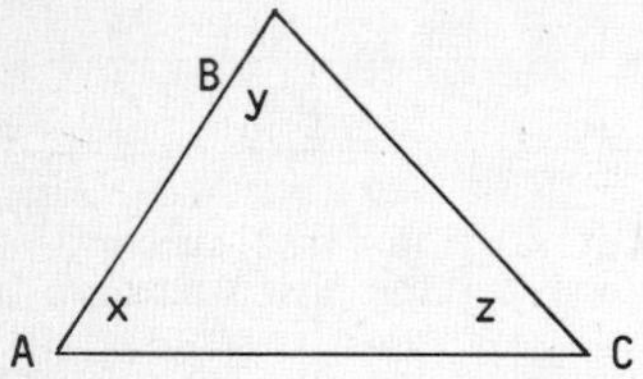

We let x, y, and z represent the measures of angles A, B, and C, respectively. The measures of the angles of a triangle add up to 180°.

Translate.

The sum of the measures is 180°.

$x + y + z = 180$

The measure of angle B is 2° more than three times the measure of angle A.

$y = 3x + 2$

The measure of angle C is 8° more than the measure of angle A.

$z = x + 8$

We now have a system of equations.

$x + y + z = 180,$

$y = 3x + 2,$

$z = x + 8$

Carry out. Solving the system we get (34,104,42).

Check. The sum of the numbers is 180, so that checks. Three times the measure of angle A is 3·34, or 102°, and 2° added to 102° is 104°, the measure of angle B. The measure of angle C, 42°, is 8° more than 34°, the measure of angle A. These values check.

State. Angles A, B, and C measure 34°, 104°, and 42°, respectively.

6. 30°, 90°, 60°

7. Familiarize. We first make a drawing.

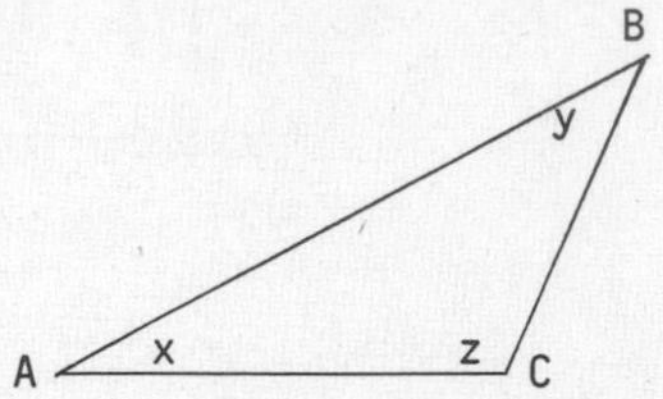

We let x, y, and z represent the measures of angles A, B, and C, respectively. The measures of the angles of a triangle add up to 180°.

Translate.

The sum of the measures is 180°.

$x + y + z = 180$

The measure of angle B is twice the measure of angle A.

$y = 2 \cdot x$

The measure of angle C is 80° more than the measure of angle A.

$z = x + 80$

We now have a system of equations.

$x + y + z = 180,$

$y = 2x,$

$z = x + 80$

Carry out. Solving the system we get (25,50,105).

Check. The sum of the numbers is 180, so that checks. The measure of angle B, 50°, is twice 25°, the measure of angle A. The measure of angle C, 105°, is 80° more than 25°, the measure of angle A. The values check.

State. Angles A, B, and C measure 25°, 50°, and 105°, respectively.

8. 32°, 96°, 52°

9. Familiarize. Let x, y, and z represent the amount spent on newspaper, television, and radio ads, respectively, in billions of dollars.

Translate.

The total expenditure was $84.8 billion.

$x + y + z = 84.8$

The total amount spent on television and radio ads was $2.6 billion more than the amount spent on newspaper ads.

$y + z = 2.6 + x$

The amount spent on newspaper ads was $5.1 billion more than the amount spent on television ads.

$x = 5.1 + y$

We have a system of equations.

$x + y + z = 84.8,$

$y + z = 2.6 + x,$

$x = 5.1 + y$

Carry out. Solving the system we get (41.1,36,7.7).

Check. The sum of the numbers is 84.8, so that checks. Also, 36 + 7.7 = 43.7 which is 2.6 + 41.1, and 41.1 = 5.1 + 36. These values check.

State. $41.1 billion was spent on newspaper ads, $36 billion was spent on television ads, and $7.7 billion was spent on radio ads.

10. Automatic transmission: $865, power door locks: $520, air conditioning: $375

11. Familiarize. Let s = the number of servings of steak, p = the number of baked potatoes, and b = the number of servings of broccoli. Then s servings of steak contain 300s Calories, 20s g of protein, and no vitamin C. In p baked potatoes there are 100p Calories, 5p g of protein, and 20p mg of vitamin C. And b servings of broccoli contain 50b Calories, 5b g of protein, and 100b mg of vitamin C. The patient requires 800 Calories, 55 g of protein, and 220 mg of vitamin C.

Translate. Write equations for the total number of calories, the total amount of protein, and the total amount of vitamin C.

$300s + 100p + 50b = 800$ (calories)

$20s + 5p + 5b = 55$ (protein)

$20p + 100b = 220$ (vitamin C)

We now have a system of equations.

Carry out. Solving the system we get (2,1,2).

Check. Two servings of steak provide 600 Calories, 40 g of protein, and no vitamin C. One baked potato provides 100 Calories, 5 g of protein, and 20 mg of vitamin C. And 2 servings of broccoli provide 100 Calories, 10 g of protein, and 200 mg of vitamin C. Together, then, they provide 800 Calories, 55 g of protein, and 220 mg of vitamin C. The values check.

State. The dietician should prepare 2 servings of steak, 1 baked potato, and 2 servings of broccoli.

12. $1\frac{1}{8}$ servings of steak, $2\frac{3}{4}$ baked potatoes, $3\frac{3}{4}$ servings of asparagus

13. Familiarize. Let x, y, and z represent the number of fraternal twin births for Orientals, African-Americans and Caucasians in the U.S., respectively, out of every 15,400 births.

Translate. Out of every 15,400 births, we have the following statistics:

The total number of fraternal twin births	is	739.
$x + y + z$	=	739

The number of fraternal twin births for Orientals	is	185	more than	the number for African-Americans.
x	=	185	+	y

The number of fraternal twin births for Orientals	is	231	more than	the number for Caucasians.
x	=	231	+	z

We have a system of equations.

$x + y + z = 739,$

$x = 185 + y,$

$x = 231 + z$

Carry out. Solving the system we get (385,200,154).

Check. The total of the numbers is 739. Also 385 is 185 more than 200, and it is 231 more than 154.

State. Out of every 15,400 births, there are 385 births of fraternal twins for Orientals, 200 for African-Americans, and 154 for Caucasians.

14. A man: 3.6 times, a woman: 18.1 times, a one-year old child: 50 times

15. Familiarize. It helps to orgainze the information in a table.

Machines Working	A	B	C	A + B	B + C	A, B, & C
Weekly Production	x	y	z	3400	4200	5700

We let x, y,and z represent the weekly productions of the individual machines.

Translate. From the table, we obtain three equations.

$x + y + z = 5700$ (All three machines working)

$x + y = 3400$ (A and B working)

$y + z = 4200$ (B and C working)

Carry out. Solving the system we get (1500,1900,2300).

Check. The sum of the weekly productions of machines A, B & C is 1500 + 1900 + 2300, or 5700. The sum of the weekly productions of machines A and B is 1500 + 1900, or 3400. The sum of the weekly productions of machines B and C is 1900 + 2300, or 4200. The numbers check.

State. In a week Machine A can polish 1500 lenses, Machine B can polish 1900 lenses, and Machine C can polish 2300 lenses.

16. A: 2200, B: 2500, C: 2700

17. Familiarize. It helps to organize the information in a table.

Pumps Working	A	B	C	A + B	A + C	A, B, & C
Gallons Per hour	x	y	z	2200	2400	3700

We let x, y, and z represent the number of gallons per hour which can be pumped by pumps A, B, and C, respectively.

Translate. From the table, we obtain three equations.

$x + y + z = 3700$ (All three pumps working)

$x + y = 2200$ (A and B working)

$x + z = 2400$ (A and C working)

Carry out. Solving the system we get (900,1300,1500).

Check. The sum of the gallons per hour pumped when all three are pumping is 900 + 1300 + 1500, or 3700. The sum of the gallons per hour pumped when only pump A and pump B are pumping is 900 + 1300, or 2200. The sum of the gallons per hour pumped when only pump A and pump C are pumping is 900 + 1500, or 2400. The numbers check.

State. The pumping capacities of pumps A, B, and C are respectively 900, 1300, and 1500 gallons per hour.

18. Pat: 10, Chris: 12, Jean: 15

19. Familiarize. Let x, y, and z represent the amount invested at 8%, 6%, and 9%, respectively. The interest earned is 0.08x, 0.06y, and 0.09z.

Translate.

The total invested was $80,000.

$x + y + z = 80{,}000$

The total interest was $6300.

$0.08x + 0.06y + 0.09z = 6300$

The interest at 8% was 4 times the interest at 6%.

$0.08x = 4 \cdot 0.06y$

We have a system of equations.

$x + y + z = 80{,}000,$

$0.08x + 0.06y + 0.09z = 6300,$

$0.08x = 4(0.06y)$

Carry out. Solving the system we get (45,000, 15,000, 20,000).

Check. The numbers add up to 80,000. Also, 0.08(45,000) + 0.06(15,000) + 0.09(20,000) = 6300. In addition, 0.08(45,000) = 3600 which is 4[0.6(15,000)]. The values check.

State. $45,000 was invested at 8%, $15,000 at 6%, and $20,000 at 9%.

20. 1869

21. $3(5 - x) + 7 = 5(x + 3) - 9$

$15 - 3x + 7 = 5x + 15 - 9$

$-3x + 22 = 5x + 6$

$22 = 8x + 6$

$16 = 8x$

$2 = x$

22. 8

23. $\frac{(a^2b^3)^5}{a^7b^{16}} = \frac{a^{10}b^{15}}{a^7b^{16}} = a^{10-7}b^{15-16} = a^3b^{-1}$, or $\frac{a^3}{b}$

24. $y = -\frac{3}{5}x - 7$

25. Familiarize. We let w,x, y, and z represent the ages of Tammy, Carmen, Dennis, and Mark, respectively.

Translate.

Tammy's age is Carmen's age plus Dennis's age.

$w = x + y$

Carmen's age is 2 plus Dennis's age plus Mark's age.

$x = 2 + y + z$

Dennis's age is four times Mark's age.

$y = 4 \cdot z$

The sum of all ages is 42.

$w + x + y + z = 42$

We now have a system of equations.

$w = x + y,$

$x = 2 + y + z,$

$y = 4z,$

$w + x + y + z = 42$

Carry out. Solving the system we get (20,12,8,2).

Check. The sum of all four numbers is 42. Tammy's age, 20, is the sum of the ages of Carmen and Dennis, 12 + 8. Carmen's age, 12, is 2 more than the sum of the ages of Dennis and Mark, 8 + 2. Dennis's age, 8, is four times Mark's age, 2. The numbers check.

State. Tammy is 20 years old.

26. 464

27. Familiarize. We first make a drawing with additional labels.

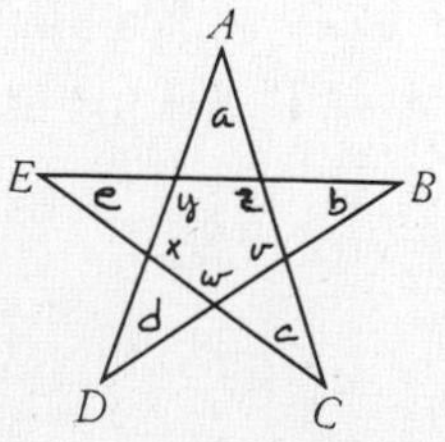

We let a, b, c, d, and e represent the angle measures at the tips of the star. We also label the interior angles of the pentagon v, w, x, y, and z. We must recall the following geometric fact:

The sum of the measures of the interior angles of a polygon of n sides is given by $(n - 2)180°$.

Using this fact we know:

1. The sum of the angle measures of a triangle is $(3 - 2)180°$, or $180°$.
2. The sum of the angle measures of a pentagon is $(5 - 2)180°$, or $3(180°)$.

Translate. Using fact (1) listed above we obtain a system of 5 equations.

$$a + v + d = 180$$
$$b + w + e = 180$$
$$c + x + a = 180$$
$$d + y + b = 180$$
$$e + z + c = 180$$

Carry out. Adding we obtain

$$2a + 2b + 2c + 2d + 2e + v + w + x + y + z = 5(180)$$

$$2(a + b + c + d + e) + (v + w + x + y + z) = 5(180)$$

Using fact (2) listed above we substitute 3(180) for $(v + w + x + y + z)$ and solve for $(a + b + c + d + e)$.

$$2(a + b + c + d + e) + 3(180) = 5(180)$$
$$2(a + b + c + d + e) = 2(180)$$
$$a + b + c + d + e = 180$$

Check. We should repeat the above calculations.

State. The sum of the angle measures at the tips of the star is 180°.

28. 5 adults, 1 student, 94 children

29. Familiarize. Let T, G, and H represent the number of tickets Tom, Gary, and Hal begin with, respectively. After Hal gives tickets to Tom and Gary, each has the following number of tickets:

Tom: $T + T$, or $2T$,

Gary: $G + G$, or $2G$,

Hal: $H - T - G$.

After Tom gives tickets to Gary and Hal, each has the following number of tickets:

Gary: $2G + 2G$, or $4G$,

Hal: $(H - T - G) + (H - T - G)$, or $2(H - T - G)$

Tom: $2T - 2G - (H - T - G)$, or $3T - H - G$

After Gary gives tickets to Hal and Tom, each has the following number of tickets:

Hal: $2(H - T - G) + 2(H - T - G)$, or $4(H - T - G)$,

Tom: $(3T - H - G) + (3T - H - G)$, or $2(3T - H - G)$,

Gary: $4G - 2(H - T - G) - (3T - H - G)$, or $7G - H - T$.

Translate. Since Hal, Tom, and Gary each finish with 40 tickets, we write the following system of equations:

$$4(H - T - G) = 40,$$
$$2(3T - H - G) = 40,$$
$$7G - H - T = 40$$

Carry out. Solving the system we find that $T = 35$.

Check. The check is left to the student.

State. Tom started with 35 tickets.

Exercise Set 3.6

1. $4x + 2y = 11,$
$3x - y = 2$

Write a matrix using only the constants.

$$\left[\begin{array}{cc|c} 4 & 2 & 11 \\ 3 & -1 & 2 \end{array}\right]$$

Multiply row 2 by 4 to make the first number in row 2 a multiple of 4.

$$\left[\begin{array}{cc|c} 4 & 2 & 11 \\ 12 & -4 & 8 \end{array}\right] \quad \text{New Row 2} = 4(\text{Row 2})$$

Multiply row 1 by -3 and add it to row 2.

$$\left[\begin{array}{cc|c} 4 & 2 & 11 \\ 0 & -10 & -25 \end{array}\right] \quad \text{New Row 2} = -3(\text{Row 1}) + \text{Row 2}$$

Reinserting the variables, we have

$$4x + 2y = 11, \quad (1)$$
$$-10y = -25 \quad (2)$$

Solve (2) for y.

$$-10y = -25$$
$$y = \frac{5}{2}$$

Back-substitute $\frac{5}{2}$ for y in (1) and solve for x.

$$4x + 2y = 11$$
$$4x + 2 \cdot \frac{5}{2} = 11$$
$$4x + 5 = 11$$
$$4x = 6$$
$$x = \frac{3}{2}$$

The solution is $\left(\frac{3}{2}, \frac{5}{2}\right)$.

2. $\left(-\frac{1}{3}, -4\right)$

3. $x + 4y = 8,$
$3x + 5y = 3$

We first write a matrix using only the constants.

$$\left[\begin{array}{cc|c} 1 & 4 & 8 \\ 3 & 5 & 3 \end{array}\right]$$

$$\left[\begin{array}{cc|c} 1 & 4 & 8 \\ 0 & -7 & -21 \end{array}\right] \quad \text{New Row 2} = -3(\text{Row 1}) + (\text{Row 2})$$

Reinserting the variables, we have

$x + 4y = 8,$ (1)
$-7y = -21$ (2)

Solve (2) for y.

$-7y = -21$
$y = 3$

Back-substitute 3 for y in (1) and solve for x.

$x + 4\cdot 3 = 8$
$x + 12 = 8$
$x = -4$

The solution is (-4,3).

4. (-3,2)

5. $5x - 3y = -2,$
$4x + 2y = 5$

Write a matrix using only the constants.

$$\left[\begin{array}{cc|c} 5 & -3 & -2 \\ 4 & 2 & 5 \end{array}\right]$$

$$\left[\begin{array}{cc|c} 5 & -3 & -2 \\ 20 & 10 & 25 \end{array}\right] \quad \text{New Row 2} = 5(\text{Row 2})$$

$$\left[\begin{array}{cc|c} 5 & -3 & -2 \\ 0 & 22 & 33 \end{array}\right] \quad \text{New Row 2} = -4(\text{Row 1}) + \text{Row 2}$$

Reinserting the variables, we have

$5x - 3y = -2,$ (1)
$22y = 33$ (2)

Solve (2) for y.

$22y = 33$
$y = \frac{3}{2}$

Back-substitute $\frac{3}{2}$ for y in (1) and solve for x.

$5x - 3y = -2$
$5x - 3 \cdot \frac{3}{2} = -2$
$5x - \frac{9}{2} = -\frac{4}{2}$
$5x = \frac{5}{2}$
$x = \frac{1}{2}$

The solution is $\left(\frac{1}{2}, \frac{3}{2}\right)$.

6. $\left(-1, \frac{5}{2}\right)$

7. $4x - y - 3z = 1,$
$8x + y - z = 5,$
$2x + y + 2z = 5$

Write a matrix using only the constants.

$$\left[\begin{array}{ccc|c} 4 & -1 & -3 & 1 \\ 8 & 1 & -1 & 5 \\ 2 & 1 & 2 & 5 \end{array}\right]$$

First interchange rows 1 and 3 so that each number below the first number in the first row is a multiple of that number.

$$\left[\begin{array}{ccc|c} 2 & 1 & 2 & 5 \\ 8 & 1 & -1 & 5 \\ 4 & -1 & -3 & 1 \end{array}\right]$$

Multiply row 1 by -4 and add it to row 2.
Multiply row 1 by -2 and add it to row 3.

$$\left[\begin{array}{ccc|c} 2 & 1 & 2 & 5 \\ 0 & -3 & -9 & -15 \\ 0 & -3 & -7 & -9 \end{array}\right]$$

Multiply row 2 by -1 and add it to row 3.

$$\left[\begin{array}{ccc|c} 2 & 1 & 2 & 5 \\ 0 & -3 & -9 & -15 \\ 0 & 0 & 2 & 6 \end{array}\right]$$

Reinserting the variables, we have

$2x + y + 2z = 5,$ (1)
$-3y - 9z = -15,$ (2)
$2z = 6$ (3)

Solve (3) for z.

$2z = 6$
$z = 3$

Back-substitute 3 for z in (2) and solve for y.

$$-3y - 9z = -15$$
$$-3y - 9(3) = -15$$
$$-3y - 27 = -15$$
$$-3y = 12$$
$$y = -4$$

Back-substitute 3 for z and -4 for y in (1) and solve for x.

$$2x + y + 2z = 5$$
$$2x + (-4) + 2(3) = 5$$
$$2x - 4 + 6 = 5$$
$$2x = 3$$
$$x = \frac{3}{2}$$

The solution is $\left(\frac{3}{2}, -4, 3\right)$.

8. $\left(2, \frac{1}{2}, -2\right)$

9. $p + q + r = 1,$
$p - 2q - 3r = 3,$
$4p + 5q + 6r = 4$

We first write a matrix using only the constants.

$$\left[\begin{array}{ccc|c} 1 & 1 & 1 & 1 \\ 1 & -2 & -3 & 3 \\ 4 & 5 & 6 & 4 \end{array}\right]$$

$$\left[\begin{array}{ccc|c} 1 & 1 & 1 & 1 \\ 0 & -3 & -4 & 2 \\ 0 & 1 & 2 & 0 \end{array}\right]$$ New Row 2 = -1(Row 1) + (Row 2)
New Row 3 = -4(Row 1) + (Row 3)

$$\left[\begin{array}{ccc|c} 1 & 1 & 1 & 1 \\ 0 & -3 & -4 & 2 \\ 0 & 3 & 6 & 0 \end{array}\right]$$ New Row 3 = 3(Row 3)

$$\left[\begin{array}{ccc|c} 1 & 1 & 1 & 1 \\ 0 & -3 & -4 & 2 \\ 0 & 0 & 2 & 2 \end{array}\right]$$ New Row 3 = (Row 2) + (Row 3)

Reinstating the variables, we have

$p + q + r = 1,$ (1)
$-3q - 4r = 2,$ (2)
$2r = 2$ (3)

Solve (3) for r.

$$2r = 2$$
$$r = 1$$

Back-substitute 1 for r in (2) and solve for q.

$$-3q - 4\cdot 1 = 2$$
$$-3q - 4 = 2$$
$$-3q = 6$$
$$q = -2$$

Back-substitute -2 for q and 1 for r in (1) and solve for p.

$$p + (-2) + 1 = 1$$
$$p - 1 = 1$$
$$p = 2$$

The solution is (2,-2,1).

10. (-1,2,-2)

11. $3p + 2r = 11,$
$q - 7r = 4,$
$p - 6q = 1$

We first write a matrix using only the constants.

$$\left[\begin{array}{ccc|c} 3 & 0 & 2 & 11 \\ 0 & 1 & -7 & 4 \\ 1 & -6 & 0 & 1 \end{array}\right]$$

$$\left[\begin{array}{ccc|c} 1 & -6 & 0 & 1 \\ 0 & 1 & -7 & 4 \\ 3 & 0 & 2 & 11 \end{array}\right]$$ Interchange Row 1 and Row 3

$$\left[\begin{array}{ccc|c} 1 & -6 & 0 & 1 \\ 0 & 1 & -7 & 4 \\ 0 & 18 & 2 & 8 \end{array}\right]$$ New Row 3 = -3(Row 1) + (Row 3)

$$\left[\begin{array}{ccc|c} 1 & -6 & 0 & 1 \\ 0 & 1 & -7 & 4 \\ 0 & 0 & 128 & -64 \end{array}\right]$$ New Row 3 = -18·(Row 2) + Row 3

Reinserting the variables, we have

$p - 6q = 1,$ (1)
$q - 7r = 4,$ (2)
$128r = -64$ (3)

Solve (3) for r.

$$128r = -64$$
$$r = -\frac{1}{2}$$

Back-substitute $-\frac{1}{2}$ for r in (2) and solve for q.

$$q - 7r = 4$$
$$q - 7\left(-\frac{1}{2}\right) = 4$$
$$q + \frac{7}{2} = 4$$
$$q = \frac{1}{2}$$

Back-substitute $\frac{1}{2}$ for q in (1) and solve for p.

$$p - 6 \cdot \frac{1}{2} = 1$$
$$p - 3 = 1$$
$$p = 4$$

The solution is $\left(4,\frac{1}{2},-\frac{1}{2}\right)$.

12. $\left(\frac{1}{2},\frac{2}{3},-\frac{5}{6}\right)$

13. We will rewrite the equations with the variables in alphabetical order:

$$-2w + 2x + 2y - 2z = -10,$$
$$w + x + y + z = -5,$$
$$3w + x - y + 4z = -2,$$
$$w + 3x - 2y + 2z = -6$$

Write a matrix using only the constants.

$$\left[\begin{array}{cccc|c} -2 & 2 & 2 & -2 & -10 \\ 1 & 1 & 1 & 1 & -5 \\ 3 & 1 & -1 & 4 & -2 \\ 1 & 3 & -2 & 2 & -6 \end{array}\right]$$

$$\left[\begin{array}{cccc|c} -1 & 1 & 1 & -1 & -5 \\ 1 & 1 & 1 & 1 & -5 \\ 3 & 1 & -1 & 4 & -2 \\ 1 & 3 & -2 & 2 & -6 \end{array}\right]$$ New Row 1 = $\frac{1}{2}$(Row 1)

$$\left[\begin{array}{cccc|c} -1 & 1 & 1 & -1 & -5 \\ 0 & 2 & 2 & 0 & -10 \\ 0 & 4 & 2 & 1 & -17 \\ 0 & 4 & -1 & 1 & -11 \end{array}\right]$$ New Row 2 = (Row 1) + (Row 2)
New Row 3 = 3(Row 1) + (Row 3)
New Row 4 = (Row 1) + (Row 4)

$$\left[\begin{array}{cccc|c} -1 & 1 & 1 & -1 & -5 \\ 0 & 2 & 2 & 0 & -10 \\ 0 & 0 & -2 & 1 & 3 \\ 0 & 0 & -5 & 1 & 9 \end{array}\right]$$ New Row 3 = -2(Row 2) + (Row 3)
New Row 4 = -2(Row 2) + (Row 4)

$$\left[\begin{array}{cccc|c} -1 & 1 & 1 & -1 & -5 \\ 0 & 2 & 2 & 0 & -10 \\ 0 & 0 & -2 & 1 & 3 \\ 0 & 0 & -10 & 2 & 18 \end{array}\right]$$ New Row 4 = 2(Row 4)

$$\left[\begin{array}{cccc|c} -1 & 1 & 1 & -1 & -5 \\ 0 & 2 & 2 & 0 & -10 \\ 0 & 0 & -2 & 1 & 3 \\ 0 & 0 & 0 & -3 & 3 \end{array}\right]$$ New Row 4 = -5(Row 3) + (Row 4)

Reinserting the variables, we have

$$-w + x + y - z = -5, \quad (1)$$
$$2x + 2y = -10, \quad (2)$$
$$-2y + z = 3, \quad (3)$$
$$-3z = 3, \quad (4)$$

Solve (4) for z.

$$-3z = 3$$
$$z = -1$$

Back-substitute -1 for z in (3) and solve for y.

$$-2y + (-1) = 3$$
$$-2y = 4$$
$$y = -2$$

Back-substitute -2 for y in (2) and solve for x.

$$2x + 2(-2) = -10$$
$$2x - 4 = -10$$
$$2x = -6$$
$$x = -3$$

Back-substitute -3 for x, -2 for y, and -1 for z in (1) and solve for w.

$$-w + (-3) + (-2) - (-1) = -5$$
$$-w - 3 - 2 + 1 = -5$$
$$-w - 4 = -5$$
$$-w = -1$$
$$w = 1$$

The solution is (1,-3,-2,-1).

14. (7,4,5,6)

15. Familiarize. Let d = the number of dimes and n = the number of nickels. The value of d dimes is \$0.10d, and the value of n nickels is \$0.05n.

Translate.

Total number of coins is 34.

$$d + n = 34$$

Total value of coins is \$1.90.

$$0.10d + 0.05n = 1.90$$

After clearing decimals we have this system.

$$d + n = 34,$$
$$10d + 5n = 190$$

Carry out. Solve using matrices.

$$\left[\begin{array}{cc|c} 1 & 1 & 34 \\ 10 & 5 & 190 \end{array}\right]$$

$$\left[\begin{array}{cc|c} 1 & 1 & 34 \\ 0 & -5 & -150 \end{array}\right] \quad \text{New Row 2} = -10(\text{Row 1}) + (\text{Row 2})$$

Reinserting the variables, we have

$$d + n = 34, \quad (1)$$
$$-5n = -150 \quad (2)$$

Solve (2) for n.

$$-5n = -150$$
$$n = 30$$

$$d + 30 = 34 \quad \text{Back-substituting}$$
$$d = 4$$

Check. The sum of the two numbers is 34. The total value is $0.10(4) + $0.05(30) = $0.40 + $1.50 = $1.90. The numbers check.

State. There are 4 dimes and 30 nickels.

16. 21 dimes, 22 quarters

17. Familiarize. We let x represent the number of pounds of the $4.05 kind and y represent the number of pounds of the $2.70 kind of granola. We organize the information in a table.

Granola	Number of pounds	Price per pound	Value
$4.05 kind	x	$4.05	$4.05x
$2.70 kind	y	$2.70	$2.70y
Mixture	15	$3.15	$3.15 × 15 or $47.25

Translate.

Total number of pounds is 15.

$$x + y = 15$$

Total value of mixture is $47.25.

$$4.05x + 2.70y = 47.25$$

After clearing decimals we have this system:

$$x + y = 15,$$
$$405x + 270y = 4725$$

Carry out. Solve using matrices.

$$\left[\begin{array}{cc|c} 1 & 1 & 15 \\ 405 & 270 & 4725 \end{array}\right]$$

$$\left[\begin{array}{cc|c} 1 & 1 & 15 \\ 0 & -135 & -1350 \end{array}\right] \quad \text{New Row 2} = -405(\text{Row 1}) + (\text{Row 2})$$

Reinserting the variables, we have

$$x + y = 15, \quad (1)$$
$$-135y = -1350 \quad (2)$$

Solve (2) for y.

$$-135y = -1350$$
$$y = 10$$

Back-substitute 10 for y in (1) and solve for x.

$$x + 10 = 15$$
$$x = 5$$

Check. The sum of the numbers is 15. The total value is $4.05(5) + $2.70(10), or $20.25 + $27.00, or $47.25. The numbers check.

State. 5 pounds of the $4.05 per lb kind and 10 pounds of the $2.70 per lb kind should be used.

18. Candy: 14 lb, nuts: 6 lb

19. Familiarize. Let x, y, and z represent the amounts invested at 7%, 8%, and 9%, respectively. Recall the formula for simple interest:

Interest = Principal × Rate × Time

Translate. We organize the information in a table.

	First Investment	Second Investment	Third Investment	Total
Principal	x	y	z	$2500
Interest Rate	7%	8%	9%	
Time	1 yr	1 yr	1 yr	
Interest	0.07x	0.08y	0.09z	$212

The first row gives us one equation:

$$x + y + z = 2500$$

The last row gives a second equation:

$$0.07x + 0.08y + 0.09z = 212$$

Amount invested at 9%	is	$1100	more than	amount invested at 8%
z	=	$1100	+	y

After clearing decimals we have this system:

$$x + y + z = 2500,$$
$$7x + 8y + 9z = 21{,}200,$$
$$-y + z = 1100$$

Carry out. Solve using matrices.

$$\left[\begin{array}{ccc|c} 1 & 1 & 1 & 2500 \\ 7 & 8 & 9 & 21,200 \\ 0 & -1 & 1 & 1100 \end{array}\right]$$

$$\left[\begin{array}{ccc|c} 1 & 1 & 1 & 2500 \\ 0 & 1 & 2 & 3700 \\ 0 & -1 & 1 & 1100 \end{array}\right]$$ New Row 2 = -7(Row 1) + (Row 2)

$$\left[\begin{array}{ccc|c} 1 & 1 & 1 & 2500 \\ 0 & 1 & 2 & 3700 \\ 0 & 0 & 3 & 4800 \end{array}\right]$$ New Row 3 = (Row 2) + (Row 3)

Reinserting the variables, we have

$x + y + z = 2500,$ (1)

$y + 2z = 3700,$ (2)

$3z = 4800$ (3)

Solve (3) for z.

$3z = 4800$

$z = 1600$

Back-substitute 1600 for z in 2 and solve for y.

$y + 2\cdot 1600 = 3700$

$y + 3200 = 3700$

$y = 500$

Back-substitute 500 for y and 1600 for z in (1) and solve for x.

$x + 500 + 1600 = 2500$

$x + 2100 = 2500$

$x = 400$

Check. The total investment is \$400 + \$500 + \$1600, or \$2500. The total interest is 0.07(\$400) + 0.08(\$500) + 0.09(\$1600) = \$28 + \$40 + \$144 = \$212. The amount invested at 9%, \$1600, is \$1100 more than the amount invested at 8%, \$500. The numbers check.

State. \$400 is invested at 7%, \$500 is invested at 8%, and \$1600 is invested at 9%.

20. \$500 at 8%, \$400 at 9%, \$2300 at 10%

21. $0.1x - 12 = 3.6x - 2.34 - 4.9x$

$10x - 1200 = 360x - 234 - 490x$ Multiplying by 100 to clear decimals

$10x - 1200 = -130x - 234$

$140x - 1200 = -234$

$140x = 966$

$x = \frac{966}{140}$, or $\frac{69}{10}$, or 6.9

22. -20

23. $4(9 - x) - 6(8 - 3x) = 5(3x + 4)$

$36 - 4x - 48 + 18x = 15x + 20$

$-12 + 14x = 15x + 20$

$-12 = x + 20$

$-32 = x$

24. $b = \frac{c}{5 + a}$

25. ◈

26. ◈

27. Familiarize. Let w, x, y, and z represent the thousand's, hundred's, ten's, and one's digits, respectively.

Translate.

The sum of the digits is 10.

$w + x + y + z = 10$

Twice the sum of the thousand's and ten's digits is the sum of the hundred's and one's digits less one.

$2(w + y) = x + z - 1$

The ten's digit is twice the thousand's digit.

$y = 2 \cdot w$

The one's digit equals the sum of the thousand's and hundred's digits.

$z = w + x$

We have a system of equations which can be written as

$w + x + y + z = 10,$

$2w - x + 2y - z = -1,$

$-2w + y = 0,$

$w + x - z = 0.$

Carry out. We can use matrices to solve the system. We get (1,3,2,4).

Check. The sum of the digits is 10. Twice the sum of 1 and 2 is 6. This is one less than the sum of 3 and 4. The ten's digit, 2, is twice the thousand's digit, 1. The one's digit, 4, equals 1 + 3. The numbers check.

State. The number is 1324.

28. $x = \frac{ce - bf}{ae - bd}$, $y = \frac{af - cd}{ae - bd}$

Exercise Set 3.7

1. $\begin{vmatrix} 2 & 7 \\ 1 & 5 \end{vmatrix} = 2\cdot 5 - 1\cdot 7 = 10 - 7 = 3$

2. -13

3. $\begin{vmatrix} 6 & -9 \\ 2 & 3 \end{vmatrix} = 6\cdot 3 - 2\cdot(-9) = 18 + 18 = 36$

4. 29

5. $\begin{vmatrix} 0 & 2 & 0 \\ 3 & -1 & 1 \\ 1 & -2 & 2 \end{vmatrix}$

$= 0\begin{vmatrix} -1 & 1 \\ -2 & 2 \end{vmatrix} - 3\begin{vmatrix} 2 & 0 \\ -2 & 2 \end{vmatrix} + 1\begin{vmatrix} 2 & 0 \\ -1 & 1 \end{vmatrix}$

$= 0 - 3[2\cdot 2 - (-2)\cdot 0] + 1[2\cdot 1 - (-1)\cdot 0]$

$= 0 - 3\cdot 4 + 1\cdot 2$

$= 0 - 12 + 2$

$= -10$

6. 1

7. $\begin{vmatrix} -1 & -2 & -3 \\ 3 & 4 & 2 \\ 0 & 1 & 2 \end{vmatrix}$

$= -1\begin{vmatrix} 4 & 2 \\ 1 & 2 \end{vmatrix} - 3\begin{vmatrix} -2 & -3 \\ 1 & 2 \end{vmatrix} + 0\begin{vmatrix} -2 & -3 \\ 4 & 2 \end{vmatrix}$

$= -1[4\cdot 2 - 1\cdot 2] - 3[-2\cdot 2 - 1(-3)] + 0$

$= -1\cdot 6 - 3\cdot(-1) + 0$

$= -6 + 3 + 0$

$= -3$

8. 3

9. $\begin{vmatrix} 3 & 2 & -2 \\ -2 & 1 & 4 \\ -4 & -3 & 3 \end{vmatrix}$

$= 3\begin{vmatrix} 1 & 4 \\ -3 & 3 \end{vmatrix} - (-2)\begin{vmatrix} 2 & -2 \\ -3 & 3 \end{vmatrix} + (-4)\begin{vmatrix} 2 & -2 \\ 1 & 4 \end{vmatrix}$

$= 3[1\cdot 3 - (-3)\cdot 4] + 2[2\cdot 3 - (-3)(-2)] - 4[2\cdot 4 - 1(-2)]$

$= 3\cdot 15 + 2\cdot 0 - 4\cdot 10$

$= 45 + 0 - 40$

$= 5$

10. -6

11. $3x - 4y = 6,$
$5x + 9y = 10$

We compute D, D_x, and D_y.

$D = \begin{vmatrix} 3 & -4 \\ 5 & 9 \end{vmatrix} = 27 - (-20) = 47$

$D_x = \begin{vmatrix} 6 & -4 \\ 10 & 9 \end{vmatrix} = 54 - (-40) = 94$

$D_y = \begin{vmatrix} 3 & 6 \\ 5 & 10 \end{vmatrix} = 30 - 30 = 0$

Then,

$x = \frac{D_x}{D} = \frac{94}{47} = 2$

and

$y = \frac{D_y}{D} = \frac{0}{47} = 0$

The solution is (2,0).

12. (-3,2)

13. $-2x + 4y = 3,$
$3x - 7y = 1$

We compute D, D_x, and D_y.

$D = \begin{vmatrix} -2 & 4 \\ 3 & -7 \end{vmatrix} = 14 - 12 = 2$

$D_x = \begin{vmatrix} 3 & 4 \\ 1 & -7 \end{vmatrix} = -21 - 4 = -25$

$D_y = \begin{vmatrix} -2 & 3 \\ 3 & 1 \end{vmatrix} = -2 - 9 = -11$

Then,

$x = \frac{D_x}{D} = \frac{-25}{2} = -\frac{25}{2}$

and

$y = \frac{D_y}{D} = \frac{-11}{2} = -\frac{11}{2}$

The solution is $\left(-\frac{25}{2}, -\frac{11}{2}\right)$.

14. $\left(\frac{9}{19}, \frac{51}{38}\right)$

15. $3x + 2y - z = 4$
$3x - 2y + z = 5$
$4x - 5y - z = -1$

We compute D, D_x, and D_y.

$D = \begin{vmatrix} 3 & 2 & -1 \\ 3 & -2 & 1 \\ 4 & -5 & -1 \end{vmatrix}$

$= 3\begin{vmatrix} -2 & 1 \\ -5 & -1 \end{vmatrix} - 3\begin{vmatrix} 2 & -1 \\ -5 & -1 \end{vmatrix} + 4\begin{vmatrix} 2 & -1 \\ -2 & 1 \end{vmatrix}$

$= 3(7) - 3(-7) + 4(0)$

$= 21 + 21 + 0$

$= 42$

$$D_x = \begin{vmatrix} 4 & 2 & -1 \\ 5 & -2 & 1 \\ -1 & -5 & -1 \end{vmatrix}$$

$$= 4\begin{vmatrix} -2 & 1 \\ -5 & -1 \end{vmatrix} - 5\begin{vmatrix} 2 & -1 \\ -5 & -1 \end{vmatrix} + (-1)\begin{vmatrix} 2 & -1 \\ -2 & 1 \end{vmatrix}$$

$= 4(7) - 5(-7) - 1(0)$
$= 28 + 35 - 0$
$= 63$

$$D_y = \begin{vmatrix} 3 & 4 & -1 \\ 3 & 5 & 1 \\ 4 & -1 & -1 \end{vmatrix}$$

$$= 3\begin{vmatrix} 5 & 1 \\ -1 & -1 \end{vmatrix} - 3\begin{vmatrix} 4 & -1 \\ -1 & -1 \end{vmatrix} + 4\begin{vmatrix} 4 & -1 \\ 5 & 1 \end{vmatrix}$$

$= 3(-4) - 3(-5) + 4(9)$
$= -12 + 15 + 36$
$= 39$

Then,

$$x = \frac{D_x}{D} = \frac{63}{42} = \frac{3}{2}$$

and

$$y = \frac{D_y}{D} = \frac{39}{42} = \frac{13}{14}.$$

We substitute in the second equation to find z.

$$3 \cdot \frac{3}{2} - 2 \cdot \frac{13}{14} + z = 5$$
$$\frac{9}{2} - \frac{13}{7} + z = 5$$
$$\frac{37}{14} + z = 5$$
$$z = \frac{33}{14}$$

The solution is $\left(\frac{3}{2}, \frac{13}{14}, \frac{33}{14}\right)$.

16. $\left(-1, -\frac{6}{7}, \frac{11}{7}\right)$

17. $2x - 3y + 5z = 27$
$x + 2y - z = -4$
$5x - y + 4z = 27$
We compute D, D_x, and D_y.

$$D = \begin{vmatrix} 2 & -3 & 5 \\ 1 & 2 & -1 \\ 5 & -1 & 4 \end{vmatrix}$$

$$= 2\begin{vmatrix} 2 & -1 \\ -1 & 4 \end{vmatrix} - 1\begin{vmatrix} -3 & 5 \\ -1 & 4 \end{vmatrix} + 5\begin{vmatrix} -3 & 5 \\ 2 & -1 \end{vmatrix}$$

$= 2(7) - 1(-7) + 5(-7)$
$= 14 + 7 - 35$
$= -14$

$$D_x = \begin{vmatrix} 27 & -3 & 5 \\ -4 & 2 & -1 \\ 27 & -1 & 4 \end{vmatrix}$$

$$= 27\begin{vmatrix} 2 & -1 \\ -1 & 4 \end{vmatrix} - (-4)\begin{vmatrix} -3 & 5 \\ -1 & 4 \end{vmatrix} + 27\begin{vmatrix} -3 & 5 \\ 2 & -1 \end{vmatrix}$$

$= 27(7) + 4(-7) + 27(-7)$
$= 189 - 28 - 189$
$= -28$

$$D_y = \begin{vmatrix} 2 & 27 & 5 \\ 1 & -4 & -1 \\ 5 & 27 & 4 \end{vmatrix}$$

$$= 2\begin{vmatrix} -4 & -1 \\ 27 & 4 \end{vmatrix} - 1\begin{vmatrix} 27 & 5 \\ 27 & 4 \end{vmatrix} + 5\begin{vmatrix} 27 & 5 \\ -4 & -1 \end{vmatrix}$$

$= 2(11) - 1(-27) + 5(-7)$
$= 22 + 27 - 35$
$= 14$

Then,

$$x = \frac{D_x}{D} = \frac{-28}{-14} = 2,$$

and

$$y = \frac{D_y}{D} = \frac{14}{-14} = -1.$$

We substitute in the second equation to find z.

$$2 + 2(-1) - z = -4$$
$$2 - 2 - z = -4$$
$$-z = -4$$
$$z = 4$$

The solution is (2,-1,4).

18. (-3,2,1)

19. $r - 2s + 3t = 6$
$2r - s - t = -3$
$r + s + t = 6$
We compute D, D_r, and D_s.

$$D = \begin{vmatrix} 1 & -2 & 3 \\ 2 & -1 & -1 \\ 1 & 1 & 1 \end{vmatrix}$$

$$= 1\begin{vmatrix} -1 & -1 \\ 1 & 1 \end{vmatrix} - 2\begin{vmatrix} -2 & 3 \\ 1 & 1 \end{vmatrix} + 1\begin{vmatrix} -2 & 3 \\ -1 & -1 \end{vmatrix}$$

$= 1(0) - 2(-5) + 1(5)$
$= 0 + 10 + 5 = 15$

$$D_r = \begin{vmatrix} 6 & -2 & 3 \\ -3 & -1 & -1 \\ 6 & 1 & 1 \end{vmatrix}$$

$$= 6\begin{vmatrix} -1 & -1 \\ 1 & 1 \end{vmatrix} - (-3)\begin{vmatrix} -2 & 3 \\ 1 & 1 \end{vmatrix} + 6\begin{vmatrix} -2 & 3 \\ -1 & -1 \end{vmatrix}$$

$= 6(0) + 3(-5) + 6(5)$

$= 0 - 15 + 30$

$= 15$

$$D_s = \begin{vmatrix} 1 & 6 & 3 \\ 2 & -3 & -1 \\ 1 & 6 & 1 \end{vmatrix}$$

$$= 1\begin{vmatrix} -3 & -1 \\ 6 & 1 \end{vmatrix} - 2\begin{vmatrix} 6 & 3 \\ 6 & 1 \end{vmatrix} + 1\begin{vmatrix} 6 & 3 \\ -3 & -1 \end{vmatrix}$$

$= 1(3) - 2(-12) + 1(3)$

$= 3 + 24 + 3$

$= 30$

Then,

$$r = \frac{D_r}{D} = \frac{15}{15} = 1,$$

and

$$s = \frac{D_s}{D} = \frac{30}{15} = 2.$$

Substitute in the third equation to find t.

$1 + 2 + t = 6$

$3 + t = 6$

$t = 3$

The solution is (1,2,3).

0. (3,4,-1)

1. $0.5x - 2.34 + 2.4x = 7.8x - 9$

$2.9x - 2.34 = 7.8x - 9$

$6.66 = 4.9x$

$\frac{6.66}{4.9} = x$

$\frac{666}{490} = x$

$\frac{333}{245} = x$

2. -12

3. Familiarize. We first make a drawing.

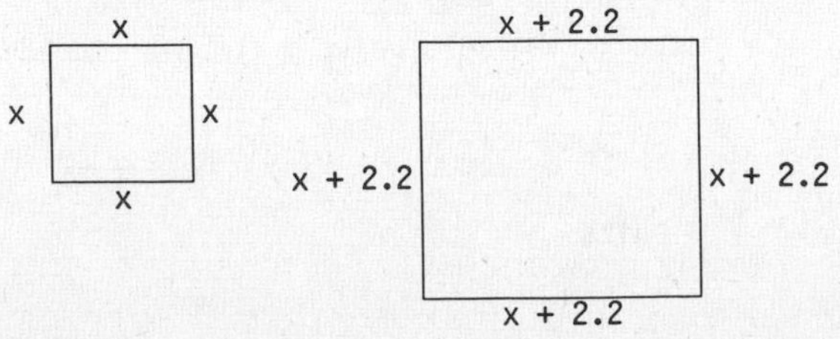

Let x represent the length of a side of the smaller square and x + 2.2 the length of a side of the larger square. The perimeter of the smaller square is 4x. The perimeter of the larger square is 4(x + 2.2).

Translate. The sum of the perimeters is 32.8 ft.

$$4x + 4(x + 2.2) = 32.8$$

Carry out. We solve the equation.

$4x + 4x + 8.8 = 32.8$

$8x = 24$

$x = 3$

Check. If x = 3 ft, then x + 2.2 = 5.2 ft. The perimeters are 4·3, or 12 ft, and 4(5.2), or 20.8 ft. The sum of the two perimeters is 12 + 20.8, or 32.8 ft. The values check.

State. The wire should be cut into two pieces measuring 12 ft and 20.8 ft.

24. 12

25. $\begin{vmatrix} 2 & x & -1 \\ -1 & 3 & 2 \\ -2 & 1 & 1 \end{vmatrix} = -12$

$$2\begin{vmatrix} 3 & 2 \\ 1 & 1 \end{vmatrix} - (-1)\begin{vmatrix} x & -1 \\ 1 & 1 \end{vmatrix} + (-2)\begin{vmatrix} x & -1 \\ 3 & 2 \end{vmatrix} = -12$$

$2(1) + 1(x + 1) - 2(2x + 3) = -12$

$2 + x + 1 - 4x - 6 = -12$

$-3x - 3 = -12$

$-3x = -9$

$x = 3$

26. 10

27. $\begin{vmatrix} x & y & 1 \\ x_1 & y_1 & 1 \\ x_2 & y_2 & 1 \end{vmatrix} = 0$

is equivalent to

$$x\begin{vmatrix} y_1 & 1 \\ y_2 & 1 \end{vmatrix} - x_1\begin{vmatrix} y & 1 \\ y_2 & 1 \end{vmatrix} + x_2\begin{vmatrix} y & 1 \\ y_1 & 1 \end{vmatrix} = 0$$

or

$$x(y_1 - y_2) - x_1(y - y_2) + x_2(y - y_1) = 0$$

or

$$xy_1 - xy_2 - x_1y + x_1y_2 + x_2y - x_2y_1 = 0. \quad (1)$$

Since the slope of the line through (x_1,y_1) and (x_2,y_2) is $\frac{y_2 - y_1}{x_2 - x_1}$, an equation of the line through (x_1,y_1) and (x_2,y_2) is

$$y - y_1 = \frac{y_2 - y_1}{x_2 - x_1}(x - x_1)$$

which is equivalent to

$$(x_2 - x_1)(y - y_1) = (y_2 - y_1)(x - x_1)$$

or

$x_2y - x_2y_1 - x_1y + x_1y_1 = y_2x - y_2x_1 - y_1x + y_1x_1$

or

$x_2y - x_2y_1 - x_1y - xy_2 + x_1y_2 + xy_1 = 0.$ (2)

Equations (1) and (2) are equivalent.

28.

Exercise Set 3.8

1. $C(x) = 45x + 600{,}000$ $R(x) = 65x$

a) $P(x) = R(x) - C(x)$
$= 65x - (45x + 600{,}000)$
$= 65x - 45x - 600{,}000$
$= 20x - 600{,}000$

b) To find the break-even point we solve the system
$R(x) = 65x,$
$C(x) = 45x + 600{,}000.$
Since $R(x) = C(x)$ at the break-even point, we can rewrite the system:
$R(x) = 65x,$ (1)
$R(x) = 45x + 600{,}000$ (2)
We solve using substitution.
$65x = 45x + 600{,}000$ Substituting 65x for R(x) in (2)
$20x = 600{,}000$
$x = 30{,}000$
Thus, 30,000 units must be produced and sold in order to break even.

2. a) $P(x) = 45x - 360{,}000$

b) 8000 units

3. $C(x) = 10x + 120{,}000$ $R(x) = 60x$

a) $P(x) = R(x) - C(x)$
$= 60x - (10x + 120{,}000)$
$= 60x - 10x - 120{,}000$
$= 50x - 120{,}000$

b) Solve the system
$R(x) = 60x,$
$C(x) = 10x + 120{,}000.$
Since $R(x) = C(x)$ at the break-even point, we can rewrite the system:
$R(x) = 60x,$ (1)
$R(x) = 10x + 120{,}000$ (2)
We solve using substitution.
$60x = 10x + 120{,}000$ Substituting 60x for R(x) in (2)
$50x = 120{,}000$
$x = 2400$
Thus, 2400 units must be produced and sold in order to break even.

4. a) $P(x) = 55x - 49{,}500$

b) 900 units

5. $C(x) = 20x + 10{,}000$ $R(x) = 100x$

a) $P(x) = R(x) - C(x)$
$= 100x - (20x + 10{,}000)$
$= 100x - 20x - 10{,}000$
$= 80x - 10{,}000$

b) Solve the system
$R(x) = 100x,$
$C(x) = 20x + 10{,}000.$
Since $R(x) = C(x)$ at the break-even point, we can rewrite the system:
$R(x) = 100x,$ (1)
$R(x) = 20x + 10{,}000$ (2)
We solve using substitution.
$100x = 20x + 10{,}000$ Substituting 100x for R(x) in (2)
$80x = 10{,}000$
$x = 125$
Thus, 125 units must be produced and sold in order to break even.

6. a) $P(x) = 45x - 22{,}500$

b) 500 units

7. $C(x) = 15x + 75{,}000$ $R(x) = 55x$

a) $P(x) = R(x) - C(x)$
$= 55x - (15x + 75{,}000)$
$= 55x - 15x - 75{,}000$
$= 40x - 75{,}000$

b) Solve the system
$R(x) = 55x,$
$C(x) = 15x + 75{,}000.$
Since $R(x) = C(x)$ at the break-even point, we can rewrite the system:
$R(x) = 55x,$ (1)
$R(x) = 15x + 75{,}000$ (2)
We solve using substitution.
$55x = 15x + 75{,}000$ Substituting 55x for R(x) in (2)
$40x = 75{,}000$
$x = 1875$
To break even 1875 units must be produced and sold.

8. a) $P(x) = 18x - 16{,}000$

b) 889 units

9. $C(x) = 50x + 195{,}000$ $R(x) = 125x$

a) $P(x) = R(x) - C(x)$
$= 125x - (50x + 195{,}000)$
$= 125x - 50x - 195{,}000$
$= 75x - 195{,}000$

b) Solve the system
$R(x) = 125x,$
$C(x) = 50x + 195{,}000.$

Since $R(x) = C(x)$ at the break-even point, we can rewrite the system:
$R(x) = 125x,$ (1)
$R(x) = 50x + 195{,}000$ (2)

We solve using substitution.
$125x = 50x + 195{,}000$ Substituting 125x for R(x) in (2)
$75x = 195{,}000$
$x = 2600$

To break even 2600 units must be produced and sold.

10. a) $P(x) = 94x - 928{,}000$

b) 9873 units

11. $D(p) = 2000 - 60p,$
$S(p) = 460 + 94p$

Since both demand and supply are quantities, the system can be rewritten:
$q = 2000 - 60p,$ (1)
$q = 460 + 94p$ (2)

Substitute 2000 - 60p for q in (2) and solve.
$2000 - 60p = 460 + 94p$
$1540 = 154p$
$10 = p$

The equilibrium price is \$10 per unit. To find the equilibrium quantity we substitute \$10 into either D(p) or S(p).
$D(10) = 2000 - 60(10) = 2000 - 600 = 1400$
The equilibrium quantity is 1400 units.
The equilibrium point is (\$10,1400).

12. (\$50,500)

13. $D(p) = 760 - 13p,$
$S(p) = 430 + 2p$

Rewrite the system:
$q = 760 - 13p,$ (1)
$q = 430 + 2p$ (2)

Substitute 760 - 13p for q in (2) and solve.
$760 - 13p = 430 + 2p$
$330 = 15p$
$22 = p$

The equilibrium price is \$22 per unit.

To find the equilibrium quantity we substitute \$22 into either D(p) or S(p).
$S(22) = 430 + 2(22) = 430 + 44 = 474$
The equilibrium quantity is 474 units.
The equilibrium point is (\$22,474).

14. (\$10,370)

15. $D(p) = 7500 - 25p,$
$S(p) = 6000 + 5p$

Rewrite the system:
$q = 7500 - 25p,$ (1)
$q = 6000 + 5p$ (2)

Substitute 7500 - 25p for q in (2) and solve.
$7500 - 25p = 6000 + 5p$
$1500 = 30p$
$50 = p$

The equilibrium price is \$50 per unit.
To find the equilibrium quantity we substitute \$50 into either D(p) or S(p).
$D(50) = 7500 - 25(50) = 7500 - 1250 = 6250$
The equilibrium quantity is 6250 units.
The equilibrium point is (\$50,6250).

16. (\$40,7600)

17. $D(p) = 1600 - 53p,$
$S(p) = 320 + 75p$

Rewrite the system:
$q = 1600 - 53p,$ (1)
$q = 320 + 75p$ (2)

Substitute 1600 - 53p for q in (2) and solve.
$1600 - 53p = 320 + 75p$
$1280 = 128p$
$10 = p$

The equilibrium price is \$10 per unit.
To find the equilibrium quantity we substitute \$10 into either D(p) or S(p).
$S(10) = 320 + 75(10) = 320 + 750 = 1070$
The equilibrium quantity is 1070 units.
The equilibrium point is (\$10,1070).

18. (\$36,4060)

19. a) $C(x)$ = Fixed costs + Variable costs
$C(x) = 125{,}000 + 750x,$
where x is the number of computers produced.

b) Each computer sells for \$1050. The total revenue is 1050 times the number of computers sold. We assume that all computers produced are sold.
$R(x) = 1050x$

c) $P(x) = R(x) - C(x)$

$P(x) = 1050x - (125{,}000 + 750x)$

$= 1050x - 125{,}000 - 750x$

$= 300x - 125{,}000$

d) $P(x) = 300x - 125{,}000$

$P(400) = 300(400) - 125{,}000$

$= 120{,}000 - 125{,}000$

$= -5000$

The company will realize a $5000 loss when 400 computers are produced and sold.

$P(700) = 300(700) - 125{,}000$

$= 210{,}000 - 125{,}000$

$= 85{,}000$

The company will realize a profit of $85,000 from the production and sale of 700 computers.

e) Solve the system

$R(x) = 1050x,$

$C(x) = 125{,}000 + 750x.$

Since $R(x) = C(x)$ at the break-even point, we can rewrite the system:

$R(x) = 1050x,$ (1)

$R(x) = 125{,}000 + 750x$ (2)

We solve using substitution.

$1050x = 125{,}000 + 750x$ Substituting $1050x$ for $R(x)$ in (2)

$300x = 125{,}000$

$x \approx 417$

To break even 417 units must be produced and sold.

20. a) $C(x) = 22{,}500 + 40x$

b) $R(x) = 85x$

c) $P(x) = 45x - 22{,}500$

d) A profit of $112,500, a loss of $4500

e) 500 units

21. a) $C(x)$ = Fixed costs + Variable costs

$C(x) = 10{,}000 + 20x,$

where x is the number of sport coats produced.

b) Each sport coat sells for $100. The total revenue is 100 times the number of coats sold. We assume that all coats produced are sold.

$R(x) = 100x$

c) $P(x) = R(x) - C(x)$

$P(x) = 100x - (10{,}000 + 20x)$

$= 100x - 10{,}000 - 20x$

$= 80x - 10{,}000$

d) $P(x) = 80x - 10{,}000$

$P(2000) = 80(2000) - 10{,}000$

$= 160{,}000 - 10{,}000$

$= 150{,}000$

The clothing firm will realize a $150,000 profit when 2000 sport coats are produced and sold.

$P(50) = 80(50) - 10{,}000$

$= 4000 - 10{,}000$

$= -6000$

The company will realize a $6000 loss when 50 coats are produced and sold.

e) Solve the system

$R(x) = 100x,$

$C(x) = 10{,}000 + 20x$

Since $R(x) = C(x)$ at the break-even point, we can rewrite the system:

$R(x) = 100x,$ (1)

$R(x) = 10{,}000 + 20x$ (2)

We solve using substitution.

$100x = 10{,}000 + 20x$ Substituting $100x$ for $R(x)$ in (2)

$80x = 10{,}000$

$x = 125$

To break even 125 coats must be produced and sold.

22. a) $C(x) = 16{,}400 + 6x$

b) $R(x) = 18x$

c) $P(x) = 12x - 16{,}400$

d) A profit of $19,600, a loss of $4400

e) 1367 units

23. $y - 3 = \frac{2}{5}(x - 1)$

The equation of the line is in point-slope form. We see that the line has slope $\frac{2}{5}$ and contains the point (1,3). Plot (1,3). Then go up two units and right 5 units to find another point on the line, (6,5). A third point can be found as a check.

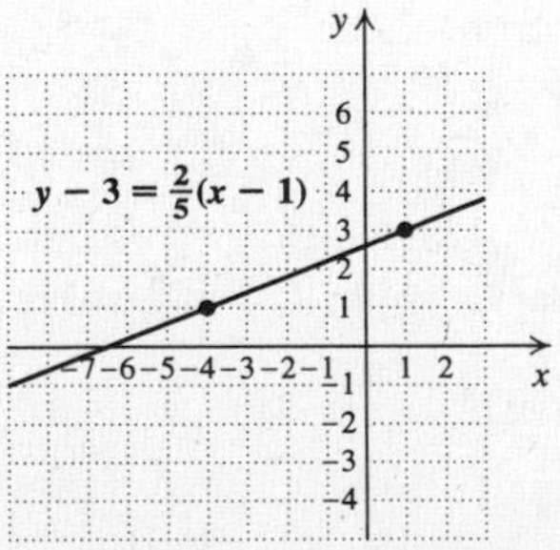

24. 5 and 6

25. $9x = 5x - \{3(2x - 7) - 4\}$

$9x = 5x - \{6x - 21 - 4\}$

$9x = 5x - \{6x - 25\}$

$9x = 5x - 6x + 25$

$9x = -x + 25$

$10x = 25$

$x = \frac{25}{10}$

$x = \frac{5}{2}$, or 2.5

26. $t = \frac{rw - v}{-s}$, or $\frac{v - rw}{s}$

27. The supply function contains the points ($2,100) and ($8,500). We find its equation:

$m = \frac{500 - 100}{8 - 2} = \frac{400}{6} = \frac{200}{3}$

$y - y_1 = m(x - x_1)$ Point-slope form

$y - 100 = \frac{200}{3}(x - 2)$

$y - 100 = \frac{200}{3}x - \frac{400}{3}$

$y = \frac{200}{3}x - \frac{100}{3}$

We can equivalently express supply S as a function of price p:

$S(p) = \frac{200}{3}p - \frac{100}{3}$

The demand function contains the points ($1,500) and ($9,100). We find its equation:

$m = \frac{100 - 500}{9 - 1} = \frac{-400}{8} = -50$

$y - y_1 = m(x - x_1)$

$y - 500 = -50(x - 1)$

$y - 500 = -50x + 50$

$y = -50x + 550$

We can equivalently express demand D as a function of price p:

$D(p) = -50p + 550$

We have a system of equations

$S(p) = \frac{200}{3}p - \frac{100}{3}$,

$D(p) = -50p + 550$.

Rewrite the system:

$q = \frac{200}{3}p - \frac{100}{3}$, (1)

$q = -50p + 550$ (2)

Substitute $\frac{200}{3}p - \frac{100}{3}$ for q in (2) and solve.

$\frac{200}{3}p - \frac{100}{3} = -50p + 550$

$200p - 100 = -150p + 1650$ Multiplying by 3 to clear fractions

$350p - 100 = 1650$

$350p = 1750$

$p = 5$

The equilibrium price is $5 per unit.

To find the equilibrium quantity, we substitute $5 into either S(p) or D(p).

$D(5) = -50(5) + 550 = -250 + 550 = 300$

The equilibrium quantity is 300 units.

The equilibrium point is ($5,300).

28. 308 pairs

29. ◈

30. ◈

31. a) 4526 units

b) $870

32. a) $8.74

b) 24,509 units

Exercise Set 4.1

1. $x - 2 \geq 6$

   -4: We substitute and get $-4 - 2 \geq 6$, or $-6 \geq 6$, a false sentence. Therefore, -4 is not a solution.

   0: We substitute and get $0 - 2 \geq 6$, or $-2 \geq 6$, a false sentence. Therefore, 0 is not a solution.

   4: We substitute and get $4 - 2 \geq 6$, or $2 \geq 6$, a false sentence. Therefore, 4 is not a solution.

   8: We substitute and get $8 - 2 \geq 6$, or $6 \geq 6$, a true sentence. Therefore, 8 is a solution.

2. Yes; yes; no; no

3. $t - 8 > 2t - 3$

   0: We substitute and get $0 - 8 > 2\cdot 0 - 3$, or $-8 > -3$, a false sentence. Therefore, 0 is not a solution.

   -8: We substitute and get $-8 - 8 > 2(-8) - 3$, or $-16 > -19$, a true sentence. Therefore, -8 is a solution.

   -9: We substitute and get $-9 - 8 > 2(-9) - 3$, or $-17 > -21$, a true sentence. Therefore, -9 is a solution.

   -3: We substitute and get $-3 - 8 > 2(-3) - 3$, or $-11 > -9$, a false sentence. Therefore, -3 is not a solution.

4. No; yes; yes; no

5. $x > 4$

   Graph: The solutions consist of all real numbers greater than 4, so we shade all numbers to the right of 4 and use an open circle at 4 to indicate that it is not a solution.

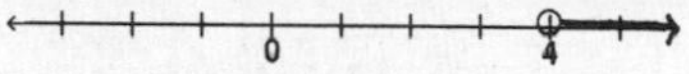

   Set-builder notation: $\{x|x > 4\}$

   Interval notation: $(4,\infty)$

6. 

   $\{y|y < 5\}$, $(-\infty,5)$

7. $t \leq 6$

   Graph: We shade all numbers to the left of 6 and use a solid endpoint at 6 to indicate that it is also a solution.

   Set-builder notation: $\{t|t \leq 6\}$

   Interval notation: $(-\infty,6]$

8. 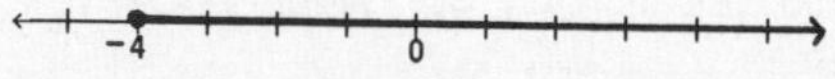

   $\{x|x \geq -4\}$, $[-4,\infty)$

9. $y < -3$

   Graph: We shade all numbers to the left of -3 and use an open circle at -3 to indicate that it is not a solution.

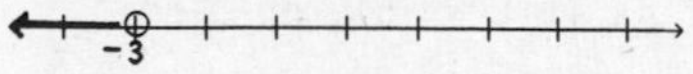

   Set-builder notation: $\{y|y < -3\}$

   Interval notation: $(-\infty,-3)$

10. 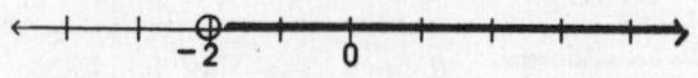

   $\{t|t > -2\}$, $(-2,\infty)$

11. $x \geq -6$

   Graph: We shade all numbers to the right of -6 and use a solid endpoint at -6 to indicate that it is also a solution.

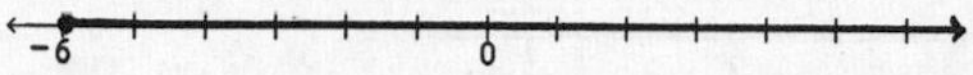

   Set-builder notation: $\{x|x \geq -6\}$

   Interval notation: $[-6,\infty)$

12. 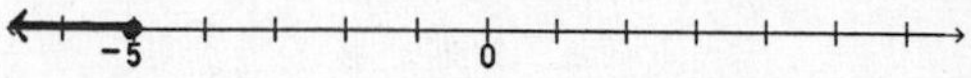

   $\{x|x \leq -5\}$, $(-\infty,-5]$

13. $$x + 8 > 3$$
   $$x + 8 + (-8) > 3 + (-8) \quad \text{Adding } -8$$
   $$x > -5$$

   The solution set is $\{x|x > -5\}$, or $(-5,\infty)$.

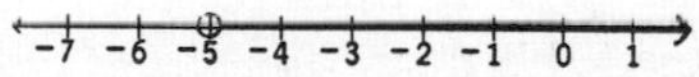

14. $\{x|x > -3\}$, or $(-3,\infty)$

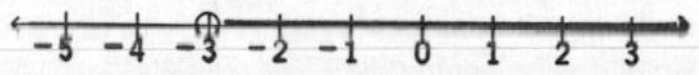

15. $$y + 3 < 9$$
   $$y + 3 + (-3) < 9 + (-3) \quad \text{Adding } -3$$
   $$y < 6$$

   The solution set is $\{y|y < 6\}$, or $(-\infty,6)$.

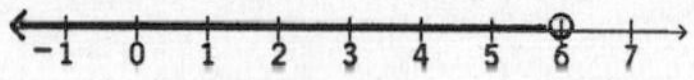

16. $\{y|y < 6\}$, or $(-\infty,6)$

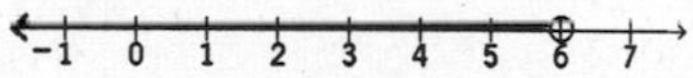

17. $$a + 9 \leq -12$$
   $$a + 9 + (-9) \leq -12 + (-9) \quad \text{Adding } -9$$
   $$a \leq -21$$

   The solution set is $\{a|a \leq -21\}$, or $(-\infty,-21]$.

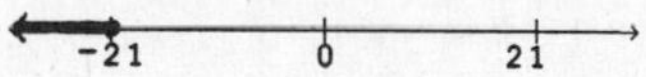

18. $\{a|a \leq -20\}$, or $(-\infty,-20]$

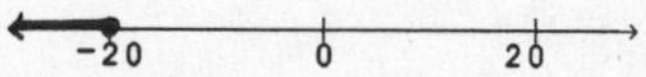

19. $t + 14 \geqslant 9$

$t + 14 + (-14) \geqslant 9 + (-14)$ Adding -14

$t \geqslant -5$

The solution set is $\{t|t \geqslant -5\}$, or $[-5,\infty)$.

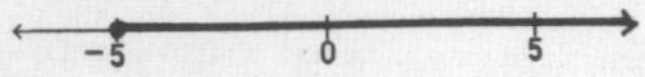

20. $\{x|x \leqslant 19\}$, or $(-\infty,19]$

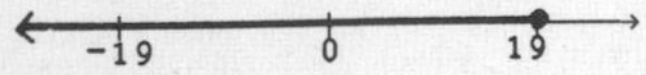

21. $y - 8 > -14$

$y - 8 + 8 > -14 + 8$ Adding 8

$y > -6$

The solution set is $\{y|y > -6\}$, or $(-6,\infty)$.

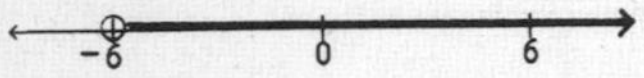

22. $\{y|y > -9\}$, or $(-9,\infty)$

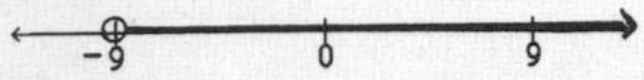

23. $x - 11 \leqslant -2$

$x - 11 + 11 \leqslant -2 + 11$ Adding 11

$x \leqslant 9$

The solution set is $\{x|x \leqslant 9\}$, or $(-\infty,9]$.

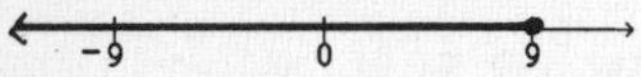

24. $\{y|y \leqslant 14\}$, or $(-\infty,14]$

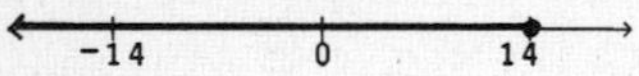

25. $8x \geqslant 24$

$\frac{1}{8} \cdot 8x \geqslant \frac{1}{8} \cdot 24$ Multiplying by $\frac{1}{8}$

$x \geqslant 3$

The solution set is $\{x|x \geqslant 3\}$, or $[3,\infty)$.

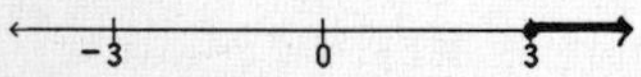

26. $\{t|t < -9\}$, or $(-\infty,-9)$

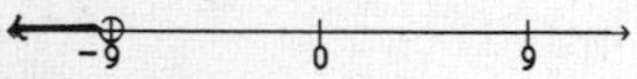

27. $0.3x < -18$

$\frac{1}{0.3}(0.3x) < \frac{1}{0.3}(-18)$ Multiplying by $\frac{1}{0.3}$

$x < -\frac{18}{0.3}$

$x < -60$

The solution set is $\{x|x < -60\}$, or $(-\infty,-60)$.

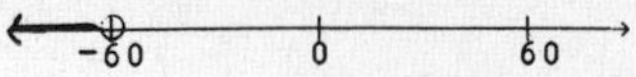

28. $\{x|x < 50\}$, or $(-\infty,50)$

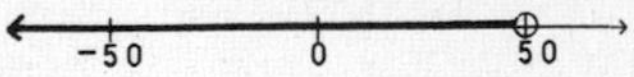

29. $-9x \geqslant -8.1$

$-\frac{1}{9}(-9x) \leqslant -\frac{1}{9}(-8.1)$ Multiplying by $-\frac{1}{9}$ and reversing the inequality sign

$x \leqslant \frac{8.1}{9}$

$x \leqslant 0.9$

The solution set is $\{x|x \leqslant 0.9\}$, or $(-\infty,0.9]$.

30. $\{y|y \geqslant -0.4\}$, or $[-0.4,\infty)$

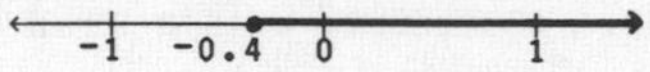

31. $-\frac{3}{4}x \geqslant -\frac{5}{8}$

$-\frac{4}{3}\left(-\frac{3}{4}x\right) \leqslant -\frac{4}{3}\left(-\frac{5}{8}\right)$ Multiplying by $-\frac{4}{3}$ and reversing the inequality sign

$x \leqslant \frac{20}{24}$

$x \leqslant \frac{5}{6}$

The solution set is $\left\{x\middle|x \leqslant \frac{5}{6}\right\}$, or $\left(-\infty,\frac{5}{6}\right]$.

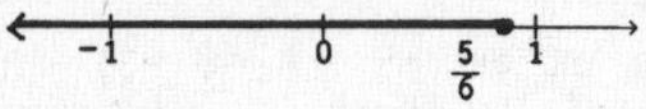

32. $\left\{y\middle|y \geqslant \frac{9}{10}\right\}$, or $\left[\frac{9}{10},\infty\right)$

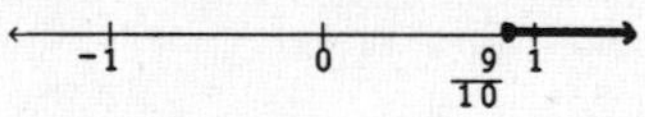

33. $2x + 7 < 19$

$2x + 7 + (-7) < 19 + (-7)$ Adding -7

$2x < 12$

$\frac{1}{2} \cdot 2x < \frac{1}{2} \cdot 12$ Multiplying by $\frac{1}{2}$

$x < 6$

The solution set is $\{x|x < 6\}$, or $(-\infty,6)$.

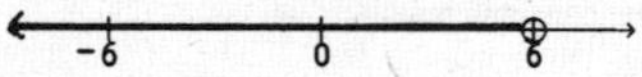

34. $\{y|y > 3\}$, or $(3,\infty)$

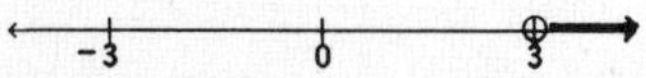

35. $5y + 2y \leqslant -21$

$7y \leqslant -21$ Collecting like terms

$\frac{1}{7}(7y) \leqslant \frac{1}{7}(-21)$ Multiplying by $\frac{1}{7}$

$y \leqslant -3$

The solution set is $\{y|y \leqslant -3\}$, or $(-\infty,-3]$.

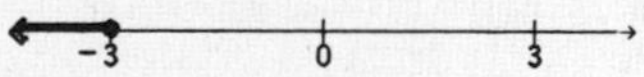

36. $\{x|x \leqslant 4\}$, or $(-\infty,4]$

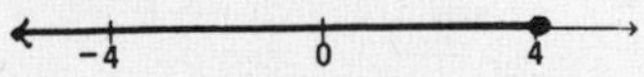

37. $2y - 7 < 5y - 9$

$-5y + 2y - 7 < -5y + 5y - 9$ Adding $-5y$

$-3y - 7 < -9$

$-3y - 7 + 7 < -9 + 7$ Adding 7

$-3y < -2$

$-\frac{1}{3}(-3y) > -\frac{1}{3}(-2)$ Multiplying by $-\frac{1}{3}$ and reversing the inequality sign

$y > \frac{2}{3}$

The solution set is $\left\{y\middle|y > \frac{2}{3}\right\}$, or $\left(\frac{2}{3},\infty\right)$.

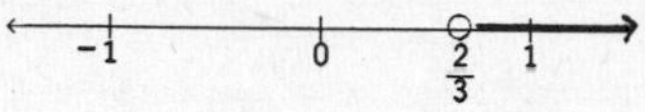

38. $\left\{x\middle|x < -\frac{2}{5}\right\}$, or $\left(-\infty,-\frac{2}{5}\right)$

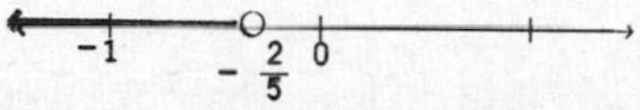

39. $0.4x + 5 \leqslant 1.2x - 4$

$-1.2x + 0.4x + 5 \leqslant -1.2x + 1.2x - 4$ Adding $-1.2x$

$-0.8x + 5 \leqslant -4$

$-0.8x + 5 + (-5) \leqslant -4 + (-5)$ Adding -5

$-0.8x \leqslant -9$

$\left(-\frac{1}{0.8}\right)(-0.8x) \geqslant \left(-\frac{1}{0.8}\right)(-9)$

Multiplying by $-\frac{1}{0.8}$ and reversing the inequality sign

$x \geqslant \frac{9}{0.8}$

$x \geqslant 11.25$

The solution set is $\{x|x \geqslant 11.25\}$, or $[11.25,\infty)$.

40. $\{y|y < 5\}$, or $(-\infty,5)$

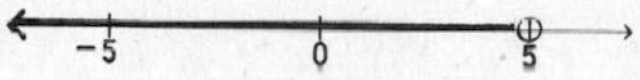

41. $3x - \frac{1}{8} \leqslant \frac{3}{8} + 2x$

$-2x + 3x - \frac{1}{8} \leqslant -2x + \frac{3}{8} + 2x$ Adding $-2x$

$x - \frac{1}{8} \leqslant \frac{3}{8}$

$x - \frac{1}{8} + \frac{1}{8} \leqslant \frac{3}{8} + \frac{1}{8}$ Adding $\frac{1}{8}$

$x \leqslant \frac{4}{8}$

$x \leqslant \frac{1}{2}$

The solution set is $\left\{x\middle|x \leqslant \frac{1}{2}\right\}$, or $\left(-\infty,\frac{1}{2}\right]$.

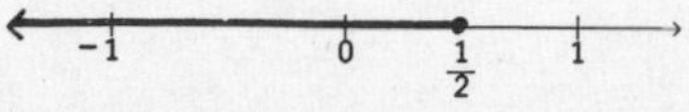

42. $\left\{x\middle|x < \frac{13}{3}\right\}$, or $\left(-\infty, \frac{13}{3}\right)$

43. $4(3y - 2) \geqslant 9(2y + 5)$

$12y - 8 \geqslant 18y + 45$

$-6y - 8 \geqslant 45$

$-6y \geqslant 53$

$y \leqslant -\frac{53}{6}$

The solution set is $\left\{y\middle|y \leqslant -\frac{53}{6}\right\}$, or $\left(-\infty,-\frac{53}{6}\right]$.

This can be expressed equivalently as

$\left\{y\middle|y \leqslant -8\frac{5}{6}\right\}$, or $\left(-\infty,-8\frac{5}{6}\right]$.

44. $\{m|m \leqslant 3.3\}$, or $(-\infty,3.3]$

45. $3(2 - 5x) + 2x < 2(4 + 2x)$

$6 - 15x + 2x < 8 + 4x$

$6 - 13x < 8 + 4x$

$6 - 17x < 8$

$-17x < 2$

$x > -\frac{2}{17}$

The solution set is $\left\{x\middle|x > -\frac{2}{17}\right\}$, or $\left(-\frac{2}{17},\infty\right)$.

46. $\left\{y\middle|y < \frac{13}{185}\right\}$, or $\left(-\infty,\frac{13}{185}\right)$

47. $5[3m - (m + 4)] > -2(m - 4)$

$5(3m - m - 4) > -2(m - 4)$

$5(2m - 4) > -2(m - 4)$

$10m - 20 > -2m + 8$

$12m - 20 > 8$

$12m > 28$

$m > \frac{28}{12}$

$m > \frac{7}{3}$

The solution set is $\left\{m\middle|m > \frac{7}{3}\right\}$, or $\left(\frac{7}{3},\infty\right)$.

48. $\left\{x\middle|x \leqslant -\frac{23}{2}\right\}$, or $\left(-\infty,-\frac{23}{2}\right]$

49. $3(r - 6) + 2 > 4(r + 2) - 21$

$3r - 18 + 2 > 4r + 8 - 21$

$3r - 16 > 4r - 13$

$-r - 16 > -13$

$-r > 3$

$r < -3$

The solution set is $\{r|r < -3\}$, or $(-\infty,-3)$.

50. $\{t|t < -12\}$, or $(-\infty,-12)$

51. $19 - (2x + 3) \leq 2(x + 3) + x$

$$19 - 2x - 3 \leq 2x + 6 + x$$
$$16 - 2x \leq 3x + 6$$
$$16 - 5x \leq 6$$
$$-5x \leq -10$$
$$x \geq 2$$

The solution set is $\{x|x \geq 2\}$, or $[2,\infty)$.

52. $\{c|c \leq 1\}$, or $(-\infty,1]$

53. $\frac{1}{4}(8y + 4) - 17 < -\frac{1}{2}(4y - 8)$

$$2y + 1 - 17 < -2y + 4$$
$$2y - 16 < -2y + 4$$
$$4y - 16 < 4$$
$$4y < 20$$
$$y < 5$$

The solution set is $\{y|y < 5\}$, or $(-\infty,5)$.

54. $\{x|x > 6\}$, or $(6,\infty)$

55. $2[4 - 2(3 - x)] - 1 \geq 4[2(4x - 3) + 7] - 25$

$$2[4 - 6 + 2x] - 1 \geq 4[8x - 6 + 7] - 25$$
$$2[-2 + 2x] - 1 \geq 4[8x + 1] - 25$$
$$-4 + 4x - 1 \geq 32x + 4 - 25$$
$$4x - 5 \geq 32x - 21$$
$$-28x - 5 \geq -21$$
$$-28x \geq -16$$
$$x \leq \frac{-16}{-28}, \text{ or } \frac{4}{7}$$

The solution set is $\left\{x\middle|x \leq \frac{4}{7}\right\}$, or $\left[-\infty,\frac{4}{7}\right]$.

56. $\left\{t\middle|t \geq -\frac{27}{19}\right\}$, or $\left[-\frac{27}{19},\infty\right]$

57. Familiarize. List the information in a table. Let x represent the score on the fourth test.

Test	Score
Test 1	89
Test 2	92
Test 3	95
Test 4	x
Total	360 or more

Translate. We can easily get an inequality from the table.

$$88 + 92 + 95 + x \geq 360$$

Carry out.

$276 + x \geq 360$ Collecting like terms

$x \geq 84$ Adding -276

Check. If you get 84 on the fourth test, your total score will be 89 + 92 + 95 + 84, or 360. Any higher score will also give you an A.

State. A score of 84 or better will give you an A.

58. 77 or better

59. Familiarize. We let x represent the number of miles traveled in a day. Then the total rental cost for a day is $30 + 0.20x$.

Translate. The total cost for a day must be less than or equal to \$96. This translates to the following inequality:

$$30 + 0.20x \leq 96$$

Carry out.

$0.20x \leq 66$ Adding -30

$x \leq \frac{66}{0.20}$ Multiplying by $\frac{1}{0.20}$

$x \leq 330$

Check. If you travel 330 miles, the total cost is $30 + 0.20(330)$, or $30 + 66 = \$96$. Any mileage less than 330 will also stay within the budget.

State. Mileage less than or equal to 330 miles allows you to stay within budget.

60. More than 36.8 miles per day

61. Familiarize. Let m = the length of a telephone call, in minutes. Then the number of minutes after the first minute of the call is $m - 1$. Using Down East Calling, the cost of the call, in cents, is $20 + 16(m - 1)$. Using Long Call Systems, the cost of the call, in cents, is $19 + 18(m - 1)$.

Translate.

Cost using Down East Calling	is less than	cost using Long Call Systems.
$20 + 16(m - 1)$	$<$	$19 + 18(m - 1)$

Carry out.

$$20 + 16(m - 1) < 19 + 18(m - 1)$$
$$20 + 16m - 16 < 19 + 18m - 18$$
$$4 + 16m < 1 + 18m$$
$$3 < 2m$$
$$1.5 < m$$

Check. We substitute a value of m greater than 1.5. We choose 2. For m = 2, the cost of the Down East call is $20 + 16(2 - 1)$, or 36¢. The cost of the Long Call call is $19 + 18(2 - 1)$, or 37¢, so the Down East call is less expensive. Since we cannot check all possible values, we stop here.

State. For calls longer than 1.5 minutes, Down East calling is less expensive.

62. Times greater than 4.25 hours

63. Familiarize. We make a table of information.

Plan A: Monthly Income	Plan B: Monthly Income
\$500 salary 4% of sales Total: 500 + 4% of sales	\$750 salary 5% of sales over \$8000 Total: 750 + 5% of sales over 8000

Suppose your gross sales were \$14,000 in one month. Compute the income from each plan.

Plan A:	Plan B:
500 + 4%(14,000)	750 + 5%(14,000 - 8000)
500 + 0.04(14,000)	750 + 0.05(6000)
500 + 560	750 + 300
\$1060	\$1050

When gross sales are \$14,000, Plan A is better.

Compute the income from each plan for gross monthly sales of \$16,000.

Plan A:	Plan B:
500 + 4%(16,000)	750 + 5%(16,000 - 8000)
500 + 0.04(16,000)	750 + 0.05(8000)
500 + 640	750 + 400
\$1140	\$1150

When gross sales are \$16,000, Plan B is better.

To determine all values for which Plan B is better, we solve an inequality.

Translate. We write an inequality stating that the income from Plan B is greater than the income from Plan A. We let S represent gross sales. Then S - 8000 represents gross sales over 8000.

$$750 + 5\%(S - 8000) > 500 + 4\%S$$

Carry out.

$$750 + 0.05S - 400 > 500 + 0.04S$$

$$350 + 0.05S > 500 + 0.04S \quad \text{Collecting like terms}$$

$$0.01S > 150 \quad \text{Adding -350 and -0.04S}$$

$$S > \frac{150}{0.01} \quad \text{Multiplying by } \frac{1}{0.01}$$

$$S > 15,000$$

Check. We calculate for S = \$15,000 and for some amount greater than \$15,000 and some amount less than \$15,000.

Plan A:	Plan B:
500 + 4%(15,000)	750 + 5%(15,000 - 8000)
500 + 0.04(15,000)	750 + 0.05(7000)
500 + 600	750 + 350
\$1100	\$1100

When S = \$15,000, income from Plan A is equal to the income from Plan B.

In the Familiarize step we saw that Plan A is better for sales of \$14,000 and Plan B is better for sales of \$16,000. We cannot check all possible values for S, so we will stop here.

State. Plan B is better than Plan A when gross sales are greater than \$15,000.

64. Less than \$116,666.67 per year

65. Familiarize. We let n represent the number of hours it will take the painter to complete the job. Plan A will pay the painter 500 + 5n, while Plan B will pay the painter 8n.

Suppose the job takes 90 hr. Compute the income from each plan.

Plan A:	Plan B:
500 + 6(90)	11(90)
500 + 540	\$990
\$1040	

If the job takes 90 hr, Plan A is better.

Computer the income from each plan if the job takes 120 hr.

Plan A:	Plan B:
500 + 6(120)	11(120)
500 + 720	\$1320
\$1220	

If the job takes 120 hr, Plan B is better.

To determine all values for which Plan A is better, we solve an inequality.

Translate. We write an inequality stating that the income from Plan A is greater than the income from Plan B.

Carry out. $500 + 6n > 11n$

$$500 > 5n \quad \text{Adding -6n}$$

$$100 > n \quad \text{Multiplying by } \frac{1}{5}$$

Check.

Plan A:	Plan B:
500 + 6(100)	11(100)
500 + 600	\$1100
\$1100	

When n = 100, the income from Plan A is equal to the income from Plan B.

In the Familiarize step we saw that Plan A is better for a 90 hr job and Plan B is better for a 120 hr job. We cannot check all possible values for n, so we stop here.

State. Plan A is better for the painter for n < 100.

66. n > 85.7

67. Familiarize. Organize the information in a table. Let x = the amount invested at 7%. Then 25,000 - x = the amount invested at 8%.

Amount invested		Rate of interest	Time	Interest (I = Prt)
x		7%	1 yr	0.07x
25,000 - x		8%	1 yr	0.08(25,000 - x)
Total	$25,000			$1800 or more

Translate. Use the information in the table to write an inequality:

$0.07x + 0.08(25,000 - x) \geq 1800$

Carry out.

$0.07x + 2000 - 0.08x \geq 1800$

$2000 - 0.01x \geq 1800$ Collecting like terms

$-0.01x \geq -200$ Adding -2000

$x \leq \frac{200}{0.01}$ Multiplying by $-\frac{1}{0.01}$ and reversing the inequality sign

$x \leq 20,000$

Check. For x = $20,000, 7%(20,000) = 0.07(20,000), or $1400, and 8%(25,000 - 20,000) = 0.08(5000), or $400. The total interest earned is 1400 + 400, or $1800. We also calculate for some amount less than $20,000 and for some amount greater than $20,000. For x = $15,000, 7%(15,000) = 0.07(15,000), or $1050, and 8%(25,000 - 15,000) = 0.08(10,000), or $800. The total interest earned is 1050 + 800, or $1850. For x = $22,000, 7%(22,000) = 0.07(22,000), or $1540, and 8%(25,000 - 22,000) = 0.08(3000), or $240. The total interest earned is 1540 + 240, or $1780. For these values the inequality, $x \leq 20,000$, gives correct results.

State. To make at least $1800 interest per year, $20,000 is the most that can be invested at 7%.

68. $5000

69. Familiarize. We let x represent the ticket price. Then 300x represents the total receipts from the ticket sales assuming 300 people will attend. The first band will play for $250 + 50%(300x). The second band will play for $550.

Translate. For school profit to be greater when the first band plays, the amount the first band charges must be less than the amount the second band charges. We now have an inequality.

$250 + 50\%(300x) < 550$

Carry out.

$250 + 0.5(300x) < 550$

$250 + 150x < 550$

$150x < 300$ Adding -250

$x < 2$ Multiplying by $\frac{1}{150}$

Check. For x = $2, the total receipts are 300(2), or $600. The first band charges 250 + 50%(600), or $550. The second band also charges $550. The school profit is the same using either band. For x = $1.99, the total receipts are 300(1.99), or $597. The first band charges 250 + 50%(597), or $548.50. Using the first band, the school profit would be 597 - 548.50, or $48.50. Using the second band, the profit would be 597 - 550, or $47. Thus, the first band produces more profit. For x = $2.01, the total receipts are 300(2.01), or $603. The first band charges 250 + 50%(603), or $551.50. Using the first band, the school profit would be 603 - 551.50, or $51.50. Using the second band, the school profit would be 603 - 550, or $53. Thus, the second band produces more profit. For these values, the inequality x < 2 gives correct results.

State. The ticket price must be less than $2. The highest price, rounded to the nearest cent, less than $2 is $1.99.

70. More than 33

71. Familiarize. We make a table listing the information. We let x represent the total medical bill.

	Plan A	Plan B
You pay the first	100	250
Insurance pays	80%(x - 100)	90%(x - 250)
You also pay	20%(x - 100)	10%(x - 250)
Total you pay	100+20%(x-100)	250+10%(x-250)

Translate. We write an inequality stating that the amount you pay is less when you choose Plan B. This gives us an inequality.

$250 + 10\%(x - 250) < 100 + 20\%(x - 100)$

Carry out.

$250 + 0.1x - 25 < 100 + 0.2x - 20$

$225 + 0.1x < 80 + 0.2x$ Collecting like terms

$145 < 0.1x$ Adding -80 and -0.1x

$1450 < x$ Multiplying by $\frac{1}{0.1}$

or

$x > 1450$

Check. We calculate for x = $1450 and also for some amount greater than $1450 and some amount less than $1450.

Plan A:	Plan B:
100 + 20%(1450 - 100)	250 + 10%(1450 - 250)
100 + 0.2(1350)	250 + 0.1(1200)
100 + 270	250 + 120
$370	$370

When x = $1450, you pay the same with either plan.

Plan A:	Plan B:
100 + 20%(1500 - 100)	250 + 10%(1500 - 250)
100 + 0.2(1400)	250 + 0.1(1250)
100 + 280	250 + 125
\$380	\$375

When x = \$1500, you pay less with Plan B.

Plan A:	Plan B:
100 + 20%(1400 - 100)	250 + 10%(1400 - 100)
100 + 0.2(1300)	250 + 0.1(1300)
100 + 260	250 + 130
\$360	\$380

When x = \$1400, you pay more with Plan B.

For these values, the inequality x > \$1450 gives correct results.

State. For Plan B to save you money, the total of the medical bills must be greater than \$1450.

72. More than 8.75 pounds

73. a) Familiarize. Find the values of F for which C is less than 1063°.

Translate. $\frac{5}{9}(F - 32) < 1063°$.

Carry out. $5(F - 32) < 9567$

$5F - 160 < 9567$

$5F < 9727$

$F < 1945.4$

Check. When F = 1945.4, $C = \frac{5}{9}(1945.4 - 32) = 1063$. Then values of F less than 1945.4 will give values of C less than 1063.

State. Gold is solid at temperatures less than 1945.4° F.

b) Familiarize. Find the values of F for which C is less than 960.8.

Translate. $\frac{5}{9}(F - 32) < 960.8$

Carry out. $5(F - 32) < 8647.2$

$5F - 160 < 8647.2$

$5F < 8807.2$

$F < 1761.44$

Check. When F = 1761.44, $C = \frac{5}{9}(1761.44 - 32) = 960.8$. Then values of F less than 1761.44 will give values of C less than 960.8.

State. Silver is solid at temperatures less than 1761.44° F.

74. a) 44,000, 165,978, 275,758

b) From 1987 on

75. a) Familiarize. Find the values of x for which $R(x) < C(x)$.

Translate. $26x < 90{,}000 + 15x$

Carry out.

$11x < 90{,}000$

$x < 8181\frac{9}{11}$

Check. $R\left(8181\frac{9}{11}\right) = \$212{,}727.27 = C\left(8181\frac{9}{11}\right)$. Calculate R(x) and C(x) for some x greater than $8181\frac{9}{11}$ and for some x less than $8181\frac{9}{11}$.

Suppose x = 8200:

$R(x) = 26(8200) = 213{,}200$ and

$C(x) = 90{,}000 + 15(8200) = 213{,}000$.

In this case $R(x) > C(x)$.

Suppose x = 8000:

$R(x) = 26(8000) = 208{,}000$ and

$C(x) = 90{,}000 + 15(8000) = 210{,}000$.

In this case $R(x) < C(x)$.

Then for $x < 8181\frac{9}{11}$, $R(x) < C(x)$.

State. We will state the result in terms of integers, since the company cannot sell a fraction of a radio. For 8181 or fewer radios the company loses money.

b) Our check in part a) shows that for $x > 8181\frac{9}{11}$, $R(x) > C(x)$ and the company makes a profit. Again, we will state the result in terms of an integer. For 8182 or more radios the company makes money.

76. a) $p < 10$

b) $p > 10$

77. $5y - 10 = 2x$

x	y
0	2
-5	0
5	4

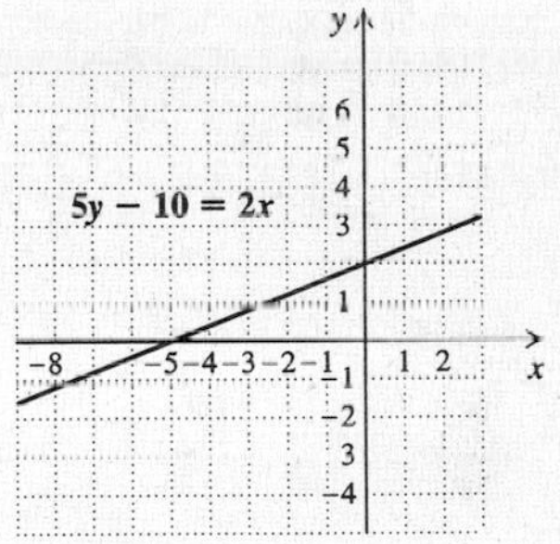

78. -2

79. $|-16| = 16$ -16 is 16 units from 0

80. -4

81.

82.

83. $3ax + 2x \geq 5ax - 4$

$2x - 2ax \geq -4$

$2x(1 - a) \geq -4$

$x(1 - a) \geq -2 \quad (a > 1,\ 1 - a < 0)$

$x \leq -\frac{2}{1 - a}$, or $\frac{2}{a - 1}$

The solution set is $\left\{x \middle| x \leq \frac{2}{a - 1}\right\}$.

84. $\left\{y \middle| y \geq -\frac{10}{b + 4}\right\}$

85. $a(by - 2) \geq b(2y + 5)$

$aby - 2a \geq 2by + 5b$

$aby - 2by \geq 2a + 5b$

$y(ab - 2b) \geq 2a + 5b \quad (a > 2,\ ab - 2b > 0)$

$y \geq \frac{2a + 5b}{ab - 2b}$, or $\frac{2a + 5b}{b(a - 2)}$

The solution set is $\left\{y \middle| y \geq \frac{2a + 5b}{b(a - 2)}\right\}$.

86. $\left\{x \middle| x < \frac{4c + 3d}{6c - 2d}\right\}$

87. $c(2 - 5x) + dx > m(4 + 2x)$

$2c - 5cx + dx > 4m + 2mx$

$-5cx + dx - 2mx > 4m - 2c$

$x(-5c + d - 2m) > 4m - 2c$

$x[d - (5c + 2m)] > 4m - 2c \quad (5c + 2m < d,\ d - (5c + 2m) > 0)$

$x > \frac{4m - 2c}{d - (5c + 2m)}$

The solution set is $\left\{x \middle| x > \frac{4m - 2c}{d - (5c + 2m)}\right\}$.

88. $\left\{x \middle| x < \frac{-3a + 2d}{c - (4a + 5d)}\right\}$

89. False. If $a = 2$, $b = 3$, $c = 4$, $d = 5$, then $2 - 4 = 3 - 5$.

90. False, because $-3 < -2$, but $9 > 4$.

91.

92.

93. $x + 5 \leq 5 + x$

$5 \leq 5$ Subtracting x

We get an inequality that is true for all real numbers x. Thus the solution is all real numbers.

94. No solution

95. $0^2 = 0$, $x^2 > 0$ for $x \neq 0$

The solution is all real numbers except 0.

96. All real numbers

Exercise Set 4.2

1. $\{5,6,7,8\} \cap \{4,6,8,10\}$

The numbers 6 and 8 are common to the two sets, so the intersection is $\{6,8\}$.

2. $\{10\}$

3. $\{2,4,6,8\} \cap \{1,3,5\}$

There are no numbers common to the two sets, so the intersection is the empty set, $\emptyset$.

4. $\{0\}$

5. $\{1,2,3,4\} \cap \{1,2,3,4\}$

The numbers 1, 2, 3, and 4 are common to the two sets, so the intersection is $\{1,2,3,4\}$.

6. $\emptyset$

7. $1 < x < 6$; $(1,6)$

8. $0 \leq y \leq 3$; $[0,3]$

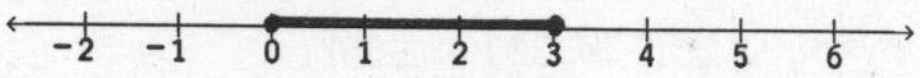

9. $-7 \leq y \leq -3$; $[-7,-3]$

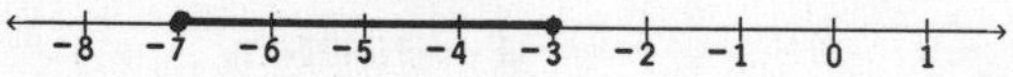

10. $-9 \leq x < -5$; $[-9,-5)$

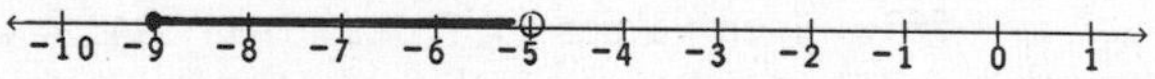

11. $-4 \leq -x < 3$

$4 \geq x > -3$ Multiplying by -1

$(-3,4]$

12. $x > -8$ and $x < -3$; $(-8,-3)$

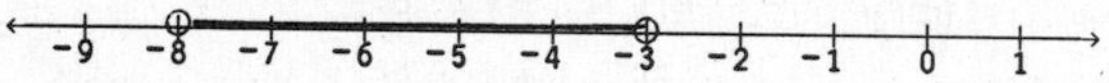

13. $6 > -x \geq -2$

$-6 < x \leq 2$ Multiplying by -1

$(-6,2]$

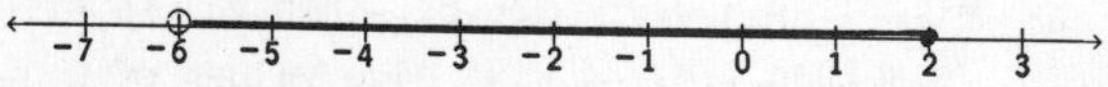

14. $x > -4$ and $x < 2$; $(-4,2)$

15. $5 > x \geq -2$; $[-2,5)$

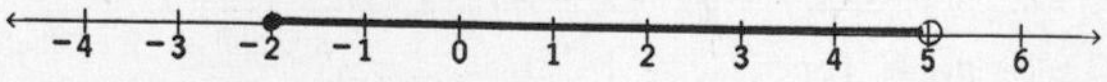

16. $3 > x \geq 0$; $[0,3)$

17. $x < 5$ and $x \geqslant 1$; $[1,5)$

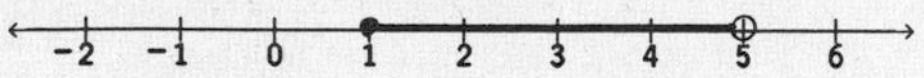

18. $x \geqslant -2$ and $x < 2$; $[-2,2)$

19. $-2 < x + 2 < 8$

$-2 + (-2) < x + 2 + (-2) < 8 + (-2)$ Adding -2

$-4 < x < 6$

The solution set is $\{x \mid -4 < x < 6\}$, or $(-4,6)$.

20. $\{x \mid -2 < x \leqslant 5\}$, or $(-2,5]$

21. $1 < 2y + 5 \leqslant 9$

$1 + (-5) < 2y + 5 + (-5) \leqslant 9 + (-5)$ Adding -5

$-4 < 2y \leqslant 4$

$\frac{1}{2} \cdot (-4) < \frac{1}{2} \cdot 2y \leqslant \frac{1}{2} \cdot 4$ Multiplying by $\frac{1}{2}$

$-2 < y \leqslant 2$

The solution set is $\{y \mid -2 < y \leqslant 2\}$, or $(-2,2]$.

22. $\{x \mid 0 \leqslant x \leqslant 1\}$, or $[0,1]$

23. $-10 \leqslant 3x - 5 \leqslant -1$

$-10 + 5 \leqslant 3x - 5 + 5 \leqslant -1 + 5$ Adding 5

$-5 \leqslant 3x \leqslant 4$

$\frac{1}{3} \cdot (-5) \leqslant \frac{1}{3} \cdot 3x \leqslant \frac{1}{3} \cdot 4$ Multiplying by $\frac{1}{3}$

$-\frac{5}{3} \leqslant x \leqslant \frac{4}{3}$

The solution set is $\left\{x \middle| -\frac{5}{3} \leqslant x \leqslant \frac{4}{3}\right\}$, or $\left[-\frac{5}{3},\frac{4}{3}\right]$.

24. $\left\{x \middle| -\frac{7}{2} < x \leqslant \frac{11}{2}\right\}$, or $\left(-\frac{7}{2},\frac{11}{2}\right]$

25. $2 < x + 3 \leqslant 9$

$2 + (-3) < x + 3 + (-3) \leqslant 9 + (-3)$ Adding -3

$-1 < x \leqslant 6$

The solution set is $\{x \mid -1 < x \leqslant 6\}$, or $(-1,6]$.

26. $\{x \mid -7 \leqslant x < 8\}$, or $[-7,8)$

27. $-6 \leqslant 2x - 3 < 6$

$-6 + 3 \leqslant 2x - 3 + 3 < 6 + 3$

$-3 \leqslant 2x < 9$

$\frac{1}{2} \cdot (-3) \leqslant \frac{1}{2} \cdot 2x < \frac{1}{2} \cdot 9$

$-\frac{3}{2} \leqslant x < \frac{9}{2}$

The solution set is $\left\{x \middle| -\frac{3}{2} \leqslant x < \frac{9}{2}\right\}$, or $\left[-\frac{3}{2},\frac{9}{2}\right)$.

28. $\left\{m \middle| -\frac{11}{3} < m \leqslant -3\right\}$, or $\left(-\frac{11}{3},-3\right]$

29. $-\frac{1}{2} < \frac{1}{4}x - 3 \leqslant \frac{1}{2}$

$-\frac{1}{2} + 3 < \frac{1}{4}x - 3 + 3 \leqslant \frac{1}{2} + 3$

$\frac{5}{2} < \frac{1}{4}x \leqslant \frac{7}{2}$

$4 \cdot \frac{5}{2} < 4 \cdot \frac{1}{4}x \leqslant 4 \cdot \frac{7}{2}$

$10 < x \leqslant 14$

The solution set is $\{x \mid 10 < x \leqslant 14\}$, or $(10,14]$.

30. $\left\{x \middle| \frac{40}{3} < x \leqslant \frac{56}{3}\right\}$, or $\left(\frac{40}{3},\frac{56}{3}\right]$

31. $\{4,5,6,7,8\} \cup \{1,4,6,11\}$

The numbers in either or both sets are 1, 4, 5, 6, 7, 8, and 11, so the union is $\{1,4,5,6,7,8,11\}$.

32. $\{2,8,9,27\}$

33. $\{2,4,6,8\} \cup \{1,3,5\}$

The numbers in either or both sets are 1, 2, 3, 4, 5, 6, and 8, so the union is $\{1,2,3,4,5,6,8\}$.

34. $\{8,9,10\}$

35. $\{4,8,11\} \cup \emptyset$

The numbers in either or both sets are 4, 8, and 11, so the union is $\{4,8,11\}$.

36. $\emptyset$

37. $x < -1$ or $x > 2$

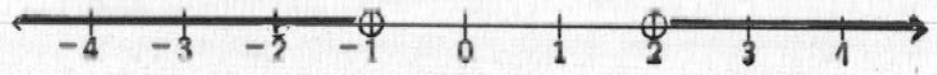

38. $x < -2$ or $x > 0$

39. $x \leqslant -3$ or $x > 1$

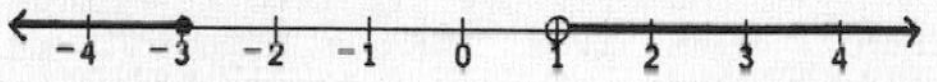

40. $x \leqslant -1$ or $x > 3$

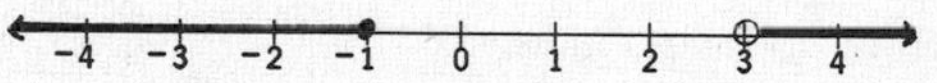

41. $x + 7 < -2$ or $x + 7 > 2$

$x + 7 + (-7) < -2 + (-7)$ or $x + 7 + (-7) > 2 + (-7)$

$x < -9$ or $x > -5$

The solution set is $\{x \mid x < -9 \text{ or } x > -5\}$, or $(-\infty,-9) \cup (-5,\infty)$.

42. $\{x \mid x < -13 \text{ or } x > -5\}$, or $(-\infty,-13) \cup (-5,\infty)$

43. $2x - 8 \leqslant -3$ or $x - 8 \geqslant 3$
$2x - 8 + 8 \leqslant -3 + 8$ or $x - 8 + 8 \geqslant 3 + 8$
$2x \leqslant 5$ or $x \geqslant 11$
$\frac{1}{2} \cdot 2x \leqslant \frac{1}{2} \cdot 5$ or $x \geqslant 11$
$x \leqslant \frac{5}{2}$ or $x \geqslant 11$

The solution set is $\left\{x \middle| x \leqslant \frac{5}{2} \text{ or } x \geqslant 11\right\}$, or $\left[-\infty, \frac{5}{2}\right] \cup [11,\infty)$.

44. $\{x \mid x \leqslant -9 \text{ or } x \geqslant 3\}$, or $(-\infty,-9] \cup [3,\infty)$

45. $7x + 4 \geqslant -17$ or $6x + 5 \geqslant -7$
$7x \geqslant -21$ or $6x \geqslant -12$
$x \geqslant -3$ or $x \geqslant -2$

The solution set is $\{x \mid x \geqslant -3\} \cup \{x \mid x \geqslant -2\}$. This is $\{x \mid x \geqslant -3\}$, or $[-3,\infty)$.

46. $\{x \mid x < 4\}$, or $(-\infty,4)$

47. $7 > -4x + 5$ or $10 \leqslant -4x + 5$
$7 + (-5) > -4x + 5 + (-5)$ or $10 + (-5) \leqslant -4x + 5 + (-5)$
$2 > -4x$ or $5 \leqslant -4x$
$-\frac{1}{4} \cdot 2 < -\frac{1}{4}(-4x)$ or $-\frac{1}{4} \cdot 5 \geqslant -\frac{1}{4}(-4x)$
$-\frac{1}{2} < x$ or $-\frac{5}{4} \geqslant x$

The solution set is $\left\{x \middle| x \leqslant -\frac{5}{4} \text{ or } x > -\frac{1}{2}\right\}$, or $\left[-\infty, -\frac{5}{4}\right] \cup \left[-\frac{1}{2},\infty\right]$.

48. The set of all real numbers

49. $3x - 7 > -10$ or $5x + 2 \leqslant 22$
$3x > -3$ or $5x \leqslant 20$
$x > -1$ or $x \leqslant 4$

Since all real numbers are greater than -1 or less than or equal to 4, the two sets fill up the entire number line. Thus the solution set is the set of all real numbers.

50. The set of all real numbers

51. $-2x - 2 < -6$ or $-2x - 2 > 6$
$-2x - 2 + 2 < -6 + 2$ or $-2x - 2 + 2 > 6 + 2$
$-2x < -4$ or $-2x > 8$
$-\frac{1}{2}(-2x) > -\frac{1}{2}(-4)$ or $-\frac{1}{2}(-2x) < -\frac{1}{2} \cdot 8$
$x > 2$ or $x < -4$

The solution set is $\{x \mid x < -4 \text{ or } x > 2\}$, or $(-\infty,-4) \cup (2,\infty)$.

52. $\left\{m \middle| m < -4 \text{ or } m > -\frac{2}{3}\right\}$, or $(-\infty,-4) \cup \left[-\frac{2}{3},\infty\right]$

53. $\frac{2}{3}x - 14 < -\frac{5}{6}$ or $\frac{2}{3}x - 14 > \frac{5}{6}$
$6\left[\frac{2}{3}x - 14\right] < 6\left[-\frac{5}{6}\right]$ or $6\left[\frac{2}{3}x - 14\right] > 6 \cdot \frac{5}{6}$
$4x - 84 < -5$ or $4x - 84 > 5$
$4x - 84 + 84 < -5 + 84$ or $4x - 84 + 84 > 5 + 84$
$4x < 79$ or $4x > 89$
$\frac{1}{4} \cdot 4x < \frac{1}{4} \cdot 79$ or $\frac{1}{4} \cdot 4x > \frac{1}{4} \cdot 89$
$x < \frac{79}{4}$ or $x > \frac{89}{4}$

The solution set is $\left\{x \middle| x < \frac{79}{4} \text{ or } x > \frac{89}{4}\right\}$, or $\left[-\infty, \frac{79}{4}\right] \cup \left[\frac{89}{4},\infty\right]$.

54. $\left\{x \middle| x \leqslant -\frac{91}{100} \text{ or } x \geqslant \frac{79}{60}\right\}$, or $\left[-\infty, -\frac{91}{100}\right] \cup \left[\frac{79}{60},\infty\right]$

55. $\frac{2x - 5}{6} \leqslant -3$ or $\frac{2x - 5}{6} \geqslant 4$
$6\left[\frac{2x - 5}{6}\right] \leqslant 6(-3)$ or $6\left[\frac{2x - 5}{6}\right] \geqslant 6 \cdot 4$
$2x - 5 \leqslant -18$ or $2x - 5 \geqslant 24$
$2x - 5 + 5 \leqslant -18 + 5$ or $2x - 5 + 5 \geqslant 24 + 5$
$2x \leqslant -13$ or $2x \geqslant 29$
$\frac{1}{2} \cdot 2x \leqslant \frac{1}{2}(-13)$ or $\frac{1}{2} \cdot 2x \geqslant \frac{1}{2} \cdot 29$
$x \leqslant -\frac{13}{2}$ or $x \geqslant \frac{29}{2}$

The solution set is $\left\{x \middle| x \leqslant -\frac{13}{2} \text{ or } x \geqslant \frac{29}{2}\right\}$, or $\left[-\infty, -\frac{13}{2}\right] \cup \left[\frac{29}{2},\infty\right]$.

56. $\left\{x \middle| x < -\frac{13}{3} \text{ or } x > 9\right\}$, or $\left[-\infty, -\frac{13}{3}\right] \cup (9,\infty)$

57. $5x - 7 \leqslant 13$ or $2x - 1 \geqslant -7$
$5x \leqslant 20$ or $2x \geqslant -6$
$x \leqslant 4$ or $x \geqslant -3$

Since all real numbers are greater than or equal to -3 or less than or equal to 4, the two sets fill up the entire number line. Thus, the solution set is the set of all real numbers.

58. $\{x \mid x \leqslant 2\}$, or $(-\infty,2]$

59. $2x - 3y = 7$, (1)
$3x + 2y = -10$ (2)

We will use the elimination method.

$4x - 6y =$	14	Multiplying (1) by 2
$9x + 6y =$	-30	Multiplying (2) by 3
$13x$	$= -16$	Adding

$x = -\frac{16}{13}$

Substitute $-\frac{16}{13}$ for x in (1) and solve for y.

$$2\left(-\frac{16}{13}\right) - 3y = 7$$
$$-\frac{32}{13} - 3y = 7$$
$$\frac{32}{13} - \frac{32}{13} - 3y = \frac{32}{13} + \frac{91}{13}$$
$$-3y = \frac{123}{13}$$
$$y = -\frac{41}{13}$$

These numbers check, so the solution is $\left(-\frac{16}{13}, -\frac{41}{13}\right)$.

60. $-\frac{8}{3}$

61. $5(2x + 3) = 3(x - 4)$
$$10x + 15 = 3x - 12$$
$$7x + 15 = -12$$
$$7x = -27$$
$$x = -\frac{27}{7}$$

62.

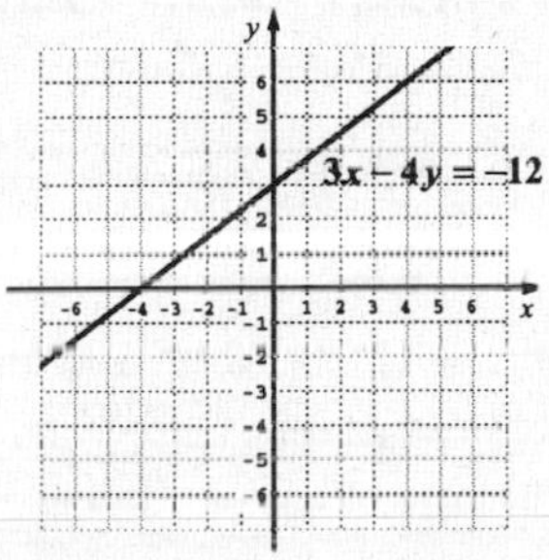

63.

64.

65. a) Substitute $\frac{5}{9}(F - 32)$ for C in the given inequality.
$$1063 \leqslant \frac{5}{9}(F - 32) < 2660$$
$$9\cdot 1063 \leqslant 9 \cdot \frac{5}{9}(F - 32) < 9\cdot 2660$$
$$9567 \leqslant 5(F - 32) < 23{,}940$$
$$9567 \leqslant 5F - 160 < 23{,}940$$
$$9727 \leqslant 5F < 24{,}100$$
$$1945.4 \leqslant F < 4820$$
The inequality for Fahrenheit temperatures is $1945.4° \leqslant F < 4820°$.

b) Substitute $\frac{5}{9}(F - 32)$ for C in the given inequality.
$$960.8 \leqslant \frac{5}{9}(F - 32) < 2180$$
$$9(960.8) \leqslant 9 \cdot \frac{5}{9}(F - 32) < 9\cdot 2180$$
$$8647.2 \leqslant 5(F - 32) < 19{,}620$$
$$8647.2 \leqslant 5F - 160 < 19{,}620$$
$$8807.2 \leqslant 5F < 19{,}780$$
$$1761.44 \leqslant F < 3956$$
The inequality for Fahrenheit temperatures is $1761.44° \leqslant F < 3956°$.

66. All the numbers between -5 and 11

67. Solve $32 < f(x) < 46$, or $32 < 2(x + 10) < 46$.
$$32 < 2(x + 10) < 46$$
$$32 < 2x + 20 < 46$$
$$12 < 2x < 26$$
$$6 < x < 13$$
For U.S. dress sizes between 6 and 13, dress sizes in Italy will be between 32 and 46.

68. $0 \text{ ft} \leqslant d \leqslant 198 \text{ ft}$

69. Solve $50{,}000 \leqslant N \leqslant 250{,}000$ where $N = 12{,}197.8t + 44{,}000$. Keep in mind that t is the number of years since 1971.
$$50{,}000 \leqslant 12{,}197.8t + 44{,}000 \leqslant 250{,}000$$
$$6000 \leqslant 12{,}197.8t \leqslant 206{,}000$$
$$0.49 \leqslant t \leqslant 16.89$$
0.49 years after the end of 1971 is in 1972 and 16.89 years after the end of 1971 is in 1988. We give the result in terms of <u>entire</u> years that satisfy the inequality. Therefore, there will be at least 50,000 and at most 250,000 women in the active duty military force from 1973 to 1987.

70. $1966 \leqslant y \leqslant 1980$

71. $4a - 2 \leqslant a + 1 \leqslant 3a + 4$
$$4a - 2 \leqslant a + 1 \quad \text{and} \quad a + 1 \leqslant 3a + 4$$
$$3a \leqslant 3 \quad \text{and} \quad -3 \leqslant 2a$$
$$a \leqslant 1 \quad \text{and} \quad -\frac{3}{2} \leqslant a$$
The solution set is $\left\{a \middle| -\frac{3}{2} \leqslant a \leqslant 1\right\}$, or $\left[-\frac{3}{2}, 1\right]$.

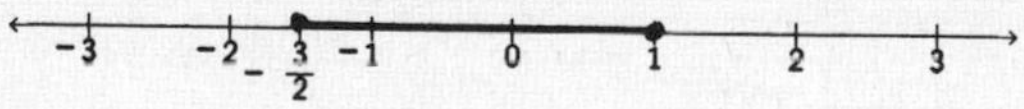

72. $\left\{m \middle| m < \frac{6}{5}\right\}$, or $\left(-\infty, \frac{6}{5}\right)$

73. $x - 10 < 5x + 6 \leqslant x + 10$

$-10 < 4x + 6 \leqslant 10$

$-16 < 4x \leqslant 4$

$-4 < x \leqslant 1$

The solution set is $\{x|-4 < x \leqslant 1\}$, or $(-4,1]$.

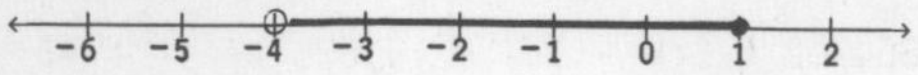

74. $\left\{y \middle| y < \frac{34}{15}\right\}$, or $\left(-\infty,\frac{34}{15}\right)$

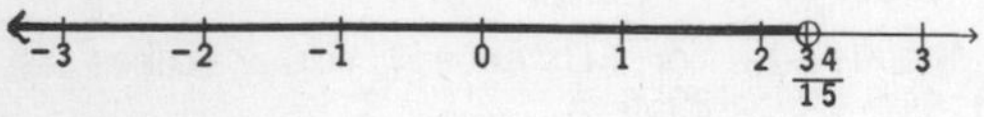

75. $3x < 4 - 5x < 5 + 3x$

$0 < 4 - 8x < 5$

$-4 < -8x < 1$

$\frac{1}{2} > x > -\frac{1}{8}$

The solution set is $\left\{x \middle| -\frac{1}{8} < x < \frac{1}{2}\right\}$, or $\left(-\frac{1}{8},\frac{1}{2}\right)$.

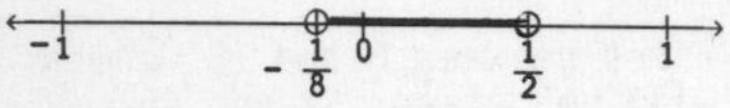

76. $\left\{x \middle| x > \frac{21}{4}\right\}$, or $\left(\frac{21}{4},\infty\right)$

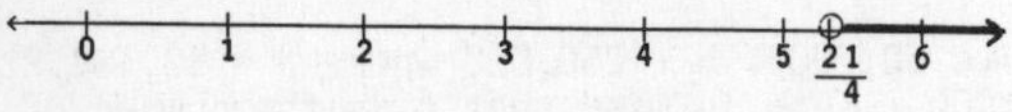

77. Given any two numbers b and c either $b < c$, $b = c$, or $b > c$. If $b > c$ is given, we know that $b \nless c$ and $b \neq c$. Thus $b \nless c$ is true.

78. True

79. Let $c = 6$ and $a = 10$. Then $c \neq a$, but $a \nless c$. The given statement is false.

80. False

81. Let $a = 5$, $c = 12$, and $b = 2$. Then $a < c$ and $b < c$, but $a \nless b$. The given statement is false.

82. False

83. $[4x - 2 < 8$ or $3(x - 1) < -2]$ and $-2 \leqslant 5x \leqslant 10$

$[4x - 2 < 8$ or $3x - 3 < -2]$ and $-\frac{2}{5} \leqslant x \leqslant 2$

$[4x < 10$ or $3x < 1]$ and $-\frac{2}{5} \leqslant x \leqslant 2$

$\left[x < \frac{5}{2}\right.$ or $\left.x < \frac{1}{3}\right]$ and $-\frac{2}{5} \leqslant x \leqslant 2$

$\left[x < \frac{5}{2}\right]$ and $-\frac{2}{5} \leqslant x \leqslant 2$

The solution set is $\left\{x \middle| -\frac{2}{5} \leqslant x \leqslant 2\right\}$, or $\left[-\frac{2}{5},2\right]$.

84. $\emptyset$

Exercise Set 4.3

1. $|x| = 3$

$x = -3$ or $x = 3$ Using the absolute-value principle

The solution set is $\{-3,3\}$.

2. $\{-5,5\}$

3. $|x| = -3$

The absolute value of a number is always nonnegative. Therefore, the solution set is $\emptyset$.

4. $\emptyset$

5. $|p| = 0$

The only number whose absolute value is 0 is 0. The solution set is $\{0\}$.

6. $\{-8.6,8.6\}$

7. $|t| = 5.5$

$t = -5.5$ or $t = 5.5$ Using the absolute-value principle

The solution set is $\{-5.5,5.5\}$.

8. $\{0\}$

9. $|x - 3| = 12$

$x - 3 = -12$ or $x - 3 = 12$ Absolute-value principle

$x = -9$ or $x = 15$

The solution set is $\{-9,15\}$.

10. $\left\{-\frac{4}{3},\frac{8}{3}\right\}$

11. $|2x - 3| = 4$

$2x - 3 = -4$ or $2x - 3 = 4$ Absolute-value principle

$2x = -1$ or $2x = 7$

$x = -\frac{1}{2}$ or $x = \frac{7}{2}$

The solution set is $\left\{-\frac{1}{2},\frac{7}{2}\right\}$.

12. $\left\{-1,\frac{1}{5}\right\}$

13. $|2y - 7| = 10$

$2y - 7 = -10$ or $2y - 7 = 10$

$2y = -3$ or $2y = 17$

$y = -\frac{3}{2}$ or $y = \frac{17}{2}$

The solution set is $\left\{-\frac{3}{2},\frac{17}{2}\right\}$.

14. $\left\{-\frac{4}{3},4\right\}$

15. $|3x - 10| = -8$

Absolute value is always nonnegative, so the equation has no solution. The solution set is $\emptyset$.

16. $\emptyset$

17. $|x| + 7 = 18$

$|x| + 7 - 7 = 18 - 7$ Adding -7

$|x| = 11$

$x = -11$ or $x = 11$ Absolute-value principle

The solution set is $\{-11,11\}$.

18. $\{-8.3,8.3\}$

19. $|5x| - 3 = 37$

$|5x| = 40$ Adding 3

$5x = -40$ or $5x = 40$

$x = -8$ or $x = 8$

The solution set is $\{-8,8\}$.

20. $\{-9,9\}$

21. $5|q| - 2 = 9$

$5|q| = 11$ Adding 2

$|q| = \frac{11}{5}$ Multiplying by $\frac{1}{5}$

$q = -\frac{11}{5}$ or $q = \frac{11}{5}$

The solution set is $\left\{-\frac{11}{5},\frac{11}{5}\right\}$.

22. $\{-2,2\}$

23. $\left|\frac{2x - 1}{3}\right| = 5$

$\frac{2x - 1}{3} = -5$ or $\frac{2x - 1}{3} = 5$

$2x - 1 = -15$ or $2x - 1 = 15$

$2x = -14$ or $2x = 16$

$x = -7$ or $x = 8$

The solution set is $\{-7,8\}$.

24. $\left\{-\frac{38}{5},\frac{46}{5}\right\}$

25. $|m + 5| + 9 = 16$

$|m + 5| = 7$ Adding -9

$m + 5 = -7$ or $m + 5 = 7$

$m = -12$ or $m = 2$

The solution set is $\{-12,2\}$.

26. $\{6,8\}$

27. $\left|\frac{2x - 1}{3}\right| = 1$

$\frac{2x - 1}{3} = -1$ or $\frac{2x - 1}{3} = 1$

$2x - 1 = -3$ or $2x - 1 = 3$

$2x = -2$ or $2x = 4$

$x = -1$ or $x = 2$

The solution set is $\{-1,2\}$.

28. $\left\{-\frac{8}{3},4\right\}$

29. $5 - 2|3x - 4| = -5$

$-2|3x - 4| = -10$

$|3x - 4| = 5$

$3x - 4 = -5$ or $3x - 4 = 5$

$3x = -1$ or $3x = 9$

$x = -\frac{1}{3}$ or $x = 3$

The solution set is $\left\{-\frac{1}{3},3\right\}$.

30. $\left\{\frac{3}{2},\frac{7}{2}\right\}$

31. The distance between 13 and 17 is $|17 - 13| = |4| = 4$. The distance is also given by $|13 - 17| = |-4| = 4$.

32. 6

33. The distance between 25 and 14 is $|25 - 14| = |11| = 11$, or $|14 - 25| = |-11| = 11$.

34. 15

35. The distance between -9 and 24 is $|24 - (-9)| = |24 + 9| = |33| = 33$, or $|-9 - 24| = |-33| = 33$.

36. 19

37. The distance between -8 and -42 is $|-8 - (-42)| = |-8 + 42| = |34| = 34$, or $|-42 - (-8)| = |-42 + 8| = |-34| = 34$.

38. 27

39. $|3x + 4| = |x - 7|$

$3x + 4 = x - 7$ or $3x + 4 = -(x - 7)$

$2x + 4 = -7$ or $3x + 4 = -x + 7$

$2x = -11$ or $4x + 4 = 7$

$x = -\frac{11}{2}$ or $4x = 3$

$x = \frac{3}{4}$

The solution set is $\left\{-\frac{11}{2},\frac{3}{4}\right\}$.

40. $\left\{11,\frac{5}{3}\right\}$

41. $|x + 5| = |x - 2|$

$x + 5 = x - 2$ or $x + 5 = -(x - 2)$

$5 = -2$ or $x + 5 = -x + 2$

False - yields no solution $\quad 2x + 5 = 2$

$2x = -3$

$x = -\frac{3}{2}$

The solution set is $\left\{-\frac{3}{2}\right\}$.

42. $\left\{-\frac{1}{2}\right\}$

43. $|2a + 4| = |3a - 1|$

$2a + 4 = 3a - 1$ or $2a + 4 = -(3a - 1)$

$-a + 4 = -1$ or $2a + 4 = -3a + 1$

$-a = -5$ or $5a + 4 = 1$

$a = 5$ or $5a = -3$

$a = -\frac{3}{5}$

The solution set is $\left\{5, -\frac{3}{5}\right\}$.

44. $\left\{-4, -\frac{10}{9}\right\}$

45. $|y - 3| = |3 - y|$

$y - 3 = 3 - y$ or $y - 3 = -(3 - y)$

$2y - 3 = 3$ or $y - 3 = -3 + y$

$2y = 6$ or $-3 = -3$

$y = 3$ $\quad$ True for all real values of y

The solution set is $\{y|y$ is a real number$\}$.

46. $\{m|m$ is a real number$\}$

47. $|5 - p| = |p + 8|$

$5 - p = p + 8$ or $5 - p = -(p + 8)$

$5 - 2p = 8$ or $5 - p = -p - 8$

$-2p = 3$ or $5 = -8$

$p = -\frac{3}{2}$ $\quad$ False

The solution set is $\left\{-\frac{3}{2}\right\}$.

48. $\left\{-\frac{11}{2}\right\}$

49. $\left|\frac{2x - 3}{6}\right| = \left|\frac{4 - 5x}{8}\right|$

$\frac{2x - 3}{6} = \frac{4 - 5x}{8}$ or $\frac{2x - 3}{6} = -\left(\frac{4 - 5x}{8}\right)$

$24\left(\frac{2x - 3}{6}\right) = 24\left(\frac{4 - 5x}{8}\right)$ or $\frac{2x - 3}{6} = \frac{-4 + 5x}{8}$

$8x - 12 = 12 - 15x$ or $24\left(\frac{2x - 3}{6}\right) = 24\left(\frac{-4 + 5x}{8}\right)$

$23x - 12 = 12$ or $8x - 12 = -12 + 15x$

$23x = 24$ or $-7x - 12 = -12$

$x = \frac{24}{23}$ or $-7x = 0$

$x = 0$

The solution set is $\left\{\frac{24}{23}, 0\right\}$.

50. $\left\{-\frac{23}{31}, 47\right\}$

51. $\left|\frac{1}{2}x - 5\right| = \left|\frac{1}{4}x + 3\right|$

$\frac{1}{2}x - 5 = \frac{1}{4}x + 3$ or $\frac{1}{2}x - 5 = -\left(\frac{1}{4}x + 3\right)$

$\frac{1}{4}x - 5 = 3$ or $\frac{1}{2}x - 5 = -\frac{1}{4}x - 3$

$\frac{1}{4}x = 8$ or $\frac{3}{4}x - 5 = -3$

$x = 32$ or $\frac{3}{4}x = 2$

$x = \frac{8}{3}$

The solution set is $\left\{32, \frac{8}{3}\right\}$.

52. $\left\{-\frac{48}{37}, -\frac{144}{5}\right\}$

53. $|x| < 3$

$-3 < x < 3$ Part (b)

The solution set is $\{x|-3 < x < 3\}$, or $(-3,3)$.

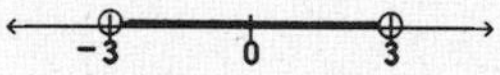

54. $\{x|-5 \leqslant x \leqslant 5\}$, or $[-5,5]$;

55. $|x| \geqslant 2$

$x \leqslant -2$ or $2 \leqslant x$ Part (c)

The solution set is $\{x|x \leqslant -2$ or $x \geqslant 2\}$, or $(-\infty,-2] \cup [2,\infty)$.

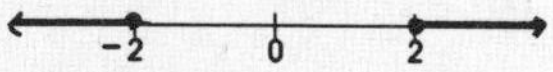

56. $\{y|y < -8$ or $y > 8\}$, or $(-\infty,-8) \cup (8,\infty)$

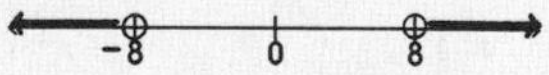

57. $|t| \geqslant 5.5$

$t \leqslant -5.5$ or $5.5 \leqslant t$ Part (c)

The solution set is $\{t|t \leqslant -5.5$ or $t \geqslant 5.5\}$, or $(-\infty,-5.5] \cup [5.5,\infty)$.

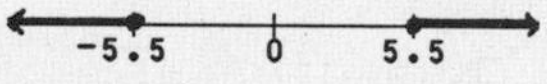

58. $\{m|m \neq 0\}$, or $(-\infty,0) \cup (0,\infty)$

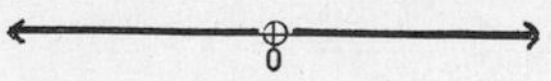

59. $|x - 3| < 1$

$-1 < x - 3 < 1$ Part (b)

$2 < x < 4$ Adding 3

The solution set is $\{x|2 < x < 4\}$, or $(2,4)$.

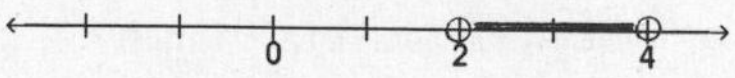

60. $\{x|-4 < x < 8\}$, or $(-4,8)$

61. $|x + 2| \leqslant 5$

$-5 \leqslant x + 2 \leqslant 5$ Part (b)

$-7 \leqslant x \leqslant 3$ Adding -2

The solution set is $\{x|-7 \leqslant x \leqslant 3\}$, or $[-7,3]$.

62. $\{x|-5 \leqslant x \leqslant -3\}$, or $[-5,-3]$

63. $|x - 3| > 1$

$x - 3 < -1$ or $1 < x - 3$ Part (c)

$x < 2$ or $4 < x$ Adding 3

The solution set is $\{x|x < 2 \text{ or } x > 4\}$, or $(-\infty,2) \cup (4,\infty)$.

64. $\{x|x < -4 \text{ or } x > 8\}$, or $(-\infty,-4) \cup (8,\infty)$

65. $|2x - 3| \leqslant 4$

$-4 \leqslant 2x - 3 \leqslant 4$ Part (b)

$-1 \leqslant 2x \leqslant 7$ Adding 3

$-\frac{1}{2} \leqslant x \leqslant \frac{7}{2}$ Multiplying by $\frac{1}{2}$

The solution set is $\left\{x\middle|-\frac{1}{2} \leqslant x \leqslant \frac{7}{2}\right\}$, or $\left[-\frac{1}{2},\frac{7}{2}\right]$.

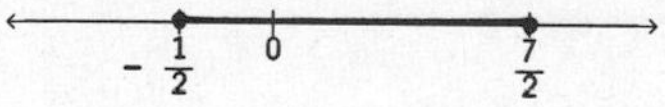

66. $\left\{x\middle|-1 \leqslant x \leqslant \frac{1}{5}\right\}$, or $\left[-1,\frac{1}{5}\right]$

67. $|2y - 7| > -1$

Since absolute value is never negative, any value of $2y - 7$, and hence any value of y, will satisfy the inequality. The solution set is the set of all real numbers.

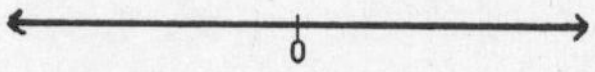

68. $\left\{y\middle|y < -\frac{4}{3} \text{ or } y > 4\right\}$, or $\left(-\infty,-\frac{4}{3}\right) \cup (4,\infty)$

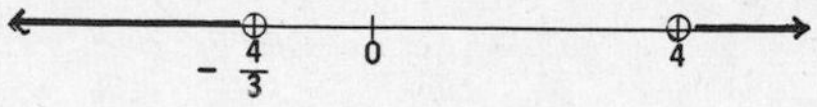

69. $|4x - 9| \geqslant 14$

$4x - 9 \leqslant -14$ or $14 \leqslant 4x - 9$ Part (c)

$4x \leqslant -5$ or $23 \leqslant 4x$

$x \leqslant -\frac{5}{4}$ or $\frac{23}{4} \leqslant x$

The solution set is $\left\{x\middle|x \leqslant -\frac{5}{4} \text{ or } x \geqslant \frac{23}{4}\right\}$, or $\left(-\infty,-\frac{5}{4}\right] \cup \left[\frac{23}{4},\infty\right)$.

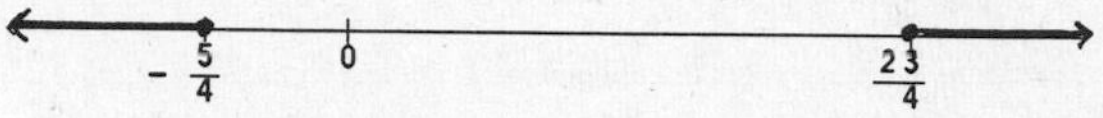

70. The set of all real numbers

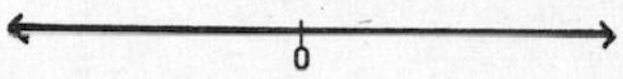

71. $|y - 3| < 12$

$-12 < y - 3 < 12$ Part (b)

$-9 < y < 15$ Adding 3

The solution set is $\{y|-9 < y < 15\}$, or $(-9,15)$.

72. $\{p|-1 < p < 5\}$, or $(-1,5)$

73. $|2x + 3| \leqslant 4$

$-4 \leqslant 2x + 3 \leqslant 4$ Part (b)

$-7 \leqslant 2x \leqslant 1$ Adding -3

$-\frac{7}{2} \leqslant x \leqslant \frac{1}{2}$ Multiplying by $\frac{1}{2}$

The solution set is $\left\{x\middle|-\frac{7}{2} \leqslant x \leqslant \frac{1}{2}\right\}$, or $\left[-\frac{7}{2},\frac{1}{2}\right]$.

74. $\left\{x\middle|-\frac{1}{5} \leqslant x \leqslant 1\right\}$, or $\left[-\frac{1}{5},1\right]$

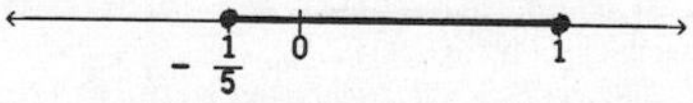

75. $|4 - 3y| > 8$

$4 - 3y < -8$ or $8 < 4 - 3y$ Part (c)

$-3y < -12$ or $4 < -3y$ Adding -4

$y > 4$ or $-\frac{4}{3} > y$ Multiplying by $-\frac{1}{3}$

The solution set is $\left\{y\middle|y < -\frac{4}{3} \text{ or } y > 4\right\}$, or $\left(-\infty,-\frac{4}{3}\right) \cup (4,\infty)$.

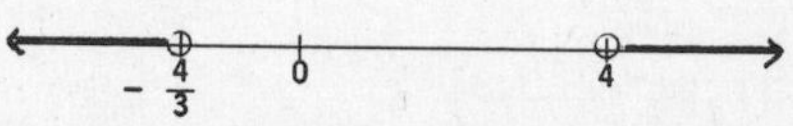

76. $\emptyset$

77. $|9 - 4x| \leqslant 14$

$-14 \leqslant 9 - 4x \leqslant 14$ Part (b)

$-23 \leqslant -4x \leqslant 5$ Adding -9

$\frac{23}{4} \geqslant x \geqslant -\frac{5}{4}$ Multiplying by $-\frac{1}{4}$

The solution set is $\left\{x \middle| -\frac{5}{4} \leqslant x \leqslant \frac{23}{4}\right\}$, or

$\left[-\frac{5}{4}, \frac{23}{4}\right]$.

78. $\left\{p \middle| p \leqslant -\frac{5}{3} \text{ or } p \geqslant \frac{19}{9}\right\}$, or $\left(-\infty, -\frac{5}{3}\right] \cup \left[\frac{19}{9}, \infty\right)$

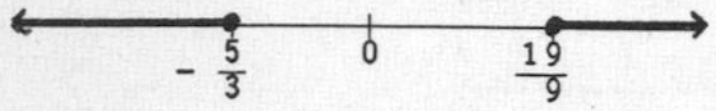

79. $|3 - 4x| < -5$

Absolute value is always nonnegative, so the inequality has no solution. The solution set is $\emptyset$.

80. $\left\{x \middle| -5 \leqslant x \leqslant \frac{25}{7}\right\}$, or $\left[-5, \frac{25}{7}\right]$

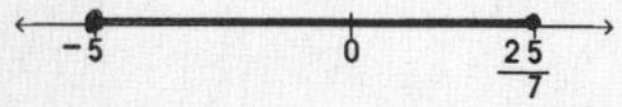

81. $7 + |2x - 1| > 16$

$|2x - 1| > 9$

$2x - 1 < -9$ or $9 < 2x - 1$ Part (c)

$2x < -8$ or $10 < 2x$ Adding 1

$x < -4$ or $5 < x$ Multiplying by $\frac{1}{2}$

The solution set is $\{x|x < -4 \text{ or } x > 5\}$, or $(-\infty,-4) \cup (5,\infty)$.

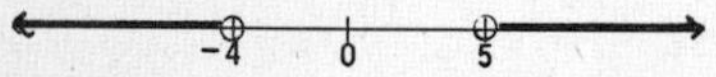

82. $\left\{x \middle| x < -\frac{16}{3} \text{ or } x > 4\right\}$, or $\left(-\infty, -\frac{16}{3}\right) \cup (4,\infty)$

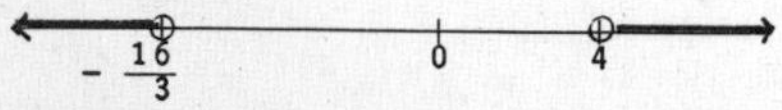

83. $\left|\frac{x - 7}{3}\right| < 4$

$-4 < \frac{x - 7}{3} < 4$ Part (b)

$-12 < x - 7 < 12$ Multiplying by 3

$-5 < x < 19$ Adding 7

The solution set is $\{x|-5 < x < 19\}$, or $(-5,19)$.

84. $\{x|-13 \leqslant x \leqslant 3\}$, or $[-13,3]$

85. $\left|\frac{2 - 5x}{4}\right| \geqslant \frac{2}{3}$

$\frac{2 - 5x}{4} \leqslant -\frac{2}{3}$ or $\frac{2}{3} \leqslant \frac{2 - 5x}{4}$ Part (c)

$2 - 5x \leqslant -\frac{8}{3}$ or $\frac{8}{3} \leqslant 2 - 5x$ Multiplying by 4

$-5x \leqslant -\frac{14}{3}$ or $\frac{2}{3} \leqslant -5x$ Adding -2

$x \geqslant \frac{14}{15}$ or $-\frac{2}{15} \geqslant x$ Multiplying by $-\frac{1}{5}$

The solution set is $\left\{x \middle| x \leqslant -\frac{2}{15} \text{ or } x \geqslant \frac{14}{15}\right\}$, or $\left(-\infty, -\frac{2}{15}\right] \cup \left[\frac{14}{15}, \infty\right)$.

86. $\left\{x \middle| x < -\frac{43}{24} \text{ or } x > \frac{9}{8}\right\}$, or $\left(-\infty, -\frac{43}{24}\right) \cup \left(\frac{9}{8}, \infty\right)$

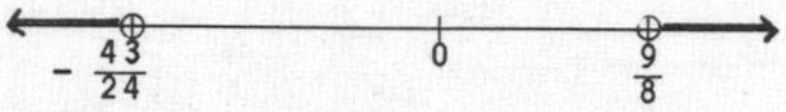

87. $|m + 5| + 9 \leqslant 16$

$|m + 5| \leqslant 7$ Adding -9

$-7 \leqslant m + 5 \leqslant 7$

$-12 \leqslant m \leqslant 2$

The solution set is $\{m|-12 \leqslant m \leqslant 2\}$, or $[-12,2]$.

88. $\{t|t \leqslant 6 \text{ or } t \geqslant 8\}$, or $(-\infty,6] \cup [8,\infty)$

89. $|g + 7| + 13 > 9$

$|g + 7| > -4$ Adding -13

Since absolute value is never negative, any value of $g + 7$, and hence any value of g, will satisfy the inequality. The solution set is the set of all real numbers.

90. The set of all real numbers

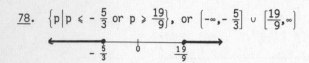

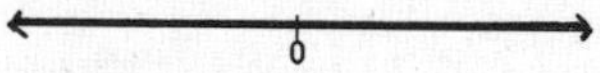

91. $\left|\frac{2x - 1}{3}\right| \leqslant 1$

$-1 \leqslant \frac{2x - 1}{3} \leqslant 1$

$-3 \leqslant 2x - 1 \leqslant 3$

$-2 \leqslant 2x \leqslant 4$

$-1 \leqslant x \leqslant 2$

The solution set is $\{x|-1 \leqslant x \leqslant 2\}$, or $[-1,2]$.

92. $\left\{x \middle| -\frac{8}{3} \leqslant x \leqslant 4\right\}$, or $\left[-\frac{8}{3}, 4\right]$

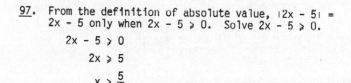

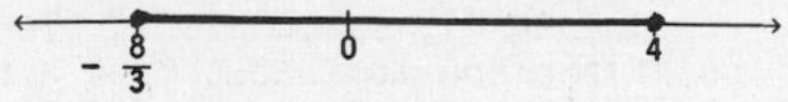

93. Familiarize. Let ℓ represent the length and w represent the width. Recall that perimeter $P = 2\ell + 2w$, and area $A = \ell w$.

Translate.

The perimeter is 628 m.

$$2\ell + 2w = 628$$

The length is 6 m more than the width.

$$\ell = 6 + w$$

We have a system of equations:

$$2\ell + 2w = 628, \quad (1)$$
$$\ell = 6 + w \quad (2)$$

Carry out. Use the substitution method to solve the system of equations and then compute the area. Substitute $6 + w$ for ℓ in (1).

$$2(6 + w) + 2w = 628$$
$$12 + 2w + 2w = 628$$
$$12 + 4w = 628$$
$$4w = 616$$
$$w = 154$$

Substitute 154 for w in (2).

$$\ell = 6 + 154 = 160$$
$$A = \ell w = 160(154) = 24{,}640$$

Check. $2\ell + 2w = 2 \cdot 160 + 2 \cdot 154 = 320 + 308 = 628$. 160 is 6 more than 154. The numbers check.

State. The area is 24,640 m^2.

94. 132 adult plates, 118 children's plates

95.

96.

97. From the definition of absolute value, $|2x - 5| = 2x - 5$ only when $2x - 5 \geqslant 0$. Solve $2x - 5 \geqslant 0$.

$$2x - 5 \geqslant 0$$
$$2x \geqslant 5$$
$$x \geqslant \frac{5}{2}$$

The solution set is $\left\{x \middle| x \geqslant \frac{5}{2}\right\}$.

98. {-31,-33}

99. $|x + 5| = x + 5$

From the definition of absolute value, $|x + 5| = x + 5$ only when $x + 5 \geqslant 0$, or $x \geqslant -5$. The solution set is $\{x|x \geqslant -5\}$.

100. $\{x|x \geqslant 1\}$

101. $|7x - 2| = x + 4$

From the definition of absolute value, we know $x + 4 \geqslant 0$, or $x \geqslant -4$. So we have $x \geqslant -4$ and

$$7x - 2 = x + 4 \quad \text{or} \quad 7x - 2 = -(x + 4)$$
$$6x = 6 \quad \text{or} \quad 7x - 2 = -x - 4$$
$$x = 1 \quad \text{or} \quad 8x = -2$$
$$x = -\frac{1}{4}$$

The solution set is

$\left\{x \middle| x \geqslant -4 \text{ and } x = 1 \text{ or } x = -\frac{1}{4}\right\}$, or $\left\{1, -\frac{1}{4}\right\}$.

102. {x|x is a real number}

103. $\left|5.2x - \frac{6}{7}\right| \leqslant -8$

From the definition of absolute value we know that $\left|5.2x - \frac{6}{7}\right| \geqslant 0$. Thus $\left|5.2x - \frac{6}{7}\right| \leqslant -8$ is false for all x. The solution set is Ø.

104. Ø

105. $|x + 5| > x$

The inequality is true for all $x < 0$ (because absolute value must be nonnegative). The solution set in this case is $\{x|x < 0\}$. If $x = 0$, we have $|0 + 5| > 0$, which is true. The solution set in this case is {0}. If $x > 0$, we have the following:

$$x + 5 < -x \quad \text{or} \quad x < x + 5$$
$$2x < -5 \quad \text{or} \quad 0 < 5$$
$$x < -\frac{5}{2}$$

Although $x > 0$ and $x < -\frac{5}{2}$ yields no solution, $x > 0$ and $5 > 0$ (true for all x) yield the solution set $\{x|x > 0\}$ in this case. The solution set for the inequality is $\{x|x < 0\} \cup \{0\} \cup \{x|x > 0\}$, or {x|x is a real number}.

106. $\{x|-4 \leqslant x \leqslant -1 \text{ or } 3 \leqslant x \leqslant 6\}$

107. Using part (b), we find that $-3 < x < 3$ is equivalent to $|x| < 3$.

108. $|y| \leqslant 5$

109. $x \leqslant -6$ or $6 \leqslant x$

$|x| \geqslant 6$ Using part (c)

110. $|x| > 4$

111. $x < -8$ or $2 < x$

$x + 3 < -5$ or $5 < x + 3$ Adding 3

$|x + 3| > 5$ Using part (c)

112. $|x + 2| < 3$

113. We first write an inequality in terms of feet. Convert $\frac{1}{8}$ in. to feet:

$$\frac{1}{8}\text{ in.} = \frac{1}{8}\text{ in.} \times \frac{1\text{ ft}}{12\text{ in.}} = \frac{1}{96}\text{ ft}$$

Then we have:

$$5 - \frac{1}{96} \leqslant p \leqslant 5 + \frac{1}{96}$$

$$-\frac{1}{96} \leqslant p - 5 \leqslant \frac{1}{96} \quad \text{Adding } -5$$

$$|p - 5| \leqslant \frac{1}{96} \quad \text{Part (b)}$$

We could also express the difference between 5 ft and p as $|5 - p|$, so the result can also be stated as $|5 - p| \leqslant \frac{1}{96}$.

We could also write an inequality in terms of inches. Convert 5 ft to inches:

$$5\text{ ft} = 5\text{ ft} \times \frac{12\text{ in.}}{1\text{ ft}} = 60\text{ in.}$$

Then we have:

$$60 - \frac{1}{8} \leqslant p \leqslant 60 + \frac{1}{8}$$

$$-\frac{1}{8} \leqslant p - 60 \leqslant \frac{1}{8} \quad \text{Adding } -60$$

$$|p - 60| \leqslant \frac{1}{8} \quad \text{Part (b)}$$

We could also express the difference between 60 inches and p as $|60 - p|$, so the result can also be stated as $|60 - p| \leqslant \frac{1}{8}$.

114. $\left\{d \middle| 5\frac{1}{2}\text{ ft} \leqslant d \leqslant 6\frac{1}{2}\text{ ft}\right\}$, or $\left[5\frac{1}{2}\text{ ft}, 6\frac{1}{2}\text{ ft}\right]$

115. 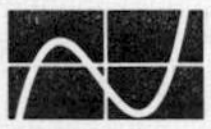

Exercise Set 4.4

1. We replace x by -4 and y by 2.

$$\begin{array}{l|l} 2x + y < -5 & \\ \hline 2(-4) + 2 \ ? \ -5 & \\ -8 + 2 & \\ -6 & \text{TRUE} \end{array}$$

Since $-6 < -5$ is true, $(-4,2)$ is a solution.

2. Yes

3. We replace x by 8 and y by 14.

$$\begin{array}{l|l} 2y - 3x > 5 & \\ \hline 2 \cdot 14 - 3 \cdot 8 \ ? \ 5 & \\ 28 - 24 & \\ 4 & \text{FALSE} \end{array}$$

Since $4 > 5$ is false, $(8,14)$ is not a solution.

4. Yes

5. Graph: $y > 2x$

We first graph the line $y = 2x$. We draw the line dashed since the inequality symbol is $>$. To determine which half-plane to shade, test a point not on the line. We try (1,1) and substitute:

$$\begin{array}{l|l} y > 2x & \\ \hline 1 \ ? \ 2 \cdot 1 & \\ & 2 \quad \text{FALSE} \end{array}$$

Since $1 > 2$ is false, (1,1) is not a solution, nor are any points in the half-plane containing (1,1). The points in the other half-plane are solutions, so we shade that half-plane and obtain the graph.

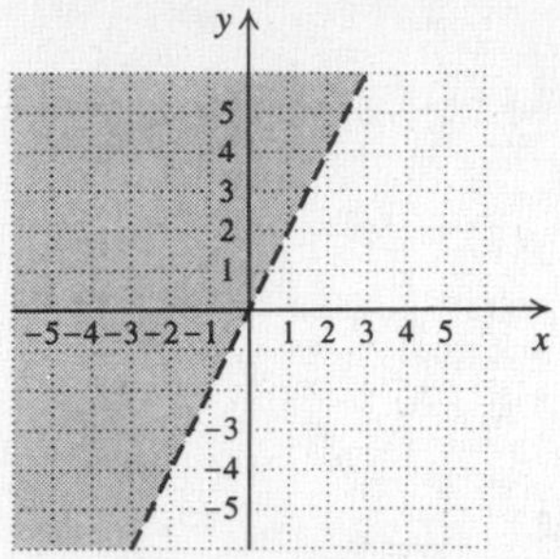

6.

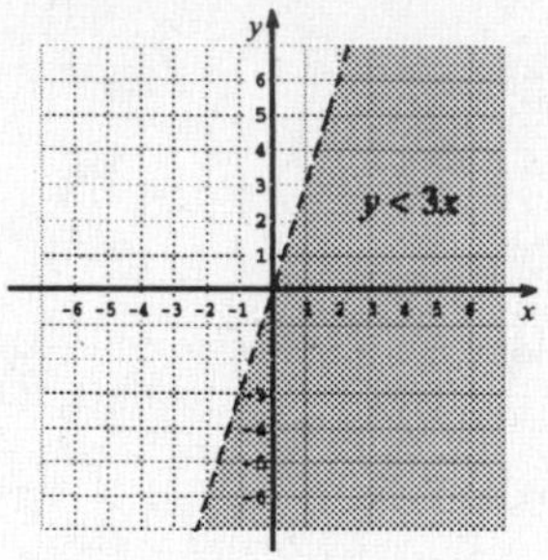

7. Graph: $y < x + 1$

First graph the line $y = x + 1$. Draw it dashed since the inequality symbol is $<$. Test the point (0,0) to determine if it is a solution.

$$\begin{array}{l|l} y < x + 1 & \\ \hline 0 \ ? \ 0 + 1 & \\ & 1 \quad \text{TRUE} \end{array}$$

Since $0 < 1$ is true, we shade the half-plane containing (0,0) and obtain the graph.

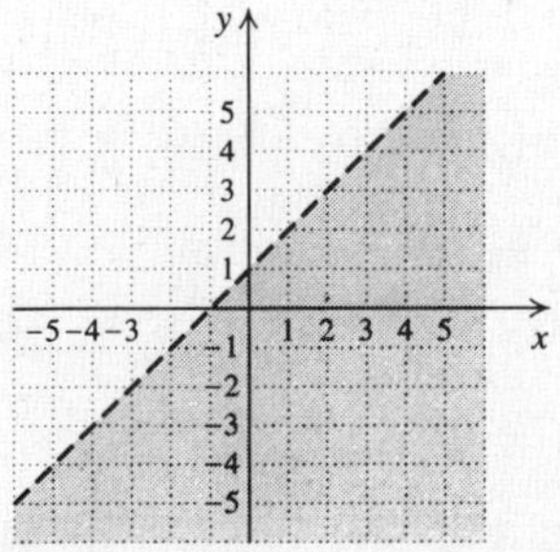

8.

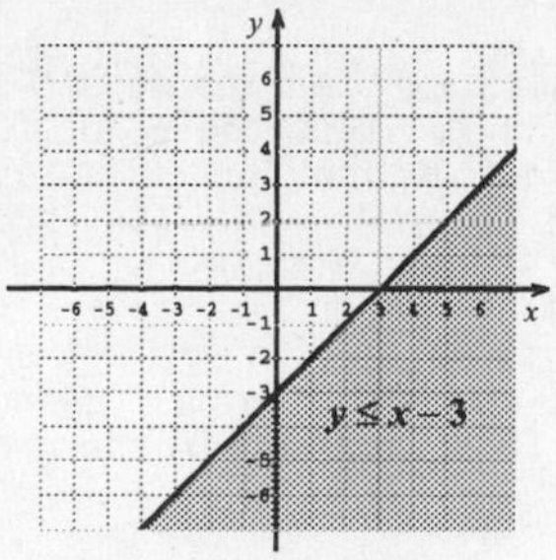

9. Graph: $y > x - 2$

We first graph $y = x - 2$. Draw a dashed line since the inequality symbol is >. Test the point (0,0) to determine if it is a solution.

$$\begin{array}{c|c} y & > x - 2 \\ \hline 0 & ?\ 0 - 2 \\ & -2 \quad \text{TRUE} \end{array}$$

Since $0 > -2$ is true, we shade the half-plane containing (0,0) and obtain the graph.

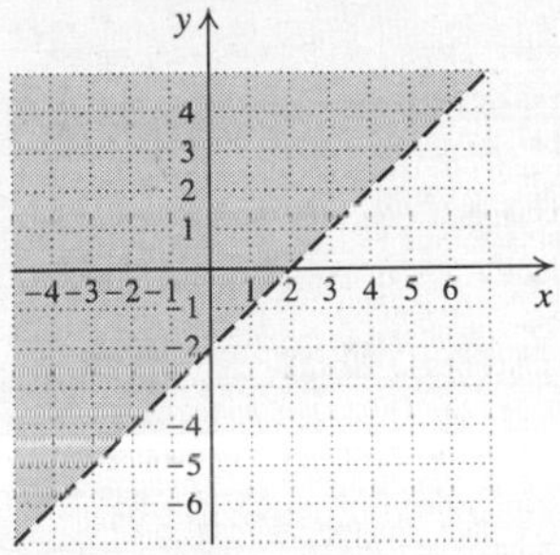

10.

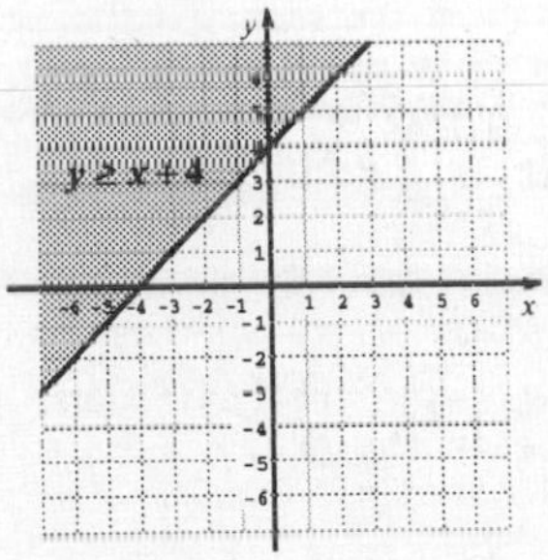

11. Graph: $x + y < 4$

First graph $x + y = 4$. Draw the line dashed since the inequality symbol is <. Test the point (0,0) to determine if it is a solution.

$$\begin{array}{c|c} x + y & < 4 \\ \hline 0 + 0 & ?\ 4 \\ 0 & \quad \text{TRUE} \end{array}$$

Since $0 < 4$ is true, we shade the half-plane containing (0,0) and obtain the graph.

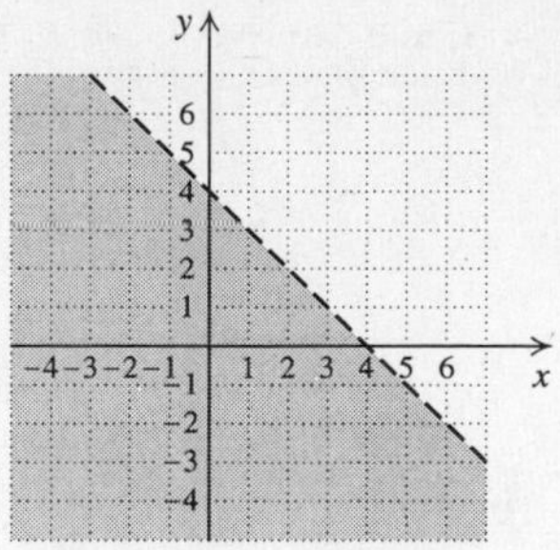

12.

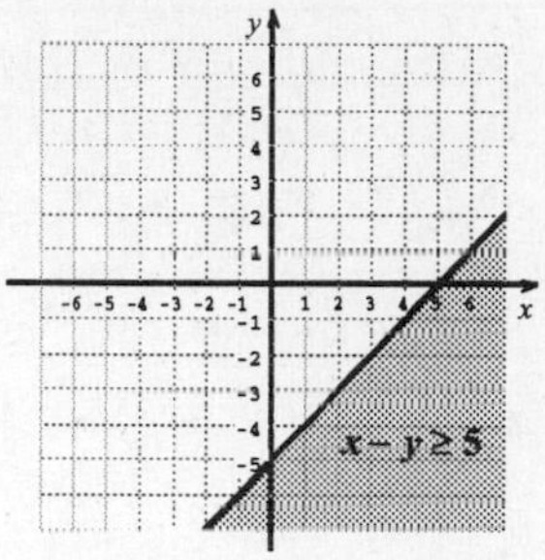

13. Graph: $3x + 4y \leqslant 12$

We first graph $3x + 4y = 12$. Draw the line solid since the inequality symbol is $\leqslant$. Test the point (0,0) to determine if it is a solution.

$$\begin{array}{c|c} 3x + 4y & \leqslant 12 \\ \hline 3 \cdot 0 + 4 \cdot 0 & ?\ 12 \\ 0 & \quad \text{TRUE} \end{array}$$

Since $0 \leqslant 12$ is true, we shade the half-plane containing (0,0) and obtain the graph.

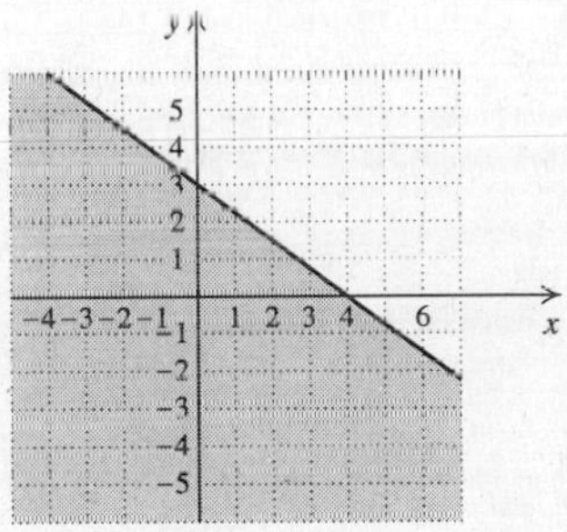

14.

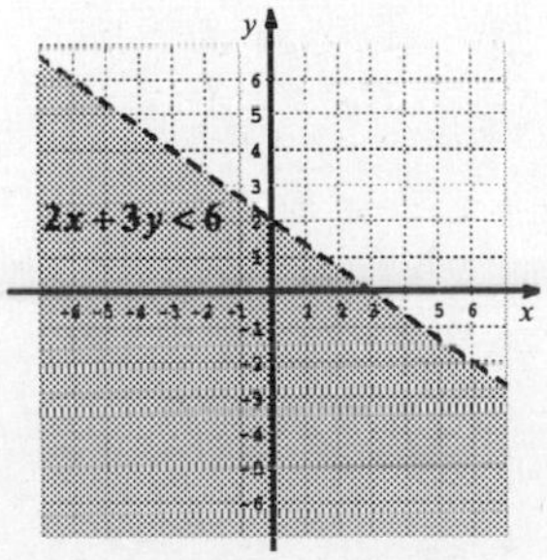

15. Graph: $2y - 3x > 6$

We first graph $2y - 3x = 6$. Draw the line dashed since the inequality symbol is >. Test the point (0,0) to determine if it is a solution.

$$\begin{array}{c|c} 2y - 3x & > 6 \\ \hline 2 \cdot 0 - 3 \cdot 0 & ?\ 6 \\ 0 & \quad \text{FALSE} \end{array}$$

Since 0 > 6 is false, we shade the half-plane that does not contain (0,0) and obtain the graph.

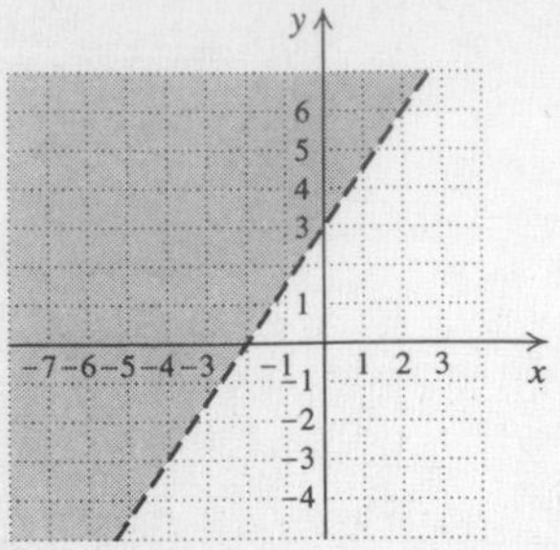

16.

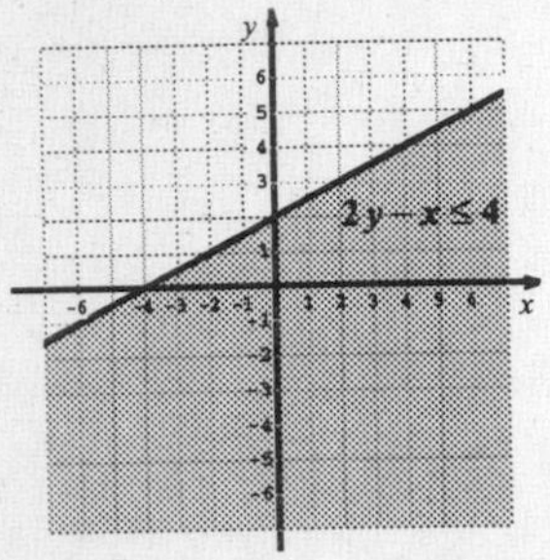

17. Graph: $3x - 2 \leqslant 5x + y$

$-2 \leqslant 2x + y$

We first graph $-2 = 2x + y$. Draw the line solid since the inequality symbol is $\leqslant$. Test the point (0,0) to determine if it is a solution.

$$\begin{array}{l|l} \multicolumn{2}{l}{-2 \leqslant 2x + y} \\ \hline -2 \;?\; 2\cdot 0 + 0 & \\ & 0 \quad \text{TRUE} \end{array}$$

Since $-2 \leqslant 0$ is true, we shade the half-plane containing (0,0) and obtain the graph.

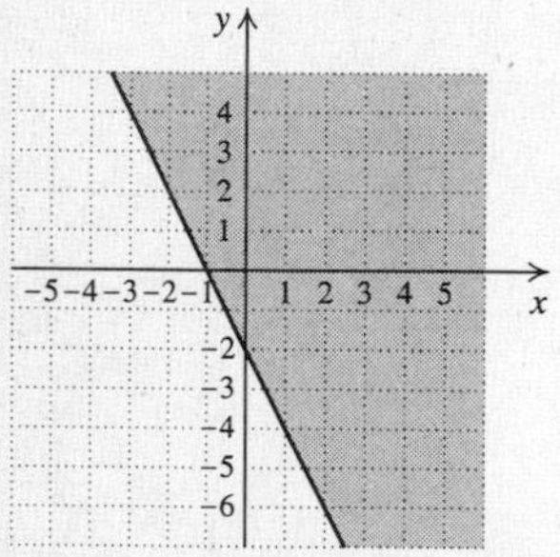

18.

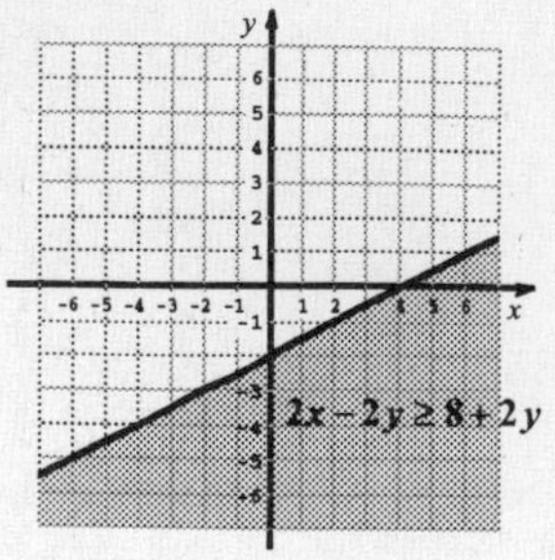

19. Graph: $x < -4$

We first graph $x = -4$. Draw the line dashed since the inequality symbol is <. Test the point (0,0) to determine if it is a solution.

$$\begin{array}{l} x < -4 \\ \hline 0 \;?\; -4 \quad \text{FALSE} \end{array}$$

Since $0 < -4$ is false, we shade the half-plane that does not contain (0,0) and obtain the graph.

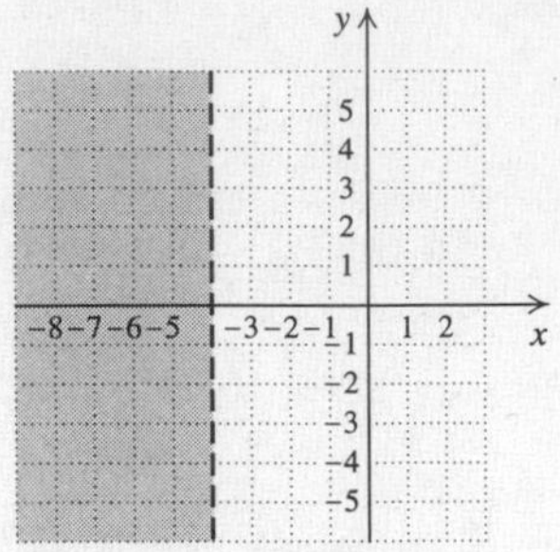

20.

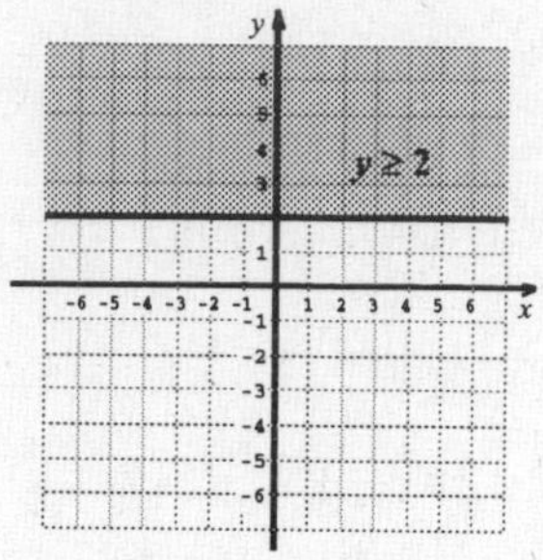

21. Graph: $y > -2$

We first graph $y = -2$. We draw the line dashed since the inequality symbol is >. Test the point (0,0) to determine if it is a solution.

$$\begin{array}{l} y > -2 \\ \hline 0 \;?\; -2 \quad \text{TRUE} \end{array}$$

Since $0 > -2$ is true, we shade the half-plane containing (0,0) and obtain the graph.

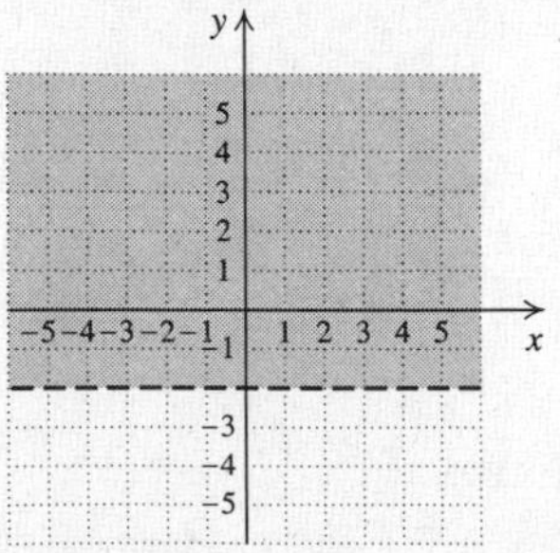

22.

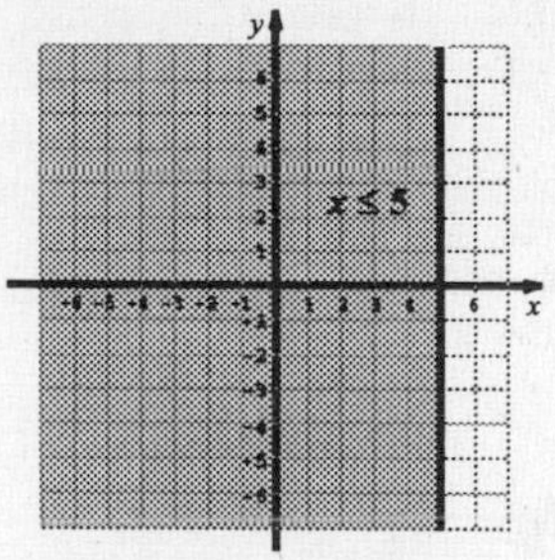

23. Graph: $-4 < y < -1$

This is a system of inequalities:

$-4 < y$,

$y < -1$

We graph the equation $-4 = y$ and see that the graph of $-4 < y$ is the half-plane above the line $-4 = y$. We also graph $y = -1$ and see that the graph of $y < -1$ is the half-plane below the line $y = -1$.

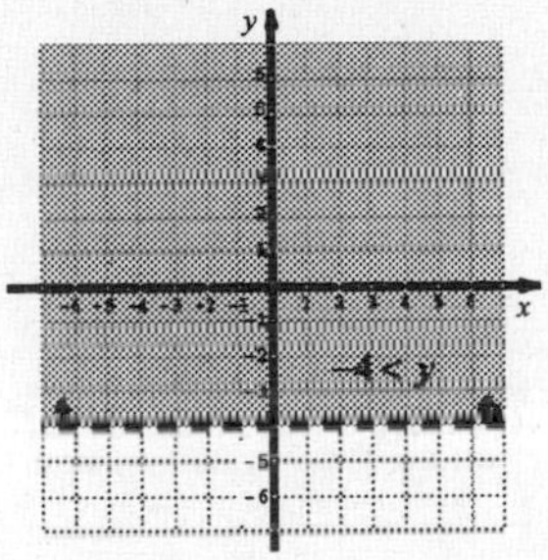

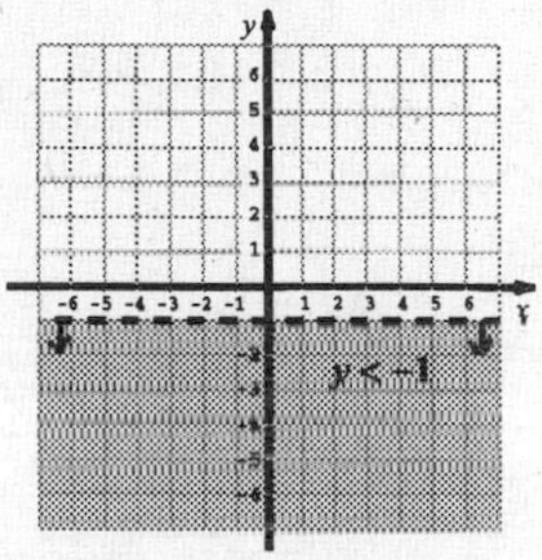

Finally, we shade the intersection of these graphs.

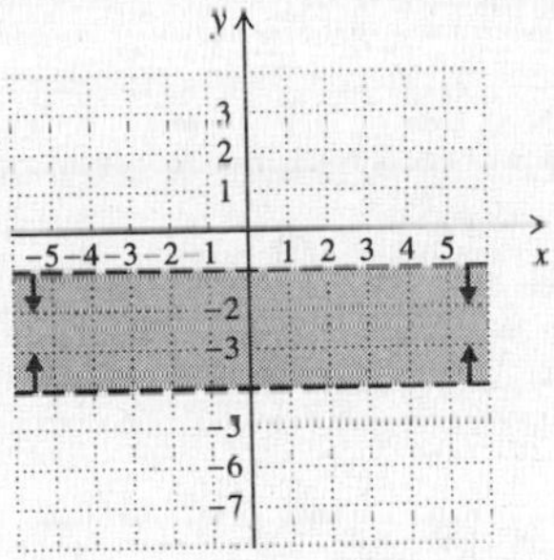

24.

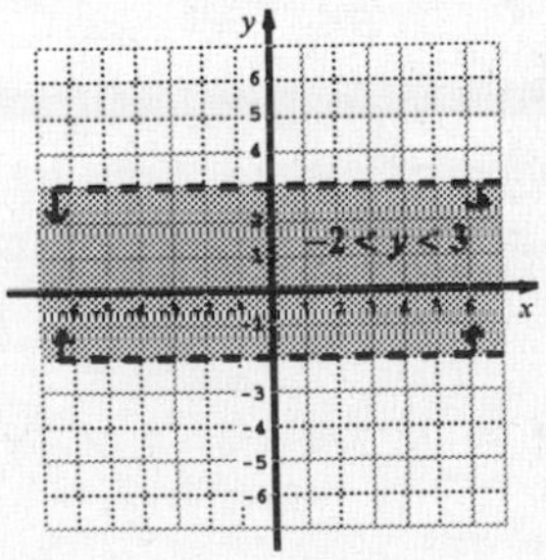

25. Graph: $-3 \leqslant x \leqslant 3$

This is a system of inequalities:

$-3 \leqslant x$,

$x \leqslant 3$

Graph $-3 \leqslant x$ and $x \leqslant 3$.

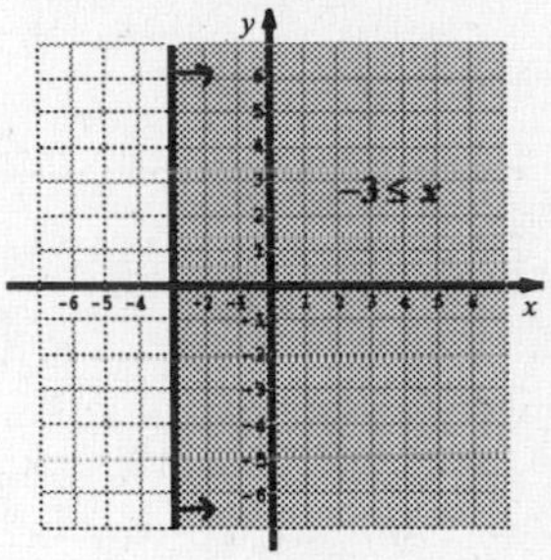

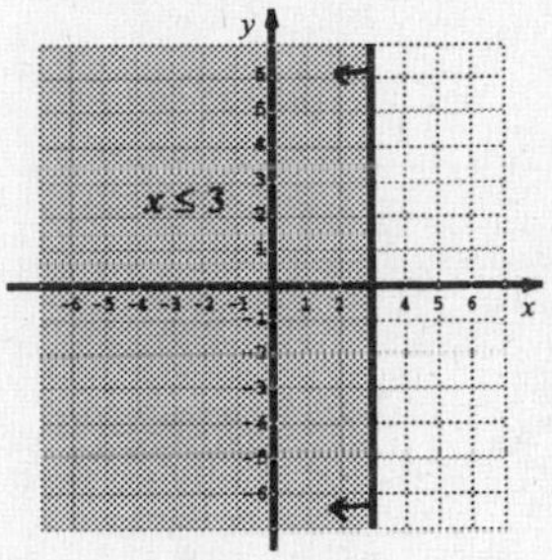

Then we shade the intersection of these graphs.

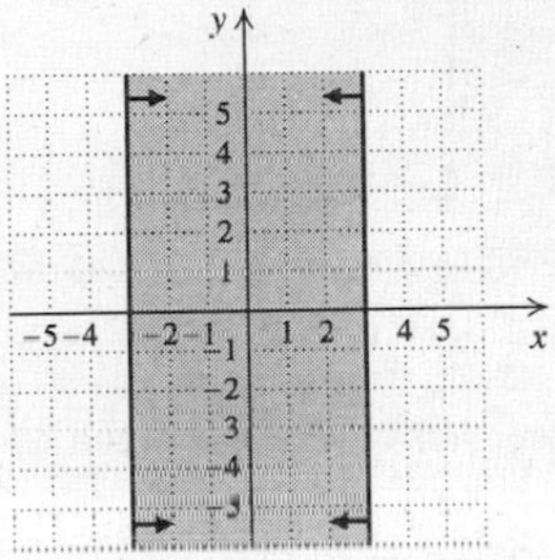

26.

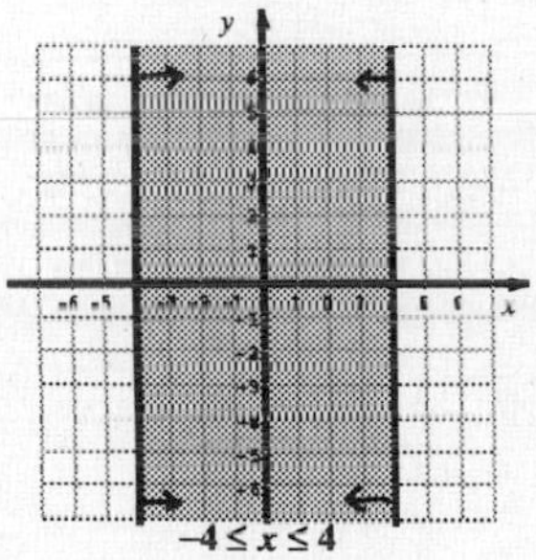

27. Graph: $0 \leqslant x \leqslant 5$

This is a system of inequalities:

$0 \leqslant x$,

$x \leqslant 5$

Graph $0 \leqslant x$ and $x \leqslant 5$.

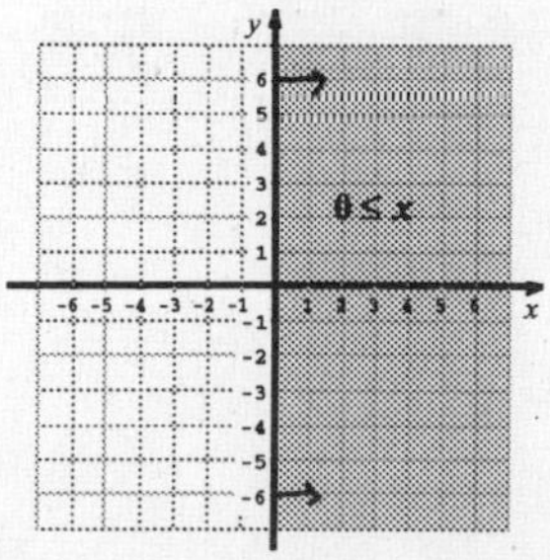

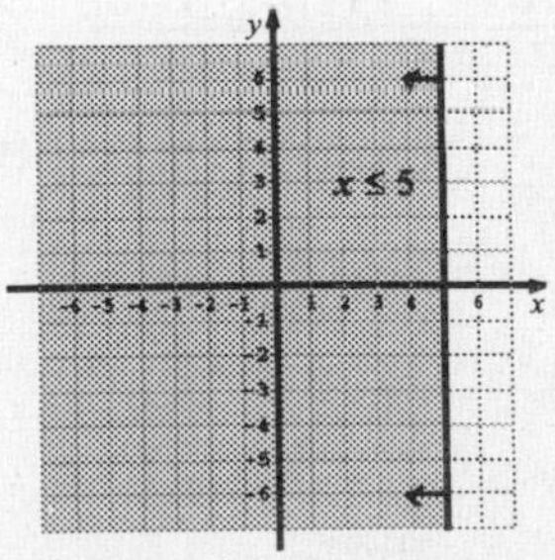

Then we shade the intersection of these graphs.

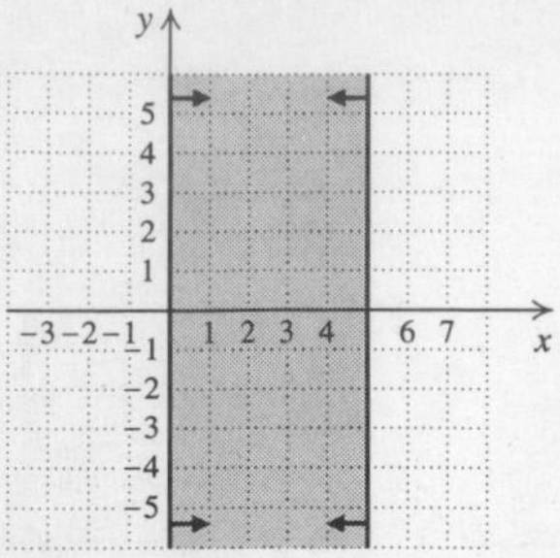

28.

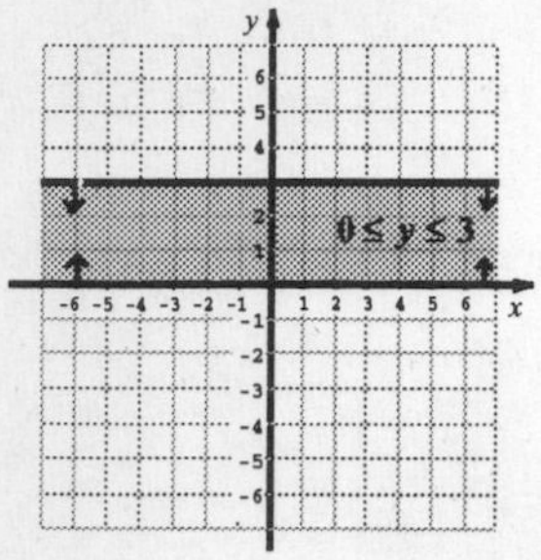

29. Graph: $y < x$, (1)

$y > -x + 3$ (2)

We graph the lines $y = x$ and $y = -x + 3$, using dashed lines. We indicate the region for each inequality by the arrows at the ends of the lines. Note where the regions overlap and shade the region of solutions.

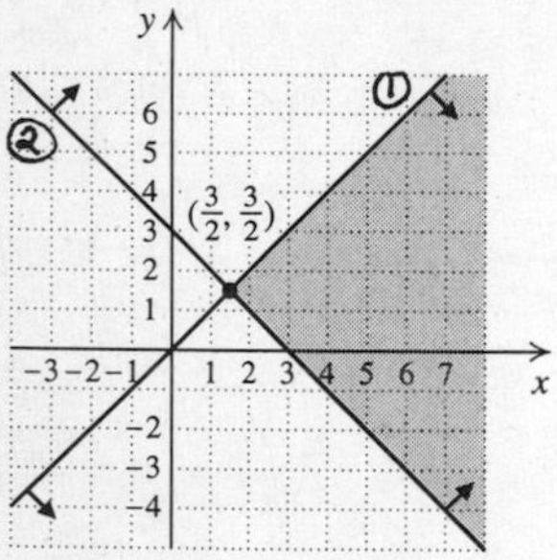

To find the vertex we solve the system of related equations:

$y = x$,

$y = -x + 3$

Solving, we obtain the vertex $\left(\frac{3}{2},\frac{3}{2}\right)$.

30.

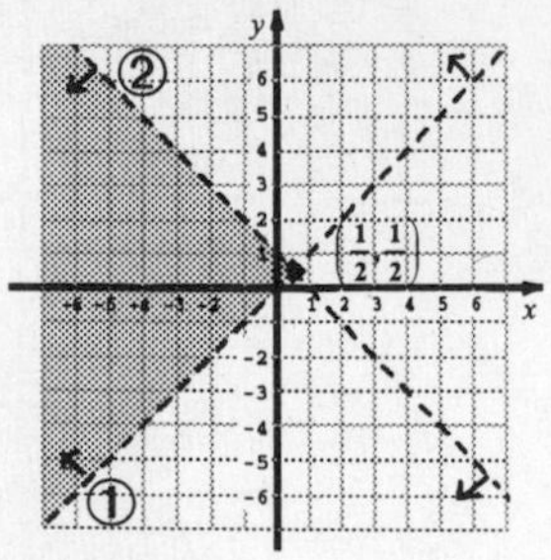

31. Graph: $y \geqslant x$, (1)

$y \leqslant -x + 4$ (2)

We graph the lines $y = x$ and $y = -x + 4$, using solid lines. We indicate the region for each inequality by the arrows at the ends of the lines. Note where the regions overlap, and shade the region of solutions.

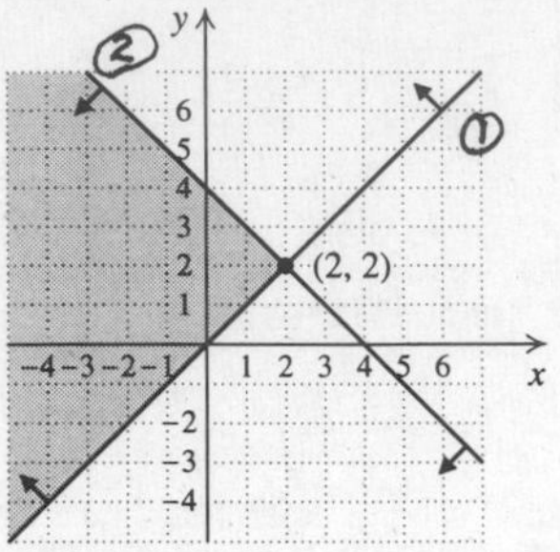

To find the vertex we solve the system of related equations:

$y = x$,

$y = -x + 4$

Solving, we obtain the vertex (2,2).

32.

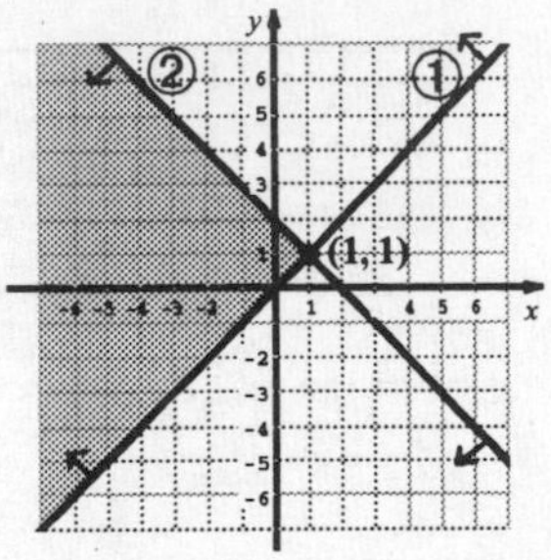

33. Graph: $y \geqslant -2$, (1)

$x \geqslant 1$ (2)

We graph the lines $y = -2$ and $x = 1$, using solid lines. We indicate the region for each inequality by arrows. Shade the region where they overlap.

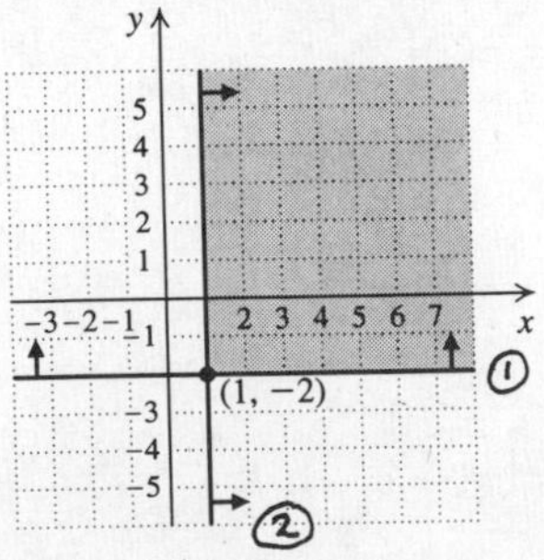

To find the vertex, we solve the system of related equations:

$y = -2$,

$x = 1$

Solving, we obtain the vertex (1,-2).

34.

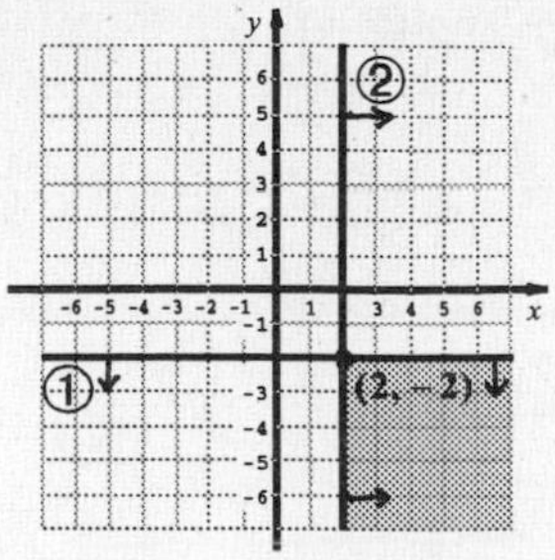

35. Graph: $x < 3$, (1)

$y > -3x + 2$ (2)

Graph the lines $x = 3$ and $y = -3x + 2$, using dashed lines. Indicate the region for each inequality by arrows, and shade the region where they overlap.

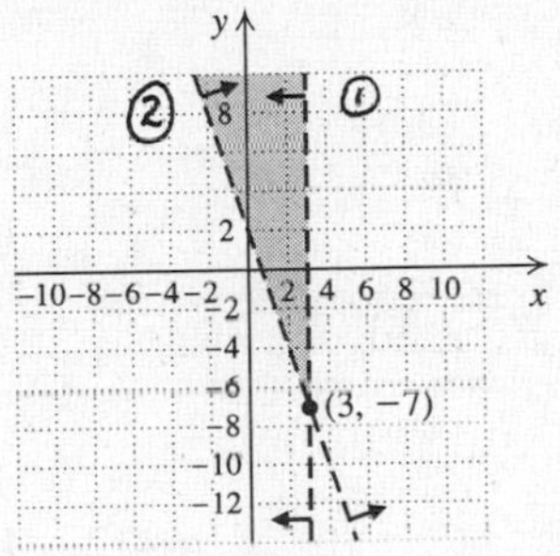

To find the vertex we solve the system of related equations:

$x = 3$,

$y = -3x + 2$

Solving, we obtain the vertex (3,-7).

36.

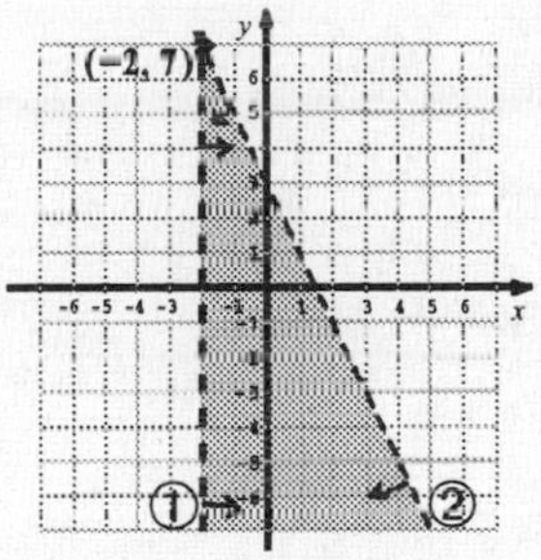

37. Graph: $y \geqslant -2$, (1)

$y \geqslant x + 3$ (2)

Graph the lines $y = -2$ and $y = x + 3$, using solid lines. Indicate the region for each inequality by arrows, and shade the region where they overlap.

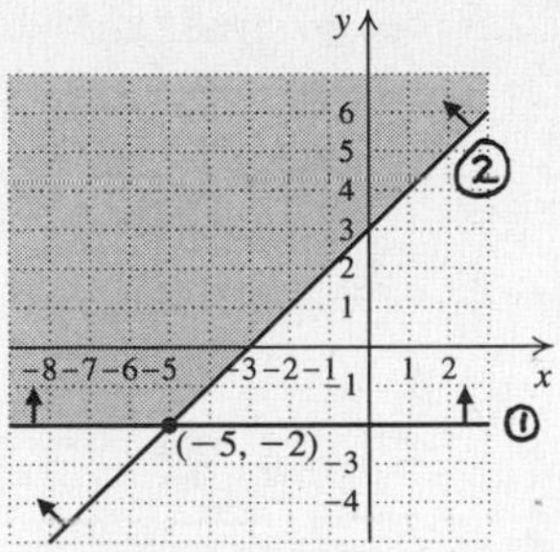

To find the vertex we solve the system of related equations:

$y = -2$,

$y = x + 3$

The vertex is (-5,-2).

38.

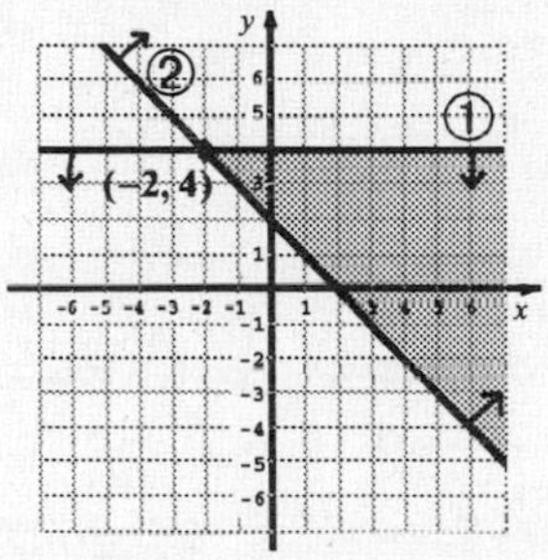

39. Graph: $x + y < 1$, (1)

$x - y < 2$ (2)

Graph the lines $x + y = 1$ and $x - y = 2$, using dashed lines. Indicate the region for each inequality by arrows, and shade the region where they overlap.

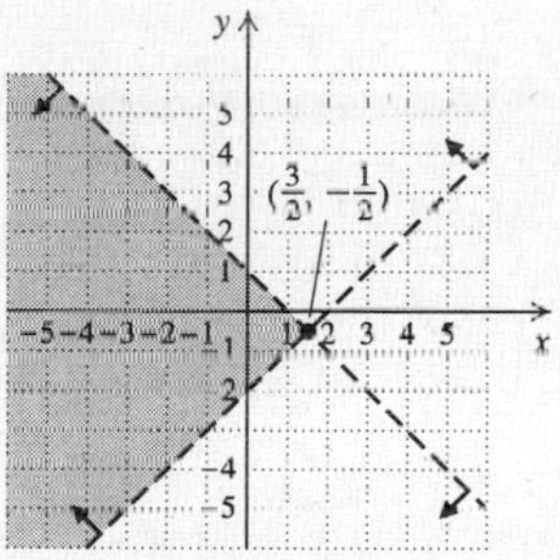

To find the vertex we solve the system of related equations:

$x + y = 1$,

$x - y = 2$

The vertex is $\left(\frac{3}{2}, -\frac{1}{2}\right)$.

40.

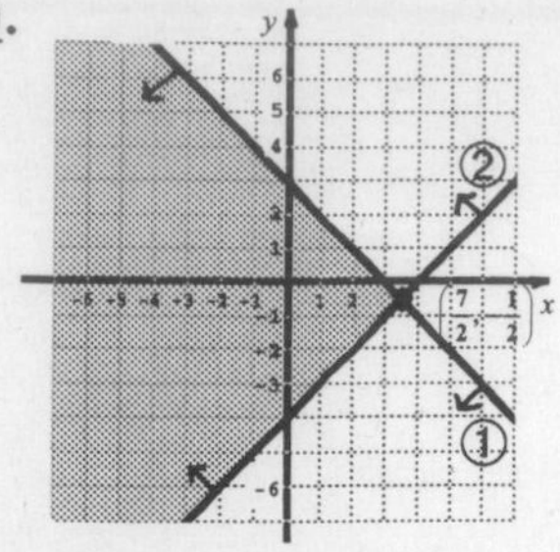

41. Graph: $y - 2x \geqslant 1$, (1)

$y - 2x \leqslant 3$ (2)

Graph the lines $y - 2x = 1$ and $y - 2x = 3$, using solid lines. Indicate the region for each inequality by arrows, and shade the region where they overlap.

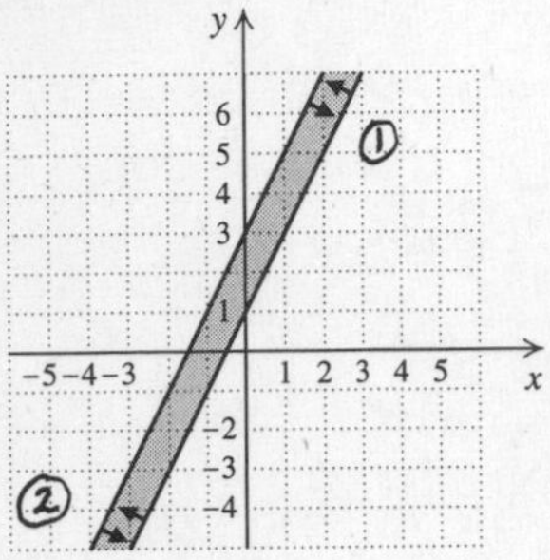

We can see from the graph that the lines are parallel. Hence there are no vertices.

42.

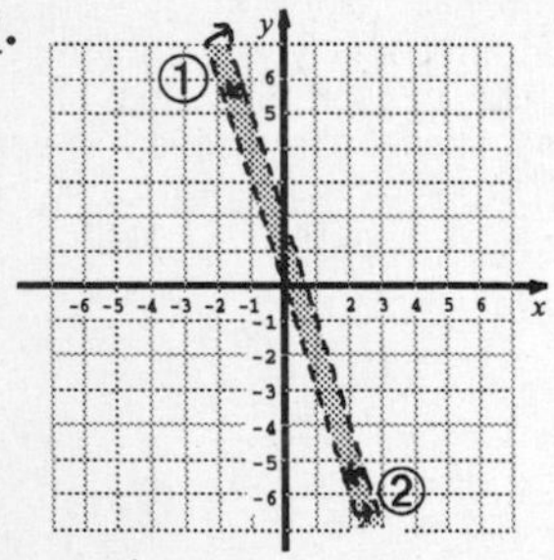

43. Graph: $2y - x \leqslant 2$, (1)

$y - 3x \geqslant -1$ (2)

Graph the lines $2y - x = 2$ and $y - 3x = -1$, using solid lines. Indicate the region for each inequality by arrows, and shade the region where they overlap.

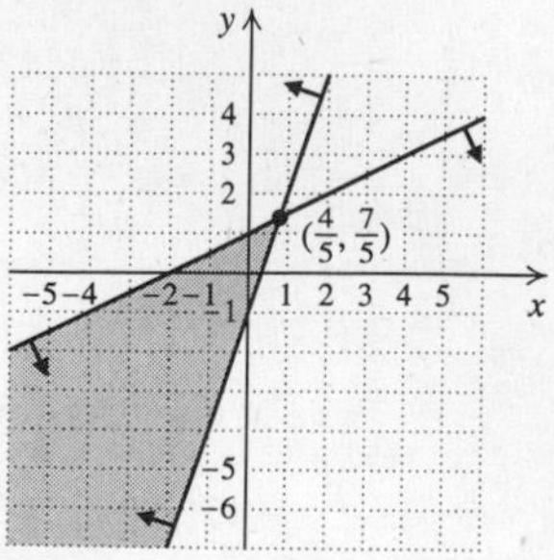

To find the vertex we solve the system of related equations:

$2y - x = 2$,

$y - 3x = -1$

The vertex is $\left[\frac{4}{5},\frac{7}{5}\right]$.

44.

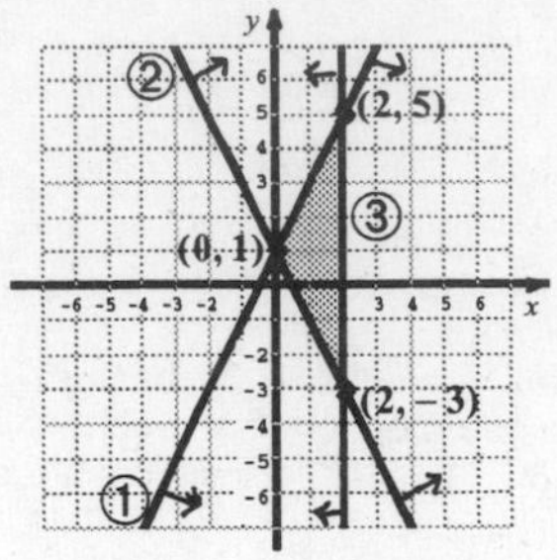

45. Graph: $x - y \leqslant 2$, (1)

$x + 2y \geqslant 8$, (2)

$y \leqslant 4$ (3)

Graph the lines $x - y = 2$, $x + 2y = 8$, and $y = 4$, using solid lines. Indicate the region for each inequality by arrows, and shade the region where they overlap.

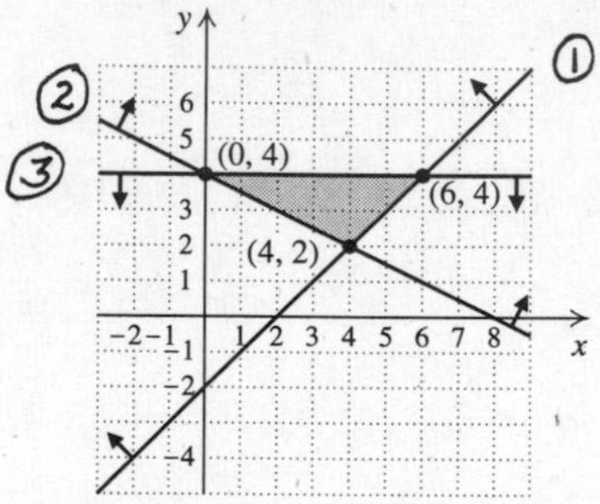

To find the vertices we solve three different systems of equations.

From (1) and (2) we have $x - y = 2$,

$x + 2y = 8$.

Solving, we obtain the vertex (4,2).

From (1) and (3) we have $x - y = 2$,

$y = 4$.

Solving, we obtain the vertex (6,4).

From (2) and (3) we have $x + 2y = 8$,

$y = 4$.

Solving, we obtain the vertex (0,4).

46.

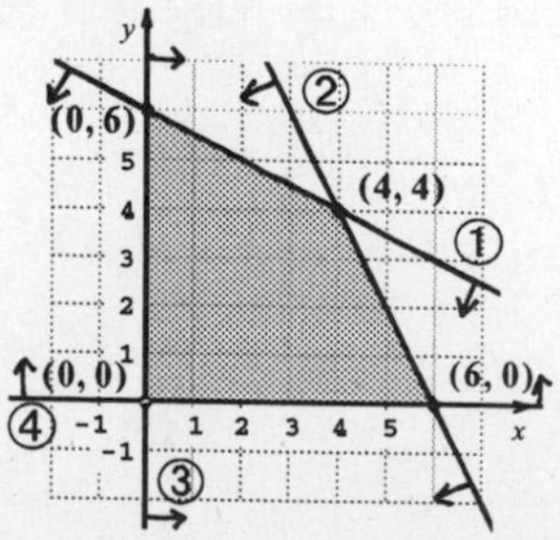

47. Graph: $4y - 3x \geqslant -12$, (1)

$4y + 3x \geqslant -36$, (2)

$y \leqslant 0$, (3)

$x \leqslant 0$ (4)

Graph the lines $4y - 3x = -12$, $4y + 3x = -36$, $y = 0$, and $x = 0$, using solid lines. Indicate the region for each inequality by arrows, and shade the region where they overlap.

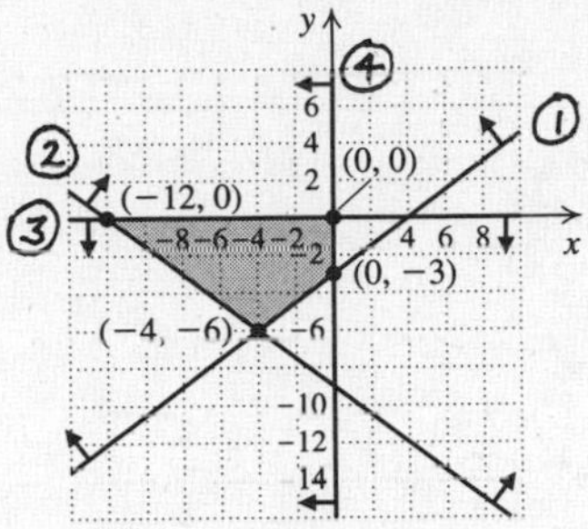

To find the vertices we solve four different systems of equations.

From (1) and (2) we have $4y - 3x = -12$,

$4y + 3x = -36$.

Solving we obtain the vertex $(-4,-6)$.

From (1) and (4) we have $4y - 3x = -12$,

$x = 0$.

Solving, we obtain the vertex $(0,-3)$.

From (2) and (3) we have $4y + 3x = -36$,

$y = 0$.

Solving, we obtain the vertex $(-12,0)$.

48.

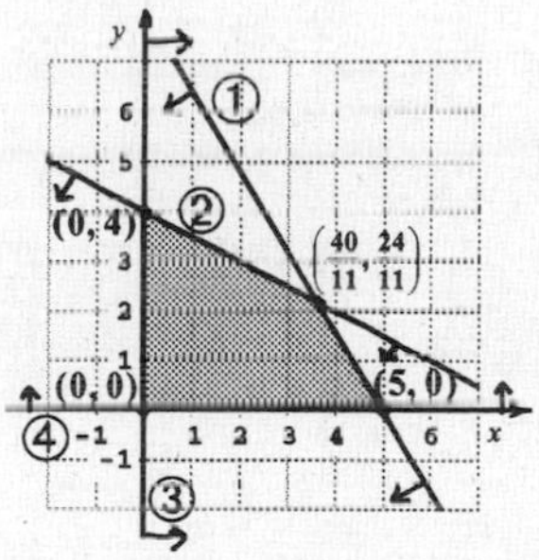

49. Graph: $3x + 4y \geqslant 12$, (1)

$5x + 6y \leqslant 30$, (2)

$1 \leqslant x \leqslant 3$ (3)

Think of (3) as two inequalities:

$1 \leqslant x$, (4)

$x \leqslant 3$ (5)

Graph the lines $3x + 4y = 12$, $5x + 6y = 30$, $x = 1$, and $x = 3$, using solid lines. Indicate the region for each inequality by arrows, and shade the region where they overlap.

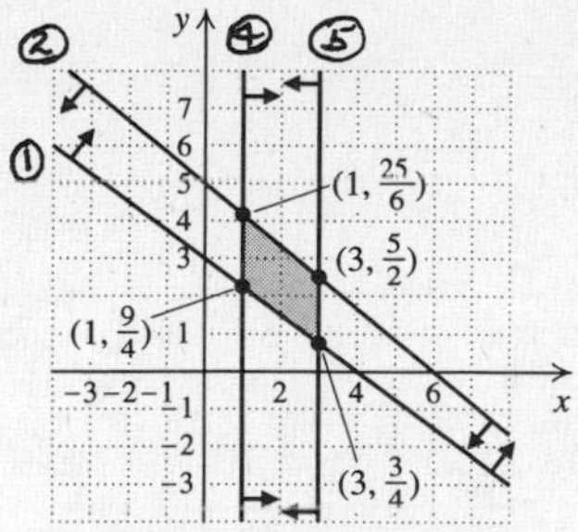

To find the vertices we solve four different systems of equations.

From (1) and (4) we have $3x + 4y = 12$,

$x = 1$.

Solving, we obtain the vertex $\left(1,\frac{9}{4}\right)$.

From (1) and (5) we have $3x + 4y = 12$,

$x = 3$.

Solving, we obtain the vertex $\left(3,\frac{3}{4}\right)$.

From (2) and (4) we have $5x + 6y = 30$,

$x = 1$.

Solving, we obtain the vertex $\left(1,\frac{25}{6}\right)$.

From (2) and (5) we have $5x + 6y = 30$,

$x = 3$.

Solving, we obtain the vertex $\left(3,\frac{5}{2}\right)$.

50.

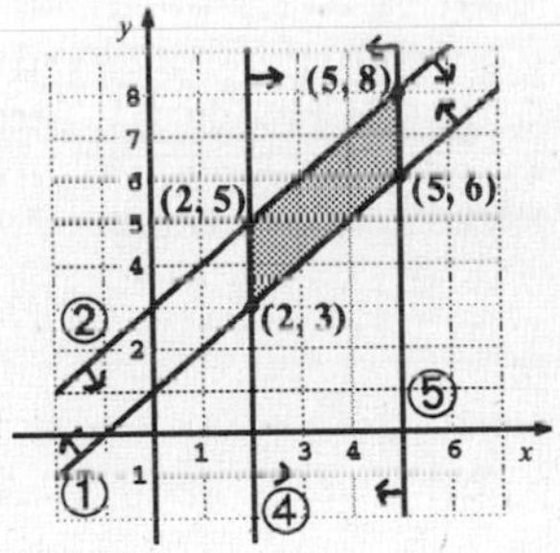

51. Familiarize. We first make a drawing. We let x represent the length of a side of the equilateral triangle. Then $x - 5$ represents the length of a side of the square.

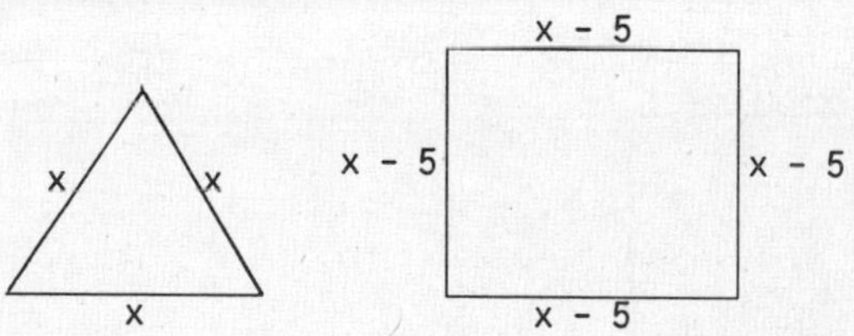

Translate.

Perimeter of triangle = Perimeter of square

$3x = 4(x - 5)$

Carry out.

$3x = 4(x - 5)$

$3x = 4x - 20$

$20 = x$

Then $x - 5 = 20 - 5 = 15$.

Check. If the length of a side of the triangle is 20 and the length of a side of the square is 15, the perimeter of the triangle is $3 \cdot 20$, or 60 and the perimeter of the square is $4 \cdot 15$, or 60. The values check.

State. The length of a side of the square is 15, and the length of a side of the triangle is 20.

52. $\left[-\frac{44}{23}, \frac{13}{23}\right]$

53. $5(3x - 4) = -2(x + 5)$

$15x - 20 = -2x - 10$

$17x - 20 = -10$

$17x = 10$

$x = \frac{10}{17}$

54. $-\frac{14}{13}$

55.

56.

57. Graph: $x + y \geqslant 5$, (1)

$x + y \leqslant -3$ (2)

Graph the lines $x + y = 5$ and $x + y = -3$, using solid lines, and indicate the region for each inequality by arrows. The regions do not overlap (the solution set is $\emptyset$), so we do not shade any portion of the graph.

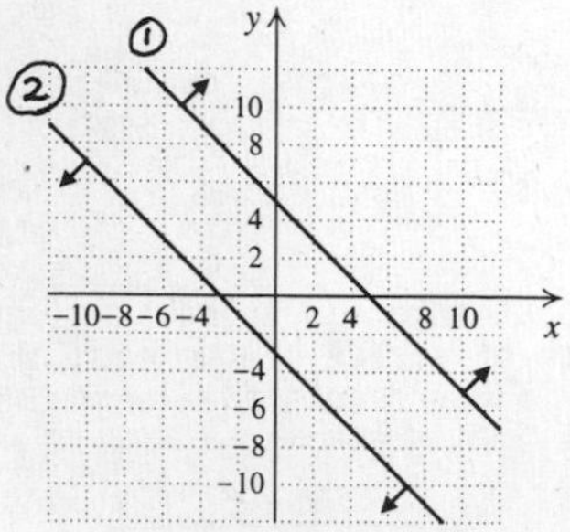

58.

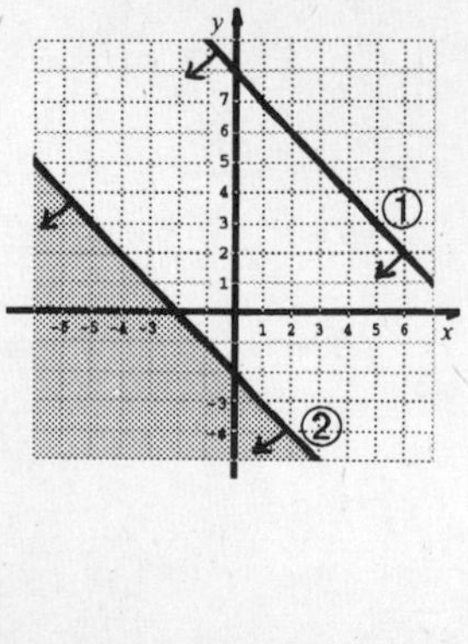

59. Graph: $x - 2y \leqslant 0$, (1)

$-2x + y \leqslant 2$, (2)

$x \leqslant 2$, (3)

$y \leqslant 2$, (4)

$x + y \leqslant 4$ (5)

Graph the five inequalities above, and shade the region where they overlap.

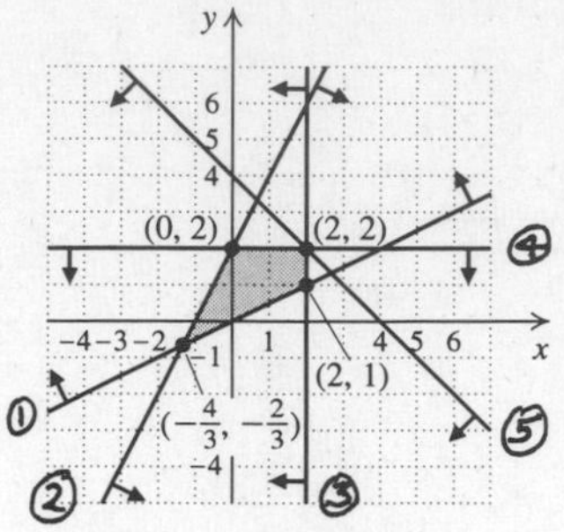

60.

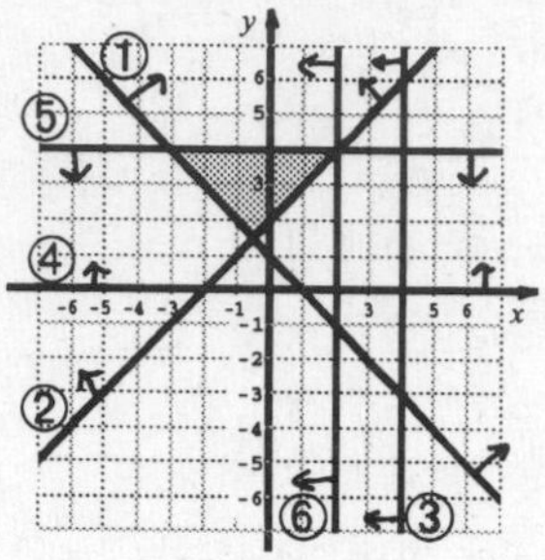

61. We graph the following system of inequalities:

$0 \leqslant L \leqslant 94$,

$0 \leqslant W \leqslant 50$

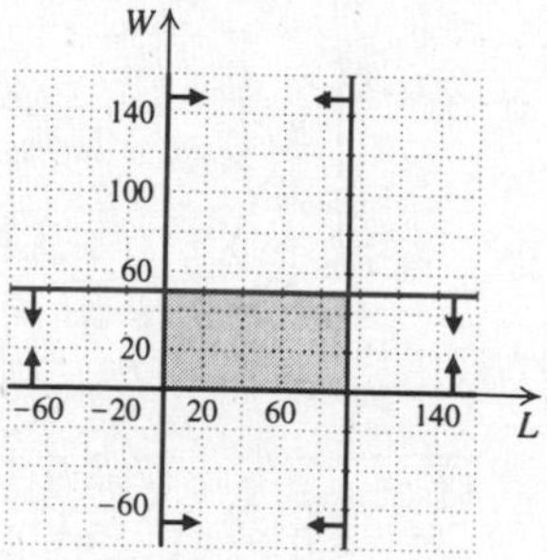

62. $2w + t \geqslant 60$,

$w \geqslant 0$,

$t \geqslant 0$

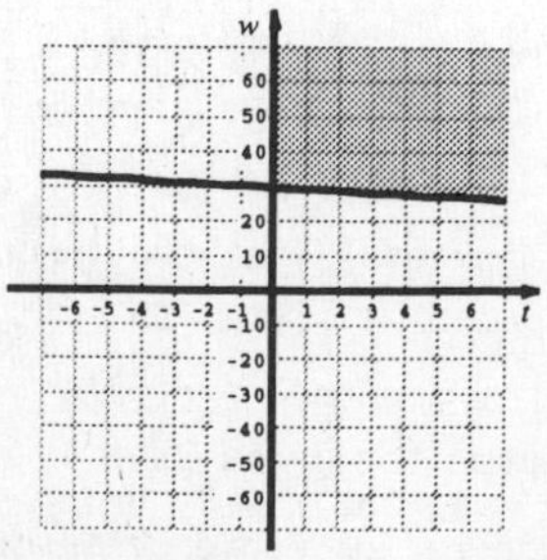

63. Let c = the number of children and a = the number of adults. Graph: $35c + 75a > 1000$,

$c \geqslant 0$,

$a \geqslant 0$

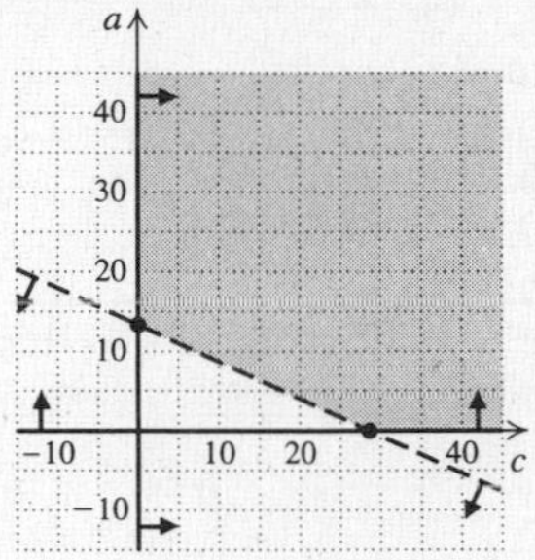

64. a) $3x + 6y > 2$

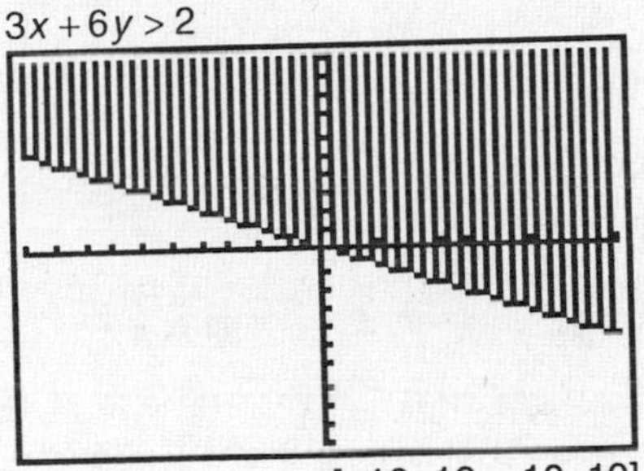

[−10, 10, −10, 10]

b) $x - 5y \leq 10$

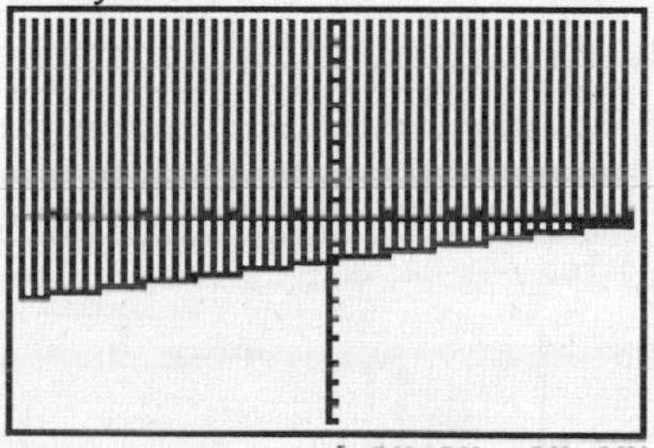

[−10, 10, −10, 10]

c) $13x - 25y + 10 \leq 0$

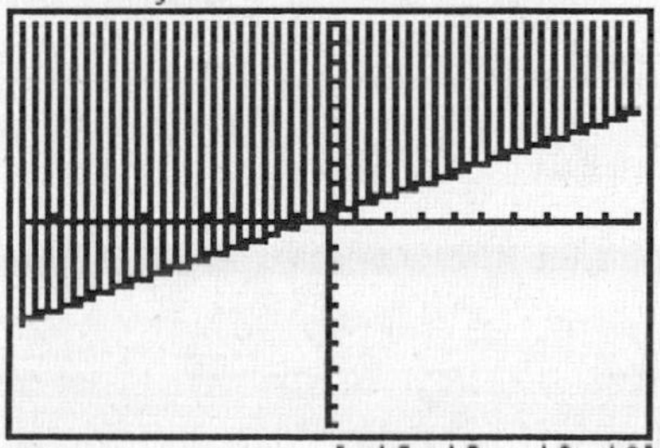

[−10, 10, −10, 10]

d) $2x + 5y > 0$

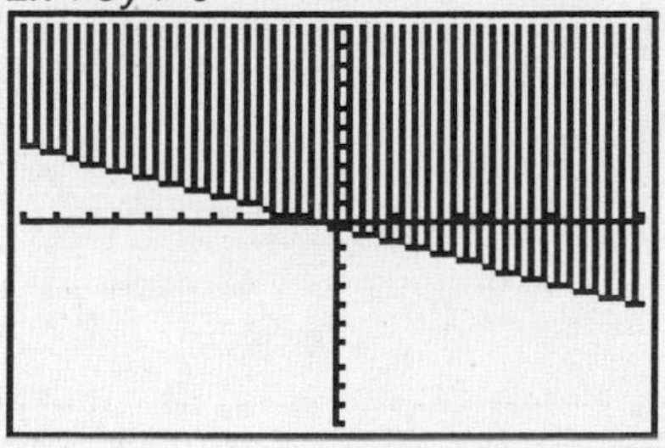

[−10, 10, −10, 10]

65. 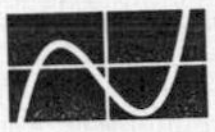

Exercise Set 4.5

1. Find the maximum and minimum values of $F = 4x + 28y$,

subject to

$5x + 3y \leqslant 34$, (1)

$3x + 5y \leqslant 30$, (2)

$x \geqslant 0$, (3)

$y \geqslant 0$. (4)

Graph the system of inequalities and find the coordinates of the vertices.

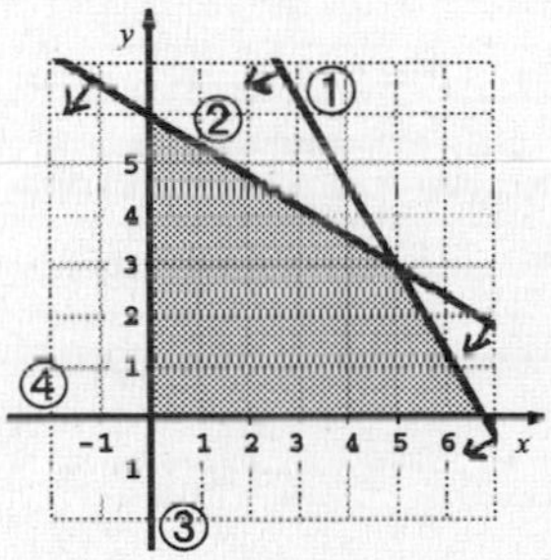

To find vertex A we solve the system

$x = 0$,

$y = 0$.

Vertex A is (0,0).

To find vertex B we solve the system

$5x + 3y = 34$,

$y = 0$.

Vertex B is $\left(\frac{34}{5},0\right)$.

To find vertex C we solve the system

$5x + 3y = 34$,

$3x + 5y = 30$.

Vertex C is (5,3).

To find vertex D we solve the system

$3x + 5y = 30,$

$x = 0.$

Vertex D is (0,6).

Now the value of F at each of these points.

Vertex (x,y)	F = 4x + 28y	
A(0,0)	$4\cdot0 + 28\cdot0 = 0 + 0 = 0$	⟵ Minimum
$B\left(\frac{34}{5},0\right)$	$4\cdot\frac{34}{5} + 28\cdot0 = \frac{136}{5} + 0 = 27\frac{1}{5}$	
C(5,3)	$4\cdot5 + 28\cdot3 = 20 + 84 = 104$	
D(0,6)	$4\cdot0 + 28\cdot6 = 0 + 168 = 168$	⟵ Maximum

The maximum value of F is 168 when x = 0 and y = 6. The minimum value of F is 0 when x = 0 and y = 0.

2. The maximum is 92.8 when x = 0 and y = 5.8. The minimum is 0 when x = 0 and y = 0.

3. Find the maximum and minimum values of $P = 16x - 2y + 40,$

subject to

$6x + 8y \leqslant 48,$ (1)

$0 \leqslant y \leqslant 4,$ (2)

$0 \leqslant x \leqslant 7.$ (3)

Think of (2) as $0 \leqslant y,$ (4)

$y \leqslant 4.$ (5)

Think of (3) as $0 \leqslant x,$ (6)

$x \leqslant 7.$ (7)

Graph the system of inequalities.

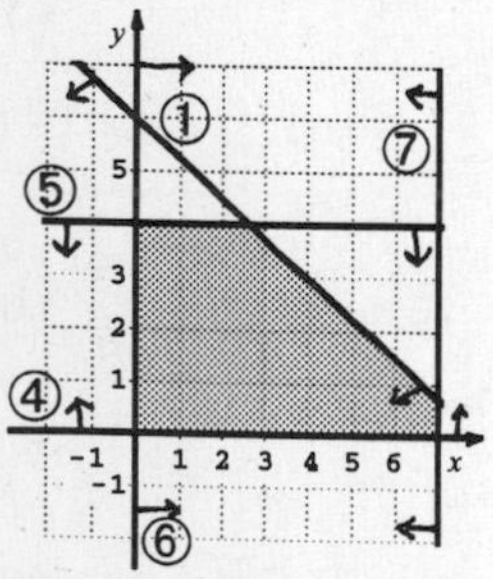

To determine the coordinates of the vertices, we solve the following systems:

$x = 0,$ $y = 0;$ $\quad$ $x = 7,$ $y = 0;$ $\quad$ $6x + 8y = 48,$ $x = 7;$

$6x + 8y = 48,$ $y = 4;$ $\quad$ $x = 0,$ $y = 4$

The vertices are (0,0), (7,0) $\left(7,\frac{3}{4}\right)$, $\left(\frac{8}{3},4\right)$, and (0,4), respectively. Compute the value of P at each of these points.

The vertices are (0,0), (7,0) $\left(7,\frac{3}{4}\right)$, $\left(\frac{8}{3},4\right)$, and (0,4), respectively. Compute the value of P at each of these points.

Vertex (x,y)	P = 16x - 2y + 40	
(0,0)	$16\cdot0 - 2\cdot0 + 40 = 0 - 0 + 40 = 40$	
(7,0)	$16\cdot7 - 2\cdot0 + 40 = 112 - 0 + 40 = 152$	⟵ Maximum
$\left(7,\frac{3}{4}\right)$	$16\cdot7 - 2\cdot\frac{3}{4} + 40 = 112 - \frac{3}{2} + 40 = 150\frac{1}{2}$	
$\left(\frac{8}{3},4\right)$	$16\cdot\frac{8}{3} - 2\cdot4 + 40 = \frac{128}{3} - 8 + 40 = 74\frac{2}{3}$	
(0,4)	$16\cdot0 - 2\cdot4 + 40 = 0 - 8 + 40 = 32$	⟵ Minimum

The maximum is 152 when x = 7 and y = 0. The minimum is 32 when x = 0 and y = 4.

4. The maximum is 124 when x = 3 and y = 0. The minimum is 40 when x = 0 and y = 4.

5. Find the maximum and minimum values of $F = 5x + 2y + 3,$

subject to

$y \leqslant 2x + 1,$ (1)

$x \leqslant 5,$ (2)

$y \geqslant 1$ (3)

Graph the system of inequalities and find the coordinates of the vertices.

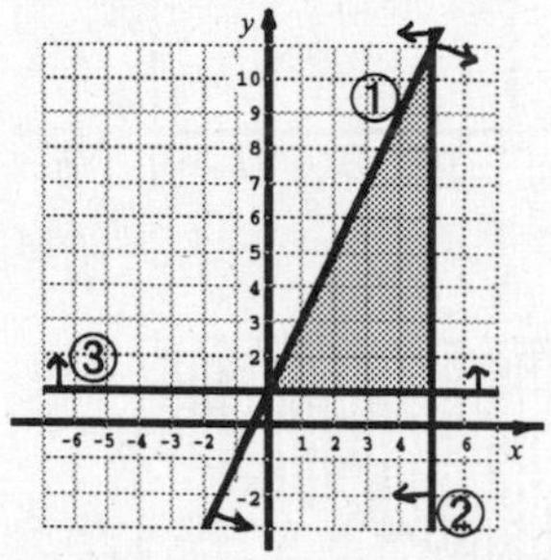

To determine the coordinates of the vertices, we solve the following systems:

$y = 2x + 1,$ $y = 1;$ $\quad$ $x = 5,$ $y = 1;$ $\quad$ $y = 2x + 1,$ $x = 5$

The solutions of the systems are (0,1), (5,1), and (5,11), respectively. Now find the value of F at each of these points.

Vertex (x,y)	F = 5x + 2y + 3
(0,1)	$5\cdot0 + 2\cdot1 + 3 = 5$ ⟵ Minimum
(5,1)	$5\cdot5 + 2\cdot1 + 3 = 30$
(5,11)	$5\cdot5 + 2\cdot11 + 3 = 50$ ⟵ Maximum

The maximum is 50 when x = 5 and y = 11. The minimum value is 5 when x = 0 and y = 1.

6. The maximum is 4 when x = 2 and y = 5. The minimum is -12 when x = 2 and y = -3.

7. Familiarize. Let x = the number of Biscuit Jumbos and y = the number of Mitimite Biscuits to be made per day.

Translate. The income I is given by

$I = \$0.10x + \$0.08y.$

We wish to maximize I subject to these facts (constraints) about x and y:

$$x + y \leqslant 200,$$
$$2x + y \leqslant 300,$$
$$x \geqslant 0,$$
$$y \geqslant 0.$$

Carry out. We graph the system of inequalities, determine the vertices, and evaluate I at each vertex.

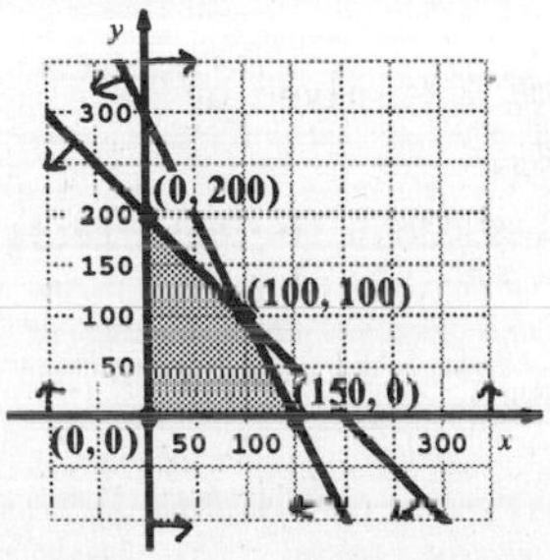

Vertex	I = \$0.10x + \$0.08y
(0, 0)	\$0.10(0) + \$0.08(0) = \$0
(0, 200)	\$0.10(0) + \$0.08(200) = \$16
(100, 100)	\$0.10(100) + \$0.08(100) = \$18
(150, 0)	\$0.10(150) + \$0.08(0) = \$15

The greatest income in the table is \$18, obtained when 100 Biscuit Jumbos and 100 Mitimite Biscuits are made.

Check. Go over the algebra and arithmetic.

State. The company will have a maximum income of \$18 when 100 of each type of biscuit is made.

8. The maximum number of miles is 480 when the car uses 9 gal and the moped uses 3 gal.

9. Familiarize. We organize the information in a table. Let x = the number of matching questions and y = the number of essay questions you answer.

Type	Number of points for each	Number answered	Total points for type
Matching	10	$3 \leqslant x \leqslant 12$	10x
Essay	25	$4 \leqslant y \leqslant 15$	25y
Total		$x + y \leqslant 20$	10x + 25y

No more than 20 questions can be answered (Number answered, Total). This is the total score on the test (Total points for type, Total).

Translate. The score S is given by

$S = 10x + 25y.$

We wish to maximize S subject to these facts (constraints) about x and y.

$$3 \leqslant x \leqslant 12,$$
$$4 \leqslant y \leqslant 15,$$
$$x + y \leqslant 20.$$

Carry out. We graph the system of inequalities, determine the vertices, and evaluate S at each vertex.

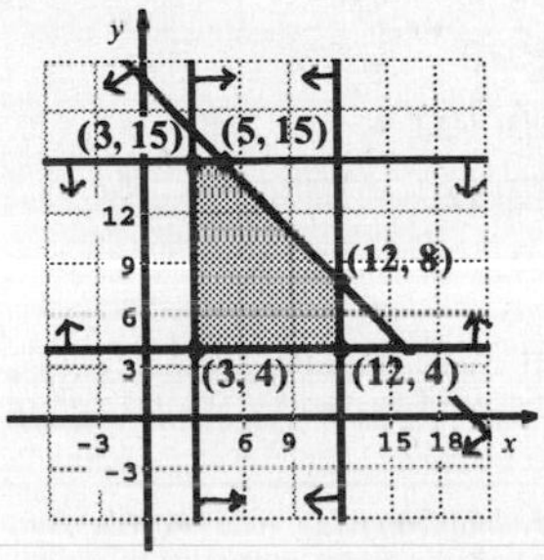

Vertex	S = 10x + 25y
(3, 4)	$10\cdot3 + 25\cdot4 = 130$
(3, 15)	$10\cdot3 + 25\cdot15 = 405$
(5, 15)	$10\cdot5 + 25\cdot15 = 425$
(12, 8)	$10\cdot12 + 25\cdot8 = 320$
(12, 4)	$10\cdot12 + 25\cdot4 = 220$

The greatest score in the table is 425, obtained when 5 matching and 15 essay questions are answered correctly.

Check. Go over the algebra and arithmetic.

State. The maximum score is 425 points when 5 matching questions and 15 essay questions are answered correctly.

10. The maximum test score is 102 when 8 short-answer questions and 10 word problems are answered correctly.

11. Familiarize. Let x = the number of motorcycles manufactured and y = the number of bicycles manufactured each month.

Translate. The profit P is given by

$P = \$134x + \$20y.$

We wish to maximize P subject to these facts (constraints) about x and y:

$10 \le x \le 60$	From 10 to 60 motorcycles will be produced.
$0 \le y \le 120$	No more than 120 bicycles can be produced.
$x + y \le 160$	Total production cannot exceed 160.

Carry out. We graph the system of inequalities, determine the vertices, and evaluate P at each vertex.

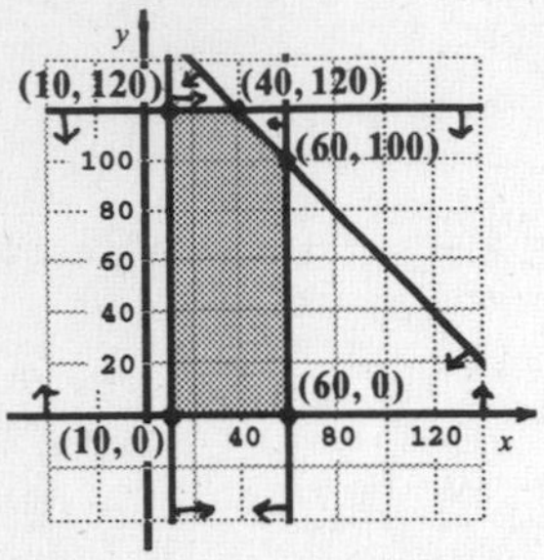

Vertex	P = \$134x + \$20y
(10,0)	$134\cdot10 + 20\cdot0 = \$1340$
(60,0)	$134\cdot60 + 20\cdot0 = \$8040$
(60,100)	$134\cdot60 + 20\cdot100 = \$10,040$
(40,120)	$134\cdot40 + 20\cdot120 = \$7760$
(10,120)	$134\cdot10 + 20\cdot120 = \$3740$

The greatest profit in the table is \$10,040, obtained when 60 motorcycles and 100 bicycles are manufactured.

Check. Go over the algebra and arithmetic.

State. The maximum profit occurs when 60 motorcycles and 100 bicycles are manufactured.

12. The maximum profit occurs when 40 hamburgers and 50 hot dogs are sold.

13. Familiarize. Let x = the number of units of lumber and y = the number of units of plywood produced per week.

Translate. The profit P is given by

$P = \$20x + \$30y.$

We wish to maximize P subject to these facts (constraints) about x and y:

$x + y \le 400,$

$x \ge 100,$

$y \ge 150.$

Carry out. We graph the system of inequalities, determine the vertices and evaluate P at each vertex.

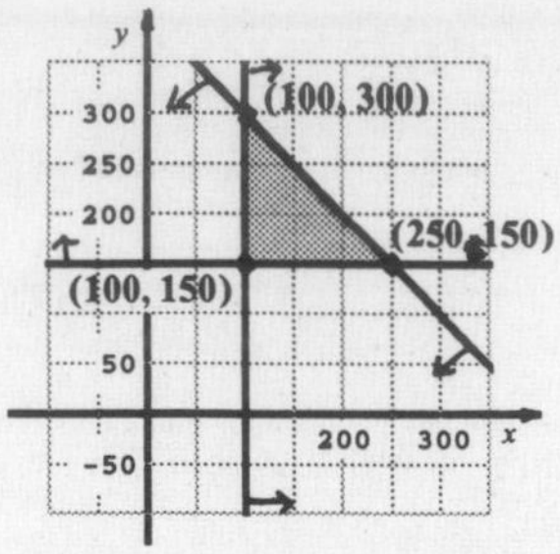

Vertex	P = \$20x + \$30y
(100, 150)	$\$20\cdot100 + \$30\cdot150 = \$6500$
(100, 300)	$\$20\cdot100 + \$30\cdot300 = \$11,000$
(250, 150)	$\$20\cdot250 + \$30\cdot150 = \$9500$

The greatest profit in the table is \$11,000, obtained when 100 units of lumber and 300 units of plywood are produced.

Check. Go over the algebra and arithmetic.

State. The maximum profit is achieved by producing 100 units of lumber and 300 units of plywood.

14. The maximum profit occurs by planting 80 acres of corn and 160 acres of oats.

15. Familiarize. Let x = the amount invested in corporate bonds and y = the amount invested in municipal bonds.

Translate. The income I is given by

$I = 0.08x + 0.075y.$

We wish to maximize I subject to these facts (constraints) about x and y:

$x + y \le \$40,000,$

$\$6000 \le x \le \$22,000,$

$0 \le y \le \$30,000.$

Carry out. We graph the system of inequalities, determine the vertices, and evaluate I at each vertex.

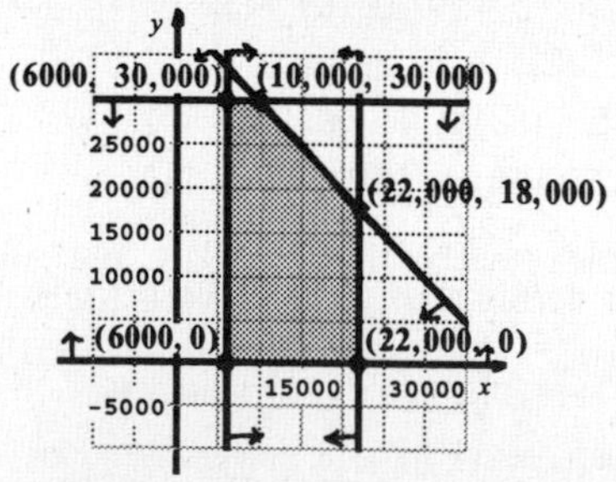

Vertex	I = 0.08x + 0.075y
(\$6000, \$0)	\$480
(\$6000, \$30,000)	\$2730
(\$10,000, \$30,000)	\$3050
(\$22,000, \$18,000)	\$3110
(\$22,000, \$0)	\$1760

The greatest profit in the table is \$3110, obtained when \$22,000 is invested in corporate bonds and \$18,000 in municipal bonds.

Check. Go over the algebra and arithmetic.

State. The maximum income of \$3110 occurs when \$22,000 is invested in corporate bonds and \$18,000 is invested in municipal bonds.

16. The maximum interest income is \$1395 when \$7000 is invested in City Bank and \$15,000 is invested in State Bank.

17. Familiarize. Let x = the number of batches of Smello and y = the number of batches of Roppo to be made. We organize the information in a table.

	Composition		Number of lbs Available
	Smello	Roppo	
Number of Batches	x	y	
English tobacco	12	8	3000
Virginia tobacco	0	8	2000
Latakia tobacco	4	0	500
Profit per Batch	\$10.56	\$6.40	

Translate. The profit P is given by

$$P = \$10.56x + \$6.40y.$$

We wish to maximize P subject to these facts (constraints) about x and y:

$$12x + 8y \leqslant 3000,$$
$$8y \leqslant 2000,$$
$$4x \leqslant 500,$$
$$x \geqslant 0,$$
$$y \geqslant 0.$$

Carry out. We graph the system of inequalities, determine the vertices, and evaluate P at each vertex.

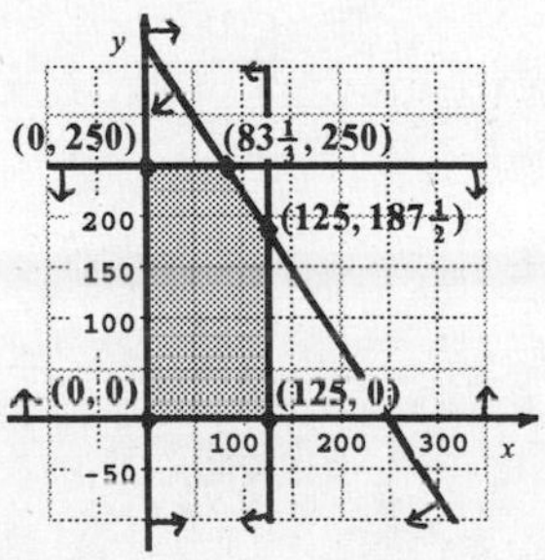

Vertex	P = \$10.56x + \$6.40y
(0, 0)	\$0
(125, 0)	\$1320
(125, 187.5)	\$2520
$\left(83\frac{1}{3}, 250\right)$	\$2480
(0, 250)	\$1600

The greatest profit in the table is \$2520, obtained when 125 batches of Smello and 187.5 batches of Roppo are produced.

Check. Go over the algebra and arithmetic.

State. The maximum profit of \$2520 occurs when 125 batches of Smello and 187.5 batches of Roppo are made.

18. The maximum profit per day is \$192 when 2 knit suits and 4 worsted suits are made.

19. Familiarize. Let x represent the number of P-1 planes and y represent the number of P-2 planes. Organize the information in a table.

Plane	Number of planes	Passengers			Cost per mile
		First	Tourist	Economy	
P-1	x	40x	40x	120x	12,000x
P-2	y	80y	30y	40y	10,000y

Translate. Suppose C is the total cost per mile. Then C = 12,000x + 10,000y. We wish to minimize C subject to these facts (constraints) about x and y:

$$40x + 80y \geqslant 2000, \quad (1)$$
$$40x + 30y \geqslant 1500, \quad (2)$$
$$120x + 40y \geqslant 2400, \quad (3)$$
$$x \geqslant 0, \quad (4)$$
$$y \geqslant 0 \quad (5)$$

Carry out. Graph the system of inequalities, determine the vertices, and evaluate C at each vertex.

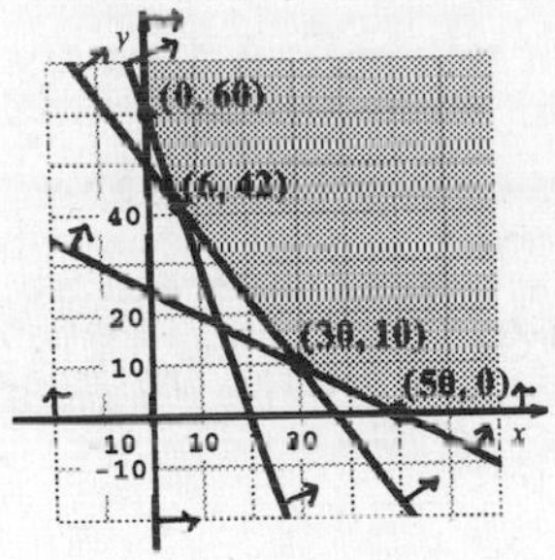

Vertex	C = 12,000x + 10,000y
(0,60)	12,000(0) + 10,000(60) = 600,000
(6,42)	12,000(6) + 10,000(42) = 492,000
(30,10)	12,000(30) + 10,000(10) = 460,000
(50,0)	12,000(50) + 10,000(0) = 600,000

The smallest value of C in the table is 460,000, obtained when x = 30 and y = 10.

Check. Go over the algebra and arithmetic.

State. In order to minimize the operating cost, 30 P-1 planes and 10 P-2 planes should be used.

20. The airline should use 30 P-2's and 15 P-3's.

21. Familiarize. Let x = the number of chairs and y = the number of sofas produced.
Translate. Find the maximum value of

$$I = \$20x + \$300y$$

subject to

$$\begin{aligned} 20x + 100y &\leqslant 1900, \\ x + 50y &\leqslant 500, \\ 2x + 20y &\leqslant 240, \\ x &\geqslant 0, \\ y &\geqslant 0. \end{aligned}$$

Carry out. Graph the system of inequalities, determine the vertices, and evaluate I at each vertex.

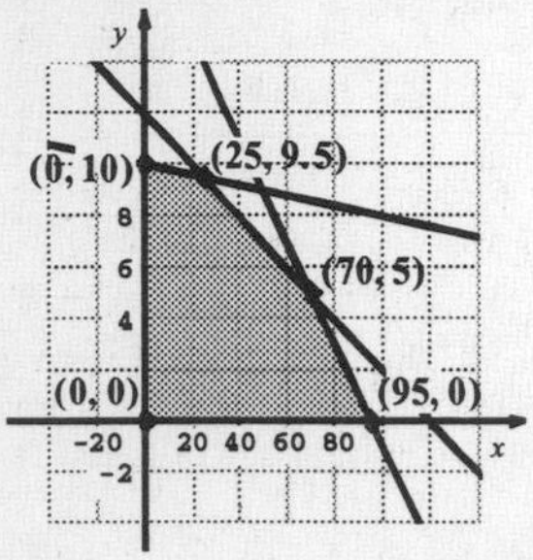

Vertex	I = \$20x + \$300y
(0, 0)	\$0
(0, 10)	\$3000
(25, 9.5)	\$3350
(70, 5)	\$2900
(95, 0)	\$1900

The greatest income in the table is \$3350, obtained when 25 chairs and 9.5 sofas are made.

Check. Go over the algebra and arithmetic.

State. The maximum income of \$3350 occurs when 25 chairs and 9.5 sofas are made. A more practical answer is that the maximum income of \$3200 is achieved when 25 chairs and 9 sofas are made.

Exercise Set 5.1

1. $-11x^4 - x^3 + x^2 + 3x - 9$

Term	$-11x^4$	$-x^3$	x^2	$3x$	-9
Degree	4	3	2	1	0
Degree of polynomial	4				

2. 3, 2, 1, 0; 3

3. $y^3 + 2y^7 + x^2y^4 - 8$

Term	y^3	$2y^7$	x^2y^4	-8
Degree	3	7	6	0
Degree of polynomial	7			

4. 2, 5, 7, 0; 7

5. $a^5 + 4a^2b^4 + 6ab + 4a - 3$

Term	a^5	$4a^2b^4$	$6ab$	$4a$	-3
Degree	5	6	2	1	0
Degree of polynomial	6				

6. 6, 8, 4, 2, 0; 8

7. $-4y^3 - 6y^2 + 7y + 23$; $-4y^3$; -4

8. $-18y^4 + 11y^3 + 6y^2 - 8y + 5$; $-18y^4$; -18

9. $3x^7 + 5x^2 - x + 12$; $3x^7$; 3

10. $-10x^4 + 7x^2 - 3x + 9$; $-10x^4$; -10

11. $-a^7 + 8a^5 + 5a^3 - 19a^2 + a$; $-a^7$; -1

12. $a^9 + 11a^4 + a^3 - 5a^2 - 7$; a^9; 1

13. $12 + 4x - 5x^2 + 3x^4$

14. $5 + 10x - 5x^2$

15. $3xy^3 + x^2y^2 - 9x^3y + 2x^4$

16. $-9xy + 5x^2y^2 + 8x^3y^2 - 5x^4$

17. $-7ab + 4ax - 7ax^2 + 4x^6$

18. $-12a + 5xy^8 + 4ax^3 - 3ax^5 + 5x^5$

19. $P(x) = 4x^2 - 3x + 2$

$$P(4) = 4\cdot 4^2 - 3\cdot 4 + 2 = 64 - 12 + 2 = 54$$

$$P(0) = 4\cdot 0^2 - 3\cdot 0 + 2 = 0 - 0 + 2 = 2$$

20. -84, 0

21. $P(y) = 8y^3 - 12y - 5$

$$P(-2) = 8(-2)^3 - 12(-2) - 5 = -64 + 24 - 5 = -45$$

$$P\left(\frac{1}{3}\right) = 8\left(\frac{1}{3}\right)^3 - 12\cdot\frac{1}{3} - 5 = 8\cdot\frac{1}{27} - 4 - 5 = \frac{8}{27} - 9 = \frac{8}{27} - \frac{243}{27} = -\frac{235}{27}, \text{ or } -8\frac{19}{27}$$

22. -168, -9

23. $-5x + 2 = -5\cdot 4 + 2 = -20 + 2 = -18$

24. -11

25. $2x^2 - 5x + 7 = 2\cdot 4^2 - 5\cdot 4 + 7 = 2\cdot 16 - 20 + 7 = 32 - 20 + 7 = 19$

26. 59

27. $x^3 - 5x^2 + x = 4^3 - 5\cdot 4^2 + 4 = 64 - 5\cdot 16 + 4 = 64 - 80 + 4 = -12$

28. 51

29. $f(x) = x^2 - 2x + 1$

$$f(-1) = (-1)^2 - 2(-1) + 1 = 1 + 2 + 1 = 4$$

30. -10

31. $g(x) = -3x^3 + 7x^2 - 3x - 2$

$$g(-1) = -3(-1)^3 + 7(-1)^2 - 3(-1) - 2 = -3(-1) + 7\cdot 1 + 3 - 2 = 3 + 7 + 3 - 2 = 11$$

32. -4

33. $P(a) = 0.4a^2 - 40a + 1039$
$P(18) = 0.4(18)^2 - 40(18) + 1039$
$= 0.4(324) - 720 + 1039$
$= 129.6 - 720 + 1039$
$= 448.6$
There are approximately 449 accidents daily involving an 18-year-old driver.

34. 399

35. Evaluate the function for $t = 8$:
$s(t) = 16t^2$
$s(8) = 16(8)^2 = 16 \cdot 64 = 1024$
The cliff is 1024 ft high.

36. 144 ft

37. Evaluate the polynomial function for $x = 75$:
$R(x) = 280x - 0.4x^2$
$R(75) = 280 \cdot 75 - 0.4(75)^2$
$= 21{,}000 - 0.4(5625)$
$= 21{,}000 - 2250 = 18{,}750$
The total revenue is $18,750.

38. $24,000

39. Evaluate the polynomial function for $x = 75$:
$C(x) = 5000 + 0.6x^2$
$C(75) = 5000 + 0.6(75)^2$
$= 5000 + 0.6(5625)$
$= 5000 + 3375$
$= 8375$
The total cost is $8375.

40. $11,000

41. We evaluate the polynomial for $r = 1.2$ and $h = 4.7$:
$2\pi rh + 2\pi r^2 = 2(3.14)(1.2)(4.7) + 2(3.14)(1.2)^2 = 44.4624 \approx 44.46$, rounded to the nearest hundredth.
The surface area is approximately 44.46 in^2.

42. 56.52 in^2

43. $6x^2 - 7x^2 + 3x^2 = (6 - 7 + 3)x^2 = 2x^2$

44. $-4y^2$

45. $5a + 7 - 4 + 2a - 6a + 3$
$= (5 + 2 - 6)a + (7 - 4 + 3)$
$= a + 6$

46. $7x + 14$

47. $3a^2b + 4b^2 - 9a^2b - 6b^2$
$= (3 - 9)a^2b + (4 - 6)b^2$
$= -6a^2b - 2b^2$

48. $-8x^3 - 3x^2y^2$

49. $8x^2 - 3xy + 12y^2 + x^2 - y^2 + 5xy + 4y^2$
$= (8 + 1)x^2 + (-3 + 5)xy + (12 - 1 + 4)y^2$
$= 9x^2 + 2xy + 15y^2$

50. $11a^2 + 3ab - 3b^2$

51. $(3x^2 + 5y^2 + 6) + (2x^2 - 3y^2 - 1)$
$= (3 + 2)x^2 + (5 - 3)y^2 + (6 - 1)$
$= 5x^2 + 2y^2 + 5$

52. $21y^2 + 3y + 4$

53. $(2a + 3b - c) + (4a - 2b + 2c)$
$= (2 + 4)a + (3 - 2)b + (-1 + 2)c$
$= 6a + b + c$

54. $14x + 8y - 6z$

55. $(a^2 - 3b^2 + 4c^2) + (-5a^2 + 2b^2 - c^2)$
$= (1 - 5)a^2 + (-3 + 2)b^2 + (4 - 1)c^2$
$= -4a^2 - b^2 + 3c^2$

56. $-5x^2 + 4y^2 - 11z^2$

57. $(x^2 + 2x - 3xy - 7) + (-3x^2 - x + 2xy + 6)$
$= (1 - 3)x^2 + (2 - 1)x + (-3 + 2)xy + (-7 + 6)$
$= -2x^2 + x - xy - 1$

58. $2a^2 + 3b - 4ab + 4$

59. $(7x^2y - 3xy^2 + 4xy) + (-2x^2y - xy^2 + xy)$
$= (7 - 2)x^2y + (-3 - 1)xy^2 + (4 + 1)xy$
$= 5x^2y - 4xy^2 + 5xy$

60. $20ab - 18ac - 3bc$

61. $(2r^2 + 12r - 11) + (6r^2 - 2r + 4) + (r^2 - r - 2)$
$= (2 + 6 + 1)r^2 + (12 - 2 - 1)r + (-11 + 4 - 2)$
$= 9r^2 + 9r - 9$

62. $-3x^2 - x - 3$

63. $\left(\frac{2}{3}xy + \frac{5}{6}xy^2 + 5.1x^2y\right) + \left(-\frac{4}{5}xy + \frac{3}{4}xy^2 - 3.4x^2y\right)$
$= \left(\frac{2}{3} - \frac{4}{5}\right)xy + \left(\frac{5}{6} + \frac{3}{4}\right)xy^2 + (5.1 - 3.4)x^2y$
$= \left(\frac{10}{15} - \frac{12}{15}\right)xy + \left(\frac{10}{12} + \frac{9}{12}\right)xy^2 + 1.7x^2y$
$= -\frac{2}{15}xy + \frac{19}{12}xy^2 + 1.7x^2y$

64. $-\frac{5}{24}xy - \frac{27}{20}x^3y^2 + 1.4y^3$

65. $5x^3 - 7x^2 + 3x - 6$

a) $-(5x^3 - 7x^2 + 3x - 6)$ Writing the opposite of P as -P

b) $-5x^3 + 7x^2 - 3x + 6$ Changing the sign of every term

66. $-(-8y^4 - 18y^3 + 4y - 9)$, $8y^4 + 18y^3 - 4y + 9$

67. $-12y^5 + 4ay^4 - 7by^2$

a) $-(-12y^5 + 4ay^4 - 7by^2)$

b) $12y^5 - 4ay^4 + 7by^2$

68. $-(7ax^3y^2 - 8by^4 - 7abx - 12ay)$,
$-7ax^3y^2 + 8by^4 + 7abx + 12ay$

69. $(8x - 4) - (-5x + 2)$
$= (8x - 4) + (5x - 2)$
$= 13x - 6$

70. $13y + 5$

71. $(-3x^2 + 2x + 9) - (x^2 + 5x - 4)$
$= (-3x^2 + 2x + 9) + (-x^2 - 5x + 4)$
$= -4x^2 - 3x + 13$

72. $-13y^2 + 2y + 11$

73. $(5a - 2b + c) - (3a + 2b - 2c)$
$= (5a - 2b + c) + (-3a - 2b + 2c)$
$= 2a - 4b + 3c$

74. $4x - 10y + 4z$

75. $(3x^2 - 2x - x^3) - (5x^2 - 8x - x^3)$
$= (3x^2 - 2x - x^3) + (-5x^2 + 8x + x^3)$
$= -2x^2 + 6x$

76. $5y^2 + 6y + 3y^3$

77. $(5a^2 + 4ab - 3b^2) - (9a^2 - 4ab + 2b^2)$
$= (5a^2 + 4ab - 3b^2) + (-9a^2 + 4ab - 2b^2)$
$= -4a^2 + 8ab - 5b^2$

78. $-3y^2 - 6yz - 12z^2$

79. $(6ab - 4a^2b + 6ab^2) - (3ab^2 - 10ab - 12a^2b)$
$= (6ab - 4a^2b + 6ab^2) + (-3ab^2 + 10ab + 12a^2b)$
$= 8a^2b + 16ab + 3ab^2$

80. $17xy + 5x^2y^2 - 7y^3$

81. $(0.09y^4 - 0.052y^3 + 0.93) -$
$(0.03y^4 - 0.084y^3 + 0.94y^2)$
$= (0.09y^4 - 0.052y^3 + 0.93) +$
$(-0.03y^4 + 0.084y^3 - 0.94y^2)$
$= 0.06y^4 + 0.032y^3 - 0.94y^2 + 0.93$

82. $0.44x^4 + 5.612x^3 - 0.04x^2 - 5.6x$

83. $\left[\frac{5}{8}x^4 - \frac{1}{4}x^2 - \frac{1}{2}\right] - \left[-\frac{3}{8}x^4 + \frac{3}{4}x^2 + \frac{1}{2}\right]$
$= \left[\frac{5}{8}x^4 - \frac{1}{4}x^2 - \frac{1}{2}\right] + \left[\frac{3}{8}x^4 - \frac{3}{4}x^2 - \frac{1}{2}\right]$
$= x^4 - x^2 - 1$

84. $\frac{29}{24}y^4 - \frac{5}{4}y^2 - 11.2y + \frac{8}{15}$

85. $P(x) = R(x) - C(x)$
$P(x) = (280x - 0.4x^2) - (5000 + 0.6x^2)$
$P(x) = (280x - 0.4x^2) + (-5000 - 0.6x^2)$
$P(x) = 280x - x^2 - 5000$

86. $P(x) = 280x - 1.2x^2 - 8000$

87. $P(x) = 280x - x^2 - 5000$
$P(70) = 280 \cdot 70 - 70^2 - 5000 = 19{,}600 - 4900 - 5000 = 9700$
The profit is $9700.

88. $8000

89. $3(y - 2) = 3y - 6$

90. $-2x - 92$

91. 

92. 

93. $2[P(x)] = 2(13x^5 - 22x^4 - 36x^3 + 40x^2 - 16x + 75)$
$= 26x^5 - 44x^4 - 72x^3 + 80x^2 - 32x + 150$
Use columns to add:

$$\begin{array}{r} 26x^5 - 44x^4 - 72x^3 + 80x^2 - 32x + 150 \\ \underline{42x^5 - 37x^4 + 50x^3 - 28x^2 + 34x + 100} \\ 68x^5 - 81x^4 - 22x^3 + 52x^2 + 2x + 250 \end{array}$$

94. $-3x^5 - 29x^4 - 158x^3 + 148x^2 - 82x + 125$

95. $2[Q(x)] = 2(42x^5 - 37x^4 + 50x^3 - 28x^2 + 34x + 100)$
$= 84x^5 - 74x^4 + 100x^3 - 56x^2 + 68x + 200$

$3[P(x)] = 3(13x^5 - 22x^4 - 36x^3 + 40x^2 - 16x + 75)$
$= 39x^5 - 66x^4 - 108x^3 + 120x^2 - 48x + 225$

Use columns to subtract:

$$\begin{array}{r} 84x^5 - 74x^4 + 100x^3 - 56x^2 + 68x + 200 \\ -39x^5 + 66x^4 + 108x^3 - 120x^2 + 48x - 225 \\ \hline 45x^5 - 8x^4 + 208x^3 - 176x^2 + 116x - 25 \end{array}$$

Adding the inverse of $3[P(x)]$

96. $178x^5 - 199x^4 + 6x^3 + 76x^2 + 38x + 600$

97. The area of the base is $x \cdot x$, or x^2.
The area of each side is $x \cdot (x - 2)$.
The total area of all four sides is $4x(x - 2)$.

The surface area of this box can be expressed as a polynomial function.

$S(x) = x^2 + 4x(x - 2)$
$= x^2 + 4x^2 - 8x$
$= 5x^2 - 8x$

98. $8x^{2a} + 7x^a + 7$

99. Writing in columns is helpful.

Add:

$$\begin{array}{r} 47x^{4a} + 3x^{3a} + 22x^{2a} + x^a + 1 \\ 37x^{3a} + 8x^{2a} \qquad + 3 \\ \hline 47x^{4a} + 40x^{3a} + 30x^{2a} + x^a + 4 \end{array}$$

100. $x^{6a} - 5x^{5a} - 4x^{4a} + x^{3a} - 2x^{2a} + 8$

101. Writing in columns is helpful.

Subtract:

$$\begin{array}{l} 2x^{5b} + 4x^{4b} + 3x^{3b} \qquad\qquad\qquad + 8 \\ \underline{-(x^{5b} \qquad\qquad + 2x^{3b} + 6x^{2b} + 9x^b + 8)} \end{array}$$

We rewrite as an addition problem.

Add:

$$\begin{array}{l} 2x^{5b} + 4x^{4b} + 3x^{3b} \qquad\qquad\qquad + 8 \\ -x^{5b} \qquad\qquad - 2x^{3b} - 6x^{2b} - 9x^b - 8 \\ \hline x^{5b} + 4x^{4b} + x^{3b} - 6x^{2b} - 9x^b \end{array}$$

Exercise Set 5.2

1. $2y^2(5y) = (2 \cdot 5)(y^2 \cdot y) = 10y^3$

2. $-6x^3y$

3. $5x(-4x^2y) = 5(-4)(x \cdot x^2)y = -20x^3y$

4. $-6a^3b^4$

5. $2x^3y^2(-5x^2y^4) = 2(-5)(x^3 \cdot x^2)(y^2 \cdot y^4) = -10x^5y^6$

6. $-56a^3b^4c^6$

7. $2x(3 - x)$
$= 2x \cdot 3 - 2x \cdot x$
$= 6x - 2x^2$

8. $4a^3 - 20a^2$

9. $3ab(a + b)$
$= 3ab \cdot a + 3ab \cdot b$
$= 3a^2b + 3ab^2$

10. $4x^2y - 6xy^2$

11. $5cd(3c^2d - 5cd^2)$
$= 5cd \cdot 3c^2d - 5cd \cdot 5cd^2$
$= 15c^3d^2 - 25c^2d^3$

12. $2a^4 - 5a^5$

13. $(2x + 3)(3x - 4)$
$= 6x^2 - 8x + 9x - 12$ FOIL
$= 6x^2 + x - 12$

14. $8a^2 - 14ab + 3b^2$

15. $(s + 3t)(s - 3t)$
$= s^2 - (3t)^2$ $(A + B)(A - B) = A^2 - B^2$
$= s^2 - 9t^2$

16. $y^2 - 16$

17. $(x - y)(x - y)$
$= x^2 - 2xy + y^2$ $(A - B)^2 = A^2 - 2AB + B^2$

18. $a^2 + 4ab + 4b^2$

19. $(y + 8x)(2y - 7x)$
$= 2y^2 - 7xy + 16xy - 56x^2$ FOIL
$= 2y^2 + 9xy - 56x^2$

20. $x^2 - xy - 2y^2$

21. $(a^2 - 2b^2)(a^2 - 3b^2)$
$= a^4 - 3a^2b^2 - 2a^2b^2 + 6b^4$ FOIL
$= a^4 - 5a^2b^2 + 6b^4$

22. $6m^4 - 13m^2n^2 + 5n^4$

23. $(x - 4)(x^2 + 4x + 16)$
$= (x - 4)(x^2) + (x - 4)(4x) + (x - 4)(16)$ Distributive law
$= x(x^2) - 4(x^2) + x(4x) - 4(4x) + x(16) - 4(16)$ Distributive law
$= x^3 - 4x^2 + 4x^2 - 16x + 16x - 64$ Multiplying monomials
$= x^3 - 64$ Collecting like terms

24. $y^3 + 27$

25. $(x + y)(x^2 - xy + y^2)$
$= (x + y)x^2 + (x + y)(-xy) + (x + y)(y^2)$
$= x(x^2) + y(x^2) + x(-xy) + y(-xy) + x(y^2) + y(y^2)$
$= x^3 + x^2y - x^2y - xy^2 + xy^2 + y^3$
$= x^3 + y^3$

26. $a^3 - b^3$

27.
$$\begin{array}{rrrrrl}
 & & a^2 & + a & - 1 & \\
 & & a^2 & + 4a & - 5 & \\ \hline
 & & -5a^2 & - 5a & + 5 & \text{Multiplying by } -5 \\
 & 4a^3 & + 4a^2 & - 4a & & \text{Multiplying by } 4a \\
a^4 & + a^3 & - a^2 & & & \text{Multiplying by } a^2 \\ \hline
a^4 & + 5a^3 & - 2a^2 & - 9a & + 5 & \text{Adding}
\end{array}$$

28. $x^4 - x^3 + x^2 - 3x + 2$

29.
$$\begin{array}{r}
4a^2b - 2ab + 3b^2 \\
ab - 2b + a \\ \hline
4a^3b - 2a^2b + 3ab^2 \quad ① \\
-6b^3 \qquad + 4ab^2 - 8a^2b^2 \quad ② \\
3ab^3 \qquad - 2a^2b^2 + 4a^3b^2 \quad ③ \\ \hline
3ab^3 - 6b^3 + 4a^3b - 2a^2b + 7ab^2 - 10a^2b^2 + 4a^3b^2 \quad ④
\end{array}$$

① Multiplying by a
② Multiplying by -2b
③ Multiplying by ab
④ Adding

30. $-4x^3y + 3xy^3 - x^2y^2 - 2y^4 + 2x^4$

31. $\left(x - \frac{1}{2}\right)\left(x - \frac{1}{4}\right)$
$= x^2 - \frac{1}{4}x - \frac{1}{2}x + \frac{1}{8}$ FOIL
$= x^2 - \frac{1}{4}x - \frac{2}{4}x + \frac{1}{8}$
$= x^2 - \frac{3}{4}x + \frac{1}{8}$

32. $b^2 - \frac{2}{3}b + \frac{1}{9}$

33. $(1.3x - 4y)(2.5x + 7y)$
$= 3.25x^2 + 9.1xy - 10xy - 28y^2$ FOIL
$= 3.25x^2 - 0.9xy - 28y^2$

34. $12a^2 + 399.928ab - 2.4b^2$

35. $P(x)\cdot Q(x) = (3x^2 - 5)(4x^2 - 7x + 2)$
$= (3x^2 - 5)(4x^2) + (3x^2 - 5)(-7x) + (3x^2 - 5)(2)$
$= 12x^4 - 20x^2 - 21x^3 + 35x + 6x^2 - 10$
$= 12x^4 - 21x^3 - 14x^2 + 35x - 10$

36. $x^5 + x^3 + 2x^2 - x + 2$

37. $(a + 2)(a + 3)$
$= a^2 + 3a + 2a + 6$ FOIL
$= a^2 + 5a + 6$

38. $x^2 + 13x + 40$

39. $(y + 3)(y - 2)$
$= y^2 - 2y + 3y - 6$ FOIL
$= y^2 + y - 6$

40. $y^2 + 3y - 28$

41. $(x + 3)^2$
$= x^2 + 2\cdot x\cdot 3 + 3^2$ $(A + B)^2 = A^2 + 2AB + B^2$
$= x^2 + 6x + 9$

42. $y^2 - 14y + 49$

43. $(x - 2y)^2$
$= x^2 - 2(x)(2y) + (2y)^2$ $(A - B)^2 = A^2 - 2AB + B^2$
$= x^2 - 4xy + 4y^2$

44. $4s^2 + 12st + 9t^2$

45. $\left(b - \frac{1}{3}\right)\left(b - \frac{1}{2}\right)$
$= b^2 - \frac{1}{2}b - \frac{1}{3}b + \frac{1}{6}$ FOIL
$= b^2 - \frac{3}{6}b - \frac{2}{6}b + \frac{1}{6}$
$= b^2 - \frac{5}{6}b + \frac{1}{6}$

46. $x^2 - \frac{13}{6}x + 1$

47. $(2x + 9)(x + 2)$
$= 2x^2 + 4x + 9x + 18$ FOIL
$= 2x^2 + 13x + 18$

48. $6b^2 - 11b - 10$

49. $(10a - 0.12b)^2$
$= (10a)^2 - 2(10a)(0.12b) + (0.12b)^2$
$(A - B)^2 = A^2 - 2AB + B^2$
$= 100a^2 - 2.4ab + 0.0144b^2$

50. $100p^4 + 46p^2q + 5.29q^2$

51. $(2x - 3y)(2x + y)$
$= 4x^2 + 2xy - 6xy - 3y^2$ FOIL
$= 4x^2 - 4xy - 3y^2$

52. $4a^2 - 8ab + 3b^2$

53. $\left(2a + \frac{1}{3}\right)^2$

$= (2a)^2 + 2(2a)\left(\frac{1}{3}\right) + \left(\frac{1}{3}\right)^2$

$(A + B)^2 = A^2 + 2AB + B^2$

$= 4a^2 + \frac{4}{3}a + \frac{1}{9}$

54. $9c^2 - 3c + \frac{1}{4}$

55. $(2x^3 - 3y^2)^2$

$= (2x^3)^2 - 2(2x^3)(3y^2) + (3y^2)^2$

$(A - B)^2 = A^2 - 2AB + B^2$

$= 4x^6 - 12x^3y^2 + 9y^4$

56. $9s^4 + 24s^2t^3 + 16t^6$

57. $(a^2b^2 + 1)^2$

$= (a^2b^2)^2 + 2(a^2b^2)\cdot 1 + 1^2$

$(A + B)^2 = A^2 + 2AB + B^2$

$= a^4b^4 + 2a^2b^2 + 1$

58. $x^4y^2 - 2x^3y^3 + x^2y^4$

59. $(20a^5 - 0.16b)^2$

$= (20a^5)^2 - 2(20a^5)(0.16b) + (0.16b)^2$

$(A - B)^2 = A^2 - 2AB + B^2$

$= 400a^{10} - 6.4a^5b + 0.0256b^2$

60. $4p^{10} + 12.4p^5q + 9.61q^2$

61. $P(x)\cdot P(x) = (3x - 4)(3x - 4) = (3x - 4)^2$

$= (3x)^2 - 2(3x)(4) + (4)^2$

$(A - B)^2 = A^2 - 2AB + B^2$

$= 9x^2 - 24x + 16$

62. $25x^4 - 110x^2 + 121$

63. $(c + 2)(c - 2) = c^2 - 2^2$

$(A + B)(A - B) = A^2 - B^2$

$= c^2 - 4$

64. $x^2 - 9$

65. $(2a + 1)(2a - 1) = (2a)^2 - 1^2$

$(A + B)(A - B) = A^2 - B^2$

$= 4a^2 - 1$

66. $9 - 4x^2$

67. $(3m - 2n)(3m + 2n) = (3m)^2 - (2n)^2$

$(A + B)(A - B) = A^2 - B^2$

$= 9m^2 - 4n^2$

68. $9x^2 - 25y^2$

69. $(x^3 + yz)(x^3 - yz) = (x^3)^2 - (yz)^2$

$(A + B)(A - B) = A^2 - B^2$

$= x^6 - y^2z^2$

70. $4a^6 - 25a^2b^2$

71. $(-mn + m^2)(mn + m^2) = (m^2 - mn)(m^2 + mn)$

$= (m^2)^2 - (mn)^2$ $\quad (A + B)(A - B) = A^2 - B^2$

$= m^4 - m^2n^2$

72. $p^2q^2 - 2.56$

73. $\left(\frac{1}{2}p^4 - \frac{2}{3}q\right)\left(\frac{1}{2}p^4 + \frac{2}{3}q\right)$

$= \left(\frac{1}{2}p^4\right)^2 - \left(\frac{2}{3}q\right)^2 = \frac{1}{4}p^8 - \frac{4}{9}q^2$

74. $\frac{9}{25}a^2b^2 - 16c^{10}$

75. $(x + 1)(x - 1)(x^2 + 1) = (x^2 - 1^2)(x^2 + 1)$

$= (x^2 - 1)(x^2 + 1)$

$= (x^2)^2 - 1^2$

$= x^4 - 1$

76. $y^4 - 16$

77. $(a - b)(a + b)(a^2 - b^2) = (a^2 - b^2)(a^2 - b^2)$

$= (a^2 - b^2)^2$

$= (a^2)^2 - 2(a^2)(b^2) + (b^2)^2$

$= a^4 - 2a^2b^2 + b^4$

78. $16x^4 - 8x^2y^2 + y^4$

79. $(a + b + 1)(a + b - 1)$

$= [(a + b) + 1][(a + b) - 1]$

$= (a + b)^2 - 1^2$

$= a^2 + 2ab + b^2 - 1$

80. $m^2 + 2mn + n^2 - 4$

81. $(2x + 3y + 4)(2x + 3y - 4)$

$= [(2x + 3y) + 4][(2x + 3y) - 4]$

$= (2x + 3y)^2 - 4^2$

$= 4x^2 + 12xy + 9y^2 - 16$

82. $9a^2 - 12ab + 4b^2 - c^2$

83. $A = P(1 + i)^2$

$A = P(1 + 2i + i^2)$

$A = P + 2Pi + Pi^2$

84. $A = P + Pi + \frac{Pi^2}{4}$

85. a) Replace each occurrence of x by t - 1.

$f(t - 1) = 5(t - 1) + (t - 1)^2$
$= 5t - 5 + t^2 - 2t + 1$
$= t^2 + 3t - 4$

b) $f(a + h) - f(a)$
$= [5(a + h) + (a + h)^2] - [5a + a^2]$
$= 5a + 5h + a^2 + 2ah + h^2 - 5a - a^2$
$= 5h + 2ah + h^2$

86. a) $-p^2 + p + 6$

b) $3h - 2ah - h^2$

87. Familiarize. We let x represent the number of days worked during the week and y represent the number of days worked during the weekend. We organize the information in a table.

	Days worked	Pay per day	Total earned
During the week	x	\$25	25x
During the weekend	y	\$35	35y
Total	17		\$485

Translate.

Total number of days worked is 17.
$x + y = 17$

Total earned is \$485.
$25x + 35y = 485$

We now have a system of equations.

$x + y = 17,$
$25x + 35y = 485$

Carry out. Solving the system we get (11,6).

Check. The total number of days is 11 + 6, or 17. The total earned is 25(11) + 35(6), or 275 + 210, or \$485. The values check.

State. Takako worked 11 weekdays.

88. 56, 58, and 60

89. Familiarize. Let a, b, and c represent the daily production of machines A, B, and C, respectively.

Translate. Rewording, we have:

Production of A, B, and C is 222 per day.
$a + b + c = 222$

Production of A and B alone is 159.
$a + b = 159$

Production of B and C alone is 147.
$b + c = 147$

We have a system of equations.

$a + b + c = 222,$
$a + b = 159,$
$b + c = 147$

Carry out. Solving the system we get (75,84,63).

Check. The daily production of the three machines together is 75 + 84 + 63, or 222. The daily production of A and B alone is 75 + 84, or 159. The daily production of B and C alone is 84 + 63, or 147. The numbers check.

State. The daily production of suitcases by machines A, B, and C is 75, 84, and 63, respectively.

90. Dimes: 5, nickels: 2, quarters: 6

91.

92.

93. $(6y)^2\left[-\frac{1}{3}x^2y^3\right]^3$
$= (36y^2)\left[-\frac{1}{27}x^6y^9\right]$
$= -\frac{36}{27}x^6y^{11}$
$= -\frac{4}{3}x^6y^{11}$

94. a^{16n^2}

95. $(-r^6s^2)^3\left[-\frac{r^2}{6}\right]^2(9s^4)^2$
$= (-r^{18}s^6)\left[\frac{r^4}{36}\right](81s^8)$
$= -\frac{81}{36}r^{22}s^{14}$
$= -\frac{9}{4}r^{22}s^{14}$

96. $\frac{8}{21}x^8y^5$

97. $(z^{n^2})^{n^3}(z^{4n^3})^{n^2}$
$= z^{n^5}\cdot z^{4n^5}$ Multiplying exponents
$= z^{5n^5}$ Adding exponents

98. $\frac{1}{4}a^{7x}b^{2y+2}$

99. $[(2x - 1)^2 - 1]^2$
$= [(4x^2 - 4x + 1) - 1]^2$
$= [4x^2 - 4x]^2$
$= (4x^2)^2 - 2(4x^2)(4x) + (4x)^2$
$= 16x^4 - 32x^3 + 16x^2$

100. $y^{3n+3}z^{n+3} - 4y^4z^{3n}$

101. $[(a + b)(a - b)][5 - (a + b)][5 + (a + b)]$
$= [a^2 - b^2][5^2 - (a + b)^2]$
$= [a^2 - b^2][25 - (a^2 + 2ab + b^2)]$
$= [a^2 - b^2][25 - a^2 - 2ab - b^2]$
$= 25a^2 - a^4 - 2a^3b - a^2b^2 - 25b^2 + a^2b^2 + 2ab^3 + b^4$
$= -a^4 - 2a^3b + 25a^2 + 2ab^3 - 25b^2 + b^4$

102. $x^3 + y^3 + 3y^2 + 3y + 1$

103. $(r^2 + s^2)^2(r^2 + 2rs + s^2)(r^2 - 2rs + s^2)$
$= (r^2 + s^2)^2(r + s)^2(r - s)^2$
$= (r^2 + s^2)^2[(r + s)(r - s)]^2$
$= (r^2 + s^2)^2(r^2 - s^2)^2$
$= [(r^2 + s^2)(r^2 - s^2)]^2$
$= (r^4 - s^4)^2$
$= r^8 - 2r^4s^4 + s^8$

104. $y^{12} - 6y^{10} + 15y^8 - 20y^6 + 15y^4 - 6y^2 + 1$

105. $(a - b + c - d)(a + b + c + d)$
$= [(a + c) - (b + d)][(a + c) + (b + d)]$
$= (a + c)^2 - (b + d)^2$
$= (a^2 + 2ac + c^2) - (b^2 + 2bd + d^2)$
$= a^2 + 2ac + c^2 - b^2 - 2bd - d^2$

106. $\frac{4}{9}x^2 - \frac{1}{9}y^2 - \frac{2}{3}y - 1$

107. $[2(y - 3) - 6(x + 4)][5(y - 3) - 4(x + 4)]$
$= [2y - 6 - 6x - 24][5y - 15 - 4x - 16]$
$= (2y - 6x - 30)(5y - 4x - 31)$

$$\begin{array}{rrrrrr}
 & & & 5y & -\ 4x & -\ 31 \\
 & & & 2y & -\ 6x & -\ 30 \\ \hline
 & & & -150y & +\ 120x & +\ 930 \\
 & -30xy & +\ 24x^2 & & +\ 186x & \\
10y^2 & -\ 8xy & & -\ 62y & & \\ \hline
10y^2 & -\ 38xy & +\ 24x^2 & -\ 212y & +\ 306x & +\ 930
\end{array}$$

(Note: This exercise could also be done using FOIL.)

108. $x^3 - \frac{1}{343}$

109. $(4x^2 + 2xy + y^2)(4x^2 - 2xy + y^2)$
$= [(4x^2 + y^2) + 2xy][(4x^2 + y^2) - 2xy]$
$= (4x^2 + y^2)^2 - (2xy)^2$
$= 16x^4 + 8x^2y^2 + y^4 - 4x^2y^2$
$= 16x^4 + 4x^2y^2 + y^4$

110. $x^4 - 25x^2 + 144$

111. $(x^a + y^b)(x^a - y^b)(x^{2a} + y^{2b})$
$= (x^{2a} - y^{2b})(x^{2a} + y^{2b})$
$= x^{4a} - y^{4b}$

112. $\frac{1}{25}x^4 + \frac{4}{25}x^3 + \frac{16}{25}x^2 + \frac{24}{25}x + \frac{36}{25}$

113. $[a - (b - 1)][(b - 1)^2 + a(b - 1) + a^2]$
$= (a - b + 1)(b^2 - 2b + 1 + ab - a + a^2)$
$= ab^2 - 2ab + a + a^2b - a^2 + a^3 - b^3 + 2b^2 -$
$b - ab^2 + ab - a^2b + b^2 - 2b + 1 + ab - a + a^2$
$= a^3 - b^3 + 3b^2 - 3b + 1$

114. $x^6 - 1$

115. $\left[\left(\frac{1}{3}x^3 - \frac{2}{3}y^2\right)\left(\frac{1}{3}x^3 + \frac{2}{3}y^2\right)\right]^2$
$= \left[\left(\frac{1}{3}x^3\right)^2 - \left(\frac{2}{3}y^2\right)^2\right]^2$
$= \left[\frac{1}{9}x^6 - \frac{4}{9}y^4\right]^2$
$= \left[\frac{1}{9}x^6\right]^2 - 2\left[\frac{1}{9}x^6\right]\left[\frac{4}{9}y^4\right] + \left[\frac{4}{9}y^4\right]^2$
$= \frac{1}{81}x^{12} - \frac{8}{81}x^6y^4 + \frac{16}{81}y^8$

116. $0.1x^8y^2 - 0.02x^5y^5 + 0.001x^2y^8$

117. $(x^{a-b})^{a+b} = x^{(a-b)(a+b)} = x^{a^2-b^2}$

118. $M^{x^2+2xy+y^2}$

Exercise Set 5.3

1. $4a^2 + 2a$
$= 2a\cdot 2a + 2a\cdot 1$
$= 2a(2a + 1)$

2. $3y(2y + 1)$

3. $y^2 - 5y$
$= y\cdot y - 5\cdot y$
$= y(y - 5)$

4. $x(x + 9)$

5. $y^3 + 9y^2$
$= y\cdot y^2 + 9\cdot y^2$
$= y^2(y + 9)$

6. $x^2(x + 8)$

7. $6x^2 - 3x^4$
$= 3x^2\cdot 2 - 3x^2\cdot x^2$
$= 3x^2(2 - x^2)$

8. $4y^2(2 + y^2)$

9. $4x^2y - 12xy^2$
$= 4xy\cdot x - 4xy\cdot 3y$
$= 4xy(x - 3y)$

10. $5x^2y^2(y + 3x)$

11. $3y^2 - 3y - 9$
$= 3 \cdot y^2 - 3 \cdot y - 3 \cdot 3$
$= 3(y^2 - y - 3)$

12. $5(x^2 - x + 3)$

13. $4ab - 6ac + 12ad$
$= 2a \cdot 2b - 2a \cdot 3c + 2a \cdot 6d$
$= 2a(2b - 3c + 6d)$

14. $2x(4y + 5z - 7w)$

15. $10a^4 + 15a^2 - 25a - 30$
$= 5 \cdot 2a^4 + 5 \cdot 3a^2 - 5 \cdot 5a - 5 \cdot 6$
$= 5(2a^4 + 3a^2 - 5a - 6)$

16. $4(3t^5 - 5t^4 + 2t^2 - 4)$

17. $-3x + 12 = -3(x - 4)$

18. $-5(x + 8)$

19. $-6y - 72 = -6(y + 12)$

20. $-8(t - 9)$

21. $-2x^2 + 4x - 12 = -2(x^2 - 2x + 6)$

22. $-2(x^2 - 6x - 20)$

23. $-3y^2 + 24x = -3(y^2 - 8x)$

24. $-7x^2 - 56y = -7(x^2 + 8y)$

25. $-3y^3 + 12y^2 - 15y = -3y(y^2 - 4y + 5)$

26. $-4m(m^3 + 8m^2 - 16)$

27. $-x^2 + 3x - 7 = -(x^2 - 3x + 7)$

28. $-(p^3 + 4p^2 - 11)$

29. $-a^4 + 2a^3 - 13a = -a(a^3 - 2a^2 + 13)$

30. $-(m^3 + m^2 - m + 2)$

31. a) $h(t) = -16t^2 + 80t$
$h(t) = -16t(t - 5)$ Factoring out $-16t$

b) Using $h(t) = -16t^2 + 80t$:
$h(3) = -16 \cdot 3^2 + 80 \cdot 3 = -16 \cdot 9 + 240 =$
$-144 + 240 = 96$
Using $h(t) = -16t(t - 5)$:
$h(3) = -16(3)(3 - 5) = -16(3)(-2) = 96$

32. a) $h(t) = -16t(t - 6)$

b) 128

33. $N(x) = \frac{1}{6}x^3 + \frac{1}{2}x^2 + \frac{1}{3}x$

$N(x) = \frac{1}{6}x(x^2 + 3x + 2)$ Factoring out $\frac{1}{6}x$

34. $f(n) = \frac{1}{2}n(n - 1)$

35. $P(n) = \frac{1}{2}n^2 - \frac{3}{2}n$

$P(n) = \frac{1}{2}n(n - 3)$ Factoring out $\frac{1}{2}n$

36. $2\pi r(h + r)$

37. $R(x) = 280x - 0.4x^2$
$R(x) = 0.4x(700 - x)$

38. $C(x) = 0.6x(0.3 + x)$

39. $a(b - 2) + c(b - 2)$
$= (b - 2)(a + c)$

40. $(x^2 - 3)(a - 2)$

41. $(x - 2)(x + 5) + (x - 2)(x + 8)$
$= (x - 2)[(x + 5) + (x + 8)]$
$= (x - 2)(2x + 13)$

42. $2m(m - 4)$

43. $a^2(x - y) + a^2(y - x)$
$= a^2(x - y) + a^2(-1)(x - y)$ Factoring out -1 to reverse the subtraction
$= a^2(x - y) - a^2(x - y)$ Simplifying
$= (x - y)(a^2 - a^2)$ Factoring out $x - y$
$= (x - y)(0)$
$= 0$
[You might have recognized that $a^2(x - y) - a^2(x - y) = 0$ and gone directly to the result.]

44. 0

45. $ac + ad + bc + bd$
$= a(c + d) + b(c + d)$
$= (c + d)(a + b)$

46. $(y + z)(x + w)$

47. $b^3 - b^2 + 2b - 2$
$= b^2(b - 1) + 2(b - 1)$
$= (b - 1)(b^2 + 2)$

48. $(y - 1)(y^2 + 3)$

49. $a^3 - 3a^2 + 6 - 2a$
$= a^2(a - 3) + 2(3 - a)$
$= a^2(a - 3) + 2(-1)(a - 3)$ Factoring out -1 to reverse the subtraction
$= a^2(a - 3) - 2(a - 3)$
$= (a - 3)(a^2 - 2)$

50. $(t + 6)(t^2 - 2)$

51. $24x^3 - 36x^2 + 72x$
$= 12x\cdot 2x^2 - 12x\cdot 3x + 12x\cdot 6$
$= 12x(2x^2 - 3x + 6)$

52. $3a(4a^3 - 7a^2 - 2)$

53. $x^6 - x^5 - x^3 + x^4$
$= x^3(x^3 - x^2 - 1 + x)$
$= x^3[x^2(x - 1) + x - 1]$ $(-1 + x = x - 1)$
$= x^3(x - 1)(x^2 + 1)$

54. $y(y - 1)(y^2 + 1)$

55. $2y^4 + 6y^2 + 5y^2 + 15$
$= 2y^2(y^2 + 3) + 5(y^2 + 3)$
$= (y^2 + 3)(2y^2 + 5)$

56. $(2 - x)(xy - 3)$

57. Familiarize. Let x, y, and z represent the number of nickels, dimes, and quarters, respectively. The value of x nickels is 0.05x, of y dimes is 0.10y, and of z quarters is 0.25z.

Translate.

The total value is \$18.10.
$0.05x + 0.10y + 0.25z = 18.10$

The number of quarters is two more than twice the number of dimes.
$z = 2 + 2\cdot y$

The total number of coins is 120.
$x + y + z = 120$

We have a system of equations.
$0.05x + 0.10y + 0.25z = 18.10,$
$z = 2 + 2y,$
$x + y + z = 120$

Carry out. Solving the system we get (40,26,54).

Check. The total value of the coins is \$0.05(40) + \$0.10(26) + \$0.25(54), or \$18.10. The number of quarters, 54, is 2 more than twice the number of dimes, 26. The total number of coins is 40 + 26 + 54, or 120. The values check.

State. There are 40 nickels, 26 dimes, and 54 quarters.

58. 21°, 84°

59.

60.

61. $P(x) = x^4 - 3x^3 - 5x^2 + 4x + 2$
$= x(x^3 - 3x^2 - 5x + 4) + 2$
$= x(x(x^2 - 3x - 5) + 4) + 2$
$= x(x(x(x - 3) - 5) + 4) + 2$

62. $P(x) = x(x(x(x + 5) + 7) - 2) - 4$

63. $P(x) = 5x^5 + 0x^4 - 3x^3 + 4x^2 - 5x + 1$
$= x(5x^4 + 0x^3 - 3x^2 + 4x - 5) + 1$
$= x(x(5x^3 + 0x^2 - 3x + 4) - 5) + 1$
$= x(x(x(5x^2 + 0x - 3) + 4) - 5) + 1$
$= x(x(x(x(5x + 0) - 3) + 4) - 5) + 1$

64. $Q(x) = x(x(x(x(4x + 2) + 0) - 5) + 0) - 7$

65. $P(x) = 2x^4 - 3x^3 + 5x^2 + 6x - 4$
$= x(2x^3 - 3x^2 + 5x + 6) - 4$
$= x(x(2x^2 - 3x + 5) + 6) - 4$
$= x(x(x(2x - 3) + 5) + 6) - 4$

$P(5) = 5(5(5(2\cdot 5 - 3) + 5) + 6) - 4$
$= 5(5(5\cdot 7 + 5) + 6) - 4$
$= 5(5\cdot 40 + 6) - 4$
$= 5\cdot 206 - 4$
$= 1030 - 4$
$= 1026$

$P(-2) = -2(-2(-2(2(-2) - 3) + 5) + 6) - 4$
$= -2(-2(-2\cdot(-7) + 5) + 6) - 4$
$= -2(-2\cdot 19 + 6) - 4$
$= -2\cdot(-32) - 4$
$= 64 - 4$
$= 60$

$P(10) = 10(10(10(2\cdot 10 - 3) + 5) + 6) - 4$
$= 10(10(10\cdot 17 + 5) + 6) - 4$
$= 10(10\cdot 175 + 6) - 4$
$= 10\cdot 1756 - 4$
$= 17{,}560 - 4$
$= 17{,}556$

66. $Q(x) = x(x(x(x(2x - 3) + 0) + 5) - 7) + 12;$
-2696; 26,619; 278,719

67. $4y^{4a} + 12y^{2a} + 10y^{2a} + 30$
$= 2(2y^{4a} + 6y^{2a} + 5y^{2a} + 15)$
$= 2[2y^{2a}(y^{2a} + 3) + 5(y^{2a} + 3)]$
$= 2(y^{2a} + 3)(2y^{2a} + 5)$

68. $(x^{2p} + 3)(2x^{2p} + 5)$

69. $4x^{a+b} + 7x^{a-b}$
$= 4 \cdot x^a \cdot x^b + 7 \cdot x^a \cdot x^{-b}$
$= x^a(4x^b + 7x^{-b})$

70. $y^{a+b}(7y^a - 5 + 3y^b)$

Exercise Set 5.4

1. $x^2 + 9x + 20$

We look for two numbers whose product is 20 and whose sum is 9. Since both 20 and 9 are positive, we need only consider positive factors.

Pairs of Factors	Sums of Factors
1, 20	21
2, 10	12
4, 5	9

The numbers we need are 4 and 5. The factorization is $(x + 4)(x + 5)$.

2. $(x + 3)(x + 5)$

3. $t^2 - 8t + 15$

Since the constant term is positive and the coefficient of the middle term is negative, we look for a factorization of 15 in which both factors are negative. Their sum must be -8.

Pairs of Factors	Sums of Factors
-1, -15	-16
-3, -5	-8

The numbers we need are -3 and -5. The factorization is $(t - 3)(t - 5)$.

4. $(y - 3)(y - 9)$

5. $x^2 - 27 - 6x = x^2 - 6x - 27$

Since the constant term is negative, we look for a factorization of -27 in which one factor is positive and one factor is negative. Their sum must be -6, so the negative factor must have the larger absolute value. Thus we consider only pairs of factors in which the negative factor has the larger absolute value.

Pairs of Factors	Sums of Factors
-27, 1	-26
-9, 3	-6

The numbers we need are -9 and 3. The factorization is $(x - 9)(x + 3)$.

6. $(t - 5)(t + 3)$

7. $2y^2 - 16y + 32$
$= 2(y^2 - 8y + 16)$ Removing the common factor

We now factor $y^2 - 8y + 16$. We look for two numbers whose product is 16 and whose sum is -8. Since the constant term is positive and the coefficient of the middle term is negative, we look for a factorization of 16 in which both factors are negative.

Pairs of Factors	Sums of Factors
-1, -16	-17
-2, -8	-10
-4, -4	-8

The numbers we want are -4 and -4.

$y^2 - 8y + 16 = (y - 4)(y - 4)$

We must not forget to include the common factor 2.

$2y^2 - 16y + 32 = 2(y - 4)(y - 4)$.

8. $2(a - 5)(a - 5)$

9. $p^3 + 3p^2 - 54p$
$= p(p^2 + 3p - 54)$ Removing the common factor

We now factor $p^2 + 3p - 54$. Since the constant term is negative, we look for a factorization of -54 in which one factor is positive and one factor is negative. We consider only pairs of factors in which the positive factor has the larger absolute value, since the sum of the factors, 3, is positive.

Pairs of Factors	Sums of Factors
54, -1	53
27, -2	25
18, -3	15
9, -6	3

The numbers we need are 9 and -6.

$p^2 + 3p - 54 = (p + 9)(p - 6)$

We must not forget to include the common factor p.

$p^3 + 3p^2 - 54p = p(p + 9)(p - 6)$

10. $m(m + 9)(m - 8)$

11. $14x + x^2 + 45 = x^2 + 14x + 45$

Since the constant term and the middle term are both positive, we look for a factorization of 45 in which both factors are positive. Their sum must be 14.

Pairs of Factors	Sums of Factors
45, 1	46
15, 3	18
9, 5	14

The numbers we need are 9 and 5. The factorization is $(x + 9)(x + 5)$.

12. $(y + 8)(y + 4)$

13. $y^2 + 2y - 63$

Since the constant term is negative, we look for a factorization of -63 in which one factor is positive and one factor is negative. We consider only pairs of factors in which the positive factor has the larger absolute value, since the sum of the factors, 2, is positive.

Pairs of Factors	Sums of Factors
63, -1	62
21, -3	18
9, -7	2

The numbers we need are 9 and -7. The factorization is $(y + 9)(y - 7)$.

14. $(p + 8)(p - 5)$

15. $t^2 - 11t + 28$

Since the constant term is positive and the coefficient of the middle term is negative, we look for a factorization of 28 in which both factors are negative. Their sum must be -11.

Pairs of Factors	Sums of Factors
-28, -1	-29
-14, -2	-16
-7, -4	-11

The numbers we need are -7 and -4. The factorization is $(t - 7)(t - 4)$.

16. $(y - 9)(y - 5)$

17. $3x + x^2 - 10 = x^2 + 3x - 10$

Since the constant term is negative, we look for a factorization of -10 in which one factor is positive and one factor is negative. We consider only pairs of factors in which the positive factor has the larger absolute value, since the sum of the factors, 3, is positive.

Pairs of Factors	Sums of Factors
10, -1	9
5, -2	3

The numbers we need are 5 and -2. The factorization is $(x + 5)(x - 2)$.

18. $(x + 3)(x - 2)$

19. $x^2 + 5x + 6$

We look for two numbers whose product is 6 and whose sum is 5. Since 6 and 5 are both positive, we need consider only positive factors.

Pairs of Factors	Sums of Factors
1, 6	7
2, 3	5

The numbers we need are 2 and 3. The factorization is $(x + 2)(x + 3)$.

20. $(y + 7)(y + 1)$

21. $56 + x - x^2 = -x^2 + x + 56 = -(x^2 - x - 56)$

We now factor $x^2 - x - 56$. Since the constant term is negative, we look for a factorization of -56 in which one factor is positive and one factor is negative. We consider only pairs of factors in which the negative factor has the larger absolute value, since the sum of the factors, -1, is negative.

Pairs of Factors	Sums of Factors
-56, 1	-55
-28, 2	-26
-14, 4	-10
-8, 7	-1

The numbers we need are -8 and 7. Thus, $x^2 - x - 56 = (x - 8)(x + 7)$. We must not forget to include the factor that was factored out earlier:

$56 + x - x^2 = -(x - 8)(x + 7)$, or

$(-x + 8)(x + 7)$, or $(8 - x)(7 + x)$

22. $(8 - y)(4 + y)$

23. $32y + 4y^2 - y^3$

There is a common factor, y. We also factor out -1 in order to make the leading coefficient positive.

$$32y + 4y^2 - y^3 = -y(-32 - 4y + y^2)$$
$$= -y(y^2 - 4y - 32)$$

Now we factor $y^2 - 4y - 32$. Since the constant term is negative, we look for a factorization of -32 in which one factor is positive and one factor is negative. We consider only pairs of factors in which the negative factor has the larger absolute value, since the sum of the factors, -4, is negative.

Pairs of Factors	Sums of Factors
-32, 1	-31
-16, 2	-14
-8, 4	-4

The numbers we need are -8 and 4. Thus, $y^2 - 4y - 32 = (y - 8)(y + 4)$. We must not forget to include the common factor:

$32y + 4y^2 - y^3 = -y(y - 8)(y + 4)$, or

$y(-y + 8)(y + 4)$, or $y(8 - y)(4 + y)$

24. $x(8 - x)(7 + x)$

25. $x^4 + 11x^2 - 80$

First make a substitution. We let $u = x^2$, so $u^2 = x^4$. Then we consider $u^2 + 11u - 80$. We look for pairs of factors of -80, one positive and one negative, such that the positive factor has the larger absolute value and the sum of the factors is 11.

Pairs of Factors	Sums of Factors
80, -1	79
40, -2	38
20, -4	16
16, -5	11
10, -8	2

The numbers we need are 16 and -5. Then $u^2 + 11u - 80 = (u + 16)(u - 5)$. Replacing u by x^2 we obtain the factorization of the original trinomial: $(x^2 + 16)(x^2 - 5)$

26. $(y^2 + 12)(y^2 - 7)$

27. $x^2 - 3x + 7$

There are no factors of 7 whose sum is -3. This trinomial is not factorable into binomials with integer coefficients.

28. Not factorable

29. $x^2 + 12xy + 27y^2$

We look for numbers p and q such that $x^2 + 12xy + 27y^2 = (x + py)(x + qy)$. Our thinking is much the same as if we were factoring $x^2 + 12x + 27$. We look for factors of 27 whose sum is 12. Those factors are 9 and 3. Then

$$x^2 + 12xy + 27y^2 = (x + 9y)(x + 3y).$$

30. $(p - 8q)(p + 3q)$

31. $x^2 - 14xy + 49y^2$

We look for numbers p and q such that

$x^2 - 14xy + 49y^2 = (x + py)(x + qy)$.

Our thinking is much the same as if we factor $x^2 - 14x + 49$. Since the constant term is positive and the coefficient of the middle term is negative, we look for a factorization of 49 in which both factors are negative. Their sum must be -14.

Pairs of Factors	Sums of Factors
-49, -1	-50
-7, -7	-14

The numbers we need are -7 and -7. The factorization is $(x - 7y)(x - 7y)$.

32. $(y + 4z)(y + 4z)$

33. $x^4 + 50x^2 + 49$

Substitute u for x^2 (and hence u^2 for x^4). Consider $u^2 + 50u + 49$. We look for a pair of positive factors of 49 whose sum is 50.

Pairs of Factors	Sums of Factors
7, 7	14
1, 49	50

The numbers we need are 1 and 49. Then $u^2 + 50 + 49 = (u + 1)(u + 49)$. Replacing u by x^2 we have

$x^4 + 50x^2 + 49 = (x^2 + 1)(x^2 + 49)$.

34. $(p^2 + 79)(p^2 + 1)$

35. $x^6 + 2x^3 - 63$

Substitute u for x^3 (and hence u^2 for x^6). Consider $u^2 + 2u - 63$. This is the same trinomial as in Exercise 13, so we know $u^2 + 2u - 63 = (u + 9)(u - 7)$. Replacing u by x^3 we have

$x^6 + 2x^3 - 63 = (x^3 + 9)(x^3 - 7)$.

36. $(x^4 - 5)(x^4 - 2)$

37. $3x^2 - 16x - 12$

We will use the FOIL method.

1. There is no common factor (other than 1 or -1).

2. Factor the first term, $3x^2$. The factors are 3x and x. We have this possibility:

   $(3x \quad)(x \quad)$

3. Factor the last term, -12. The possibilities are 12(-1), (-12)(1), 6(-2), (-6)2, 4(-3), and (-4)3 as well as (-1)12, (1)(-12), (-2)6, 2(-6), (-3)4, and 3(-4). We look for factors such that the sum of their products is the middle term -16x. Try some possibilities and check by multiplying.

   $(3x - 3)(x + 4) = 3x^2 + 9x - 12$

   This gives a middle term with a positive coefficient. But, more significantly, the expression 3x - 3 has a common factor of 3. But we know there is no common factor (other than 1 or -1) so we can eliminate (3x - 3) and all other factors containing a common factor, such as (3x + 6), (3x - 12), and so on.

   We try another possibility:

   $(3x + 2)(x - 6) = 3x^2 - 16x - 12$

   The factorization is $(3x + 2)(x - 6)$.

38. $(3x + 5)(2x - 5)$

39. $6x^3 - 15x - x^2 = 6x^3 - x^2 - 15x$

We will use the grouping method.

1. Look for a common factor. We factor out x:

   $x(6x^2 - x - 15)$

2. Factor the trinomial $6x^2 - x - 15$. Multiply the leading coefficient, 6, and the constant, -15.

   $6(-15) = -90$

3. Try to factor -90 so the sum of the factors is -1. We need only consider pairs of factors in which the negative term has the larger absolute value, since their sum is negative.

Pairs of Factors	Sums of Factors
-90, 1	-89
-45, 2	-43
-30, 3	-27
-16, 5	-11
-10, 9	-1

4. We split the middle term, -x, using the results of step (3):

   $-x = -10x + 9x$

5. Factor by grouping:

$$6x^2 - x - 15 = 6x^2 - 10x + 9x - 15$$
$$= 2x(3x - 5) + 3(3x - 5)$$
$$= (3x - 5)(2x + 3)$$

We must include the common factor to get a factorization of the original trinomial:

$6x^3 - 15x - x^2 = x(3x - 5)(2x + 3)$

40. $y(5y + 4)(2y - 3)$

41. $3a^2 - 10a + 8$

We will use the FOIL method.

1. There is no common factor (other than 1 or -1).

2. Factor the first term, $3a^2$. The factors are 3a and a. We have this possibility: $(3a \quad)(a \quad)$

3. Factor the last term, 8. The possibilities are $8\cdot1$, (-8)(-1), $4\cdot2$, and (-4)(-2) as well as $1\cdot8$, (-1)(-8), $2\cdot4$, and (-2)(-4). Look for factors such that the sum of the products is the middle term, -10a. Try some possibilities and check by multiplying. Trial and error leads us to the correct factorization, $(3a - 4)(a - 2)$.

42. $(4a - 1)(3a - 1)$

43. $35y^2 + 34y + 8$

We will use the grouping method.

1. There is no common factor (other than 1 or -1).

2. Multiply the leading coefficient, 35, and the constant, 8: $35(8) = 280$

3. Try to factor 280 so the sum of the factors is 34. We need only consider pairs of positive factors since 280 and 34 are both positive.

Pairs of Factors	Sums of Factors
280, 1	281
140, 2	142
70, 4	74
56, 5	61
40, 7	47
28, 10	38
20, 14	34

4. Split 34y using the results of step (3):

$34y = 20y + 14y$

5. Factor by grouping:

$$\begin{aligned}35y^2 + 34y + 8 &= 35y^2 + 20y + 14y + 8\\ &= 5y(7y + 4) + 2(7y + 4)\\ &= (7y + 4)(5y + 2)\end{aligned}$$

44. $(3a + 2)(3a + 4)$

45. $4t + 10t^2 - 6 = 10t^2 + 4t - 6$

We will use the FOIL method.

1. Factor out the common factor, 2:

$2(5t^2 + 2t - 3)$

2. Now we factor the trinomial $5t^2 + 2t - 3$.

Factor the first term, $5t^2$. The factors are 5t and t. We have this possibility: $(5t \quad)(t \quad)$

3. Factor the last term, -3. The possibilities are (1)(-3) and (-1)3 as well as (-3)(1) and 3(-1). Look for factors such that the sum of the products is the middle term, 2t. Trial and error leads us to the correct factorization:
$5t^2 + 2t - 3 = (5t - 3)(t + 1)$

We must include the common factor to get a factorization of the original trinomial:

$4t + 10t^2 - 6 = 2(5t - 3)(t + 1)$

46. $2(5x + 3)(3x - 1)$

47. $8x^2 - 16 - 28x = 8x^2 - 28x - 16$

We will use the grouping method.

1. Factor out the common factor, 4:

$4(2x^2 - 7x - 4)$

2. Now we factor the trinomial $2x^2 - 7x - 4$. Multiply the leading coefficient, 2, and the constant, -4: $2(-4) = -8$

3. Factor -8 so the sum of the factors is -7. We need only consider pairs of factors in which the negative factor has the larger absolute value, since their sum is negative.

Pairs of Factors	Sums of Factors
-4, 2	-2
-8, 1	-7

4. Split -7x using the results of step (3):

$-7x = -8x + x$

5. Factor by grouping:

$$\begin{aligned}2x^2 - 7x - 4 &= 2x^2 - 8x + x - 4\\ &= 2x(x - 4) + (x - 4)\\ &= (x - 4)(2x + 1)\end{aligned}$$

We must include the common factor to get a factorization of the original trinomial:

$8x^2 - 16 - 28x = 4(x - 4)(2x + 1)$

48. $6(3x - 4)(x + 1)$

49. $12x^3 - 31x^2 + 20x$

We will use the FOIL method.

1. Factor out the common factor, x:

$x(12x^2 - 31x + 20)$

2. We now factor the trinomial $12x^2 - 31x + 20$. Factor the first term, $12x^2$. The factors are 12x, x and 6x, 2x and 4x, 3x. We have these possibilities: $(12x \quad)(x \quad)$, $(6x \quad)(2x \quad)$, $(4x \quad)(3x \quad)$

3. Factor the last term, 20. The possibilities are $20\cdot1$, $(-20)(-1)$, $10\cdot2$, $(-10)(-2)$, $5\cdot4$, and $(-5)(-4)$ as well as $1\cdot20$, $(-1)(-20)$, $2\cdot10$, $(-2)(-10)$, $4\cdot5$, and $(-4)(-5)$. Look for factors such that the sum of the products is the middle term, -31x. Trial and error leads us to the correct factorization:
$12x^2 - 31x + 20 = (4x - 5)(3x - 4)$

We must include the common factor to get a factorization of the original trinomial:

$12x^3 - 31x^2 + 20x = x(4x - 5)(3x - 4)$

50. $x(5x + 2)(3x - 5)$

51. $14x^4 - 19x^3 - 3x^2$

We will use the grouping method.

1. Factor out the common factor, x^2:

$x^2(14x^2 - 19x - 3)$

2. Now we factor the trinomial $14x^2 - 19x - 3$. Multiply the leading coefficient, 14, and the constant, -3: $14(-3) = -42$

3. Factor -42 so the sum of the factors is -19. We need only consider pairs of factors in which the negative factor has the larger absolute value, since the sum is negative.

Pairs of Factors	Sums of Factors
-42, 1	-41
-21, 2	-19
-14, 3	-11
-7, 6	-1

4. Split $-19x$ using the results of step (3):

   $-19x = -21x + 2x$

5. Factor by grouping:

   $$14x^2 - 19x - 3 = 14x^2 - 21x + 2x - 3$$
   $$= 7x(2x - 3) + 2x - 3$$
   $$= (2x - 3)(7x + 1)$$

   We must include the common factor to get a factorization of the original trinomial:

   $$14x^4 - 19x^3 - 3x^2 = x^2(2x - 3)(7x + 1)$$

52. $2x^2(5x - 2)(7x - 4)$

53. $3a^2 - a - 4$

We will use the FOIL method.

1. There is no common factor (other than 1 or -1).

2. Factor the first term, $3a^2$. The factors are $3a$ and a. We have this possibility: $(3a \quad)(a \quad)$

3. Factor the last term, -4. The possibilities are $4(-1)$, $(-4)(1)$, and $2(-2)$ as well as $(-1)4$, $(1)(-4)$, and $(-2)2$. Look for factors such that the sum of the products is the middle term, $-a$. Trial and error leads us to the correct factorization: $(3a - 4)(a + 1)$

54. $(6a + 5)(a - 2)$

55. $9x^2 + 15x + 4$

We will use the grouping method.

1. There is no common factor (other than 1 or -1).

2. Multiply the leading coefficient and the constant: $9(4) = 36$

3. Factor 36 so the sum of the factors is 15. We need only consider pairs of positive factors since 36 and 15 are both positive.

Pairs of Factors	Sums of Factors
36, 1	37
18, 2	20
12, 3	15
9, 4	13
6, 6	12

4. Split $15x$ using the results of step (3):

   $15x = 12x + 3x$

5. Factor by grouping:

   $$9x^2 + 15x + 4 = 9x^2 + 12x + 3x + 4$$
   $$= 3x(3x + 4) + 3x + 4$$
   $$= (3x + 4)(3x + 1)$$

56. $(3y - 2)(2y + 1)$

57. $3 + 35z - 12z^2 = -12z^2 + 35z + 3$

We will use the FOIL method.

1. Factor out -1 so the leading coefficient is positive: $-(12z^2 - 35z - 3)$

2. Now we factor the trinomial $12z^2 - 35z - 3$. Factor the first term, $12z^2$. The factors are $12z$, z and $6z$, $2z$ and $4z$, $3z$. We have these possibilities: $(12z \quad)(z \quad)$, $(6z \quad)(2z \quad)$, $(4z \quad)(3z \quad)$

3. Factor the last term, -3. The possibilites are $3(-1)$ and $(-3)(1)$ as well as $(-1)3$ and $(1)(-3)$. Look for factors such that the sum of the products is the middle term, $-35z$. Trial and error leads us to the correct factorization: $(12z + 1)(z - 3)$

We must include the common factor to get a factorization of the original trinomial:

$3 + 35z - 12z^2 = -(12z + 1)(z - 3)$, or $(12z + 1)(-z + 3)$, or $(1 + 12z)(3 - z)$

58. $(2 - 3a)(4 + 3a)$

59. $-8t^2 - 8t + 30$

We will use the grouping method.

1. Factor out -2: $-2(4t^2 + 4t - 15)$

2. Now we factor the trinomial $4t^2 + 4t - 15$. Multiply the leading coefficient and the constant: $4(-15) = -60$

3. Factor -60 so the sum of the factors is 4. The desired factorization is $10(-6)$.

4. Split $4t$ using the results of step (3):

   $4t = 10t - 6t$

5. Factor by grouping:

   $$4t^2 + 4t - 15 = 4t^2 + 10t - 6t - 15$$
   $$= 2t(2t + 5) - 3(2t + 5)$$
   $$= (2t + 5)(2t - 3)$$

We must include the common factor to get a factorization of the original trinomial:

$$-8t^2 - 8t + 30 = -2(2t + 5)(2t - 3)$$

60. $-3(4a - 1)(3a - 1)$

61. $3x^3 - 5x^2 - 2x$

We will use the FOIL method.

1. Factor out the common factor, x: $x(3x^2 - 5x - 2)$

2. Now we factor the trinomial $3x^2 - 5x - 2$. Factor the first term, $3x^2$. The factors are $3x$ and x. We have this possibility: $(3x \quad)(x \quad)$

3. Factor the last term, -2. The possibilities are $2(-1)$ and $(-2)(1)$ as well as $(-1)2$ and $(1)(-2)$. Look for factors such that the sum of the products is the middle term, $-5x$. Trial and error leads us to the correct factorization: $(3x + 1)(x - 2)$

We must include the common factor to get a factorization of the original trinomial:

$$3x^3 - 5x^2 - 2x = x(3x + 1)(x - 2)$$

62. $y(6y - 5)(3y + 2)$

63. $24x^2 - 2 - 47x = 24x^2 - 47x - 2$

We will use the grouping method.

1. There is no common factor (other than 1 or -1).
2. Multiply the leading coefficient and the constant: $24(-2) = -48$
3. Factor -48 so the sum of the factors is -47. The desired factorization is $-48 \cdot 1$.
4. Split $-47x$ using the results of step (3):

   $-47x = -48x + x$

5. Factor by grouping:

$$\begin{aligned} 24x^2 - 47x - 2 &= 24x^2 - 48x + x - 2 \\ &= 24x(x - 2) + (x - 2) \\ &= (x - 2)(24x + 1) \end{aligned}$$

64. $(5y + 1)(3y - 10)$

65. $63x^3 + 111x^2 + 36x$

We will use the FOIL method.

1. Factor out the common factor, $3x$.

   $3x(21x^2 + 37x + 12)$

   Now we will factor the trinomial $21x^2 + 37x + 12$.

2. Factor the first term $21x^2$. The factors are $21x$, x and $7x$, $3x$. We have these possibilities: $(21x \quad)(x \quad)$ and $(7x \quad)(3x \quad)$.
3. Factor the last term, 12. The possibilites are $12 \cdot 1$, $(-12)(-1)$, $6 \cdot 2$, $(-6)(-2)$, $4 \cdot 3$, and $(-4)(-3)$ as well as $1 \cdot 12$, $(-1)(-12)$, $2 \cdot 6$, $(-2)(-6)$, $3 \cdot 4$, and $(-3)(-4)$. Look for factors such that the sum of the products is the middle term, $37x$. Trial and error leads us to the correct factorization: $(7x + 3)(3x + 4)$

   We must include the common factor to get a factorization of the original trinomial:

   $$63x^3 + 111x^2 + 36x = 3x(7x + 3)(3x + 4)$$

66. $5y(5y + 4)(2y + 3)$

67. $40x^4 + 16x^2 - 12$

We will use the grouping method.

1. Factor out the common factor, 4.

   $4(10x^4 + 4x^2 - 3)$

   Now we will factor the trinomial $10x^4 + 4x^2 - 3$. Substitute u for x^2 (and u^2 for x^4), and factor $10u^2 + 4u - 3$.

2. Multiply the leading coefficient and the constant: $10(-3) = -30$
3. Factor -30 so the sum of the factors is 4. This cannot be done. The trinomial $10u^2 + 4u - 3$ cannot be factored into binomials with integer coefficients. We have

   $$40x^4 + 16x^2 - 12 = 4(10x^4 + 4x^2 - 3)$$

68. $(6y^2 + 5)(4y^2 - 3)$

69. $12a^2 - 17ab + 6b^2$

We will use the FOIL method. (Our thinking is much the same as if we were factoring $12a^2 - 17a + 6$.)

1. There is no common factor (other than 1 or -1).
2. Factor the first term, $12a^2$. The factors are $12a$, a and $6a$, $2a$ and $4a$, $3a$. We have these possibilities: $(12a \quad)(a \quad)$ and $(6a \quad)(2a \quad)$ and $(4a \quad)(3a \quad)$.
3. Factor the last term, $6b^2$. The possibilities are $6b \cdot b$, $(-6b)(-b)$, $3b \cdot 2b$, and $(-3b)(-2b)$ as well as $b \cdot 6b$, $(-b)(-6b)$, $2b \cdot 3b$, and $(-2b)(-3b)$. Look for factors such that the sum of the products is the middle term, $-17ab$. Trial and error leads us to the correct factorization: $(4a - 3b)(3a - 2b)$

70. $(4p - 3q)(5p - 2q)$

71. $2x^2 + xy - 6y^2$

We will use the grouping method.

1. There is no common factor (other than 1 or -1).
2. Multiply the coefficients of the first and last terms: $2(-6) = -12$
3. Factor -12 so the sum of the factors is 1. The desired factorization is $4(-3)$.
4. Split xy using the results of step (3):

   $xy = 4xy - 3xy$

5. Factor by grouping:

$$\begin{aligned} 2x^2 + xy - 6y^2 &= 2x^2 + 4xy - 3xy - 6y^2 \\ &= 2x(x + 2y) - 3y(x + 2y) \\ &= (x + 2y)(2x - 3y) \end{aligned}$$

72. $(4m + 3n)(2m - 3n)$

73. $6x^2 - 29xy + 28y^2$

We will use the FOIL method.

1. There is no common factor (other than 1 or -1).
2. Factor the first term, $6x^2$. The factors are $6x$, x and $3x$, $2x$. We have these possibilities: $(6x \quad)(x \quad)$ and $(3x \quad)(2x \quad)$.
3. Factor the last term, $28y^2$. The possibilities are $28y \cdot y$, $(-28y)(-y)$, $14y \cdot 2y$, $(-14y)(-2y)$, $7y \cdot 4y$, and $(-7y)(-4y)$ as well as $y \cdot 28y$, $(-y)(-28y)$, $2y \cdot 14y$, $(-2y)(-14y)$, $4y \cdot 7y$, and $(-4y)(-7y)$. Look for factors such that the sum of the products is the middle term, $-29xy$. Trial and error leads us to the correct factorization: $(3x - 4y)(2x - 7y)$

74. $(2p + 3q)(5p - 4q)$

75. $9x^2 - 30xy + 25y^2$

We will use the grouping method.

1. There are no common factors (other than 1 or -1).
2. Multiply the coefficients of the first and last terms: $9(25) = 225$
3. Factor 225 so the sum of the factors is -30. The desired factorization is $-15(-15)$.
4. Split $-30xy$ using the results of step (3):

   $-30xy = -15xy - 15xy$
5. Factor by grouping:

   $$\begin{aligned} 9x^2 - 30xy + 25y^2 &= 9x^2 - 15xy - 15xy + 25y^2 \\ &= 3x(3x - 5y) - 5y(3x - 5y) \\ &= (3x - 5y)(3x - 5y) \end{aligned}$$

76. $(2p + 3q)(2p + 3q)$

77. $6x^6 + x^3 - 2$

We will use the FOIL method.

1. There is no common factor (other than 1 or -1). Substitute u for x^3 (and u^2 for x^6). We factor $6u^2 + u - 2$.
2. Factor the first term, $6u^2$. The factors are 6u, u and 3u, 2u. We have these possibilities: (6u   )(u   ) and (3u   )(2u   )
3. Factor the last term, -2. The possibilities are 2(-1) and (-2)(1) as well as (-1)2 and (1)(-2). Look for factors such that the sum of the products is the middle term, u. Trial and error leads us to the correct factorization: $(3u + 2)(2u - 1)$

   Replacing u by x^3 we have

   $6x^6 + x^3 - 2 = (3x^3 + 2)(2x^3 - 1)$.

78. $(2p^4 + 5)(p^4 + 3)$

79. $h(0) = -16(0)^2 + 80(0) + 224 = 224$ ft

$h(1) = -16(1)^2 + 80(1) + 224 = 288$ ft

$h(3) = -16(3)^2 + 80(3) + 224 = 320$ ft

$h(4) = -16(4)^2 + 80(4) + 224 = 288$ ft

$h(6) = -16(6)^2 + 80(6) + 224 = 128$ ft

80. 880 ft, 960 ft, 1024 ft, 624 ft, 240 ft

81.

82.

83. $p^2q^2 + 7pq + 12$

The factorization will be of the form (pq   )(pq   ). We look for factors of 12 whose sum is 7. The factors we need are 4 and 3. The factorization is $(pq + 4)(pq + 3)$.

84. $(2x^2y^3 + 5)(x^2y^3 - 4)$

85. $x^2 - \frac{4}{25} + \frac{3}{5}x = x^2 + \frac{3}{5}x - \frac{4}{25}$

We look for factors of $-\frac{4}{25}$ whose sum is $\frac{3}{5}$. The factors are $\frac{4}{5}$ and $-\frac{1}{5}$. The factorization is $\left(x + \frac{4}{5}\right)\left(x - \frac{1}{5}\right)$.

86. $\left(y + \frac{4}{7}\right)\left(y - \frac{2}{7}\right)$

87. $y^2 + 0.4y - 0.05$

We look for factors of -0.05 whose sum is 0.4. The factors are -0.1 and 0.5. The factorization is $(y - 0.1)(y + 0.5)$.

88. $(t + 0.9)(t - 0.3)$

89. $7a^2b^2 + 6 + 13ab = 7a^2b^2 + 13ab + 6$

We will use the grouping method. There is no common factor (other than 1 or -1). Multiply the leading coefficient and the constant: $7(6) = 42$. Factor 42 so the sum of the factors is 13. The desired factorization is $6 \cdot 7$. Split the middle term and factor by grouping.

$$\begin{aligned} 7a^2b^2 + 13ab + 6 &= 7a^2b^2 + 6ab + 7ab + 6 \\ &= ab(7ab + 6) + 7ab + 6 \\ &= (7ab + 6)(ab + 1) \end{aligned}$$

90. $(9xy - 4)(xy + 1)$

91. $216x + 78x^2 + 6x^3 = 6x^3 + 78x^2 + 216x$

Factor out the common factor, 6x.

$6x(x^2 + 13x + 36)$

Now factor $x^2 + 13x + 36$. Look for factors of 36 whose sum is 13. The factors are 9 and 4. Then $x^2 + 13x + 36 = (x + 9)(x + 4)$, and $6x^3 + 78x^2 + 216x = 6x(x + 9)(x + 4)$.

92. $(x^a + 8)(x^a - 3)$

93. $4x^{2a} - 4x^a - 3$

Substitute u for x^a (and u^2 for x^{2a}). We factor $4u^2 - 4u - 3$.

We will use the grouping method. There is no common factor (other than 1 or -1). Multiply the leading coefficient and the constant: $4(-3) = -12$. Factor -12 so the sum of the factors is -4. The desired factorization is $-6 \cdot 2$. Split the middle term and factor by grouping.

$$\begin{aligned} 4u^2 - 4u - 3 &= 4u^2 - 6u + 2u - 3 \\ &= 2u(2u - 3) + 2u - 3 \\ &= (2u - 3)(2u + 1) \end{aligned}$$

Replacing u by x^a we have

$4x^{2a} - 4x^a - 3 = (2x^a - 3)(2x^a + 1)$.

94. $(x + a)(x + b)$

95. $bdx^2 + adx + bcx + ac$

$= dx(bx + a) + c(bx + a)$ Factoring by grouping

$= (bx + a)(dx + c)$

96. $a^2(p^a + 2)(p^a - 1)$

97. $2ar^2 + 4asr + as^2 - asr$

$= 2ar^2 + 3asr + as^2$ Combining like terms

$= a(2r^2 + 3sr + s^2)$ Factoring out a

Factor $2r^2 + 3sr + s^2$ using the grouping method. Multiply the first and last coefficients: $2 \cdot 1 = 2$. Factor 2 so the sum of the factors is 3. The desired factorization is $2 \cdot 1$. Split the middle term and factor by grouping.

$$2r^2 + 3sr + s^2 = 2r^2 + 2sr + sr + s^2$$
$$= 2r(r + s) + s(r + s)$$
$$= (r + s)(2r + s)$$

We must include the common factor to get a factorization of the original expression. The factorization is $a(r + s)(2r + s)$.

98. $(x - 4)(x + 8)$

99. $6(x - 7)^2 + 13(x - 7) - 5$

Let $u = x - 7$, make the substitutions and factor using the grouping method. We have $6u^2 + 13u - 5$. Multiply the leading coefficient and the constant: $6(-5) = -30$. Factor -30 so the sum of the factors is 13. The desired factorization is $15(-2)$.

Split the middle term and factor by grouping.

$$6u^2 + 13u - 5 = 6u^2 + 15u - 2u - 5$$
$$= 3u(2u + 5) - (2u + 5)$$
$$= (2u + 5)(3u - 1)$$

Replace u by $x - 7$. The factorization is $[2(x - 7) + 5][3(x - 7) - 1]$, or $(2x - 9)(3x - 22)$.

100. 76, -76, 28, -28, 20, or -20

101. $x^2 + qx - 32$

All such q are the sums of the factors of -32.

Pairs of Factors	Sums of Factors
32, -1	31
-32, 1	-31
16, -2	14
-16, 2	-14
8, -4	4
-8, 4	-4

q can be 31, -31, 14, -14, 4, or -4.

102. $x - 365$

103. See the answer section in the text.

Exercise Set 5.5

1. $y^2 - 6y + 9 = (y - 3)^2$ Find the square terms and write the quantitites that were squared with a minus sign between them.

2. $(x - 4)^2$

3. $x^2 + 14x + 49 = (x + 7)^2$ Find the square terms and write the quantities that were squared with a plus sign between them.

4. $(x + 8)^2$

5. $x^2 + 1 + 2x = x^2 + 2x + 1$ Changing order

$= (x + 1)^2$ Factoring the trinomial square

6. $(x - 1)^2$

7. $2a^2 + 8a + 8 = 2(a^2 + 4a + 4)$ Factoring out the common factor

$= 2(a + 2)^2$ Factoring the trinomial square

8. $4(a - 2)^2$

9. $y^2 + 36 - 12y = y^2 - 12y + 36$ Changing order

$= (y - 6)^2$ Factoring the trinomial square

10. $(y + 6)^2$

11. $-18y^2 + y^3 + 81y = y^3 - 18y^2 + 81y$ Changing order

$= y(y^2 - 18y + 81)$ Factoring out the common factor

$= y(y - 9)^2$ Factoring the trinomial square

12. $a(a + 12)^2$

13. $12a^2 + 36a + 27 = 3(4a^2 + 12a + 9)$ Factoring out the common factor

$= 3(2a + 3)^2$ Factoring the trinomial square

14. $5(2y + 5)^2$

15. $2x^2 - 40x + 200 = 2(x^2 - 20x + 100)$

$= 2(x - 10)^2$

16. $2(4x + 3)^2$

17. $1 - 8d + 16d^2 = (1 - 4d)^2$ Find the square terms and write the quantities that were squared with a minus sign between them.

18. $(5y - 8)^2$

19. $y^4 + 8y^2 + 16 = (y^2 + 4)^2$ Note that $y^4 = (y^2)^2$.

20. $(a^2 - 5)^2$

21. $0.25x^2 + 0.30x + 0.09 = (0.5x + 0.3)^2$ Find the square terms and write the quantities that were squared with a plus sign between them.

22. $(0.2x - 0.7)^2$

23. $p^2 - 2pq + q^2 = (p - q)^2$

24. $(m + n)^2$

25. $a^2 + 4ab + 4b^2 = (a + 2b)^2$

26. $(7p - q)^2$

27. $25a^2 - 30ab + 9b^2 = (5a - 3b)^2$

28. $(7p - 6q)^2$

29. $x^4 + 2x^2y^2 + y^4 = (x^2 + y^2)^2$ Note that $x^4 = (x^2)^2$ and $y^4 = (y^2)^2$.

30. $(p^4 + q^4)^2$

31. $x^2 - 16 = x^2 - 4^2 = (x + 4)(x - 4)$

32. $(y + 5)(y - 5)$

33. $p^2 - 49 = p^2 - 7^2 = (p + 7)(p - 7)$

34. $(m + 8)(m - 8)$

35. $p^2q^2 - 25 = (pq)^2 - 5^2 = (pq + 5)(pq - 5)$

36. $(ab + 9)(ab - 9)$

37. $6x^2 - 6y^2 = 6(x^2 - y^2)$ Factoring out the common factor

$= 6(x + y)(x - y)$ Factoring the difference of squares

38. $8(x + y)(x - y)$

39. $4xy^4 - 4xz^4 = 4x(y^4 - z^4)$ Factoring out the common factor

$= 4x[(y^2)^2 - (z^2)^2]$

$= 4x(y^2 + z^2)(y^2 - z^2)$ Factoring the difference of squares

$= 4x(y^2 + z^2)(y + z)(y - z)$ Factoring $y^2 - z^2$

40. $25a(b^2 + z^2)(b + z)(b - z)$

41. $4a^3 - 49a = a(4a^2 - 49)$

$= a[(2a)^2 - 7^2]$

$= a(2a + 7)(2a - 7)$

42. $x(3x + 5)(3x - 5)$

43. $3x^8 - 3y^8 = 3(x^8 - y^8)$

$= 3[(x^4)^2 - (y^4)^2]$

$= 3(x^4 + y^4)(x^4 - y^4)$

$= 3(x^4 + y^4)[(x^2)^2 - (y^2)^2]$

$= 3(x^4 + y^4)(x^2 + y^2)(x^2 - y^2)$

$= 3(x^4 + y^4)(x^2 + y^2)(x + y)(x - y)$

44. $a^2(3a + b)(3a - b)$

45. $9a^4 - 25a^2b^4 = a^2(9a^2 - 25b^4)$

$= a^2[(3a)^2 - (5b^2)^2]$

$= a^2(3a + 5b^2)(3a - 5b^2)$

46. $x^2(4x^2 + 11y^2)(4x^2 - 11y^2)$

47. $\frac{1}{25} - x^2 = \left(\frac{1}{5}\right)^2 - x^2$

$= \left(\frac{1}{5} + x\right)\left(\frac{1}{5} - x\right)$

48. $\left(\frac{1}{4} + y\right)\left(\frac{1}{4} - y\right)$

49. $0.04x^2 - 0.09y^2 = (0.2x)^2 - (0.3y)^2$

$= (0.2x + 0.3y)(0.2x - 0.3y)$

50. $(0.1x + 0.2y)(0.1x - 0.2y)$

51. $m^3 - 7m^2 - 4m + 28 = m^2(m - 7) - 4(m - 7)$ Factoring by grouping

$= (m - 7)(m^2 - 4)$

$= (m - 7)(m + 2)(m - 2)$ Factoring the difference of squares

52. $(x + 8)(x + 1)(x - 1)$

53. $a^3 - ab^2 - 2a^2 + 2b^2 = a(a^2 - b^2) - 2(a^2 - b^2)$ Factoring by grouping

$= (a^2 - b^2)(a - 2)$

$= (a + b)(a - b)(a - 2)$ Factoring the difference of squares

54. $(p + 5)(p - 5)(q + 3)$

55. $(a + b)^2 - 100 = (a + b)^2 - 10^2$

$= (a + b + 10)(a + b - 10)$

56. $(p + 5)(p - 19)$

57. $a^2 + 2ab + b^2 - 9 = (a^2 + 2ab + b^2) - 9$ Grouping as a difference of squares

$= (a + b)^2 - 3^2$

$= (a + b + 3)(a + b - 3)$

58. $(x - y + 5)(x - y - 5)$

59. $r^2 - 2r + 1 - 4s^2 = (r^2 - 2r + 1) - 4s^2$ Grouping as a difference of squares

$= (r - 1)^2 - (2s)^2$

$= (r - 1 + 2s)(r - 1 - 2s)$

60. $(c + 2d + 3p)(c + 2d - 3p)$

61. $50a^2 - 2m^2 - 4mn - 2n^2$

$= 2(25a^2 - m^2 - 2mn - n^2)$ Factoring out the common factor

$= 2[25a^2 - (m^2 + 2mn + n^2)]$ Factoring out -1 and rewriting as a subtraction

$= 2[25a^2 - (m + n)^2]$ Factoring the trinomial square

$= 2[5a + (m + n)][5a - (m + n)]$ Factoring a difference of squares

$= 2(5a + m + n)(5a - m - n)$ Removing parentheses

62. $3(x + 2y + 1)(x - 2y - 1)$

63. $9 - a^2 + 2ab - b^2$

$= 9 - (a^2 - 2ab + b^2)$ Factoring out -1 and rewriting as a subtraction

$= 9 - (a - b)^2$ Factoring the trinomial square

$= [3 + (a - b)][3 - (a - b)]$ Factoring a difference of squares

$= (3 + a - b)(3 - a + b)$ Removing parentheses

64. $(4 + x - y)(4 - x + y)$

65. $x - y + z = 6,$ (1)

$2x + y - z = 0,$ (2)

$x + 2y + z = 3$ (3)

Add (1) and (2).

$x - y + z = 6$ (1)

$2x + y - z = 0$ (2)

$3x \quad = 6$ Adding

$x = 2$

Add (2) and (3).

$2x + y - z = 0$ (2)

$x + 2y + z = 3$ (3)

$3x + 3y \quad = 3$ (4)

Substitute 2 for x in (4).

$3(2) + 3y = 3$

$6 + 3y = 3$

$3y = -3$

$y = -1$

Substitute 2 for x and -1 for y in (1).

$2 - (-1) + z = 6$

$3 + z = 6$

$z = 3$

The solution is (2,-1,3).

66. $\left\{x \middle| x \leqslant -\frac{4}{7} \text{ or } x \geqslant 2\right\}$, or $\left[-\infty, -\frac{4}{7}\right] \cup [2, \infty)$

67. $|5 - 7x| \leqslant 9$

$-9 \leqslant 5 - 7x \leqslant 9$

$-14 \leqslant -7x \leqslant 4$

$2 \geqslant x \geqslant -\frac{4}{7}$

The solution is $\left\{x \middle| -\frac{4}{7} \leqslant x \leqslant 2\right\}$, or $\left[-\frac{4}{7}, 2\right]$.

68. $\left\{x \middle| x < \frac{14}{19}\right\}$, or $\left(-\infty, \frac{14}{19}\right)$

69.

70.

71. $-225x + x^3 = x^3 - 225x$

$= x(x^2 - 225)$

$= x(x + 15)(x - 15)$

72. $4y(y - 12)^2$

73. $3xy^2 - 150xy + 1875x = 3x(y^2 - 50y + 625)$

$= 3x(y - 25)^2$

74. $7(x + 25y)(x - 25y)$

75. $12x^2 - 72xy + 108y^2 = 12(x^2 - 6xy + 9y^2)$

$= 12(x - 3y)^2$

76. $-3\left(\frac{1}{2}p - \frac{2}{5}\right)^2$

77. $-\frac{8}{27}r^2 - \frac{10}{9}rs - \frac{1}{6}s^2 + \frac{2}{3}rs$

$= -\frac{8}{27}r^2 - \frac{4}{9}rs - \frac{1}{6}s^2$ Collecting like terms

$= -\frac{1}{54}(16r^2 + 24rs + 9s^2)$

$= -\frac{1}{54}(4r + 3s)^2$

78. $(x^2y^2 - 4)^2$

79. $-24ab + 16a^2 + 9b^2 = 16a^2 - 24ab + 9b^2$
$= (4a - 3b)^2$

80. $\left(\frac{1}{6}x^4 + \frac{2}{3}\right)^2$

81. $0.09x^2 + 0.48x + 0.64 = (0.3x + 0.8)^2$

82. $(x^a + y)(x^a - y)$

83. $x^{4a} - y^{2b} = (x^{2a})^2 - (y^b)^2 = (x^{2a} + y^b)(x^{2a} - y^b)$

84. $4(y^{2a} + 5)$

85. $25y^{2a} - (x^{2b} - 2x^b + 1)$
$= (5y^a)^2 - (x^b - 1)^2$
$= [5y^a + (x^b - 1)][5y^a - (x^b - 1)]$
$= (5y^a + x^b - 1)(5y^a - x^b + 1)$

86. $8(a - 7)^2$

87. $3(x + 1)^2 + 12(x + 1) + 12$
$= 3[(x + 1)^2 + 4(x + 1) + 4]$
$= 3[(x + 1) + 2]^2$
$= 3(x + 3)^2$

88. $5(c^{50} + 4d^{50})(c^{25} + 2d^{25})(c^{25} - 2d^{25})$

89. $9x^{2n} - 6x^n + 1 = (3x^n)^2 - 6x^n + 1$
$= (3x^n - 1)^2$

90. $c(c^w + 1)^2$

91. If $P(x) = x^2$, then
$P(a + h) - P(a) = (a + h)^2 - a^2$
$= [(a + h) + a][a + h) - a]$
$= (2a + h)h$, or $h(2a + h)$

92. a) $\pi h(R + r)(R - r)$

b) 3,014,400 cm^3

Exercise Set 5.6

1. $x^3 + 8 = x^3 + 2^3$
$= (x + 2)(x^2 - 2x + 4)$
$A^3 + B^3 = (A + B)(A^2 - AB + B^2)$

2. $(c + 3)(c^2 - 3c + 9)$

3. $y^3 - 64 = y^3 - 4^3$
$= (y - 4)(y^2 + 4y + 16)$
$A^3 - B^3 = (A - B)(A^2 + AB + B^2)$

4. $(z - 1)(z^2 + z + 1)$

5. $w^3 + 1 = w^3 + 1^3$
$= (w + 1)(w^2 - w + 1)$
$A^3 + B^3 = (A + B)(A^2 - AB + B^2)$

6. $(x + 5)(x^2 - 5x + 25)$

7. $8a^3 + 1 = (2a)^3 + 1^3$
$= (2a + 1)(4a^2 - 2a + 1)$
$A^3 + B^3 = (A + B)(A^2 - AB + B^2)$

8. $(3x + 1)(9x^2 - 3x + 1)$

9. $y^3 - 8 = y^3 - 2^3$
$= (y - 2)(y^2 + 2y + 4)$
$A^3 - B^3 = (A - B)(A^2 + AB + B^2)$

10. $(p - 3)(p^2 + 3p + 9)$

11. $8 - 27b^3 = 2^3 - (3b)^3$
$= (2 - 3b)(4 + 6b + 9b^2)$

12. $(4 - 5x)(16 + 20x + 25x^2)$

13. $64y^3 + 1 = (4y)^3 + 1^3$
$= (4y + 1)(16y^2 - 4y + 1)$

14. $(5x + 1)(25x^2 - 5x + 1)$

15. $8x^3 + 27 = (2x)^3 + 3^3$
$= (2x + 3)(4x^2 - 6x + 9)$

16. $(3y + 4)(9y^2 - 12y + 16)$

17. $a^3 - b^3 = (a - b)(a^2 + ab + b^2)$

18. $(x - y)(x^2 + xy + y^2)$

19. $a^3 + \frac{1}{8} = a^3 + \left(\frac{1}{2}\right)^3$
$= \left(a + \frac{1}{2}\right)\left(a^2 - \frac{1}{2}a + \frac{1}{4}\right)$

20. $\left(b + \frac{1}{3}\right)\left(b^2 - \frac{1}{3}b + \frac{1}{9}\right)$

21. $2y^3 - 128 = 2(y^3 - 64)$
$= 2(y^3 - 4^3)$
$= 2(y - 4)(y^2 + 4y + 16)$

22. $3(z - 1)(z^2 + z + 1)$

23. $24a^3 + 3 = 3(8a^3 + 1)$
$= 3[(2a)^3 + 1^3]$
$= 3(2a + 1)(4a^2 - 2a + 1)$

24. $2(3x + 1)(9x^2 - 3x + 1)$

25. $rs^3 + 64r = r(s^3 + 64)$
$= r(s^3 + 4^3)$
$= r(s + 4)(s^2 - 4s + 16)$

26. $a(b + 5)(b^2 - 5b + 25)$

27. $5x^3 - 40z^3 = 5(x^3 - 8z^3)$
$= 5[x^3 - (2z)^3]$
$= 5(x - 2z)(x^2 + 2xz + 4z^2)$

28. $2(y - 3z)(y^2 + 3yz + 9z^2)$

29. $x^3 + 0.001 = x^3 + (0.1)^3$
$= (x + 0.1)(x^2 - 0.1x + 0.01)$

30. $(y + 0.5)(y^2 - 0.5y + 0.25)$

31. $64x^6 - 8t^6 = 8(8x^6 - t^6)$
$= 8[(2x^2)^3 - (t^2)^3]$
$= 8(2x^2 - t^2)(4x^4 + 2x^2t^2 + t^4)$

32. $(5c^2 - 2d^2)(25c^4 + 10c^2d^2 + 4d^4)$

33. $2y^4 - 128y = 2y(y^3 - 64)$
$= 2y(y^3 - 4^3)$
$= 2y(y - 4)(y^2 + 4y + 16)$

34. $3z^2(z - 1)(z^2 + z + 1)$

35. $z^6 - 1 = (z^3)^2 - 1^2$ Writing as a difference of squares
$= (z^3 + 1)(z^3 - 1)$ Factoring a difference of squares
$= (z + 1)(z^2 - z + 1)(z - 1)(z^2 + z + 1)$ Factoring a sum and a difference of cubes

36. $(t^2 + 1)(t^4 - t^2 + 1)$

37. $t^6 + 64y^6 = (t^2)^3 + (4y^2)^3$
$= (t^2 + 4y^2)(t^4 - 4t^2y^2 + 16y^4)$

38. $(p + q)(p^2 - pq + q^2)(p - q)(p^2 + pq + q^2)$

39. Familiarize. Let w represent the width and ℓ represent the length of the rectangle. If the width is increased by 2 ft, the new width is w + 2. Also, recall the formulas for the perimeter and area of a rectangle:

$P = 2\ell + 2w$
$A = \ell w$

Translate.

Width is length less 7 ft.
$w = \ell - 7$

When the length is ℓ and the width is w + 2, the perimeter is 66 ft, so we have

$66 = 2\ell + 2(w + 2).$

We have a system of equations:

$w = \ell - 7,$
$66 = 2\ell + 2(w + 2)$

Carry out. Solving the system we get (19,12).

Check. The width, 12 ft, is 7 ft less than the length, 19 ft. When the width is increased by 2 ft it becomes 14 ft, and the perimeter of the rectangle is $2 \cdot 19 + 2 \cdot 14 = 38 + 28 = 66$ ft. The numbers check.

State. The area of the original rectangle is 19 ft·12 ft, or 228 ft².

40. $\{27, -27\}$

41. $|5x - 6| \leqslant 39$
$-39 \leqslant 5x - 6 \leqslant 39$
$-33 \leqslant 5x \leqslant 45$
$-\frac{33}{5} \leqslant x \leqslant 9$

The solution set is $\left\{x \middle| -\frac{33}{5} \leqslant x \leqslant 9\right\}$, or $\left[-\frac{33}{5}, 9\right]$.

42. $\left\{x \middle| x < -\frac{33}{5} \text{ or } x > 9\right\}$, or $\left(-\infty, -\frac{33}{5}\right] \cup (9, \infty)$

43.

44.

45. $x^{6a} + y^{3b} = (x^{2a})^3 + (y^b)^3$
$= (x^{2a} + y^b)(x^{4a} - x^{2a}y^b + y^{2b})$

46. $(ax - by)(a^2x^2 + axby + b^2y^2)$

47. $3x^{3a} + 24y^{3b} = 3(x^{3a} + 8y^{3b})$
$= 3[(x^a)^3 + (2y^b)^3]$
$= 3(x^a + 2y^b)(x^{2a} - 2x^ay^b + 4y^{2b})$

48. $\left(\frac{2}{3}x + \frac{1}{4}y\right)\left(\frac{4}{9}x^2 - \frac{1}{6}xy + \frac{1}{16}y^2\right)$

49. $\frac{1}{24}x^3y^3 + \frac{1}{3}z^3 = \frac{1}{3}\left(\frac{1}{8}x^3y^3 + z^3\right)$
$= \frac{1}{3}\left[\left(\frac{1}{2}xy\right)^3 + z^3\right]$
$= \frac{1}{3}\left(\frac{1}{2}xy + z\right)\left(\frac{1}{4}x^2y^2 - \frac{1}{2}xyz + z^2\right)$

50. $\frac{1}{2}\left(\frac{1}{2}x^a + y^{2a}z^{3b}\right)\left(\frac{1}{4}x^{2a} - \frac{1}{2}x^ay^{2a}z^{3b} + y^{4a}z^{6b}\right)$

51. $7x^3 + \frac{7}{8} = 7\left(x^3 + \frac{1}{8}\right)$
$= 7\left[x^3 + \left(\frac{1}{2}\right)^3\right]$
$= 7\left(x + \frac{1}{2}\right)\left(x^2 - \frac{1}{2}x + \frac{1}{4}\right)$

52. $(c - 2d)^2(c^2 - cd + d^2)^2$

53. $(x + y)^3 - x^3$
$= [(x + y) - x][(x + y)^2 + x(x + y) + x^2]$
$= (x + y - x)(x^2 + 2xy + y^2 + x^2 + xy + x^2)$
$= y(3x^2 + 3xy + y^2)$

54. $(x - 1)^3(x - 2)(x^2 - x + 1)$

55. $(a + 2)^3 - (a - 2)^3$
$= [(a + 2) - (a - 2)][(a + 2)^2 + (a + 2)(a - 2) + (a - 2)^2]$
$= (a + 2 - a + 2)(a^2 + 4a + 4 + a^2 - 4 + a^2 - 4a + 4)$
$= 4(3a^2 + 4)$

56. $(y - 8)(y - 1)(y^2 + y + 1)$

57. If $P(x) = x^3$, then
$P(a + h) - P(a) = (a + h)^3 - a^3$
$= [(a+h)-a][(a+h)^2+(a+h)(a)+a^2]$
$= (a+h-a)(a^2+2ah+h^2+a^2+ah+a^2)$
$= h(3a^2 + 3ah + h^2)$

58. $h(2a + h)(a^2 + ah + h^2)(3a^2 + 3ah + h^2)$

59.

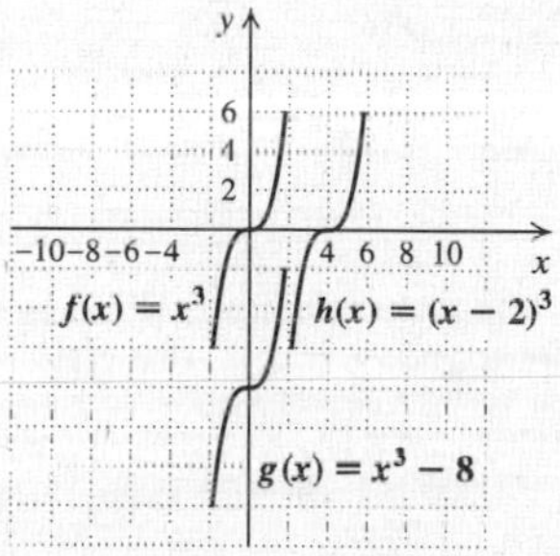

Exercise Set 5.7

1. $x^2 - 144$
$= x^2 - 12^2$ Difference of squares
$= (x + 12)(x - 12)$

2. $(y + 9)(y - 9)$

3. $2x^2 + 11x + 12$
$= (2x + 3)(x + 4)$ FOIL or grouping method

4. $(4a - 1)(2a + 5)$

5. $3x^4 - 12$
$= 3(x^4 - 4)$ Difference of squares
$= 3(x^2 + 2)(x^2 - 2)$

6. $2x(y + 5)(y - 5)$

7. $a^2 + 25 + 10a$
$= a^2 + 10a + 25$ Trinomial square
$= (a + 5)^2$

8. $(p + 8)^2$

9. $2x^2 - 10x - 132$
$= 2(x^2 - 5x - 66)$
$= 2(x - 11)(x + 6)$ Trial and error

10. $3(y - 12)(y + 7)$

11. $9x^2 - 25y^2$
$= (3x)^2 - (5y)^2$ Difference of squares
$= (3x + 5y)(3x - 5y)$

12. $(4a + 9b)(4a - 9b)$

13. $m^6 - 1$
$= (m^3)^2 - 1^2$ Difference of squares
$= (m^3 + 1)(m^3 - 1)$ Sum and difference of cubes
$= (m + 1)(m^2 - m + 1)(m - 1)(m^2 + m + 1)$

14. $(2t + 1)(4t^2 - 2t + 1)(2t - 1)(4t^2 + 2t + 1)$

15. $x^2 + 6x - y^2 + 9$
$= x^2 + 6x + 9 - y^2$
$= (x + 3)^2 - y^2$ Difference of squares
$= [(x + 3) + y][(x + 3) - y]$
$= (x + y + 3)(x - y + 3)$

16. $(t + p + 5)(t - p + 5)$

17. $250x^3 - 128y^3$
$= 2(125x^3 - 64y^3)$
$= 2[(5x)^3 - (4y)^3]$ Difference of cubes
$= 2(5x - 4y)(25x^2 + 20xy + 16y^2)$

18. $(3a - 7b)(9a^2 + 21ab + 49b^2)$

19. $8m^3 + m^6 - 20$
$= (m^3)^2 + 8m^3 - 20$
$= (m^3 - 2)(m^3 + 10)$ Trial and error

20. $(x + 6)(x - 6)(x + 1)(x - 1)$

21. $ac + cd - ab - bd$
$= c(a + d) - b(a + d)$ Factoring by grouping
$= (a + d)(c - b)$

22. $(x - y)(w + z)$

23. $4c^2 - 4cd + d^2$ Trinomial square
$= (2c - d)^2$

24. $(10b + a)(7b - a)$

25. $-7x^2 + 2x^3 + 4x - 14$
$= 2x^3 - 7x^2 + 4x - 14$
$= x^2(2x - 7) + 2(2x - 7)$ Factoring by grouping
$= (2x - 7)(x^2 + 2)$

26. $(m + 3)(3m^2 + 8)$

27. $2x^3 + 6x^2 - 8x - 24$
$= 2(x^3 + 3x^2 - 4x - 12)$
$= 2[x^2(x + 3) - 4(x + 3)]$ Factoring by grouping
$= 2(x + 3)(x^2 - 4)$ Difference of squares
$= 2(x + 3)(x + 2)(x - 2)$

28. $3(x + 2)(x + 3)(x - 3)$

29. $16x^3 + 54y^3$
$= 2(8x^3 + 27y^3)$
$= 2[(2x)^3 + (3y)^3]$ Sum of cubes
$= 2(2x + 3y)(4x^2 - 6xy + 9y^2)$

30. $2(5a + 3b)(25a^2 - 15ab + 9b^2)$

31. $36y^2 - 35 + 12y$
$= 36y^2 + 12y - 35$
$= (6y - 5)(6y + 7)$ FOIL or grouping method

32. $-2b(7a + 1)(2a - 1)$

33. $a^8 - b^8$ Difference of squares
$= (a^4 + b^4)(a^4 - b^4)$ Difference of squares
$= (a^4 + b^4)(a^2 + b^2)(a^2 - b^2)$ Difference of squares
$= (a^4 + b^4)(a^2 + b^2)(a + b)(a - b)$

34. $2(x^2 + 4)(x + 2)(x - 2)$

35. $a^3b - 16ab^3$
$= ab(a^2 - 16b^2)$ Difference of squares
$= ab(a + 4b)(a - 4b)$

36. $xy(x + 5y)(x - 5y)$

37. $a(b - 2) + c(b - 2)$
$= (b - 2)(a + c)$ Removing a common factor

38. $(x - 2)(2x + 13)$

39. $7a^4 - 14a^3 + 21a^2 - 7a$
$= 7a(a^3 - 2a^2 + 3a - 1)$ Removing a common factor

40. $(a + b)^2(a - b)$

41. $42ab + 27a^2b^2 + 8$
$= 27a^2b^2 + 42ab + 8$
$= (9ab + 2)(3ab + 4)$ FOIL or grouping method

42. $(4xy - 3)(5xy - 2)$

43. $8y^4 - 125y$
$= y(8y^3 - 125)$
$= y[(2y)^3 - 5^3]$ Difference of cubes
$= y(2y - 5)(4y^2 + 10y + 25)$

44. $p(4p - 1)(16p^2 + 4p + 1)$

45. $a^2 - b^2 - 6b - 9$
$= a^2 - (b^2 + 6b + 9)$ Factoring out -1
$= a^2 - (b + 3)^2$ Difference of squares
$= [a + (b + 3)][a - (b + 3)]$
$= (a + b + 3)(a - b - 3)$

46. $(m + n + 4)(m - n - 4)$

47. Familiarize. It is helpful to organize the information in a table. We let x represent the number of correct answers and y the number of wrong answers.

	Number of questions	Point value
Correct answers	x	$2x$
Wrong answers	y	$-\frac{1}{2}y$
Total	75	100

Translate. The total number of questions is 75. This gives us one statement.

$$x + y = 75$$

The total point value (score) of 100 gives us a second statement.

$$2x - \frac{1}{2}y = 100$$

We now have a system of equations.

$$x + y = 75 \quad \text{or} \quad x + y = 75$$
$$2x - \frac{1}{2}y = 100 \quad \text{or} \quad 4x - y = 200$$

Carry out. Solving the system we get (55,20).

Check. The total number of questions is 55 + 20, or 75. The total score is $2(55) - \frac{1}{2}(20)$, or 110 - 10, or 100. The numbers check.

State. There were 55 correct answers and 20 wrong answers.

48. $\frac{80}{7}$

49. ◆

50. ◆

51. $20x^2 - 4x - 39$
$= (10x + 13)(2x - 3)$ FOIL or grouping method

52. $(5y^2 - 12x)(6y^2 - 5x)$

53. $72a^2 + 284a - 160$
$= 4(18a^2 + 71a - 40)$
$= 4(9a + 40)(2a - 1)$ FOIL or grouping method

54. $z(3xy + 4z)(xy + 7z)$

55. $7x^3 - \frac{7}{8}$
$= 7\left(x^3 - \frac{1}{8}\right)$ Difference of cubes
$= 7\left(x - \frac{1}{2}\right)\left(x^2 + \frac{1}{2}x + \frac{1}{4}\right)$

56. $(11 + y^2)(2 + y)(2 - y)$

57. $[a^3 - (a + b)^3]$ Difference of cubes
$= [a - (a + b)][a^2 + a(a + b) + (a + b)^2]$
$= (a - a - b)(a^2 + a^2 + ab + a^2 + 2ab + b^2)$ Simplifying
$= -b(3a^2 + 3ab + b^2)$

58. $x(x - 2p)$

59. $x^4 - 50x^2 + 49$
$= (x^2)^2 - 50x^2 + 49$
$= (x^2 - 1)(x^2 - 49)$ Trial and error
$= (x + 1)(x - 1)(x + 7)(x - 7)$ Difference of squares

60. $y(y - 1)^2(y - 2)$

61. $27x^{6s} + 64y^{3t}$
$= (3x^{2s})^3 + (4y^t)^3$ Sum of cubes
$= (3x^{2s} + 4y^t)(9x^{4s} - 12x^{2s}y^t + 16y^{2t})$

62. $(x - 1)^3(x^2 + 1)(x + 1)$

63. $4x^2 + 4xy + y^2 - r^2 + 6rs - 9s^2$
$= (4x^2 + 4xy + y^2) - (r^2 - 6rs + 9s^2)$ Grouping
$= (2x + y)^2 - (r - 3s)^2$ Difference of squares
$= [(2x + y) + (r - 3s)][(2x + y) - (r - 3s)]$
$= (2x + y + r - 3s)(2x + y - r + 3s)$

64. $-(x - 1)^3(x)(x^2 - 3x + 3)$, or
$x(1 - x)^3(x^2 - 3x + 3)$

65. $c^{2w+1} - 2c^{w+1} + c$
$= c^{2w}\cdot c - 2c^w\cdot c + c$
$= c(c^{2w} - 2c^w + 1)$
$= c[(c^w)^2 - 2(c^w) + 1]$ Trinomial square
$= c(c^w - 1)^2$

66. $6(2x^a + 1)(2x^a - 1)$

67. $y^9 - y$
$= y(y^8 - 1)$ Difference of squares
$= y(y^4 + 1)(y^4 - 1)$ Difference of squares
$= y(y^4 + 1)(y^2 + 1)(y^2 - 1)$ Difference of squares
$= y(y^4 + 1)(y^2 + 1)(y + 1)(y - 1)$

68. $\left(1 - \frac{x^9}{10}\right)\left(1 + \frac{x^9}{10} + \frac{x^{18}}{100}\right)$

69. $3a^2 + 3b^2 - 3c^2 - 3d^2 + 6ab - 6cd$
$= 3(a^2 + b^2 - c^2 - d^2 + 2ab - 2cd)$
$= 3[(a^2 + 2ab + b^2) - (c^2 + 2cd + d^2)]$ Grouping
$= 3[(a + b)^2 - (c + d)^2]$ Difference of squares
$= 3(a + b + c + d)(a + b - c - d)$

70. $3x(x + 5)$

71. $(m - 1)^3 + (m + 1)^3$ Sum of cubes
$= [(m - 1) + (m + 1)][(m - 1)^2 - (m - 1)(m + 1) + (m + 1)^2]$
$= (m - 1 + m + 1)(m^2 - 2m + 1 - m^2 + 1 + m^2 + 2m + 1)$
$= 2m(m^2 + 3)$

72. $3(a - 7)^2$

73. $\left(x + \frac{2}{x}\right)^2 = 6$
$x^2 + 4 + \frac{4}{x^2} = 6$
$x^2 - 2 + \frac{4}{x^2} = 0$

Now consider $x^3 + \frac{8}{x^3}$.

$x^3 + \frac{8}{x^3}$ Sum of cubes
$= \left(x + \frac{2}{x}\right)\left(x^2 - 2 + \frac{4}{x^2}\right)$
$= \left(x + \frac{2}{x}\right)(0)$ Substituting 0 for $x^2 - 2 + \frac{4}{x^2}$
$= 0$

Exercise Set 5.8

1. $x^2 + 3x = 28$
$x^2 + 3x - 28 = 0$ Getting 0 on one side
$(x + 7)(x - 4) = 0$ Factoring
$x + 7 = 0$ or $x - 4 = 0$ Principle of zero products
$x = -7$ or $x = 4$

The solutions are -7 and 4. The solution set is $\{-7,4\}$.

2. $\{9,-5\}$

3. $y^2 + 16 = 8y$
$y^2 - 8y + 16 = 0$ Getting 0 on one side
$(y - 4)(y - 4) = 0$ Factoring
$y - 4 = 0$ or $y - 4 = 0$ Principle of zero products
$y = 4$ or $y = 4$ We have a repeated root.
The solution is 4. The solution set is {4}.

4. {1}

5. $x^2 - 12x + 36 = 0$
$(x - 6)(x - 6) = 0$ Factoring
$x - 6 = 0$ or $x - 6 = 0$ Principle of zero products
$x = 6$ or $x = 6$ We have a repeated root.
The solution is 6. The solution set is {6}.

6. {-8}

7. $9x + x^2 + 20 = 0$
$x^2 + 9x + 20 = 0$ Changing order
$(x + 5)(x + 4) = 0$ Factoring
$x + 5 = 0$ or $x + 4 = 0$ Principle of zero products
$x = -5$ or $x = -4$
The solutions are -5 and -4. The solution set is {-5,-4}.

8. {-5,-3}

9. $x^2 + 8x = 0$
$x(x + 8) = 0$ Factoring
$x = 0$ or $x + 8 = 0$ Principle of zero products
$x = 0$ or $x = -8$
The solutions are 0 and -8. The solution set is {0,-8}.

10. {0,-9}

11. $x^2 - 9 = 0$
$(x + 3)(x - 3) = 0$
$x + 3 = 0$ or $x - 3 = 0$
$x = -3$ or $x = 3$
The solutions are -3 and 3. The solution set is {-3,3}.

12. {-4,4}

13. $z^2 = 36$
$z^2 - 36 = 0$
$(z + 6)(z - 6) = 0$
$z + 6 = 0$ or $z - 6 = 0$
$z = -6$ or $z = 6$
The solutions are -6 and 6. The solution set is {-6,6}.

14. {-9,9}

15. $x^2 + 14x + 45 = 0$
$(x + 9)(x + 5) = 0$
$x + 9 = 0$ or $x + 5 = 0$
$x = -9$ or $x = -5$
The solutions are -9 and -5. The solution set is {-9,-5}.

16. {-8,-4}

17. $y^2 + 2y = 63$
$y^2 + 2y - 63 = 0$
$(y + 9)(y - 7) = 0$
$y + 9 = 0$ or $y - 7 = 0$
$y = -9$ or $y = 7$
The solutions are -9 and 7. The solution set is {-9,7}.

18. {-8,5}

19. $p^2 - 11p = -28$
$p^2 - 11p + 28 = 0$
$(p - 7)(p - 4) = 0$
$p - 7 = 0$ or $p - 4 = 0$
$p = 7$ or $p = 4$
The solutions are 7 and 4. The solution set is {7,4}.

20. {9,5}

21. $32 + 4x - x^2 = 0$
$0 = x^2 - 4x - 32$
$0 = (x - 8)(x + 4)$
$x - 8 = 0$ or $x + 4 = 0$
$x = 8$ or $x = -4$
The solutions are 8 and -4. The solution set is {8,-4}.

22. {-9,-3}

23. $3b^2 + 8b + 4 = 0$
$(3b + 2)(b + 2) = 0$
$3b + 2 = 0$ or $b + 2 = 0$
$3b = -2$ or $b = -2$
$b = -\frac{2}{3}$ or $b = -2$
The solutions are $-\frac{2}{3}$ and -2. The solution set is $\left\{-\frac{2}{3},-2\right\}$.

24. $\left\{-\frac{4}{3},-\frac{1}{3}\right\}$

25. $8y^2 - 10y + 3 = 0$
$(4y - 3)(2y - 1) = 0$
$4y - 3 = 0$ or $2y - 1 = 0$
$4y = 3$ or $2y = 1$
$y = \frac{3}{4}$ or $y = \frac{1}{2}$
The solutions are $\frac{3}{4}$ and $\frac{1}{2}$. The solution set is $\left\{\frac{3}{4},\frac{1}{2}\right\}$.

26. $\left\{-\frac{3}{4},-2\right\}$

27. $6z - z^2 = 0$
$0 = z^2 - 6z$
$0 = z(z - 6)$
$z = 0$ or $z - 6 = 0$
$z = 0$ or $z = 6$
The solutions are 0 and 6. The solution set is $\{0,6\}$.

28. $\{0,8\}$

29. $12z^2 + z = 6$
$12z^2 + z - 6 = 0$
$(4z + 3)(3z - 2) = 0$
$4z + 3 = 0$ or $3z - 2 = 0$
$4z = -3$ or $3z = 2$
$z = -\frac{3}{4}$ or $z = \frac{2}{3}$
The solutions are $-\frac{3}{4}$ and $\frac{2}{3}$. The solution set is $\left\{-\frac{3}{4},\frac{2}{3}\right\}$.

30. $\left\{-\frac{5}{6},2\right\}$

31. $5x^2 - 20 = 0$
$5(x^2 - 4) = 0$
$5(x + 2)(x - 2) = 0$
$x + 2 = 0$ or $x - 2 = 0$
$x = -2$ or $x = 2$
The solutions are -2 and 2. The solution set is $\{-2,2\}$.

32. $\{-3,3\}$

33. $2x^2 - 15x = -7$
$2x^2 - 15x + 7 = 0$
$(2x - 1)(x - 7) = 0$
$2x - 1 = 0$ or $x - 7 = 0$
$2x = 1$ or $x = 7$
$x = \frac{1}{2}$ or $x = 7$
The solutions are $\frac{1}{2}$ and 7. The solution set is $\left\{\frac{1}{2},7\right\}$.

34. $\{8,1\}$

35. $21r^2 + r - 10 = 0$
$(3r - 2)(7r + 5) = 0$
$3r - 2 = 0$ or $7r + 5 = 0$
$3r = 2$ or $7r = -5$
$r = \frac{2}{3}$ or $r = -\frac{5}{7}$
The solutions are $\frac{2}{3}$ and $-\frac{5}{7}$. The solution set is $\left\{\frac{2}{3},-\frac{5}{7}\right\}$.

36. $\left\{\frac{7}{4},-\frac{4}{3}\right\}$

37. $15y^2 = 3y$
$15y^2 - 3y = 0$
$3y(5y - 1) = 0$
$3y = 0$ or $5y - 1 = 0$
$y = 0$ or $5y = 1$
$y = 0$ or $y = \frac{1}{5}$
The solutions are 0 and $\frac{1}{5}$. The solution set is $\left\{0,\frac{1}{5}\right\}$.

38. $\left\{0,\frac{1}{2}\right\}$

39. $x^2 - \frac{1}{25} = 0$
$\left(x + \frac{1}{5}\right)\left(x - \frac{1}{5}\right) = 0$
$x + \frac{1}{5} = 0$ or $x - \frac{1}{5} = 0$
$x = -\frac{1}{5}$ or $x = \frac{1}{5}$
The solutions are $-\frac{1}{5}$ and $\frac{1}{5}$. The solution set is $\left\{-\frac{1}{5},\frac{1}{5}\right\}$.

40. $\left\{-\frac{1}{8},\frac{1}{8}\right\}$

41. $2x^3 - 2x^2 = 12x$
$2x^3 - 2x^2 - 12x = 0$
$2x(x^2 - x - 6) = 0$
$2x(x - 3)(x + 2) = 0$
$2x = 0$ or $x - 3 = 0$ or $x + 2 = 0$
$x = 0$ or $x = 3$ or $x = -2$
The solutions are 0, 3, and -2. The solution set is $\{0,3,-2\}$.

42. $\{0,5,2\}$

43. Familiarize. Let x represent the number.

Translate.

Square of number	plus	number	is	132.
x^2	+	x	=	132

Carry out. We solve the equation:

$$x^2 + x = 132$$
$$x^2 + x - 132 = 0$$
$$(x - 11)(x + 12) = 0$$
$$x - 11 = 0 \quad \text{or} \quad x + 12 = 0$$
$$x = 11 \quad \text{or} \quad x = -12$$

Check. The square of 11, which is 121, plus 11 is 132. The square of -12, which is 144, plus -12 is 132. Both numbers check.

State. The number is 11 or -12.

44. -13 or 12

45. Familiarize. We let w represent the width and w + 5 represent the length. We make a drawing and label it.

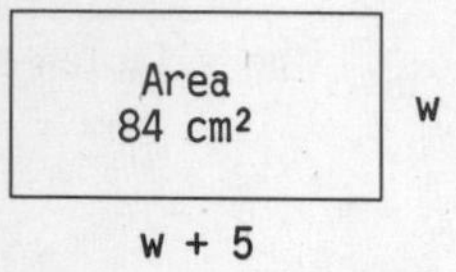

Recall that the formula for the area of a rectangle is A = length × width.

Translate.

Area	is	84 cm².
$w(w + 5)$	=	84

Carry out. We solve the equation:

$$w(w + 5) = 84$$
$$w^2 + 5w = 84$$
$$w^2 + 5w - 84 = 0$$
$$(w + 12)(w - 7) = 0$$
$$w + 12 = 0 \quad \text{or} \quad w - 7 = 0$$
$$w = -12 \quad \text{or} \quad w = 7$$

Check. The number -12 is not a solution, because width cannot be negative. If the width is 7 cm and the length is 5 cm more, or 12 cm, then the area is 12·7, or 84 cm². This is a solution.

State. The length is 12 cm, and the width is 7 cm.

46. Length: 12 cm, width: 8 cm

47. Familiarize. We make a drawing and label it. We let x represent the length of a side of the original square.

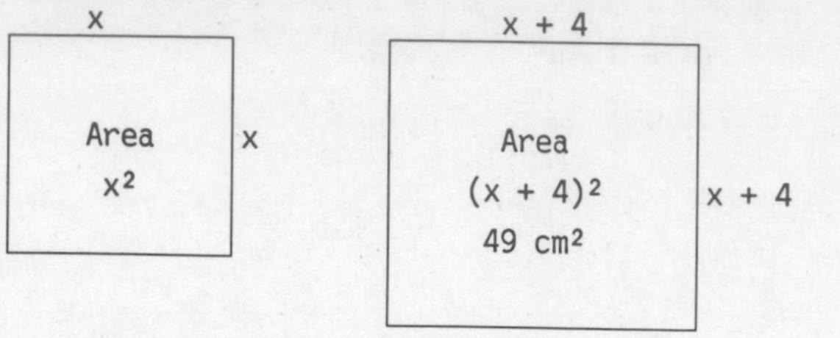

Translate.

Area of new square	is	49 cm².
$(x + 4)^2$	=	49

Carry out. We solve the equation:

$$(x + 4)^2 = 49$$
$$x^2 + 8x + 16 = 49$$
$$x^2 + 8x - 33 = 0$$
$$(x + 11)(x - 3) = 0$$
$$x + 11 = 0 \quad \text{or} \quad x - 3 = 0$$
$$x = -11 \quad \text{or} \quad x = 3$$

Check. We only check 3 since the length of a side cannot be negative. If we increase the length by 4, the new length is 3 + 4, or 7 cm. The new area is 7·7, or 49 cm². We have a solution.

State. The length of a side of the original square is 3 cm.

48. 6 m

49. Familiarize. We make a drawing and label it with both known and unknown information. We let x represent the width of the frame.

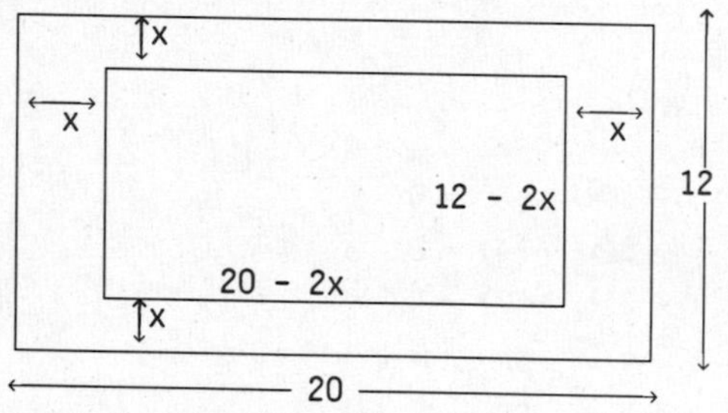

The length and width of the picture that shows are represented by 20 - 2x and 12 - 2x. The area of the picture that shows is 84 cm².

Translate. Using the formula for the area of a rectangle, $A = \ell \cdot w$, we have

$$84 = (20 - 2x)(12 - 2x).$$

Carry out. We solve the equation.

$$84 = 240 - 64x + 4x^2$$
$$84 = 4(60 - 16x + x^2).$$
$$21 = 60 - 16x + x^2$$
$$0 = x^2 - 16x + 39$$
$$0 = (x - 3)(x - 13)$$

$$x - 3 = 0 \quad \text{or} \quad x - 13 = 0$$
$$x = 3 \quad \text{or} \quad x = 13$$

Check. We see that 13 is not a solution because when $x = 13$, $20 - 2x = -6$ and $12 - 2x = -14$, and the length and width of the frame cannot be negative. We check 3. When $x = 3$, $20 - 2x = 14$ and $12 - 2x = 6$ and $14 \cdot 6 = 84$. The area is 84. The value checks.

State. The width of the frame is 3 cm.

50. 2 cm

51. Familiarize. We make a drawing and label it. We let x represent the length of the rectangle. Then $x - 5$ represents the width.

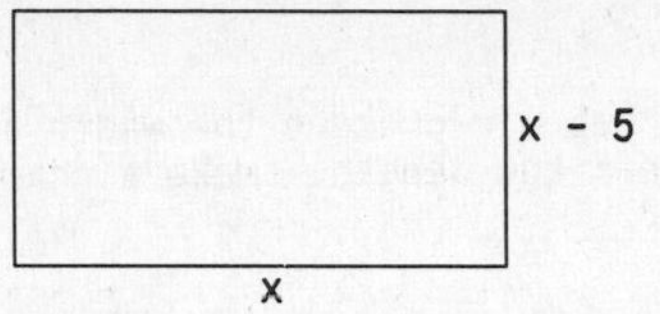

Translate. We use the formula for the area of a rectangle, $A = \ell \cdot w$.

$$24 = x(x - 5)$$

Carry out. We solve the equation.

$$24 = x^2 - 5x$$
$$0 = x^2 - 5x - 24$$
$$0 = (x - 8)(x + 3)$$

$$x - 8 = 0 \quad \text{or} \quad x + 3 = 0$$
$$x = 8 \quad \text{or} \quad x = -3$$

Check. We only check 8 since the length of a rectangle cannot be negative. If $x = 8$, then $x - 5 = 3$ and the area is $8 \cdot 3$, or 24. The value checks.

State. The length is 8 m, and the width is 3 m.

52. Length: 6 m, width: 2 m

53. Familiarize. We make a drawing and label it with both known and unknown information. We let x represent the width of the sidewalk.

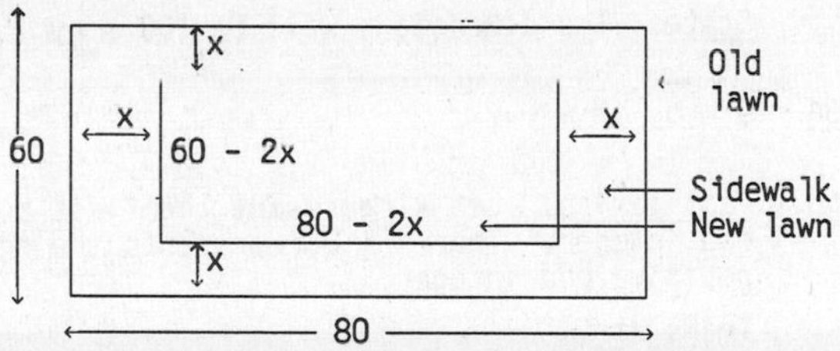

The area of the old lawn is $80 \cdot 60$, or 4800 ft^2. Then the area of the new lawn is $\frac{1}{6} \cdot 4800$, or 800 ft^2.

Translate. Rewording, we have

Area of new lawn is 800 ft^2.

$$(80 - 2x)(60 - 2x) = 800$$

Carry out. We solve the equation.

$$4800 - 280x + 4x^2 = 800$$
$$4x^2 - 280x + 4000 = 0$$
$$x^2 - 70x + 1000 = 0$$
$$(x - 20)(x - 50) = 0$$

$$x = 20 \quad \text{or} \quad x = 50$$

Check. If the sidewalk is 20 ft wide, the new lawn will have length $80 - 2 \cdot 20$, or 40 ft. The width will be $60 - 2 \cdot 20$, or 20 ft. The area of the new lawn will be $40 \cdot 20$, or 800 ft^2. This is $\frac{1}{6}$ of 4800 ft^2, the area of the old lawn, so this answer checks.

If the sidewalk is 50 ft wide, the new lawn will have length $80 - 2 \cdot 50$, or -20. Since length cannot be negative, this is not a solution.

State. The sidewalk is 20 ft wide.

54. 5 ft

55. Familiarize. We make a drawing.

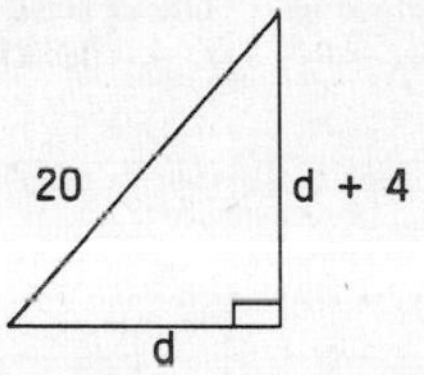

Since we have a right triangle, we can use the Pythagorean theorem:

$$a^2 + b^2 = c^2$$

Translate. Substituting, we have

$$d^2 + (d + 4)^2 = 20^2.$$

Carry out. We solve the equation.

$$d^2 + d^2 + 8d + 16 = 400$$
$$2d^2 + 8d - 384 = 0$$
$$2(d^2 + 4d - 192) = 0$$
$$2(d + 16)(d - 12) = 0$$

$$d + 16 = 0 \quad \text{or} \quad d - 12 = 0$$
$$d = -16 \quad \text{or} \quad d = 12$$

Check. We only check 12 since the length cannot be negative. If $d = 12$, then $d + 4 = 16$; $12^2 + 16^2 = 144 + 256 = 400 = 20^2$, so the answer checks.

State. The distance d is 12 ft, and the height of the tower is 16 ft.

56. 150 ft by 200 ft

57. Familiarize. Let h represent the height of the sail. Then $h + 9$ represents the base. Recall that the formula for the area of a triangle is $A = \frac{1}{2} \times \text{base} \times \text{height}$.

Translate.

Area is 56 m^2.

$$\frac{1}{2}(h + 9)h = 56$$

Carry out. We solve the equation:

$$\frac{1}{2}(h + 9)h = 56$$
$$(h + 9)h = 112 \quad \text{Multiplying by 2}$$
$$h^2 + 9h = 112$$
$$h^2 + 9h - 112 = 0$$
$$(h + 16)(h - 7) = 0$$
$$h + 16 = 0 \quad \text{or} \quad h - 7 = 0$$
$$h = -16 \quad \text{or} \quad h = 7$$

Check. We only check 7, since height cannot be negative. If the height is 7 m, the base is 7 + 9, or 16 m and the area is $\frac{1}{2} \cdot 16 \cdot 7$, or 56 m^2. We have a solution.

State. The height is 7 m, and the base is 16 m.

58. Height: 6 ft, base: 4 ft

59. Familiarize. We let x represent the first integer, x + 2 the second, and x + 4 the third.

Translate.

Square of the first	+	Square of the third	=	136
x^2	+	$(x + 4)^2$	=	136

Carry out. We solve the equation:

$$x^2 + (x + 4)^2 = 136$$
$$x^2 + x^2 + 8x + 16 = 136$$
$$2x^2 + 8x - 120 = 0$$
$$2(x^2 + 4x - 60) = 0$$
$$2(x + 10)(x - 6) = 0$$
$$x + 10 = 0 \quad \text{or} \quad x - 6 = 0$$
$$x = -10 \quad \text{or} \quad x = 6$$

Check. If x = -10, the consecutive even integers are -10, -8, and -6. The square of -10 is 100; the square of -6 is 36. The sum of 100 and 36 is 136. If x = 6, the consecutive even integers are 6, 8, and 10. The square of 6 is 36; the square of 10 is 100. The sum of 36 and 100 is 136. Both sets check.

State. The three consecutive even integers can be -10, -8, and -6 or 6, 8, and 10.

60. 16, 18, 20

61. Familiarize. We make a drawing. Let h = the height the ladder reaches on the wall. Then the length of the ladder is h + 1.

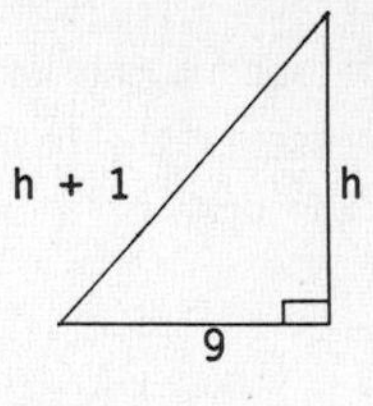

Translate. We use the Pythagorean theorem.

$$9^2 + h^2 = (h + 1)^2$$

Carry out. We solve the equation.

$$81 + h^2 = h^2 + 2h + 1$$
$$80 = 2h$$
$$40 = h$$

Check. If h = 40, then h + 1 = 41; $9^2 + 40^2 = 81 + 1600 = 1681 = 41^2$, so the answer checks.

State. The ladder is 41 ft long.

62. 24 ft

63. Familiarize. Let w represent the width and w + 25 represent the length. Make a drawing.

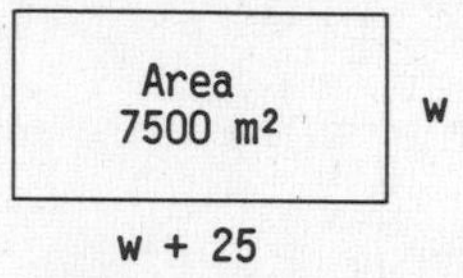

Recall that the formula for the area of a rectangle is A = length × width.

Translate.

Area	is	7500 m^2.
$w(w + 25)$	=	7500

Carry out. We solve the equation:

$$w(w + 25) = 7500$$
$$w^2 + 25w = 7500$$
$$w^2 + 25w - 7500 = 0$$
$$(w + 100)(w - 75) = 0$$
$$w + 100 = 0 \quad \text{or} \quad w - 75 = 0$$
$$w = -100 \quad \text{or} \quad w = 75$$

Check. The number -100 is not a solution because width cannot be negative. If the width is 75 m and the length is 25 m more, or 100 m, then the area will be 100·75, or 7500 m^2. This is a solution.

State. The dimensions will be 100 m by 75 m.

64. 9 m by 12 m

65. Familiarize. Let x represent the first positive odd integer. Then x + 2 represents the next positive odd integer.

Translate.

Square of first integer	plus	Square of second integer	is	202.
x^2	+	$(x + 2)^2$	=	202

Carry out. We solve the equation:

$$x^2 + (x + 2)^2 = 202$$
$$x^2 + x^2 + 4x + 4 = 202$$
$$2x^2 + 4x - 198 = 0$$
$$2(x^2 + 2x - 99) = 0$$
$$2(x + 11)(x - 9) = 0$$
$$x + 11 = 0 \quad \text{or} \quad x - 9 = 0$$
$$x = -11 \quad \text{or} \quad x = 9$$

Check. We only check 9 since the problem asks for consecutive positive odd integers. If $x = 9$, then $x + 2 = 11$, and 9 and 11 are consecutive positive odd integers. The sum of the squares of 9 and 11 is 81 + 121, or 202. The numbers check.

State. The integers are 9 and 11.

66. 13 and 15

67. Familiarize. We will use the given function $h(t) = -16t^2 + 80t + 224$.

Translate.

$$\text{Height} \quad \text{is} \quad 0 \text{ ft.}$$
$$-16t^2 + 80t + 224 = 0$$

Carry out. We solve the equation:

$$-16(t^2 - 5t - 14) = 0$$
$$-16(t - 7)(t + 2) = 0$$
$$t - 7 = 0 \quad \text{or} \quad t + 2 = 0$$
$$t = 7 \quad \text{or} \quad t = -2$$

Check. The number -2 is not a solution, since time cannot be negative. When $t = 7$, $h(7) = -16\cdot 7^2 + 80\cdot 7 + 224 = 0$. We have a solution.

State. The object reaches the ground after 7 sec.

68. 11 sec

69. Familiarize. Recall that in order for a firm to break even, its revenue must equal its production costs.

Translate.

$$\text{Revenue} \quad \text{equals} \quad \text{cost.}$$
$$\frac{4}{9}x^2 + x = \frac{1}{3}x^2 + x + 1$$

Carry out. We solve the equation:

$$\frac{4}{9}x^2 + x = \frac{1}{3}x^2 + x + 1$$
$$4x^2 + 9x = 3x^2 + 9x + 9 \quad \text{Multiplying by 9}$$
$$x^2 - 9 = 0$$
$$(x + 3)(x - 3) = 0$$
$$x + 3 = 0 \quad \text{or} \quad x - 3 = 0$$
$$x = -3 \quad \text{or} \quad x = 3$$

Check. The number -3 cannot be a solution, since the number of clocks cannot be negative. If 3 clocks are produced and sold the revenue is $R(3) = \frac{4}{9}\cdot 3^2 + 3 = 7$ hundred dollars and the cost is $C(3) = \frac{1}{3}\cdot 3^2 + 3 + 1 = 7$ hundred dollars. We have a solution.

State. The firm breaks even when 3 clocks are produced and sold.

70. 6

71. Familiarize. Let r represent the speed of the faster car and d represent its distance. Then r - 15 and 651 - d represent the speed and distance of the slower car, respectively. We organize the information in a table.

	Speed	Time	Distance
Faster car	r	7	d
Slower car	r - 15	7	651 - d

Translate. We use the formula $rt = d$. Each row of the table gives us an equation.

$$7r = d$$
$$7(r - 15) = 651 - d$$

Carry out. We use the substitution method substituting 7r for d in the second equation and solving for r.

$$7(r - 15) = 651 - 7r \quad \text{Substituting}$$
$$7r - 105 = 651 - 7r$$
$$14r = 756$$
$$r = 54$$

Check. If $r = 54$, then the speed of the faster car is 54 mph and the speed of the slower car is 54 - 15, or 39 mph. The distance the faster car travels is $54\cdot 7$, or 378 miles. The distance the slower car travels is $39\cdot 7$, or 273 miles. The total of the two distances is 378 + 273, or 651 miles. The values check.

State. The speed of the faster car is 54 mph. The speed of the slower car is 39 mph.

72. 36 color, 42 black and white

73. $2x - 14 + 9x > -8x + 16 + 10x$

$$11x - 14 > 2x + 16 \quad \text{Collecting like terms}$$
$$9x - 14 > 16 \quad \text{Adding } -2x$$
$$9x > 30 \quad \text{Adding 14}$$
$$x > \frac{10}{3} \quad \text{Multiplying by } \frac{1}{9}$$

The solution set is $\left\{x \middle| x > \frac{10}{3}\right\}$, or $\left(\frac{10}{3},\infty\right)$.

74. (2,-2,-4)

75. $(3x^2 - 7x - 20)(x - 5) = 0$

$$(3x + 5)(x - 4)(x - 5) = 0$$
$$3x + 5 = 0 \quad \text{or} \quad x - 4 = 0 \quad \text{or} \quad x - 5 = 0$$
$$3x = -5 \quad \text{or} \quad x = 4 \quad \text{or} \quad x = 5$$
$$x = -\frac{5}{3} \quad \text{or} \quad x = 4 \quad \text{or} \quad x = 5$$

The solutions are $-\frac{5}{3}$, 4, and 5. The solution set is $\left\{-\frac{5}{3},4,5\right\}$.

76. $\left\{-\frac{11}{8},\frac{2}{3},-\frac{1}{4}\right\}$

77. $(x + 1)^3 = (x - 1)^3 + 26$

$x^3 + 3x^2 + 3x + 1 = (x^3 - 3x^2 + 3x - 1) + 26$

$x^3 + 3x^2 + 3x + 1 = x^3 - 3x^2 + 3x + 25$

$6x^2 - 24 = 0$

$6(x^2 - 4) = 0$

$6(x + 2)(x - 2) = 0$

$x + 2 = 0$ or $x - 2 = 0$

$x = -2$ or $x = 2$

The solutions are -2 and 2. The solution set is $\{-2,2\}$.

78. $\{1\}$

79. $3x^3 + 6x^2 - 27x - 54 = 0$

$3(x^3 + 2x^2 - 9x - 18) = 0$

$3[x^2(x + 2) - 9(x + 2)] = 0$

$3(x + 2)(x^2 - 9) = 0$

$3(x + 2)(x + 3)(x - 3) = 0$

$x + 2 = 0$ or $x + 3 = 0$ or $x - 3 = 0$

$x = -2$ or $x = -3$ or $x = 3$

The solutions are -2, -3, and 3. The solution set is $\{-2,-3,3\}$.

80. $\{-3,-2,2\}$

81. Familiarize. Let x represent the length of a side of the square and y represent the length of a side of the triangle.

Translate.

Area of the square is 9 cm².

$x^2 = 9$

Perimeter of the square equals perimeter of the triangle.

$4x = 3y$

We have a system of equations:

$x^2 = 9$, (1)

$4x = 3y$ (2)

Carry out. Solve (1).

$x^2 = 9$

$x^2 - 9 = 0$

$(x + 3)(x - 3) = 0$

$x + 3 = 0$ or $x - 3 = 0$

$x = -3$ or $x = 3$

Length cannot be negative, so we only consider $x = 3$. Substitute 3 for x in (2).

$4x = 3y$

$4\cdot 3 = 3y$

$4 = y$

Check. The perimeter of a square with a side of 3 cm is $4\cdot 3$, or 12 cm, and the area is 3^2, or 9 cm². The perimeter of an equilateral triangle with a side of length 4 cm is $3\cdot 4$, or 12 cm. The numbers check.

State. The length of a side of the triangle is 4 cm.

82. 3 and 14

83. Familiarize. It is helpful to list the information in a table and to make a drawing. We let r represent the speed, in km/h, of the tugboat. Then r - 7 represents the speed of the freighter. Recall that Distance = Rate × Time.

Boat	Speed	Time	Distance
Tugboat	r	4	4r
Freighter	r - 7	4	4(r - 7)

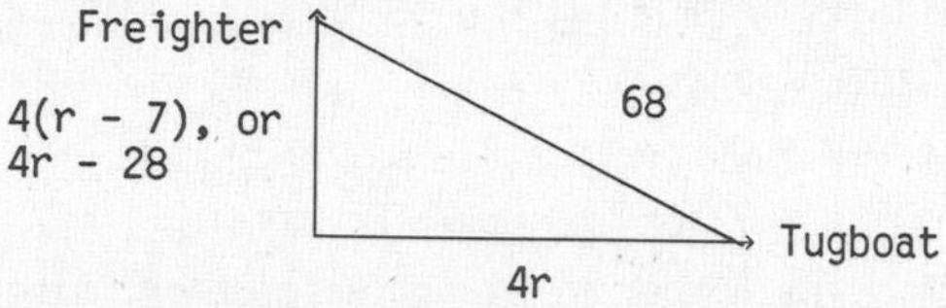

We will use the Pythagorean property:

$a^2 + b^2 = c^2$

Translate. Substituting, we have

$(4r - 28)^2 + (4r)^2 = 68^2.$

Carry out. We solve the equation.

$16r^2 - 224r + 784 + 16r^2 = 4624$

$32r^2 - 224r - 3840 = 0$

$r^2 - 7r - 120 = 0$

$(r + 8)(r - 15) = 0$

$r + 8 = 0$ or $r - 15 = 0$

$r = -8$ or $r = 15$

Check. We only check 15 since the speeds of the boats cannot be negative. If the speed of the tugboat is 15 km/h, then the speed of the freighter is 15 - 7, or 8 km/h, and the distances they travel are $4\cdot 15$, or 60 km, and $4\cdot 8$, or 32 km, respectively. $60^2 + 32^2 = 3600 + 1024 = 4624$, and $4624 = 68^2$, so the answer checks.

State. The speed of the tugboat is 15 km/h, and the speed of the freighter is 8 km/h.

84. 14 cm by 28 cm

85. $\{-1.91, 4.32\}$

86. No real-number solutions

87. $\{6.90\}$

88. $\{-3.33, 5.15\}$

89. $\{3.48\}$

90. $\{-1.77, 0.85, 2.72\}$

Exercise Set 6.1

1. $v(t) = \frac{4t^2 - 5t + 2}{t + 3}$

$v(0) = \frac{4\cdot0^2 - 5\cdot0 + 2}{0 + 3} = \frac{0 - 0 + 2}{0 + 3} = \frac{2}{3}$

$v(3) = \frac{4\cdot3^2 - 5\cdot3 + 2}{3 + 3} = \frac{36 - 15 + 2}{3 + 3} = \frac{23}{6}$

$v(7) = \frac{4\cdot7^2 - 5\cdot7 + 2}{7 + 3} = \frac{196 - 35 + 2}{7 + 3} = \frac{163}{10}$

2. 42, $-\frac{11}{7}$, 15

3. $r(y) = \frac{3y^3 - 2y}{y - 5}$

$r(0) = \frac{3\cdot0^3 - 2\cdot0}{0 - 5} = \frac{0 - 0}{0 - 5} = \frac{0}{-5} = 0$

$r(4) = \frac{3\cdot4^3 - 2\cdot4}{4 - 5} = \frac{192 - 8}{-1} = \frac{184}{-1} = -184$

$r(5) = \frac{3\cdot5^3 - 2\cdot5}{5 - 5} = \frac{375 - 10}{0}$

Since division by zero is not defined, r(5) does not exist.

4. 25.12, 27.475, does not exist

5. $g(x) = \frac{2x^3 - 9}{x^2 - 4x + 4}$

$g(0) = \frac{2\cdot0^3 - 9}{0^2 - 4\cdot0 + 4} = \frac{0 - 9}{0 - 0 + 4} = -\frac{9}{4}$

$g(2) = \frac{2\cdot2^3 - 9}{2^2 - 4\cdot2 + 4} = \frac{16 - 9}{4 - 8 + 4} = \frac{7}{0}$

Since division by zero is not defined, g(2) does not exist.

$g(-1) = \frac{2(-1)^3 - 9}{(-1)^2 - 4(-1) + 4} = \frac{-2 - 9}{1 + 4 + 4} = -\frac{11}{9}$

6. 0, $\frac{2}{5}$, $\frac{40}{7}$

7. $f(t) = \frac{9 - t^2}{5 - 6t + t^2}$

$f(-3) = \frac{9 - (-3)^2}{5 - 6(-3) + (-3)^2} = \frac{9 - 9}{5 + 18 + 9} = \frac{0}{32} = 0$

$f(0) = \frac{9 - 0^2}{5 - 6\cdot0 + 0^2} = \frac{9 - 0}{5 - 0 + 0} = \frac{9}{5}$

$f(1) = \frac{9 - 1^2}{5 - 6\cdot1 + 1^2} = \frac{9 - 1}{5 - 6 + 1} = \frac{8}{0}$

Since division by zero is not defined, f(1) does not exist.

8. 0, does not exist, $-\frac{4}{9}$

9. $f(x) = \frac{7}{3 - x}$

To avoid division by zero, we must determine which x-value causes 3 - x to be zero. We set

$3 - x = 0$

$3 = x.$ Adding x on both sides.

The domain of f = {x|x is a real number and $x \neq 3$}.

10. {t|t is a real number and $t \neq 5$}

11. $v(t) = \frac{t - 7}{t^2 - 4t}$

We set

$t^2 - 4t = 0$

$t(t - 4) = 0$

$t = 0$ or $t - 4 = 0$

$t = 0$ or $t = 4.$

The domain of v = {t|t is a real number and $t \neq 0$ and $t \neq 4$}.

12. {x|x is a real number and $x \neq 0$ and $x \neq 3$}

13. $s(x) = \frac{5}{x^2 - 4}$

We set

$x^2 - 4 = 0$

$(x + 2)(x - 2) = 0$

$x + 2 = 0$ or $x - 2 = 0$

$x = -2$ or $x = 2.$

The domain of s = {x|x is a real number and $x \neq -2$ and $x \neq 2$}.

14. {y|y is a real number and $y \neq -5$ and $y \neq 5$}

15. $F(x) = \frac{x^2 - 4}{x^2 - 8x + 12}$

We set

$x^2 - 8x + 12 = 0$

$(x - 6)(x - 2) = 0$

$x - 6 = 0$ or $x - 2 = 0$

$x = 6$ or $x = 2.$

The domain of F = {x|x is a real number and $x \neq 6$ and $x \neq 2$}.

16. {x|x is a real number and $x \neq 4$ and $x \neq 2$}

17. $\frac{3x}{3x} \cdot \frac{x + 1}{x + 3} = \frac{3x(x + 1)}{3x(x + 3)}$

18. $\frac{(4 - y^2)(-1)}{(6 - y)(-1)}$

19. $\frac{t - 3}{t + 2} \cdot \frac{t + 3}{t + 3} = \frac{(t - 3)(t + 3)}{(t + 2)(t + 3)}$

20. $\frac{(p - 4)(p + 5)}{(p - 5)(p + 5)}$

21. $\frac{x^2 - 3}{x - 6} \cdot \frac{x + 6}{x + 6} = \frac{(x^2 - 3)(x + 6)}{(x - 6)(x + 6)}$

22. $\frac{(t^2 - 9)(t + 2)}{(3 - t)(t + 2)}$

23. $\frac{9y^2}{15y} = \frac{3y\cdot3y}{3y\cdot5} = \frac{3y}{3y} \cdot \frac{3y}{5} = \frac{3y}{5}$

24. $\frac{x}{3}$

25. $\frac{8t^3}{4t^7} = \frac{4t^3\cdot2}{4t^3\cdot t^4} = \frac{4t^3}{4t^3} \cdot \frac{2}{t^4} = \frac{2}{t^4}$

26. $\frac{3}{2y^2}$

27. $\frac{2a - 6}{2} = \frac{2(a - 3)}{2\cdot 1} = \frac{2}{2} \cdot \frac{a - 3}{1} = a - 3$

28. $a - 2$

29. $\frac{6x - 9}{12} = \frac{3(2x - 3)}{3\cdot 4} = \frac{3}{3} \cdot \frac{2x - 3}{4} = \frac{2x - 3}{4}$

30. $\frac{5a - 6}{3}$

31. $\frac{4y - 12}{4y + 12} = \frac{4(y - 3)}{4(y + 3)} = \frac{4}{4} \cdot \frac{y - 3}{y + 3} = \frac{y - 3}{y + 3}$

32. $\frac{x + 2}{x - 2}$

33. $\frac{6x - 12}{5x - 10} = \frac{6(x - 2)}{5(x - 2)} = \frac{6}{5} \cdot \frac{x - 2}{x - 2} = \frac{6}{5}$

34. $\frac{7}{3}$

35. $\frac{12 - 6x}{5x - 10} = \frac{-6(-2 + x)}{5(x - 2)} = \frac{-6(x - 2)}{5(x - 2)} = \frac{-6}{5} \cdot \frac{x - 2}{x - 2}$

$= -\frac{6}{5}$

36. $-\frac{7}{3}$

37. $\frac{t^2 - 16}{t^2 - 8t + 16} = \frac{(t + 4)(t - 4)}{(t - 4)(t - 4)} = \frac{t + 4}{t - 4} \cdot \frac{t - 4}{t - 4}$

$= \frac{t + 4}{t - 4}$

38. $\frac{p - 5}{p + 5}$

39. $\frac{x^2 + 9x + 8}{x^2 - 3x - 4} = \frac{(x + 1)(x + 8)}{(x + 1)(x - 4)} = \frac{x + 1}{x + 1} \cdot \frac{x + 8}{x - 4}$

$= \frac{x + 8}{x - 4}$

40. $\frac{t - 9}{t + 4}$

41. $\frac{16 - t^2}{t^2 - 8t + 16} = \frac{16 - t^2}{16 - 8t + t^2} = \frac{(4 + t)(4 - t)}{(4 - t)(4 - t)}$

$= \frac{4 + t}{4 - t} \cdot \frac{4 - t}{4 - t} = \frac{4 + t}{4 - t}$

42. $\frac{5 - p}{5 + p}$

43. $\frac{5x^2}{3t^5} \cdot \frac{9t^8}{25x} = \frac{5x^2\cdot 9t^8}{3t^5\cdot 25x}$

$= \frac{5x\cdot x\cdot 3t^5\cdot 3t^3}{3t^5\cdot 5x\cdot 5}$

$= \frac{5x\cdot 3t^5}{5x\cdot 3t^5} \cdot \frac{x\cdot 3t^3}{5}$

$= \frac{3t^3x}{5}$

44. $\frac{7a^2}{6b^4}$

45. $\frac{3x - 6}{5x} \cdot \frac{x^3}{5x - 10} = \frac{(3x - 6)(x^3)}{5x(5x - 10)}$

$= \frac{3(x - 2)(x)(x^2)}{5\cdot x\cdot 5(x - 2)}$

$= \frac{(x - 2)(x)}{(x - 2)(x)} \cdot \frac{3\cdot x^2}{5\cdot 5}$

$= \frac{3x^2}{25}$

46. $\frac{3t^2}{4}$

47. $\frac{y^2 - 16}{2y + 6} \cdot \frac{y + 3}{y - 4} = \frac{(y^2 - 16)(y + 3)}{(2y + 6)(y - 4)}$

$= \frac{(y + 4)(y - 4)(y + 3)}{2(y + 3)(y - 4)}$

$= \frac{(y - 4)(y + 3)}{(y - 4)(y + 3)} \cdot \frac{y + 4}{2}$

$= \frac{y + 4}{2}$

48. $\frac{m + n}{4}$

49. $\frac{x^2 - 16}{x^2} \cdot \frac{x^2 - 4x}{x^2 - x - 12} = \frac{(x^2 - 16)(x^2 - 4x)}{x^2(x^2 - x - 12)}$

$= \frac{(x + 4)(x - 4)(x)(x - 4)}{x\cdot x(x - 4)(x + 3)}$

$= \frac{x(x - 4)}{x(x - 4)} \cdot \frac{(x + 4)(x - 4)}{x(x + 3)}$

$= \frac{(x + 4)(x - 4)}{x(x + 3)}$

50. $\frac{y(y + 5)}{y - 3}$

51. $\frac{6 - 2t}{t^2 + 4t + 4} \cdot \frac{t^3 + 2t^2}{t^8 - 9t^6}$

$= \frac{(6 - 2t)(t^3 + 2t^2)}{(t^2 + 4t + 4)(t^8 - 9t^6)}$

$= \frac{-2(-3 + t)(t^2)(t + 2)}{(t + 2)(t + 2)(t^6)(t + 3)(t - 3)}$

$= -\frac{t^2(t - 3)(t + 2)}{t^2(t - 3)(t + 2)} \cdot \frac{-2}{t^4(t + 2)(t + 3)}$

$= \frac{-2}{t^4(t + 2)(t + 3)}$

52. $\frac{x^2(x + 3)(x - 3)}{-4}$

53. $\frac{x^2 - 2x - 35}{2x^3 - 3x^2} \cdot \frac{4x^3 - 9x}{7x - 49}$

$= \frac{(x^2 - 2x - 35)(4x^3 - 9x)}{(2x^3 - 3x^2)(7x - 49)}$

$= \frac{(x - 7)(x + 5)(x)(2x + 3)(2x - 3)}{x\cdot x(2x - 3)(7)(x - 7)}$

$= \frac{(x - 7)(x)(2x - 3)}{(x - 7)(x)(2x - 3)} \cdot \frac{(x + 5)(2x + 3)}{x\cdot 7}$

$= \frac{(x + 5)(2x + 3)}{7x}$

54. $\frac{1}{y + 1}$

55. $\dfrac{c^3 + 8}{c^5 - 4c^3} \cdot \dfrac{c^6 - 4c^5 + 4c^4}{c^2 - 2c + 4}$

$= \dfrac{(c^3 + 8)(c^6 - 4c^5 + 4c^4)}{(c^5 - 4c^3)(c^2 - 2c + 4)}$

$= \dfrac{(c + 2)(c^2 - 2c + 4)(c^4)(c - 2)(c - 2)}{c^3(c + 2)(c - 2)(c^2 - 2c + 4)}$

$= \dfrac{c^3(c + 2)(c^2 - 2c + 4)(c - 2)}{c^3(c + 2)(c^2 - 2c + 4)(c - 2)} \cdot \dfrac{c(c - 2)}{1}$

$= c(c - 2)$

56. $\dfrac{x(x - 3)(x - 3)}{x + 3}$

57. $\dfrac{a^3 - b^3}{3a^2 + 9ab + 6b^2} \cdot \dfrac{a^2 + 2ab + b^2}{a^2 - b^2}$

$= \dfrac{(a^3 - b^3)(a^2 + 2ab + b^2)}{(3a^2 + 9ab + 6b^2)(a^2 - b^2)}$

$= \dfrac{(a - b)(a^2 + ab + b^2)(a + b)(a + b)}{3(a + b)(a + 2b)(a + b)(a - b)}$

$= \dfrac{(a - b)(a + b)(a + b)}{(a - b)(a + b)(a + b)} \cdot \dfrac{a^2 + ab + b^2}{3(a + 2b)}$

$= \dfrac{a^2 + ab + b^2}{3(a + 2b)}$

58. $\dfrac{x^2 - xy + y^2}{3(x + 3y)}$

59. $\dfrac{4x^2 - 9y^2}{8x^3 - 27y^3} \cdot \dfrac{4x^2 + 6xy + 9y^2}{4x^2 + 12xy + 9y^2}$

$= \dfrac{(4x^2 - 9y^2)(4x^2 + 6xy + 9y^2)}{(8x^3 - 27y^3)(4x^2 + 12xy + 9y^2)}$

$= \dfrac{(2x + 3y)(2x - 3y)(4x^2 + 6xy + 9y^2)\cdot 1}{(2x - 3y)(4x^2 + 6xy + 9y^2)(2x + 3y)(2x + 3y)}$

$= \dfrac{(2x + 3y)(2x - 3y)(4x^2 + 6xy + 9y^2)}{(2x + 3y)(2x - 3y)(4x^2 + 6xy + 9y^2)} \cdot \dfrac{1}{2x + 3y}$

$= \dfrac{1}{2x + 3y}$

60. $\dfrac{(x - y)(2x + 3y)}{2(x + y)(9x^2 + 6xy + 4y^2)}$

61. $\dfrac{16a^7}{3b^5} \div \dfrac{8a^3}{6b} = \dfrac{16a^7}{3b^5} \cdot \dfrac{6b}{8a^3}$ Multiplying by the reciprocal of the divisor

$= \dfrac{16a^7(6b)}{3b^5(8a^3)}$

$= \dfrac{8a^3\cdot 2a^4\cdot 3b\cdot 2}{3b\cdot b^4\cdot 8a^3}$

$= \dfrac{8a^3\cdot 3b}{8a^3\cdot 3b} = \dfrac{2a^4\cdot 2}{b^4}$

$= \dfrac{4a^4}{b^4}$

62. $6x^4y^7$

63. $\dfrac{3y + 15}{y^7} \div \dfrac{y + 5}{y^2} = \dfrac{3y + 15}{y^7} \cdot \dfrac{y^2}{y + 5}$

$= \dfrac{(3y + 15)(y^2)}{y^7(y + 5)}$

$= \dfrac{3(y + 5)(y^2)}{y^7(y + 5)}$

$= \dfrac{(y + 5)(y^2)}{(y + 5)(y^2)} \cdot \dfrac{3}{y^5}$

$= \dfrac{3}{y^5}$

64. $\dfrac{6}{x^5}$

65. $\dfrac{y^2 - 9}{y^2} \div \dfrac{y^5 + 3y^4}{y + 2} = \dfrac{y^2 - 9}{y^2} \cdot \dfrac{y + 2}{y^5 + 3y^4}$

$= \dfrac{(y^2 - 9)(y + 2)}{y^2(y^5 + 3y^4)}$

$= \dfrac{(y + 3)(y - 3)(y + 2)}{y^2(y^4)(y + 3)}$

$= \dfrac{y + 3}{y + 3} \cdot \dfrac{(y - 3)(y + 2)}{y^6}$

$= \dfrac{(y - 3)(y + 2)}{y^6}$

66. $\dfrac{(x + 2)(x + 4)}{x^7}$

67. $\dfrac{4a^2 - 1}{a^2 - 4} \div \dfrac{2a - 1}{a - 2} = \dfrac{4a^2 - 1}{a^2 - 4} \cdot \dfrac{a - 2}{2a - 1}$

$= \dfrac{(4a^2 - 1)(a - 2)}{(a^2 - 4)(2a - 1)}$

$= \dfrac{(2a + 1)(2a - 1)(a - 2)}{(a + 2)(a - 2)(2a - 1)}$

$= \dfrac{(2a - 1)(a - 2)}{(2a - 1)(a - 2)} \cdot \dfrac{2a + 1}{a + 2}$

$= \dfrac{2a + 1}{a + 2}$

68. $\dfrac{5x + 2}{x - 3}$

69. $\dfrac{x^2 - y^2}{4x + 4y} \div \dfrac{3y - 3x}{12x^2} = \dfrac{x^2 - y^2}{4x + 4y} \cdot \dfrac{12x^2}{3y - 3x}$

$= \dfrac{(x^2 - y^2)(12x^2)}{(4x + 4y)(3y - 3x)}$

$= \dfrac{(x + y)(x - y)(3)(4)(x^2)}{4(x + y)(-1)(3)(-y + x)}$

$= \dfrac{(x + y)(x - y)(3)(4)}{(x + y)(x - y)(3)(4)} \cdot \dfrac{x^2}{-1}$

$= -x^2$

70. $-\dfrac{1}{y^3}$

71. $\dfrac{x^2 - 16}{x^2 - 10x + 25} \div \dfrac{3x - 12}{x^2 - 3x - 10}$

$= \dfrac{x^2 - 16}{x^2 - 10x + 25} \cdot \dfrac{x^2 - 3x - 10}{3x - 12}$

$= \dfrac{(x^2 - 16)(x^2 - 3x - 10)}{(x^2 - 10x + 25)(3x - 12)}$

$= \dfrac{(x + 4)(x - 4)(x - 5)(x + 2)}{(x - 5)(x - 5)(3)(x - 4)}$

$= \dfrac{(x - 4)(x - 5)}{(x - 4)(x - 5)} \cdot \dfrac{(x + 4)(x + 2)}{(x - 5)(3)}$

$= \dfrac{(x + 4)(x + 2)}{3(x - 5)}$

72. $\dfrac{(y + 6)(y + 3)}{3(y - 4)}$

73. $\dfrac{y^3 + 3y}{y^2 - 9} \div \dfrac{y^2 + 5y - 14}{y^2 + 4y - 21}$

$= \dfrac{y^3 + 3y}{y^2 - 9} \cdot \dfrac{y^2 + 4y - 21}{y^2 + 5y - 14}$

$= \dfrac{(y^3 + 3y)(y^2 + 4y - 21)}{(y^2 - 9)(y^2 + 5y - 14)}$

$= \dfrac{y(y^2 + 3)(y + 7)(y - 3)}{(y + 3)(y - 3)(y + 7)(y - 2)}$

$= \dfrac{(y + 7)(y - 3)}{(y + 7)(y - 3)} \cdot \dfrac{y(y^2 + 3)}{(y + 3)(y - 2)}$

$= \dfrac{y(y^2 + 3)}{(y + 3)(y - 2)}$

74. $\dfrac{a(a^2 + 4)}{(a + 4)(a + 3)}$

75. $\dfrac{x^3 - 64}{x^3 + 64} \div \dfrac{x^2 - 16}{x^2 - 4x + 16}$

$= \dfrac{x^3 - 64}{x^3 + 64} \cdot \dfrac{x^2 - 4x + 16}{x^2 - 16}$

$= \dfrac{(x^3 - 64)(x^2 - 4x + 16)}{(x^3 + 64)(x^2 - 16)}$

$= \dfrac{(x - 4)(x^2 + 4x + 16)(x^2 - 4x + 16)}{(x + 4)(x^2 - 4x + 16)(x + 4)(x - 4)}$

$= \dfrac{(x - 4)(x^2 - 4x + 16)}{(x - 4)(x^2 - 4x + 16)} \cdot \dfrac{x^2 + 4x + 16}{(x + 4)(x + 4)}$

$= \dfrac{x^2 + 4x + 16}{(x + 4)(x + 4)}$, or $\dfrac{x^2 + 4x + 16}{(x + 4)^2}$

76. $\dfrac{4y^2 + 6y + 9}{(4y - 1)(2y + 3)}$

77. $\dfrac{8a^3 + b^3}{2a^2 + 3ab + b^2} \div \dfrac{8a^2 - 4ab + 2b^2}{4a^2 + 4ab + b^2}$

$= \dfrac{8a^3 + b^3}{2a^2 + 3ab + b^2} \cdot \dfrac{4a^2 + 4ab + b^2}{8a^2 - 4ab + 2b^2}$

$= \dfrac{(8a^3 + b^3)(4a^2 + 4ab + b^2)}{(2a^2 + 3ab + b^2)(8a^2 - 4ab + 2b^2)}$

$= \dfrac{(2a + b)(4a^2 - 2ab + b^2)(2a + b)(2a + b)}{(2a + b)(a + b)(2)(4a^2 - 2ab + b^2)}$

$= \dfrac{(2a + b)(4a^2 - 2ab + b^2)}{(2a + b)(4a^2 - 2ab + b^2)} \cdot \dfrac{(2a + b)(2a + b)}{(a + b)(2)}$

$= \dfrac{(2a + b)(2a + b)}{2(a + b)}$, or $\dfrac{(2a + b)^2}{2(a + b)}$

78. $\dfrac{2(2x - y)}{x}$

79. $3x + y = 13,$ (1)

$x = y + 1$ (2)

$3(y + 1) + y = 13$ Substituting $y + 1$ for x in (1)

$3y + 3 + y = 13$

$4y + 3 = 13$

$4y = 10$

$y = \dfrac{10}{4}$, or $\dfrac{5}{2}$

$x = \dfrac{5}{2} + 1$ Substituting $\dfrac{5}{2}$ for y in (2)

$x = \dfrac{7}{2}$

The solution is $\left(\dfrac{7}{2}, \dfrac{5}{2}\right)$.

80. 29

81. $\dfrac{2}{3}(3x - 4) = 8$

$\dfrac{3}{2} \cdot \dfrac{2}{3}(3x - 4) = \dfrac{3}{2} \cdot 8$ Multiplying by $\dfrac{3}{2}$

$3x - 4 = 12$

$3x = 16$

$x = \dfrac{16}{3}$

82. The full price is $8.

83.

84.

85. To find the domain of f we set

$x - 3 = 0$

$x = 3.$

The domain of $f = \{x \mid x \text{ is a real number and } x \neq 3\}$.

Simplify: $f(x) = \dfrac{x^2 - 9}{x - 3} = \dfrac{(x + 3)(x - 3)}{(x - 3) \cdot 1}$

$= \dfrac{x - 3}{x - 3} \cdot \dfrac{x + 3}{1} = x + 3$

Graph $f(x) = x + 3$ using the domain found above.

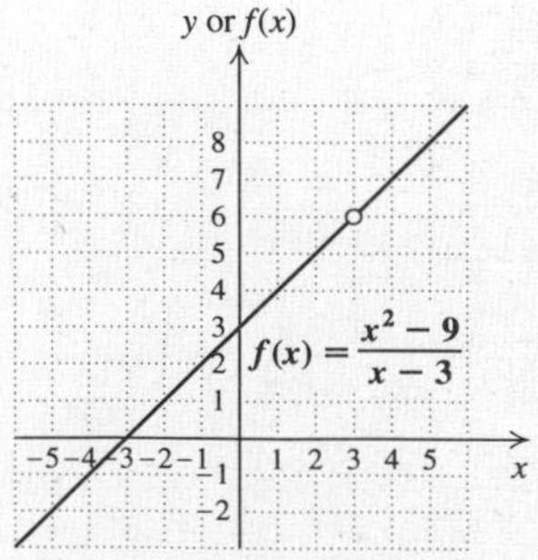

86. a) $\dfrac{2x + 2h + 3}{4x + 4h - 1}$

b) $\dfrac{2x + 3}{8x - 9}$

c) $\dfrac{x + 5}{4x - 1}$

87. $\left[\dfrac{r^2 - 4s^2}{r + 2s} \div (r + 2s)\right] \cdot \dfrac{2s}{r - 2s}$

$= \left[\dfrac{r^2 - 4s^2}{r + 2s} \cdot \dfrac{1}{r + 2s}\right] \cdot \dfrac{2s}{r - 2s}$

$= \dfrac{(r + 2s)(r - 2s)(2s)}{(r + 2s)(r + 2s)(r - 2s)}$

$= \dfrac{(r + 2s)(r - 2s)}{(r + 2s)(r - 2s)} \cdot \dfrac{2s}{r + 2s}$

$= \dfrac{2s}{r + 2s}$

88. $\dfrac{(d - 1)(d - 5)}{5d(d + 5)}$

89. $\dfrac{x(x + 1) - 2(x + 3)}{(x + 1)(x + 2)(x + 3)} = \dfrac{x^2 + x - 2x - 6}{(x + 1)(x + 2)(x + 3)}$

$= \dfrac{x^2 - x - 6}{(x + 1)(x + 2)(x + 3)}$

$= \dfrac{(x - 3)(x + 2)}{(x + 1)(x + 2)(x + 3)}$

$= \dfrac{x + 2}{x + 2} \cdot \dfrac{x - 3}{(x + 1)(x + 3)}$

$= \dfrac{x - 3}{(x + 1)(x + 3)}$

90. $-\dfrac{4}{x - 2}$, or $\dfrac{4}{2 - x}$

91. $\dfrac{m^2 - t^2}{m^2 + t^2 + m + t + 2mt} = \dfrac{m^2 - t^2}{(m^2 + 2mt + t^2) + (m + t)}$

$= \dfrac{(m + t)(m - t)}{(m + t)^2 + (m + t)}$

$= \dfrac{(m + t)(m - t)}{(m + t)[(m + t) + 1]}$

$= \dfrac{m + t}{m + t} \cdot \dfrac{m - t}{m + t + 1}$

$= \dfrac{m - t}{m + t + 1}$

92. $\dfrac{a^2 + 2}{a^2 - 3}$

93. $\dfrac{x^3 + x^2 - y^3 - y^2}{x^2 - 2xy + y^2}$

$= \dfrac{(x^3 - y^3) + (x^2 - y^2)}{x^2 - 2xy + y^2}$

$= \dfrac{(x - y)(x^2 + xy + y^2) + (x + y)(x - y)}{(x - y)^2}$

$= \dfrac{(x - y)[(x^2 + xy + y^2) + (x + y)]}{(x - y)(x - y)}$

$= \dfrac{x - y}{x - y} \cdot \dfrac{x^2 + xy + y^2 + x + y}{x - y}$

$= \dfrac{x^2 + xy + y^2 + x + y}{x - y}$

94. $\dfrac{(u^2 - uv + v^2)^2}{u - v}$

95. $\dfrac{x^5 - x^3 + x^2 - 1 - (x^3 - 1)(x + 1)^2}{(x^2 - 1)^2}$

$= \dfrac{x^5 - x^3 + (x^2 - 1) - [(x^3 - 1)(x + 1)^2]}{(x^2 - 1)^2}$

$= \dfrac{x^3(x^2-1) + (x^2-1) - [(x-1)(x^2+x+1)(x+1)(x+1)]}{(x^2 - 1)^2}$

$= \dfrac{x^3(x^2-1) + (x^2-1) - [(x^2-1)(x+1)(x^2+x+1)]}{(x^2 - 1)^2}$

$= \dfrac{(x^2 - 1)[x^3 + 1 - (x + 1)(x^2 + x + 1)]}{(x^2 - 1)^2}$

$= \dfrac{-2x^2 - 2x}{x^2 - 1}$

$= \dfrac{-2x(x + 1)}{(x + 1)(x - 1)}$

$= \dfrac{-2x\cancel{(x + 1)}}{\cancel{(x + 1)}(x - 1)}$

$= \dfrac{-2x}{x - 1}$

96. a) $\dfrac{16(x + 1)}{(x - 1)^2(x^2 + x + 1)}$

(Note that $x \neq -1$ is an additional restriction, since -1 is not in the domain of f.)

b) $\dfrac{x^2 + x + 1}{(x + 1)^3}$

(Note that $x \neq 1$ is an additional restriction, since 1 is not in the domain of g.)

c) $\dfrac{(x + 1)^3}{x^2 + x + 1}$

(Note that $x \neq -1$ and $x \neq 1$ are restrictions, since -1 is not in the domain of f and 1 is not in the domain of either f and g.)

97. $\{x \mid x$ is a real number and $x \neq 0$ and $x \neq -8.70\}$

98. $\{x \mid x$ is a real number and $x \neq -4.28$ and $x \neq -2.62\}$

99. $\{x \mid x$ is a real number and $x \neq 1.25$ and $x \neq 5.32\}$

100. $\{x \mid x$ is a real number and $x \neq -4.55$ and $x \neq 2.10$ and $x \neq 3.55\}$

Exercise Set 6.2

1. $\dfrac{3}{2a} + \dfrac{5}{2a} = \dfrac{8}{2a} = \dfrac{2\cdot 4}{2a} = \dfrac{\cancel{2}\cdot 4}{\cancel{2}\cdot a} = \dfrac{4}{a}$

2. $\dfrac{4}{y}$

3. $\dfrac{3}{4a^2b} - \dfrac{7}{4a^2b} = \dfrac{-4}{4a^2b} = \dfrac{-1\cdot 4}{4a^2b} = \dfrac{-1\cdot\cancel{4}}{\cancel{4}a^2b} = -\dfrac{1}{a^2b}$

4. $\dfrac{1}{3m^2n^2}$

5. $\frac{a - 3b}{a + b} + \frac{a + 5b}{a + b} = \frac{a - 3b + a + 5b}{a + b}$

$= \frac{2a + 2b}{a + b} = \frac{2(a + b)}{a + b}$

$= \frac{2\cancel{(a + b)}}{1\cdot\cancel{(a + b)}} = 2$

6. 2

7. $\frac{4y + 2}{y - 2} - \frac{y - 3}{y - 2} = \frac{4y + 2 - (y - 3)}{y - 2}$

$= \frac{4y + 2 - y + 3}{y - 2}$

$= \frac{3y + 5}{y - 2}$

8. $\frac{2t + 4}{t - 4}$

9. $\frac{3x - 4}{x^2 - 5x + 4} + \frac{3 - 2x}{x^2 - 5x + 4} = \frac{3x - 4 + 3 - 2x}{x^2 - 5x + 4}$

$= \frac{x - 1}{(x - 4)(x - 1)}$

$= \frac{1\cdot\cancel{(x - 1)}}{(x - 4)\cancel{(x - 1)}} = \frac{1}{x - 4}$

10. $\frac{1}{x - 7}$

11. $\frac{3a - 8}{a^2 - 9} - \frac{2a - 5}{a^2 - 9} = \frac{3a - 8 - (2a - 5)}{a^2 - 9}$

$= \frac{3a - 8 - 2a + 5}{a^2 - 9}$

$= \frac{a - 3}{(a + 3)(a - 3)}$

$= \frac{1\cdot\cancel{(a - 3)}}{(a + 3)\cancel{(a - 3)}} = \frac{1}{a + 3}$

12. $\frac{1}{a + 5}$

13. $\frac{a^2}{a - b} + \frac{b^2}{b - a} = \frac{a^2}{a - b} + \frac{-1}{-1}\cdot\frac{b^2}{b - a}$

$= \frac{a^2}{a - b} + \frac{-b^2}{a - b}$

$= \frac{a^2 - b^2}{a - b} = \frac{(a + b)(a - b)}{a - b}$

$= \frac{(a + b)\cancel{(a - b)}}{1\cdot\cancel{(a - b)}} = a + b$

14. $-(s + r)$

15. $\frac{3}{x} - \frac{8}{-x} = \frac{3}{x} + \frac{1}{-1}\cdot\frac{8}{-x} = \frac{3}{x} + \frac{8}{x} = \frac{11}{x}$

16. $\frac{7}{a}$

17. $\frac{2x - 9}{x^2 - 25} - \frac{4 - x}{25 - x^2} = \frac{2x - 9}{x^2 - 25} + \frac{1}{-1}\cdot\frac{4 - x}{25 - x^2}$

$= \frac{2x - 9}{x^2 - 25} + \frac{4 - x}{x^2 - 25}$

$= \frac{x - 5}{x^2 - 25} = \frac{x - 5}{(x + 5)(x - 5)}$

$= \frac{1\cdot\cancel{(x - 5)}}{(x + 5)\cancel{(x - 5)}} = \frac{1}{x + 5}$

18. $\frac{-2}{y^2 - 16}$, or $-\frac{2}{y^2 - 16}$

19. $\frac{t^2 + 3}{t^4 - 16} + \frac{7}{16 - t^4} = \frac{t^2 + 3}{t^4 - 16} + \frac{-1}{-1}\cdot\frac{7}{16 - t^4}$

$= \frac{t^2 + 3}{t^4 - 16} + \frac{-7}{t^4 - 16}$

$= \frac{t^2 - 4}{t^4 - 16}$

$= \frac{(t + 2)(t - 2)}{(t^2 + 4)(t + 2)(t - 2)}$

$= \frac{1\cdot\cancel{(t + 2)(t - 2)}}{(t^2 + 4)\cancel{(t + 2)(t - 2)}}$

$= \frac{1}{t^2 + 4}$

20. $\frac{1}{y^2 + 9}$

21. $\frac{m - 3n}{m^3 - n^3} - \frac{2n}{n^3 - m^3} = \frac{m - 3n}{m^3 - n^3} + \frac{1}{-1}\cdot\frac{2n}{n^3 - m^3}$

$= \frac{m - 3n}{m^3 - n^3} + \frac{2n}{m^3 - n^3}$

$= \frac{m - n}{m^3 - n^3}$

$= \frac{m - n}{(m - n)(m^2 + mn + n^2)}$

$= \frac{1\cdot\cancel{(m - n)}}{\cancel{(m - n)}(m^2 + mn + n^2)}$

$= \frac{1}{m^2 + mn + n^2}$

22. $\frac{1}{r^2 + rs + s^2}$

23. $\frac{y - 2}{y + 4} + \frac{y + 3}{y - 5}$

[LCD is $(y + 4)(y - 5)$.]

$= \frac{y - 2}{y + 4}\cdot\frac{y - 5}{y - 5} + \frac{y + 3}{y - 5}\cdot\frac{y + 4}{y + 4}$

$= \frac{(y^2 - 7y + 10) + (y^2 + 7y + 12)}{(y + 4)(y - 5)}$

$= \frac{2y^2 + 22}{(y + 4)(y - 5)}$

24. $\frac{2x^2 - x + 14}{(x + 3)(x - 4)}$

25. $2 + \frac{x - 3}{x + 1} = \frac{2}{1} + \frac{x - 3}{x + 1}$

[LCD is $x + 1$.]

$= \frac{2}{1}\cdot\frac{x + 1}{x + 1} + \frac{x - 3}{x + 1}$

$= \frac{(2x + 2) + (x - 3)}{x + 1}$

$= \frac{3x - 1}{x + 1}$

26. $\frac{4y - 13}{y - 5}$

27. $\dfrac{4xy}{x^2 - y^2} + \dfrac{x - y}{x + y}$

$= \dfrac{4xy}{(x + y)(x - y)} + \dfrac{x - y}{x + y}$

[LCD is $(x + y)(x - y)$.]

$= \dfrac{4xy}{(x + y)(x - y)} + \dfrac{x - y}{x + y} \cdot \dfrac{x - y}{x - y}$

$= \dfrac{4xy + x^2 - 2xy + y^2}{(x + y)(x - y)}$

$= \dfrac{x^2 + 2xy + y^2}{(x + y)(x - y)} = \dfrac{(x + y)(x + y)}{(x + y)(x - y)}$

$= \dfrac{\cancel{(x + y)}(x + y)}{\cancel{(x + y)}(x - y)} = \dfrac{x + y}{x - y}$

28. $\dfrac{a^2 + 7ab + b^2}{(a + b)(a - b)}$

29. $\dfrac{9x + 2}{3x^2 - 2x - 8} + \dfrac{7}{3x^2 + x - 4}$

$= \dfrac{9x + 2}{(3x + 4)(x - 2)} + \dfrac{7}{(3x + 4)(x - 1)}$

[LCD is $(3x + 4)(x - 2)(x - 1)$.]

$= \dfrac{9x + 2}{(3x + 4)(x - 2)} \cdot \dfrac{x - 1}{x - 1} + \dfrac{7}{(3x + 4)(x - 1)} \cdot \dfrac{x - 2}{x - 2}$

$= \dfrac{9x^2 - 7x - 2 + 7x - 14}{(3x + 4)(x - 2)(x - 1)}$

$= \dfrac{9x^2 - 16}{(3x + 4)(x - 2)(x - 1)} = \dfrac{(3x + 4)(3x - 4)}{(3x + 4)(x - 2)(x - 1)}$

$= \dfrac{\cancel{(3x + 4)}(3x - 4)}{\cancel{(3x + 4)}(x - 2)(x - 1)} = \dfrac{3x - 4}{(x - 2)(x - 1)}$

30. $\dfrac{3y^2 + 7y + 14}{(2y - 5)(y + 2)(y - 1)}$

31. $\dfrac{4}{x + 1} + \dfrac{x + 2}{x^2 - 1} + \dfrac{3}{x - 1}$

$= \dfrac{4}{x + 1} + \dfrac{x + 2}{(x + 1)(x - 1)} + \dfrac{3}{x - 1}$

[LCD is $(x + 1)(x - 1)$.]

$= \dfrac{4}{x + 1} \cdot \dfrac{x - 1}{x - 1} + \dfrac{x + 2}{(x + 1)(x - 1)} + \dfrac{3}{x - 1} \cdot \dfrac{x + 1}{x + 1}$

$= \dfrac{4x - 4 + x + 2 + 3x + 3}{(x + 1)(x - 1)}$

$= \dfrac{8x + 1}{(x + 1)(x - 1)}$

32. $\dfrac{4y + 17}{(y + 2)(y - 2)}$

33. $\dfrac{x - 1}{3x + 15} - \dfrac{x + 3}{5x + 25}$

$= \dfrac{x - 1}{3(x + 5)} - \dfrac{x + 3}{5(x + 5)}$

[LCD is $3 \cdot 5(x + 5)$, or $15(x + 5)$.]

$= \dfrac{x - 1}{3(x + 5)} \cdot \dfrac{5}{5} - \dfrac{x + 3}{5(x + 5)} \cdot \dfrac{3}{3}$

$= \dfrac{5x - 5 - (3x + 9)}{15(x + 5)}$

$= \dfrac{5x - 5 - 3x - 9}{15(x + 5)}$

$= \dfrac{2x - 14}{15(x + 5)}$

34. $\dfrac{y - 34}{20(y + 2)}$

35. $\dfrac{5ab}{a^2 - b^2} - \dfrac{a - b}{a + b}$

$= \dfrac{5ab}{(a + b)(a - b)} - \dfrac{a - b}{a + b}$

[LCD is $(a + b)(a - b)$.]

$= \dfrac{5ab}{(a + b)(a - b)} - \dfrac{a - b}{a + b} \cdot \dfrac{a - b}{a - b}$

$= \dfrac{5ab - (a^2 - 2ab + b^2)}{(a + b)(a - b)}$

$= \dfrac{5ab - a^2 + 2ab - b^2}{(a + b)(a - b)}$

$= \dfrac{-a^2 + 7ab - b^2}{(a + b)(a - b)}$

36. $\dfrac{-x^2 + 4xy - y^2}{(x + y)(x - y)}$

37. $\dfrac{x}{x^2 + 9x + 20} - \dfrac{4}{x^2 + 7x + 12}$

$= \dfrac{x}{(x + 5)(x + 4)} - \dfrac{4}{(x + 3)(x + 4)}$

[LCD is $(x + 5)(x + 4)(x + 3)$.]

$= \dfrac{x}{(x + 5)(x + 4)} \cdot \dfrac{x + 3}{x + 3} - \dfrac{4}{(x + 3)(x + 4)} \cdot \dfrac{x + 5}{x + 5}$

$= \dfrac{x^2 + 3x - (4x + 20)}{(x + 5)(x + 4)(x + 3)}$

$= \dfrac{x^2 + 3x - 4x - 20}{(x + 5)(x + 4)(x + 3)}$

$= \dfrac{x^2 - x - 20}{(x + 5)(x + 4)(x + 3)}$

$= \dfrac{(x - 5)(x + 4)}{(x + 5)(x + 4)(x + 3)}$

$= \dfrac{(x - 5)\cancel{(x + 4)}}{(x + 5)\cancel{(x + 4)}(x + 3)}$

$= \dfrac{x - 5}{(x + 5)(x + 3)}$

38. $\dfrac{x - 6}{(x + 6)(x + 4)}$

39. $\dfrac{3y}{y^2 - 7y + 10} - \dfrac{2y}{y^2 - 8y + 15}$

$= \dfrac{3y}{(y - 5)(y - 2)} - \dfrac{2y}{(y - 5)(y - 3)}$

[LCD is $(y - 5)(y - 2)(y - 3)$.]

$= \dfrac{3y}{(y - 5)(y - 2)} \cdot \dfrac{y - 3}{y - 3} - \dfrac{2y}{(y - 5)(y - 3)} \cdot \dfrac{y - 2}{y - 2}$

$= \dfrac{3y^2 - 9y - (2y^2 - 4y)}{(y - 5)(y - 2)(y - 3)}$

$= \dfrac{3y^2 - 9y - 2y^2 + 4y}{(y - 5)(y - 2)(y - 3)}$

$= \dfrac{y^2 - 5y}{(y - 5)(y - 2)(y - 3)} = \dfrac{y(y - 5)}{(y - 5)(y - 2)(y - 3)}$

$= \dfrac{y\cancel{(y - 5)}}{\cancel{(y - 5)}(y - 2)(y - 3)} = \dfrac{y}{(y - 2)(y - 3)}$

40. $\dfrac{2x^2 + 21x}{(x - 4)(x - 2)(x + 3)}$

41. $\dfrac{y}{y^2 - y - 20} + \dfrac{2}{y + 4}$

$= \dfrac{y}{(y - 5)(y + 4)} + \dfrac{2}{y + 4}$

[LCD is (y - 5)(y + 4).]

$= \dfrac{y}{(y - 5)(y + 4)} + \dfrac{2}{y + 4} \cdot \dfrac{y - 5}{y - 5}$

$= \dfrac{y + 2y - 10}{(y - 5)(y + 4)}$

$= \dfrac{3y - 10}{(y - 5)(y + 4)}$

42. $\dfrac{3}{t - 3}$

43. $\dfrac{3y + 2}{y^2 + 5y - 24} + \dfrac{7}{y^2 + 4y - 32}$

$= \dfrac{3y + 2}{(y + 8)(y - 3)} + \dfrac{7}{(y + 8)(y - 4)}$

[LCD is (y + 8)(y - 3)(y - 4).]

$= \dfrac{3y + 2}{(y + 8)(y - 3)} \cdot \dfrac{y - 4}{y - 4} + \dfrac{7}{(y + 8)(y - 4)} \cdot \dfrac{y - 3}{y - 3}$

$= \dfrac{3y^2 - 10y - 8 + 7y - 21}{(y + 8)(y - 3)(y - 4)}$

$= \dfrac{3y^2 - 3y - 29}{(y + 8)(y - 3)(y - 4)}$

44. $\dfrac{5x^2 - 11x - 6}{(x - 5)(x - 2)(x - 3)}$

45. $\dfrac{3x - 1}{x^2 + 2x - 3} - \dfrac{x + 4}{x^2 - 9}$

$= \dfrac{3x - 1}{(x + 3)(x - 1)} - \dfrac{x + 4}{(x + 3)(x - 3)}$

[LCD is (x + 3)(x - 1)(x - 3).]

$= \dfrac{3x - 1}{(x + 3)(x - 1)} \cdot \dfrac{x - 3}{x - 3} - \dfrac{x + 4}{(x + 3)(x - 3)} \cdot \dfrac{x - 1}{x - 1}$

$= \dfrac{3x^2 - 10x + 3 - (x^2 + 3x - 4)}{(x + 3)(x - 1)(x - 3)}$

$= \dfrac{3x^2 - 10x + 3 - x^2 - 3x + 4}{(x + 3)(x - 1)(x - 3)}$

$= \dfrac{2x^2 - 13x + 7}{(x + 3)(x - 1)(x - 3)}$

46. $\dfrac{2p^2 + 7p + 10}{(p + 6)(p - 4)(p + 4)}$

47. $\dfrac{2}{a^2 - 5a + 4} + \dfrac{-2}{a^2 - 4}$

$= \dfrac{2}{(a - 4)(a - 1)} + \dfrac{-2}{(a + 2)(a - 2)}$

[LCD is (a - 4)(a - 1)(a + 2)(a - 2).]

$= \dfrac{2}{(a - 4)(a - 1)} \cdot \dfrac{(a + 2)(a - 2)}{(a + 2)(a - 2)} +$

$\dfrac{-2}{(a + 2)(a - 2)} \cdot \dfrac{(a - 4)(a - 1)}{(a - 4)(a - 1)}$

$= \dfrac{2(a^2 - 4) - 2(a^2 - 5a + 4)}{(a - 4)(a - 1)(a + 2)(a - 2)}$

$= \dfrac{2a^2 - 8 - 2a^2 + 10a - 8}{(a - 4)(a - 1)(a + 2)(a - 2)}$

$= \dfrac{10a - 16}{(a - 4)(a - 1)(a + 2)(a - 2)}$, or

$\dfrac{10a - 16}{(a^2 - 5a + 4)(a^2 - 4)}$

48. $\dfrac{21a - 45}{(a - 6)(a - 1)(a + 3)(a - 3)}$, or

$\dfrac{21a - 45}{(a^2 - 7a + 6)(a^2 - 9)}$

49. $3 + \dfrac{t}{t + 2} - \dfrac{2}{t^2 - 4} = \dfrac{3}{1} + \dfrac{t}{t + 2} - \dfrac{2}{(t + 2)(t - 2)}$

[LCD is (t + 2)(t - 2).]

$= \dfrac{3}{1} \cdot \dfrac{(t+2)(t-2)}{(t+2)(t-2)} + \dfrac{t}{t+2} \cdot \dfrac{t-2}{t-2} - \dfrac{2}{(t+2)(t-2)}$

$= \dfrac{3t^2 - 12 + t^2 - 2t - 2}{(t + 2)(t - 2)}$

$= \dfrac{4t^2 - 2t - 14}{(t + 2)(t - 2)}$

50. $\dfrac{3t^2 + 3t - 21}{(t + 3)(t - 3)}$

51. $\dfrac{1}{x + 1} - \dfrac{x}{x - 2} + \dfrac{x^2 + 2}{x^2 - x - 2}$

$= \dfrac{1}{x + 1} - \dfrac{x}{x - 2} + \dfrac{x^2 + 2}{(x - 2)(x + 1)}$

[LCD is (x + 1)(x - 2).]

$= \dfrac{1}{x + 1} \cdot \dfrac{x - 2}{x - 2} - \dfrac{x}{x - 2} \cdot \dfrac{x + 1}{x + 1} + \dfrac{x^2 + 2}{(x - 2)(x + 1)}$

$= \dfrac{x - 2 - (x^2 + x) + x^2 + 2}{(x + 1)(x - 2)}$

$= \dfrac{x - 2 - x^2 - x + x^2 + 2}{(x + 1)(x - 2)}$

$= \dfrac{0}{(x + 1)(x - 2)} = 0$

52. $\dfrac{-y}{(y + 3)(y - 1)}$, or $-\dfrac{y}{(y + 3)(y - 1)}$

53. $\dfrac{4x}{x^2 - 1} + \dfrac{3x}{1 - x} - \dfrac{4}{x - 1}$

$= \dfrac{4x}{x^2 - 1} + \dfrac{-1}{-1} \cdot \dfrac{3x}{1 - x} - \dfrac{4}{x - 1}$

$= \dfrac{4x}{(x + 1)(x - 1)} + \dfrac{-3x}{x - 1} - \dfrac{4}{x - 1}$

[LCD is $(x + 1)(x - 1)$.]

$= \dfrac{4x}{(x + 1)(x - 1)} + \dfrac{-3x}{x - 1} \cdot \dfrac{x + 1}{x + 1} - \dfrac{4}{x - 1} \cdot \dfrac{x + 1}{x + 1}$

$= \dfrac{4x - 3x^2 - 3x - 4x - 4}{(x + 1)(x - 1)}$

$= \dfrac{-3x^2 - 3x - 4}{(x + 1)(x - 1)}$

54. $\dfrac{-14y^2 - 3y + 3}{(2y + 1)(2y - 1)}$

55. $\dfrac{1}{t^2 + 5t + 6} - \dfrac{2}{t^2 + 3t + 2} + \dfrac{1}{t^2 - 3t - 4}$

$= \dfrac{1}{(t + 3)(t + 2)} - \dfrac{2}{(t + 2)(t + 1)} + \dfrac{1}{(t - 4)(t + 1)}$

[LCD is $(t + 3)(t + 2)(t + 1)(t - 4)$.]

$= \dfrac{1}{(t + 3)(t + 2)} \cdot \dfrac{(t + 1)(t - 4)}{(t + 1)(t - 4)} -$

$\dfrac{2}{(t + 2)(t + 1)} \cdot \dfrac{(t + 3)(t - 4)}{(t + 3)(t - 4)} +$

$\dfrac{1}{(t - 4)(t + 1)} \cdot \dfrac{(t + 3)(t + 2)}{(t + 3)(t + 2)}$

$= \dfrac{(t^2 - 3t - 4) - 2(t^2 - t - 12) + (t^2 + 5t + 6)}{(t + 3)(t + 2)(t + 1)(t - 4)}$

$= \dfrac{t^2 - 3t - 4 - 2t^2 + 2t + 24 + t^2 + 5t + 6}{(t + 3)(t + 2)(t + 1)(t - 4)}$

$= \dfrac{4t + 26}{(t + 3)(t + 2)(t + 1)(t - 4)}$

56. $\dfrac{-6x + 42}{(x - 3)(x - 2)(x + 1)(x + 3)}$

57. $\dfrac{15x^{-7}y^{12}z^4}{35x^{-2}y^6z^{-3}} = \dfrac{15}{35}x^{-7-(-2)}y^{12-6}z^{4-(-3)}$

$= \dfrac{3}{7}x^{-5}y^6z^7 = \dfrac{3}{7} \cdot \dfrac{1}{x^5} \cdot y^6z^7$

$= \dfrac{3y^6z^7}{7x^5}$

58. $y = \dfrac{5}{4}x + \dfrac{11}{2}$

59. Familiarize. We let x, y, and z represent the number of rolls of dimes, nickels, and quarters, respectively.

Coins	Number of rolls	Value per roll	Total Value
Dimes	x	50 × 0.10, or 5.00	5x
Nickels	y	40 × 0.05, or 2.00	2y
Quarters	z	40 × 0.25, or 10.00	10z
Total	12		$70.00

Translate. The number of rolls of nickels is three more than the number of rolls of dimes. This gives us one equation.

$y = x + 3$

From the table we get two more equations.

$x + y + z = 12$

$5x + 2y + 10z = 70$

Carry out. Solving the system we get the ordered triple (2,5,5).

Check. 2 rolls of dimes = 2 × \$5, or \$10

5 rolls of nickels = 5 × \$2, or \$10

5 rolls of quarters = 5 × \$10, or \$50

The total value is \$10 + \$10 + \$50 = \$70.

The total number of rolls of coins is 2 + 5 + 5, or 12, and the number of rolls of nickels, 5, is three more than the number of rolls of dimes, 2. The numbers check.

State. Robert has 2 rolls of dimes, 5 rolls of nickels, and 5 rolls of quarters.

60. 4 30-min tapes, 8 60-min tapes

61.

62.

63. $x^8 - x^4 = x^4(x^2 + 1)(x + 1)(x - 1)$

$x^5 - x^2 = x^2(x - 1)(x^2 + x + 1)$

$x^5 - x^3 = x^3(x + 1)(x - 1)$

$x^5 + x^2 = x^2(x + 1)(x^2 - x + 1)$

The LCM is $x^4(x^2 + 1)(x + 1)(x - 1)(x^2 + x + 1)(x^2 - x + 1)$.

64. $2ab(a^2+ab+b^2)(a+b)^2(a^2-ab+b^2)(a-b)(2b+3a)$

65. The LCM is $8a^4b^7$.

One expression is $2a^3b^7$.

Then the other expression must contain 8, a^4, and one of the following:

no factor of b, b, b^2, b^3, b^4, b^5, b^6, or b^7.

Thus, all the possibilities for the other expression are $8a^4$, $8a^4b$, $8a^4b^2$, $8a^4b^3$, $8a^4b^4$, $8a^4b^5$, $8a^4b^6$, $8a^4b^7$.

66. Every 60 years

67. Find the LCM of the number of years it takes the planets to revolve around the sun.

Earth: 1

Jupiter: $12 = 2 \cdot 2 \cdot 3$

Saturn: $30 = 2 \cdot 3 \cdot 5$

Uranus: $84 = 2 \cdot 2 \cdot 3 \cdot 7$

The LCM is $2 \cdot 2 \cdot 3 \cdot 5 \cdot 7$, or 420.

It will take 420 years.

68. $\dfrac{x^4 + 6x^3 + 2x^2}{(x + 2)(x - 2)(x + 5)}$

69. $(f - g)(x) = \frac{x^3}{x^2 - 4} - \frac{x^2}{x^2 + 3x - 10}$

$= \frac{x^3}{(x + 2)(x - 2)} - \frac{x^2}{(x + 5)(x - 2)}$

$= \frac{x^3(x + 5) - x^2(x + 2)}{(x + 2)(x - 2)(x + 5)}$

$= \frac{x^4 + 5x^3 - x^3 - 2x^2}{(x + 2)(x - 2)(x + 5)}$

$= \frac{x^4 + 4x^3 - 2x^2}{(x + 2)(x - 2)(x + 5)}$

70. $\frac{x^5}{(x^2 - 4)(x^2 + 3x - 10)}$

71. $(f/g)(x) = \frac{x^3}{x^2 - 4} \div \frac{x^2}{x^2 + 3x - 10}$

$= \frac{x^2 \cdot x}{(x + 2)(x - 2)} \cdot \frac{(x + 5)(x - 2)}{x^2}$

$= \frac{x^2(x - 2)}{x^2(x - 2)} \cdot \frac{x(x + 5)}{x + 2}$

$= \frac{x(x + 5)}{x + 2}$

(Note that $x \neq 0$, $x \neq -5$, and $x \neq 2$ are additional restrictions, since $g(0) = 0$, -5 is not in the domain of g, and 2 is not in the domain of either f or g.)

72. $\{x|x$ is a real number and $x \neq -2$ and $x \neq 2$ and $x \neq -5\}$

73. $2x^{-2} + 3x^{-2}y^{-2} - 7xy^{-1}$

$= \frac{2}{x^2} + \frac{3}{x^2y^2} - \frac{7x}{y}$

[LCD is x^2y^2.]

$= \frac{2}{x^2} \cdot \frac{y^2}{y^2} + \frac{3}{x^2y^2} - \frac{7x}{y} \cdot \frac{x^2y}{x^2y}$

$= \frac{2y^2 + 3 - 7x^3y}{x^2y^2}$

74. $\frac{9x^2 + 28x + 15}{(x - 3)(x + 3)^2}$

75. $4(y-1)(2y-5)^{-1} + 5(2y+3)(5-2y)^{-1} + (y-4)(2y-5)^{-1}$

$= \frac{4(y - 1)}{2y - 5} + \frac{-1}{-1} \cdot \frac{5(2y + 3)}{5 - 2y} + \frac{y - 4}{2y - 5}$

$= \frac{4y - 4 - 10y - 15 + y - 4}{2y - 5}$

$= \frac{-5y - 23}{2y - 5}$, or $\frac{5y + 23}{5 - 2y}$

76. $\frac{1}{2x(x - 5)}$

77. $\frac{x^2 - 7x + 12}{x^2 - x - 29/3}\left[\frac{3x + 2}{x^2 + 5x - 24} + \frac{7}{x^2 + 4x - 32}\right]$

$= \frac{x^2 - 7x + 12}{x^2 - x - 29/3}\left[\frac{3x + 2}{(x + 8)(x - 3)} + \frac{7}{(x + 8)(x - 4)}\right]$

$= \frac{x^2-7x+12}{x^2-x-29/3}\left[\frac{3x+2}{(x+8)(x-3)} \cdot \frac{x-4}{x-4} + \frac{7}{(x+8)(x-4)} \cdot \frac{x-3}{x-3}\right]$

$= \frac{x^2 - 7x + 12}{x^2 - x - 29/3}\left[\frac{3x^2 - 10x - 8 + 7x - 21}{(x + 8)(x - 3)(x - 4)}\right]$

$= \frac{x^2 - 7x + 12}{x^2 - x - 29/3}\left[\frac{3x^2 - 3x - 29}{(x + 8)(x - 3)(x - 4)}\right]$

$= \frac{(x - 4)(x - 3)(3)(x^2 - x - 29/3)}{(x^2 - x - 29/3)(x + 8)(x - 3)(x - 4)}$

$= \frac{\cancel{(x - 4)}\cancel{(x - 3)}(3)\cancel{(x^2 - x - 29/3)}}{\cancel{(x^2 - x - 29/3)}(x + 8)\cancel{(x - 3)}\cancel{(x - 4)}}$

$= \frac{3}{x + 8}$

78. $-4t^4$

79. $\frac{9t^3}{3t^3 - 12t^2 + 9t} \div \left[\frac{t + 4}{t^2 - 9} - \frac{3t - 1}{t^2 + 2t - 3}\right]$

$= \frac{9t^3}{3t^3 - 12t^2 + 9t} \div \left[\frac{t + 4}{(t + 3)(t - 3)} - \frac{3t - 1}{(t + 3)(t - 1)}\right]$

$= \frac{9t^3}{3t^3-12t^2+9t} \div \left[\frac{t+4}{(t+3)(t-3)} \cdot \frac{t-1}{t-1} - \frac{3t-1}{(t+3)(t-1)} \cdot \frac{t-3}{t-3}\right]$

$= \frac{9t^3}{3t^3 - 12t^2 + 9t} \div \left[\frac{t^2 + 3t - 4 - (3t^2 - 10t + 3)}{(t + 3)(t - 3)(t - 1)}\right]$

$= \frac{9t^3}{3t^3 - 12t^2 + 9t} \div \left[\frac{t^2 + 3t - 4 - 3t^2 + 10t - 3}{(t + 3)(t - 3)(t - 1)}\right]$

$= \frac{9t^3}{3t(t^2 - 4t + 3)} \div \frac{-2t^2 + 13t - 7}{(t + 3)(t - 3)(t - 1)}$

$= \frac{3t \cdot 3t^2}{3t(t - 3)(t - 1)} \cdot \frac{(t + 3)(t - 3)(t - 1)}{-2t^2 + 13t - 7}$

$= \frac{3t \cdot 3t^2(t + 3)\cancel{(t - 3)}\cancel{(t - 1)}}{3t\cancel{(t - 3)}\cancel{(t - 1)}(-2t^2 + 13t - 7)}$

$= \frac{3t^2(t + 3)}{-2t^2 + 13t - 7}$

Exercise Set 6.3

1. $\frac{\frac{1}{x} + 4}{\frac{1}{x} - 3} = \frac{\frac{1}{x} + 4}{\frac{1}{x} - 3} \cdot \frac{x}{x}$ Using the LCD

$= \frac{\frac{1}{x} \cdot x + 4 \cdot x}{\frac{1}{x} \cdot x - 3 \cdot x}$

$= \frac{1 + 4x}{1 - 3x}$

2. $\frac{1 + 7y}{1 - 5y}$

3. $\dfrac{x - x^{-1}}{x + x^{-1}} = \dfrac{x - \frac{1}{x}}{x + \frac{1}{x}}$ Rewriting with positive exponents

$= \dfrac{x - \frac{1}{x}}{x + \frac{1}{x}} \cdot \dfrac{x}{x}$ Using the LCD

$= \dfrac{x \cdot x - \frac{1}{x} \cdot x}{x \cdot x + \frac{1}{x} \cdot x}$

$= \dfrac{x^2 - 1}{x^2 + 1}$

4. $\dfrac{y^2 + 1}{y^2 - 1}$

5. $\dfrac{\frac{3}{x} + \frac{4}{y}}{\frac{4}{x} - \frac{3}{y}} = \dfrac{\frac{3}{x} + \frac{4}{y}}{\frac{4}{x} - \frac{3}{y}} \cdot \dfrac{xy}{xy}$ Using the LCD

$= \dfrac{\frac{3}{x} \cdot xy + \frac{4}{y} \cdot xy}{\frac{4}{x} \cdot xy - \frac{3}{y} \cdot xy}$

$= \dfrac{3y + 4x}{4y - 3x}$

6. $\dfrac{2z + 5y}{z - 4y}$

7. $\dfrac{\frac{x^2 - y^2}{xy}}{\frac{x - y}{y}} = \dfrac{x^2 - y^2}{xy} \cdot \dfrac{y}{x - y}$ Multiplying by the reciprocal of the divisor

$= \dfrac{(x + y)(x - y) \cdot y}{xy(x - y)}$

$= \dfrac{(x + y)\cancel{(x - y)} \cdot \cancel{y}}{x\cancel{y}\cancel{(x - y)}}$

$= \dfrac{x + y}{x}$

8. $\dfrac{a + b}{a}$

9. $\dfrac{a - \frac{3a}{b}}{b - \frac{b}{a}} = \dfrac{a - \frac{3a}{b}}{b - \frac{b}{a}} \cdot \dfrac{ab}{ab}$ Using the LCD

$= \dfrac{a(ab) - \frac{3a}{b} \cdot ab}{b(ab) - \frac{b}{a} \cdot ab}$

$= \dfrac{a^2b - 3a^2}{ab^2 - b^2}$

$= \dfrac{a^2(b - 3)}{b^2(a - 1)}$

10. $\dfrac{3}{3x + 2}$

11. $\dfrac{a^{-1} + b^{-1}}{\frac{a^2 - b^2}{ab}} = \dfrac{\frac{1}{a} + \frac{1}{b}}{\frac{a^2 - b^2}{ab}}$

$= \dfrac{\frac{1}{a} + \frac{1}{b}}{\frac{a^2 - b^2}{ab}} \cdot \dfrac{ab}{ab}$ Using the LCD

$= \dfrac{\frac{1}{a} \cdot ab + \frac{1}{b} \cdot ab}{\frac{a^2 - b^2}{ab} \cdot ab}$

$= \dfrac{b + a}{a^2 - b^2} = \dfrac{b + a}{(a + b)(a - b)}$

$= \dfrac{\cancel{(a + b)} \cdot (1)}{\cancel{(a + b)}(a - b)}$ $(b + a = a + b)$

$= \dfrac{1}{a - b}$

12. $\dfrac{1}{x - y}$

13. $\dfrac{\frac{1}{x + h} - \frac{1}{x}}{h} = \dfrac{\frac{1}{x + h} \cdot \frac{x}{x} - \frac{1}{x} \cdot \frac{x + h}{x + h}}{h}$ Adding in the numerator

$= \dfrac{\frac{x - x - h}{x(x + h)}}{h} = \dfrac{\frac{-h}{x(x + h)}}{h}$

$= \dfrac{-h}{x(x + h)} \cdot \dfrac{1}{h}$ Multiplying by the reciprocal of the divisor

$= \dfrac{-1 \cdot \cancel{h} \cdot 1}{x(x + h)\cancel{(h)}}$ $(-h = -1 \cdot h)$

$= -\dfrac{1}{x(x + h)}$

14. $\dfrac{1}{a(a - h)}$

15. $\dfrac{\frac{y^2 - y - 6}{y^2 - 5y - 14}}{\frac{y^2 + 6y + 5}{y^2 - 6y - 7}} = \dfrac{y^2 - y - 6}{y^2 - 5y - 14} \cdot \dfrac{y^2 - 6y - 7}{y^2 + 6y + 5}$

$= \dfrac{(y - 3)\cancel{(y + 2)}\cancel{(y - 7)}\cancel{(y + 1)}}{\cancel{(y - 7)}\cancel{(y + 2)}(y + 5)\cancel{(y + 1)}}$

$= \dfrac{(y - 3)(y + 2)(y - 7)(y + 1)}{(y - 7)(y + 2)(y + 5)(y + 1)}$

$= \dfrac{y - 3}{y + 5}$

16. $\dfrac{(x - 4)(x - 7)}{(x - 5)(x + 6)}$

17. $$\frac{\frac{1}{x-2}+\frac{3}{x-1}}{\frac{2}{x-1}+\frac{5}{x-2}}$$

$$= \frac{\frac{1}{x-2}+\frac{3}{x-1}}{\frac{2}{x-1}+\frac{5}{x-2}}\cdot\frac{(x-2)(x-1)}{(x-2)(x-1)}$$ Using the LCD

$$= \frac{\frac{1}{x-2}\cdot(x-2)(x-1)+\frac{3}{x-1}\cdot(x-2)(x-1)}{\frac{2}{x-1}\cdot(x-2)(x-1)+\frac{5}{x-2}\cdot(x-2)(x-1)}$$

$$= \frac{x-1+3(x-2)}{2(x-2)+5(x-1)}$$

$$= \frac{x-1+3x-6}{2x-4+5x-5} = \frac{4x-7}{7x-9}$$

18. $\frac{3y-1}{7y-5}$

19. $$\frac{a(a+3)^{-1}-2(a-1)^{-1}}{a(a+3)^{-1}-(a-1)^{-1}}$$

$$= \frac{\frac{a}{a+3}-\frac{2}{a-1}}{\frac{a}{a+3}-\frac{1}{a-1}}$$

$$= \frac{\frac{a}{a+3}-\frac{2}{a-1}}{\frac{a}{a+3}-\frac{1}{a-1}}\cdot\frac{(a+3)(a-1)}{(a+3)(a-1)}$$ Using the LCD

$$= \frac{\frac{a}{a+3}\cdot(a+3)(a-1)-\frac{2}{a-1}\cdot(a+3)(a-1)}{\frac{a}{a+3}\cdot(a+3)(a-1)-\frac{1}{a-1}\cdot(a+3)(a-1)}$$

$$= \frac{a(a-1)-2(a+3)}{a(a-1)-(a+3)}$$

$$= \frac{a^2-a-2a-6}{a^2-a-a-3} = \frac{a^2-3a-6}{a^2-2a-3}$$

20. $\frac{a^2-6a-6}{a^2-4a-2}$

21. $$\frac{\frac{x}{x^2+3x-4}-\frac{1}{x^2+3x-4}}{\frac{x}{x^2+6x+8}+\frac{3}{x^2+6x+8}} = \frac{\frac{x-1}{x^2+3x-4}}{\frac{x+3}{x^2+6x+8}}$$

Adding in the numerator and the denominator

$$= \frac{x-1}{x^2+3x-4}\cdot\frac{x^2+6x+8}{x+3}$$

$$= \frac{(x-1)(x+4)(x+2)}{(x+4)(x-1)(x+3)}$$

$$= \frac{\cancel{(x-1)}\cancel{(x+4)}(x+2)}{\cancel{(x+4)}\cancel{(x-1)}(x+3)} = \frac{x+2}{x+3}$$

22. $\frac{x-4}{x-2}$

23. $$\frac{\frac{y}{y^2-1}+\frac{3}{1-y^2}}{\frac{y^2}{y^2-1}+\frac{9}{1-y^2}} = \frac{\frac{y}{y^2-1}+\frac{-1}{-1}\cdot\frac{3}{1-y^2}}{\frac{y^2}{y^2-1}+\frac{-1}{-1}\cdot\frac{9}{1-y^2}}$$

$$= \frac{\frac{y}{y^2-1}-\frac{3}{y^2-1}}{\frac{y^2}{y^2-1}-\frac{9}{y^2-1}}$$

$$= \frac{\frac{y-3}{y^2-1}}{\frac{y^2-9}{y^2-1}}$$ Adding in the numerator and the denominator

$$= \frac{y-3}{y^2-1}\cdot\frac{y^2-1}{y^2-9}$$

$$= \frac{(y-3)(y^2-1)}{(y^2-1)(y+3)(y-3)}$$

$$= \frac{\cancel{(y-3)}\cancel{(y^2-1)}}{\cancel{(y^2-1)}(y+3)\cancel{(y-3)}}$$

$$= \frac{1}{y+3}$$

24. $\frac{1}{y+5}$

25. $$\frac{\frac{2}{a^2-1}+\frac{1}{a+1}}{\frac{3}{a^2-1}+\frac{2}{a-1}} = \frac{\frac{2}{(a+1)(a-1)}+\frac{1}{a+1}}{\frac{3}{(a+1)(a-1)}+\frac{2}{a-1}}$$

$$= \frac{\frac{2}{(a+1)(a-1)}+\frac{1}{a+1}}{\frac{3}{(a+1)(a-1)}+\frac{2}{a-1}}\cdot\frac{(a+1)(a-1)}{(a+1)(a-1)}$$ Using the LCD

$$= \frac{\frac{2}{(a+1)(a-1)}\cdot(a+1)(a-1)+\frac{1}{a+1}\cdot(a+1)(a-1)}{\frac{3}{(a+1)(a-1)}\cdot(a+1)(a-1)+\frac{2}{a-1}\cdot(a+1)(a-1)}$$

$$= \frac{2+a-1}{3+2(a+1)} = \frac{a+1}{3+2a+2} = \frac{a+1}{2a+5}$$

26. $\frac{2a-3}{a+1}$

27. $$\frac{\frac{5}{x^2-4}-\frac{3}{x-2}}{\frac{4}{x^2-4}-\frac{2}{x+2}} = \frac{\frac{5}{(x+2)(x-2)}-\frac{3}{x-2}}{\frac{4}{(x+2)(x-2)}-\frac{2}{x+2}}$$

$$= \frac{\frac{5}{(x+2)(x-2)}-\frac{3}{x-2}}{\frac{4}{(x+2)(x-2)}-\frac{2}{x+2}}\cdot\frac{(x+2)(x-2)}{(x+2)(x-2)}$$ Using the LCD

$$= \frac{\frac{5}{(x+2)(x-2)}\cdot(x+2)(x-2)-\frac{3}{x-2}\cdot(x+2)(x-2)}{\frac{4}{(x+2)(x-2)}\cdot(x+2)(x-2)-\frac{2}{x+2}\cdot(x+2)(x-2)}$$

$$= \frac{5-3(x+2)}{4-2(x-2)} = \frac{5-3x-6}{4-2x+4} = \frac{-1-3x}{8-2x}$$

28. $\frac{7-3x}{3-2x}$

29. $$\frac{\dfrac{y^2}{y^2-9} - \dfrac{y}{y+3}}{\dfrac{y}{y^2-9} - \dfrac{1}{y-3}} = \frac{\dfrac{y^2}{(y+3)(y-3)} - \dfrac{y}{y+3}}{\dfrac{y}{(y+3)(y-3)} - \dfrac{1}{y-3}}$$

$$= \frac{\dfrac{y^2}{(y+3)(y-3)} - \dfrac{y}{y+3}}{\dfrac{y}{(y+3)(y-3)} - \dfrac{1}{y-3}} \cdot \frac{(y+3)(y-3)}{(y+3)(y-3)}$$ Using the LCD

$$= \frac{\dfrac{y^2}{(y+3)(y-3)} \cdot (y+3)(y-3) - \dfrac{y}{y+3} \cdot (y+3)(y-3)}{\dfrac{y}{(y+3)(y-3)} \cdot (y+3)(y-3) - \dfrac{1}{y-3} \cdot (y+3)(y-3)}$$

$$= \frac{y^2 - y(y-3)}{y-(y+3)} = \frac{y^2 - y^2 + 3y}{y - y - 3} = \frac{3y}{-3} = \frac{\cancel{3}y}{-1\cdot\cancel{3}} = -y$$

30. $-y$

31. $$\frac{\dfrac{a}{a+3} + \dfrac{4}{5a}}{\dfrac{a}{2a+6} + \dfrac{3}{a}} = \frac{\dfrac{a}{a+3} + \dfrac{4}{5a}}{\dfrac{a}{2(a+3)} + \dfrac{3}{a}}$$

$$= \frac{\dfrac{a}{a+3} + \dfrac{4}{5a}}{\dfrac{a}{2(a+3)} + \dfrac{3}{a}} \cdot \frac{10a(a+3)}{10a(a+3)}$$ Using the LCD

$$= \frac{\dfrac{a}{a+3} \cdot 10a(a+3) + \dfrac{4}{5a} \cdot 10a(a+3)}{\dfrac{a}{2(a+3)} \cdot 10a(a+3) + \dfrac{3}{a} \cdot 10a(a+3)}$$

$$= \frac{10a^2 + 8(a+3)}{5a^2 + 30(a+3)} = \frac{10a^2 + 8a + 24}{5a^2 + 30a + 90}$$

$$= \frac{2(5a^2 + 4a + 12)}{5(a^2 + 6a + 18)}$$

32. $$\frac{6(a^2 + 5a + 10)}{3a^2 + 2a + 4}$$

33. $$\frac{\dfrac{x}{x+y} + \dfrac{x}{y}}{\dfrac{x}{3x+3y} + \dfrac{y}{x}} = \frac{\dfrac{x}{x+y} + \dfrac{x}{y}}{\dfrac{x}{3(x+y)} + \dfrac{y}{x}}$$

$$= \frac{\dfrac{x}{x+y} + \dfrac{x}{y}}{\dfrac{x}{3(x+y)} + \dfrac{y}{x}} \cdot \frac{3xy(x+y)}{3xy(x+y)}$$ Using the LCD

$$= \frac{\dfrac{x}{x+y} \cdot 3xy(x+y) + \dfrac{x}{y} \cdot 3xy(x+y)}{\dfrac{x}{3(x+y)} \cdot 3xy(x+y) + \dfrac{y}{x} \cdot 3xy(x+y)}$$

$$= \frac{3x^2y + 3x^2(x+y)}{x^2y + 3y^2(x+y)}$$

$$= \frac{3x^2y + 3x^3 + 3x^2y}{x^2y + 3xy^2 + 3y^3} = \frac{3x^3 + 6x^2y}{x^2y + 3xy^2 + 3y^3}$$

$$= \frac{3x^2(x+2y)}{y(x^2 + 3xy + 3y^2)}$$

34. $$\frac{5y(x^2 + xy + y^2)}{x^2(6y + 5x)}$$

35. $$\frac{\dfrac{1}{x^2-1} + \dfrac{1}{x^2+4x+3}}{\dfrac{1}{x^2-1} + \dfrac{1}{x^2-3x+2}} = \frac{\dfrac{1}{(x+1)(x-1)} + \dfrac{1}{(x+3)(x+1)}}{\dfrac{1}{(x+1)(x-1)} + \dfrac{1}{(x-2)(x-1)}}$$

$$= \frac{\dfrac{1}{(x+1)(x-1)} + \dfrac{1}{(x+3)(x+1)}}{\dfrac{1}{(x+1)(x-1)} + \dfrac{1}{(x-2)(x-1)}} \cdot \frac{(x+1)(x-1)(x+3)(x-2)}{(x+1)(x-1)(x+3)(x-2)}$$

Using the LCD

$$= \frac{(x+3)(x-2) + (x-1)(x-2)}{(x+3)(x-2) + (x+1)(x+3)}$$

$$= \frac{x^2 + x - 6 + x^2 - 3x + 2}{x^2 + x - 6 + x^2 + 4x + 3}$$

$$= \frac{2x^2 - 2x - 4}{2x^2 + 5x - 3}$$

$$= \frac{2(x^2 - x - 2)}{(2x-1)(x+3)}$$

$$= \frac{2(x-2)(x+1)}{(2x-1)(x+3)}$$

36. $$\frac{2x^2 - 5x + 2}{2x^2 + 2x}, \text{ or } \frac{(2x-1)(x-2)}{2x(x+1)}$$

37. $$\frac{\dfrac{y}{y^2-4} - \dfrac{2y}{y^2+y-6}}{\dfrac{2y}{y^2-4} - \dfrac{y}{y^2+5y+6}} = \frac{\dfrac{y}{(y+2)(y-2)} - \dfrac{2y}{(y+3)(y-2)}}{\dfrac{2y}{(y+2)(y-2)} - \dfrac{y}{(y+3)(y+2)}}$$

$$= \frac{\dfrac{y}{(y+2)(y-2)} - \dfrac{2y}{(y+3)(y-2)}}{\dfrac{2y}{(y+2)(y-2)} - \dfrac{y}{(y+3)(y+2)}} \cdot \frac{(y+2)(y-2)(y+3)}{(y+2)(y-2)(y+3)}$$ Using the LCD

$$= \frac{y(y+3) - 2y(y+2)}{2y(y+3) - y(y-2)}$$

$$= \frac{y^2 + 3y - 2y^2 - 4y}{2y^2 + 6y - y^2 + 2y} = \frac{-y^2 - y}{y^2 + 8y}$$

$$= \frac{y(-y-1)}{y(y+8)} = \frac{y(-y-1)}{y(y+8)} = \frac{-y-1}{y+8}$$

38. $$\frac{-2y^2 + 13y - 21}{2(y^2 - y - 20)}, \text{ or } \frac{-(2y-7)(y-3)}{2(y-5)(y+4)}$$

39. $$\frac{\dfrac{1}{a^2+7a+12} + \dfrac{1}{a^2+a-6}}{\dfrac{1}{a^2+2a-8} + \dfrac{1}{a^2+5a+4}} = \frac{\dfrac{1}{(a+3)(a+4)} + \dfrac{1}{(a+3)(a-2)}}{\dfrac{1}{(a+4)(a-2)} + \dfrac{1}{(a+4)(a+1)}}$$

$$= \frac{\dfrac{1}{(a+3)(a+4)} + \dfrac{1}{(a+3)(a-2)}}{\dfrac{1}{(a+4)(a-2)} + \dfrac{1}{(a+4)(a+1)}} \cdot \frac{(a+3)(a+4)(a-2)(a+1)}{(a+3)(a+4)(a-2)(a+1)}$$

Using the LCD

$$= \frac{(a-2)(a+1) + (a+4)(a+1)}{(a+3)(a+1) + (a+3)(a-2)}$$

$$= \frac{a^2 - a - 2 + a^2 + 5a + 4}{a^2 + 4a + 3 + a^2 + a - 6}$$

$$= \frac{2a^2 + 4a + 2}{2a^2 + 5a - 3}$$

$$= \frac{2(a^2 + 2a + 1)}{(2a-1)(a+3)}$$

$$= \frac{2(a+1)^2}{(2a-1)(a+3)}$$

40. $$\frac{2a^2 + 7a - 15}{2a^2 - 2a - 12}, \text{ or } \frac{(2a-3)(a+5)}{2(a-3)(a+2)}$$

41. $$\frac{\frac{2}{x^2-7x+12}-\frac{1}{x^2+7x+10}}{\frac{2}{x^2-x-6}-\frac{1}{x^2+x-20}} = \frac{\frac{2}{(x-3)(x-4)}-\frac{1}{(x+5)(x+2)}}{\frac{2}{(x-3)(x+2)}-\frac{1}{(x+5)(x-4)}}$$

$$= \frac{\frac{2}{(x-3)(x-4)}-\frac{1}{(x+5)(x+2)}}{\frac{2}{(x-3)(x+2)}-\frac{1}{(x+5)(x-4)}} \cdot \frac{(x-3)(x-4)(x+5)(x+2)}{(x-3)(x-4)(x+5)(x+2)}$$

Using the LCD

$$= \frac{2(x+5)(x+2) - (x-3)(x-4)}{2(x-4)(x+5) - (x-3)(x+2)}$$

$$= \frac{2(x^2+7x+10) - (x^2-7x+12)}{2(x^2+x-20) - (x^2-x-6)}$$

$$= \frac{2x^2+14x+20-x^2+7x-12}{2x^2+2x-40-x^2+x+6}$$

$$= \frac{x^2+21x+8}{x^2+3x-34}$$

42. $\frac{2x^2-11x-27}{2x^2+21x+13}$

43. $\frac{a}{x+y} = b$

$(x+y)\frac{a}{x+y} = b(x+y)$ Multiplying by $x + y$

$a = bx + by$ Simplifying

$a - by = bx$ Adding $-by$

$\frac{1}{b}(a - by) = \frac{1}{b}\cdot bx$ Multiplying by $\frac{1}{b}$

$\frac{1}{b}\cdot a - \frac{1}{b}\cdot by = x$ Using the distributive law

$\frac{a}{b} - y = x$ Simplifying

44. 11

45. Familiarize. We let ℓ and w represent the length and width of the second rectangle. Then $\ell - 3$ and $w - 4$ represent the length and width of the first rectangle. The perimeter of the second rectangle is $2\ell + 2w$; the perimeter of the first rectangle is $2(\ell - 3) + 2(w - 4)$, or $2\ell + 2w - 14$.

Translate.

Perimeter of second	=	2	$\cdot$	Perimeter of first	$-$	1
$2\ell + 2w$	=	2	$\cdot$	$(2\ell + 2w - 14)$	$-$	1

Carry out. We first solve for $2\ell + 2w$.

$2\ell + 2w = 2(2\ell + 2w - 14) - 1$

$2\ell + 2w = 4\ell + 4w - 28 - 1$

$2\ell + 2w = 4\ell + 4w - 29$

$29 = 2\ell + 2w$

If $2\ell + 2w = 29$, then $2\ell + 2w - 14 = 29 - 14$, or 15.

Check. The perimeter of the second rectangle is 1 less than twice the perimeter of the first rectangle:

$29 = 2\cdot 15 - 1$

The values check.

State. The perimeter of the first rectangle is 15; the perimeter of the second rectangle is 29.

46. \$17

47. 

48. 

49. $$\frac{5x^{-1}-5y^{-1}+10x^{-1}y^{-1}}{6x^{-1}-6y^{-1}+12x^{-1}y^{-1}} = \frac{\frac{5}{x}-\frac{5}{y}+\frac{10}{xy}}{\frac{6}{x}-\frac{6}{y}+\frac{12}{xy}}$$

$$= \frac{\frac{5}{x}-\frac{5}{y}+\frac{10}{xy}}{\frac{6}{x}-\frac{6}{y}+\frac{12}{xy}} \cdot \frac{xy}{xy}$$

$$= \frac{5y-5x+10}{6y-6x+12} = \frac{5(y-x+2)}{6(y-x+2)}$$

$$= \frac{5\cancel{(y-x+2)}}{6\cancel{(y-x+2)}} = \frac{5}{6}$$

50. $\frac{x^4}{81}$

51. $$(a^2-ab+b^2)^{-1}(a^2b^{-1}+b^2a^{-1})\cdot (a^{-2}-b^{-2})(a^{-2}+2a^{-1}b^{-1}+b^{-2})^{-1}$$

$$= \frac{(a^2b^{-1}+b^2a^{-1})(a^{-2}-b^{-2})}{(a^2-ab+b^2)(a^{-2}+2a^{-1}b^{-1}+b^{-2})}$$

$$= \frac{\left(\frac{a^2}{b}+\frac{b^2}{a}\right)\left(\frac{1}{a^2}-\frac{1}{b^2}\right)}{(a^2-ab+b^2)\left(\frac{1}{a^2}+\frac{2}{ab}+\frac{1}{b^2}\right)}$$

$$= \frac{\left(\frac{a^3+b^3}{ab}\right)\left(\frac{b^2-a^2}{a^2b^2}\right)}{(a^2-ab+b^2)\left(\frac{b^2+2ab+a^2}{a^2b^2}\right)}$$

$$= \frac{(a+b)(a^2-ab+b^2)(b+a)(b-a)}{a^3b^3} \cdot \frac{a^2b^2}{(a^2-ab+b^2)(b+a)^2}$$

$$= \frac{b-a}{ab}$$

52. $\frac{x}{x^3-1}$

53. First simplify the given expression.

$$1+\cfrac{1}{1+\cfrac{1}{1+\cfrac{1}{1+\cfrac{1}{x}}}} = 1+\cfrac{1}{1+\cfrac{1}{1+\cfrac{1}{\frac{x+1}{x}}}}$$

$$= 1+\cfrac{1}{1+\cfrac{1}{1+\frac{x}{x+1}}} = 1+\cfrac{1}{1+\cfrac{1}{\frac{2x+1}{x+1}}}$$

$$= 1+\cfrac{1}{1+\frac{x+1}{2x+1}} = 1+\cfrac{1}{\frac{3x+2}{2x+1}}$$

$$= 1+\frac{2x+1}{3x+2} = \frac{5x+3}{3x+2}$$

The reciprocal is $\frac{3x+2}{5x+3}$.

54. x

55. $f(x) = \frac{3}{x^2}$, $f(x + h) = \frac{3}{(x + h)^2}$

$$\frac{f(x + h) - f(x)}{h} = \frac{\frac{3}{(x + h)^2} - \frac{3}{x^2}}{h}$$

$$= \frac{3x^2 - 3(x + h)^2}{x^2(x + h)^2} \cdot \frac{1}{h}$$

$$= \frac{3x^2 - 3x^2 - 6xh - 3h^2}{x^2(x + h)^2} \cdot \frac{1}{h}$$

$$= \frac{-6xh - 3h^2}{x^2(x + h)^2h}$$

$$= \frac{-3h(2x + h)}{x^2(x + h)^2h}$$

$$= \frac{-3\cancel{h}(2x + h)}{x^2(x + h)^2\cancel{h}}$$

$$= \frac{-3(2x + h)}{x^2(x + h)^2}$$

56. $\frac{-5}{x(x + h)}$

57. $f(x) = \frac{1}{1 - x}$, $f(x + h) = \frac{1}{1 - x - h}$

$$\frac{f(x + h) - f(x)}{h} = \frac{\frac{1}{1 - x - h} - \frac{1}{1 - x}}{h}$$

$$= \frac{1 - x - (1 - x - h)}{(1 - x - h)(1 - x)} \cdot \frac{1}{h}$$

$$= \frac{1 - x - 1 + x + h}{(1 - x - h)(1 - x)} \cdot \frac{1}{h}$$

$$= \frac{h}{(1 - x - h)(1 - x)h}$$

$$= \frac{\cancel{h} \cdot 1}{(1 - x - h)(1 - x)\cancel{h}}$$

$$= \frac{1}{(1 - x - h)(1 - x)}$$

58. $\frac{1}{(1 + x + h)(1 + x)}$

59. To avoid division by zero in $\frac{1}{x}$ and $\frac{8}{x^2}$ we must exclude 0 from the domain of F($x \neq 0$). To avoid division by zero in the complex fraction we set

$$2 - \frac{8}{x^2} = 0$$

$$2x^2 - 8 = 0$$

$$2(x^2 - 4) = 0$$

$$2(x + 2)(x - 2) = 0$$

$$x + 2 = 0 \quad \text{or} \quad x - 2 = 0$$

$$x = -2 \quad \text{or} \quad x = 2.$$

The domain of F = {x|x is a real number and $x \neq 0$ and $x \neq -2$ and $x \neq 2$}.

60. {x|x is a real number and $x \neq 1$ and $x \neq -1$ and $x \neq 4$ and $x \neq -4$ and $x \neq 5$ and $x \neq -5$}

61. 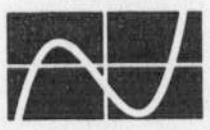

Exercise Set 6.4

1. $\frac{2}{5} + \frac{7}{8} = \frac{y}{20}$, LCD is 40

$$40\left[\frac{2}{5} + \frac{7}{8}\right] = 40 \cdot \frac{y}{20}$$

$$40 \cdot \frac{2}{5} + 40 \cdot \frac{7}{8} = 40 \cdot \frac{y}{20}$$

$$16 + 35 = 2y$$

$$51 = 2y$$

$$\frac{51}{2} = y$$

Check:

$\frac{2}{5} + \frac{7}{8}$	$= \frac{y}{20}$	
$\frac{16}{40} + \frac{35}{40}$?	$\frac{\frac{51}{2}}{20}$	
$\frac{51}{40}$	$\frac{51}{2} \cdot \frac{1}{20}$	
	$\frac{51}{40}$	TRUE

The solution is $\frac{51}{2}$.

2. $\frac{51}{5}$

3. $\frac{x}{3} - \frac{x}{4} = 12$, LCD is 12

$$12\left[\frac{x}{3} - \frac{x}{4}\right] = 12 \cdot 12$$

$$12 \cdot \frac{x}{3} - 12 \cdot \frac{x}{4} = 12 \cdot 12$$

$$4x - 3x = 144$$

$$x = 144$$

Check:

$\frac{x}{3} - \frac{x}{4}$	$= 12$	
$\frac{144}{3} - \frac{144}{4}$?	12	
$48 - 36$		
12		TRUE

The solution is 144.

4. $-\frac{225}{2}$

5. $\frac{1}{3} - \frac{5}{6} = \frac{1}{x}$, LCD is 6x

$$6x\left[\frac{1}{3} - \frac{5}{6}\right] = 6x \cdot \frac{1}{x}$$

$$6x \cdot \frac{1}{3} - 6x \cdot \frac{5}{6} = 6x \cdot \frac{1}{x}$$

$$2x - 5x = 6$$

$$-3x = 6$$

$$x = -2$$

Check:

$\frac{1}{3} - \frac{5}{6}$	$= \frac{1}{x}$	
$\frac{2}{6} - \frac{5}{6}$?	$\frac{1}{-2}$	
$-\frac{3}{6}$	$-\frac{1}{2}$	
$-\frac{1}{2}$		TRUE

The solution is -2.

6. $\frac{40}{9}$

7. $\frac{2}{3} - \frac{1}{5} = \frac{7}{3x}$, LCD is $3\cdot5\cdot x$

$$3\cdot5\cdot x\left(\frac{2}{3} - \frac{1}{5}\right) = 3\cdot5\cdot x \cdot \frac{7}{3x}$$
$$3\cdot5\cdot x \cdot \frac{2}{3} - 3\cdot5\cdot x \cdot \frac{1}{5} = 3\cdot5\cdot x \cdot \frac{7}{3x}$$
$$10x - 3x = 35$$
$$7x = 35$$
$$x = 5$$

Check:

$$\frac{2}{3} - \frac{1}{5} = \frac{7}{3x}$$
$$\frac{2}{3} - \frac{1}{5} \;?\; \frac{7}{3\cdot5}$$
$$\frac{10}{15} - \frac{3}{15} \;\Big|\; \frac{7}{15}$$
$$\frac{7}{15} \;\Big|\; \quad \text{TRUE}$$

The solution is 5.

8. 7

9. $\frac{2}{6} + \frac{1}{2x} = \frac{1}{3}$, LCD is $2\cdot3\cdot x$

$$2\cdot3\cdot x\left(\frac{2}{6} + \frac{1}{2x}\right) = 2\cdot3\cdot x \cdot \frac{1}{3}$$
$$2\cdot3\cdot x \cdot \frac{2}{6} + 2\cdot3\cdot x \cdot \frac{1}{2x} = 2\cdot3\cdot x \cdot \frac{1}{3}$$
$$2x + 3 = 2x$$
$$3 = 0$$

We get a false equation. The given equation has no solution.

10. No solution

11. $\frac{4}{z} + \frac{2}{z} = 3$, LCD is z

$$z\left(\frac{4}{z} + \frac{2}{z}\right) = z\cdot3$$
$$z \cdot \frac{4}{z} + z \cdot \frac{2}{z} = 3z$$
$$4 + 2 = 3z$$
$$6 = 3z$$
$$2 = z$$

Check:

$$\frac{4}{z} + \frac{2}{z} = 3$$
$$\frac{4}{2} + \frac{2}{2} \;?\; 3$$
$$2 + 1 \;\Big|$$
$$3 \;\Big|\; \quad \text{TRUE}$$

The solution is 2.

12. $-\frac{1}{2}$

13. $y + \frac{5}{y} = -6$, LCD is y

$$y\left(y + \frac{5}{y}\right) = y(-6)$$
$$y\cdot y + y \cdot \frac{5}{y} = -6y$$
$$y^2 + 5 = -6y$$
$$y^2 + 6y + 5 = 0$$
$$(y + 5)(y + 1) = 0$$
$$y + 5 = 0 \quad \text{or} \quad y + 1 = 0$$
$$y = -5 \quad \text{or} \quad y = -1$$

Check:

For -5:

$$y + \frac{5}{y} = -6$$
$$-5 + \frac{5}{-5} \;?\; -6$$
$$-5 - 1 \;\Big|$$
$$-6 \;\Big|\; \quad \text{TRUE}$$

For -1:

$$y + \frac{5}{y} = -6$$
$$-1 + \frac{5}{-1} \;?\; -6$$
$$-1 - 5 \;\Big|$$
$$-6 \;\Big|\; \quad \text{TRUE}$$

The solutions are -5 and -1.

14. -4, -1

15. $2x - \frac{6}{x} = 1$, LCD is x

$$x\left(2x - \frac{6}{x}\right) = x\cdot1$$
$$x\cdot2x - x \cdot \frac{6}{x} = x$$
$$2x^2 - 6 = x$$
$$2x^2 - x - 6 = 0$$
$$(x - 2)(2x + 3) = 0$$
$$x - 2 = 0 \quad \text{or} \quad 2x + 3 = 0$$
$$x = 2 \quad \text{or} \quad 2x = -3$$
$$x = 2 \quad \text{or} \quad x = -\frac{3}{2}$$

Check:

For 2:

$$2x - \frac{6}{x} = 1$$
$$2\cdot2 - \frac{6}{2} \;?\; 1$$
$$4 - 3 \;\Big|$$
$$1 \;\Big|\; \quad \text{TRUE}$$

For $-\frac{3}{2}$:

$$2x - \frac{6}{x} = 1$$
$$2\left(-\frac{3}{2}\right) - \frac{6}{-\frac{3}{2}} \;?\; 1$$
$$-3 + 4 \;\Big|$$
$$1 \;\Big|\; \quad \text{TRUE}$$

The solutions are 2 and $-\frac{3}{2}$.

16. $-\frac{5}{2}$, 3

17. $\frac{y - 1}{y - 3} = \frac{2}{y - 3}$, LCD is $y - 3$

$$(y - 3) \cdot \frac{y - 1}{y - 3} = (y - 3) \cdot \frac{2}{y - 3}$$
$$y - 1 = 2$$
$$y = 3$$

Check:

$$\begin{array}{c|c} \frac{y-1}{y-3} & = \frac{2}{y-3} \\ \hline \frac{3-1}{3-3} & ?\ \frac{2}{3-3} \\ \frac{2}{0} & \frac{2}{0} \end{array} \quad \text{UNDEFINED}$$

We know that 3 is not a solution of the original equation because it results in division by 0. The equation has no solution.

18. No solution

19. $\frac{x+1}{x} = \frac{3}{2}$, LCD is $2x$

$$2x \cdot \frac{x+1}{x} = 2x \cdot \frac{3}{2}$$
$$2(x+1) = x \cdot 3$$
$$2x + 2 = 3x$$
$$2 = x$$

Check:

$$\begin{array}{c|c} \frac{x+1}{x} & = \frac{3}{2} \\ \hline \frac{2+1}{2} & ?\ \frac{3}{2} \\ \frac{3}{2} & \end{array} \quad \text{TRUE}$$

The solution is 2.

20. 3

21. $\frac{x-3}{x+2} = \frac{1}{5}$, LCD is $5(x + 2)$

$$5(x+2) \cdot \frac{x-3}{x+2} = 5(x+2) \cdot \frac{1}{5}$$
$$5(x-3) = x + 2$$
$$5x - 15 = x + 2$$
$$4x = 17$$
$$x = \frac{17}{4}$$

Check:

$$\begin{array}{c|c} \frac{x-3}{x+2} & = \frac{1}{5} \\ \hline \frac{\frac{17}{4} - \frac{12}{4}}{\frac{17}{4} + \frac{8}{4}} & ?\ \frac{1}{5} \\ \frac{5}{4} \cdot \frac{4}{25} & \\ \frac{1}{5} & \end{array} \quad \text{TRUE}$$

The solution is $\frac{17}{4}$.

22. 14

23. $\frac{3}{y+1} = \frac{2}{y-3}$, LCD is $(y + 1)(y - 3)$

$$(y+1)(y-3) \cdot \frac{3}{y+1} = (y+1)(y-3) \cdot \frac{2}{y-3}$$
$$3(y-3) = 2(y+1)$$
$$3y - 9 = 2y + 2$$
$$y = 11$$

Check:

$$\begin{array}{c|c} \frac{3}{y+1} & = \frac{2}{y-3} \\ \hline \frac{3}{11+1} & ?\ \frac{2}{11-3} \\ \frac{3}{12} & \frac{2}{8} \\ \frac{1}{4} & \frac{1}{4} \end{array} \quad \text{TRUE}$$

The solution is 11.

24. -11

25. $\frac{7}{5x-2} = \frac{5}{4x}$, LCD is $4x(5x - 2)$

$$4x(5x-2) \cdot \frac{7}{5x-2} = 4x(5x-2) \cdot \frac{5}{4x}$$
$$4x \cdot 7 = 5(5x-2)$$
$$28x = 25x - 10$$
$$3x = -10$$
$$x = -\frac{10}{3}$$

Since $-\frac{10}{3}$ checks, it is the solution.

26. 11

27. $\frac{2}{x} - \frac{3}{x} + \frac{4}{x} = 5$, LCD is x

$$x\left(\frac{2}{x} - \frac{3}{x} + \frac{4}{x}\right) = x \cdot 5$$
$$2 - 3 + 4 = 5x$$
$$3 = 5x$$
$$\frac{3}{5} = x$$

Check:

$$\begin{array}{c|c} \frac{2}{x} - \frac{3}{x} + \frac{4}{x} & = 5 \\ \hline \frac{2}{\frac{3}{5}} - \frac{3}{\frac{3}{5}} + \frac{4}{\frac{3}{5}} & ?\ \frac{15}{3} \\ \frac{10}{3} - \frac{15}{3} + \frac{20}{3} & \\ \frac{15}{3} & \end{array} \quad \text{TRUE}$$

The solution is $\frac{3}{5}$.

28. $\frac{3}{4}$

29. $\frac{1}{2} - \frac{4}{9x} = \frac{4}{9} - \frac{1}{6x}$, LCD is $18x$

$$18x\left(\frac{1}{2} - \frac{4}{9x}\right) = 18x\left(\frac{4}{9} - \frac{1}{6x}\right)$$
$$9x - 8 = 8x - 3$$
$$x = 5$$

Since 5 checks, it is the solution.

30. -1

31. $$\frac{z}{z-1} = \frac{6}{z+1},\ \text{LCD is } (z-1)(z+1)$$

$$(z-1)(z+1)\cdot\frac{z}{z-1} = (z-1)(z+1)\cdot\frac{6}{z+1}$$

$$(z+1)\cdot z = (z-1)\cdot 6$$

$$z^2 + z = 6z - 6$$

$$z^2 - 5z + 6 = 0$$

$$(z-3)(z-2) = 0$$

$$z - 3 = 0 \quad\text{or}\quad z - 2 = 0$$

$$z = 3 \quad\text{or}\quad z = 2$$

Both 3 and 2 check. The solutions are 3 and 2.

32. 3, 2

33. $$\frac{60}{x} - \frac{60}{x-5} = \frac{2}{x},\ \text{LCD is } x(x-5)$$

$$x(x-5)\left[\frac{60}{x} - \frac{60}{x-5}\right] = x(x-5)\cdot\frac{2}{x}$$

$$60(x-5) - 60x = 2(x-5)$$

$$60x - 300 - 60x = 2x - 10$$

$$-300 = 2x - 10$$

$$-290 = 2x$$

$$-145 = x$$

Since -145 checks, it is the solution.

34. -23

35. $$\frac{x}{x-2} + \frac{x}{x^2-4} = \frac{x+3}{x+2}$$

$$\frac{x}{x-2} + \frac{x}{(x+2)(x-2)} = \frac{x+3}{x+2},\ \text{LCD is } (x+2)(x-2)$$

$$(x+2)(x-2)\left[\frac{x}{x-2} + \frac{x}{(x+2)(x-2)}\right] = (x+2)(x-2)\cdot\frac{x+3}{x+2}$$

$$x(x+2) + x = (x-2)(x+3)$$

$$x^2 + 2x + x = x^2 + x - 6$$

$$3x = x - 6$$

$$2x = -6$$

$$x = -3$$

Since -3 checks, it is the solution.

36. 4

37. $$\frac{a}{2a-6} - \frac{3}{a^2-6a+9} = \frac{a-2}{3a-9}$$

$$\frac{a}{2(a-3)} - \frac{3}{(a-3)(a-3)} = \frac{a-2}{3(a-3)}$$

LCD is $2\cdot 3(a-3)(a-3)$

$$6(a-3)(a-3)\left[\frac{a}{2(a-3)} - \frac{3}{(a-3)(a-3)}\right] = 6(a-3)(a-3)\cdot\frac{a-2}{3(a-3)}$$

$$3a(a-3) - 6\cdot 3 = 2(a-3)(a-2)$$

$$3a^2 - 9a - 18 = 2(a^2 - 5a + 6)$$

$$3a^2 - 9a - 18 = 2a^2 - 10a + 12$$

$$a^2 + a - 30 = 0$$

$$(a+6)(a-5) = 0$$

$$a + 6 = 0 \quad\text{or}\quad a - 5 = 0$$

$$a = -6 \quad\text{or}\quad a = 5$$

Both -6 and 5 check. The solutions are -6 and 5.

38. 3

39. $$\frac{2x+3}{x-1} = \frac{10}{x^2-1} + \frac{2x-3}{x+1}$$

$$\frac{2x+3}{x-1} = \frac{10}{(x-1)(x+1)} + \frac{2x-3}{x+1}$$

LCD is $(x-1)(x+1)$

$$(x-1)(x+1)\cdot\frac{2x+3}{x-1} = (x-1)(x+1)\left[\frac{10}{(x-1)(x+1)} + \frac{2x-3}{x+1}\right]$$

$$(x+1)(2x+3) = 10 + (x-1)(2x-3)$$

$$2x^2 + 5x + 3 = 10 + 2x^2 - 5x + 3$$

$$5x + 3 = 13 - 5x$$

$$10x = 10$$

$$x = 1$$

We know that 1 is not a solution of the original equation because it results in division by 0. The equation has no solution.

40. No solution

41. $81x^4 - y^4 = (9x^2 + y^2)(9x^2 - y^2)$

$\quad = (9x^2 + y^2)(3x + y)(3x - y)$

42. a) Inconsistent

b) Consistent

43. Familiarize. Let x, y, and z represent the number of multiple-choice, true-false and fill-in questions, respectively.

Translate. The total number of questions is 70.

$$x + y + z = 70$$

Number of true-false is twice number of fill-ins.

$$y = 2z$$

Number of multiple-choice is 5 less than number of true-false.

$$x = y - 5$$

Carry out. Solving the system of three equations we get (25,30,15).

Check. The sum of 25, 30, and 15 is 70. The number of true-false, 30, is twice the number of fill-ins, 15. The number of multiple-choice, 25, is 5 less than the number of true-false, 30.

State. On the test there are 25 multiple-choice, 30 true-false, and 15 fill-in questions.

44. 16 and 18 or -18 and -16

45.

46.

47. Set f(a) equal to g(a) and solve for a.

$$\frac{a - \frac{3}{2}}{a + \frac{2}{3}} = \frac{a + \frac{1}{2}}{a - \frac{2}{3}}$$

$$\frac{\frac{2a - 3}{2}}{\frac{3a + 2}{3}} = \frac{\frac{2a + 1}{2}}{\frac{3a - 2}{3}}$$

$$\frac{2a - 3}{2} \cdot \frac{3}{3a + 2} = \frac{2a + 1}{2} \cdot \frac{3}{3a - 2},$$

LCD is 2(3a + 2)(3a - 2)

$$2(3a + 2)(3a - 2) \cdot \frac{2a - 3}{2} \cdot \frac{3}{3a + 2} =$$
$$2(3a + 2)(3a - 2) \cdot \frac{2a + 1}{2} \cdot \frac{3}{3a - 2}$$

$$3(3a - 2)(2a - 3) = 3(3a + 2)(2a + 1)$$
$$(3a - 2)(2a - 3) = (3a + 2)(2a + 1)$$
$$6a^2 - 13a + 6 = 6a^2 + 7a + 2$$
$$-13a + 6 = 7a + 2$$
$$-20a = -4$$
$$a = \frac{1}{5}$$

The result checks. Thus, when $a = \frac{1}{5}$, f(a) = g(a).

48. -8

49.

$$\left[\frac{1}{1 + x} + \frac{x}{1 - x}\right] \div \left[\frac{x}{1 + x} - \frac{1}{1 - x}\right] = -1$$

$$\frac{1\cdot(1 - x) + x(1 + x)}{(1 + x)(1 - x)} \div \frac{x(1 - x) - 1\cdot(1 + x)}{(1 + x)(1 - x)} = -1$$

$$\frac{x^2 + 1}{(1 + x)(1 - x)} \cdot \frac{(1 + x)(1 - x)}{-x^2 - 1} = -1$$

$$\frac{x^2 + 1}{-x^2 - 1} = -1$$

$$-\frac{x^2 + 1}{x^2 + 1} = -1$$

$$-1 = -1$$

We know that 1 and -1 cannot be solutions of the original equation, because they result in division by 0. Since -1 = -1 is true for all values of x, all real numbers except 1 and -1 are solutions.

50. $-\frac{7}{2}$

51.

$$\frac{2.315}{y} - \frac{12.6}{17.4} = \frac{6.71}{7} + 0.763$$

LCD is 7(17.4)y

$$7(17.4)y\left[\frac{2.315}{y} - \frac{12.6}{17.4}\right] = 7(17.4)y\left[\frac{6.71}{7} + 0.763\right]$$

$$7(17.4)(2.315)-7(12.6)y=(17.4)(6.71)y+7(17.4)(0.763)y$$
$$281.967 - 88.2y = 116.754y + 92.9334y$$
$$281.967 = 88.2y + 116.754y + 92.9334y$$
$$281.967 = 297.8874y$$
$$0.9465556 \approx y$$

Since 0.9465556 checks, it is the solution.

52. 0.0854697

53.

$$\frac{x^2 + 6x - 16}{x - 2} = x + 8, \; x \neq 2$$

$$\frac{(x + 8)(x - 2)}{x - 2} = x + 8$$

$$\frac{(x + 8)\cancel{(x - 2)}}{\cancel{x - 2}} = x + 8$$

$$x + 8 = x + 8$$
$$8 = 8$$

Since 8 = 8 is true for all values of x, the original equation is true for any possible replacements of the variable. It is an identity.

54. Identity

55. 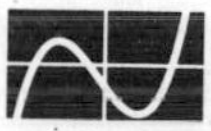

Exercise Set 6.5

1. Familiarize. We let x = the number.

Translate.

The reciprocal of 5	plus	the reciprocal of 7	is	the reciprocal of the number.
$\frac{1}{5}$	+	$\frac{1}{7}$	=	$\frac{1}{x}$

Carry out. We solve the equation.

$$\frac{1}{5} + \frac{1}{7} = \frac{1}{x}, \text{ LCD} = 35x$$

$$35x\left[\frac{1}{5} + \frac{1}{7}\right] = 35x \cdot \frac{1}{x}$$

$$7x + 5x = 35$$
$$12x = 35$$
$$x = \frac{35}{12}$$

Check. The reciprocal of $\frac{35}{12}$ is $\frac{12}{35}$. The sum of $\frac{1}{5} + \frac{1}{7}$ is $\frac{7}{35} + \frac{5}{35}$, or $\frac{12}{35}$, so the value checks.

State. The number is $\frac{35}{12}$.

2. 2

3. Familiarize. We let x = the number.

Translate.

A number plus 6 times its reciprocal is -5.

$$x \quad + \quad 6 \quad \cdot \quad \frac{1}{x} \quad = \quad -5$$

Carry out. We solve the equation.

$$x + \frac{6}{x} = -5,\ \text{LCD is } x$$
$$x\left(x + \frac{6}{x}\right) = x(-5)$$
$$x^2 + 6 = -5x$$
$$x^2 + 5x + 6 = 0$$
$$(x + 3)(x + 2) = 0$$
$$x + 3 = 0 \quad \text{or} \quad x + 2 = 0$$
$$x = -3 \quad \text{or} \quad x = -2$$

Check. The possible solutions are -3 and -2. We check -3 in the conditions of the problem.

Number:	-3
6 times the reciprocal of the number:	$6\left(-\frac{1}{3}\right) = -2$
Sum of the number and 6 times its reciprocal:	$-3 + (-2) = -5$

The number -3 checks. So does -2, but that check is left to the student.

State. The numbers -3 and -2 both satisfy the conditions of the problem.

4. -3, -7

5. Familiarize. We let x = the first integer. Then x + 1 = the second, and their product = x(x + 1).

Translate.

Reciprocal of the product is $\frac{1}{72}$.

$$\frac{1}{x(x + 1)} = \frac{1}{72}$$

Carry out. We solve the equation.

$$72x(x + 1) \cdot \frac{1}{x(x + 1)} = 72x(x + 1) \cdot \frac{1}{72}$$
$$72 = x(x + 1)$$
$$72 = x^2 + x$$
$$0 = x^2 + x - 72$$
$$0 = (x + 9)(x - 8)$$
$$x + 9 = 0 \quad \text{or} \quad x - 8 = 0$$
$$x = -9 \quad \text{or} \quad x = 8$$

Check. When x = -9, x + 1 = -8, so -9 and -8 is a possible solution. When x = 8, x + 1 = 9, so 8 and 9 is also a possible solution, We check -9 and -8. Their product is -9(-8), or 72, and the reciprocal of the product is $\frac{1}{72}$. These numbers check. Now we check 8 and 9. Their product is 8·9, or 72, and the reciprocal of the product is $\frac{1}{72}$. Both possible solutions check.

State. The numbers are -9 and -8 or 8 and 9.

6. 6 and 7 or -7 and -6

7. Familiarize. The job takes Sam 5 hours working alone and Willy 9 hours working alone. Then in 1 hour, Sam does $\frac{1}{5}$ of the job and Willy does $\frac{1}{9}$ of the job. Working together, they can do $\frac{1}{5} + \frac{1}{9}$ of the job in 1 hour. Let t represent the time required for Sam and Willy, working together, to do the job.

Translate. We want to find t such that

$$t\left(\frac{1}{5}\right) + t\left(\frac{1}{9}\right) = 1, \text{ or } \frac{t}{5} + \frac{t}{9} = 1,$$

where 1 represents one entire job.

Carry out. We solve the equation.

$$45\left(\frac{t}{5} + \frac{t}{9}\right) = 45\cdot 1$$
$$9t + 5t = 45$$
$$14t = 45$$
$$t = \frac{45}{14}$$

Check. The possible solution is $\frac{45}{14}$ hours. If Sam works $\frac{45}{14}$ hours, he will do $\frac{1}{5} \cdot \frac{45}{14}$, or $\frac{9}{14}$ of the job. If Willy works $\frac{45}{14}$ hours, he will do $\frac{1}{9} \cdot \frac{45}{14}$, or $\frac{5}{14}$ of the job. Together, they will do $\frac{9}{14} + \frac{5}{14}$ of the job, or one complete job.

State. Working together it will take them $\frac{45}{14}$, or $3\frac{3}{14}$ hours.

8. $1\frac{5}{7}$ hr

9. Familiarize. The pool can be filled in 12 hours with only the pipe and in 30 hours with only the hose. Then in 1 hour, the pipe fills $\frac{1}{12}$ of the pool, and the hose fills $\frac{1}{30}$ of the pool. Using both the pipe and the hose, $\frac{1}{12} + \frac{1}{30}$ of the pool can be filled in 1 hour. Suppose that it takes t hours to fill the pool using both the pipe and hose.

Translate. We want to find t such that

$$t\left(\frac{1}{12}\right) + t\left(\frac{1}{30}\right) = 1, \text{ or } \frac{t}{12} + \frac{t}{30} = 1,$$

where 1 represents one entire job.

Carry out. We solve the equation.

$$60\left(\frac{t}{12} + \frac{t}{30}\right) = 60\cdot 1$$
$$5t + 2t = 60$$
$$7t = 60$$
$$t = \frac{60}{7}$$

Check. The possible solution is $\frac{60}{7}$ hours. If the pipe is used $\frac{60}{7}$ hours, it fills $\frac{1}{12} \cdot \frac{60}{7}$, or $\frac{5}{7}$ of the pool. If the hose is used $\frac{60}{7}$ hours, it fills $\frac{1}{30} \cdot \frac{60}{7}$, or $\frac{2}{7}$ of the pool. Using both, $\frac{5}{7} + \frac{2}{7}$ of the pool, or all of it, will be filled in $\frac{60}{7}$ hours.

State. Using both the pipe and the hose, it will take $\frac{60}{7}$, or $8\frac{4}{7}$ hours, to fill the pool.

10. 9.9 hr

11. Familiarize. The job takes Bill 5.5 hours working alone and his partner 7.5 hours working alone. Then in 1 hour, Bill does $\frac{1}{5.5}$ of the job and his partner does $\frac{1}{7.5}$ of the job. Working together, they can do $\frac{1}{5.5} + \frac{1}{7.5}$ of the job in 1 hour. Suppose it takes them t hours working together.

Translate. We want to find t such that

$$t\left(\frac{1}{5.5}\right) + t\left(\frac{1}{7.5}\right) = 1, \text{ or } \frac{t}{5.5} + \frac{t}{7.5} = 1.$$

Carry out. We solve the equation.

$$5.5(7.5)\left(\frac{t}{5.5} + \frac{t}{7.5}\right) = 5.5(7.5)(1)$$
$$7.5t + 5.5t = 41.25$$
$$13t = 41.25$$
$$t = \frac{41.25}{13}, \text{ or } \frac{41.25}{13} \cdot \frac{4}{4}$$
$$t = \frac{165}{52}, \text{ or } 3\frac{9}{52}$$

Check. The possible solution is $3\frac{9}{52}$ hours. If Bill works $3\frac{9}{52}$ hours, he will do $\frac{1}{5.5}\left(\frac{165}{52}\right) = \frac{2}{11} \cdot \frac{165}{52}$, or $\frac{15}{26}$ of the job. If his partner works $3\frac{9}{52}$ hours, he will do $\frac{1}{7.5}\left(\frac{165}{52}\right) = \frac{2}{15} \cdot \frac{165}{52}$, or $\frac{11}{26}$ of the job. Together they will do $\frac{15}{26} + \frac{11}{26}$ of the job, or all of it.

State. Working together, it will take them $3\frac{9}{52}$ hours.

12. 2.475 hr

13. Familiarize.
Let t = the time it takes Claudia to paint the house alone. Jan takes 4 times as long as Claudia, or 4t. Thus in 1 day Claudia does $\frac{1}{t}$ of the job and Jan does $\frac{1}{4t}$ of the job.

Translate. In 8 days Claudia and Jan will complete one entire job, so we have

$$8\left(\frac{1}{t}\right) + 8\left(\frac{1}{4t}\right) = 1, \text{ or } \frac{8}{t} + \frac{2}{t} = 1.$$

Carry out. We solve the equation.

$$t\left(\frac{8}{t} + \frac{2}{t}\right) = t \cdot 1$$
$$8 + 2 = t$$
$$10 = t$$

Check. In 1 day, Claudia will do $\frac{1}{10}$ of the job and Jan will do $\frac{1}{40}$ of the job. Together they will do $\frac{1}{10} + \frac{1}{40} = \frac{4}{40} + \frac{1}{40} = \frac{5}{40} = \frac{1}{8}$ of the job. Thus, in 8 days they will do $8 \cdot \frac{1}{8} = 1$ job. The answer checks.

State. It would take Claudia 10 days and Jan 40 days working alone.

14. Zsuzanne: $\frac{4}{3}$ hr, Stan: 4 hr

15. Familiarize. Working alone, Rosita does $\frac{1}{2}$ of the job in 1 hr. Let t = the time it takes Helga to wax the car, working alone. Then in 1 hr she does $\frac{1}{t}$ of the job. Represent 45 min as $\frac{3}{4}$ hr.

Translate. In $\frac{3}{4}$ hr they do 1 entire job, working together, so we have

$$\frac{3}{4}\left(\frac{1}{2}\right) + \frac{3}{4}\left(\frac{1}{t}\right) = 1, \text{ or } \frac{3}{8} + \frac{3}{4t} = 1.$$

Carry out. We solve the equation.

$$8t\left(\frac{3}{8} + \frac{3}{4t}\right) = 8t \cdot 1$$
$$3t + 6 = 8t$$
$$6 = 5t$$
$$\frac{6}{5} = t$$

Check. In $\frac{3}{4}$ hr, Rosita will do $\frac{3}{4} \cdot \frac{1}{2}$, or $\frac{3}{8}$, of the job, and Helga will do $\frac{3}{4}\left(\frac{1}{\frac{6}{5}}\right)$, or $\frac{3}{4} \cdot \frac{5}{6} = \frac{5}{8}$, of the job. Together they do $\frac{3}{8} + \frac{5}{8} = 1$ job. The answer checks.

State. It would take Helga $\frac{6}{5}$, or $1\frac{1}{5}$ hr, working alone.

16. 6 hr

17. Familiarize. Let t represent the time it takes Jake, working alone. Since it takes Jake half the time it takes Skyler, 2t represents Skyler's time, working alone. In 1 hr, Jake does $\frac{1}{t}$ and Skyler does $\frac{1}{2t}$ of the job.

Translate. Working together, they can do the entire job in 4 hr, so we want to find t such that

$$4\left(\frac{1}{t}\right) + 4\left(\frac{1}{2t}\right) = 1, \text{ or } \frac{4}{t} + \frac{2}{t} = 1.$$

Carry out. We solve the equation.

$$t\left(\frac{4}{t} + \frac{2}{t}\right) = t\cdot 1$$
$$4 + 2 = t$$
$$6 = t$$

Check. If Jake does the job in 6 hr, then in 4 hr he does $4 \cdot \frac{1}{6}$, or $\frac{2}{3}$, of the job. It takes Skyler $2\cdot 6$, or 12 hr, so in 4 hr he does $4 \cdot \frac{1}{12}$, or $\frac{1}{3}$, of the job. Working together, they do $\frac{2}{3} + \frac{1}{3} = 1$ job in 4 hr. The answer checks.

State. It takes Jake 6 hr and Skyler 12 hr working alone.

18. Sara: 22.5 min, Kate: 45 min

19. Familiarize. We will convert hours to minutes:

2 hr = $2\cdot 60$ min = 120 min

2 hr 55 min = 120 min + 55 min = 175 min

Let t = the time it takes Deb to do the job alone. Then t + 120 = the time it takes John alone. In 1 hour (60 minutes) Deb does $\frac{1}{t}$ and John does $\frac{1}{t + 120}$ of the job.

Translate. In 175 min John and Deb will complete one entire job, so we have

$$175\left(\frac{1}{t}\right) + 175\left(\frac{1}{t + 120}\right) = 1, \text{ or } \frac{175}{t} + \frac{175}{t + 120} = 1.$$

Carry out. We solve the equation.

$$t(t + 120)\left[\frac{175}{t} + \frac{175}{t + 120}\right] = t(t + 120)(1)$$
$$175(t + 120) + 175t = t^2 + 120t$$
$$175t + 21{,}000 + 175t = t^2 + 120t$$
$$0 = t^2 - 230t - 21{,}000$$
$$0 = (t - 300)(t + 70)$$

$$t - 300 = 0 \quad \text{or} \quad t + 70 = 0$$
$$t = 300 \quad \text{or} \quad t = -70$$

Check. Since negative time has no meaning in this problem, -70 is not a solution of the original problem. If the job takes Deb 300 min and it takes John 300 + 120 = 420 min, then in 175 min they would complete

$$175\left(\frac{1}{300}\right) + 175\left(\frac{1}{420}\right) = \frac{7}{12} + \frac{5}{12} = 1 \text{ job.}$$

The result checks.

State. It would take Deb 300 min, or 5 hr, to do the job alone.

20. 8 hr

21. Familiarize. Let t = the time it takes the new machine to do the job alone. Then 2t = the time it takes the old machine to do the job alone. In 1 hr the new machine does $\frac{1}{t}$ of the job and the old machine does $\frac{1}{2t}$ of the job.

Translate. In 15 hr the two machines will complete one entire job, so we have

$$15\left(\frac{1}{t}\right) + 15\left(\frac{1}{2t}\right) = 1, \text{ or } \frac{15}{t} + \frac{15}{2t} = 1.$$

Carry out. We solve the equation.

$$2t\left(\frac{15}{t} + \frac{15}{2t}\right) = 2t\cdot 1$$
$$30 + 15 = 2t$$
$$45 = 2t$$
$$\frac{45}{2} = t$$

Check. In 1 hr, the new machine will do $\frac{1}{\frac{45}{2}}$, or $\frac{2}{45}$ of the job and the old machine will do $\frac{1}{2 \cdot \frac{45}{2}}$, or $\frac{1}{45}$ of the job. Together they will do $\frac{2}{45} + \frac{1}{45} = \frac{3}{45} = \frac{1}{15}$ of the job. Then in 15 hr they will do $15 \cdot \frac{1}{15} = 1$ job. The answer checks.

State. Working alone, the job would take the new machine $\frac{45}{2}$, or $22\frac{1}{2}$ hr, and it would take the old machine $2 \cdot \frac{45}{2}$, or 45 hr.

22. 2 hr

23. Familiarize. We first make a drawing. We let r = the speed of the boat in still water. Then r - 3 = the speed upstream and r + 3 = the speed downstream.

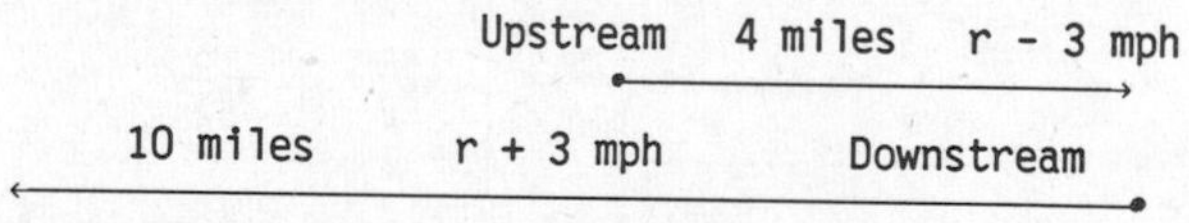

We organize the information in a table. The time is the same both upstream and downstream so we use t for each time.

	Distance	Speed	Time
Upstream	4	r - 3	t
Downstream	10	r + 3	t

Translate. Using $t = \frac{d}{r}$ we get two different equations from the rows of the table.

$$t = \frac{4}{r - 3} \text{ and } t = \frac{10}{r + 3}$$

Carry out. Since both rational expressions represent the same time, t, we can set them equal to each other and solve.

$$\frac{4}{r - 3} = \frac{10}{r + 3}, \text{ LCD is } (r - 3)(r + 3)$$
$$(r - 3)(r + 3) \cdot \frac{4}{r - 3} = (r - 3)(r + 3) \cdot \frac{10}{r + 3}$$
$$4(r + 3) = 10(r - 3)$$
$$4r + 12 = 10r - 30$$
$$42 = 6r$$
$$7 = r$$

Check. If $r = 7$ mph, then $r - 3$ is 4 mph and $r + 3$ is 10 mph. The time upstream is $\frac{4}{4}$, or 1 hour. The time downstream is $\frac{10}{10}$, or 1 hour. The times are the same. The values check.

State. The speed of the boat in still water is 7 mph.

24. 12 mph

25. Familiarize. We first make a drawing. We let r = the person's rate on a nonmoving sidewalk. Then $r + 7$ = the rate walking forward on the moving sidewalk, and $r - 7$ = the rate walking in the opposite direction.

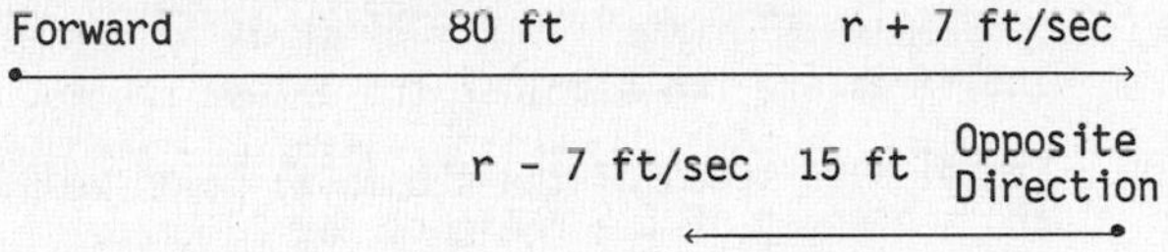

We organize the information in a table. Both distances are covered in the same amount of time. We let t represent the time.

	Distance	Rate	Time
Forward	80	$r + 7$	t
Opposite Direction	15	$r - 7$	t

Translate. Using $t = \frac{d}{r}$ we get two equations.

$$t = \frac{80}{r + 7} \text{ and } t = \frac{15}{r - 7}$$

Carry out. Set the rational expressions equal to each other and solve.

$$\frac{80}{r + 7} = \frac{15}{r - 7}, \text{ LCD is } (r + 7)(r - 7)$$

$$(r + 7)(r - 7) \cdot \frac{80}{r + 7} = (r + 7)(r - 7) \cdot \frac{15}{r - 7}$$

$$80(r - 7) = 15(r + 7)$$

$$80r - 560 = 15r + 105$$

$$65r = 665$$

$$r = \frac{665}{65} = \frac{133}{13}$$

Check. If the answer checks, the rate on a nonmoving sidewalk is $\frac{133}{13}$ ft/sec. We calculate:

Walking forward: Rate is $\frac{133}{13} + 7 = \frac{224}{13}$ ft/sec

In opposite direction: Rate is $\frac{133}{13} - 7 = \frac{42}{13}$ ft/sec

Now we calculate time, using $t = \frac{d}{r}$.

At $\frac{224}{13}$ ft/sec: $t = \frac{80}{\frac{224}{13}} = \frac{80}{1} \cdot \frac{13}{224} = \frac{65}{14}$ sec

At $\frac{42}{13}$ ft/sec: $t = \frac{15}{\frac{42}{13}} = \frac{15}{1} \cdot \frac{13}{42} = \frac{65}{14}$ sec

The times are the same, so the answer checks.

State. The person would walk $\frac{133}{13}$, or $10\frac{3}{13}$ ft/sec, on a nonmoving sidewalk.

26. 10.8 ft/sec

27. Familiarize. We first make a drawing. Let r = Rosanna's speed. Then $r + 2$ = Simone's speed.

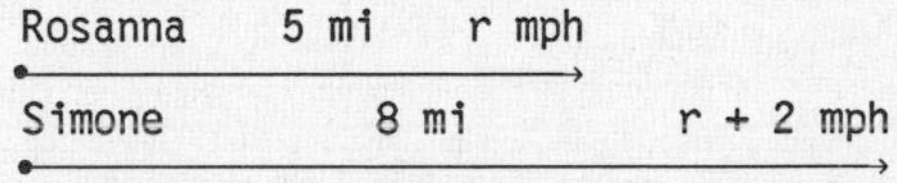

We organize the information in a table. Both distances are covered in the same amount of time. We let t represent the time.

	Distance	Rate	Time
Rosanna	5	r	t
Simone	8	$r + 2$	t

Translate. Using $t = \frac{d}{r}$ we get two equations.

$$t = \frac{5}{r} \text{ and } t = \frac{8}{r + 2}$$

Carry out. Set the rational expressions equal to each other and solve.

$$\frac{5}{r} = \frac{8}{r + 2}, \text{ LCD is } r(r + 2)$$

$$r(r + 2) \cdot \frac{5}{r} = r(r + 2) \cdot \frac{8}{r + 2}$$

$$5(r + 2) = 8r$$

$$5r + 10 = 8r$$

$$10 = 3r$$

$$\frac{10}{3} = r$$

Check. If the answer checks, Rosanna's speed is $\frac{10}{3}$ mph, and Simone's speed is $\frac{10}{3} + 2$, or $\frac{16}{3}$ mph. We calculate time, using $t = \frac{d}{r}$.

Rosanna: $t = \frac{5}{\frac{10}{3}} = \frac{5}{1} \cdot \frac{3}{10} = \frac{3}{2}$ hr

Simone: $t = \frac{8}{\frac{16}{3}} = \frac{8}{1} \cdot \frac{3}{16} = \frac{3}{2}$ hr

The times are the same, so the answer checks.

State. Rosanna's speed is $\frac{10}{3}$, or $3\frac{1}{3}$ mph, and Simone's speed is $\frac{16}{3}$, or $5\frac{1}{3}$ mph.

28. Local: 35 mph, express: 42 mph

29. Familiarize. We let r = the speed of the B train. Then r - 12 = the speed of the A train. The times are the same. We use t for each. We organize the information in a table.

	Distance	Rate	Time
A train	230	r - 12	t
B train	290	r	t

Translate. Using $t = \frac{d}{r}$, we get two equations.

$$t = \frac{230}{r - 12} \text{ and } t = \frac{290}{r}$$

Carry out. Set the rational expressions equal to each other and solve.

$$\frac{230}{r - 12} = \frac{290}{r}, \text{ LCD is } r(r - 12)$$

$$r(r - 12) \cdot \frac{230}{r - 12} = r(r - 12) \cdot \frac{290}{r}$$

$$230r = 290(r - 12)$$

$$230r = 290r - 3480$$

$$-60r = -3480$$

$$r = 58$$

Check. If the speed of the B train is 58 mph, then the speed of the A train is 58 - 12, or 46 mph. The time for the A train is $\frac{230}{46}$, or 5 hours. The time for the B train is $\frac{290}{58}$, or 5 hours. The time are the same. The values check.

State. The speed of the A train is 46 mph; the speed of the B train is 58 mph.

30. Freight: 66 mph, passenger: 80 mph

31. Familiarize. Let r = the speed of the freighter. Then r + 10 = the speed of the steamboat. Both distances are covered in the same amount of time. We let t = the time. We organize the information in a table.

	Distance	Rate	Time
Steamboat	75	r + 10	t
Freighter	50	r	t

Translate. Using $t = \frac{d}{r}$, we get two equations.

$$t = \frac{75}{r + 10} \text{ and } t = \frac{50}{r}$$

Carry out. We set the rational expressions equal to each other and solve.

$$\frac{75}{r + 10} = \frac{50}{r}, \text{ LCD is } r(r + 10)$$

$$r(r + 10) \cdot \frac{75}{r + 10} = r(r + 10) \cdot \frac{50}{r}$$

$$75r = 50(r + 10)$$

$$75r = 50r + 500$$

$$25r = 500$$

$$r = 20$$

Check. If the speed of the freighter is 20 km/h, then the speed of the steamboat is 20 + 10, or 30 km/h. The time for the steamboat is $\frac{75}{30}$, or $\frac{5}{2}$ hr. The time for the freighter is $\frac{50}{20}$, or $\frac{5}{2}$ hr. The times are the same, so the answer checks.

State. The speed of the steamboat is 30 km/h, and the speed of the freighter is 20 km/h.

32. Jaime: 23 km/h, Mara: 15 km/h

33. Familiarize. We first make a drawing. We let r = the speed of the river. Then 15 + r = her speed downstream and 15 - r = her speed upstream.

140 km   15 + r   t hours   Downstream
←————————————————————

35 km   15 - r   t hours   Upstream
————————————→

The times are the same. Let t represent the time. We organize the information in a table.

	Distance	Speed	Time
Downstream	140	15 + r	t
Upstream	35	15 - r	t

Translate. Using $t = \frac{d}{r}$, we get two equations from the table.

$$t = \frac{140}{15 + r} \text{ and } t = \frac{35}{15 - r}$$

Carry out. Set the rational expressions equal to each other and solve.

$$\frac{140}{15 + r} = \frac{35}{15 - r}, \text{ LCD is } (15 + r)(15 - r)$$

$$(15 + r)(15 - r) \cdot \frac{140}{15 + r} = (15 + r)(15 - r) \cdot \frac{35}{15 - r}$$

$$140(15 - r) = 35(15 + r)$$

$$2100 - 140r = 525 + 35r$$

$$1575 = 175r$$

$$9 = r$$

Check. If r = 9, then the speed downstream is 15 + 9, or 24 km/h and the speed upstream is 15 - 9, or 6 km/h. The time for the trip downstream is $\frac{140}{24}$, or $5\frac{5}{6}$ hours. The time for the trip upstream is $\frac{35}{6}$, or $5\frac{5}{6}$ hours. The times are same. The values check.

State. The speed of the river is 9 km/h.

34. $1\frac{1}{5}$ km/h

35. Familiarize. We make a drawing. Let c = the speed of the current.

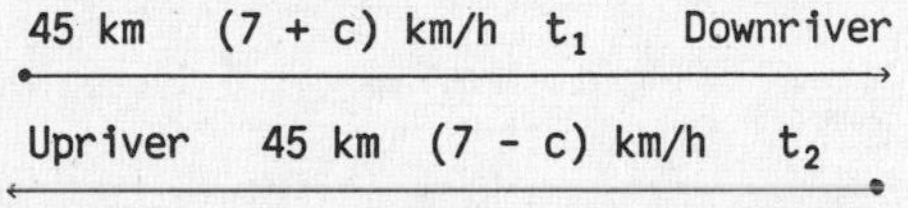

We organize the information in a table.

	Distance	Speed	Time
Downriver	45	$7 + c$	t_1
Upriver	45	$7 - c$	t_2

Translate. We use the information that $t_1 + t_2 = 14$. Since $t = \frac{d}{r}$ we have $t_1 = \frac{45}{7 + c}$ and $t_2 = \frac{45}{7 - c}$. Substituting, we have

$$\frac{45}{7 + c} + \frac{45}{7 - c} = 14.$$

Carry out. We solve the equation.

$$(7 + c)(7 - c)\left[\frac{45}{7 + c} + \frac{45}{7 - c}\right] = (7 + c)(7 - c)14$$
$$45(7 - c) + 45(7 + c) = 14(49 - c^2)$$
$$315 - 45c + 315 + 45c = 686 - 14c^2$$
$$14c^2 - 56 = 0$$
$$14(c + 2)(c - 2) = 0$$

$c + 2 = 0$ or $c - 2 = 0$

$c = -2$ or $c = 2$

Check. Since speed cannot be negative in this problem, 2 cannot be a solution of the original problem. If the speed of the current is 2 km/h, the barge travels upriver at 7 - 2, or 5 km/h. At this rate it takes $\frac{45}{5}$, or 9 hr, to travel 45 km. The barge travels downriver at 7 + 2, or 9 km/h. At this rate it takes $\frac{45}{9}$, or 5 hr, to travel 45 km. The total travel time is 9 + 5, or 14 hr. The answer checks.

State. The speed of the current is 2 km/h.

36. 3 mph

37. Familiarize. Let w = the wind speed. We organize the information in a table.

	Distance	Speed	Time
Into the wind	240	$100 - w$	t_1
With the wind	240	$100 + w$	t_2

Translate. We use the information that $t_1 + t_2 = 5$. Since $t = \frac{d}{r}$, we have $t_1 = \frac{240}{100 - w}$ and $t_2 = \frac{240}{100 + w}$. Substituting, we have

$$\frac{240}{100 - w} + \frac{240}{100 + w} = 5.$$

Carry out. We solve the equation.

$$(100-w)(100+w)\left[\frac{240}{100-w} + \frac{240}{100+w}\right] = (100-w)(100+w)(5)$$
$$240(100+w) + 240(100-w) = 5(10{,}000-w^2)$$
$$24{,}000 + 240w + 24{,}000 - 240w = 50{,}000 - 5w^2$$
$$5w^2 - 2000 = 0$$
$$5(w + 20)(w - 20) = 0$$

$w + 20 = 0$ or $w - 20 = 0$

$w = -20$ or $w = 20$

Check. Since speed cannot be negative in this problem, -20 is not a solution of the original problem. If the wind speed is 20 mph, the plane travels at 100 - 20, or 80 mph, into the wind. At this rate it takes $\frac{240}{80}$, or 3 hr, to travel 240 mi. The plane travels 100 + 20, or 120 mph, with the wind. At this rate it takes $\frac{240}{120}$, or 2 hr, to travel 240 mi. The total travel time is 3 + 2, or 5 hr. The answer checks.

State. The wind speed is 20 mph.

38. 5 m per minute

39. Familiarize. Let r = the speed at which the car actually traveled, and let t = the actual travel time. We organize the information in a table.

	Distance	Speed	Time
Actual speed	120	r	t
Faster speed	120	$r + 10$	$t - 2$

Translate. From the first row of the table we have $120 = rt$, and from the second row we have $120 = (r + 10)(t - 2)$. Solving the first equation for t, we have $t = \frac{120}{r}$. Substituting for t in the second equation, we have

$$120 = (r + 10)\left[\frac{120}{r} - 2\right].$$

Carry out. We solve the equation.

$$120 = (r + 10)\left[\frac{120}{r} - 2\right]$$
$$120 = 120 - 2r + \frac{1200}{r} - 20$$
Multiplying
$$20 = -2r + \frac{1200}{r}$$
$$r \cdot 20 = r\left[-2r + \frac{1200}{r}\right]$$

$20r = -2r^2 + 1200$

$2r^2 + 20r - 1200 = 0$

$2(r^2 + 10r - 600) = 0$

$2(r + 30)(r - 20) = 0$

$r + 30 = 0$ or $r - 20 = 0$

$r = -30$ or $r = 20$

Check. Since speed cannot be negative in this problem, -30 cannot be a solution of the original problem. If the speed is 20 mph, it takes $\frac{120}{20}$, or 6 hr, to travel 120 mi. If the speed is 10 mph faster, or 30 mph, it takes $\frac{120}{30}$, or 4 hr, to travel 120 mi. Since 4 hr is 2 hr less time than 6 hr, the answer checks.

State. The speed was 20 mph.

40. 12 mph

41. Familiarize. Let n = the number.

Translate.

8% of what number is 480.

$0.08 \cdot n = 480$

Carry out. We solve the equation.

$0.08n = 480$

$n = 6000$

Check. 8% of 6000 is 0.08(6000), or 480. The answer checks.

State. 8% of 6000 is 480. The number is 6000.

42. $\{x | x \text{ is a real number and } x \neq 5 \text{ and } x \neq -1\}$

43. $|x - 2| = 9$

$x - 2 = 9$ or $x - 2 = -9$

$x = 11$ or $x = -7$

The solutions are 11 and -7.

44. $2y - 8xy^2 + 6xy$

45.

46.

47. Familiarize. We first make a drawing. We let x = the speed of the boat in still water and r = the speed of the stream. Then x + r = the speed downstream and x - r = the speed upstream.

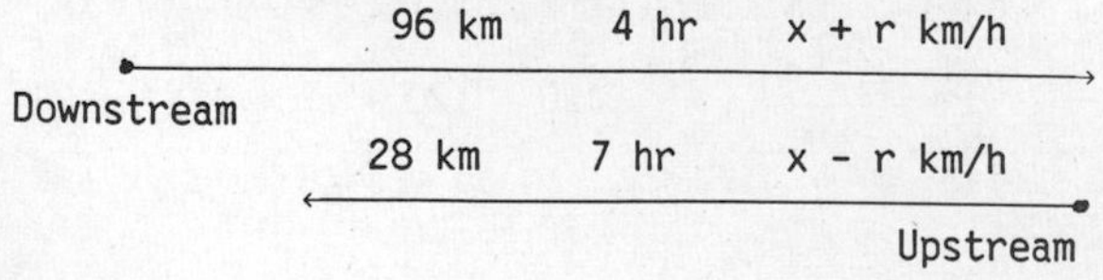

We organize the information in a table.

	Distance	Speed	Time
Downstream	96	x + r	4
Upstream	28	x - r	7

Translate. Using d = rt, we get a system of equations from the table.

$96 = (x + r)4$ or $x + r = 24$

$28 = (x - r)7$ or $x - r = 4$

Carry out. Solving the system we get (10,14).

Check. The speed downstream is 14 + 10, or 24 km/h. The distance downstream is $24 \cdot 4$, or 96 km. The speed upstream is 14 - 10, or 4 km/h. The distance upstream is $4 \cdot 7$, or 28 km. The values check.

State. The speed of the boat is 14 km/h; the speed of the stream is 10 km/h.

48. The airplane can fly 700 mi away from the airport and return to the airport without refueling.

49. Familiarize. We let x = the speed of the current and 3x = the speed of the boat. Then the speed up the river is 3x - x, or 2x, and the speed down the river is 3x + x, or 4x. The total distance is 100 km; thus the distance each way is 50 km. Using $t = \frac{d}{r}$, we can use $\frac{50}{2x}$ for the time up the river and $\frac{50}{4x}$ for the time down the river.

Translate. Since the total of the times is 10 hours, we have the following equation.

$\frac{50}{2x} + \frac{50}{4x} = 10$

Carry out. We solve the equation. The LCD is 4x.

$4x\left(\frac{50}{2x} + \frac{50}{4x}\right) = 4x \cdot 10$

$100 + 50 = 40x$

$150 = 40x$

$\frac{15}{4} = x$

or $x = 3\frac{3}{4}$

Check. If the speed of the current is $\frac{15}{4}$ km/h, then the speed of the boat is $3 \cdot \frac{15}{4}$, or $\frac{45}{4}$. The speed up the river is $\frac{45}{4} - \frac{15}{4}$, or $\frac{15}{2}$ km/h, and the time traveling up the river is $50 \div \frac{15}{2}$, or $6\frac{2}{3}$ hr. The speed down the river is $\frac{45}{4} + \frac{15}{4}$, or 15 km/h, and the time traveling down the river is $50 \div 15$, or $3\frac{1}{3}$ hr. The total time for the trip is $6\frac{2}{3} + 3\frac{1}{3}$, or 10 hr. The value checks.

State. The speed of the current is $3\frac{3}{4}$ km/h.

50. 30 mi

51. Familiarize. If the drain is closed, $\frac{1}{9}$ of the tank is filled in 1 hr. If the tank is not being filled, $\frac{1}{11}$ of the tank is drained in 1 hr. If the tank is being filled with the drain left open, $\frac{1}{9} - \frac{1}{11}$ of the tank is filled in 1 hr. Let t = the time it takes to fill the tank with the drain left open.

Translate. We want to find t such that

$$t\left(\frac{1}{9} - \frac{1}{11}\right) = 1, \text{ or } \frac{t}{9} - \frac{t}{11} = 1.$$

Carry out. We solve the equation.

$$9\cdot 11\left(\frac{t}{9} - \frac{t}{11}\right) = 9\cdot 11\cdot 1$$
$$11t - 9t = 99$$
$$2t = 99$$
$$t = \frac{99}{2}$$

Check. In $\frac{99}{2}$ hr, we have $\frac{99}{2}\left(\frac{1}{9} - \frac{1}{11}\right) = \frac{11}{2} - \frac{9}{2} = \frac{2}{2} = 1$ full tank.

State. It will take $\frac{99}{2}$, or $49\frac{1}{2}$ hr, to fill the tank.

52. 40 min

53. Familiarize. It helps to first make a drawing.

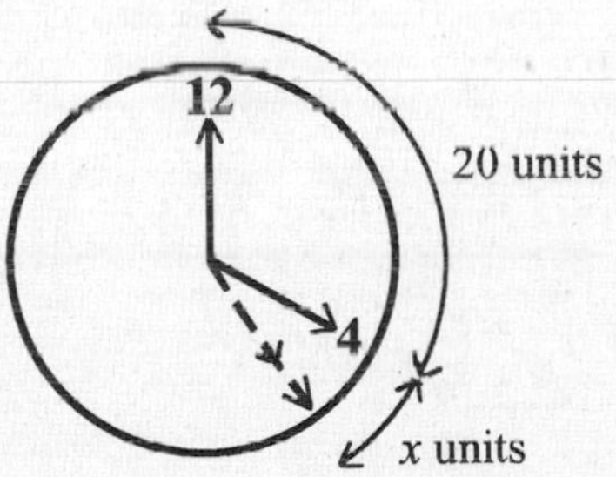

The minute hand moves 60 units per hour while the hour hand moves 5 units per hour, where one unit represents one minute on the face of the clock. When the hands are in the same position the first time, the hour hand will have moved x units and the minute hand will have moved x + 20 units. The times are the same. We use t for time.

	Distance	Speed	Time
Minute	x + 20	60	t
Hour	x	5	t

Translate. Using $t = \frac{d}{r}$, we get two equations from the table.

$$t = \frac{x + 20}{60} \text{ and } t = \frac{x}{5}$$

Carry out. Set the rational expressions equal to each other and solve.

$$\frac{x + 20}{60} = \frac{x}{5}, \text{ LCD is } 60$$
$$60 \cdot \frac{x + 20}{60} = 60 \cdot \frac{x}{5}$$
$$x + 20 = 12x$$
$$20 = 11x$$
$$\frac{20}{11} = x$$
$$\text{or } x = 1\frac{9}{11}$$

Check. If the hour hand moves $1\frac{9}{11}$ units, then the minute hand moves $1\frac{9}{11} + 20$, or $21\frac{9}{11}$ units $\left(21\frac{9}{11}\text{ minutes after 4}\right)$. The time for the hour hand is $\frac{20}{11} \div 5$, or $\frac{4}{11}$ hour. The time for the minute hand is $\frac{240}{11} \div 60$, or $\frac{4}{11}$ hour. The times are the same. The values check.

State. At $21\frac{9}{11}$ minutes after 4:00, the hands will be in the same position.

54. $8\frac{2}{11}$ min after 10:30

55. Familiarize. We make a drawing and organize the information in a table. Remember: $t = \frac{d}{r}$.

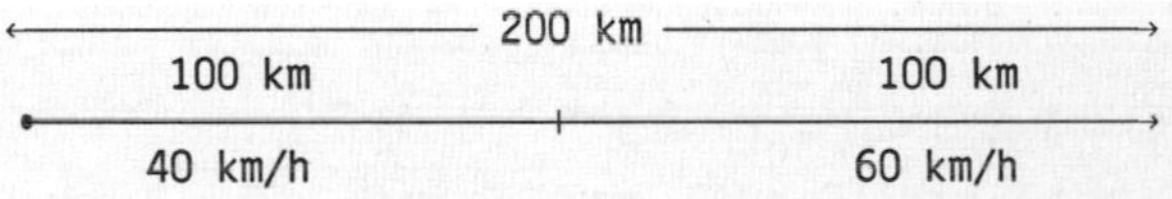

	Distance	Speed	Time
1st part	100	40	$\frac{100}{40}$, or $\frac{5}{2}$
2nd part	100	60	$\frac{100}{60}$, or $\frac{5}{3}$

The total distance is 200 km.

The total time is $\frac{5}{2} + \frac{5}{3}$, or $\frac{25}{6}$ hr.

Translate. $\text{Average speed} = \frac{\text{Total distance}}{\text{Total time}}$

$$= \frac{200 \text{ km}}{\frac{25}{6} \text{ hr}}$$

Carry out. We simplify.

$$\text{Average speed} = \frac{200 \text{ km}}{1} \cdot \frac{6}{25 \text{ hr}} = 48 \text{ km/h}$$

Check. Calculate the total time at 48 km/h. $\frac{200}{48} = \frac{25}{6}$ hr. The answer checks.

State. The average speed was 48 km/h.

56. $51\frac{3}{7}$ mph

Exercise Set 6.6

1. $\dfrac{30x^8 - 15x^6 + 40x^4}{5x^4}$

$= \dfrac{30x^8}{5x^4} - \dfrac{15x^6}{5x^4} + \dfrac{40x^4}{5x^4}$

$= 6x^4 - 3x^2 + 8$

2. $4y^4 + 3y^3 - 6$

3. $\dfrac{-14a^3 + 28a^2 - 21a}{7a}$

$= \dfrac{-14a^3}{7a} + \dfrac{28a^2}{7a} - \dfrac{21a}{7a}$

$= -2a^2 + 4a - 3$

4. $-8x^3 - 6x^2 - 3x$

5. $(9y^4 - 18y^3 + 27y^2) \div 9y$

$= \dfrac{9y^4}{9y} - \dfrac{18y^3}{9y} + \dfrac{27y^2}{9y}$

$= y^3 - 2y^2 + 3y$

6. $12a^2 + 14a - 10$

7. $(36x^6 - 18x^4 - 12x^2) \div -6x$

$= \dfrac{36x^6}{-6x} - \dfrac{18x^4}{-6x} - \dfrac{12x^2}{-6x}$

$= -6x^5 - (-3x^3) - (-2x)$

$= -6x^5 + 3x^3 + 2x$

8. $-6y^5 + 9y^2 + 1$

9. $(a^2b - a^3b^3 - a^5b^5) \div a^2b$

$= \dfrac{a^2b}{a^2b} - \dfrac{a^3b^3}{a^2b} - \dfrac{a^5b^5}{a^2b}$

$= 1 - ab^2 - a^3b^4$

10. $x - xy - x^2$

11. $(6p^2q^2 - 9p^2q + 12pq^2) \div -3pq$

$= \dfrac{6p^2q^2}{-3pq} - \dfrac{9p^2q}{-3pq} + \dfrac{12pq^2}{-3pq}$

$= -2pq - (-3p) + (-4q)$

$= -2pq + 3p - 4q$

12. $4z - 2y^2z^3 + 3y^4z^2$

13.
$$\begin{array}{rl}
 & x + 7 \\
x + 3 & \overline{)\,x^2 + 10x + 21} \\
 & \underline{x^2 + 3x} \\
 & \quad 7x + 21 \qquad (x^2 + 10x) - (x^2 + 3x) = 7x \\
 & \quad \underline{7x + 21} \\
 & \qquad 0
\end{array}$$

The answer is $x + 7$.

14. $y - 4$

15.
$$\begin{array}{rl}
 & a - 12 \\
a + 4 & \overline{)\,a^2 - 8a - 16} \\
 & \underline{a^2 + 4a} \\
 & \quad -12a - 16 \qquad (a^2 - 8a) - (a^2 + 4a) = -12a \\
 & \quad \underline{-12a - 48} \\
 & \qquad 32 \qquad (-12a - 16) - (-12a - 48) = 32
\end{array}$$

The answer is $a - 12$, R 32, or $a - 12 + \dfrac{32}{a + 4}$.

16. $y - 5$, R -50, or $y - 5 + \dfrac{-50}{y - 5}$

17.
$$\begin{array}{rl}
 & x - 6 \\
x - 5 & \overline{)\,x^2 - 11x + 23} \\
 & \underline{x^2 - 5x} \\
 & \quad -6x + 23 \\
 & \quad \underline{-6x + 30} \\
 & \qquad -7
\end{array}$$

The answer is $x - 6$, R -7, or $x - 6 + \dfrac{-7}{x - 5}$.

18. $x - 4$, R -5, or $x - 4 + \dfrac{-5}{x - 7}$

19.
$$\begin{array}{rl}
 & y - 5 \\
y + 5 & \overline{)\,y^2 + 0y - 25} \qquad \text{Writing in the missing term} \\
 & \underline{y^2 + 5y} \\
 & \quad -5y - 25 \\
 & \quad \underline{-5y - 25} \\
 & \qquad 0
\end{array}$$

The answer is $y - 5$.

20. $a + 9$

21.
$$\begin{array}{rl}
 & y^2 - 2y - 1 \\
y - 2 & \overline{)\,y^3 - 4y^2 + 3y - 6} \\
 & \underline{y^3 - 2y^2} \\
 & \quad -2y^2 + 3y \\
 & \quad \underline{-2y^2 + 4y} \\
 & \qquad -y - 6 \\
 & \qquad \underline{-y + 2} \\
 & \qquad\quad -8
\end{array}$$

The answer is $y^2 - 2y - 1$, R -8, or $y^2 - 2y - 1 + \dfrac{-8}{y - 2}$.

22. $x^2 - 2x - 2$, R -13, or $x^2 - 2x - 2 + \dfrac{-13}{x - 3}$

23.
$$\begin{array}{rl}
 & 2x^2 - x + 1 \\
x + 2 & \overline{)\,2x^3 + 3x^2 - x - 3} \\
 & \underline{2x^3 + 4x^2} \\
 & \quad -x^2 - x \\
 & \quad \underline{-x^2 - 2x} \\
 & \qquad x - 3 \\
 & \qquad \underline{x + 2} \\
 & \qquad\quad -5
\end{array}$$

The answer is $2x^2 - x + 1$, R -5, or $2x^2 - x + 1 + \dfrac{-5}{x + 2}$.

24. $3x^2 + x - 1$, R -4, or $3x^2 + x - 1 + \frac{-4}{x - 2}$

25.
$$\begin{array}{r} a^2 + 4a + 15 \\ a - 4\overline{)a^3 + 0a^2 - a + 12} \\ \underline{a^3 - 4a^2} \\ 4a^2 - a \\ \underline{4a^2 - 16a} \\ 15a + 12 \\ \underline{15a - 60} \\ 72 \end{array}$$

The answer is $a^2 + 4a + 15$, R 72, or $a^2 + 4a + 15 + \frac{72}{a - 4}$.

26. $x^2 - 2x + 3$

27.
$$\begin{array}{r} 4x^2 - 6x + 9 \\ 2x + 3\overline{)8x^3 + 0x^2 + 0x + 27} \\ \underline{8x^3 + 12x^2} \\ -12x^2 + 0x \\ \underline{-12x^2 - 18x} \\ 18x + 27 \\ \underline{18x + 27} \\ 0 \end{array}$$

The answer is $4x^2 - 6x + 9$.

28. $16y^2 + 8y + 4$

29.
$$\begin{array}{r} x^2 + 6 \\ x^2 - 7\overline{)x^4 - x^2 - 42} \\ \underline{x^4 - 7x^2} \\ 6x^2 - 42 \\ \underline{6x^2 - 42} \\ 0 \end{array}$$

The answer is $x^2 + 6$.

30. $y^2 + 2$, R -48, or $y^2 + 2 + \frac{-48}{y^2 - 3}$

31.
$$\begin{array}{r} x^3 + x^2 - 1 \\ x - 1\overline{)x^4 + 0x^3 - x^2 - x + 2} \\ \underline{x^4 - x^3} \\ x^3 - x^2 \\ \underline{x^3 - x^2} \\ 0 - x + 2 \\ \underline{-x + 1} \\ 1 \end{array}$$

The answer is $x^3 + x^2 - 1$, R 1, or $x^3 + x^2 - 1 + \frac{1}{x - 1}$.

32. $y^3 - y^2 - 1$, R 4, or $y^3 - y^2 - 1 + \frac{4}{y + 1}$

33.
$$\begin{array}{r} 2y^2 + 2y - 1 \\ 5y - 2\overline{)10y^3 + 6y^2 - 9y + 10} \\ \underline{10y^3 - 4y^2} \\ 10y^2 - 9y \\ \underline{10y^2 - 4y} \\ -5y + 10 \\ \underline{-5y + 2} \\ 8 \end{array}$$

The answer is $2y^2 + 2y - 1$, R 8, or $2y^2 + 2y - 1 + \frac{8}{5y - 2}$.

34. $3x^2 - x + 4$, R 10, or $3x^2 - x + 4 + \frac{10}{2x - 3}$

35.
$$\begin{array}{r} 2x^2 - x - 9 \\ x^2 + 2\overline{)2x^4 - x^3 - 5x^2 + x - 6} \\ \underline{2x^4 + 4x^2} \\ -x^3 - 9x^2 + x \\ \underline{-x^3 - 2x} \\ -9x^2 + 3x - 6 \\ \underline{-9x^2 - 18} \\ 3x + 12 \end{array}$$

The answer is $2x^2 - x - 9$, R $3x + 12$, or $2x^2 - x - 9 + \frac{3x + 12}{x^2 + 2}$.

36. $3x^2 + 2x - 5$, R $2x - 5$, or $3x^2 + 2x - 5 + \frac{2x - 5}{x^2 - 2}$

37.
$$\begin{array}{r} 2x^3 + x^2 - 1 \\ x^2 + 1\overline{)2x^5 + x^4 + 2x^3 + 0x^2 + x} \\ \underline{2x^5 + 2x^3} \\ x^4 + 0x^2 \\ \underline{x^4 + x^2} \\ -x^2 + x \\ \underline{-x^2 - 1} \\ x + 1 \end{array}$$

The answer is $2x^3 + x^2 - 1$, R $x + 1$, or $2x^3 + x^2 - 1 + \frac{x + 1}{x^2 + 1}$.

38. $2x^3 - x + 1$, R $-x + 5$, or $2x^3 - x + 1 + \frac{-x + 5}{x^2 - 1}$

39. $x^2 - 5x = 0$

$x(x - 5) = 0$

$x = 0$ or $x - 5 = 0$ Principle of zero products

$x = 0$ or $x = 5$

The solutions are 0 and 5.

40. $-\frac{8}{5}, \frac{8}{5}$

41. Familiarize. Let x, $x + 1$, and $x + 2$ represent the three consecutive positive integers.

Translate. Rewording, we write an equation.

Product of first and second	is	product of second and third	less	26.
$x(x + 1)$	$=$	$(x + 1)(x + 2)$	$-$	26

Carry out. We solve the equation.

$$x^2 + x = x^2 + 3x + 2 - 26$$
$$x = 3x - 24$$
$$24 = 2x$$
$$12 = x$$

If the first integer is 12, the next two are 13 and 14.

Check. The product of 12 and 13 is 156. The product of 13 and 14 is 182, and 182 - 26 = 156. The numbers check.

State. The three consecutive positive integers are 12, 13, and 14.

42. $-54a^3$

43.

44.

45.
$$\begin{array}{r} x^2 + 2y \\ x^2 - xy + y^2 \overline{)\, x^4 - x^3y + x^2y^2 + 2x^2y - 2xy^2 + 2y^3} \\ \underline{x^4 - x^3y + x^2y^2} \qquad\qquad\qquad\qquad \\ 0 + 2x^2y - 2xy^2 + 2y^3 \\ \underline{2x^2y - 2xy^2 + 2y^3} \\ 0 \end{array}$$

The answer is $x^2 + 2y$.

46. $a^2 + ab$

47.
$$\begin{array}{r} x^3 + x^2y + xy^2 + y^3 \\ x - y \overline{)\, x^4 \qquad\qquad\qquad\qquad - y^4} \\ \underline{x^4 - x^3y} \qquad\qquad\qquad\qquad \\ x^3y \qquad\qquad\qquad \\ \underline{x^3y - x^2y^2} \qquad\qquad \\ x^2y^2 \qquad\qquad \\ \underline{x^2y^2 - xy^3} \qquad \\ xy^3 - y^4 \\ \underline{xy^3 - y^4} \\ 0 \end{array}$$

The answer is $x^3 + x^2y + xy^2 + y^3$.

48. $a^6 - a^5b + a^4b^2 - a^3b^3 + a^2b^4 - ab^5 + b^6$

49.
$$\begin{array}{r} x^2 + (-k - 2)x + (2k + 7) \\ x + 2 \overline{)\, x^3 - kx^2 + 3x + 7k} \\ \underline{x^3 + 2x^2} \qquad\qquad\qquad \\ (-k - 2)x^2 + 3x \qquad \\ \underline{(-k - 2)x^2 + (-2k - 4)x} \qquad \\ (2k + 7)x + 7k \\ \underline{(2k + 7)x + (4k + 14)} \end{array}$$

The remainder must be 0.

Thus, we solve the following equation for k.

$$7k - (4k + 14) = 0$$
$$7k - 4k - 14 = 0$$
$$3k = 14$$
$$k = \frac{14}{3}$$

50. $-\frac{3}{2}$

51. a), b)

$$\begin{array}{r} 3 \\ x + 2 \overline{)\, 3x + 7} \\ \underline{3x + 6} \\ 1 \end{array}$$

$$f(x) = 3 + \frac{1}{x + 2}$$

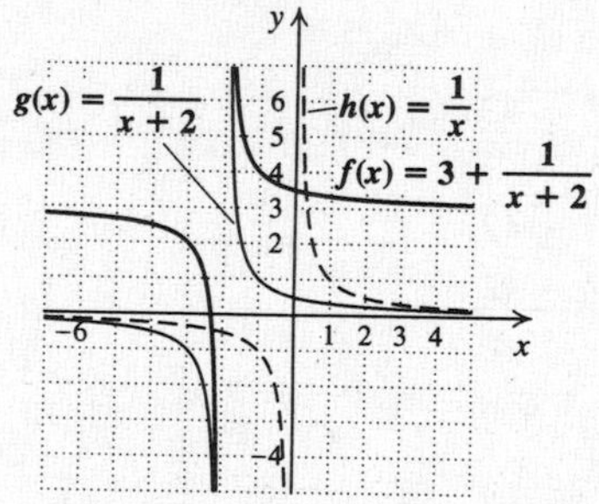

c) The graph of f looks like the graph of g, shifted up 3 units. The graph of g looks like the graph of h, shifted to the left 2 units.

Exercise Set 6.7

1. $(x^3 - 2x^2 + 2x - 5) \div (x - 1)$

$$\begin{array}{r|rrrr} 1 & 1 & -2 & 2 & -5 \\ & & 1 & -1 & 1 \\ \hline & 1 & -1 & 1 & -4 \end{array}$$

The answer is $x^2 - x + 1$ with R-4, or $x^2 - x + 1 + \frac{-4}{x - 1}$.

2. $x^2 - 3x + 5$, R-10, or $x^2 - 3x + 5 + \frac{-10}{x + 1}$

3. $(a^2 + 11a - 19) \div (a + 4) =$
$(a^2 + 11a - 19) \div [a - (-4)]$

$$\begin{array}{r|rrr} -4 & 1 & 11 & -19 \\ & & -4 & -28 \\ \hline & 1 & 7 & -47 \end{array}$$

The answer is $a + 7$ with R-47, or
$a + 7 + \frac{-47}{a + 4}$.

4. $a + 15$, R41, or $a + 15 + \frac{41}{a - 4}$

5. $(x^3 - 7x^2 - 13x + 3) \div (x - 2)$

$$\begin{array}{r|rrrr} 2 & 1 & -7 & -13 & 3 \\ & & 2 & -10 & -46 \\ \hline & 1 & -5 & -23 & -43 \end{array}$$

The answer is $x^2 - 5x - 23$ with R-43, or
$x^2 - 5x - 23 + \frac{-43}{x - 2}$.

6. $x^2 - 9x + 5$, R-7, or $x^2 - 9x + 5 + \frac{-7}{x + 2}$

7. $(3x^3 + 7x^2 - 4x + 3) \div (x + 3) =$
$(3x^3 + 7x^2 - 4x + 3) \div [x - (-3)]$

$$\begin{array}{r|rrrr} -3 & 3 & 7 & -4 & 3 \\ & & -9 & 6 & -6 \\ \hline & 3 & -2 & 2 & -3 \end{array}$$

The answer is $3x^2 - 2x + 2$ with R-3, or
$3x^2 -2x + 2 + \frac{-3}{x + 3}$.

8. $3x^2 + 16x + 44$, R135, or $3x^2 + 16x + 44 + \frac{135}{x - 3}$

9. $(y^3 - 3y + 10) \div (y - 2) =$
$(y^3 + 0y^2 - 3y + 10) \div (y - 2)$

$$\begin{array}{r|rrrr} 2 & 1 & 0 & -3 & 10 \\ & & 2 & 4 & 2 \\ \hline & 1 & 2 & 1 & 12 \end{array}$$

The answer is $y^2 + 2y + 1$ with R12, or
$y^2 + 2y + 1 + \frac{12}{y - 2}$.

10. $x^2 - 4x + 8$, R-8, or $x^2 - 4x + 8 + \frac{-8}{x + 2}$

11. $(3x^4 - 25x^2 - 18) \div (x - 3) =$
$(3x^4 + 0x^3 - 25x^2 + 0x - 18) \div (x - 3)$

$$\begin{array}{r|rrrrr} 3 & 3 & 0 & -25 & 0 & -18 \\ & & 9 & 27 & 6 & 18 \\ \hline & 3 & 9 & 2 & 6 & 0 \end{array}$$

The answer is $3x^3 + 9x^2 + 2x + 6$ with R0, or
$3x^3 + 9x^2 + 2x + 6$.

12. $6y^3 - 3y^2 + 9y + 1$, R3, or
$6y^3 - 3y^2 + 9y + 1 + \frac{3}{y + 3}$

13. $(x^3 - 27) \div (x - 3) =$
$(x^3 + 0x^2 + 0x - 27) \div (x - 3)$

$$\begin{array}{r|rrrr} 3 & 1 & 0 & 0 & -27 \\ & & 3 & 9 & 27 \\ \hline & 1 & 3 & 9 & 0 \end{array}$$

The answer is $x^2 + 3x + 9$ with R0, or
$x^2 + 3x + 9$.

14. $y^2 - 3y + 9$

15. $(y^5 - 1) \div (y - 1) =$
$(y^5 + 0y^4 + 0y^3 + 0y^2 + 0y - 1) \div (y - 1)$

$$\begin{array}{r|rrrrrr} 1 & 1 & 0 & 0 & 0 & 0 & -1 \\ & & 1 & 1 & 1 & 1 & 1 \\ \hline & 1 & 1 & 1 & 1 & 1 & 0 \end{array}$$

The answer is $y^4 + y^3 + y^2 + y + 1$ with R0, or
$y^4 + y^3 + y^2 + y + 1$.

16. $x^4 + 2x^3 + 4x^2 + 8x + 16$

17. $(3x^4 + 8x^3 + 2x^2 - 7x - 4) \div (x + 2) =$
$(3x^4 + 8x^3 + 2x^2 - 7x - 4) \div [(x - (-2)]$

$$\begin{array}{r|rrrrr} -2 & 3 & 8 & 2 & -7 & -4 \\ & & -6 & -4 & 4 & 6 \\ \hline & 3 & 2 & -2 & -3 & 2 \end{array}$$

The answer is $3x^3 + 2x^2 - 2x - 3$ with R2, or
$3x^3 + 2x^2 - 2x - 3 + \frac{2}{x + 2}$.

18. $2x^3 - 3x^2 - 2x + 3$, R4, or
$2x^3 - 3x^2 - 2x + 3 + \frac{4}{x + 1}$

19. $(3x^3 + 7x^2 - x + 1) \div \left[x + \frac{1}{3}\right] =$
$(3x^3 + 7x^2 - x + 1) \div \left[x - \left[-\frac{1}{3}\right]\right]$

$$\begin{array}{r|rrrr} -\frac{1}{3} & 3 & 7 & -1 & 1 \\ & & -1 & -2 & 1 \\ \hline & 3 & 6 & -3 & 2 \end{array}$$

The answer is $3x^2 + 6x - 3$ with R2, or
$3x^2 + 6x - 3 + \frac{2}{x + \frac{1}{3}}$.

20. $8x^2 - 2x + 6$, R2, or $8x^2 - 2x + 6 + \frac{2}{x - \frac{1}{2}}$

21. Graph: $2x - 3y < 6$

First graph the line $2x - 3y = 6$. The intercepts are (0,-2) and (3,0). We draw the line dashed since the inequality is <. Since the ordered pair (0,0) is a solution of the inequality ($2\cdot0 - 3\cdot0 < 6$ is true), we shade the upper half-plane.

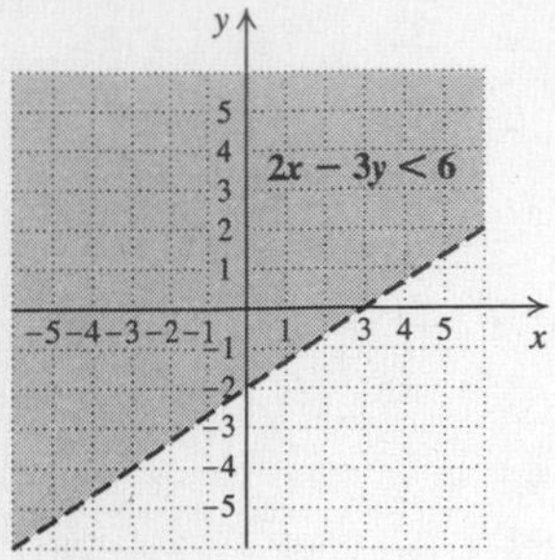

22.

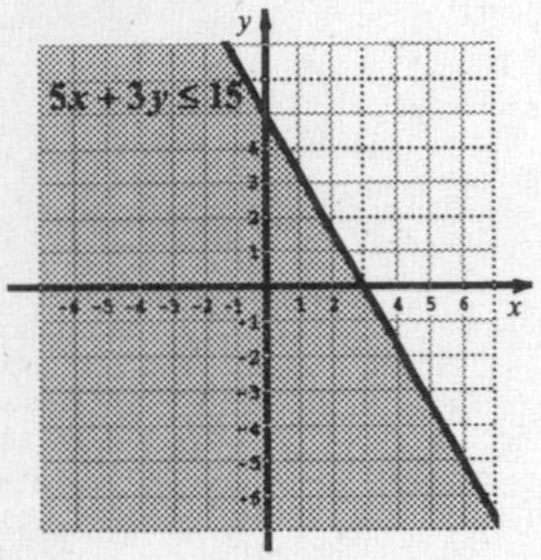

23. Graph: $y > 4$

First graph the line $y = 4$. The line is parallel to the x-axis with y-intercept (0,4). We draw the line dashed since the inequality is >. Since the ordered pair (0,0) is not a solution of the inequality (0 > 4 is false), we shade the upper half-plane.

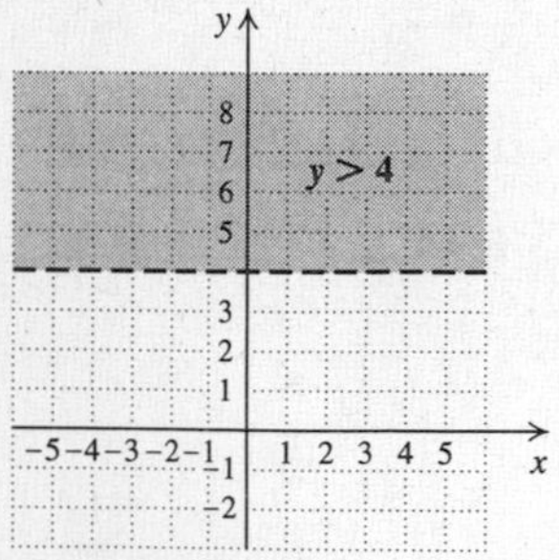

24.

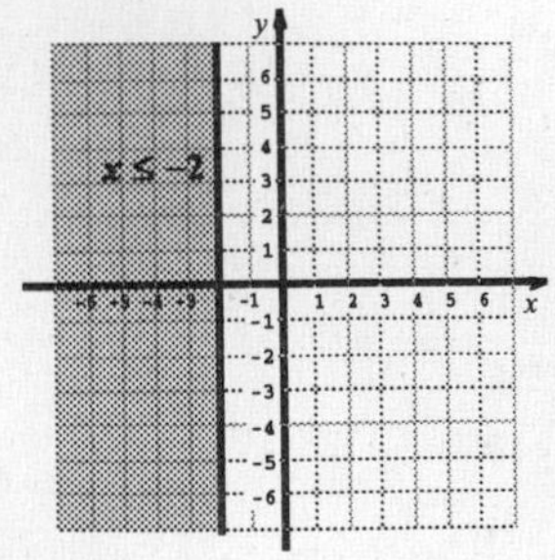

25.

26.

27. a)

4	1	-5	5	-4
		4	-4	4
	1	-1	1	0

The remainder is 0. Therefore, we know that $x^3 - 5x^2 + 5x - 4 = (x - 4)(x^2 - x + 1)$.

b) $f(x) = (x - 4)(x^2 - x + 1)$, so
$f(4) = (4 - 4)(4^2 - 4 + 1) = 0$

c) $f(4) = 4^3 - 5\cdot4^2 + 5\cdot4 - 4$
$= 64 - 80 + 20 - 4$
$= 0$

Exercise Set 6.8

1. $\frac{W_1}{W_2} = \frac{d_1}{d_2}$

$\frac{d_2W_1}{W_2} = d_1$ Multiplying by d_2

2. $W_1 = \frac{d_1W_2}{d_2}$

3. $s = \frac{(v_1 + v_2)t}{2}$

$2s = (v_1 + v_2)t$ Multiplying by 2

$\frac{2s}{v_1 + v_2} = t$ Multiplying by $\frac{1}{v_1 + v_2}$

4. $v_1 = \frac{2s}{t} - v_2$

5. $\frac{1}{R} = \frac{1}{r_1} + \frac{1}{r_2}$

$Rr_1r_2 \cdot \frac{1}{R} = Rr_1r_2\left(\frac{1}{r_1} + \frac{1}{r_2}\right)$ Multiplying by Rr_1r_2

$r_1r_2 = Rr_2 + Rr_1$

$r_1r_2 - Rr_1 = Rr_2$ Adding $-Rr_1$

$r_1(r_2 - R) = Rr_2$ Factoring

$r_1 = \frac{Rr_2}{r_2 - R}$ Multiplying by $\frac{1}{r_2 - R}$

6. $R = \frac{r_1r_2}{r_2 + r_1}$

7. $R = \frac{gs}{g + s}$

$R(g + s) = gs$ Multiplying by $g + s$

$Rg + Rs = gs$ Removing parentheses

$Rg = gs - Rs$ Adding $-Rs$

$Rg = s(g - R)$ Factoring

$\frac{Rg}{g - R} = s$ Multiplying by $\frac{1}{g - R}$

8. $g = \frac{Rs}{s - R}$

9. $$I = \frac{2V}{R + 2r}$$

$I(R + 2r) = 2V$ Multiplying by $R + 2r$

$IR + 2Ir = 2V$ Removing parentheses

$2Ir = 2V - IR$ Adding $-IR$

$r = \frac{2V - IR}{2I}$ Multiplying by $\frac{1}{2I}$

10. $R = \frac{2V}{I} - 2r$

11. $$\frac{1}{p} + \frac{1}{q} = \frac{1}{f}$$

$pqf\left[\frac{1}{p} + \frac{1}{q}\right] = pqf \cdot \frac{1}{f}$ Multiplying by pqf

$qf + pf = pq$

$pf = pq - qf$ Adding $-qf$

$pf = q(p - f)$ Factoring

$\frac{pf}{p - f} = q$ Multiplying by $\frac{1}{p - f}$

12. $p = \frac{qf}{q - f}$

13. $$I = \frac{nE}{R + nr}$$

$I(R + nr) = nE$ Multiplying by $R + nr$

$IR + Inr = nE$ Removing parentheses

$Inr = nE - IR$ Adding $-IR$

$r = \frac{nE - IR}{In}$ Multiplying by $\frac{1}{In}$

14. $n = \frac{IR}{E - Ir}$

15. $$S = \frac{H}{m(t_1 - t_2)}$$

$m(t_1 - t_2)S = H$ Multiplying by $m(t_1 - t_2)$

16. $t_1 = \frac{H}{Sm} + t_2$

17. $$\frac{E}{e} = \frac{R + r}{r}$$

$er \cdot \frac{E}{e} = er \cdot \frac{R + r}{r}$ Multiplying by er

$rE = e(R + r)$

$\frac{rE}{R + r} = e$ Multiplying by $\frac{1}{R + r}$

18. $r = \frac{er}{E - e}$

19. $$S = \frac{a - ar^n}{1 - r}$$

$S(1 - r) = a - ar^n$ Multiplying by $1 - r$

$S(1 - r) = a(1 - r^n)$ Factoring

$\frac{S - Sr}{1 - r^n} = a$ Multiplying by $\frac{1}{1 - r^n}$

20. $r = 1 - \frac{a}{S}$

21. Familiarize and Translate. We use the formula given.

Carry out. First solve the formula for R.

$$A = \frac{9R}{I}$$

$$AI = 9R$$

$$\frac{AI}{9} = R$$

Then substitute 2.4 for A and 45 for I.

$$\frac{2.4(45)}{9} = R$$

$$12 = R$$

Check. We substitute in the original formula.

$$\begin{array}{c|c} A & = 9R/I \\ \hline 2.4 \ ? & \frac{9\cdot 12}{45} \\ & \frac{108}{45} \\ & 2.4 \quad \text{TRUE} \end{array}$$

The answer checks.

State. 12 earned runs were given up.

22. $5\frac{5}{23}$ ohms

23. Familiarize. We want to find r_2 using the formula $\frac{1}{R} = \frac{1}{r_1} + \frac{1}{r_2}$. We know R = 5 ohms and r_1 = 50 ohms.

Translate. Using a result from Example 4, we have

$$r_2 = \frac{Rr_1}{r_1 - R}.$$

Carry out. We substitute and compute.

$$r_2 = \frac{5\cdot 50}{50 - 5}$$

$$= \frac{250}{45}$$

$$= \frac{50}{9}, \text{ or } 5\frac{5}{9}$$

Check. We substitute into the original formula.

$$\begin{array}{c|c} \frac{1}{R} & = \frac{1}{r_1} + \frac{1}{r_2} \\ \hline \frac{1}{5} \ ? & \frac{1}{50} + \frac{1}{\frac{50}{9}} \\ & \frac{1}{50} + \frac{9}{50} \\ & \frac{10}{50} \\ & \frac{1}{5} \qquad \text{TRUE} \end{array}$$

The answer checks.

State. A resistor with a resistance of $5\frac{5}{9}$ ohms should be used.

24. 6 cm

25. $\frac{1}{t} = \frac{1}{u} + \frac{1}{v}$

$tuv \cdot \frac{1}{t} = tuv\left(\frac{1}{u} + \frac{1}{v}\right)$ Multiplying by tuv

$uv = tv + tu$

$uv = t(v + u)$ Factoring

$\frac{uv}{v + u} = t$ Multiplying by $\frac{1}{v + u}$

26. $Q = \frac{2Tt - 2AT}{A - q}$

27. $v = \frac{d_2 - d_1}{t_2 - t_1}$

$(t_2 - t_1)v = (t_2 - t_1) \cdot \frac{d_2 - d_1}{t_2 - t_1}$ Multiplying by $(t_2 - t_1)$

$(t_2 - t_1)v = d_2 - d_1$

$t_2 - t_1 = \frac{d_2 - d_1}{v}$ Multiplying by $\frac{1}{v}$

$t_2 = \frac{d_2 - d_1}{v} + t_1$ Adding t_1

28. $r = \frac{A}{P} - 1$

29. Familiarize. We want to find r using the formula $P = \frac{A}{1 + r}$. We know P = \$1600 and A = \$1712.

Translate. Using the result of Exercise 28, we have $r = \frac{A}{P} - 1$.

Carry out. Substitute and compute.

$r = \frac{1712}{1600} - 1$

$r = 1.07 - 1$

$r = 0.07$

Check. We substitute in the original formula.

$P = \frac{A}{1 + r}$

$1600 \ ? \ \frac{1712}{1 + 0.07}$

$\frac{1712}{1.07}$

1600 TRUE

The answer checks.

State. The interest rate is 0.07, or 7%.

30. 3:45 A.M.

31. $I_t = \frac{I_f}{1 - T}$

$(1 - T)I_t = I_f$ Multiplying by (1 - T)

$1 - T = \frac{I_f}{I_t}$ Multiplying by $\frac{1}{I_t}$

$-T = \frac{I_f}{I_t} - 1$ Adding -1

$-1\cdot(-T) = -1\cdot\left(\frac{I_f}{I_t} - 1\right)$ Multiplying by -1

$T = -\frac{I_f}{I_t} + 1$

32. $d = \frac{LD}{R + L}$

33. $\frac{V^2}{R^2} = \frac{2g}{R + h}$

$(R + h) \cdot \frac{V^2}{R^2} = (R + h) \cdot \frac{2g}{R + h}$ Multiplying by (R + h)

$\frac{(R + h)V^2}{R^2} = 2g$

$R + h = \frac{2gR^2}{V^2}$ Multiplying by $\frac{R^2}{V^2}$

$h = \frac{2gR^2}{V^2} - R$ Adding -R

34. $t_1 = t_2 - \frac{v_2 - v_1}{a}$

35. Familiarize and Translate. We want to find T using the formula $A = \frac{2Tt + Qq}{2T + Q}$. We know t = 79, q = 90, A = 84, and Q = 5.

Carry out. We first solve the formula for T.

$(2T + Q)A = 2Tt + Qq$

$2AT + AQ = 2Tt + Qq$

$2AT - 2Tt = Qq - AQ$

$T(2A - 2t) = Qq - AQ$

$T = \frac{Qq - AQ}{2A - 2t}$

Now we substitute and compute.

$T = \frac{(5\cdot90) - (84\cdot5)}{(2\cdot84) - (2\cdot79)}$

$T = \frac{30}{10}$

$T = 3$

Check. We substitute in the original formula.

$A = \frac{2Tt + Qq}{2T + Q}$

$84 \ ? \ \frac{(2\cdot3\cdot79) + (5\cdot90)}{(2\cdot3) + 5}$

$\frac{924}{11}$

84 TRUE

The answer checks.

State. The student took 3 tests.

36. (-15,2), (-30,8)

37. Graph: $6x - y < 6$

First graph the line 6x - y = 6. The intercepts are (0,-6) and (1,0). We draw the line dashed since the inequality is <. Since the ordered pair (0,0) is a solution of the inequality (6·0 - 0 < 6 is true), we shade the half-plane containing (0,0).

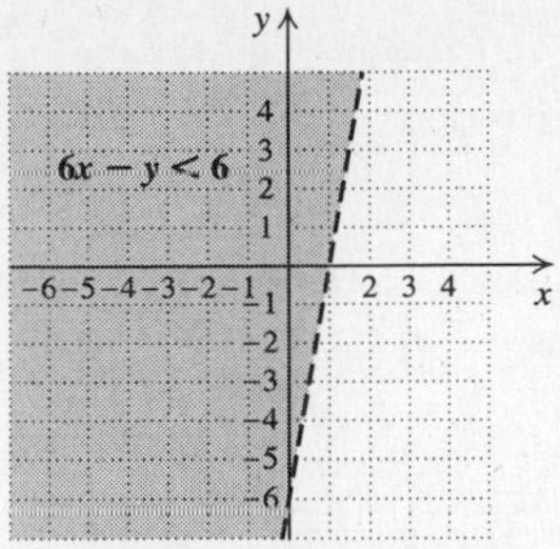

38. $8a^3 - 2a$

39. $t^3 + 8b^3 = t^3 + (2b)^3 = (t + 2b)(t^2 - 2tb + 4b^2)$

40. $-\frac{5}{3}, \frac{7}{2}$

41. Familiarize. We can use the formula in Exercise 33 to find h, the satellite's height above the earth. Then to find the satellite's distance from the center of the earth, we will compute R + h, the sum of the earth's radius and the satellite's height above earth. We know V = 6.5 mi/sec, R = 3960 mi, and g = 32.2 ft/sec^2.

Translate. Using the result of Exercise 33, we have $h = \frac{2gR^2}{V^2} - R$. Then $R + h = R + \frac{2gR^2}{V^2} - R = \frac{2gR^2}{V^2}$.

Carry out. We must convert 32.2 ft/sec^2 to mi/sec^2 so all units of length are the same.

$$32.2 \frac{ft}{sec^2} \cdot \frac{1 \text{ mi}}{5280 \text{ ft}} \approx 0.0060984 \frac{mi}{sec^2}$$

Now we substitute and compute.

$$R + h = \frac{2(0.0060984)(3960)^2}{(6.5)^2}$$

$$R + h \approx 4527$$

Check. We go over our calculations. The answer checks.

State. The satellite is 4527 mi from the center of the earth.

42. $M = \frac{2ab}{b + a}$

43. $$x^2\left(1 - \frac{2pq}{x}\right) = \frac{2p^2q^3 - pq^2x}{-q}$$

$$x^2 - 2pqx = -2p^2q^2 + pqx$$ Multiplying on the left; dividing on the right

$$x^2 - 3pqx + 2p^2q^2 = 0$$

$$(x - pq)(x - 2pq) = 0$$

$$x - pq = 0 \quad \text{or} \quad x - 2pq = 0$$

$$x = pq \quad \text{or} \quad x = 2pq$$

The solution set is {pq, 2pq}.

44. $t_1 = t_2 + \frac{(d_2 - d_1)(t_4 - t_3)}{a(t_4 - t_2)(t_4 - t_3) + d_3 - d_4}$

Exercise Set 7.1

1. The square roots of 16 are 4 and -4, because $4^2 = 16$ and $(-4)^2 = 16$.

2. 15, -15

3. The square roots of 144 are 12 and -12, because $12^2 = 144$ and $(-12)^2 = 144$.

4. 3, -3

5. The square roots of 400 are 20 and -20, because $20^2 = 400$ and $(-20)^2 = 400$.

6. 9, -9

7. The square roots of 49 are 7 and -7, because $7^2 = 49$ and $(-7)^2 = 49$.

8. 30, -30

9. $-\sqrt{\frac{49}{36}} = -\frac{7}{6}$ Since $\sqrt{\frac{49}{36}} = \frac{7}{6}$, $-\sqrt{\frac{49}{36}} = -\frac{7}{6}$.

10. $-\frac{19}{3}$

11. $\sqrt{196} = 14$ Remember, $\sqrt{}$ indicates the principal square root.

12. 21

13. $-\sqrt{\frac{16}{81}} = -\frac{4}{9}$ Since $\sqrt{\frac{16}{81}} = \frac{4}{9}$, $-\sqrt{\frac{16}{81}} = -\frac{4}{9}$.

14. $-\frac{9}{12}$, or $-\frac{3}{4}$

15. $\sqrt{0.09} = 0.3$

16. 0.6

17. $-\sqrt{0.0049} = -0.07$

18. 0.12

19. $5\sqrt{p^2 + 4}$

    The radicand is the expression written under the radical sign, $p^2 + 4$.

20. $y^2 - 8$

21. $x^2y^2 \sqrt{\frac{x}{y + 4}}$

    The radicand is the expression written under the radical sign, $\frac{x}{y + 4}$.

22. $\frac{a}{a^2 - b}$

23. $f(y) = \sqrt{5y - 10}$

    $f(6) = \sqrt{5\cdot 6 - 10} = \sqrt{20}$

    $f(2) = \sqrt{5\cdot 2 - 10} = \sqrt{0} = 0$

    $f(0) = \sqrt{5\cdot 0 - 10} = \sqrt{-10}$

    Since negative numbers do not have real-number square roots, 0 is not in the domain of f.

    $f(-3) = \sqrt{5(-3) - 10} = \sqrt{-25}$

    Since negative numbers do not have real-number square roots, -3 is not in the domain of f.

24. $g(x) = \sqrt{x^2 - 25}$

    $g(-5) = \sqrt{(-5)^2 - 25} = \sqrt{0} = 0$

    $g(5) = \sqrt{5^2 - 25} = \sqrt{0} = 0$

    $g(0) = \sqrt{0^2 - 25} = \sqrt{-25}$; 0 is not in the domain of g.

    $g(6) = \sqrt{6^2 - 25} = \sqrt{11}$

    $g(-6) = \sqrt{(-6)^2 - 25} = \sqrt{11}$

25. $t(x) = -\sqrt{2x + 1}$

    $t(4) = -\sqrt{2\cdot 4 + 1} = -\sqrt{9} = -3$

    $t(-4) = -\sqrt{2(-4) + 1} = -\sqrt{-7}$; -4 is not in the domain of t.

    $t(0) = -\sqrt{2\cdot 0 + 1} = -\sqrt{1} = -1$

    $t(12) = -\sqrt{2\cdot 12 + 1} = -\sqrt{25} = -5$

24. 0, 0, 0 is not in the domain of g, $\sqrt{11}$, $\sqrt{11}$

    $g(6) = \sqrt{6^2 - 25} = \sqrt{11}$

    $g(-6) = \sqrt{(-6)^2 - 25} = \sqrt{11}$

25. $t(x) = -\sqrt{2x + 1}$

    $t(4) = -\sqrt{2\cdot 4 + 1} = -\sqrt{9} = -3$

    $t(-4) = -\sqrt{2(-4) + 1} = -\sqrt{-7}$; -4 is not in the domain of t.

    $t(0) = -\sqrt{2\cdot 0 + 1} = -\sqrt{1} = -1$

    $t(12) = -\sqrt{2\cdot 12 + 1} = -\sqrt{25} = -5$

26. 0 is not in the domain of p, $\sqrt{30}$, $\sqrt{30}$, $\sqrt{180}$, $\sqrt{180}$

27. $f(t) = \sqrt{t^2 + 1}$

    $f(5) = \sqrt{5^2 + 1} = \sqrt{26}$

    $f(-5) = \sqrt{(-5)^2 + 1} = \sqrt{26}$

    $f(0) = \sqrt{0^2 + 1} = \sqrt{1} = 1$

    $f(10) = \sqrt{10^2 + 1} = \sqrt{101}$

    $f(-10) = \sqrt{(-10)^2 + 1} = \sqrt{101}$

28. -2, 0, -4, -2, -6, -4

29. $g(x) = \sqrt{x^3 + 9}$

$g(2) = \sqrt{2^3 + 9} = \sqrt{17}$

$g(-2) = \sqrt{(-2)^3 + 9} = \sqrt{1} = 1$

$g(3) = \sqrt{3^3 + 9} = \sqrt{36} = 6$

$g(-3) = \sqrt{(-3)^3 + 9} = \sqrt{-18}$; -3 is not in the domain of g.

30. 2 is not in the domain of f, -2 is not in the domain of f, $\sqrt{17}$, -3 is not in the domain of f

31. $\sqrt{16x^2} = \sqrt{(4x)^2} = |4x| = 4|x|$

Since x might be negative, absolute-value notation is necessary.

32. $5|t|$

33. $\sqrt{(-7c)^2} = |-7c| = |-7|\cdot|c| = 7|c|$

Since c might be negative, absolute-value notation is necessary.

34. $6|b|$

35. $\sqrt{(a + 1)^2} = |a + 1|$

Since a + 1 might be negative, absolute-value notation is necessary.

36. $|5 - b|$

37. $\sqrt{x^2 - 4x + 4} = \sqrt{(x - 2)^2} = |x - 2|$

Since x - 2 might be negative, absolute-value notation is necessary.

38. $|y + 8|$

39. $\sqrt{4x^2 + 28x + 49} = \sqrt{(2x + 7)^2} = |2x + 7|$

Since 2x + 7 might be negative, absolute-value notation is necessary.

40. $|3x - 5|$

41. $\sqrt[4]{625} = 5$ Since $5^4 = 625$

42. -4

43. $\sqrt[5]{-1} = -1$ Since $(-1)^5 = -1$

44. 2

45. $\sqrt[5]{-\frac{32}{243}} = -\frac{2}{3}$ Since $\left(-\frac{2}{3}\right)^5 = -\frac{32}{243}$

46. $-\frac{1}{2}$

47. $\sqrt[6]{x^6} = |x|$

The index is even. Use absolute-value notation since x could have a negative value.

48. $|y|$

49. $\sqrt[4]{(5a)^4} = |5a| = 5|a|$

The index is even. Use absolute-value notation since a could have a negative value.

50. $7|b|$

51. $\sqrt[10]{(-6)^{10}} = |-6| = 6$

52. 10

53. $\sqrt[414]{(a + b)^{414}} = |a + b|$

The index is even. Use absolute-value notation since a + b could have a negative value.

54. $|2a + b|$

55. $\sqrt[7]{y^7} = y$

We do not use absolute-value notation when the index is odd.

56. -6

57. $\sqrt[5]{(x - 2)^5} = x - 2$

We do not use absolute-value notation when the index is odd.

58. 2xy

59. $\sqrt{16x^2} = \sqrt{(4x)^2} = 4x$ Assuming x is nonnegative

60. 5t

61. $\sqrt{(-6b)^2} = \sqrt{36b^2} = 6b$ Assuming b is nonnegative

62. 7c

63. $\sqrt{(a + 1)^2} = a + 1$ Assuming a + 1 is nonnegative

64. 5 + b

65. $\sqrt{4x^2 + 8x + 4} = \sqrt{4(x^2 + 2x + 1)} =$

$\sqrt{[2(x + 1)]^2} = 2(x + 1)$, or $2x + 2$

66. $3(x + 2)$, or $3x + 6$

67. $\sqrt{9t^2 - 12t + 4} = \sqrt{(3t - 2)^2} = 3t - 2$

68. 5t - 2

69. $\sqrt[3]{27} = 3$ $[3^3 = 27]$

70. -4

71. $\sqrt[4]{16x^4} = \sqrt[4]{(2x)^4} = 2x$

72. $3x$

73. $\sqrt[3]{-216} = -6 \qquad [(-6)^3 = -216]$

74. 10

75. $-\sqrt[3]{-125y^3} = -(-5y) \qquad [(-5y)^3 = -125y^3]$
$= 5y$

76. $4x$

77. $\sqrt[5]{0.00032(x + 1)^5} = 0.2(x + 1)$
$[0.00032(x + 1)^5 = (0.2(x + 1))^5]$

78. $0.02(y - 2)$

79. $\sqrt[6]{64x^{12}} = 2x^2 \qquad [64x^{12} = (2x^2)^6]$

80. $3y^3$

81. $f(x) = \sqrt[3]{x + 1}$
$f(7) = \sqrt[3]{7 + 1} = \sqrt[3]{8} = 2$
$f(26) = \sqrt[3]{26 + 1} = \sqrt[3]{27} = 3$
$f(-9) = \sqrt[3]{-9 + 1} = \sqrt[3]{-8} = -2$
$f(-65) = \sqrt[3]{-65 + 1} = \sqrt[3]{-64} = -4$

82. 1, 5, 3, -5

83. $g(t) = \sqrt[4]{t - 3}$
$g(19) = \sqrt[4]{19 - 3} = \sqrt[4]{16} = 2$
$g(-13) = \sqrt[4]{-13 - 3} = \sqrt[4]{-16}$; -13 is not in the domain of g.
$g(1) = \sqrt[4]{1 - 3} = \sqrt[4]{-2}$; 1 is not in the domain of g.
$g(84) = \sqrt[4]{84 - 3} = \sqrt[4]{81} = 3$

84. 1, 2, -82 is not in the domain of f, 3

85. $f(x) = \sqrt{x + 7}$
We need to find all x-values such that $x + 7$ is nonnegative. We solve the inequality:
$x + 7 \geqslant 0$
$x \geqslant -7$ Adding -7 on both sides
Domain of $f = \{x|x \geqslant -7\}$

86. $\{x|x \geqslant 10\}$

87. $g(t) = \sqrt[4]{2t - 9}$
Since the index is even, the radicand must be nonnegative. We solve the inequality:
$2t - 9 \geqslant 0$
$2t \geqslant 9$ Adding 9 on both sides
$t \geqslant \frac{9}{2}$ Multiplying by $\frac{1}{2}$ on both sides
Domain of $g = \left\{t \middle| t \geqslant \frac{9}{2}\right\}$

88. $\{x|x \geqslant -1\}$

89. $g(x) = \sqrt[4]{5 - x}$
Since the index is even, the radicand must be nonnegative. We solve the inequality:
$5 - x \geqslant 0$
$5 \geqslant x$
Domain of $g = \{x|x \leqslant 5\}$

90. $\{t|t$ is a real number$\}$

91. $f(t) = \sqrt[5]{3t + 7}$
Since the index is odd, the radicand can be any real number.
Domain of $f = \{t|t$ is a real number$\}$

92. $\left\{t \middle| t \geqslant -\frac{5}{2}\right\}$

93. $h(z) = -\sqrt[6]{5z + 3}$
Since the index is even, the radicand must be nonnegative. We solve the inequality:
$5z + 3 \geqslant 0$
$5z \geqslant -3$
$z \geqslant -\frac{3}{5}$
Domain of $h = \left\{z \middle| z \geqslant -\frac{3}{5}\right\}$

94. $\left\{x \middle| x \geqslant \frac{5}{7}\right\}$

95. $f(t) = 7 + 2\sqrt[8]{3t - 5}$
Since the index is even, the radicand must be nonnegative. We solve the inequality:
$3t - 5 \geqslant 0$
$3t \geqslant 5$
$t \geqslant \frac{5}{3}$
Domain of $f = \left\{t \middle| t \geqslant \frac{5}{3}\right\}$

96. $\left\{t \middle| t \geqslant \frac{4}{5}\right\}$

97. $(a^3b^2c^5)^3 = a^{3\cdot3}b^{2\cdot3}c^{5\cdot3} = a^9b^6c^{15}$

98. $10a^{10}b^9$

99. $(x - 3)(x + 3) = x^2 - 3^2 = x^2 - 9$

100. $a^2 - b^2x^2$

101.

102.

103. $N = 2.5\sqrt{A}$

a) $N = 2.5\sqrt{25} = 2.5(5) = 12.5 \approx 13$

b) $N = 2.5\sqrt{36} = 2.5(6) = 15$

c) $N = 2.5\sqrt{49} = 2.5(7) = 17.5 \approx 18$

d) $N = 2.5\sqrt{64} = 2.5(8) = 20$

104.

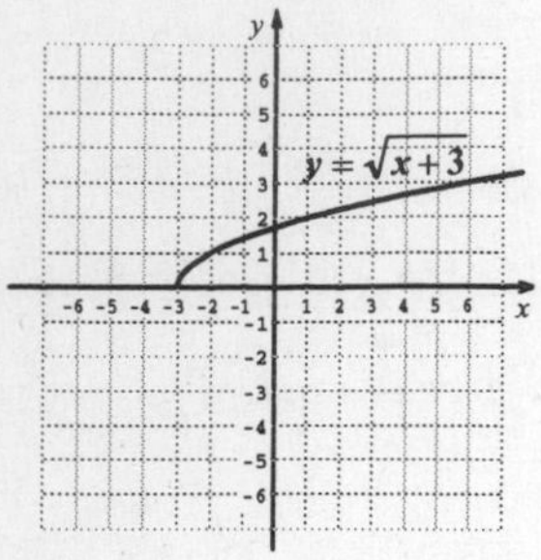

105. $y = \sqrt{x} + 3$

Make a table of values. Note that x must be nonnegative. Plot these points and draw the graph.

x	y
0	3
1	4
2	4.4
3	4.7
4	5
5	5.2

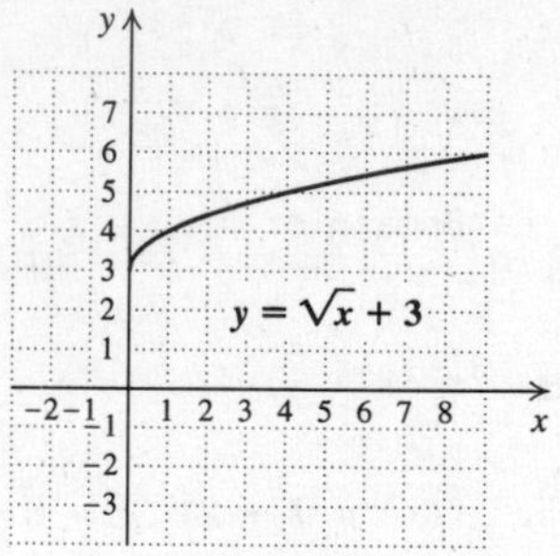

106.

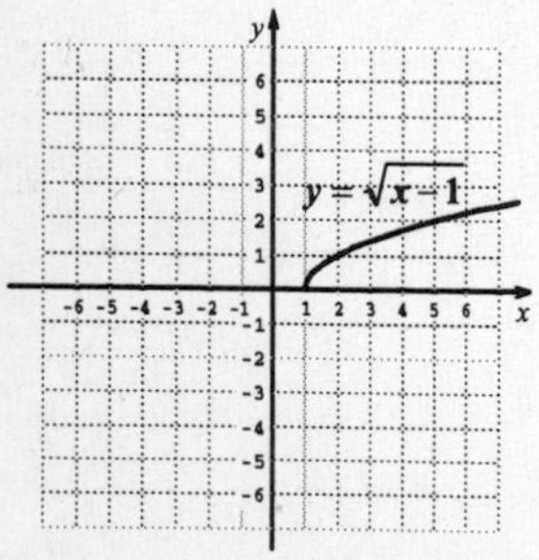

107. $y = \sqrt{x} - 2$

Make a table of values. Note that x must be nonnegative. Plot these points and draw the graph.

x	y
0	-2
1	-1
3	-0.3
4	0
5	0.3

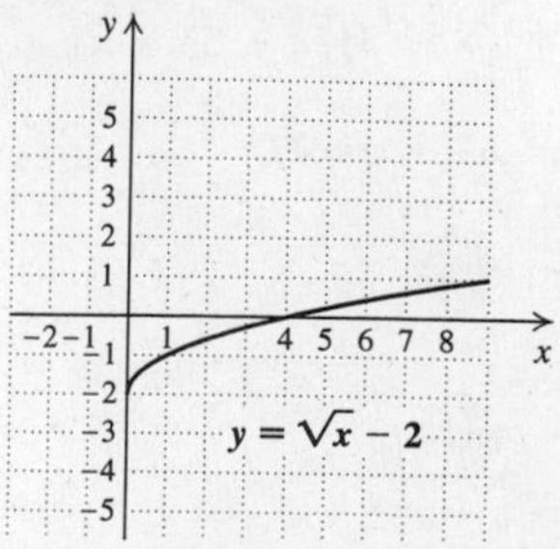

Exercise Set 7.2

1. $x^{1/4} = \sqrt[4]{x}$

2. $\sqrt[5]{y}$

3. $(8)^{1/3} = \sqrt[3]{8} = 2$

4. 4

5. $81^{1/4} = \sqrt[4]{81} = 3$

6. 4

7. $9^{1/2} = \sqrt{9} = 3$

8. 5

9. $(xyz)^{1/3} = \sqrt[3]{xyz}$

10. $\sqrt[4]{ab}$

11. $(a^2b^2)^{1/5} = \sqrt[5]{a^2b^2}$

12. $\sqrt[4]{x^3y^3}$

13. $a^{2/3} = \sqrt[3]{a^2}$

14. $\sqrt{b^3}$, or $b\sqrt{b}$

15. $16^{3/4} = \sqrt[4]{16^3} = (\sqrt[4]{16})^3 = 2^3 = 8$

16. 128

17. $49^{3/2} = \sqrt{49^3} = (\sqrt{49})^3 = 7^3 = 343$

18. 81

19. $9^{5/2} = \sqrt{9^5} = (\sqrt{9})^5 = 3^5 = 243$

20. 729

21. $(81x)^{3/4} = \sqrt[4]{(81x)^3} = \sqrt[4]{81^3x^3}$, or $\sqrt[4]{81^3} \cdot \sqrt[4]{x^3} = (\sqrt[4]{81})^3 \cdot \sqrt[4]{x^3} = 3^3\,\sqrt[4]{x^3} = 27\sqrt[4]{x^3}$

22. $25\sqrt[3]{a^2}$

23. $(25x^4)^{3/2} = \sqrt{(25x^4)^3} = \sqrt{25^3 \cdot x^{12}} = \sqrt{25^3} \cdot \sqrt{x^{12}} = (\sqrt{25})^3x^6 = 5^3x^6 = 125x^6$

24. $27y^9$

25. $(8a^2b^4)^{2/3} = \sqrt[3]{(8a^2b^4)^2} = \sqrt[3]{64a^4b^8}$, or $\sqrt[3]{64a^3b^6 \cdot ab^2} = 4ab^2\,\sqrt[3]{ab^2}$

26. $9a^3\,\sqrt[3]{ab^2}$

27. $\sqrt[3]{20} = 20^{1/3}$

28. $19^{1/3}$

29. $\sqrt{17} = 17^{1/2}$

30. $6^{1/2}$

31. $\sqrt{x^3} = x^{3/2}$

32. $a^{5/2}$

33. $\sqrt[5]{m^2} = m^{2/5}$

34. $n^{4/5}$

35. $\sqrt[4]{cd} = (cd)^{1/4}$ Parentheses are required.

36. $(xy)^{1/5}$

37. $\sqrt[5]{xy^2z} = (xy^2z)^{1/5}$

38. $(x^3y^2z^2)^{1/7}$

39. $(\sqrt{3mn})^3 = (3mn)^{3/2}$

40. $(7xy)^{4/3}$

41. $(\sqrt[7]{8x^2y})^5 = (8x^2y)^{5/7}$

42. $(2a^5b)^{7/6}$

43. $x^{-1/3} = \dfrac{1}{x^{1/3}}$

44. $\dfrac{1}{y^{1/4}}$

45. $(2rs)^{-3/4} = \dfrac{1}{(2rs)^{3/4}}$

46. $\dfrac{1}{(5xy)^{5/6}}$

47. $\left(\dfrac{1}{10}\right)^{-2/3} = \dfrac{1}{\left(\dfrac{1}{10}\right)^{2/3}} = 1\cdot\left(\dfrac{10}{1}\right)^{2/3} = 10^{2/3}$

48. $8^{3/4}$

49. $\dfrac{1}{x^{-2/3}} = x^{2/3}$

50. $x^{5/6}$

51. $5x^{-1/4} = 5\cdot x^{-1/4} = 5 \cdot \dfrac{1}{x^{1/4}} = \dfrac{5}{x^{1/4}}$

52. $\dfrac{8}{a^{1/3}}$

53. $\dfrac{3}{5m^{-1/2}} = \dfrac{3}{5} \cdot \dfrac{1}{m^{-1/2}} = \dfrac{3}{5} \cdot m^{1/2} = \dfrac{3m^{1/2}}{5}$

54. $\dfrac{2x^{1/3}}{7}$

55. $5^{3/4}\cdot 5^{1/8} = 5^{3/4+1/8} = 5^{6/8+1/8} = 5^{7/8}$

We added exponents after finding a common denominator.

56. $11^{7/6}$

57. $\dfrac{7^{5/8}}{7^{3/8}} = 7^{5/8-3/8} = 7^{2/8} = 7^{1/4}$

We subtracted exponents and simplified using arithmetic.

58. $9^{2/11}$

59. $\dfrac{8.3^{3/4}}{8.3^{2/5}} = 8.3^{3/4-2/5} = 8.3^{15/20-8/20} = 8.3^{7/20}$

We subtracted exponents after finding a common denominator.

60. $3.9^{7/20}$

61. $(10^{3/5})^{2/5} = 10^{3/5\cdot 2/5} = 10^{6/25}$

We multiplied exponents.

62. $5^{15/28}$

63. $a^{2/3}\cdot a^{5/4} = a^{2/3+5/4} = a^{8/12+15/12} = a^{23/12}$

We add exponents after finding a common denominator.

64. $x^{17/12}$

65. $(x^{2/3})^{3/7} = x^{2/3\cdot 3/7} = x^{6/21} = x^{2/7}$

We multiplied exponents and simplified using arithmetic.

66. $a^{3/5}$

67. $(m^{2/3}n^{1/2})^{1/4} = m^{2/3\cdot 1/4}\cdot n^{1/2\cdot 1/4} = m^{2/12}n^{1/8} =$
$m^{1/6}n^{1/8}$

We raised each factor to the power 1/4, multiplied exponents, and simplified using arithmetic.

68. $x^{1/12}y^{1/10}$

69. $(a^{-2/3}b^{-1/4})^{-6} = a^{-2/3\cdot(-6)}\cdot b^{-1/4\cdot(-6)} =$
$a^{12/3}b^{6/4} = a^4b^{3/2}$

We raised each factor to the power -6, multiplied exponents, and simplified using arithmetic.

70. $m^2n^{25/3}$

71. $\sqrt[6]{a^4} = a^{4/6}$ Converting to an exponential expression

$= a^{2/3}$ Using arithmetic to simplify the exponent

$= \sqrt[3]{a^2}$ Converting back to radical notation

72. $\sqrt[3]{y}$

73. $\sqrt[3]{8y^6} = \sqrt[3]{2^3y^6}$ Recognizing that 8 is 2^3

$= (2^3y^6)^{1/3}$ Converting to an exponential expression

$= 2^{3/3}y^{6/3}$ Using the product and power rules

$= 2y^2$ Using arithmetic to simplify the exponents

74. x^2y^3

75. $\sqrt[4]{32} = \sqrt[4]{2^5} = 2^{5/4} = 2^{4/4+1/4} = 2^{4/4} \cdot 2^{1/4} = 2\sqrt[4]{2}$

76. $\sqrt{3}$

77. $\sqrt[6]{4x^2} = \sqrt[6]{2^2x^2} = \sqrt[6]{(2x)^2} = [(2x)^2]^{1/6} = (2x)^{2/6} =$
$(2x)^{1/3} = \sqrt[3]{2x}$

78. $2x\sqrt{y}$

79. $\sqrt[5]{32c^{10}d^{15}} = \sqrt[5]{2^5c^{10}d^{15}} = (2^5c^{10}d^{15})^{1/5} =$
$2^{5/5}c^{10/5}d^{15/5} = 2c^2d^3$

80. $2x^3y^4$

81. $\sqrt[6]{\dfrac{m^{12}n^{24}}{64}} = \sqrt[6]{\dfrac{m^{12}n^{24}}{2^6}} = \left(\dfrac{m^{12}n^{24}}{2^6}\right)^{1/6} = \dfrac{m^{12/6}n^{24/6}}{2^{6/6}} =$
$= \dfrac{m^2n^4}{2}$

82. $\dfrac{x^3y^4}{2}$

83. $\sqrt[8]{r^4s^2} = (r^4s^2)^{1/8} = r^{4/8} \cdot s^{2/8} = r^{2/4} \cdot s^{1/4} =$
$(r^2s)^{1/4} = \sqrt[4]{r^2s}$

84. $\sqrt{2ts}$

85. $\sqrt[3]{27a^3b^9} = \sqrt[3]{3^3a^3b^9} = (3^3a^3b^9)^{1/3} =$
$3^{3/3}a^{3/3}b^{9/3} = 3ab^3$

86. $3x^2y^2$

87.
$$x^2 - 1 = 8$$
$$x^2 - 9 = 0$$
$$(x + 3)(x - 3) = 0$$
$$x + 3 = 0 \quad \text{or} \quad x - 3 = 0$$
$$x = -3 \quad \text{or} \quad x = 3$$
Both values check. The solutions are -3 and 3.

88. $-\dfrac{11}{2}$

89.
$$\frac{1}{x} + 2 = 5$$
$$\frac{1}{x} = 3$$
$$x \cdot \frac{1}{x} = x \cdot 3$$
$$1 = 3x$$
$$\frac{1}{3} = x$$
This value checks. The solution is $\frac{1}{3}$.

90. $93,500

91.

92.

93. $\sqrt[5]{x^2y\sqrt{xy}} = \sqrt[5]{x^2y(xy)^{1/2}} = \sqrt[5]{x^2yx^{1/2}y^{1/2}} =$
$\sqrt[5]{x^{5/2}y^{3/2}} = (x^{5/2}y^{3/2})^{1/5} = x^{5/10}y^{3/10} =$
$(x^5y^3)^{1/10} = \sqrt[10]{x^5y^3}$

94. $x^3\sqrt[6]{x}$

95. $\sqrt[4]{\sqrt[3]{8x^3y^6}} = \sqrt[4]{(8x^3y^6)^{1/3}} = \sqrt[4]{(2^3x^3y^6)^{1/3}} = \sqrt[4]{2xy^2}$

96. $\sqrt[6]{p + q}$

97. a) $L = \dfrac{(0.000169)60^{2.27}}{1} \approx 1.8$ m

b) $L = \dfrac{(0.000169)75^{2.27}}{0.9906} \approx 3.1$ m

c) $L = \dfrac{(0.000169)80^{2.27}}{2.4} \approx 1.5$ m

d) $L = \dfrac{(0.000169)100^{2.27}}{1.1} \approx 5.3$ m

98.

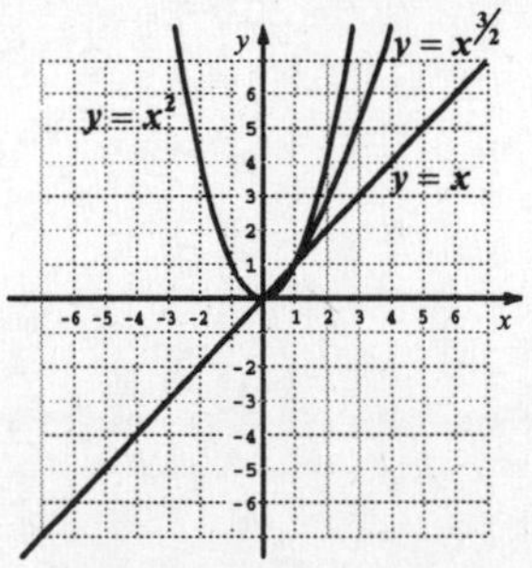

99. $r(1900) = 10^{-12^{2-1900/5700}} \approx 10^{-12}(0.7937)$, or 7.937×10^{-13}. The ratio is about 7.937×10^{-13} to 1.

Exercise Set 7.3

1. $\sqrt{3}\,\sqrt{2} = \sqrt{3\cdot 2} = \sqrt{6}$

2. $\sqrt{35}$

3. $\sqrt[3]{2}\,\sqrt[3]{5} = \sqrt[3]{2\cdot 5} = \sqrt[3]{10}$

4. $\sqrt[3]{14}$

5. $\sqrt[4]{8}\,\sqrt[4]{9} = \sqrt[4]{8\cdot 9} = \sqrt[4]{72}$

6. $\sqrt[4]{18}$

7. $\sqrt{3a}\,\sqrt{10b} = \sqrt{3a\cdot 10b} = \sqrt{30ab}$

8. $\sqrt{26xy}$

9. $\sqrt[5]{9t^2}\,\sqrt[5]{2t} = \sqrt[5]{9t^2\cdot 2t} = \sqrt[5]{18t^3}$

10. $\sqrt[5]{80y^4}$

11. $\sqrt{x - a}\,\sqrt{x + a} = \sqrt{(x - a)(x + a)} = \sqrt{x^2 - a^2}$

12. $\sqrt{y^2 - b^2}$

13. $\sqrt[3]{0.3x}\,\sqrt[3]{0.2x} = \sqrt[3]{0.3x\cdot 0.2x} = \sqrt[3]{0.06x^2}$

14. $\sqrt[3]{0.21y^2}$

15. $\sqrt[4]{x - 1}\,\sqrt[4]{x^2 + x + 1} = \sqrt[4]{(x - 1)(x^2 + x + 1)}$
$= \sqrt[4]{x^3 - 1}$

16. $\sqrt[5]{(x - 2)^3}$

17. $\sqrt{\frac{6}{x}}\,\sqrt{\frac{y}{5}} = \sqrt{\frac{6}{x}\cdot\frac{y}{5}} = \sqrt{\frac{6y}{5x}}$

18. $\sqrt{\frac{7s}{11t}}$

19. $\sqrt[7]{\frac{x - 3}{4}}\,\sqrt[7]{\frac{5}{x + 2}} = \sqrt[7]{\frac{x - 3}{4}\cdot\frac{5}{x + 2}} = \sqrt[7]{\frac{5x - 15}{4x + 8}}$

20. $\sqrt[6]{\frac{3a}{b^2 - 4}}$

21. $\sqrt[3]{7}\cdot\sqrt{2} = 7^{1/3}\cdot 2^{1/2}$ Converting to exponential notation
$= 7^{2/6}\cdot 2^{3/6}$ Rewriting so that exponents have a common denominator
$= (7^2\cdot 2^3)^{1/6}$ Using the laws of exponents
$= \sqrt[6]{7^2\cdot 2^3}$ Converting back to radical notation
$= \sqrt[6]{392}$ Multiplying under the radical

22. $\sqrt[12]{300{,}125}$

23. $\sqrt{x}\,\sqrt[3]{2x} = x^{1/2}\cdot(2x)^{1/3} = x^{1/2}\cdot 2^{1/3}\cdot x^{1/3} =$
$2^{1/3}x^{1/2+1/3} = 2^{1/3}x^{3/6+2/6} = 2^{1/3}\cdot x^{5/6} =$
$2^{2/6}x^{5/6} = (2^2\cdot x^5)^{1/6} = \sqrt[6]{2^2\cdot x^5} = \sqrt[6]{4x^5}$

24. $\sqrt[15]{27y^8}$

25. $\sqrt{x}\,\sqrt[3]{x - 2} = x^{1/2}\cdot(x - 2)^{1/3} = x^{3/6}\cdot(x - 2)^{2/6} =$
$[x^3(x - 2)^2]^{1/6} = \sqrt[6]{x^3(x - 2)^2} =$
$\sqrt[6]{x^3(x^2 - 4x + 4)} = \sqrt[6]{x^5 - 4x^4 + 4x^3}$

26. $\sqrt[4]{3xy^2 + 24xy + 48x}$

27. $\sqrt[5]{yx^2}\,\sqrt{xy} = (yx^2)^{1/5}(xy)^{1/2} = y^{1/5}x^{2/5}x^{1/2}y^{1/2} =$
$x^{2/5+1/2}y^{1/5+1/2} = x^{4/10+5/10}y^{2/10+5/10} =$
$x^{9/10}y^{7/10} = (x^9y^7)^{1/10} = \sqrt[10]{x^9y^7}$

28. $\sqrt[10]{2^2a^9b^9}$, or $\sqrt[10]{4a^9b^9}$

29. $\sqrt[4]{x(y + 1)^2}\,\sqrt[3]{x^2(y + 1)} =$
$[x(y + 1)^2]^{1/4}[x^2(y + 1)]^{1/3} =$
$x^{1/4}(y + 1)^{2/4}x^{2/3}(y + 1)^{1/3} =$
$x^{11/12}(y + 1)^{10/12} = [x^{11}(y + 1)^{10}]^{1/12} =$
$\sqrt[12]{x^{11}(y + 1)^{10}}$

30. $2\,\sqrt[10]{2^2a^9b^7}$, or $2\,\sqrt[10]{4a^9b^7}$

31. $\sqrt[5]{a^2bc^3}\,\sqrt[3]{ab^2c} = (a^2bc^3)^{1/5}(ab^2c)^{1/3} =$
$a^{2/5}b^{1/5}c^{3/5}a^{1/3}b^{2/3}c^{1/3} = a^{11/15}b^{13/15}c^{14/15} =$
$(a^{11}b^{13}c^{14})^{1/15} = \sqrt[15]{a^{11}b^{13}c^{14}}$

32. $\sqrt[12]{x^{11}(y - 1)^{10}}$

33. $\sqrt{27} = \sqrt{9\cdot 3}$ 9 is the largest perfect square factor of 27
$= \sqrt{9}\cdot\sqrt{3}$
$= 3\sqrt{3}$

34. $2\sqrt{7}$

35. $\sqrt{45} = \sqrt{9\cdot 5}$ 9 is the largest perfect square factor of 45
$= \sqrt{9}\cdot\sqrt{5}$
$= 3\sqrt{5}$

36. $2\sqrt{3}$

37. $\sqrt{8} = \sqrt{4\cdot 2} = \sqrt{4}\cdot\sqrt{2} = 2\sqrt{2}$

38. $3\sqrt{2}$

39. $\sqrt{24} = \sqrt{4\cdot 6} = \sqrt{4}\cdot\sqrt{6} = 2\sqrt{6}$

40. $2\sqrt{5}$

41. $\sqrt{180x^4} = \sqrt{36\cdot5\cdot x^4}$ Factoring the radicand

$= \sqrt{36\cdot x^4\cdot 5}$

$= \sqrt{36}\ \sqrt{x^4}\ \sqrt{5}$ Factoring into several radicals

$= 6x^2\ \sqrt{5}$ Taking square roots.

42. $5y^3\sqrt{7}$

43. $\sqrt[3]{800} = \sqrt[3]{8\cdot100}$ 8 is the largest perfect cube factor of 800

$= \sqrt[3]{8}\cdot\sqrt[3]{100}$

$= 2\sqrt[3]{100}$

44. $3\sqrt[3]{10}$

45. $\sqrt[3]{-16x^6} = \sqrt[3]{-8\cdot2\cdot x^6}$ Factoring the radicand

$= \sqrt[3]{-8\cdot x^6\cdot 2}$

$= \sqrt[3]{-8}\ \sqrt[3]{x^6}\ \sqrt[3]{2}$ Factoring into several radicals

$= -2x^2\ \sqrt[3]{2}$ Taking cube roots

46. $-2a^2\ \sqrt[3]{4}$

47. $\sqrt[3]{54x^8} = \sqrt[3]{27\cdot2\cdot x^6\cdot x^2} = \sqrt[3]{27\cdot x^6\cdot 2\cdot x^2} =$
$\sqrt[3]{27}\ \sqrt[3]{x^6}\ \sqrt[3]{2x^2} = 3x^2\ \sqrt[3]{2x^2}$

48. $2y\sqrt[3]{5}$

49. $\sqrt[3]{80x^8} = \sqrt[3]{8\cdot10\cdot x^6\cdot x^2} = \sqrt[3]{8\cdot x^6\cdot 10\cdot x^2} =$
$\sqrt[3]{8}\ \sqrt[3]{x^6}\ \sqrt[3]{10x^2} = 2x^2\ \sqrt[3]{10x^2}$

50. $3m\sqrt[3]{4m^2}$

51. $\sqrt[4]{32} = \sqrt[4]{16\cdot2} = \sqrt[4]{16}\cdot\sqrt[4]{2} = 2\sqrt[4]{2}$

52. $2\sqrt[4]{5}$

53. $\sqrt[4]{810} = \sqrt[4]{81\cdot10} = \sqrt[4]{81}\cdot\sqrt[4]{10} = 3\sqrt[4]{10}$

54. $2\sqrt[4]{10}$

55. $\sqrt[4]{96a^8} = \sqrt[4]{16\cdot6\cdot a^8} = \sqrt[4]{16\cdot a^8\cdot 6} =$
$\sqrt[4]{16}\ \sqrt[4]{a^8}\ \sqrt[4]{6} = 2a^2\ \sqrt[4]{6}$

56. $2x^2\ \sqrt[4]{15}$

57. $\sqrt[4]{162c^4d^6} = \sqrt[4]{81\cdot2\cdot c^4\cdot d^4\cdot d^2} =$
$\sqrt[4]{81\cdot c^4\cdot d^4\cdot 2\cdot d^2} = \sqrt[4]{81}\ \sqrt[4]{c^4}\ \sqrt[4]{d^4}\ \sqrt[4]{2d^2} =$
$3cd\sqrt[4]{2d^2}$

58. $3x^2y^2\ \sqrt[4]{3y^2}$

59. $\sqrt[3]{(x+y)^4} = \sqrt[3]{(x+y)^3(x+y)} =$
$\sqrt[3]{(x+y)^3}\ \sqrt[3]{x+y} = (x+y)\sqrt[3]{x+y}$

60. $(a-b)\sqrt[3]{(a-b)^2}$

61. $\sqrt[3]{8000(m+n)^8} = \sqrt[3]{8000(m+n)^6(m+n)^2} =$
$\sqrt[3]{8000}\ \sqrt[3]{(m+n)^6}\ \sqrt[3]{(m+n)^2} = 20(m+n)^2\ \sqrt[3]{(m+n)^2}$

62. $-10(x+y)^3\ \sqrt[3]{x+y}$

63. $\sqrt[5]{-a^6b^{11}c^{17}} = \sqrt[5]{-1\cdot a^5\cdot a\cdot b^{10}\cdot b\cdot c^{15}\cdot c^2} =$
$\sqrt[5]{-1\cdot a^5\cdot b^{10}\cdot c^{15}\cdot a\cdot b\cdot c^2} =$
$\sqrt[5]{-1}\ \sqrt[5]{a^5}\ \sqrt[5]{b^{10}}\ \sqrt[5]{c^{15}}\ \sqrt[5]{abc^2} = -ab^2c^3\ \sqrt[5]{abc^2}$

64. $x^2yz^4\ \sqrt[5]{x^3y^3z^2}$

65. $\sqrt{3}\ \sqrt{6} = \sqrt{3\cdot6} = \sqrt{18} = \sqrt{9\cdot2} = 3\sqrt{2}$

66. $5\sqrt{2}$

67. $\sqrt{15}\ \sqrt{12} = \sqrt{15\cdot12} = \sqrt{180} = \sqrt{36\cdot5} = 6\sqrt{5}$

68. 8

69. $\sqrt{6}\ \sqrt{8} = \sqrt{6\cdot8} = \sqrt{48} = \sqrt{16\cdot3} = 4\sqrt{3}$

70. $6\sqrt{7}$

71. $\sqrt[3]{3}\ \sqrt[3]{18} = \sqrt[3]{3\cdot18} = \sqrt[3]{54} = \sqrt[3]{27\cdot2} = 3\sqrt[3]{2}$

72. $30\sqrt{3}$

73. $\sqrt{5b^3}\ \sqrt{10c^4} = \sqrt{5b^3\cdot10c^4}$ } Multiplying radicands

$= \sqrt{50b^3c^4}$

$= \sqrt{25b^2c^4\cdot 2b}$ Factoring the radicand

$= \sqrt{25}\ \sqrt{b^2}\ \sqrt{c^4}\ \sqrt{2b}$ Factoring into radicals

$= 5bc^2\sqrt{2b}$ Taking the square root

74. $-2a\sqrt[3]{15a^2}$

75. $\sqrt[3]{10x^5}\ \sqrt[3]{-75x^2} = \sqrt[3]{10x^5\cdot(-75x^2)}$ } Multiplying radicands

$= \sqrt[3]{-750x^7}$

$= \sqrt[3]{-125x^6\cdot 6x}$ Factoring the radicand

$= \sqrt[3]{-125}\ \sqrt[3]{x^6}\ \sqrt[3]{6x}$ Factoring into radicals

$= -5x^2\ \sqrt[3]{6x}$ Taking the cube root

76. $2x^2y\sqrt{6}$

77. $\sqrt[3]{y^4}\ \sqrt[3]{16y^5} = \sqrt[3]{y^4\cdot16y^5} = \sqrt[3]{16y^9} = \sqrt[3]{8y^9\cdot2} =$
$\sqrt[3]{8}\ \sqrt[3]{y^9}\ \sqrt[3]{2} = 2y^3\ \sqrt[3]{2}$

78. $5^2t^3\sqrt[3]{t}$, or $25t^3\sqrt[3]{t}$

79. $\sqrt[3]{(b+3)^4}\sqrt[3]{(b+3)^2} = \sqrt[3]{(b+3)^4(b+3)^2} =$

$\sqrt[3]{(b+3)^6} = (b+3)^2$

80. $(x+y)^2\sqrt[3]{(x+y)^2}$

81. $\sqrt{12a^3b}\sqrt{8a^4b^2} = \sqrt{12a^3b\cdot 8a^4b^2} = \sqrt{96a^7b^3} =$

$\sqrt{16a^6b^2\cdot 6ab} = \sqrt{16}\sqrt{a^6}\sqrt{b^2}\sqrt{6ab} = 4a^3b\sqrt{6ab}$

82. $6a^2b^4\sqrt{15ab}$

83. $\sqrt[5]{a^2(b+c)^4}\sqrt[5]{a^4(b+c)^7} =$

$\sqrt[5]{a^2(b+c)^4\cdot a^4(b+c)^7} = \sqrt[5]{a^6(b+c)^{11}} =$

$\sqrt[5]{a^5(b+c)^{10}\cdot a(b+c)} =$

$\sqrt[5]{a^5}\sqrt[5]{(b+c)^{10}}\sqrt[5]{a(b+c)} = a(b+c)^2\sqrt[5]{a(b+c)}$

84. $x(y-z)^3\sqrt[5]{x^4(y-z)}$

85. $\sqrt[5]{a^3b}\sqrt{ab} = (a^3b)^{1/5}(ab)^{1/2}$ Converting to exponential notation

$= a^{3/5}b^{1/5}a^{1/2}b^{1/2}$

$= a^{3/5+1/2}b^{1/5+1/2}$ } Using the laws of exponents

$= a^{11/10}b^{7/10}$ Adding exponents

$= a^{1\frac{1}{10}}b^{\frac{7}{10}}$ Writing 11/10 as a mixed number

$= a\cdot a^{1/10}\cdot b^{7/10}$ Factoring $a^{11/10}$

$= a(ab^7)^{1/10}$

$= a\sqrt[10]{ab^7}$ Converting back to radical notation

86. $xy\sqrt[6]{xy^5}$

87. $\sqrt[3]{4xy^2}\sqrt{2x^3y^3} = (4xy^2)^{1/3}(2x^3y^3)^{1/2}$

$= (2^2xy^2)^{1/3}(2x^3y^3)^{1/2}$ Writing 4 as 2^2

$= 2^{2/3}x^{1/3}y^{2/3}2^{1/2}x^{3/2}y^{3/2}$

$= 2^{7/6}x^{11/6}y^{13/6}$

$= 2^{1\frac{1}{6}}x^{1\frac{5}{6}}y^{2\frac{1}{6}}$

$= 2\cdot 2^{1/6}\cdot x\cdot x^{5/6}\cdot y^2\cdot y^{1/6}$

$= 2xy^2(2x^5y)^{1/6}$

$= 2xy^2\sqrt[6]{2x^5y}$

88. $3a^2b\sqrt[4]{ab}$

89. $\sqrt[4]{x^3y^5}\sqrt{xy} = (x^3y^5)^{1/4}(xy)^{1/2}$

$= x^{3/4}y^{5/4}x^{1/2}y^{1/2}$

$= x^{5/4}y^{7/4}$

$= x^{1\frac{1}{4}}y^{1\frac{3}{4}}$

$= x\cdot x^{1/4}\cdot y\cdot y^{3/4}$

$= xy(xy^3)^{1/4}$

$= xy\sqrt[4]{xy^3}$

90. $a\sqrt[10]{ab^7}$

91. $\sqrt{a^4b^3c^4}\sqrt[3]{ab^2c} = (a^4b^3c^4)^{1/2}(ab^2c)^{1/3}$

$= a^{4/2}b^{3/2}c^{4/2}a^{1/3}b^{2/3}c^{1/3}$

$= a^{14/6}b^{13/6}c^{14/6}$

$= a^{2\frac{2}{6}}b^{2\frac{1}{6}}c^{2\frac{2}{6}}$

$= a^2\cdot a^{2/6}\cdot b^2\cdot b^{1/6}\cdot c^2\cdot c^{2/6}$

$= a^2b^2c^2(a^2bc^2)^{1/6}$

$= a^2b^2c^2\sqrt[6]{a^2bc^2}$

92. $xyz\sqrt[6]{x^5yz^2}$

93. $\sqrt[3]{x^2yz^2}\sqrt[4]{xy^3z^3} = (x^2yz^2)^{1/3}(xy^3z^3)^{1/4}$

$= x^{2/3}y^{1/3}z^{2/3}x^{1/4}y^{3/4}z^{3/4}$

$= x^{11/12}y^{13/12}z^{17/12}$

$= x^{\frac{11}{12}}y^{1\frac{1}{12}}z^{1\frac{5}{12}}$

$= x^{11/12}\cdot y\cdot y^{1/12}\cdot z\cdot z^{5/12}$

$= yz(x^{11}yz^5)^{1/12}$

$= yz\sqrt[12]{x^{11}yz^5}$

94. $ac\sqrt[12]{a^2b^{11}c^5}$

95. $\sqrt[3]{4a^2(b-5)^2}\sqrt{8a(b-5)^3}$

$= [4a^2(b-5)^2]^{1/3}[8a(b-5)^3]^{1/2}$

$= (2^2)^{1/3}a^{2/3}(b-5)^{2/3}(2^3)^{1/2}a^{1/2}(b-5)^{3/2}$

$(4 = 2^2,\ 8 = 2^3)$

$= 2^{2/3}a^{2/3}(b-5)^{2/3}2^{3/2}a^{1/2}(b-5)^{3/2}$

$= 2^{13/6}a^{7/6}(b-5)^{13/6}$

$= 2^{2\frac{1}{6}}a^{1\frac{1}{6}}(b-5)^{2\frac{1}{6}}$

$= 2^2\cdot 2^{1/6}\cdot a\cdot a^{1/6}(b-5)^2(b-5)^{1/6}$

$= 2^2a(b-5)^2[2a(b-5)]^{1/6}$

$= 2^2a(b-5)^2\sqrt[6]{2a(b-5)}$

96. $9x(y-2)^2\sqrt[6]{3^5x^5(y-2)}$

97. Using a calculator,

$\sqrt{180} \approx 13.41640787 \approx 13.416$

Using Table 1,

$\sqrt{180} = \sqrt{36\cdot 5} = \sqrt{36}\sqrt{5} = 6\sqrt{5} \approx 6(2.236)$

$\approx 13.416.$

98. 11.136

99. Using a calculator,

$$\frac{8 + \sqrt{480}}{4} \approx \frac{8 + 21.9089023}{4} = \frac{29.9089023}{4} \approx$$

7.477225575 ≈ 7.477.

Using Table 1,

$$\frac{8 + \sqrt{480}}{4} = \frac{8 + \sqrt{16\cdot 30}}{4} = \frac{8 + 4\sqrt{30}}{4}$$

$$= \frac{4(2 + \sqrt{30})}{4}$$

$$= 2 + \sqrt{30}$$

$$\approx 2 + 5.477$$

$$= 7.477.$$

100. -3.071

101. Using a calculator,

$$\frac{16 - \sqrt{48}}{20} \approx \frac{16 - 6.92820323}{20} = \frac{9.07179677}{20} \approx$$

0.453589838 ≈ 0.454

Using Table 1,

$$\frac{16 - \sqrt{48}}{20} \approx \frac{16 - 6.928}{20} = \frac{9.072}{20} = 0.4536 \approx 0.454$$

102. 0.919

103. Using a calculator,

$$\frac{24 + \sqrt{128}}{8} \approx \frac{24 + 11.3137085}{8} = \frac{35.3137085}{8} \approx$$

$$\approx 4.414213562 \approx 4.414$$

Using Table 1,

$$\frac{24 + \sqrt{128}}{8} = \frac{24 + \sqrt{64\cdot 2}}{8} = \frac{24 + 8\sqrt{2}}{8} =$$

$$\frac{8(3 + \sqrt{2})}{8} = 3 + \sqrt{2} \approx 3 + 1.414 = 4.414$$

104. 7.209

105. $\sqrt{24{,}500{,}000{,}000} = \sqrt{245\cdot 10^8}$ Factoring the radicand

$= \sqrt{245}\cdot\sqrt{10^8}$ Factoring the expression

$\approx 15.65247584\cdot 10^4$ Approximating $\sqrt{245}$ with a calculator and finding $\sqrt{10^8}$

$\approx 156{,}524.7584$ Multiplying

106. 128,452.3258

107. $\sqrt{468{,}200{,}000{,}000} = \sqrt{4682\cdot 10^8}$

$= \sqrt{4682}\cdot\sqrt{10^8}$

$\approx 68.42514158\cdot 10^4$

$\approx 684{,}251.4158$

108. 315,277.6554

109. $\sqrt{0.0000000395} = \sqrt{3.95\cdot 10^{-8}}$

$= \sqrt{3.95}\cdot\sqrt{10^{-8}}$

$\approx 1.987460691\cdot 10^{-4}$

≈ 0.0001987460691

110. 0.0003928103868

111. $\sqrt{0.0000005001} = \sqrt{50.01\cdot 10^{-8}}$

$= \sqrt{50.01}\cdot\sqrt{10^{-8}}$

$\approx 7.071774883\cdot 10^{-4}$

≈ 0.0007071774883

112. 0.003178207042

113. Familiarize. Let x and y represent the number of 30-sec and 60-sec commercials, respectively. Then the total number of minutes of commercial time during the show is $\frac{30x + 60y}{60}$, or $\frac{x}{2} + y$. (We divide by 60 to convert seconds to minutes.)

Translate. Rewording where necessary, we write two equations.

Total number of commercials	is	12.	
$x + y$	$=$	12	
Number of 30-sec commercials	is	total minutes of commercial time	less 6.
x	$=$	$\frac{x}{2} + y$	$- 6$

Carry out. Solving the system of equations, we get (4,8).

Check. If there are 4 30-sec and 8 60-sec commercials, the total number of commercials is 12. The total amount of commercial time is 4·30 sec + 8·60 sec = 600 sec, or 10 min. Then the number of 30-sec commercials is 6 less than the total number of minutes of commercial time. The values check.

State. 8 60-sec commercials were used.

114. $4x^2 - 9$

115. $4x^2 - 49 = (2x)^2 - 7^2 = (2x + 7)(2x - 7)$

116. $2(x - 9)(x - 4)$

117. ◆

118. ◆

119. $r = 2\sqrt{5L}$

a) $r = 2\sqrt{5\cdot 20} = 2\sqrt{100} = 2\cdot 10 = 20$ mph

b) $r = 2\sqrt{5\cdot 70} = 2\sqrt{350}$

$\approx 2 \times 18.708$ Using Table 1 or a calculator

≈ 37.4 mph Multiplying and rounding

c) $r = 2\sqrt{5\cdot 90} = 2\sqrt{450}$

$\approx 2 \times 21.213$ Using Table 1 or a calculator

≈ 42.4 mph Multiplying and rounding

120. a) -3.3° C

b) -16.6° C

c) -25.5° C

d) -54.0° C

121. $\sqrt[3]{5x^{k+1}}\,\sqrt[3]{25x^k} = 5x^7$

$\sqrt[3]{5x^{k+1}\cdot 25x^k} = 5x^7$

$\sqrt[3]{125x^{2k+1}} = 5x^7$

$\sqrt[3]{125}\,\sqrt[3]{x^{2k+1}} = 5x^7$

$5\sqrt[3]{x^{2k+1}} = 5x^7$

$\sqrt[3]{x^{2k+1}} = x^7$

$(x^{2k+1})^{1/3} = x^7$

$x^{\frac{2k+1}{3}} = x^7$

Since the base is the same, the exponents must be equal. We have:

$\frac{2k+1}{3} = 7$

$2k + 1 = 21$

$2k = 20$

$k = 10$

122. 6

123. We must assume $2x + 3$ is nonnegative.

$2x + 3 \geqslant 0$

$2x \geqslant -3$

$x \geqslant -\frac{3}{2}$

Thus we must assume $x \geqslant -\frac{3}{2}$.

124.

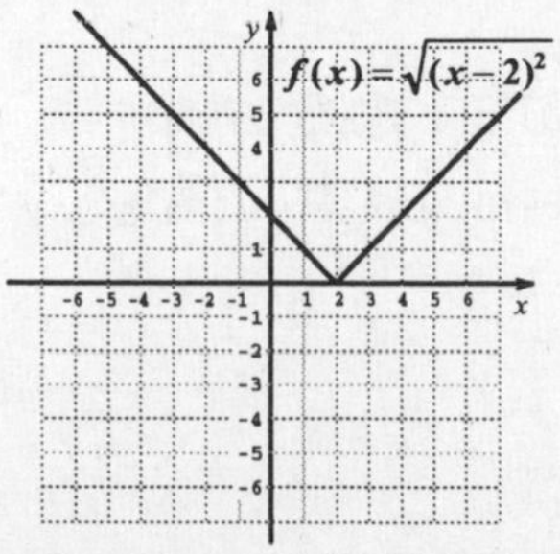

$\{x | x$ is a real number$\}$

Exercise Set 7.4

1. $\frac{\sqrt{21a}}{\sqrt{3a}} = \sqrt{\frac{21a}{3a}} = \sqrt{7}$

2. $\sqrt{7}$

3. $\frac{\sqrt[3]{54}}{\sqrt[3]{2}} = \sqrt[3]{\frac{54}{2}} = \sqrt[3]{27} = 3$

4. 2

5. $\frac{\sqrt{40xy^3}}{\sqrt{8x}} = \sqrt{\frac{40xy^3}{8x}} = \sqrt{5y^3} = \sqrt{y^2\cdot 5y} = \sqrt{y^2}\,\sqrt{5y} =$

$y\sqrt{5y}$

6. $2b\sqrt{2b}$

7. $\frac{\sqrt[3]{96a^4b^2}}{\sqrt[3]{12a^2b}} = \sqrt[3]{\frac{96a^4b^2}{12a^2b}} = \sqrt[3]{8a^2b} = \sqrt[3]{8}\,\sqrt[3]{a^2b} =$

$2\sqrt[3]{a^2b}$

8. $3xy\sqrt[3]{y^2}$

9. $\frac{\sqrt{144xy}}{2\sqrt{2}} = \frac{1}{2}\cdot\sqrt{\frac{144xy}{2}} = \frac{1}{2}\sqrt{72xy} = \frac{1}{2}\sqrt{36\cdot 2\cdot x\cdot y} =$

$\frac{1}{2}\sqrt{36}\,\sqrt{2xy} = \frac{1}{2}\cdot 6\sqrt{2xy} = 3\sqrt{2xy}$

10. $\frac{5}{3}\sqrt{ab}$

11. $\frac{\sqrt[4]{48x^9y^{13}}}{\sqrt[4]{3xy^5}} = \sqrt[4]{\frac{48x^9y^{13}}{3xy^5}} = \sqrt[4]{16x^8y^8} = 2x^2y^2$

12. $2a^2b^5\sqrt[5]{b}$

13. $\frac{\sqrt{x^3 - y^3}}{\sqrt{x - y}} = \sqrt{\frac{x^3 - y^3}{x - y}} = \sqrt{\frac{(x - y)(x^2 + xy + y^2)}{x - y}} =$

$\sqrt{\frac{\cancel{(x - y)}(x^2 + xy + y^2)}{\cancel{(x - y)}(1)}} = \sqrt{x^2 + xy + y^2}$

14. $\sqrt{r^2 - rs + s^2}$

15. $\frac{\sqrt[3]{a^2}}{\sqrt[4]{a}} = \frac{a^{2/3}}{a^{1/4}}$ Converting to exponential notation

$= a^{2/3-1/4}$ Subtracting exponents

$= a^{5/12}$
$= \sqrt[12]{a^5}$ Converting back to radical notation

16. $\sqrt[15]{x^7}$

17. $\dfrac{\sqrt[4]{x^2y^3}}{\sqrt[3]{xy^2}} = \dfrac{(x^2y^3)^{1/4}}{(xy^2)^{1/3}} = \dfrac{x^{2/4}y^{3/4}}{x^{1/3}y^{2/3}}$ Converting to exponential notation; using the product and power rules

$= x^{2/4-1/3}y^{3/4-2/3}$ Subtracting exponents

$= x^{2/12}y^{1/12}$
$= (x^2y)^{1/12}$
$= \sqrt[12]{x^2y}$
Converting back to radical notation

18. $\sqrt[15]{a^4b^2}$

19. $\dfrac{\sqrt{ab^3c}}{\sqrt[5]{a^2b^3c}} = \dfrac{(ab^3c)^{1/2}}{(a^2b^3c)^{1/5}}$

$= \dfrac{a^{1/2}b^{3/2}c^{1/2}}{a^{2/5}b^{3/5}c^{1/5}}$

$= a^{1/10}b^{9/10}c^{3/10}$ Subtracting exponents

$= (ab^9c^3)^{1/10}$

$= \sqrt[10]{ab^9c^3}$

20. $\sqrt[10]{xy^3z^3}$

21. $\dfrac{\sqrt[4]{(3x-1)^3}}{\sqrt[5]{(3x-1)^3}} = \dfrac{(3x-1)^{3/4}}{(3x-1)^{3/5}}$ Converting to exponential notation

$= (3x-1)^{3/4-3/5}$ Subtracting exponents

$= (3x-1)^{3/20}$
$= \sqrt[20]{(3x-1)^3}$
Converting back to radical notation

22. $\sqrt[12]{5-3x}$

23. $\sqrt{\dfrac{16}{25}} = \dfrac{\sqrt{16}}{\sqrt{25}} = \dfrac{4}{5}$

24. $\dfrac{10}{9}$

25. $\sqrt[3]{\dfrac{64}{27}} = \dfrac{\sqrt[3]{64}}{\sqrt[3]{27}} = \dfrac{4}{3}$

26. $\dfrac{7}{8}$

27. $\sqrt{\dfrac{49}{y^2}} = \dfrac{\sqrt{49}}{\sqrt{y^2}} = \dfrac{7}{y}$

28. $\dfrac{11}{x}$

29. $\sqrt{\dfrac{25y^3}{x^4}} = \dfrac{\sqrt{25y^3}}{\sqrt{x^4}} = \dfrac{\sqrt{25y^2\cdot y}}{\sqrt{x^4}} = \dfrac{\sqrt{25y^2}\,\sqrt{y}}{\sqrt{x^4}} = \dfrac{5y\sqrt{y}}{x^2}$

30. $\dfrac{6a^2\sqrt{a}}{b^3}$

31. $\sqrt[3]{\dfrac{8x^5}{27y^3}} = \dfrac{\sqrt[3]{8x^5}}{\sqrt[3]{27y^3}} = \dfrac{\sqrt[3]{8x^3\cdot x^2}}{\sqrt[3]{27y^3}} = \dfrac{\sqrt[3]{8x^3}\,\sqrt[3]{x^2}}{\sqrt[3]{27y^3}} = \dfrac{2x\sqrt[3]{x^2}}{3y}$

32. $\dfrac{2x^2\sqrt[3]{x}}{3y^2}$

33. $\sqrt[4]{\dfrac{16a^4}{81}} = \dfrac{\sqrt[4]{16a^4}}{\sqrt[4]{81}} = \dfrac{2a}{3}$

34. $\dfrac{3x}{y^2}$

35. $\sqrt[4]{\dfrac{a^5b^8}{c^{10}}} = \dfrac{\sqrt[4]{a^5b^8}}{\sqrt[4]{c^{10}}} = \dfrac{\sqrt[4]{a^4b^8\cdot a}}{\sqrt[4]{c^8\cdot c^2}} = \dfrac{\sqrt[4]{a^4b^8}\,\sqrt[4]{a}}{\sqrt[4]{c^8}\,\sqrt[4]{c^2}} =$

$\dfrac{ab^2\sqrt[4]{a}}{c^2\sqrt[4]{c^2}}$, or $\dfrac{ab^2}{c^2}\sqrt[4]{\dfrac{a}{c^2}}$

36. $\dfrac{x^2y^3}{z}\sqrt[4]{\dfrac{x}{z^2}}$

37. $\sqrt[5]{\dfrac{32x^6}{y^{11}}} = \dfrac{\sqrt[5]{32x^6}}{\sqrt[5]{y^{11}}} = \dfrac{\sqrt[5]{32x^5\cdot x}}{\sqrt[5]{y^{10}\cdot y}} = \dfrac{\sqrt[5]{32x^5}\cdot\sqrt[5]{x}}{\sqrt[5]{y^{10}}\,\sqrt[5]{y}} =$

$\dfrac{2x\sqrt[5]{x}}{y^2\sqrt[5]{y}}$, or $\dfrac{2x}{y^2}\sqrt[5]{\dfrac{x}{y}}$

38. $\dfrac{3a}{b^2}\sqrt[5]{\dfrac{a^4}{b^3}}$

39. $\sqrt[6]{\dfrac{x^6y^8}{z^{15}}} = \dfrac{\sqrt[6]{x^6y^8}}{\sqrt[6]{z^{15}}} = \dfrac{\sqrt[6]{x^6y^6\cdot y^2}}{\sqrt[6]{z^{12}\cdot z^3}} = \dfrac{\sqrt[6]{x^6y^6}\,\sqrt[6]{y^2}}{\sqrt[6]{z^{12}}\,\sqrt[6]{z^3}} =$

$\dfrac{xy\sqrt[6]{y^2}}{z^2\sqrt[6]{z^3}}$, or $\dfrac{xy}{z^2}\sqrt[6]{\dfrac{y^2}{z^3}}$

40. $\dfrac{ab^2}{c^2}\sqrt[6]{\dfrac{a^3}{c}}$

41. a) $\sqrt{(6a)^3} = \sqrt{6^3a^3} = \sqrt{6^2a^2}\,\sqrt{6a} = 6a\sqrt{6a}$

b) $(\sqrt{6a})^3 = \sqrt{6a}\,\sqrt{6a}\,\sqrt{6a} = 6a\sqrt{6a}$

42. $7y\sqrt{7y}$

43. a) $(\sqrt{16b^2})^3 = \sqrt{16b^2}\,\sqrt{16b^2}\,\sqrt{16b^2} =$

$16b^2\,\sqrt{16b^2} = 16b^2\cdot 4b = 64b^3$

b) $\sqrt{(16b^2)^3} = \sqrt{(2^4b^2)^3} = \sqrt{2^{12}b^6} = 2^6b^3$, or $64b^3$

44. $125r^3$

45. a) $\sqrt{(18a^2b)^3} = \sqrt{5832a^6b^3} = \sqrt{2916a^6b^2\cdot 2b} =$

$54a^3b\sqrt{2b}$

b) $(\sqrt{18a^2b})^3 = \sqrt{18a^2b}\,\sqrt{18a^2b}\,\sqrt{18a^2b} =$

$18a^2b\sqrt{18a^2b} = 18a^2b\sqrt{9a^2\cdot 2b} = 18a^2b\cdot 3a\sqrt{2b} =$

$54a^3b\sqrt{2b}$

46. $24x^3y\sqrt{3y}$

47. a) $(\sqrt[3]{3c^2d})^4 = \sqrt[3]{3c^2d}\,\sqrt[3]{3c^2d}\,\sqrt[3]{3c^2d}\,\sqrt[3]{3c^2d} =$

$3c^2d\,\sqrt[3]{3c^2d}$

b) $\sqrt[3]{(3c^2d)^4} = \sqrt[3]{81c^8d^4} = \sqrt[3]{27c^6d^3\cdot 3c^2d} =$

$3c^2d\,\sqrt[3]{3c^2d}$

48. $2x^2y\,\sqrt[3]{2x^2y}$

49. a) $\sqrt[3]{(5x^2y)^2} = \sqrt[3]{25x^4y^2} = \sqrt[3]{x^3\cdot 25xy^2} = x\sqrt[3]{25xy^2}$

b) $(\sqrt[3]{5x^2y})^2 = \sqrt[3]{5x^2y}\,\sqrt[3]{5x^2y} = \sqrt[3]{25x^4y^2} =$

$\sqrt[3]{x^3\cdot 25xy^2} = x\sqrt[3]{25xy^2}$

50. $b\sqrt[3]{36a^2b}$

51. a) $\sqrt[4]{(x^2y)^3} = \sqrt[4]{x^6y^3} = \sqrt[4]{x^4\cdot x^2y^3} = x\sqrt[4]{x^2y^3}$

b) $(\sqrt[4]{x^2y})^3 = \sqrt[4]{x^2y}\,\sqrt[4]{x^2y}\,\sqrt[4]{x^2y} = \sqrt[4]{x^6y^3} =$

$\sqrt[4]{x^4\cdot x^2y^3} = x\sqrt[4]{x^2y^3}$

52. $a^2\,\sqrt[4]{8a}$

53. a) $(\sqrt[3]{8a^4b})^2 = (\sqrt[3]{8a^3\cdot ab})^2 = (2a\sqrt[3]{ab})^2 =$

$2a\sqrt[3]{ab}\cdot 2a\sqrt[3]{ab} = 4a^2\,\sqrt[3]{a^2b^2}$

b) $\sqrt[3]{(8a^4b)^2} = \sqrt[3]{64a^8b^2} = \sqrt[3]{64a^6\cdot a^2b^2} = 4a^2\,\sqrt[3]{a^2b^2}$

54. $9y^3\,\sqrt[3]{x^2y}$

55. a) $(\sqrt[4]{16x^2y^3})^2 = (2\sqrt[4]{x^2y^3})^2 =$

$2\sqrt[4]{x^2y^3}\cdot 2\sqrt[4]{x^2y^3} = 4\sqrt[4]{x^4y^6} = 4\sqrt[4]{x^4y^4\cdot y^2} =$

$4xy\sqrt[4]{y^2}$, or $4xy(y^{2/4}) = 4xy(y^{1/2}) = 4xy\sqrt{y}$

b) $\sqrt[4]{(16x^2y^3)^2} = \sqrt[4]{256x^4y^6} = \sqrt[4]{256x^4y^4\cdot y^2} =$

$4xy\sqrt[4]{y^2}$, or $4xy(y^{2/4}) = 4xy(y^{1/2}) = 4xy\sqrt{y}$

56. $8y^3\,\sqrt[4]{x^3y^3}$

57.

$$\frac{12x}{x-4} - \frac{3x^2}{x+4} = \frac{384}{x^2-16}$$

$$\frac{12x}{x-4} - \frac{3x^2}{x+4} = \frac{384}{(x+4)(x-4)}$$

The LCD is $(x + 4)(x - 4)$.

$$(x+4)(x-4)\left[\frac{12x}{x-4} - \frac{3x^2}{x+4}\right] = (x+4)(x-4)\cdot\frac{384}{(x+4)(x-4)}$$

$$\begin{aligned}
12x(x + 4) - 3x^2(x - 4) &= 384\\
12x^2 + 48x - 3x^3 + 12x^2 &= 384\\
-3x^3 + 24x^2 + 48x - 384 &= 0\\
-3x^2(x - 8) + 48(x - 8) &= 0\\
(x - 8)(-3x^2 + 48) &= 0
\end{aligned}$$

$$\begin{array}{lll}
x - 8 = 0 & \text{or} & -3x^2 + 48 = 0\\
x = 8 & \text{or} & -3(x^2 - 16) = 0\\
x = 8 & \text{or} & -3(x + 4)(x - 4) = 0\\
x = 8 & \text{or} & x + 4 = 0 \quad \text{or} \quad x - 4 = 0\\
x = 8 & \text{or} & x = -4 \quad \text{or} \quad x = 4
\end{array}$$

Check: For $x = 8$:

$$\begin{array}{c|c}
\multicolumn{2}{c}{\dfrac{12x}{x-4} - \dfrac{3x^2}{x+4} = \dfrac{384}{x^2-16}}\\
\hline
\dfrac{12\cdot 8}{8-4} - \dfrac{3\cdot 8^2}{8+4} \;?& \dfrac{384}{8^2-16}\\
\dfrac{96}{4} - \dfrac{192}{12} & \dfrac{384}{48}\\
24 - 16 & 8\\
8 & \qquad \text{TRUE}
\end{array}$$

8 is a solution.

For $x = -4$:

$$\begin{array}{c|c}
\multicolumn{2}{c}{\dfrac{12x}{x-4} - \dfrac{3x^2}{x+4} = \dfrac{384}{x^2-16}}\\
\hline
\dfrac{12(-4)}{-4-4} - \dfrac{3(-4)^2}{-4+4} \;?& \dfrac{384}{(-4)^2-16}\\
\dfrac{-48}{-8} - \dfrac{48}{0} & \dfrac{384}{16-16} \qquad \text{UNDEFINED}
\end{array}$$

-4 is not a solution.

For $x = 4$:

$$\begin{array}{c|c}
\multicolumn{2}{c}{\dfrac{12x}{x-4} - \dfrac{3x^2}{x+4} = \dfrac{384}{x^2-16}}\\
\hline
\dfrac{12\cdot 4}{4-4} - \dfrac{3\cdot 4^2}{4+4} \;?& \dfrac{384}{4^2-16}\\
\dfrac{48}{0} - \dfrac{48}{8} & \dfrac{384}{16-16} \qquad \text{UNDEFINED}
\end{array}$$

4 is not a solution.

The solution is 8.

58. $\frac{15}{2}$

This value checks. The solution is $\frac{15}{2}$.

59. Familiarize. Let x and y represent the width and length of the rectangle, respectively.

Translate. We write two equations.

The width is one-fourth the length.

$$x \quad = \quad \frac{1}{4} \quad \cdot \quad y$$

The area is twice the perimeter.

$$xy \quad = \quad 2\cdot \quad (2x + 2y)$$

Carry out. Solving the system of equations we get (5,20).

Check. The width, 5, is one-fourth the length, 20. The area is 5·20, or 100. The perimeter is 2·5 + 2·20, or 50. Since 100 = 2·50, the area is twice the perimeter. The values check.

State. The width is 5, and the length is 20.

60.

61.

62. a) 1.62 sec

b) 1.99 sec

c) 2.20 sec

63. $\dfrac{7\sqrt{a^2b}\ \sqrt{25xy}}{5\sqrt{a^{-4}b^{-1}}\ \sqrt{49x^{-1}y^{-3}}} = \dfrac{7\sqrt{25a^2bxy}}{5\sqrt{49a^{-4}b^{-1}x^{-1}y^{-3}}} =$

$\dfrac{7}{5}\sqrt{\dfrac{25a^2bxy}{49a^{-4}b^{-1}x^{-1}y^{-3}}} = \dfrac{7}{5}\sqrt{\dfrac{25}{49}\cdot a^6b^2x^2y^4} =$

$\dfrac{7}{5}\cdot\dfrac{5}{7}\cdot a^3bxy^2 = a^3bxy^2$

64. $9\sqrt[3]{9n^2}$

65. $\dfrac{\sqrt{44x^2y^9z}\ \sqrt{22y^9z^6}}{(\sqrt{11xy^8z^2})^2} = \dfrac{\sqrt{44\cdot 22x^2y^{18}z^7}}{\sqrt{11\cdot 11x^2y^{16}z^4}} =$

$\sqrt{\dfrac{44\cdot 22x^2y^{18}z^7}{11\cdot 11x^2y^{16}z^4}} = \sqrt{4\cdot 2y^2z^3} = \sqrt{4y^2z^2\cdot 2z} = 2yz\sqrt{2z}$

Exercise Set 7.5

1. $6\sqrt{3} + 2\sqrt{3} = (6 + 2)\sqrt{3} = 8\sqrt{3}$

2. $17\sqrt{5}$

3. $9\sqrt[3]{5} - 6\sqrt[3]{5} = (9 - 6)\sqrt[3]{5} = 3\sqrt[3]{5}$

4. $8\sqrt[5]{2}$

5. $4\sqrt[3]{y} + 9\sqrt[3]{y} = (4 + 9)\sqrt[3]{y} = 13\sqrt[3]{y}$

6. $3\sqrt[4]{t}$

7. $8\sqrt{2} - 6\sqrt{2} + 5\sqrt{2} = (8 - 6 + 5)\sqrt{2} = 7\sqrt{2}$

8. $7\sqrt{6}$

9. $4\sqrt[3]{3} - \sqrt{5} + 2\sqrt[3]{3} + \sqrt{5} =$

$(4 + 2)\sqrt[3]{3} + (-1 + 1)\sqrt{5} = 6\sqrt[3]{3}$

10. $6\sqrt{7} + \sqrt[4]{11}$

11. $8\sqrt{27} - 3\sqrt{3} = 8\sqrt{9\cdot 3} - 3\sqrt{3}$ } Factoring the first radical

$= 8\sqrt{9}\cdot\sqrt{3} - 3\sqrt{3}$

$= 8\cdot 3\sqrt{3} - 3\sqrt{3}$ Taking the square root of 9

$= 24\sqrt{3} - 3\sqrt{3}$

$= (24 - 3)\sqrt{3}$ Factoring out $\sqrt{3}$ to collect like radical terms

$= 21\sqrt{3}$

12. $41\sqrt{2}$

13. $8\sqrt{45} + 7\sqrt{20} = 8\sqrt{9\cdot 5} + 7\sqrt{4\cdot 5}$ } Factoring the radicals

$= 8\sqrt{9}\cdot\sqrt{5} + 7\sqrt{4}\cdot\sqrt{5}$

$= 8\cdot 3\sqrt{5} + 7\cdot 2\sqrt{5}$ Taking the square roots

$= 24\sqrt{5} + 14\sqrt{5}$

$= (24 + 14)\sqrt{5}$ Factoring out $\sqrt{5}$ to collect like radical terms

$= 38\sqrt{5}$

14. $66\sqrt{3}$

15. $18\sqrt{72} + 2\sqrt{98} = 18\sqrt{36\cdot 2} + 2\sqrt{49\cdot 2} =$

$18\sqrt{36}\cdot\sqrt{2} + 2\sqrt{49}\cdot\sqrt{2} = 18\cdot 6\sqrt{2} + 2\cdot 7\sqrt{2} =$

$108\sqrt{2} + 14\sqrt{2} = (108 + 14)\sqrt{2} = 122\sqrt{2}$

16. $4\sqrt{5}$

17. $3\sqrt[3]{16} + \sqrt[3]{54} = 3\sqrt[3]{8\cdot 2} + \sqrt[3]{27\cdot 2} =$

$3\sqrt[3]{8}\cdot\sqrt[3]{2} + \sqrt[3]{27}\cdot\sqrt[3]{2} = 3\cdot 2\sqrt[3]{2} + 3\sqrt[3]{2} =$

$6\sqrt[3]{2} + 3\sqrt[3]{2} = (6 + 3)\sqrt[3]{2} = 9\sqrt[3]{2}$

18. -7

19. $\sqrt{5a} + 2\sqrt{45a^3} = \sqrt{5a} + 2\sqrt{9a^2\cdot 5a} =$

$\sqrt{5a} + 2\sqrt{9a^2}\cdot\sqrt{5a} = \sqrt{5a} + 2\cdot 3a\sqrt{5a} =$

$\sqrt{5a} + 6a\sqrt{5a} = (1 + 6a)\sqrt{5a}$

20. $(4x - 2)\sqrt{3x}$

21. $\sqrt[3]{24x} - \sqrt[3]{3x^4} = \sqrt[3]{8\cdot 3x} - \sqrt[3]{x^3\cdot 3x} =$

$\sqrt[3]{8}\cdot\sqrt[3]{3x} - \sqrt[3]{x^3}\cdot\sqrt[3]{3x} = 2\sqrt[3]{3x} - x\sqrt[3]{3x} =$

$(2 - x)\sqrt[3]{3x}$

22. $(3 - x)\sqrt[3]{2x}$

23. $\sqrt{8y - 8} + \sqrt{2y - 2} = \sqrt{4(2y - 2)} + \sqrt{2y - 2} =$

$\sqrt{4}\cdot\sqrt{2y - 2} + \sqrt{2y - 2} = 2\sqrt{2y - 2} + \sqrt{2y - 2} =$

$(2 + 1)\sqrt{2y - 2} = 3\sqrt{2y - 2}$

24. $3\sqrt{3t + 3}$

25. $\sqrt{x^3 - x^2} + \sqrt{9x - 9} = \sqrt{x^2(x - 1)} + \sqrt{9(x - 1)} =$

$\sqrt{x^2}\cdot\sqrt{x - 1} + \sqrt{9}\cdot\sqrt{x - 1} = x\sqrt{x - 1} + 3\sqrt{x - 1} =$

$(x + 3)\sqrt{x - 1}$

26. $(2 - x)\sqrt{x - 1}$

27. $5\sqrt[3]{32} - \sqrt[3]{108} + 2\sqrt[3]{256} =$

$5\sqrt[3]{8\cdot 4} - \sqrt[3]{27\cdot 4} + 2\sqrt[3]{64\cdot 4} =$

$5\sqrt[3]{8}\cdot\sqrt[3]{4} - \sqrt[3]{27}\cdot\sqrt[3]{4} + 2\sqrt[3]{64}\cdot\sqrt[3]{4} =$

$5\cdot 2\sqrt[3]{4} - 3\sqrt[3]{4} + 2\cdot 4\sqrt[3]{4} = 10\sqrt[3]{4} - 3\sqrt[3]{4} + 8\sqrt[3]{4} =$

$(10 - 3 + 8)\sqrt[3]{4} = 15\sqrt[3]{4}$

28. $2\sqrt[3]{x}$

29. $\sqrt{x^3 + x^2} + \sqrt{4x^3 + 4x^2} - \sqrt{9x^3 + 9x^2} =$

$\sqrt{x^2(x + 1)} + \sqrt{4x^2(x + 1)} - \sqrt{9x^2(x + 1)} =$

$\sqrt{x^2}\cdot\sqrt{x + 1} + \sqrt{4x^2}\cdot\sqrt{x + 1} - \sqrt{9x^2}\cdot\sqrt{x + 1} =$

$x\sqrt{x + 1} + 2x\sqrt{x + 1} - 3x\sqrt{x + 1} =$

$(x + 2x - 3x)\sqrt{x + 1} = 0$

30. $-8\sqrt{5x^2 + 4}$

31. $\sqrt[4]{x^5 - x^4} + 3\sqrt[4]{x^9 - x^8} = \sqrt[4]{x^4(x - 1)} + 3\sqrt[4]{x^8(x - 1)} =$

$\sqrt[4]{x^4}\cdot\sqrt[4]{x - 1} + 3\sqrt[4]{x^8}\sqrt[4]{x - 1} = x\sqrt[4]{x - 1} + 3x^2\sqrt[4]{x - 1} =$

$(x + 3x^2)\sqrt[4]{x - 1}$, or $(3x^2 + x)\sqrt[4]{x - 1}$

32. $(2a - 2a^2)\sqrt[4]{1 + a}$

33. $\sqrt{6}(2 - 3\sqrt{6}) = \sqrt{6}\cdot 2 - \sqrt{6}\cdot 3\sqrt{6} = 2\sqrt{6} - 3\cdot 6 =$

$2\sqrt{6} - 18$

34. $4\sqrt{3} + 3$

35. $\sqrt{2}(\sqrt{3} - \sqrt{5}) = \sqrt{2}\cdot\sqrt{3} - \sqrt{2}\cdot\sqrt{5} = \sqrt{6} - \sqrt{10}$

36. $5 - \sqrt{10}$

37. $\sqrt{3}(2\sqrt{5} - 3\sqrt{4}) = \sqrt{3}(2\sqrt{5} - 3\cdot 2) =$

$\sqrt{3}\cdot 2\sqrt{5} - \sqrt{3}\cdot 6 = 2\sqrt{15} - 6\sqrt{3}$

38. $6\sqrt{5} - 4$

39. $\sqrt[3]{2}(\sqrt[3]{4} - 2\sqrt[3]{32}) = \sqrt[3]{2}\cdot\sqrt[3]{4} - \sqrt[3]{2}\cdot 2\sqrt[3]{32} =$

$\sqrt[3]{8} - 2\sqrt[3]{64} = 2 - 2\cdot 4 = 2 - 8 = -6$

40. $3 - 4\sqrt[3]{63}$

41. $\sqrt[3]{a}(\sqrt[3]{2a^2} + \sqrt[3]{16a^2}) = \sqrt[3]{a}\cdot\sqrt[3]{2a^2} + \sqrt[3]{a}\cdot\sqrt[3]{16a^2} =$

$\sqrt[3]{2a^3} + \sqrt[3]{16a^3} = \sqrt[3]{a^3\cdot 2} + \sqrt[3]{8a^3\cdot 2} = a\sqrt[3]{2} + 2a\sqrt[3]{2} =$

$3a\sqrt[3]{2}$

42. $-2x\sqrt[3]{3}$

43. $\sqrt[4]{x}(\sqrt[4]{x^7} + \sqrt[4]{3x^2}) = \sqrt[4]{x}\cdot\sqrt[4]{x^7} + \sqrt[4]{x}\cdot\sqrt[4]{3x^2} =$

$\sqrt[4]{x^8} + \sqrt[4]{3x^3} = x^2 + \sqrt[4]{3x^3}$

44. $\sqrt[4]{2a^2} - a^3$

45. $(5 - \sqrt{7})(5 + \sqrt{7}) = 5^2 - (\sqrt{7})^2 = 25 - 7 = 18$

46. 4

47. $(\sqrt{5} + \sqrt{8})(\sqrt{5} - \sqrt{8}) = (\sqrt{5})^2 - (\sqrt{8})^2 =$

$5 - 8 = -3$

48. -2

49. $(3 - 2\sqrt{7})(3 + 2\sqrt{7}) = 3^2 - (2\sqrt{7})^2 = 9 - 4\cdot 7 =$

$9 - 28 = -19$

50. -2

51. $(\sqrt{a} + \sqrt{2})(\sqrt{a} + \sqrt{3})$

$= \sqrt{a}\,\sqrt{a} + \sqrt{a}\,\sqrt{3} + \sqrt{2}\,\sqrt{a} + \sqrt{2}\,\sqrt{3}$ Using FOIL

$= a + \sqrt{3a} + \sqrt{2a} + \sqrt{6}$ Multiplying radicals

52. $2 - 3\sqrt{x} + x$

53. $(3 - \sqrt[3]{5})(2 + \sqrt[3]{5})$

$= 3\cdot 2 + 3\sqrt[3]{5} - 2\sqrt[3]{5} - \sqrt[3]{5}\,\sqrt[3]{5}$ Using FOIL

$= 6 + 3\sqrt[3]{5} - 2\sqrt[3]{5} - \sqrt[3]{25}$

$= 6 + \sqrt[3]{5} - \sqrt[3]{25}$ Simplifying

54. $8 + 2\sqrt[3]{6} - \sqrt[3]{36}$

55. $(2\sqrt{7} - 4\sqrt{2})(3\sqrt{7} + 6\sqrt{2}) =$

$2\sqrt{7}\cdot 3\sqrt{7} + 2\sqrt{7}\cdot 6\sqrt{2} - 4\sqrt{2}\cdot 3\sqrt{7} - 4\sqrt{2}\cdot 6\sqrt{2} =$

$6\cdot 7 + 12\sqrt{14} - 12\sqrt{14} - 24\cdot 2 =$

$42 + 12\sqrt{14} - 12\sqrt{14} - 48 = -6$

56. $24 - 7\sqrt{15}$

57. $(2\sqrt[3]{3} + \sqrt[3]{2})(\sqrt[3]{3} - 2\sqrt[3]{2}) =$

$2\sqrt[3]{3}\cdot\sqrt[3]{3} - 2\sqrt[3]{3}\cdot 2\sqrt[3]{2} + \sqrt[3]{2}\cdot\sqrt[3]{3} - \sqrt[3]{2}\cdot 2\sqrt[3]{2} =$

$2\sqrt[3]{9} - 4\sqrt[3]{6} + \sqrt[3]{6} - 2\sqrt[3]{4} = 2\sqrt[3]{9} - 3\sqrt[3]{6} - 2\sqrt[3]{4}$

58. $6\sqrt[4]{63} - 9\sqrt[4]{42} + 2\sqrt[4]{54} - 3\sqrt[4]{36}$

59. $(1 + \sqrt{3})^2 = 1^2 + 2\sqrt{3} + (\sqrt{3})^2$ Squaring a binomial

$= 1 + 2\sqrt{3} + 3$

$= 4 + 2\sqrt{3}$

60. $6 + 2\sqrt{5}$

61. $(a + \sqrt{b})^2 = a^2 + 2a\sqrt{b} + (\sqrt{b})^2$ Squaring a binomial

$= a^2 + 2a\sqrt{b} + b$

62. $x^2 - 2x\sqrt{y} + y$

63. $(2x - \sqrt[4]{y})^2 = (2x)^2 - 2\cdot 2x\sqrt[4]{y} + (\sqrt[4]{y})^2$

$= 4x^2 - 4x\sqrt[4]{y} + y^{2/4}$

$= 4x^2 - 4x\sqrt[4]{y} + y^{1/2}$

$= 4x^2 - 4x\sqrt[4]{y} + \sqrt{y}$

64. $9a^2 + 6a\sqrt[4]{b} + \sqrt{b}$

65. $(\sqrt{m} + \sqrt{n})^2 = (\sqrt{m})^2 + 2\sqrt{m}\cdot\sqrt{n} + (\sqrt{n})^2$

$= m + 2\sqrt{mn} + n$

66. $r - 2\sqrt{rs} + s$

67. $\sqrt{a}\,(\sqrt[3]{a} + \sqrt[4]{ab})$

$= a^{1/2}[a^{1/3} + (ab)^{1/4}]$ Converting to exponential notation

$= a^{1/2}(a^{1/3} + a^{1/4}b^{1/4})$ Using the laws of exponents

$= a^{1/2}a^{1/3} + a^{1/2}a^{1/4}b^{1/4}$ Using the distributive law

$= a^{1/2+1/3} + a^{1/2+1/4}b^{1/4}$ Adding exponents

$= a^{5/6} + a^{3/4}b^{1/4}$

$= a^{5/6} + (a^3b)^{1/4}$ Using the laws of exponents

$= \sqrt[6]{a^5} + \sqrt[4]{a^3b}$ Converting back to radical notation

68. $\sqrt[6]{x^5y^2} + x\sqrt[4]{y}$

69. $\sqrt[3]{x^2y}(\sqrt{xy} - \sqrt[5]{xy^3})$

$= (x^2y)^{1/3}[(xy)^{1/2} - (xy^3)^{1/5}]$

$= x^{2/3}y^{1/3}(x^{1/2}y^{1/2} - x^{1/5}y^{3/5})$

$= x^{2/3}y^{1/3}x^{1/2}y^{1/2} - x^{2/3}y^{1/3}x^{1/5}y^{3/5}$

$= x^{2/3+1/2}y^{1/3+1/2} - x^{2/3+1/5}y^{1/3+3/5}$

$= x^{7/6}y^{5/6} - x^{13/15}y^{14/15}$

$= x^{1\frac{1}{6}}y^{\frac{5}{6}} - x^{13/15}y^{14/15}$ Writing a mixed number

$= x\cdot x^{1/6}y^{5/6} - x^{13/15}y^{14/15}$

$= x(xy^5)^{1/6} - (x^{13}y^{14})^{1/15}$

$= x\sqrt[6]{xy^5} - \sqrt[15]{x^{13}y^{14}}$

70. $a\sqrt[12]{a^2b^7} - \sqrt[20]{a^{18}b^{13}}$

71. $(m + \sqrt[3]{n^2})(2m + \sqrt[4]{n})$

$= (m + n^{2/3})(2m + n^{1/4})$ Converting to exponential notation

$= 2m^2 + mn^{1/4} + 2mn^{2/3} + n^{2/3}n^{1/4}$ Using FOIL

$= 2m^2 + mn^{1/4} + 2mn^{2/3} + n^{2/3+1/4}$ Adding exponents

$= 2m^2 + mn^{1/4} + 2mn^{2/3} + n^{11/12}$

$= 2m^2 + m\sqrt[4]{n} + 2m\sqrt[3]{n^2} + \sqrt[12]{n^{11}}$ Converting back to radical notation

72. $3r^2 - r\sqrt[5]{s} - 3r\sqrt[4]{s^3} + \sqrt[20]{s^{19}}$

73. $(a + \sqrt[4]{b})(a - \sqrt[3]{c})$

$= (a + b^{1/4})(a - c^{1/3})$

$= a^2 - ac^{1/3} + ab^{1/4} - b^{1/4}c^{1/3}$

$= a^2 - ac^{1/3} + ab^{1/4} - b^{3/12}c^{4/12}$ Finding a common denominator

$= a^2 - ac^{1/3} + ab^{1/4} - (b^3c^4)^{1/12}$

$= a^2 - a\sqrt[3]{c} + a\sqrt[4]{b} - \sqrt[12]{b^3c^4}$ Converting back to radical notation

74. $x^2 + x\sqrt[4]{z} - x\sqrt[3]{y} - \sqrt[12]{y^4z^3}$

75. $(\sqrt{x} - \sqrt[3]{yz})(\sqrt[4]{x} + \sqrt[5]{xz})$

$= [x^{1/2} - (yz)^{1/3}][x^{1/4} + (xz)^{1/5}]$

$= (x^{1/2} - y^{1/3}z^{1/3})(x^{1/4} + x^{1/5}z^{1/5})$

$= x^{1/2}x^{1/4} + x^{1/2}x^{1/5}z^{1/5} - x^{1/4}y^{1/3}z^{1/3} -$
$x^{1/5}y^{1/3}z^{1/3}z^{1/5}$

$= x^{3/4} + x^{7/10}z^{1/5} - x^{1/4}y^{1/3}z^{1/3} - x^{1/5}y^{1/3}z^{8/15}$

Adding exponents

$= x^{3/4} + x^{7/10}z^{2/10} - x^{3/12}y^{4/12}z^{4/12} -$
$x^{3/15}y^{5/15}z^{8/15}$ Finding common denominators

$= x^{3/4} + (x^7z^2)^{1/10} - (x^3y^4z^4)^{1/12} - (x^3y^5z^8)^{1/15}$

$= \sqrt[4]{x^3} + \sqrt[10]{x^7z^2} - \sqrt[12]{x^3y^4z^4} - \sqrt[15]{x^3y^5z^8}$

76. $\sqrt[6]{4a^5} - \sqrt[10]{a^5b^2c^2} + \sqrt[12]{432a^4b^3} - \sqrt[20]{243b^9c^4}$

77. $\dfrac{5}{x - 1} + \dfrac{9}{x^2 + x + 1} = \dfrac{15}{x^3 - 1}$

$\dfrac{5}{x-1} + \dfrac{9}{x^2+x+1} = \dfrac{15}{(x-1)(x^2+x+1)}$

The LCD is $(x - 1)(x^2 + x + 1)$.

$(x-1)(x^2+x+1)\left[\dfrac{5}{x-1} + \dfrac{9}{x^2+x+1}\right] =$

$(x-1)(x^2+x+1) \cdot \dfrac{15}{(x-1)(x^2+x+1)}$

$5(x^2 + x + 1) + 9(x - 1) = 15$

$5x^2 + 5x + 5 + 9x - 9 = 15$

$5x^2 + 14x - 4 = 15$

$5x^2 + 14x - 19 = 0$

$(5x + 19)(x - 1) = 0$

$5x + 19 = 0$ or $x - 1 = 0$

$5x = -19$ or $x = 1$

$x = -\dfrac{19}{5}$ or $x = 1$

Only $-\dfrac{19}{5}$ checks. The solution is $-\dfrac{19}{5}$.

78. $\dfrac{x - 2}{x + 3}$

79. $\sqrt{432} - \sqrt{6125} + \sqrt{845} - \sqrt{4800} =$

$\sqrt{144\cdot 3} - \sqrt{1225\cdot 5} + \sqrt{169\cdot 5} - \sqrt{1600\cdot 3} =$

$12\sqrt{3} - 35\sqrt{5} + 13\sqrt{5} - 40\sqrt{3} = -28\sqrt{3} - 22\sqrt{5}$

80. $(25x - 30y - 9xy)\sqrt{2xy}$

81. $\frac{1}{2}\sqrt{36a^5bc^4} - \frac{1}{2}\sqrt[3]{64a^4bc^6} + \frac{1}{6}\sqrt{144a^3bc^2} =$

$\frac{1}{2}\sqrt{36a^4c^4\cdot ab} - \frac{1}{2}\sqrt[3]{64a^3c^6\cdot ab} + \frac{1}{6}\sqrt{144a^2c^2\cdot ab} =$

$\frac{1}{2}(6a^2c^2)\sqrt{ab} - \frac{1}{2}(4ac^2)\sqrt[3]{ab} + \frac{1}{6}(12ac)\sqrt{ab} =$

$3a^2c^2\sqrt{ab} - 2ac^2\sqrt[3]{ab} + 2ac\sqrt{ab} =$

$(3a^2c^2 + 2ac)\sqrt{ab} - 2ac^2\sqrt[3]{ab}$, or

$ac[(3ac + 2)\sqrt{ab} - 2c\sqrt[3]{ab}]$

82. $(7x^2 - 2y^2)\sqrt{x + y}$

83. $\sqrt{9 + 3\sqrt{5}}\sqrt{9 - 3\sqrt{5}} = \sqrt{(9 + 3\sqrt{5})(9 - 3\sqrt{5})} =$

$\sqrt{81 - 9\cdot 5} = \sqrt{81 - 45} = \sqrt{36} = 6$

84. $2x - 2\sqrt{x^2 - 4}$

85. $(\sqrt{3} + \sqrt{5} - \sqrt{6})^2 = [(\sqrt{3} + \sqrt{5}) - \sqrt{6}]^2 =$

$(\sqrt{3} + \sqrt{5})^2 - 2(\sqrt{3} + \sqrt{5})(\sqrt{6}) + (\sqrt{6})^2 =$

$3 + 2\sqrt{15} + 5 - 2\sqrt{18} - 2\sqrt{30} + 6 =$

$14 + 2\sqrt{15} - 2\sqrt{9\cdot 2} - 2\sqrt{30} =$

$14 + 2\sqrt{15} - 6\sqrt{2} - 2\sqrt{30}$

86. $\sqrt[3]{y} - y$

87. $(\sqrt[3]{9} - 2)(\sqrt[3]{9} + 4) = \sqrt[3]{81} + 4\sqrt[3]{9} - 2\sqrt[3]{9} - 8 =$

$\sqrt[3]{27\cdot 3} + 2\sqrt[3]{9} - 8 = 3\sqrt[3]{3} + 2\sqrt[3]{9} - 8$

88. $9 + 6\sqrt{3} + 3 = 12 + 6\sqrt{3}$

Exercise Set 7.6

1. $\sqrt{\frac{6}{5}} = \sqrt{\frac{6}{5}\cdot\frac{5}{5}} = \sqrt{\frac{30}{25}} = \frac{\sqrt{30}}{\sqrt{25}} = \frac{\sqrt{30}}{5}$

2. $\frac{\sqrt{66}}{6}$

3. $\sqrt{\frac{10}{7}} = \sqrt{\frac{10}{7}\cdot\frac{7}{7}} = \sqrt{\frac{70}{49}} = \frac{\sqrt{70}}{\sqrt{49}} = \frac{\sqrt{70}}{7}$

4. $\frac{\sqrt{66}}{3}$

5. $\frac{6\sqrt{5}}{5\sqrt{3}} = \frac{6\sqrt{5}}{5\sqrt{3}}\cdot\frac{\sqrt{3}}{\sqrt{3}} = \frac{6\sqrt{15}}{5\cdot 3} = \frac{2\sqrt{15}}{5}$

6. $\frac{\sqrt{6}}{5}$

7. $\sqrt[3]{\frac{16}{9}} = \sqrt[3]{\frac{16}{9}\cdot\frac{3}{3}} = \sqrt[3]{\frac{48}{27}} = \frac{\sqrt[3]{8\cdot 6}}{\sqrt[3]{27}} = \frac{2\sqrt[3]{6}}{3}$

8. $\frac{\sqrt[3]{6}}{3}$

9. $\frac{\sqrt[3]{3a}}{\sqrt[3]{5c}} = \frac{\sqrt[3]{3a}}{\sqrt[3]{5c}}\cdot\frac{\sqrt[3]{(5c)^2}}{\sqrt[3]{(5c)^2}} = \frac{\sqrt[3]{3a\cdot 25c^2}}{\sqrt[3]{(5c)^3}} = \frac{\sqrt[3]{75ac^2}}{5c}$

10. $\frac{\sqrt[3]{63xy^2}}{3y}$

11. $\frac{\sqrt[3]{5y^4}}{\sqrt[3]{6x^4}} = \frac{\sqrt[3]{5y^4}}{\sqrt[3]{6x^4}}\cdot\frac{\sqrt[3]{36x^2}}{\sqrt[3]{36x^2}} = \frac{\sqrt[3]{y^3\cdot 180x^2y}}{\sqrt[3]{216x^6}} = \frac{y\sqrt[3]{180x^2y}}{6x^2}$

12. $\frac{a\sqrt[3]{147ab}}{7b}$

13. $\frac{1}{\sqrt[3]{xy}} = \frac{1}{\sqrt[3]{xy}}\cdot\frac{\sqrt[3]{(xy)^2}}{\sqrt[3]{(xy)^2}} = \frac{\sqrt[3]{(xy)^2}}{\sqrt[3]{(xy)^3}} = \frac{\sqrt[3]{x^2y^2}}{xy}$

14. $\frac{\sqrt[3]{a^2b^2}}{ab}$

15. $\sqrt{\frac{7a}{18}} = \sqrt{\frac{7a}{18}\cdot\frac{2}{2}} = \sqrt{\frac{14a}{36}} = \frac{\sqrt{14a}}{\sqrt{36}} = \frac{\sqrt{14a}}{6}$

16. $\frac{\sqrt{30x}}{10}$

17. $\sqrt{\frac{9}{20x^2y}} = \sqrt{\frac{9}{20x^2y}\cdot\frac{5y}{5y}} = \sqrt{\frac{9\cdot 5y}{100x^2y^2}} = \frac{\sqrt{9\cdot 5y}}{\sqrt{100x^2y^2}} =$

$\frac{3\sqrt{5y}}{10xy}$

18. $\frac{\sqrt{10a}}{8ab}$

19. $\sqrt[3]{\frac{9}{100x^2y^5}} = \sqrt[3]{\frac{9}{100x^2y^5}\cdot\frac{10xy}{10xy}} = \sqrt[3]{\frac{90xy}{1000x^3y^6}} =$

$\frac{\sqrt[3]{90xy}}{\sqrt[3]{1000x^3y^6}} = \frac{\sqrt[3]{90xy}}{10xy^2}$

20. $\frac{\sqrt[3]{42a^2b^2}}{6a^2b}$

21. $\frac{\sqrt{7}}{\sqrt{3x}} = \frac{\sqrt{7}}{\sqrt{3x}}\cdot\frac{\sqrt{7}}{\sqrt{7}} = \frac{7}{\sqrt{21x}}$

22. $\frac{6}{\sqrt{30x}}$

23. $\sqrt{\frac{14}{21}} = \sqrt{\frac{2}{3}} = \sqrt{\frac{2}{3}\cdot\frac{2}{2}} = \sqrt{\frac{4}{6}} = \frac{\sqrt{4}}{\sqrt{6}} = \frac{2}{\sqrt{6}}$

24. $\frac{2}{\sqrt{5}}$

25. $\frac{4\sqrt{13}}{3\sqrt{7}} = \frac{4\sqrt{13}}{3\sqrt{7}}\cdot\frac{\sqrt{13}}{\sqrt{13}} = \frac{4\cdot 13}{3\sqrt{91}} = \frac{52}{3\sqrt{91}}$

26. $\dfrac{105}{2\sqrt{105}}$

27. $\dfrac{\sqrt[3]{7}}{\sqrt[3]{2}} = \dfrac{\sqrt[3]{7}}{\sqrt[3]{2}} \cdot \dfrac{\sqrt[3]{49}}{\sqrt[3]{49}} = \dfrac{7}{\sqrt[3]{98}}$

28. $\dfrac{5}{\sqrt[3]{100}}$

29. $\sqrt{\dfrac{7x}{3y}} = \sqrt{\dfrac{7x}{3y} \cdot \dfrac{7x}{7x}} = \dfrac{\sqrt{49x^2}}{\sqrt{21xy}} = \dfrac{7x}{\sqrt{21xy}}$

30. $\dfrac{6a}{\sqrt{30ab}}$

31. $\dfrac{\sqrt[3]{5y^4}}{\sqrt[3]{6x^5}} = \dfrac{\sqrt[3]{5y^4}}{\sqrt[3]{6x^5}} \cdot \dfrac{\sqrt[3]{25y^2}}{\sqrt[3]{25y^2}} = \dfrac{\sqrt[3]{125y^6}}{\sqrt[3]{150x^5y^2}} =$

$\dfrac{5y^2}{\sqrt[3]{x^3 \cdot 150x^2y^2}} = \dfrac{5y^2}{x\sqrt[3]{150x^2y^2}}$

32. $\dfrac{3a^2}{\sqrt[3]{63ab^2}}$

33. $\dfrac{\sqrt{ab}}{3} = \dfrac{\sqrt{ab}}{3} \cdot \dfrac{\sqrt{ab}}{\sqrt{ab}} = \dfrac{ab}{3\sqrt{ab}}$

34. $\dfrac{xy}{5\sqrt{xy}}$

35. $\sqrt{\dfrac{x^3y}{2}} = \sqrt{\dfrac{x^3y}{2} \cdot \dfrac{xy}{xy}} = \sqrt{\dfrac{x^4y^2}{2xy}} = \dfrac{\sqrt{x^4y^2}}{\sqrt{2xy}} = \dfrac{x^2y}{\sqrt{2xy}}$

36. $\dfrac{ab^3}{\sqrt{3ab}}$

37. $\dfrac{\sqrt[3]{a^2b}}{\sqrt[3]{5}} = \dfrac{\sqrt[3]{a^2b}}{\sqrt[3]{5}} \cdot \dfrac{\sqrt[3]{ab^2}}{\sqrt[3]{ab^2}} = \dfrac{\sqrt[3]{a^3b^3}}{\sqrt[3]{5ab^2}} = \dfrac{ab}{\sqrt[3]{5ab^2}}$

38. $\dfrac{xy}{\sqrt[3]{7x^2y}}$

39. $\sqrt[3]{\dfrac{x^4y^2}{3}} = \sqrt[3]{\dfrac{x^4y^2}{3} \cdot \dfrac{x^2y}{x^2y}} = \sqrt[3]{\dfrac{x^6y^3}{3x^2y}} = \dfrac{\sqrt[3]{x^6y^3}}{\sqrt[3]{3x^2y}} = \dfrac{x^2y}{\sqrt[3]{3x^2y}}$

40. $\dfrac{a^2b}{\sqrt[3]{2ab^2}}$

41. $\dfrac{5}{8 - \sqrt{6}} = \dfrac{5}{8 - \sqrt{6}} \cdot \dfrac{8 + \sqrt{6}}{8 + \sqrt{6}} = \dfrac{5(8 + \sqrt{6})}{8^2 - (\sqrt{6})^2} =$

$\dfrac{5(8 + \sqrt{6})}{64 - 6} = \dfrac{5(8 + \sqrt{6})}{58}$

42. $\dfrac{7(9 - \sqrt{10})}{71}$

43. $\dfrac{-4\sqrt{7}}{\sqrt{5} - \sqrt{3}} = \dfrac{-4\sqrt{7}}{\sqrt{5} - \sqrt{3}} \cdot \dfrac{\sqrt{5} + \sqrt{3}}{\sqrt{5} + \sqrt{3}} =$

$\dfrac{-4\sqrt{7}(\sqrt{5} + \sqrt{3})}{5 - 3} = \dfrac{-4\sqrt{7}(\sqrt{5} + \sqrt{3})}{2} =$

$-2\sqrt{7}(\sqrt{5} + \sqrt{3})$

44. $\dfrac{3\sqrt{2}(\sqrt{3} + \sqrt{5})}{2}$

45. $\dfrac{\sqrt{5} - 2\sqrt{6}}{\sqrt{3} - 4\sqrt{5}} = \dfrac{\sqrt{5} - 2\sqrt{6}}{\sqrt{3} - 4\sqrt{5}} \cdot \dfrac{\sqrt{3} + 4\sqrt{5}}{\sqrt{3} + 4\sqrt{5}} =$

$\dfrac{\sqrt{15} + 4 \cdot 5 - 2\sqrt{18} - 8\sqrt{30}}{(\sqrt{3})^2 - (4\sqrt{5})^2} =$

$\dfrac{\sqrt{15} + 20 - 2\sqrt{9 \cdot 2} - 8\sqrt{30}}{3 - 16 \cdot 5} =$

$\dfrac{\sqrt{15} + 20 - 6\sqrt{2} - 8\sqrt{30}}{-77}$

46. $\dfrac{3\sqrt{2} + 2\sqrt{42} - 3\sqrt{15} - 6\sqrt{35}}{-25}$

47. $\dfrac{\sqrt{x} - \sqrt{y}}{\sqrt{x} + \sqrt{y}} = \dfrac{\sqrt{x} - \sqrt{y}}{\sqrt{x} + \sqrt{y}} \cdot \dfrac{\sqrt{x} - \sqrt{y}}{\sqrt{x} - \sqrt{y}} =$

$\dfrac{x - \sqrt{xy} - \sqrt{xy} + y}{x - y} = \dfrac{x - 2\sqrt{xy} + y}{x - y}$

48. $\dfrac{a + 2\sqrt{ab} + b}{a - b}$

49. $\dfrac{5\sqrt{3} - 3\sqrt{2}}{3\sqrt{2} - 2\sqrt{3}} = \dfrac{5\sqrt{3} - 3\sqrt{2}}{3\sqrt{2} - 2\sqrt{3}} \cdot \dfrac{3\sqrt{2} + 2\sqrt{3}}{3\sqrt{2} + 2\sqrt{3}} =$

$\dfrac{15\sqrt{6} + 10 \cdot 3 - 9 \cdot 2 - 6\sqrt{6}}{9 \cdot 2 - 4 \cdot 3} = \dfrac{12 + 9\sqrt{6}}{6} =$

$\dfrac{3(4 + 3\sqrt{6})}{3 \cdot 2} = \dfrac{3}{3} \cdot \dfrac{4 + 3\sqrt{6}}{2} = \dfrac{4 + 3\sqrt{6}}{2}$

50. $\dfrac{4\sqrt{6} + 9}{3}$

51. $\dfrac{3\sqrt{x} + \sqrt{y}}{2\sqrt{x} + 3\sqrt{y}} = \dfrac{3\sqrt{x} + \sqrt{y}}{2\sqrt{x} + 3\sqrt{y}} \cdot \dfrac{2\sqrt{x} - 3\sqrt{y}}{2\sqrt{x} - 3\sqrt{y}} =$

$\dfrac{6x - 9\sqrt{xy} + 2\sqrt{xy} - 3y}{4x - 9y} = \dfrac{6x - 7\sqrt{xy} - 3y}{4x - 9y}$

52. $\dfrac{6a - 7\sqrt{ab} + 2b}{9a - 4b}$

53. $\dfrac{\sqrt{3} + 5}{8} = \dfrac{\sqrt{3} + 5}{8} \cdot \dfrac{\sqrt{3} - 5}{\sqrt{3} - 5} = \dfrac{3 - 25}{8(\sqrt{3} - 5)} =$

$\dfrac{-22}{8(\sqrt{3} - 5)} = \dfrac{2(-11)}{2 \cdot 4(\sqrt{3} - 5)} = \dfrac{2}{2} \cdot \dfrac{-11}{4(\sqrt{3} - 5)} =$

$\dfrac{-11}{4(\sqrt{3} - 5)}$

54. $\dfrac{7}{5(3 + \sqrt{2})}$

55. $\dfrac{\sqrt{3} - 5}{\sqrt{2} + 5} = \dfrac{\sqrt{3} - 5}{\sqrt{2} + 5} \cdot \dfrac{\sqrt{3} + 5}{\sqrt{3} + 5} =$

$\dfrac{3 - 25}{\sqrt{6} + 5\sqrt{2} + 5\sqrt{3} + 25} = \dfrac{-22}{\sqrt{6} + 5\sqrt{2} + 5\sqrt{3} + 25}$

56. $\dfrac{-3}{3\sqrt{2} + 3\sqrt{3} + 7\sqrt{6} + 21}$

57. $\dfrac{\sqrt{x} - \sqrt{y}}{\sqrt{x} + \sqrt{y}} = \dfrac{\sqrt{x} - \sqrt{y}}{\sqrt{x} + \sqrt{y}} \cdot \dfrac{\sqrt{x} + \sqrt{y}}{\sqrt{x} + \sqrt{y}} =$

$\dfrac{x - y}{x + \sqrt{xy} + \sqrt{xy} + y} = \dfrac{x - y}{x + 2\sqrt{xy} + y}$

58. $\dfrac{x - y}{x - 2\sqrt{xy} + y}$

59. $\dfrac{4\sqrt{6} - 5\sqrt{3}}{2\sqrt{3} + 7\sqrt{6}} = \dfrac{4\sqrt{6} - 5\sqrt{3}}{2\sqrt{3} + 7\sqrt{6}} \cdot \dfrac{4\sqrt{6} + 5\sqrt{3}}{4\sqrt{6} + 5\sqrt{3}} =$

$\dfrac{16\cdot 6 - 25\cdot 3}{8\sqrt{18} + 10\cdot 3 + 28\cdot 6 + 35\sqrt{18}} = \dfrac{96 - 75}{43\sqrt{18} + 30 + 168} =$

$\dfrac{21}{43\sqrt{9\cdot 2} + 198} = \dfrac{21}{43\cdot 3\sqrt{2} + 198} = \dfrac{3\cdot 7}{3(43\sqrt{2} + 66)} =$

$\dfrac{7}{43\sqrt{2} + 66}$

60. $\dfrac{53}{75\sqrt{6} - 187}$

61. $\dfrac{\sqrt{3} + 2\sqrt{x}}{\sqrt{3} - \sqrt{x}} = \dfrac{\sqrt{3} + 2\sqrt{x}}{\sqrt{3} - \sqrt{x}} \cdot \dfrac{\sqrt{3} - 2\sqrt{x}}{\sqrt{3} - 2\sqrt{x}} =$

$\dfrac{3 - 4x}{3 - 2\sqrt{3x} - \sqrt{3x} + 2x} = \dfrac{3 - 4x}{3 - 3\sqrt{3x} + 2x}$

62. $\dfrac{5 - 9x}{5 + 4\sqrt{5x} + 3x}$

63. $\dfrac{a + b\sqrt{c}}{a + \sqrt{c}} = \dfrac{a + b\sqrt{c}}{a + \sqrt{c}} \cdot \dfrac{a - b\sqrt{c}}{a - b\sqrt{c}} =$

$\dfrac{a^2 - b^2c}{a^2 - ab\sqrt{c} + a\sqrt{c} - bc} = \dfrac{a^2 - b^2c}{a^2 + (-ab + a)\sqrt{c} - bc},$

or $\dfrac{a^2 - b^2c}{a^2 + (a - ab)\sqrt{c} - bc}$

64. $\dfrac{a^2b - c^2}{ab + (1 - a)c\sqrt{b} - c^2}$

65. $\dfrac{1}{2} - \dfrac{1}{3} = \dfrac{1}{t}$ LCD is 6t

$6t\left(\dfrac{1}{2} - \dfrac{1}{3}\right) = 6t\left(\dfrac{1}{t}\right)$

$3t - 2t = 6$

$t = 6$

Check:

$\frac{1}{2} - \frac{1}{3} = \frac{1}{t}$	
$\frac{1}{2} - \frac{1}{3}$?	$\frac{1}{6}$
$\frac{3}{6} - \frac{2}{6}$	
$\frac{1}{6}$	TRUE

The solution is 6.

66. 1

67.

68.

69. $\dfrac{a - \sqrt{a + b}}{\sqrt{a + b} - b} = \dfrac{a - \sqrt{a + b}}{\sqrt{a + b} - b} \cdot \dfrac{\sqrt{a + b} + b}{\sqrt{a + b} + b} =$

$\dfrac{a\sqrt{a + b} + ab - (a + b) - b\sqrt{a + b}}{a + b - b^2} =$

$\dfrac{ab + (a - b)\sqrt{a + b} - a - b}{a + b - b^2}$

70. $\dfrac{15y + 6\sqrt{y(z + y)} + 20y\sqrt{z} + 8\sqrt{yz(z + y)}}{21y - 4z}$

71. $\dfrac{b + \sqrt{b}}{1 + b + \sqrt{b}} = \dfrac{b + \sqrt{b}}{(1 + b) + \sqrt{b}} \cdot \dfrac{(1 + b) - \sqrt{b}}{(1 + b) - \sqrt{b}} =$

$\dfrac{b(1 + b) - b\sqrt{b} + \sqrt{b}(1 + b) - b}{(1 + b)^2 - b} =$

$\dfrac{b + b^2 - b\sqrt{b} + \sqrt{b} + b\sqrt{b} - b}{1 + 2b + b^2 - b} = \dfrac{b^2 + \sqrt{b}}{1 + b + b^2}$

72. $\dfrac{1}{\sqrt{y + 18} + \sqrt{y}}$

73. $\dfrac{\sqrt{x + 6} - 5}{\sqrt{x + 6} + 5} = \dfrac{\sqrt{x + 6} - 5}{\sqrt{x + 6} + 5} \cdot \dfrac{\sqrt{x + 6} + 5}{\sqrt{x + 6} + 5} =$

$\dfrac{x + 6 - 25}{x + 6 + 5\sqrt{x + 6} + 5\sqrt{x + 6} + 25} =$

$\dfrac{x - 19}{x + 10\sqrt{x + 6} + 31}$

74. $\dfrac{-3\sqrt{a^2 - 3}}{a^2 - 3}$

75. $5\sqrt{\frac{x}{y}} + 4\sqrt{\frac{y}{x}} - \frac{3}{\sqrt{xy}} = \frac{5\sqrt{x}}{\sqrt{y}} + \frac{4\sqrt{y}}{\sqrt{x}} - \frac{3}{\sqrt{xy}} =$

$\frac{5\sqrt{x}}{\sqrt{y}} \cdot \frac{\sqrt{x}}{\sqrt{x}} + \frac{4\sqrt{y}}{\sqrt{x}} \cdot \frac{\sqrt{y}}{\sqrt{y}} - \frac{3}{\sqrt{xy}} = \frac{5x}{\sqrt{xy}} + \frac{4y}{\sqrt{xy}} - \frac{3}{\sqrt{xy}} =$

$\frac{5x + 4y - 3}{\sqrt{xy}} = \frac{5x + 4y - 3}{\sqrt{xy}} \cdot \frac{\sqrt{xy}}{\sqrt{xy}} = \frac{(5x + 4y - 3)\sqrt{xy}}{xy}$

76. $1 - \sqrt{w}$

77. $\frac{1}{4 + \sqrt{3}} + \frac{1}{\sqrt{3}} + \frac{1}{\sqrt{3} - 4} =$

$\frac{1}{4+\sqrt{3}} \cdot \frac{\sqrt{3}(\sqrt{3}-4)}{\sqrt{3}(\sqrt{3}-4)} + \frac{1}{\sqrt{3}} \cdot \frac{(4+\sqrt{3})(\sqrt{3}-4)}{(4+\sqrt{3})(\sqrt{3}-4)} + \frac{1}{\sqrt{3}-4} \cdot \frac{\sqrt{3}(4+\sqrt{3})}{\sqrt{3}(4+\sqrt{3})} =$

$\frac{3 - 4\sqrt{3} - 16 + 3 + 4\sqrt{3} + 3}{\sqrt{3}(4 + \sqrt{3})(\sqrt{3} - 4)} = \frac{-7}{\sqrt{3}(-16 + 3)} =$

$\frac{-7}{-13\sqrt{3}} \cdot \frac{\sqrt{3}}{\sqrt{3}} = \frac{7\sqrt{3}}{39}$

Exercise Set 7.7

1. $\sqrt{2x - 3} = 1$

$(\sqrt{2x - 3})^2 = 1^2$ Principle of powers (squaring)

$2x - 3 = 1$

$2x = 4$

$x = 2$

Check:
$$\begin{array}{c|l} \sqrt{2x - 3} & = 1 \\ \hline \sqrt{2 \cdot 2 - 3} & ?\ 1 \\ \sqrt{1} & \\ 1 & \text{TRUE} \end{array}$$

The solution is 2.

2. 33

3. $\sqrt{3x} + 1 = 7$

$\sqrt{3x} = 6$ Adding to isolate the radical

$(\sqrt{3x})^2 = 6^2$ Principle of powers (squaring)

$3x = 36$

$x = 12$

Check:
$$\begin{array}{c|l} \sqrt{3x} + 1 & = 7 \\ \hline \sqrt{3 \cdot 12} + 1 & ?\ 7 \\ 6 + 1 & \\ 7 & \text{TRUE} \end{array}$$

The solution is 12.

4. 32

5. $\sqrt{y + 1} - 5 = 8$

$\sqrt{y + 1} = 13$ Adding to isolate the radical

$(\sqrt{y + 1})^2 = 13^2$ Principle of powers (squaring)

$y + 1 = 169$

$y = 168$

Check:
$$\begin{array}{c|l} \sqrt{y + 1} - 5 & = 8 \\ \hline \sqrt{168 + 1} - 5 & ?\ 8 \\ 13 - 5 & \\ 8 & \text{TRUE} \end{array}$$

The solution is 168.

6. 11

7. $\sqrt{y - 3} + 4 = 2$

$\sqrt{y - 3} = -2$

At this point we might observe that this equation has no real-number solution, because the principle square root of a number is never negative. However, we will continue with the solution.

$(\sqrt{y - 3})^2 = (-2)^2$

$y - 3 = 4$

$y = 7$

Check:
$$\begin{array}{c|l} \sqrt{y - 3} + 4 & = 2 \\ \hline \sqrt{7 - 3} + 4 & ?\ 2 \\ \sqrt{4} + 4 & \\ 6 & \text{FALSE} \end{array}$$

The number 7 does not check. There is no solution.

8. -3

9. $\sqrt[3]{x + 5} = 2$

$(\sqrt[3]{x + 5})^3 = 2^3$

$x + 5 = 8$

$x = 3$

Check:
$$\begin{array}{c|l} \sqrt[3]{x + 5} & = 2 \\ \hline \sqrt[3]{3 + 5} & ?\ 2 \\ \sqrt[3]{8} & \\ 2 & \text{TRUE} \end{array}$$

The solution is 3.

10. 29

11. $\sqrt[4]{y - 3} = 2$

$(\sqrt[4]{y - 3})^4 = 2^4$

$y - 3 = 16$

$y = 19$

Check:
$$\begin{array}{c|l} \sqrt[4]{y - 3} & = 2 \\ \hline \sqrt[4]{19 - 3} & ?\ 2 \\ \sqrt[4]{16} & \\ 2 & \text{TRUE} \end{array}$$

The solution is 19.

12. 78

13. $\sqrt{3y + 1} = 9$

$(\sqrt{3y + 1})^2 = 9^2$

$3y + 1 = 81$

$3y = 80$

$y = \frac{80}{3}$

Check: $\sqrt{3y + 1} = 9$; $\sqrt{3 \cdot \frac{80}{3} + 1}$? 9; $\sqrt{81}$; 9 TRUE

The solution is $\frac{80}{3}$.

14. 84

15. $3\sqrt{x} = 6$

$\sqrt{x} = 2$ Multiplying by $\frac{1}{3}$

$(\sqrt{x})^2 = 2^2$

$x = 4$

Check: $3\sqrt{x} = 6$; $3\sqrt{4}$? 6; $3 \cdot 2$; 6 TRUE

The solution is 4.

16. $\frac{1}{16}$

17. $2y^{1/2} - 7 = 9$

$2\sqrt{y} - 7 = 9$

$2\sqrt{y} = 16$

$\sqrt{y} = 8$

$(\sqrt{y})^2 = 8^2$

$y = 64$

Check: $2y^{1/2} - 7 = 9$; $2 \cdot 64^{1/2} - 7$? 9; $2 \cdot 8 - 7$; 9 TRUE

The solution is 64.

18. No solution

19. $\sqrt[3]{x} = -3$

$(\sqrt[3]{x})^3 = (-3)^3$

$x = -27$

Check: $\sqrt[3]{x} = -3$; $\sqrt[3]{-27}$? -3; -3 TRUE

The solution is -27.

20. -64

21. $\sqrt{y + 3} - 20 = 0$

$\sqrt{y + 3} = 20$

$(\sqrt{y + 3})^2 = 20^2$

$y + 3 = 400$

$y = 397$

Check: $\sqrt{y + 3} - 20 = 0$; $\sqrt{397 + 3} - 20$? 0; $\sqrt{400} - 20$; 0 TRUE

The solution is 397.

22. 117

23. $(x + 2)^{1/2} = -4$

$\sqrt{x + 2} = -4$

We might observe that this equation has no real-number solution, since the principal square root of a number is never negative. However, we will go through the solution process.

$(\sqrt{x + 2})^2 = (-4)^2$

$x + 2 = 16$

$x = 14$

Check: $(x + 2)^{1/2} = -4$; $(14 + 2)^{1/2}$? -4; $16^{1/2}$; 4 FALSE

The number 14 does not check. The equation has no solution.

24. No solution

25. $\sqrt{2x + 3} - 5 = -2$

$\sqrt{2x + 3} = 3$

$(\sqrt{2x + 3})^2 = 3^2$

$2x + 3 = 9$

$2x = 6$

$x = 3$

Check: $\sqrt{2x + 3} - 5 = -2$; $\sqrt{2 \cdot 3 + 3} - 5$? -2; $\sqrt{9} - 5$; -2 TRUE

The solution is 3.

26. $\frac{8}{3}$

27. $8 = x^{-1/2}$

$8 = \frac{1}{x^{1/2}}$

$8 = \frac{1}{\sqrt{x}}$

$8 \cdot \sqrt{x} = \frac{1}{\sqrt{x}} \cdot \sqrt{x}$

$8\sqrt{x} = 1$

$(8\sqrt{x})^2 = 1^2$

$64x = 1$

$x = \frac{1}{64}$

Check: $8 = x^{-1/2}$; 8 ? $\left(\frac{1}{64}\right)^{-1/2}$; $64^{1/2}$; 8 TRUE

The solution is $\frac{1}{64}$.

28. $\frac{1}{9}$

29. $\sqrt[3]{6x + 9} + 8 = 5$

$\sqrt[3]{6x + 9} = -3$

$(\sqrt[3]{6x + 9})^3 = (-3)^3$

$6x + 9 = -27$

$6x = -36$

$x = -6$

Check: $\sqrt[3]{6x + 9} + 8 = 5$; $\sqrt[3]{6(-6) + 9} + 8$? 5; $\sqrt[3]{-27} + 8$; 5 TRUE

The solution is -6.

30. $-\frac{5}{3}$

31. $\sqrt{3y + 1} = \sqrt{2y + 6}$ One radical is already isolated.

$(\sqrt{3y + 1})^2 = (\sqrt{2y + 6})^2$ Squaring both sides

$3y + 1 = 2y + 6$

$y = 5$

The number 5 checks and is the solution.

32. 2

33. $2\sqrt{1 - x} = \sqrt{5}$ One radical is already isolated.

$(2\sqrt{1 - x})^2 = (\sqrt{5})^2$ Squaring both sides

$4(1 - x) = 5$

$4 - 4x = 5$

$-4x = 1$

$x = -\frac{1}{4}$

The number $-\frac{1}{4}$ checks and is the solution.

34. 3

35. $2\sqrt{t - 1} = \sqrt{3t - 1}$

$(2\sqrt{t - 1})^2 = (\sqrt{3t - 1})^2$

$4(t - 1) = 3t - 1$

$4t - 4 = 3t - 1$

$t = 3$

The number 3 checks and is the solution.

36. -1

37. $\sqrt{y - 5} + \sqrt{y} = 5$

$\sqrt{y - 5} = 5 - \sqrt{y}$ Adding $-\sqrt{y}$; this isolates one of the radical terms

$(\sqrt{y - 5})^2 = (5 - \sqrt{y})^2$ Squaring both sides

$y - 5 = 25 - 10\sqrt{y} + y$

$-30 = -10\sqrt{y}$ Isolating the remaining radical term

$3 = \sqrt{y}$ Multiplying by $-\frac{1}{10}$

$3^2 = (\sqrt{y})^2$ Squaring both sides

$9 = y$

The number 9 checks and is the solution.

38. No solution

39. $3 + \sqrt{z - 6} = \sqrt{z + 9}$ One radical is already isolated.

$(3 + \sqrt{z - 6})^2 = (\sqrt{z + 9})^2$ Squaring both sides

$9 + 6\sqrt{z - 6} + z - 6 = z + 9$

$6\sqrt{z - 6} = 6$

$\sqrt{z - 6} = 1$ Multiplying by $\frac{1}{6}$

$(\sqrt{z - 6})^2 = 1^2$ Squaring both sides

$z - 6 = 1$

$z = 7$

The number 7 checks and is the solution.

40. 7, 3

41. $\sqrt{20 - x} + 8 = \sqrt{9 - x} + 11$

$\sqrt{20 - x} = \sqrt{9 - x} + 3$ Adding -8

$(\sqrt{20 - x})^2 = (\sqrt{9 - x} + 3)^2$ Squaring both sides

$20 - x = 9 - x + 6\sqrt{9 - x} + 9$

$2 = 6\sqrt{9 - x}$ Isolating the remaining radical

$1 = 3\sqrt{9 - x}$ Multiplying by $\frac{1}{2}$

$1^2 = (3\sqrt{9 - x})^2$ Squaring both sides

$1 = 9(9 - x)$

$1 = 81 - 9x$

$-80 = -9x$

$\frac{80}{9} = x$

The number $\frac{80}{9}$ checks and is the solution.

42. $\frac{15}{4}$

43. $\sqrt{x + 2} + \sqrt{3x + 4} = 2$

$\sqrt{x + 2} = 2 - \sqrt{3x + 4}$ Isolating one radical

$(\sqrt{x + 2})^2 = (2 - \sqrt{3x + 4})^2$

$x + 2 = 4 - 4\sqrt{3x + 4} + 3x + 4$

$-2x - 6 = -4\sqrt{3x + 4}$ Isolating the remaining radical

$x + 3 = 2\sqrt{3x + 4}$ Multiplying by $-\frac{1}{2}$

$(x + 3)^2 = (2\sqrt{3x + 4})^2$

$x^2 + 6x + 9 = 4(3x + 4)$

$x^2 + 6x + 9 = 12x + 16$

$x^2 - 6x - 7 = 0$

$(x - 7)(x + 1) = 0$

$x - 7 = 0$ or $x + 1 = 0$

$x = 7$ or $x = -1$

Check: For 7:

$$\begin{array}{c|l} \sqrt{x+2} + \sqrt{3x+4} = 2 & \\ \hline \sqrt{7+2} + \sqrt{3\cdot 7+4} \;?\; 2 & \\ \sqrt{9} + \sqrt{25} & \\ 8 & \text{FALSE} \end{array}$$

For -1:

$$\begin{array}{c|l} \sqrt{x+2} + \sqrt{3x+4} = 2 & \\ \hline \sqrt{-1+2} + \sqrt{3(-1)+4} \;?\; 2 & \\ \sqrt{1} + \sqrt{1} & \\ 2 & \text{TRUE} \end{array}$$

Since -1 checks but 7 does not, the solution is -1.

44. $\frac{1}{3}$, -1

45. $\sqrt{4y+1} - \sqrt{y-2} = 3$

$$\sqrt{4y+1} = 3 + \sqrt{y-2}$$
$$(\sqrt{4y+1})^2 = (3 + \sqrt{y-2})^2$$
$$4y + 1 = 9 + 6\sqrt{y-2} + y - 2$$
$$3y - 6 = 6\sqrt{y-2}$$
$$y - 2 = 2\sqrt{y-2}$$
$$(y-2)^2 = (2\sqrt{y-2})^2$$
$$y^2 - 4y + 4 = 4(y-2)$$
$$y^2 - 4y + 4 = 4y - 8$$
$$y^2 - 8y + 12 = 0$$
$$(y-6)(y-2) = 0$$
$$y - 6 = 0 \quad \text{or} \quad y - 2 = 0$$
$$y = 6 \quad \text{or} \quad y = 2$$

The numbers 6 and 2 check and are the solutions.

46. 1

47. $\sqrt{3x-5} + \sqrt{2x+3} + 1 = 0$

$$\sqrt{3x-5} + 1 = -\sqrt{2x+3}$$
$$(\sqrt{3x-5} + 1)^2 = (-\sqrt{2x+3})^2$$
$$3x - 5 + 2\sqrt{3x-5} + 1 = 2x + 3$$
$$2\sqrt{3x-5} = -x + 7$$
$$(2\sqrt{3x-5})^2 = (-x+7)^2$$
$$4(3x-5) = x^2 - 14x + 49$$
$$12x - 20 = x^2 - 14x + 49$$
$$0 = x^2 - 26x + 69$$
$$0 = (x-23)(x-3)$$
$$x - 23 = 0 \quad \text{or} \quad x - 3 = 0$$
$$x = 23 \quad \text{or} \quad x = 3$$

Neither number checks. There is no solution.

48. 2

49. $2\sqrt{3x+6} - \sqrt{4x+9} = 5$

$$2\sqrt{3x+6} = 5 + \sqrt{4x+9}$$
$$(2\sqrt{3x+6})^2 = (5 + \sqrt{4x+9})^2$$
$$4(3x+6) = 25 + 10\sqrt{4x+9} + 4x + 9$$
$$12x + 24 = 34 + 10\sqrt{4x+9} + 4x$$
$$8x - 10 = 10\sqrt{4x+9}$$
$$4x - 5 = 5\sqrt{4x+9}$$
$$(4x-5)^2 = (5\sqrt{4x+9})^2$$
$$16x^2 - 40x + 25 = 25(4x+9)$$
$$16x^2 - 40x + 25 = 100x + 225$$
$$16x^2 - 140x - 200 = 0$$
$$4(4x+5)(x-10) = 0$$
$$4x + 5 = 0 \quad \text{or} \quad x - 10 = 0$$
$$x = -\frac{5}{4} \quad \text{or} \quad x = 10$$

Since 10 checks and $-\frac{5}{4}$ does not, 10 is the solution.

50. 7

51. $3\sqrt{t+1} - \sqrt{2t-5} = 7$

$$3\sqrt{t+1} = 7 + \sqrt{2t-5}$$
$$(3\sqrt{t+1})^2 = (7 + \sqrt{2t-5})^2$$
$$9(t+1) = 49 + 14\sqrt{2t-5} + 2t - 5$$
$$9t + 9 = 44 + 14\sqrt{2t-5} + 2t$$
$$7t - 35 = 14\sqrt{2t-5}$$
$$t - 5 = 2\sqrt{2t-5}$$
$$(t-5)^2 = (2\sqrt{2t-5})^2$$
$$t^2 - 10t + 25 = 4(2t-5)$$
$$t^2 - 10t + 25 = 8t - 20$$
$$t^2 - 18t + 45 = 0$$
$$(t-15)(t-3) = 0$$
$$t - 15 = 0 \quad \text{or} \quad t - 3 = 0$$
$$t = 15 \qquad t = 3$$

Since 15 checks and 3 does not, the solution is 15.

52. 5

53. $\frac{3}{2x} + \frac{1}{x} = \frac{2x + 3.5}{3x}$ LCD is 6x

$$6x\left(\frac{3}{2x} + \frac{1}{x}\right) = 6x\left(\frac{2x+3.5}{3x}\right)$$
$$9 + 6 = 4x + 7$$
$$8 = 4x$$
$$2 = x$$

The number 2 checks and is the solution.

54. Height: 7 in., base: 9 in.

55.

56.

57. $V = 1.2\sqrt{h}$

$V = 1.2\sqrt{30{,}000}$

$V \approx 208$ mi   Using a calculator to compute

58. 72.25 ft

59. $\dfrac{x + \sqrt{x + 1}}{x - \sqrt{x + 1}} = \dfrac{5}{11}$

$11(x + \sqrt{x + 1}) = 5(x - \sqrt{x + 1})$

$11x + 11\sqrt{x + 1} = 5x - 5\sqrt{x + 1}$

$16\sqrt{x + 1} = -6x$

$8\sqrt{x + 1} = -3x$

$(8\sqrt{x + 1})^2 = (-3x)^2$

$64(x + 1) = 9x^2$

$64x + 64 = 9x^2$

$0 = 9x^2 - 64x - 64$

$0 = (9x + 8)(x - 8)$

$9x + 8 = 0$ or $x - 8 = 0$

$9x = -8$ or $x = 8$

$x = -\dfrac{8}{9}$ or $x = 8$

Since $-\dfrac{8}{9}$ checks but 8 does not, the solution is $-\dfrac{8}{9}$.

60. 6912

61. $\sqrt[4]{z^2 + 17} = 3$

$(\sqrt[4]{z^2 + 17})^4 = 3^4$

$z^2 + 17 = 81$

$z^2 - 64 = 0$

$(z + 8)(z - 8) = 0$

$z + 8 = 0$ or $z - 8 = 0$

$z = -8$ or $z = 8$

The numbers -8 and 8 check and are the solutions.

62. 0

63. $\sqrt[3]{x^2 + x + 15} - 3 = 0$

$(\sqrt[3]{x^2 + x + 15})^3 = 3^3$

$x^2 + x + 15 = 27$

$x^2 + x - 12 = 0$

$(x + 4)(x - 3) = 0$

$x + 4 = 0$ or $x - 3 = 0$

$x = -4$ or $x = 3$

The numbers -4 and 3 check and are the solutions.

64. 6, -1

65. $\sqrt{8 - b} = b\sqrt{8 - b}$

$(\sqrt{8 - b})^2 = (b\sqrt{8 - b})^2$

$(8 - b) = b^2(8 - b)$

$0 = b^2(8 - b) - (8 - b)$

$0 = (8 - b)(b^2 - 1)$

$0 = (8 - b)(b + 1)(b - 1)$

$8 - b = 0$ or $b + 1 = 0$ or $b - 1 = 0$

$8 = b$ or $b = -1$ or $b = 1$

Since the numbers 8 and 1 check but -1 does not, 8 and 1 are the solutions.

66. 2

67. $6\sqrt{y} + 6y^{-1/2} = 37$

$6\sqrt{y} + \dfrac{6}{\sqrt{y}} = 37$

$\sqrt{y}\left[6\sqrt{y} + \dfrac{6}{\sqrt{y}}\right] = \sqrt{y}\cdot 37$

$6y + 6 = 37\sqrt{y}$

$(6y + 6)^2 = (37\sqrt{y})^2$

$36y^2 + 72y + 36 = 1369y$

$36y^2 - 1297y + 36 = 0$

$(36y - 1)(y - 36) = 0$

$36y - 1 = 0$ or $y - 36 = 0$

$36y = 1$ or $y = 36$

$y = \dfrac{1}{36}$ or $y = 36$

The numbers $\dfrac{1}{36}$ and 36 check and are the solutions.

68. 0, $\dfrac{125}{4}$

Exercise Set 7.8

1. $a = 3, \quad b = 5$

Find c.

$c^2 = a^2 + b^2$ Pythagorean equation

$c^2 = 3^2 + 5^2$ Substituting

$c^2 = 9 + 25$

$c^2 = 34$

$c = \sqrt{34}$ Exact answer

$c \approx 5.831$ Approximation

2. $\sqrt{164}$; 12.806

3. $a = 12, \quad b = 12$

Find c.

$c^2 = a^2 + b^2$ Pythagorean equation

$c^2 = 12^2 + 12^2$ Substituting

$c^2 = 144 + 144$

$c^2 = 288$

$c = \sqrt{288}$ Exact answer

$c \approx 16.971$ Approximation

4. $\sqrt{200}$; 14.142

5. $b = 12, \quad c = 13$

Find a.

$a^2 + b^2 = c^2$

$a^2 + 12^2 = 13^2$ Substituting

$a^2 + 144 = 169$

$a^2 = 25$

$a = 5$

6. $\sqrt{119}$; 10.909

7. $c = 6, \quad a = \sqrt{5}$

Find b.

$a^2 + b^2 = c^2$

$(\sqrt{5})^2 + b^2 = 6^2$

$5 + b^2 = 36$

$b^2 = 31$

$b = \sqrt{31}$ Exact answer

$b \approx 5.568$ Approximation

8. 4

9. $b = 1, \quad c = \sqrt{13}$

Find a.

$a^2 + b^2 = c^2$

$a^2 + 1^2 = (\sqrt{13})^2$

$a^2 + 1 = 13$

$a^2 = 12$

$a = \sqrt{12}$ Exact answer

$a \approx 3.464$ Approximation

10. $\sqrt{19}$; 4.359

11. $a = 1, \quad c = \sqrt{n}$

Find b.

$a^2 + b^2 = c^2$

$1^2 + b^2 = (\sqrt{n})^2$

$1 + b^2 = n$

$b^2 = n - 1$

$b = \sqrt{n - 1}$

12. $\sqrt{4 - n}$

13. $d^2 = 10^2 + 15^2$

$d^2 = 100 + 225$

$d^2 = 325$

$d = \sqrt{325}$

$d \approx 18.028$

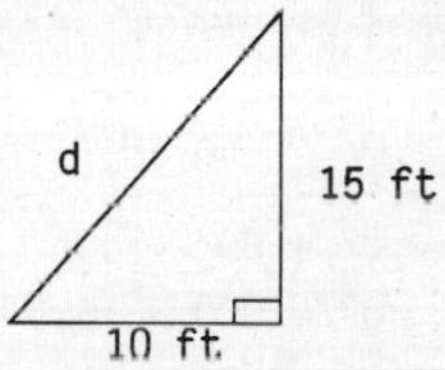

The wire is $\sqrt{325}$, or 18.028 ft long.

14. $\sqrt{8450} \approx 91.924$ ft

15. We first make a drawing and let d = the distance, in feet, to second base. A right triangle is formed in which the length of the leg from second base to third base is 90 ft. The length of the leg from third base to where the catcher fields the ball is 90 - 10, or 80 ft.

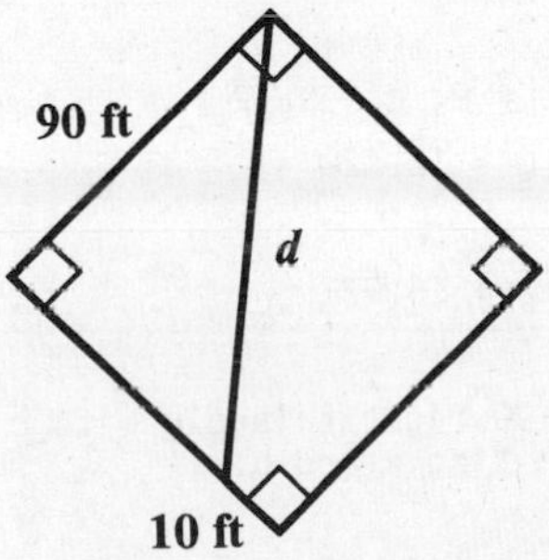

We substitute these values into the Pythagorean equation to find d.

$d^2 = 90^2 + 80^2$

$d^2 = 8100 + 6400$

$d^2 = 14{,}500$

$d = \sqrt{14{,}500}$

Exact answer: $d = \sqrt{14{,}500}$ ft

Approximation: $d \approx 120.416$ ft

16. $\sqrt{340} + 8 \approx 26.439$ ft

17.

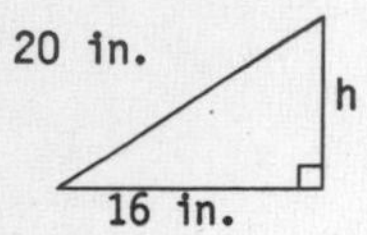

$h^2 + 16^2 = 20^2$

$h^2 + 256 = 400$

$h^2 = 144$

$h = 12$ in. Exact answer

The height is 12 in.

18. 20 in.

19.

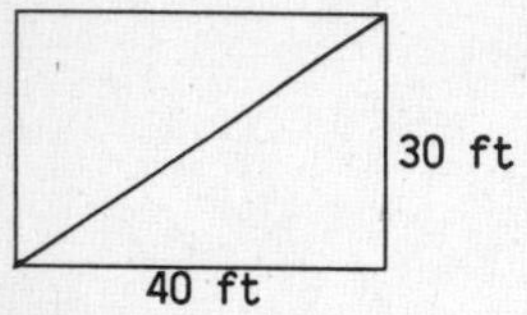

$d^2 = 30^2 + 40^2$

$d^2 = 900 + 1600$

$d^2 = 2500$

$d = 50$ ft Exact answer

20. $\sqrt{13{,}000} \approx 114.018$ m

21. Since one acute angle is 45°, this is an isosceles right triangle with a = 9. Then b = 9 also. We substitute to find c.

$c = a\sqrt{2}$

$c = 9\sqrt{2}$

Exact answer: $b = 9$, $c = 9\sqrt{2}$

Approximation: $c \approx 12.728$

22. $a = 12$; $c = 12\sqrt{2} \approx 16.971$

23. This is a 30-60-90 right triangle with c = 14. We substitute to find a and b.

$c = 2a$

$14 = 2a$

$7 = a$

$b = a\sqrt{3}$

$b = 7\sqrt{3}$

Exact answer: $a = 7$, $b = 7\sqrt{3}$

Approximation: $b \approx 12.124$

24. $a = 9$; $b = 9\sqrt{3} \approx 15.588$

25. This is a 30-60-90 right triangle with b = 5. We substitute to find a and c.

$b = a\sqrt{3}$

$5 = a\sqrt{3}$

$\frac{5}{\sqrt{3}} = a$

$\frac{5\sqrt{3}}{3} = a$ Rationalizing the denominator

$c = 2a$

$c = 2 \cdot \frac{5\sqrt{3}}{3}$

$c = \frac{10\sqrt{3}}{3}$

Exact answer: $a = \frac{5\sqrt{3}}{3}$, $c = \frac{10\sqrt{3}}{3}$

Approximation: $a \approx 2.887$, $c \approx 5.774$

26. $a = 4\sqrt{2} \approx 5.657$; $b = 4\sqrt{2} \approx 5.657$

27. This is an isosceles right triangle with c = 13. We substitute to find a.

$a = \frac{c\sqrt{2}}{2}$

$a = \frac{13\sqrt{2}}{2}$

Since a = b, we have $b = \frac{13\sqrt{2}}{2}$ also.

Exact answer: $a = \frac{13\sqrt{2}}{2}$, $b = \frac{13\sqrt{2}}{2}$

Approximation: $a \approx 9.192$, $b \approx 9.192$

28. $a = \frac{7\sqrt{3}}{3} \approx 4.041$; $c = \frac{14\sqrt{3}}{3} \approx 8.083$

29. This is a 30-60-90 right triangle with a = 12. We substitute to find b and c.

$b = a\sqrt{3}$ $\quad c = 2a$

$b = 12\sqrt{3}$ $\quad c = 2 \cdot 12$

$c = 24$

Exact answer: $b = 12\sqrt{3}$, $c = 24$

Approximation: $b \approx 20.785$

30. $b = 9\sqrt{3} \approx 15.588$; $c = 18$

31.

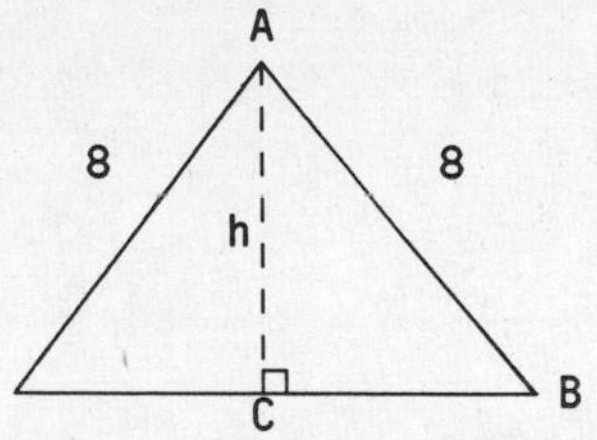

This is an equilateral triangle, so all the angles are 60°. The altitude bisects one angle and one side. Then triangle ABC is a 30-60-90 right triangle with the shorter leg of length 8/2, or 4, and hypotenuse of length 8. We substitute to find the length of the other leg.

$b = a\sqrt{3}$

$h = 4\sqrt{3}$ Substituting h for b and 4 for a

Exact answer: $h = 4\sqrt{3}$

Approximation: $h \approx 6.928$

32. $5\sqrt{3} \approx 8.660$

33.

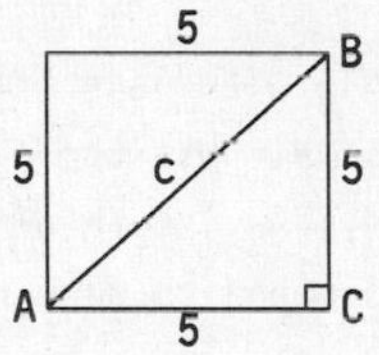

Triangle ABC is an isosceles right triangle with a = 5. We substitute to find c.

$c = a\sqrt{2}$

$c = 5\sqrt{2}$

Exact answer: $c = 5\sqrt{2}$

Approximation: $c \approx 7.071$

34. $7\sqrt{2} \approx 9.899$

35.

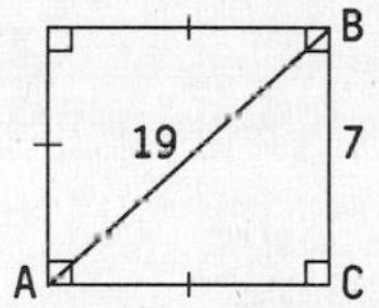

Triangle ABC is an isosceles right triangle with c = 19. We substitute to find a.

$a = \dfrac{c\sqrt{2}}{2}$

$a = \dfrac{19\sqrt{2}}{2}$

Since a = b, we have $b = \dfrac{19\sqrt{2}}{2}$ also.

Exact answer: $a = \dfrac{19\sqrt{2}}{2}$, $b = \dfrac{19\sqrt{2}}{2}$

Approximation: $a \approx 13.435$, $b \approx 13.435$

36. $\dfrac{15\sqrt{2}}{2} \approx 10.607$

37.

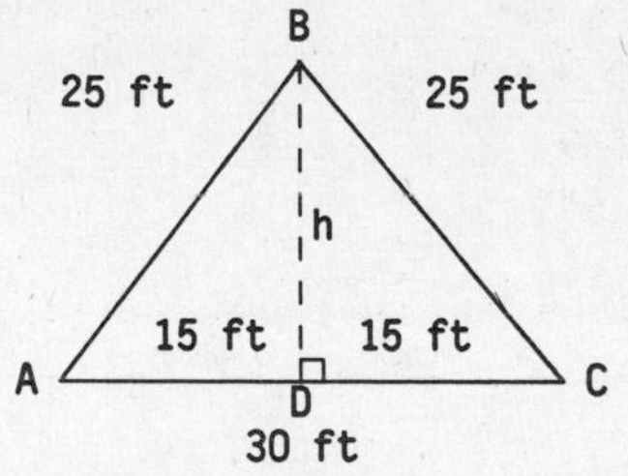

Triangle $\overline{ABC}$ is an isosceles triangle. The height from B to $\overline{AC}$ bisects $\overline{AC}$. Thus, $\overline{DC}$ measures 15 ft.

$h^2 + 15^2 = 25^2$

$h^2 + 225 = 625$

$h^2 = 400$

$h = \sqrt{400}$

$h = 20$

If the height of triangle ABC is 20 ft and the base is 30 ft, the area is $\frac{1}{2} \cdot 30 \cdot 20$, or 300 ft^2.

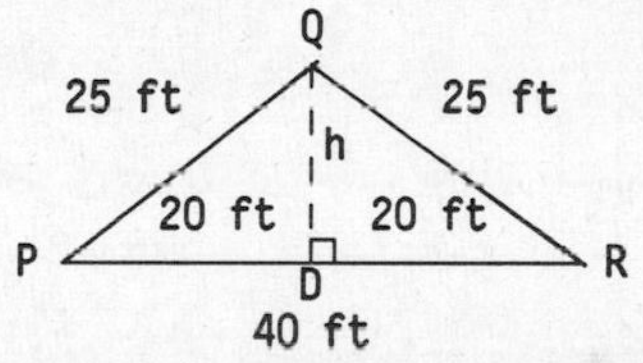

Triangle $\overline{PQR}$ is an isosceles triangle. The height from Q to $\overline{PR}$ bisects $\overline{PR}$. Thus, $\overline{SR}$ measures 20 ft.

$h^2 + 20^2 = 25^2$

$h^2 + 400 = 625$

$h^2 = 225$

$h = \sqrt{225}$

$h = 15$

If the height of triangle PQR is 15 ft and the base is 40 ft, the area is $\frac{1}{2} \cdot 40 \cdot 15$, or 300 ft^2.

The areas of the two triangles are the same.

38. 102.767 ft

39. 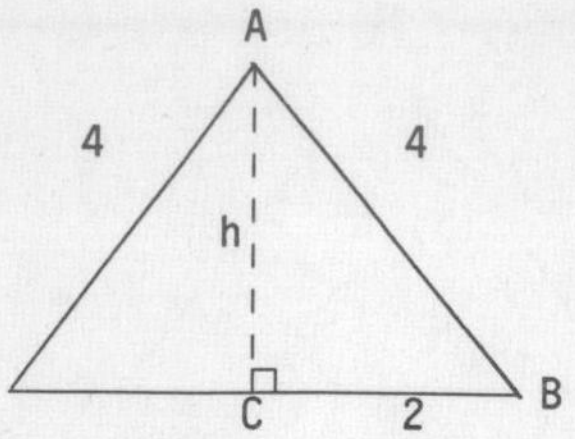

The entrance is an equilateral triangle, so all the angles are 60°. The altitude bisects one angle and one side. Then triangle ABC is a 30-60-90 right triangle with the shorter leg of length 4/2, or 2, and hypotenuse of length 4. We substitute to find h, the height of the tent.

$b = a\sqrt{3}$

$h = 2\sqrt{3}$ Substituting h for b and 2 for a

Exact answer: $h = 2\sqrt{3}$ ft

Approximation: $h \approx 3.464$ ft

40. $d = s + s\sqrt{2}$

41. 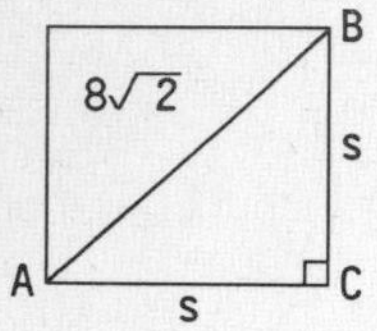

Triangle ABC is an isosceles right triangle with $c = 8\sqrt{2}$. We substitute to find a.

$$a = \frac{c\sqrt{2}}{2} = \frac{8\sqrt{2}\cdot\sqrt{2}}{2} = \frac{8\cdot 2}{2} = 8$$

The length of a side of the square is 8 ft.

42. $\sqrt{181} \approx 13.454$ cm

43. 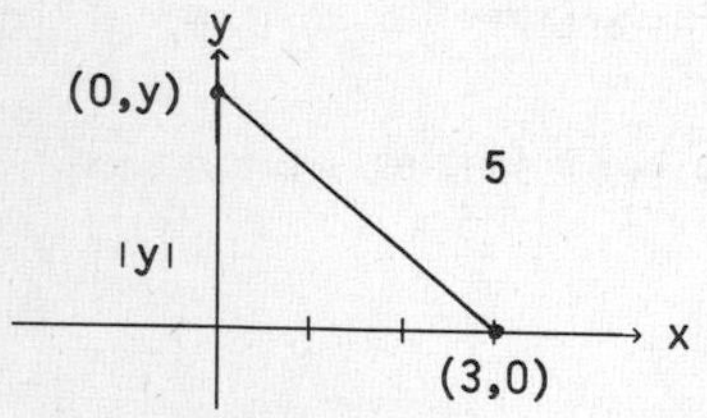

$|y|^2 + 3^2 = 5^2$

$y^2 + 9 = 25$

$y^2 = 16$

$y = \pm 4$

The points are (0,4) and (0,-4).

44. (3,0), (-3,0)

45. $x^2 - 11x + 24 = 0$

$(x - 8)(x - 3) = 0$

$x - 8 = 0$ or $x - 3 = 0$

$x = 8$ or $x = 3$

46. $\frac{3}{2}$, -7

47.

48.

49. 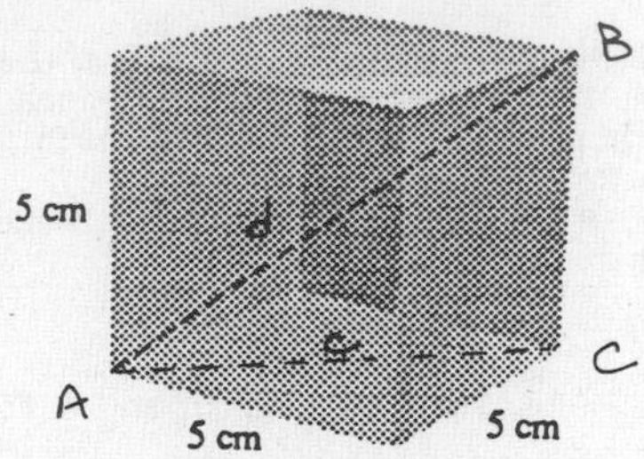

First find the length of a diagonal of the base of the cube. It is the hypotenuse of an isosceles right triangle with a = 5 cm. Then $c = a\sqrt{2} = 5\sqrt{2}$ cm.

Triangle ABC is a right triangle with legs of $5\sqrt{2}$ cm and 5 cm and hypotenuse d. Use the Pythagorean equation to find d, the length of the diagonal that connects two opposite corners of the cube.

$d^2 = (5\sqrt{2})^2 + 5^2$

$d^2 = 25\cdot 2 + 25$

$d^2 = 50 + 25$

$d^2 = 75$

$d = \sqrt{75}$

Exact answer: $d = \sqrt{75}$ cm

50. 9

Exercise Set 7.9

1. $\sqrt{-15} = \sqrt{-1\cdot 15} = \sqrt{-1}\cdot\sqrt{15} = i\sqrt{15}$, or $\sqrt{15}i$

2. $i\sqrt{17}$, or $\sqrt{17}i$

3. $\sqrt{-16} = \sqrt{-1\cdot 16} = \sqrt{-1}\cdot\sqrt{16} = i\cdot 4 = 4i$

4. 5i

5. $-\sqrt{-12} = -\sqrt{-1\cdot 4\cdot 3} = -\sqrt{-1}\cdot\sqrt{4}\cdot\sqrt{3} = -i\cdot 2\cdot\sqrt{3} =$ $-2i\sqrt{3}$, or $-2\sqrt{3}i$

6. $-2i\sqrt{5}$, or $-2\sqrt{5}i$

7. $\sqrt{-3} = \sqrt{-1\cdot 3} = \sqrt{-1}\cdot\sqrt{3} = i\sqrt{3}$, or $\sqrt{3}i$

8. 2i

9. $\sqrt{-81} = \sqrt{-1\cdot 81} = \sqrt{-1}\cdot\sqrt{81} = i\cdot 9 = 9i$

10. $3i\sqrt{3}$, or $3\sqrt{3}i$

11. $\sqrt{-98} = \sqrt{-1\cdot 98} = \sqrt{-1}\cdot\sqrt{98} = i\cdot 7\sqrt{2} = 7i\sqrt{2}$, or $7\sqrt{2}i$

12. $-3i\sqrt{2}$, or $-3\sqrt{2}i$

13. $-\sqrt{-49} = -\sqrt{-1\cdot 49} = -\sqrt{-1}\cdot\sqrt{49} = -i\cdot 7 = -7i$

14. $-5i\sqrt{5}$, or $-5\sqrt{5}i$

15. $4 - \sqrt{-60} = 4 - \sqrt{-1\cdot 60} = 4 - \sqrt{-1}\cdot\sqrt{60} =$
$4 - i\cdot 2\sqrt{15} = 4 - 2\sqrt{15}i$, or $4 - 2i\sqrt{15}$

16. $6 - 2i\sqrt{21}$, or $6 - 2\sqrt{21}i$

17. $\sqrt{-4} + \sqrt{-12} = \sqrt{-1\cdot 4} + \sqrt{-1\cdot 12} =$
$\sqrt{-1}\cdot\sqrt{4} + \sqrt{-1}\cdot\sqrt{12} = i\cdot 2 + i\cdot 2\sqrt{3} =$
$(2 + 2\sqrt{3})i$

18. $(-2\sqrt{19} + 5\sqrt{5})i$

19. $(3 + 2i) + (5 - i) = (3 + 5) + (2 - 1)i$ Collecting the real and the imaginary parts
$= 8 + i$

20. $5 + 11i$

21. $(4 - 3i) + (5 - 2i) = (4 + 5) + (-3 - 2)i$ Collecting the real and the imaginary parts
$= 9 - 5i$

22. $-1 - 8i$

23. $(9 - i) + (-2 + 5i) = (9 - 2) + (-1 + 5)i$
$= 7 + 4i$

24. $8 + i$

25. $(3 - i) - (5 + 2i) = (3 - 5) + (-1 - 2)i$
$= -2 - 3i$

26. $-9 + 5i$

27. $(4 - 2i) - (5 - 3i) = (4 - 5) + [-2 - (-3)]i$
$= -1 + i$

28. $-3 + 2i$

29. $(9 + 5i) - (-2 - i) = [9 - (-2)] + [5 - (-1)]i$
$= 11 + 6i$

30. $4 - 7i$

31. $\sqrt{-25}\,\sqrt{-36} = \sqrt{-1}\cdot\sqrt{25}\cdot\sqrt{-1}\cdot\sqrt{36}$
$= i\cdot 5\cdot i\cdot 6$
$= i^2\cdot 30$
$= -1\cdot 30$ $i^2 = -1$
$= -30$

32. -63

33. $\sqrt{-6}\,\sqrt{-5} = \sqrt{-1}\cdot\sqrt{6}\cdot\sqrt{-1}\cdot\sqrt{5} = i\cdot\sqrt{6}\cdot i\cdot\sqrt{5}$
$= i^2\cdot\sqrt{30} = -1\cdot\sqrt{30} = -\sqrt{30}$

34. $-\sqrt{70}$

35. $\sqrt{-50}\,\sqrt{-3} = \sqrt{-1\cdot 25\cdot 2}\cdot\sqrt{-1\cdot 3} =$
$\sqrt{-1}\cdot\sqrt{25}\cdot\sqrt{2}\cdot\sqrt{-1}\cdot\sqrt{3} = i\cdot 5\cdot\sqrt{2}\cdot i\cdot\sqrt{3} =$
$i^2\cdot 5\cdot\sqrt{6} = -1\cdot 5\cdot\sqrt{6} = -5\sqrt{6}$

36. $-6\sqrt{6}$

37. $\sqrt{-48}\,\sqrt{-6} = \sqrt{-1\cdot 16\cdot 3}\cdot\sqrt{-1\cdot 3\cdot 2}$
$= \sqrt{-1}\cdot\sqrt{16}\cdot\sqrt{3}\cdot\sqrt{-1}\cdot\sqrt{3}\cdot\sqrt{2}$
$= i\cdot 4\cdot 3\cdot i\cdot\sqrt{2}$ $\sqrt{3}\cdot\sqrt{3} = 3$
$= i^2\cdot 12\cdot\sqrt{2}$
$= -1\cdot 12\cdot\sqrt{2}$
$= -12\sqrt{2}$

38. $-15\sqrt{5}$

39. $5i\cdot 8i = 40\cdot i^2$
$= 40\cdot(-1)$ $i^2 = -1$
$= -40$

40. -54

41. $5i\cdot(-7i) = -35\cdot i^2$
$= -35\cdot(-1)$ $i^2 = -1$
$= 35$

42. 28

43. $5i(3 - 2i) = 5i\cdot 3 + 5i(-2i)$ Using the distributive law
$= 15i - 10i^2$
$= 15i + 10$ $i^2 = -1$
$= 10 + 15i$

44. $24 + 20i$

45. $-3i(7 - 4i) = -3i\cdot 7 + (-3i)(-4i) = -21i + 12i^2 =$
$-21i - 12 = -12 - 21i$

46. $-21 - 63i$

47. $(3 + 2i)(1 + i) = 3 + 3i + 2i + 2i^2$ Using FOIL
$= 3 + 3i + 2i - 2$ $i^2 = -1$
$= 1 + 5i$

48. $-7 + 26i$

49. $(2 + 3i)(6 - 2i) = 12 - 4i + 18i - 6i^2$ Using FOIL
$= 12 - 4i + 18i + 6$ $i^2 = -1$
$= 18 + 14i$

50. $16 + 7i$

51. $(6 - 5i)(3 + 4i) = 18 + 24i - 15i - 20i^2 =$
$18 + 24i - 15i + 20 = 38 + 9i$

52. $40 + 13i$

53. $(7 - 2i)(2 - 6i) = 14 - 42i - 4i + 12i^2 =$
$14 - 42i - 4i - 12 = 2 - 46i$

54. $8 + 31i$

55. $(5 - 3i)(4 - 5i) = 20 - 25i - 12i + 15i^2 =$
$20 - 25i - 12i - 15 = 5 - 37i$

56. $7 - 61i$

57. $(-2 + 3i)(-2 + 5i) = 4 - 10i - 6i + 15i^2 =$
$4 - 10i - 6i - 15 = -11 - 16i$

58. $-15 - 30i$

59. $(-5 - 4i)(3 + 7i) = -15 - 35i - 12i - 28i^2 =$
$-15 - 35i - 12i + 28 = 13 - 47i$

60. $39 - 37i$

61. $(3 - 2i)^2 = 3^2 - 2\cdot3\cdot2i + (2i)^2$ Squaring a binomial
$= 9 - 12i + 4i^2$
$= 9 - 12i - 4$ $i^2 = -1$
$= 5 - 12i$

62. $21 - 20i$

63. $(2 + 3i)^2 = 2^2 + 2\cdot2\cdot3i + (3i)^2$ Squaring a binomial
$= 4 + 12i + 9i^2$
$= 4 + 12i - 9$
$= -5 + 12i$

64. $12 + 16i$

65. $(-2 + 3i)^2 = 4 - 12i + 9i^2 = 4 - 12i - 9 =$
$-5 - 12i$

66. $21 + 20i$

67. $i^7 = i^6\cdot i = (i^2)^3\cdot i = (-1)^3\cdot i = -1\cdot i = -i$

68. $-i$

69. $i^{24} = (i^2)^{12} = (-1)^{12} = 1$

70. $-i$

71. $i^{42} = (i^2)^{21} = (-1)^{21} = -1$

72. 1

73. $i^9 = (i^2)^4\cdot i = (-1)^4\cdot i = 1\cdot i = i$

74. i

75. $i^6 = (i^2)^3 = (-1)^3 = -1$

76. 1

77. $(5i)^3 = 5^3\cdot i^3 = 125\cdot i^2\cdot i = 125(-1)(i) = -125i$

78. $-243i$

79. $7 + i^4 = 7 + (i^2)^2 = 7 + (-1)^2 = 7 + 1 = 8$

80. $-18 - i$

81. $i^4 - 26i = (i^2)^2 - 26i = (-1)^2 - 26i = 1 - 26i$

82. $38i$

83. $i^2 + i^4 = -1 + (i^2)^2 = -1 + (-1)^2 = -1 + 1 = 0$

84. i

85. $i^5 + i^7 = i^4\cdot i + i^6\cdot i = (i^2)^2\cdot i + (i^2)^3\cdot i =$
$(-1)^2\cdot i + (-1)^3\cdot i = 1\cdot i + (-1)i = i - i = 0$

86. 0

87. $1 + i + i^2 + i^3 + i^4 = 1 + i + i^2 + i^2\cdot i + (i^2)^2$
$= 1 + i + (-1) + (-1)\cdot i + (-1)^2$
$= 1 + i - 1 - i + 1$
$= 1$

88. i

89. $5 - \sqrt{-64} = 5 - \sqrt{-1}\cdot\sqrt{64} = 5 - i\cdot8 = 5 - 8i$

90. $(2\sqrt{3} + 36)i$

91. $\dfrac{5}{3 - i} = \dfrac{5}{3 - i}\cdot\dfrac{3 + i}{3 + i}$ Multiplying by 1, using the conjugate
$= \dfrac{15 + 5i}{10}$ Performing the multiplication
$= \dfrac{15}{10} + \dfrac{5}{10}i$, or $\dfrac{3}{2} + \dfrac{1}{2}i$ Writing in the form a + bi

92. $\frac{15}{26} - \frac{3}{26}i$

93. $\frac{2i}{7 + 3i} = \frac{2i}{7 + 3i} \cdot \frac{7 - 3i}{7 - 3i}$ Multiplying by 1, using the conjugate

$= \frac{14i + 6}{58}$ Performing the multiplication

$= \frac{6}{58} + \frac{14}{58}i$, or $\frac{3}{29} + \frac{7}{29}i$ Writing in the form a + bi

94. $-\frac{20}{29} + \frac{8}{29}i$

95. $\frac{7}{6i} = \frac{7}{6i} \cdot \frac{-6i}{-6i} = \frac{-42i}{36} = -\frac{7}{6}i$

96. $-\frac{3}{10}i$

97. $\frac{8 - 3i}{7i} = \frac{8 - 3i}{7i} \cdot \frac{-7i}{-7i} = \frac{-56i - 21}{49} = -\frac{21}{49} - \frac{56}{49}i = -\frac{3}{7} - \frac{8}{7}i$

98. $\frac{8}{5} - \frac{3}{5}i$

99. $\frac{3 + 2i}{2 + i} = \frac{3 + 2i}{2 + i} \cdot \frac{2 - i}{2 - i} = \frac{8 + i}{5} = \frac{8}{5} + \frac{1}{5}i$

100. $\frac{15}{26} + \frac{29}{26}i$

101. $\frac{5 - 2i}{2 + 5i} = \frac{5 - 2i}{2 + 5i} \cdot \frac{2 - 5i}{2 - 5i} = \frac{-29i}{29} = -i$

102. $\frac{6}{25} - \frac{17}{25}i$

103. $\frac{3 - 5i}{3 - 2i} = \frac{3 - 5i}{3 - 2i} \cdot \frac{3 + 2i}{3 + 2i} = \frac{19 - 9i}{13} = \frac{19}{13} - \frac{9}{13}i$

104. $\frac{38}{41} - \frac{27}{41}i$

105. Substitute 1 + 2i for x in the equation.

$$\begin{array}{r|l}
\multicolumn{2}{c}{x^2 - 2x + 5 = 0} \\
\hline
(1 + 2i)^2 - 2(1 + 2i) + 5 & ?\ 0 \\
1 + 4i + 4i^2 - 2 - 4i + 5 & \\
1 - 4 - 2 + 5 & \\
0 & \text{TRUE}
\end{array}$$

1 + 2i is a solution.

106. Yes

107. Substitute 1 - i for x in the equation.

$$\begin{array}{r|l}
\multicolumn{2}{c}{x^2 + 2x + 2 = 0} \\
\hline
(1 - i)^2 + 2(1 - i) + 2 & ?\ 0 \\
1 - 2i + i^2 + 2 - 2i + 2 & \\
1 - 2i - 1 + 2 - 2i + 2 & \\
4 - 4i & \text{FALSE}
\end{array}$$

1 - i is not a solution.

108. No

109. $\frac{196}{x^2 - 7x + 49} - \frac{2x}{x + 7} = \frac{2058}{x^3 + 343}$

$\frac{196}{x^2 - 7x + 49} - \frac{2x}{x + 7} = \frac{2058}{(x + 7)(x^2 - 7x + 49)}$

The LCD is $(x + 7)(x^2 - 7x + 49)$. Using the LCD to clear fractions, we have

$$196(x + 7) - 2x(x^2 - 7x + 49) = 2058$$
$$196x + 1372 - 2x^3 + 14x^2 - 98x = 2058$$
$$-2x^3 + 14x^2 + 98x - 686 = 0$$
$$-2(x^3 - 7x^2 - 49x + 343) = 0$$
$$-2[x^2(x - 7) - 49(x - 7)] = 0$$
$$-2(x - 7)(x^2 - 49) = 0$$
$$-2(x - 7)(x + 7)(x - 7) = 0$$

$x - 7 = 0$ or $x + 7 = 0$ or $x - 7 = 0$

$x = 7$ or $x = -7$ or $x = 7$

Since 7 checks and -7 does not, the solution is 7.

110. $\frac{70}{29}$

111. $28 = 3x^2 - 17x$

$0 = 3x^2 - 17x - 28$

$0 = (3x + 4)(x - 7)$

$3x + 4 = 0$ or $x - 7 = 0$

$3x = -4$ or $x = 7$

$x = -\frac{4}{3}$ or $x = 7$

Both values check. The solutions are $-\frac{4}{3}$ and 7.

112. -4 - 8i; -2 + 4i; 8 - 6i

113. $\frac{1}{\frac{1 - i}{10} - \left(\frac{1 - i}{10}\right)^2} = \frac{1}{\frac{1 - i}{10} - \left(\frac{-2i}{100}\right)} = \frac{1}{\frac{1 - i}{10} + \frac{i}{50}} =$

$\frac{1}{\frac{1 - i}{10} + \frac{i}{50}} \cdot \frac{50}{50} = \frac{50}{5 - 5i + i} = \frac{50}{5 - 4i} =$

$\frac{50}{5 - 4i} \cdot \frac{5 + 4i}{5 + 4i} = \frac{250 + 200i}{41} = \frac{250}{41} + \frac{200}{41}i$

114. 0

115. $(1 - i)^3(1 + i)^3 =$

$(1 - i)(1 + i)\cdot(1 - i)(1 + i)\cdot(1 - i)(1 + i) =$

$(1 - i^2)(1 - i^2)(1 - i^2) =$

$[1 - (-1)][1 - (-1)][1 - (-1)] =$

$(1 + 1)(1 + 1)(1 + 1) = 2\cdot2\cdot2 = 8$

116. $-1 - \sqrt{5}i$

117. $\frac{6}{1 + \frac{3}{i}} = \frac{6}{\frac{i + 3}{i}} = 6 \cdot \frac{i}{i + 3} = \frac{6i}{i - 3} =$

$\frac{6i}{i + 3} \cdot \frac{-i + 3}{-i + 3} = \frac{6 + 18i}{10} = \frac{3}{5} + \frac{9}{5}i$

118. $-\frac{2}{3}i$

119. $\frac{i - i^{38}}{1 + i} = \frac{i - (i^2)^{19}}{1 + i} = \frac{i - (-1)^{19}}{1 + i} = \frac{i + 1}{1 + i} = 1$

Exercise Set 8.1

1. $4x^2 = 20$

   $x^2 = 5$ Multiplying by $\frac{1}{4}$

   $x = \sqrt{5}$ or $x = -\sqrt{5}$ Taking square roots

   Check: $4x^2 = 20$

   $4(\pm\sqrt{5})^2 \ ? \ 20$

   $4 \cdot 5$

   20 | TRUE

   The solutions are $\sqrt{5}$ and $-\sqrt{5}$, or $\pm\sqrt{5}$.

2. $\pm\sqrt{7}$

3. $25x^2 + 4 = 0$

   $x^2 = -\frac{4}{25}$ Isolating x^2

   $x = \sqrt{-\frac{4}{25}}$ or $x = -\sqrt{-\frac{4}{25}}$ Principle of square roots

   $x = \frac{2}{5}i$ or $x = -\frac{2}{5}i$

   Check: $25x^2 + 4 = 0$

   $25\left(\pm\frac{2}{5}i\right)^2 + 4 \ ? \ 0$

   $25\left(-\frac{4}{5}\right) + 4$

   $-4 + 4$

   0 | TRUE

   The solutions are $\frac{2}{5}i$ and $-\frac{2}{5}i$, or $\pm\frac{2}{5}i$.

4. $\pm\frac{4}{3}i$

5. $2x^2 - 3 = 0$

   $x^2 = \frac{3}{2}$

   $x = \sqrt{\frac{3}{2}}$ or $x = -\sqrt{\frac{3}{2}}$ Principle of square roots

   $x = \sqrt{\frac{3}{2}\cdot\frac{2}{2}}$ or $x = -\sqrt{\frac{3}{2}\cdot\frac{2}{2}}$ Rationalizing denominators

   $x = \frac{\sqrt{6}}{2}$ or $x = -\frac{\sqrt{6}}{2}$

   Check: $2x^2 - 3 = 0$

   $2\left(\pm\frac{\sqrt{6}}{2}\right)^2 - 3 \ ? \ 0$

   $2 \cdot \frac{6}{4} - 3$

   $3 - 3$

   0 | TRUE

   The solutions are $\frac{\sqrt{6}}{2}$ and $-\frac{\sqrt{6}}{2}$, or $\pm\frac{\sqrt{6}}{2}$.

6. $\pm\frac{\sqrt{21}}{3}$

7. $(x + 2)^2 = 49$

   $x + 2 = 7$ or $x + 2 = -7$ Principle of square roots

   $x = 5$ or $x = -9$

   The solutions are 5 and -9.

8. $1 \pm \sqrt{6}$

9. $(x - 3)^2 = 21$

   $x - 3 = \sqrt{21}$ or $x - 3 = -\sqrt{21}$ Principle of square roots

   $x = 3 + \sqrt{21}$ or $x = 3 - \sqrt{21}$

   The solutions are $3 + \sqrt{21}$ and $3 - \sqrt{21}$, or $3 \pm \sqrt{21}$.

10. $-3 \pm \sqrt{6}$

11. $(x - 13)^2 = 8$

    $x - 13 = \sqrt{8}$ or $x - 13 = -\sqrt{8}$

    $x - 13 = 2\sqrt{2}$ or $x - 13 = -2\sqrt{2}$

    $x = 13 + 2\sqrt{2}$ or $x = 13 - 2\sqrt{2}$

    The solutions are $13 + 2\sqrt{2}$ and $13 - 2\sqrt{2}$, or $13 \pm 2\sqrt{2}$.

12. 21, 5

13. $(x - 7)^2 = -4$

    $x - 7 = \sqrt{-4}$ or $x - 7 = -\sqrt{-4}$

    $x - 7 = 2i$ or $x - 7 = -2i$

    $x = 7 + 2i$ or $x = 7 - 2i$

    The solutions are $7 + 2i$ and $7 - 2i$, or $7 \pm 2i$.

14. $-1 \pm 3i$

15. $(x - 9)^2 = 34$

    $x - 9 = \sqrt{34}$ or $x - 9 = -\sqrt{34}$

    $x = 9 + \sqrt{34}$ or $x = 9 - \sqrt{34}$

    The solutions are $9 + \sqrt{34}$ and $9 - \sqrt{34}$, or $9 \pm \sqrt{34}$.

16. 7, -3

17. $\left(x - \frac{3}{2}\right)^2 = \frac{7}{2}$

$x - \frac{3}{2} = \sqrt{\frac{7}{2}}$ or $x - \frac{3}{2} = -\sqrt{\frac{7}{2}}$

$x - \frac{3}{2} = \sqrt{\frac{7}{2}\cdot\frac{2}{2}}$ or $x - \frac{3}{2} = -\sqrt{\frac{7}{2}\cdot\frac{2}{2}}$

$x - \frac{3}{2} = \frac{\sqrt{14}}{2}$ or $x - \frac{3}{2} = -\frac{\sqrt{14}}{2}$

$x = \frac{3}{2} + \frac{\sqrt{14}}{2}$ or $x = \frac{3}{2} - \frac{\sqrt{14}}{2}$

$x = \frac{3 + \sqrt{14}}{2}$ or $x = \frac{3 - \sqrt{14}}{2}$

The solutions are $\frac{3 + \sqrt{14}}{2}$ and $\frac{3 - \sqrt{14}}{2}$, or $\frac{3 \pm \sqrt{14}}{2}$.

18. $\frac{-3 \pm \sqrt{17}}{4}$

19. $x^2 + 6x + 9 = 64$

$(x + 3)^2 = 64$

$x + 3 = 8$ or $x + 3 = -8$

$x = 5$ or $x = -11$

The solutions are 5 and -11.

20. 5, -15

21. $y^2 - 14y + 49 = 4$

$(y - 7)^2 = 4$

$y - 7 = 2$ or $y - 7 = -2$

$y = 9$ or $y = 5$

The solutions are 9 and 5.

22. 5, 3

23. $x^2 + 10x$

We take half the coefficient of x and square it:
Half of 10 is 5, and $5^2 = 25$. We add 25.

$x^2 + 10x + 25$, $(x + 5)^2$

24. $x^2 + 16x + 64$, $(x + 8)^2$

25. $x^2 - 8x$

We take half the coefficient of x and square it:
Half of -8 is -4, and $(-4)^2 = 16$. We add 16.

$x^2 - 8x + 16$, $(x - 4)^2$

26. $x^2 - 6x + 9$, $(x - 3)^2$

27. $x^2 - 24x$

We take half the coefficient of x and square it:
$\frac{1}{2}(-24) = -12$ and $(-12)^2 = 144$. We add 144.

$x^2 - 24x + 144$, $(x - 12)^2$

28. $x^2 - 18x + 81$, $(x - 9)^2$

29. $x^2 + 9x$

$\frac{1}{2}\cdot 9 = \frac{9}{2}$, and $\left(\frac{9}{2}\right)^2 = \frac{81}{4}$. We add $\frac{81}{4}$.

$x^2 + 9x + \frac{81}{4}$, $\left(x + \frac{9}{2}\right)^2$

30. $x^2 + 3x + \frac{9}{4}$, $\left(x + \frac{3}{2}\right)^2$

31. $x^2 - 7x$

$\frac{1}{2}(-7) = -\frac{7}{2}$, and $\left(-\frac{7}{2}\right)^2 = \frac{49}{4}$. We add $\frac{49}{4}$.

$x^2 - 7x + \frac{49}{4}$, $\left(x - \frac{7}{2}\right)^2$

32. $x^2 - 11x + \frac{121}{4}$, $\left(x - \frac{11}{2}\right)^2$

33. $x^2 + \frac{2}{3}x$

$\frac{1}{2}\cdot\frac{2}{3} = \frac{1}{3}$, and $\left(\frac{1}{3}\right)^2 = \frac{1}{9}$. We add $\frac{1}{9}$.

$x^2 + \frac{2}{3}x + \frac{1}{9}$, $\left(x + \frac{1}{3}\right)^2$

34. $x^2 + \frac{2}{5}x + \frac{1}{25}$, $\left(x + \frac{1}{5}\right)^2$

35. $x^2 - \frac{5}{6}x$

$\frac{1}{2}\left(-\frac{5}{6}\right) = -\frac{5}{12}$, and $\left(-\frac{5}{12}\right)^2 = \frac{25}{144}$. We add $\frac{25}{144}$.

$x^2 - \frac{5}{6}x + \frac{25}{144}$, $\left(x - \frac{5}{12}\right)^2$

36. $x^2 - \frac{5}{3}x + \frac{25}{36}$, $\left(x - \frac{5}{6}\right)^2$

37. $x^2 + \frac{9}{5}x$

$\frac{1}{2}\cdot\frac{9}{5} = \frac{9}{10}$, and $\left(\frac{9}{10}\right)^2 = \frac{81}{100}$. We add $\frac{81}{100}$.

$x^2 + \frac{9}{5}x + \frac{81}{100}$, $\left(x + \frac{9}{10}\right)^2$

38. $x^2 + \frac{9}{4}x + \frac{81}{64}$, $\left(x + \frac{9}{8}\right)^2$

39. $x^2 + 8x = -7$

$x^2 + 8x + 16 = -7 + 16$ Adding 16 on both sides to complete the square

$(x + 4)^2 = 9$ Factoring

$x + 4 = \pm 3$ Principle of square roots

$x = -4 \pm 3$

$x = -4 - 3$ or $x = -4 + 3$

$x = -7$ or $x = -1$

The solutions are -7 and -1.

40. -7, 1

41. $x^2 - 10x = 22$

$x^2 - 10x + 25 = 22 + 25$ Adding 25 on both sides to complete the square

$(x - 5)^2 = 47$

$x - 5 = \pm\sqrt{47}$ Principle of square roots

$x = 5 \pm \sqrt{47}$

The solutions are $5 \pm \sqrt{47}$.

42. $4 \pm \sqrt{7}$

43. $x^2 + 6x + 5 = 0$

$x^2 + 6x = -5$ Adding -5 on both sides

$x^2 + 6x + 9 = -5 + 9$ Completing the square

$(x + 3)^2 = 4$

$x + 3 = \pm 2$

$x = -3 \pm 2$

$x = -3 - 2$ or $x = -3 + 2$

$x = -5$ or $x = -1$

The solutions are -5 and -1.

44. -9, -1

45. $x^2 - 10x + 21 = 0$

$x^2 - 10x = -21$

$x^2 - 10x + 25 = -21 + 25$

$(x - 5)^2 = 4$

$x - 5 = \pm 2$

$x = 5 \pm 2$

$x = 5 - 2$ or $x = 5 + 2$

$x = 3$ or $x = 7$

The solutions are 3 and 7.

46. 4, 6

47. $x^2 + 5x + 6 = 0$

$x^2 + 5x = -6$

$x^2 + 5x + \frac{25}{4} = -6 + \frac{25}{4}$

$\left(x + \frac{5}{2}\right)^2 = \frac{1}{4}$

$x + \frac{5}{2} = \pm \frac{1}{2}$

$x = -\frac{5}{2} \pm \frac{1}{2}$

$x = -\frac{5}{2} - \frac{1}{2}$ or $x = -\frac{5}{2} + \frac{1}{2}$

$x = -3$ or $x = -2$

The solutions are -3 and -2.

48. -4, -3

49. $x^2 + 4x + 1 = 0$

$x^2 + 4x = -1$

$x^2 + 4x + 4 = -1 + 4$

$(x + 2)^2 = 3$

$x + 2 = \pm\sqrt{3}$

$x = -2 \pm \sqrt{3}$

The solutions are $-2 \pm \sqrt{3}$.

50. $-3 \pm \sqrt{2}$

51. $x^2 - 10x + 23 = 0$

$x^2 - 10x = -23$

$x^2 - 10x + 25 = -23 + 25$

$(x - 5)^2 = 2$

$x - 5 = \pm\sqrt{2}$

$x = 5 \pm \sqrt{2}$

The solutions are $5 \pm \sqrt{2}$.

52. $3 \pm \sqrt{5}$

53. $x^2 + 6x + 13 = 0$

$x^2 + 6x = -13$

$x^2 + 6x + 9 = -13 + 9$

$(x + 3)^2 = -4$

$x + 3 = \pm 2i$

$x = -3 \pm 2i$

The solutions are $-3 \pm 2i$.

54. $-4 \pm 3i$

55. $2x^2 - 5x - 3 = 0$

$2x^2 - 5x = 3$

$x^2 - \frac{5}{2}x = \frac{3}{2}$ Dividing by 2 on both sides

$x^2 - \frac{5}{2}x + \frac{25}{16} = \frac{3}{2} + \frac{25}{16}$

$\left(x - \frac{5}{4}\right)^2 = \frac{49}{16}$

$x - \frac{5}{4} = \pm \frac{7}{4}$

$x = \frac{5}{4} \pm \frac{7}{4}$

$x = \frac{5}{4} - \frac{7}{4}$ or $x = \frac{5}{4} + \frac{7}{4}$

$x = -\frac{1}{2}$ or $x = 3$

The solutions are $-\frac{1}{2}$ and 3.

56. $-2, \frac{1}{3}$

57. $4x^2 + 8x + 3 = 0$

$$4x^2 + 8x = -3$$

$$x^2 + 2x = -\frac{3}{4}$$

$$x^2 + 2x + 1 = -\frac{3}{4} + 1$$

$$(x + 1)^2 = \frac{1}{4}$$

$$x + 1 = \pm\frac{1}{2}$$

$$x = -1 \pm \frac{1}{2}$$

$x = -1 - \frac{1}{2}$ or $x = -1 + \frac{1}{2}$

$x = -\frac{3}{2}$ or $x = -\frac{1}{2}$

The solutions are $-\frac{3}{2}$ and $-\frac{1}{2}$.

58. $-\frac{4}{3}, -\frac{2}{3}$

59. $6x^2 - x - 2 = 0$

$$6x^2 - x = 2$$

$$x^2 - \frac{1}{6}x = \frac{1}{3}$$

$$x^2 - \frac{1}{6}x + \frac{1}{144} = \frac{1}{3} + \frac{1}{144}$$

$$\left(x - \frac{1}{12}\right)^2 = \frac{49}{144}$$

$$x - \frac{1}{12} = \pm\frac{7}{12}$$

$$x = \frac{1}{12} \pm \frac{7}{12}$$

$x = \frac{1}{12} - \frac{7}{12}$ or $x = \frac{1}{12} + \frac{7}{12}$

$x = -\frac{1}{2}$ or $x = \frac{2}{3}$

The solutions are $-\frac{1}{2}$ and $\frac{2}{3}$.

60. $-\frac{3}{2}, \frac{5}{3}$

61. $2x^2 + 4x + 1 = 0$

$$2x^2 + 4x = -1$$

$$x^2 + 2x = -\frac{1}{2}$$

$$x^2 + 2x + 1 = -\frac{1}{2} + 1$$

$$(x + 1)^2 = \frac{1}{2}$$

$$x + 1 = \pm\sqrt{\frac{1}{2}}$$

$x + 1 = \pm\frac{\sqrt{2}}{2}$ Rationalizing the denominator

$$x = -1 \pm \frac{\sqrt{2}}{2}$$

The solutions are $-1 \pm \frac{\sqrt{2}}{2}$.

62. $-2, -\frac{1}{2}$

63. $3x^2 - 5x - 3 = 0$

$$3x^2 - 5x = 3$$

$$x^2 - \frac{5}{3}x = 1$$

$$x^2 - \frac{5}{3}x + \frac{25}{36} = 1 + \frac{25}{36}$$

$$\left(x - \frac{5}{6}\right)^2 = \frac{61}{36}$$

$$x - \frac{5}{6} = \pm\frac{\sqrt{61}}{6}$$

$$x = \frac{5 \pm \sqrt{61}}{6}$$

The solutions are $\frac{5 \pm \sqrt{61}}{6}$.

64. $\frac{3 \pm \sqrt{13}}{4}$

65. Familiarize. We are already familiar with the compound-interest formula.

Translate. We substitute into the formula.

$$A = P(1 + r)^t$$

$$2890 = 2560(1 + r)^2$$

Carry out. We solve for r.

$$\frac{2890}{2560} = (1 + r)^2$$

$$\frac{289}{256} = (1 + r)^2$$

$\sqrt{\frac{289}{256}} = 1 + r$ or $-\sqrt{\frac{289}{256}} = 1 + r$ Principle of square roots

$\frac{17}{16} = 1 + r$ or $-\frac{17}{16} = 1 + r$

$\frac{1}{16} = r$ or $-\frac{33}{16} = r$

Check. Since the interest rate cannot be negative, we need only check $\frac{1}{16}$, or 6.25%. If \$2560 were invested at 6.25% interest, compounded annually, then in 2 years it would grow to \$2560(1.0625)², or \$2890. The number 6.25% checks.

State. The interest rate is 6.25%, or $6\frac{1}{4}$%.

66. 10%

67. Familiarize. We are already familiar with the compound-interest formula.

Translate. We substitute into the formula.

$$A = P(1 + r)^t$$

$$1440 = 1000(1 + r)^2$$

Carry out. We solve for r.

$\frac{1440}{1000} = (1 + r)^2$

$\frac{36}{25} = (1 + r)^2$ Simplifying

$\sqrt{\frac{36}{25}} = 1 + r$ or $-\sqrt{\frac{36}{25}} = 1 + r$

$\frac{6}{5} = 1 + r$ or $-\frac{6}{5} = 1 + r$

$\frac{1}{5} = r$ or $-\frac{11}{5} = r$

Check. Since the interest rate cannot be negative, we need only check $\frac{1}{5}$, or 20%. If \$1000 were invested at 20% interest, compounded annually, then in 2 years it would grow to $\$1000(1.2)^2$, or \$1440. The number 20% checks.

State. The interest rate is 20%.

68. 18.75%

69. Familiarize. We are already familiar with the compound-interest formula.

Translate. We substitute into the formula.

$A = P(1 + r)^t$

$6760 = 6250(1 + r)^2$

Carry out. We solve for r.

$\frac{6760}{6250} = (1 + r)^2$

$\frac{676}{625} = (1 + r)^2$

$\sqrt{\frac{676}{625}} = 1 + r$ or $-\sqrt{\frac{676}{625}} = 1 + r$

$\frac{26}{25} = 1 + r$ or $-\frac{26}{25} = 1 + r$

$\frac{1}{25} = r$ or $-\frac{51}{25} = r$

Check. Since the interest rate cannot be negative, we need only check $\frac{1}{25}$, or 4%. If \$6250 were invested at 4% interest, compounded annually, then in 2 years it would grow to $\$6250(1.04)^2$, or \$6760. The number 4% checks.

State. The interest rate is 4%.

70. 8%

71. Familiarize. We are already familiar with the distance formula.

Translate. We substitute into the formula.

$s = 16t^2$

$1377 = 16t^2$

Carry out. We solve for t.

$1377 = 16t^2$

$\frac{1377}{16} = t^2$

$86.0625 = t^2$

$\sqrt{86.0625} = t$ Principle of square roots; rejecting the negative square root

$9.3 \approx t$

Check. Since $16(9.3)^2 = 1383.84 \approx 1377$, our answer checks.

State. It takes about 9.3 sec. for the object to fall.

72. 8.3 sec

73. Familiarize. We are already familiar with the distance formula.

Translate. We substitute into the formula.

$s = 16t^2$

$1451 = 16t^2$

Carry out. We solve for t.

$1451 = 16t^2$

$\frac{1451}{16} = t^2$

$90.6875 = t^2$

$\sqrt{90.6875} = t$ Principle of square roots; rejecting the negative square root

$9.5 \approx t$

Check. Since $16(9.5)^2 = 1444 \approx 1451$, our answer checks.

State. It takes about 9.5 sec for the object to fall.

74. 6.3 sec

75. Graph: $y = 2x + 1$

x	y
0	1
2	5
-3	-5

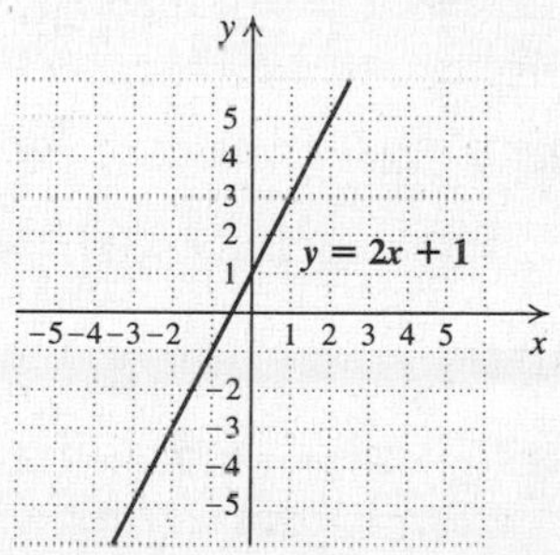

76. $2\sqrt{22}$

77. $14 - \sqrt{88} \approx 14 - 9.3808 \approx 4.6$

78. $\frac{\sqrt{10}}{5}$

79.

80. 

81. In order for $x^2 + bx + 64$ to be a trinomial square, the following must be true:

$$\left(\frac{b}{2}\right)^2 = 64$$

$$\frac{b^2}{4} = 64$$

$$b^2 = 256$$

$$b = 16 \quad \text{or} \quad b = -16$$

82. $\pm 10\sqrt{3}$

83. $x(2x^2 + 9x - 56)(3x + 10) = 0$

$x(2x - 7)(x + 8)(3x + 10) = 0$

$x = 0$ or $2x - 7 = 0$ or $x + 8 = 0$ or $3x + 10 = 0$

$x = 0$ or $x = \frac{7}{2}$ or $x = -8$ or $x = -\frac{10}{3}$

The solutions are -8, $-\frac{10}{3}$, 0, and $\frac{7}{2}$.

84. $\frac{1}{3}$, $\frac{1}{18}$

85. Familiarize. It is helpful to list information in a chart and make a drawing. Let r represent the speed of Boat A. Then r - 7 represents the speed of Boat B.

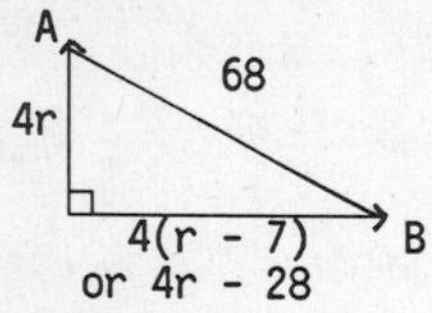

Boat	r	t	d
A	r	4	4r
B	r - 7	4	4(r - 7)

Translate. We use the Pythagorean equation: $a^2 + b^2 = c^2$

$$(4r - 28)^2 + (4r)^2 = 68^2$$

Carry out.

$$(4r - 28)^2 + (4r)^2 = 68^2$$

$$16r^2 - 224r + 784 + 16r^2 = 4624$$

$$32r^2 - 224r - 3840 = 0$$

$$r^2 - 7r - 120 = 0$$

$$(r + 8)(r - 15) = 0$$

$r + 8 = 0$ or $r - 15 = 0$

$r = -8$ or $r = 15$

Check. We only check r = 15 since the speeds of the boats cannot be negative. If the speed of Boat A is 15 km/h, then the speed of Boat B is 15 - 7, or 8 km/h, and the distances they travel are 4·15 (or 60) and 4·8 (or 32).

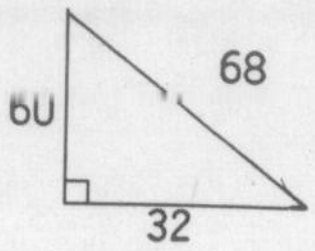

$60^2 + 32^2 = 3600 + 1024 = 4624 = 68^2$

The values check.

State. The speed of Boat A is 15 km/h, and the speed of Boat B is 8 km/h.

86. 5, 6, 7

87. 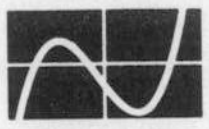

88. 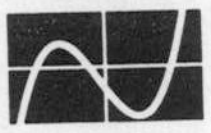

Exercise Set 8.2

1. $x^2 + 6x + 4 = 0$

$a = 1,\ b = 6,\ c = 4$

$$x = \frac{-b \pm \sqrt{b^2 - 4ac}}{2a}$$

$$x = \frac{-6 \pm \sqrt{6^2 - 4\cdot1\cdot4}}{2\cdot1} = \frac{-6 \pm \sqrt{36 - 16}}{2}$$

$$x = \frac{-6 \pm \sqrt{20}}{2} = \frac{-6 \pm 2\sqrt{5}}{2}$$

$$x = \frac{2(-3 \pm \sqrt{5})}{2} = -3 \pm \sqrt{5}$$

The solutions are $-3 + \sqrt{5}$ and $-3 - \sqrt{5}$.

2. $3 \pm \sqrt{13}$

3. $3p^2 = -8p - 5$

$3p^2 + 8p + 5 = 0$

$(3p + 5)(p + 1) = 0$

$3p + 5 = 0$ or $p + 1 = 0$

$p = -\frac{5}{3}$ or $p = -1$

The solutions are $-\frac{5}{3}$ and -1.

4. $3 \pm \sqrt{7}$

5. $x^2 - x + 1 = 0$

$a = 1,\ b = -1,\ c = 1$

$$x = \frac{-(-1) \pm \sqrt{(-1)^2 - 4\cdot1\cdot1}}{2\cdot1} = \frac{1 \pm \sqrt{1 - 4}}{2}$$

$$x = \frac{1 \pm \sqrt{-3}}{2} = \frac{1 \pm i\sqrt{3}}{2}$$

The solutions are $\frac{1 + i\sqrt{3}}{2}$ and $\frac{1 - i\sqrt{3}}{2}$.

6. $\frac{-1 \pm i\sqrt{7}}{2}$

7. $x^2 + 13 = 4x$

$x^2 - 4x + 13 = 0$ Finding standard form

$a = 1,\ b = -4,\ c = 13$

$$x = \frac{-(-4) \pm \sqrt{(-4)^2 - 4\cdot1\cdot13}}{2\cdot1} = \frac{4 \pm \sqrt{16 - 52}}{2}$$

$$x = \frac{4 \pm \sqrt{-36}}{2} = \frac{4 \pm 6i}{2} = 2 \pm 3i$$

The solutions are $2 + 3i$ and $2 - 3i$.

8. $3 \pm 2i$

9. $r^2 + 3r = 8$

$r^2 + 3r - 8 = 0$ Finding standard form

$a = 1,\ b = 3,\ c = -8$

$$r = \frac{-3 \pm \sqrt{3^2 - 4\cdot1\cdot(-8)}}{2\cdot1} = \frac{-3 \pm \sqrt{9 + 32}}{2}$$

$$r = \frac{-3 \pm \sqrt{41}}{2}$$

The solutions are $\frac{-3 + \sqrt{41}}{2}$ and $\frac{-3 - \sqrt{41}}{2}$.

10. $3 \pm \sqrt{5}$

11. $1 + \frac{2}{x} + \frac{5}{x^2} = 0$

$x^2 + 2x + 5 = 0$ Multiplying by x^2, the LCD

$a = 1,\ b = 2,\ c = 5$

$$x = \frac{-2 \pm \sqrt{2^2 - 4\cdot1\cdot5}}{2\cdot1} = \frac{-2 \pm \sqrt{4 - 20}}{2}$$

$$x = \frac{-2 \pm \sqrt{-16}}{2} = \frac{-2 \pm 4i}{2} = -1 \pm 2i$$

The solutions are $-1 + 2i$ and $-1 - 2i$.

12. $1 \pm 2i$

13. $3x + x(x - 2) = 0$

$3x + x^2 - 2x = 0$

$x^2 + x = 0$

$x(x + 1) = 0$

$x = 0$ or $x + 1 = 0$

$x = 0$ or $x = -1$

The solutions are 0 and -1.

14. 0, -1

15. $14x^2 + 9x = 0$

$x(14x + 9) = 0$

$x = 0$ or $14x + 9 = 0$

$x = 0$ or $14x = -9$

$x = 0$ or $x = -\frac{9}{14}$

The solutions are 0 and $-\frac{9}{14}$.

16. $0, -\frac{8}{19}$

17. $25x^2 - 20x + 4 = 0$

$(5x - 2)(5x - 2) = 0$

$5x - 2 = 0$ or $5x - 2 = 0$

$5x = 2$ or $5x = 2$

$x = \frac{2}{5}$ or $x = \frac{2}{5}$

The solution is $\frac{2}{5}$.

18. $-\frac{7}{6}$

19. $4x(x - 2) - 5x(x - 1) = 2$

$4x^2 - 8x - 5x^2 + 5x = 2$

$-x^2 - 3x = 2$

$-x^2 - 3x - 2 = 0$

$x^2 + 3x + 2 = 0$ Multiplying by -1

$(x + 2)(x + 1) = 0$

$x + 2 = 0$ or $x + 1 = 0$

$x = -2$ or $x = -1$

The solutions are -2 and -1.

20. $-\frac{3}{4}, -2$

21. $14(x - 4) - (x + 2) = (x + 2)(x - 4)$

$14x - 56 - x - 2 = x^2 - 2x - 8$ Removing parentheses

$13x - 58 = x^2 - 2x - 8$

$0 = x^2 - 15x + 50$

$0 = (x - 10)(x - 5)$

$x - 10 = 0$ or $x - 5 = 0$

$x = 10$ or $x = 5$

The solutions are 10 and 5.

22. 15, 1

23. $5x^2 = 13x + 17$

$5x^2 - 13x - 17 = 0$

$a = 5,\ b = -13,\ c = -17$

$$x = \frac{-(-13) \pm \sqrt{(-13)^2 - 4(5)(-17)}}{2\cdot5}$$

$$x = \frac{13 \pm \sqrt{169 + 340}}{10} = \frac{13 \pm \sqrt{509}}{10}$$

The solutions are $\frac{13 + \sqrt{509}}{10}$ and $\frac{13 - \sqrt{509}}{10}$.

24. $\frac{4}{3}, 7$

25. $x^2 + 5 = 2x$

$x^2 - 2x + 5 = 0$

$a = 1,\ b = -2,\ c = 5$

$x = \frac{-(-2) \pm \sqrt{(-2)^2 - 4\cdot1\cdot5}}{2\cdot1} = \frac{2 \pm \sqrt{4 - 20}}{2}$

$x = \frac{2 \pm \sqrt{-16}}{2} = \frac{2 \pm 4i}{2}$

$x = \frac{2(1 \pm 2i)}{2} = 1 \pm 2i$

The solutions are $1 + 2i$ and $1 - 2i$.

26. $2 \pm i$

27. $x + \frac{1}{x} = \frac{13}{6}$, LCD is $6x$

$6x\left[x + \frac{1}{x}\right] = 6x \cdot \frac{13}{6}$

$6x^2 + 6 = 13x$

$6x^2 - 13x + 6 = 0$

$(2x - 3)(3x - 2) = 0$

$2x - 3 = 0$ or $3x - 2 = 0$

$2x = 3$ or $3x = 2$

$x = \frac{3}{2}$ or $x = \frac{2}{3}$

The solutions are $\frac{3}{2}$ and $\frac{2}{3}$.

28. $\frac{3}{2}$, 6

29. $\frac{1}{x} + \frac{1}{x + 3} = \frac{1}{2}$, LCD is $2x(x + 3)$

$2x(x + 3)\left[\frac{1}{x} + \frac{1}{x + 3}\right] = 2x(x + 3) \cdot \frac{1}{2}$

$2(x + 3) + 2x = x(x + 3)$

$2x + 6 + 2x = x^2 + 3x$

$0 = x^2 - x - 6$

$0 = (x - 3)(x + 2)$

$x - 3 = 0$ or $x + 2 = 0$

$x = 3$ or $x = -2$

The solutions are 3 and -2.

30. $5 \pm \sqrt{53}$

31. $(2t - 3)^2 + 17t = 15$

$4t^2 - 12t + 9 + 17t = 15$

$4t^2 + 5t - 6 = 0$

$(4t - 3)(t + 2) = 0$

$4t - 3 = 0$ or $t + 2 = 0$

$t = \frac{3}{4}$ or $t = -2$

The solutions are $\frac{3}{4}$ and -2.

32. -3, 2

33. $(x - 2)^2 + (x + 1)^2 = 0$

$x^2 - 4x + 4 + x^2 + 2x + 1 = 0$

$2x^2 - 2x + 5 = 0$

$a = 2,\ b = -2,\ c = 5$

$x = \frac{-(-2) \pm \sqrt{(-2)^2 - 4\cdot2\cdot5}}{2\cdot2} = \frac{2 \pm \sqrt{4 - 40}}{4}$

$x = \frac{2 \pm \sqrt{-36}}{4} = \frac{2 \pm 6i}{4}$

$x = \frac{2(1 \pm 3i)}{2\cdot2} = \frac{1 \pm 3i}{2}$

The solutions are $\frac{1 + 3i}{2}$ and $\frac{1 - 3i}{2}$.

34. $-1 \pm 2i$

35. $x^3 - 8 = 0$

$x^3 - 2^3 = 0$

$(x - 2)(x^2 + 2x + 4) = 0$

$x - 2 = 0$ or $x^2 + 2x + 4 = 0$

$x = 2$ or $x = \frac{-2 \pm \sqrt{2^2 - 4\cdot1\cdot4}}{2\cdot1}$

$x = 2$ or $x = \frac{-2 \pm \sqrt{-12}}{2} = \frac{-2 \pm 2i\sqrt{3}}{2}$

$x = 2$ or $x = -1 \pm i\sqrt{3}$

The solutions are 2, $-1 + i\sqrt{3}$, and $-1 - i\sqrt{3}$.

36. -1, $\frac{1 \pm i\sqrt{3}}{2}$

37. $x^2 + 4x - 7 = 0$

$a = 1,\ b = 4,\ c = -7$

$x = \frac{-4 \pm \sqrt{4^2 - 4\cdot1\cdot(-7)}}{2\cdot1} = \frac{-4 \pm \sqrt{16 + 28}}{2}$

$x = \frac{-4 \pm \sqrt{44}}{2} = \frac{-4 \pm 2\sqrt{11}}{2} = -2 \pm \sqrt{11}$

Using a calculator we find that $\sqrt{11} \approx 3.317$.

$-2 + \sqrt{11} \approx -2 + 3.317 \approx 1.317 \approx 1.3$

$-2 - \sqrt{11} \approx -2 - 3.317 \approx -5.317 \approx -5.3$

The solutions are approximately 1.3 and -5.3.

38. -0.8, -5.2

39. $x^2 - 6x + 4 = 0$

$a = 1,\ b = -6,\ c = 4$

$x = \frac{-(-6) \pm \sqrt{(-6)^2 - 4\cdot1\cdot4}}{2\cdot1} = \frac{6 \pm \sqrt{36 - 16}}{2}$

$x = \frac{6 \pm \sqrt{20}}{2} = \frac{6 \pm 2\sqrt{5}}{2} = 3 \pm \sqrt{5}$

Using a calculator we find that $\sqrt{5} \approx 2.236$.

$3 + \sqrt{5} \approx 3 + 2.236 \approx 5.236 \approx 5.2$

$3 - \sqrt{5} \approx 3 - 2.236 \approx 0.764 \approx 0.8$

The solutions are approximately 5.2 and 0.8.

40. 3.7, 0.3

41. $2x^2 - 3x - 7 = 0$

$a = 2,\ b = -3,\ c = -7$

$$x = \frac{-(-3) \pm \sqrt{(-3)^2 - 4\cdot 2\cdot(-7)}}{2\cdot 2} = \frac{3 \pm \sqrt{9 + 56}}{4}$$

$$x = \frac{3 \pm \sqrt{65}}{4}$$

Using a calculator we find that $\sqrt{65} \approx 8.062$.

$$\frac{3 + \sqrt{65}}{4} \approx \frac{3 + 8.062}{4} \approx \frac{11.062}{4} \approx 2.7655 \approx 2.8$$

$$\frac{3 - \sqrt{65}}{4} \approx \frac{3 - 8.062}{4} \approx \frac{-5.062}{4} \approx -1.2655 \approx -1.3$$

The solutions are approximately 2.8 and -1.3.

42. 1.5, -0.5

43. $5x^2 = 3 + 8x$

$5x^2 - 8x - 3 = 0$

$a = 5,\ b = -8,\ c = -3$

$$x = \frac{-(-8) \pm \sqrt{(-8)^2 - 4\cdot 5\cdot(-3)}}{2\cdot 5} = \frac{8 \pm \sqrt{64 + 60}}{10}$$

$$x = \frac{8 \pm \sqrt{124}}{10} = \frac{8 \pm 2\sqrt{31}}{10} = \frac{4 \pm \sqrt{31}}{5}$$

Using a calculator we find that $\sqrt{31} \approx 5.568$.

$$\frac{4 + \sqrt{31}}{5} \approx \frac{4 + 5.568}{5} \approx \frac{9.568}{5} \approx 1.9136 \approx 1.9$$

$$\frac{4 - \sqrt{31}}{5} \approx \frac{4 - 5.568}{5} \approx \frac{-1.568}{5} \approx -0.3136 \approx -0.3$$

The solutions are approximately 1.9 and -0.3.

44. 0.8, -1.8

45. Familiarize. Let x = the number of pounds of Kenyan coffee and y = the number of pounds of Peruvian coffee in the mixture. We organize the information in a table.

Type of Coffee	Kenyan	Peruvian	Mixture
Price per pound	\$4.50	\$7.50	\$5.70
Number of pounds	x	y	50
Total cost	\$4.50x	\$7.50y	\$5.70 × 50, or \$285

Translate. From the last two rows of the table we get a system of equations.

$$x + y = 50,$$
$$4.50x + 7.50y = 285$$

Carry out. Solving the system of equations, we get (30,20).

Check. The total number of pounds in the mixture is 30 + 20, or 50. The total cost of the mixture is \$4.50(30) + \$7.50(20), or \$135 + \$150, or \$285. The values check.

State. The mixture should consist of 30 lb of Kenyan coffee and 20 lb of Peruvian coffee.

46. ◈

47. a) The x-intercepts occur when g(x) = 0. Solve:

$$2x^2 - 3x - 20 = 0$$
$$(2x + 5)(x - 4) = 0$$
$$2x + 5 = 0 \quad \text{or} \quad x - 4 = 0$$
$$x = -\frac{5}{2} \quad \text{or} \quad x = 4$$

The x-intercepts are $\left(-\frac{5}{2}, 0\right)$ and (4,0).

b) Solve:

$$2x^2 - 3x - 20 = -12$$
$$2x^2 - 3x - 8 = 0$$

$$x = \frac{-(-3) \pm \sqrt{(-3)^2 - 4(2)(-8)}}{2\cdot 2} = \frac{3 \pm \sqrt{73}}{4}$$

The x-values are $\frac{3 + \sqrt{73}}{4}$ and $\frac{3 - \sqrt{73}}{4}$.

48. a) $\left(\frac{-1 + \sqrt{33}}{2}, 0\right), \left(\frac{-1 - \sqrt{33}}{2}, 0\right)$

b) $\frac{-1 + \sqrt{69}}{2}, \frac{-1 - \sqrt{69}}{2}$

49. $x^2 - 0.75x - 0.5 = 0$

$100x^2 - 75x - 50 = 0$ Multiplying by 100

$4x^2 - 3x - 2 = 0$ Multiplying by $\frac{1}{25}$

$a = 4,\ b = -3,\ c = -2$

$$x = \frac{-(-3) \pm \sqrt{(-3)^2 - 4\cdot 4\cdot(-2)}}{2\cdot 4} = \frac{3 \pm \sqrt{9 + 32}}{8}$$

$$x = \frac{3 \pm \sqrt{41}}{8} \approx \frac{3 \pm 6.403124237}{8}$$

$$\frac{3 + 6.403124237}{8} \approx 1.1753905$$

$$\frac{3 - 6.403124237}{8} \approx -0.4253905$$

The solutions are approximately 1.1753905 and -0.4253905.

50. 0.3392101, -1.1792101

51. $x^2 + x - \sqrt{2} = 0$

$a = 1,\ b = 1,\ c = -\sqrt{2}$

$$x = \frac{-1 \pm \sqrt{1^2 - 4\cdot 1\cdot(-\sqrt{2})}}{2\cdot 1} = \frac{-1 \pm \sqrt{1 + 4\sqrt{2}}}{2}$$

52. $\frac{1 \pm \sqrt{1 + 4\sqrt{3}}}{2}$

53. $x^2 + \sqrt{5}x - \sqrt{3} = 0$

$a = 1,\ \ b = \sqrt{5},\ \ c = -\sqrt{3}$

$$x = \frac{-\sqrt{5} \pm \sqrt{(\sqrt{5})^2 - 4\cdot1\cdot(-\sqrt{3})}}{2\cdot1} = \frac{-\sqrt{5} \pm \sqrt{5 + 4\sqrt{3}}}{2}$$

54. $\dfrac{-5\sqrt{2} \pm \sqrt{34}}{4}$

55. $(1 + \sqrt{3})x^2 - (3 + 2\sqrt{3})x + 3 = 0$

$a = 1 + \sqrt{3},\ b = -(3 + 2\sqrt{3}),\ c = 3$

$$x = \frac{-[-(3 + 2\sqrt{3})] \pm \sqrt{[-(3 + 2\sqrt{3})]^2 - 4\cdot(1 + \sqrt{3})\cdot3}}{2(1 + \sqrt{3})}$$

$$x = \frac{3 + 2\sqrt{3} \pm \sqrt{9}}{2 + 2\sqrt{3}} = \frac{3 + 2\sqrt{3} \pm 3}{2 + 2\sqrt{3}}$$

$$x = \frac{3 + 2\sqrt{3} + 3}{2 + 2\sqrt{3}} = \frac{6 + 2\sqrt{3}}{2 + 2\sqrt{3}} = \frac{3 + \sqrt{3}}{1 + \sqrt{3}}$$

$$= \frac{3 + \sqrt{3}}{1 + \sqrt{3}}\cdot\frac{1 - \sqrt{3}}{1 - \sqrt{3}} = \frac{-2\sqrt{3}}{-2} = \sqrt{3}$$

$$x = \frac{3 + 2\sqrt{3} - 3}{2 + 2\sqrt{3}} = \frac{2\sqrt{3}}{2 + 2\sqrt{3}} = \frac{\sqrt{3}}{1 + \sqrt{3}}$$

$$= \frac{\sqrt{3}}{1 + \sqrt{3}}\cdot\frac{1 - \sqrt{3}}{1 - \sqrt{3}} = \frac{\sqrt{3} - 3}{-2} = \frac{3 - \sqrt{3}}{2}$$

56. $-i \pm i\sqrt{1 - i}$

57. $kx^2 + 3x - k = 0$

$k(-2)^2 + 3(-2) - k = 0$ Substituting -2 for x

$4k - 6 - k = 0$

$3k = 6$

$k = 2$

$2x^2 + 3x - 2 = 0$ Substituting 2 for k

$(2x - 1)(x + 2) = 0$

$2x - 1 = 0$ or $x + 2 = 0$

$x = \frac{1}{2}$ or $x = -2$

The other solution is $\frac{1}{2}$.

58.

Exercise Set 8.3

1. $\dfrac{1}{x} = \dfrac{x - 2}{24}$, LCD is 24x

$24x \cdot \dfrac{1}{x} = 24x \cdot \dfrac{x - 2}{24}$ Multiplying by the LCD

$24 = x(x - 2)$

$24 = x^2 - 2x$

$0 = x^2 - 2x - 24$ Standard form

$0 = (x - 6)(x + 4)$ Factoring

$x = 6$ or $x = -4$ Principle of zero products

Check: For 6:

$\dfrac{1}{x} = \dfrac{x - 2}{24}$

$\dfrac{1}{6} \ ? \ \dfrac{6 - 2}{24}$

$\dfrac{4}{24}$

$\dfrac{1}{6}$ TRUE

For -4:

$\dfrac{1}{x} = \dfrac{x - 2}{24}$

$\dfrac{1}{-4} \ ? \ \dfrac{-4 - 2}{24}$

$-\dfrac{1}{4}$ | $\dfrac{-6}{24}$

$-\dfrac{1}{4}$ TRUE

The solutions are 6 and -4.

2. -7, 4

3. $\dfrac{1}{2x - 1} - \dfrac{1}{2x + 1} = \dfrac{1}{4}$

LCD is $4(2x - 1)(2x + 1)$

$4(2x-1)(2x+1)\left[\dfrac{1}{2x-1} - \dfrac{1}{2x+1}\right] = 4(2x-1)(2x+1) \cdot \dfrac{1}{4}$

$4(2x + 1) - 4(2x - 1) = (2x - 1)(2x + 1)$

$8x + 4 - 8x + 4 = 4x^2 - 1$

$8 = 4x^2 - 1$

$9 = 4x^2$

$\dfrac{9}{4} = x^2$

$x = \dfrac{3}{2}$ or $x = -\dfrac{3}{2}$

Both numbers check. The solutions are $\frac{3}{2}$ and $-\frac{3}{2}$, or $\pm\frac{3}{2}$.

4. -8, 2

5. $\dfrac{50}{x} - \dfrac{50}{x - 5} = -\dfrac{1}{2}$, LCD is $2x(x - 5)$

$2x(x - 5)\left[\dfrac{50}{x} - \dfrac{50}{x - 5}\right] = 2x(x - 5)\left[-\dfrac{1}{2}\right]$

$100(x - 5) - 100x = -x(x - 5)$

$100x - 500 - 100x = -x^2 + 5x$

$x^2 - 5x - 500 = 0$

$(x - 25)(x + 20) = 0$

$x - 25 = 0$ or $x + 20 = 0$

$x = 25$ or $x = -20$

Both numbers check. The solutions are 25 and -20.

6. 2, -1

7. $\frac{x+2}{x} = \frac{x-1}{2}$, LCD is 2x

$2x \cdot \frac{x+2}{x} = 2x \cdot \frac{x-1}{2}$

$2(x+2) = x(x-1)$

$2x + 4 = x^2 - x$

$0 = x^2 - 3x - 4$

$0 = (x-4)(x+1)$

$x - 4 = 0$ or $x + 1 = 0$

$x = 4$ or $x = -1$

Both numbers check. The solutions are 4 and -1.

8. 6, -3

9. $x - 6 = \frac{1}{x+6}$, LCD is x + 6

$(x+6)(x-6) = (x+6) \cdot \frac{1}{x+6}$

$x^2 - 36 = 1$

$x^2 = 37$

$x = \sqrt{37}$ or $x = -\sqrt{37}$

Both numbers check. The solutions are $\pm\sqrt{37}$.

10. $\pm 5\sqrt{2}$

11. $\frac{2}{x} = \frac{x+3}{5}$, LCD is 5x

$5x \cdot \frac{2}{x} = 5x \cdot \frac{x+3}{5}$

$5 \cdot 2 = x(x+3)$

$10 = x^2 + 3x$

$0 = x^2 + 3x - 10$

$0 = (x+5)(x-2)$

$x + 5 = 0$ or $x - 2 = 0$

$x = -5$ or $x = 2$

Both numbers check. The solutions are -5 and 2.

12. $\frac{7 \pm \sqrt{85}}{2}$

13. $x + 5 = \frac{3}{x-5}$, LCD is x - 5

$(x-5)(x+5) = (x-5) \cdot \frac{3}{x-5}$

$x^2 - 25 = 3$

$x^2 = 28$

$x = \pm\sqrt{28}$, or $\pm 2\sqrt{7}$

Both numbers check. The solutions are $\pm 2\sqrt{7}$.

14. $\pm\sqrt{65}$

15. $\frac{40}{x} - \frac{20}{x-3} = \frac{8}{7}$, LCD = 7x(x - 3)

$7x(x-3)\left[\frac{40}{x} - \frac{20}{x-3}\right] = 7x(x-3) \cdot \frac{8}{7}$

$280(x-3) - 140x = 8x(x-3)$

$280x - 840 - 140x = 8x^2 - 24x$

$0 = 8x^2 - 164x + 840$

$0 = 2x^2 - 41x + 210$

$0 = (2x-21)(x-10)$

$2x - 21 = 0$ or $x - 10 = 0$

$x = \frac{21}{2}$ or $x = 10$

Both numbers check. The solutions are $\frac{21}{2}$ and 10.

16. $-\frac{11}{9}$, 2

17. $\frac{5}{x+2} + \frac{3x}{x+6} = 2$, LCD is (x + 2)(x + 6)

$(x+2)(x+6)\left[\frac{5}{x+2} + \frac{3x}{x+6}\right] = (x+2)(x+6)\cdot 2$

$5(x+6) + 3x(x+2) = 2(x^2 + 8x + 12)$

$5x + 30 + 3x^2 + 6x = 2x^2 + 16x + 24$

$x^2 - 5x + 6 = 0$

$(x-3)(x-2) = 0$

$x = 3$ or $x = 2$

Both numbers check. The solutions are 3 and 2.

18. $-\frac{21}{4}$, -1

19. $\frac{3}{3x+1} + \frac{6x}{11x-1} = 1$, LCD is (3x + 1)(11x - 1)

$(3x+1)(11x-1)\left[\frac{3}{3x+1} + \frac{6x}{11x-1}\right] = (3x+1)(11x-1)\cdot 1$

$3(11x-1) + 6x(3x+1) = (3x+1)(11x-1)$

$33x - 3 + 18x^2 + 6x = 33x^2 + 8x - 1$

$0 = 15x^2 - 31x + 2$

$0 = (15x-1)(x-2)$

$15x - 1 = 0$ or $x - 2 = 0$

$x = \frac{1}{15}$ or $x = 2$

Both numbers check. The solutions are $\frac{1}{15}$ and 2.

20. $\frac{3}{2}$, 1

21. $\frac{16}{5(x-2)(x+2)} + \frac{9}{5(x+2)(x+3)} = \frac{-x}{(x-2)(x+3)}$,

LCD is $5(x - 2)(x + 2)(x + 3)$

$$5(x-2)(x+2)(x+3)\left[\frac{16}{5(x-2)(x+2)} + \frac{9}{5(x+2)(x+3)}\right] =$$

$$5(x-2)(x+2)(x+3)\left[\frac{-x}{(x-2)(x+3)}\right]$$

$$16(x + 3) + 9(x - 2) = -5x(x + 2)$$

$$16x + 48 + 9x - 18 = -5x^2 - 10x$$

$$5x^2 + 35x + 30 = 0$$

$$x^2 + 7x + 6 = 0 \quad \text{Multiplying by } \frac{1}{5}$$

$$(x + 6)(x + 1) = 0$$

$x = -6$ or $x = -1$

Both numbers check. The solutions are -6 and -1.

22. 4

23. $\frac{6}{x^2 - 4x - 5} - \frac{6}{x^2 - 2x - 3} = \frac{x}{x^2 - 8x + 15}$

$\frac{6}{(x - 5)(x + 1)} - \frac{6}{(x - 3)(x + 1)} = \frac{x}{(x - 5)(x - 3)}$,

LCD is $(x - 5)(x + 1)(x - 3)$

$$(x-5)(x+1)(x-3)\left[\frac{6}{(x-5)(x+1)} - \frac{6}{(x-3)(x+1)}\right] =$$

$$(x-5)(x+1)(x-3)\left[\frac{x}{(x-5)(x-3)}\right]$$

$$6(x - 3) - 6(x - 5) = x(x + 1)$$

$$6x - 18 - 6x + 30 = x^2 + x$$

$$0 = x^2 + x - 12$$

$$0 = (x + 4)(x - 3)$$

$x = -4$ or $x = 3$

Only -4 checks. The solution is -4.

24. 5, -4

25. Familiarize. We first make a drawing, labeling it with the known and unknown information. We can also organize the information in a table. We let r represent the speed and t the time for the first part of the trip.

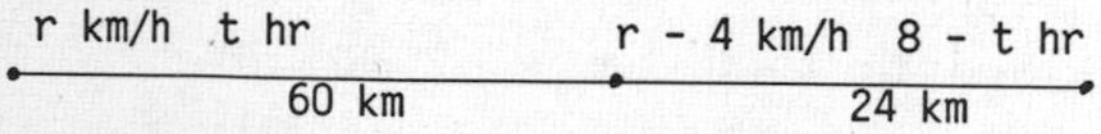

Canoe trip	Distance	Speed	Time
1st part	60	r	t
2nd part	24	r - 4	8 - t

Translate. Using $r = \frac{d}{t}$, we get two equations from the table, $r = \frac{60}{t}$ and $r - 4 = \frac{24}{8 - t}$.

Carry out. We substitute $\frac{60}{t}$ for r in the second equation and solve for t.

$\frac{60}{t} - 4 = \frac{24}{8 - t}$, LCD is $t(8 - t)$

$$t(8 - t)\left[\frac{60}{t} - 4\right] = t(8 - t) \cdot \frac{24}{8 - t}$$

$$60(8 - t) - 4t(8 - t) = 24t$$

$$4t^2 - 116t + 480 = 0 \quad \text{Standard form}$$

$$t^2 - 29t + 120 = 0 \quad \text{Multiplying by } \frac{1}{4}$$

$$(t - 24)(t - 5) = 0$$

$t = 24$ or $t = 5$

Check. Since the time cannot be negative (If $t = 24$, $8 - t = -16$.), we only check 5 hr. If $t = 5$, then $8 - t = 3$. The speed of the first part is $\frac{60}{5}$, or 12 km/h. The speed of the second part is $\frac{24}{3}$, or 8 km/h. The speed of the second part is 4 km/h slower than the first part. The value checks.

State. The speed of the first part was 12 km/h, and the speed of the second part was 8 km/h.

26. First part: 60 mph, second part: 50 mph

27. Familiarize. We first make a drawing. We also organize the information in a table. We let r represent the speed and t the time of the slower trip.

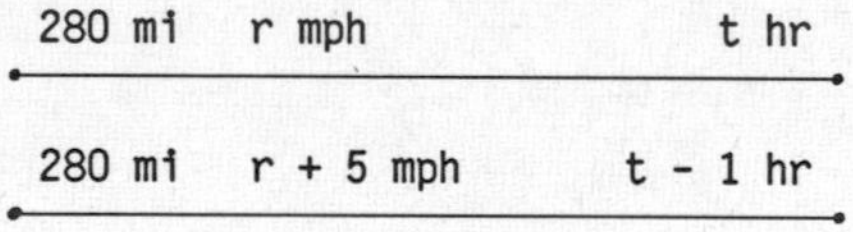

Trip	Distance	Speed	Time
Slower	280	r	t
Faster	280	r + 5	t - 1

Translate.

Using $t = \frac{d}{r}$, we get two equations from the table, $t = \frac{280}{r}$, and $t - 1 = \frac{280}{r + 5}$.

Carry out. We substitute $\frac{280}{r}$ for t in the second equation and solve for r.

$\frac{280}{r} - 1 = \frac{280}{r + 5}$, LCD is $r(r + 5)$

$$r(r + 5)\left[\frac{280}{r} - 1\right] = r(r + 5) \cdot \frac{280}{r + 5}$$

$$280(r + 5) - r(r + 5) = 280r$$

$$0 = r^2 + 5r - 1400$$

$$0 = (r - 35)(r + 40)$$

$r = 35$ or $r = -40$

Check. Since negative speed has no meaning in this problem, we only check 35. If $r = 35$, then the time for the slow trip is $\frac{280}{35}$, or 8 hours. If $r = 35$, then $r + 5 = 40$ and the time for the fast trip is $\frac{280}{40}$, or 7 hours. This is 1 hour less time than the slow trip took, so we have an answer to the problem.

State. The speed is 35 mph.

28. 40 mph

29. Familiarize. We make a drawing and then organize the information in a table. We let r represent the speed and t the time of the super-prop.

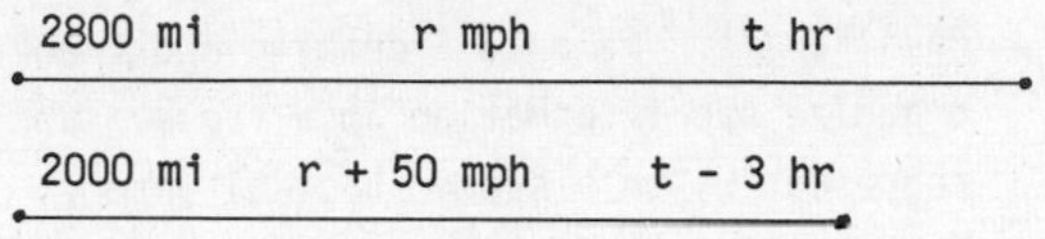

Plane	Distance	Rate	Time
Super-prop	2800	r	t
Turbo-jet	2000	r + 50	t - 3

Translate. Using $r = \frac{d}{t}$, we get two equations from the table,

$$r = \frac{2800}{t} \quad \text{and} \quad r + 50 = \frac{2000}{t - 3}.$$

Carry out. We substitute $\frac{2800}{t}$ for r in the second equation and solve for t.

$$\frac{2800}{t} + 50 = \frac{2000}{t - 3}, \text{ LCD is } t(t - 3)$$

$$t(t - 3)\left[\frac{2800}{t} + 50\right] = t(t - 3) \cdot \frac{2000}{t - 3}$$

$$2800(t - 3) + 50t(t - 3) = 2000t$$

$$50t^2 + 650t - 8400 = 0$$

$$t^2 + 13t - 168 = 0$$

$$(t + 21)(t - 8) = 0$$

$$t = -21 \quad \text{or} \quad t = 8$$

Check. Since negative time has no meaning in this problem, we only check 8 hours. If t = 8, then t - 3 = 5. The speed of the super-prop is $\frac{2800}{8}$, or 350 mph. The speed of the turbo-jet is $\frac{2000}{5}$, or 400 mph. Since the speed of the turbo-jet is 50 mph faster than the speed of the super-prop, the value checks.

State. The speed of the super-prop is 350 mph; the speed of the turbo-jet is 400 mph.

30. Cessna: 150 mph, Beechcraft: 200 mph

31. Familiarize. We make a drawing and then organize the information in a table. Let r represent the speed and t the time of the trip to Richmond.

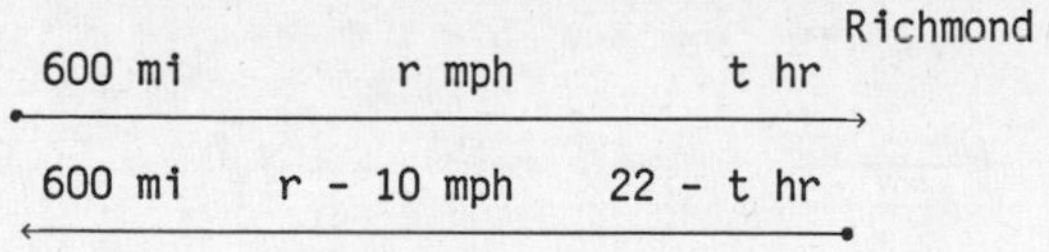

Trip	Distance	Speed	Time
To Ricmond	600	r	t
Return	600	r - 10	22 - t

Translate. Using $t = \frac{d}{r}$, we get two equations from the table,

$$t = \frac{600}{r} \quad \text{and} \quad 22 - t = \frac{600}{r - 10}.$$

Carry out. We substitute $\frac{600}{r}$ for t in the second equation and solve for r.

$$22 - \frac{600}{r} = \frac{600}{r - 10}, \text{ LCD is } r(r - 10)$$

$$r(r - 10)\left[22 - \frac{600}{r}\right] = r(r - 10) \cdot \frac{600}{r - 10}$$

$$22r(r - 10) - 600(r - 10) = 600r$$

$$22r^2 - 1420r + 6000 = 0$$

$$11r^2 - 710r + 3000 = 0$$

$$(11r - 50)(r - 60) = 0$$

$$r = \frac{50}{11} \quad \text{or} \quad r = 60$$

Check. Since negative speed has no meaning in this problem $\left[\text{If } r = \frac{50}{11}, \text{ then } r - 10 = -\frac{60}{11}.\right]$, we only check 60 mph. If r = 60, then the time of the trip to Richmond is $\frac{600}{60}$, or 10 hr. The speed of the return trip is 60 - 10, or 50 mph, and the time is $\frac{600}{50}$, or 12 hr. The total time for the round trip is 10 + 12, or 22 hr. The value checks.

State. The speed to Richmond was 60 mph, and the speed of the return trip was 50 mph.

32. To Hillsboro: 10 mph, return trip: 4 mph

33. Familiarize. We make a drawing and organize the information in a table. Let r represent the speed of the boat in still water, and let t represent the time of the trip upstream.

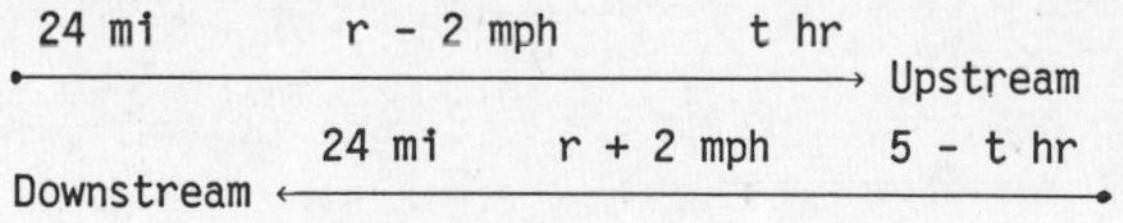

Trip	Distance	Rate	Time
Upstream	24	r - 2	t
Downstream	24	r + 2	5 - t

Translate. Using $t = \frac{d}{r}$, we get two equations from the table,

$$t = \frac{24}{r - 2} \text{ and } 5 - t = \frac{24}{r + 2}.$$

Carry out. We substitute $\frac{24}{r - 2}$ for t in the second equation and solve for r.

$$5 - \frac{24}{r - 2} = \frac{24}{r + 2}, \text{ LCD is } (r - 2)(r + 2)$$

$$(r-2)(r+2)\left[5 - \frac{24}{r-2}\right] = (r-2)(r+2) \cdot \frac{24}{r+2}$$

$$5(r-2)(r+2) - 24(r+2) = 24(r-2)$$

$$5r^2 - 48r - 20 = 0$$

$$(5r + 2)(r - 10) = 0$$

$$r = -\frac{2}{5} \text{ or } r = 10$$

Check. Since negative speed has no meaning in this problem, we only check 10 mph. If r = 10, then the speed upstream is 10 - 2, or 8 mph, and the time is $\frac{24}{8}$, or 3 hr. The speed downstream is 10 + 2, or 12 mph, and the time is $\frac{24}{12}$, or 2 hr. The total time of the round trip is 3 + 2, or 5 hr. The value checks.

State. The speed of the boat in still water is 10 mph.

34. 15 mph

35. Familiarize. Let x represent the time it takes the smaller pipe to fill the tank. Then x - 3 represents the time it takes the larger pipe to fill the tank. It takes them 2 hr to fill the tank when both pipes are working together, so they can fill $\frac{1}{2}$ of the tank in 1 hr. The smaller pipe will fill $\frac{1}{x}$ of the tank in 1 hr, and the larger pipe will fill $\frac{1}{x - 3}$ of the tank in 1 hr.

Translate. We have an equation.

$$\frac{1}{x} + \frac{1}{x - 3} = \frac{1}{2}$$

Carry out. We solve the equation.
We multiply by the LCD which is 2x(x - 3).

$$2x(x - 3)\left[\frac{1}{x} + \frac{1}{x - 3}\right] = 2x(x - 3) \cdot \frac{1}{2}$$

$$2(x - 3) + 2x = x(x - 3)$$

$$0 = x^2 - 7x + 6$$

$$0 = (x - 6)(x - 1)$$

$$x = 6 \text{ or } x = 1$$

Check. Since negative time has no meaning in this problem, 1 is not a solution (1 - 3 = -2). We only check 6 hr. This is the time it would take the smaller pipe working alone. Then the larger pipe would take 6 - 3, or 3 hr working alone. The larger pipe would fill $2\left(\frac{1}{3}\right)$, or $\frac{2}{3}$, of the tank in 2 hr, and the smaller pipe would fill $2\left(\frac{1}{6}\right)$, or $\frac{1}{3}$, or the tank in 2 hr. Thus in 2 hr they would fill $\frac{2}{3} + \frac{1}{3}$ of the tank. This is all of it, so the numbers check.

State. It takes the smaller pipe, working alone, 6 hr to fill the tank.

36. 12 hr

37. Familiarize. We make a drawing and then organize the information in a table. We let r represent Ellen's speed in still water. Then r - 2 is the speed upstream and r + 2 is the speed downstream. Using $t = \frac{d}{r}$, we let $\frac{1}{r - 2}$ represent the time upstream and $\frac{1}{r + 2}$ represent the time downstream.

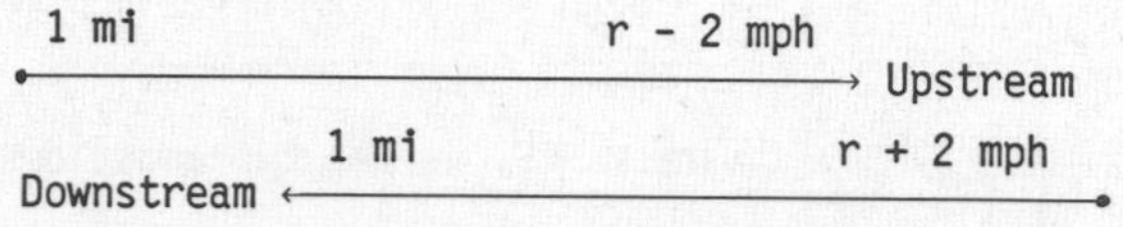

Trip	Distance	Speed	Time
Upstream	1	r - 2	$\frac{1}{r - 2}$
Downstream	1	r + 2	$\frac{1}{r + 2}$

Translate. The time for the round trip is 1 hour. We now have an equation.

$$\frac{1}{r - 2} + \frac{1}{r + 2} = 1$$

Carry out. We solve the equation. We multiply by the LCD, (r - 2)(r + 2).

$$(r - 2)(r + 2)\left[\frac{1}{r - 2} + \frac{1}{r + 2}\right] = (r - 2)(r + 2) \cdot 1$$

$$(r + 2) + (r - 2) = (r - 2)(r + 2)$$

$$2r = r^2 - 4$$

$$0 = r^2 - 2r - 4$$

$$a = 1, b = -2, c = -4$$

$$r = \frac{-(-2) \pm \sqrt{(-2)^2 - 4 \cdot 1(-4)}}{2 \cdot 1}$$

$$r = \frac{2 \pm \sqrt{4 + 16}}{2} = \frac{2 \pm \sqrt{20}}{2}$$

$$r = \frac{2 \pm 2\sqrt{5}}{2} = 1 \pm \sqrt{5}$$

$$1 + \sqrt{5} \approx 1 + 2.236 \approx 3.24$$

$$1 - \sqrt{5} \approx 1 - 2.236 \approx -1.24$$

Check. Since negative speed has no meaning in this problem, we only check 3.24 mph. If $r \approx 3.24$, then $r - 2 \approx 1.24$ and $r + 2 \approx 5.24$. The time it takes to travel upstream is approximately $\frac{1}{1.24}$, or 0.806 hr, and the time it takes to travel downstream is approximately $\frac{1}{5.24}$, or 0.191 hr. The total time is 0.997 which is approximately 1 hour. The value checks.

State. Ellen's speed in still water is approximately 3.24 mph.

38. 9.34 km/h

39. $\sqrt{3x + 1} = \sqrt{2x - 1} + 1$

$3x + 1 = 2x - 1 + 2\sqrt{2x - 1} + 1$ Squaring both sides

$x + 1 = 2\sqrt{2x - 1}$

$x^2 + 2x + 1 = 4(2x - 1)$ Squaring both sides again

$x^2 + 2x + 1 = 8x - 4$

$x^2 - 6x + 5 = 0$

$(x - 1)(x - 5) = 0$

$x = 1$ or $x = 5$

Both numbers check. The solutions are 1 and 5.

40. $\frac{1}{x - 2}$

41. $\sqrt[3]{18y^3}\,\sqrt[3]{4x^2} = \sqrt[3]{72x^2y^3} = \sqrt[3]{8y^3\cdot 9x^2} = 2y\sqrt[3]{9x^2}$

42.

43.

44. $\frac{-3 \pm \sqrt{57}}{2}$

45. $\frac{x^2}{x - 2} - \frac{x + 4}{2} + \frac{2 - 4x}{x - 2} + 1 = 0$

LCD is $2(x - 2)$

$2(x - 2)\left[\frac{x^2}{x - 2} - \frac{x + 4}{2} + \frac{2 - 4x}{x - 2} + 1\right] = 2(x - 2)\cdot 0$

$2x^2 - (x-2)(x+4) + 2(2-4x) + 2(x-2) = 0$

$2x^2 - x^2 - 2x + 8 + 4 - 8x + 2x - 4 = 0$

$x^2 - 8x + 8 = 0$

Use the quadratic formula.

$x = \frac{-(-8) \pm \sqrt{(-8)^2 - 4\cdot 1\cdot 8}}{2\cdot 1}$

$x = \frac{8 \pm \sqrt{32}}{2} = \frac{8 \pm 4\sqrt{2}}{2} = 4 \pm 2\sqrt{2}$

Both numbers check. The solutions are $4 \pm 2\sqrt{2}$.

46. $\frac{-i \pm \sqrt{23}}{4}$

47. $\frac{1}{a - 1} = a + 1$, LCD is $a - 1$

$(a - 1)\cdot\frac{1}{a - 1} = (a - 1)(a + 1)$

$1 = a^2 - 1$

$2 = a^2$

$\pm\sqrt{2} = a$

Both values check. The solution is $\pm\sqrt{2}$.

48. $2.50

49. 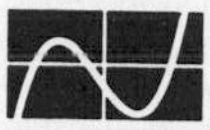

Exercise Set 8.4

1. $x^2 - 6x + 9 = 0$

$a = 1,\ b = -6,\ c = 9$

We substitute and compute the discriminant.

$b^2 - 4ac = (-6)^2 - 4\cdot 1\cdot 9$

$= 36 - 36 = 0$

Since $b^2 - 4ac = 0$, there is just one solution, and it is a real number.

2. One real

3. $x^2 + 7 = 0$

$a = 1,\ b = 0,\ c = 7$

We substitute and compute the discriminant.

$b^2 - 4ac = 0^2 - 4\cdot 1\cdot 7$

$= -28$

Since $b^2 - 4ac < 0$, there are two imaginary-number solutions.

4. Two imaginary

5. $x^2 - 2 = 0$

$a = 1,\ b = 0,\ c = -2$

We substitute and compute the discriminant.

$b^2 - 4ac = 0^2 - 4\cdot 1\cdot(-2)$

$= 8$

Since $b^2 - 4ac > 0$, there are two real solutions.

6. Two real

7. $4x^2 - 12x + 9 = 0$

$a = 4,\ b = -12,\ c = 9$

We substitute and compute the discriminant.

$b^2 - 4ac = (-12)^2 - 4\cdot 4\cdot 9$

$= 144 - 144 = 0$

Since $b^2 - 4ac = 0$, there is just one solution, and it is a real number.

8. Two real

9. $x^2 - 2x + 4 = 0$

$a = 1,\ b = -2,\ c = 4$

We substitute and compute the discriminant.

$b^2 - 4ac = (-2)^2 - 4\cdot1\cdot4$

$= 4 - 16 = -12$

Since $b^2 - 4ac < 0$, there are two imaginary-number solutions.

10. Two imaginary

11. $a^2 - 10a + 21 = 0$

$a = 1,\ b = -10,\ c = 21$

We substitute and compute the discriminant.

$b^2 - 4ac = (-10)^2 - 4\cdot1\cdot21$

$= 100 - 84 = 16$

Since $b^2 - 4ac > 0$, there are two real solutions.

12. One real

13. $6x^2 + 5x - 4 = 0$

$a = 6,\ b = 5,\ c = -4$

We substitute and compute the discriminant.

$b^2 - 4ac = 5^2 - 4\cdot6\cdot(-4)$

$= 25 + 96 = 121$

Since $b^2 - 4ac > 0$, there are two real solutions.

14. Two real

15. $9t^2 - 3t = 0$

$a = 9,\ b = -3,\ c = 0$

We substitute and compute the discriminant.

$b^2 - 4ac = (-3)^2 - 4\cdot9\cdot0$

$= 9 - 0 = 9$

Since $b^2 - 4ac > 0$, there are two real solutions.

16. Two real

17. $x^2 + 5x = 7$

$x^2 + 5x - 7 = 0$ Standard form

$a = 1,\ b = 5,\ c = -7$

We substitute and compute the discriminant.

$b^2 - 4ac = 5^2 - 4\cdot1\cdot(-7)$

$= 25 + 28 = 53$

Since $b^2 - 4ac > 0$, there are two real solutions.

18. Two imaginary

19. $y^2 = \frac{1}{2}y - \frac{3}{5}$

$y^2 - \frac{1}{2}y + \frac{3}{5} = 0$ Standard form

$a = 1,\ b = -\frac{1}{2},\ c = \frac{3}{5}$

We substitute and compute the discriminant.

$b^2 - 4ac = \left(-\frac{1}{2}\right)^2 - 4\cdot1\cdot\frac{3}{5}$

$= \frac{1}{4} - \frac{12}{5} = -\frac{43}{20}$

Since $b^2 - 4ac < 0$, there are two imaginary-number solutions.

20. Two real

21. $4x^2 - 4\sqrt{3}x + 3 = 0$

$a = 4,\ b = -4\sqrt{3},\ c = 3$

We substitute and compute the discriminant.

$b^2 - 4ac = (-4\sqrt{3})^2 - 4\cdot4\cdot3$

$= 48 - 48$

$= 0$

Since $b^2 - 4ac = 0$, there is just one solution, and it is a real number.

22. Two real

23. The solutions are -11 and 9.

$x = -11$ or $x = 9$

$x + 11 = 0$ or $x - 9 = 0$

$(x + 11)(x - 9) = 0$ Principle of zero products

$x^2 + 2x - 99 = 0$ FOIL

24. $x^2 - 16 = 0$

25. The only solution is 7. It must be a "double" solution.

$x = 7$ or $x = 7$

$x - 7 = 0$ or $x - 7 = 0$

$(x - 7)(x - 7) = 0$ Principle of zero products

$x^2 - 14x + 49 = 0$ FOIL

26. $x^2 + 10x + 25 = 0$

27. The solutions are -3 and -5.

$x = -3$ or $x = -5$

$x + 3 = 0$ or $x + 5 = 0$

$(x + 3)(x + 5) = 0$

$x^2 + 8x + 15 = 0$

28. $x^2 + 9x + 14 = 0$

29. The solutions are 4 and $\frac{2}{3}$.

$$x = 4 \quad \text{or} \quad x = \frac{2}{3}$$
$$x - 4 = 0 \quad \text{or} \quad x - \frac{2}{3} = 0$$
$$(x - 4)\left(x - \frac{2}{3}\right) = 0$$
$$x^2 - \frac{2}{3}x - 4x + \frac{8}{3} = 0$$
$$x^2 - \frac{14}{3}x + \frac{8}{3} = 0$$
$$3x^2 - 14x + 8 = 0 \quad \text{Multiplying by 3}$$

30. $4x^2 - 23x + 15 = 0$

31. The solutions are $\frac{1}{2}$ and $\frac{1}{3}$.

$$x = \frac{1}{2} \quad \text{or} \quad x = \frac{1}{3}$$
$$x - \frac{1}{2} = 0 \quad \text{or} \quad x - \frac{1}{3} = 0$$
$$\left(x - \frac{1}{2}\right)\left(x - \frac{1}{3}\right) = 0$$
$$x^2 - \frac{1}{3}x - \frac{1}{2}x + \frac{1}{6} = 0$$
$$x^2 - \frac{5}{6}x + \frac{1}{6} = 0$$
$$6x^2 - 5x + 1 = 0 \quad \text{Multiplying by 6}$$

32. $8x^2 + 6x + 1 = 0$

33. The solutions are $-\frac{2}{5}$ and $\frac{6}{5}$.

$$x = -\frac{2}{5} \quad \text{or} \quad x = \frac{6}{5}$$
$$x + \frac{2}{5} = 0 \quad \text{or} \quad x - \frac{6}{5} = 0$$
$$\left(x + \frac{2}{5}\right)\left(x - \frac{6}{5}\right) = 0$$
$$x^2 - \frac{4}{5}x - \frac{12}{25} = 0$$
$$25x^2 - 20x - 12 = 0 \quad \text{Multiplying by 25}$$

34. $49x^2 + 7x - 6 = 0$

35. The solutions are $\sqrt{2}$ and $3\sqrt{2}$.

$$x = \sqrt{2} \quad \text{or} \quad x = 3\sqrt{2}$$
$$x - \sqrt{2} = 0 \quad \text{or} \quad x - 3\sqrt{2} = 0$$
$$(x - \sqrt{2})(x - 3\sqrt{2}) = 0$$
$$x^2 - 4\sqrt{2}x + 6 = 0$$

36. $x^2 - \sqrt{3}x - 6 = 0$

37. The solutions are $-\sqrt{5}$ and $-2\sqrt{5}$.

$$x = -\sqrt{5} \quad \text{or} \quad x = -2\sqrt{5}$$
$$x + \sqrt{5} = 0 \quad \text{or} \quad x + 2\sqrt{5} = 0$$
$$(x + \sqrt{5})(x + 2\sqrt{5}) = 0$$
$$x^2 + 3\sqrt{5}x + 10 = 0$$

38. $x^2 + 4\sqrt{6}x + 18 = 0$

39. The solutions are $3i$ and $-3i$.

$$x = 3i \quad \text{or} \quad x = -3i$$
$$x - 3i = 0 \quad \text{or} \quad x + 3i = 0$$
$$(x - 3i)(x + 3i) = 0$$
$$x^2 - (3i)^2 = 0$$
$$x^2 + 9 = 0$$

40. $x^2 + 16 = 0$

41. The solutions are $5 - 2i$ and $5 + 2i$.

$$x = 5 - 2i \quad \text{or} \quad x = 5 + 2i$$
$$x - 5 + 2i = 0 \quad \text{or} \quad x - 5 - 2i = 0$$
$$[x + (-5 + 2i)][x + (-5 - 2i)] = 0$$
$$x^2 + x(-5-2i) + x(-5+2i) + (-5+2i)(-5-2i) = 0$$
$$x^2 - 5x - 2xi - 5x + 2xi + 25 - 4i^2 = 0$$
$$x^2 - 10x + 29 = 0$$
$$(i^2 = -1)$$

42. $x^2 - 4x + 53 = 0$

43. The solutions are $\frac{1 + 3i}{2}$ and $\frac{1 - 3i}{2}$.

$$x = \frac{1 + 3i}{2} \quad \text{or} \quad x = \frac{1 - 3i}{2}$$
$$x - \frac{1 + 3i}{2} = 0 \quad \text{or} \quad x - \frac{1 - 3i}{2} = 0$$
$$\left(x - \frac{1 + 3i}{2}\right)\left(x - \frac{1 - 3i}{2}\right) = 0$$
$$x^2 - \frac{x - 3xi}{2} - \frac{x + 3xi}{2} + \frac{1 - 9i^2}{4} = 0$$
$$x^2 - \frac{x}{2} + \frac{3xi}{2} - \frac{x}{2} - \frac{3xi}{2} + \frac{10}{4} = 0$$
$$x^2 - x + \frac{5}{2} = 0$$
$$2x^2 - 2x + 5 = 0$$

44. $9x^2 - 12x + 5 = 0$

45. Familiarize. Let x and y represent the number of 30-sec and 60-sec commercials, respectively. Then the amount of time for the 30-sec commercials was 30x sec, or $\frac{30x}{60} = \frac{x}{2}$ min. The amount of time for the 60-sec commercials was 60x sec, or $\frac{60x}{60} = x$ min.

Translate. Rewording, we write two equations. We will express time in minutes.

Total number of commercials is 12.

$$x + y = 12$$

Time for 30-sec commercials is total commercial time less 6 min.

$$\frac{x}{2} = \frac{x}{2} + x - 6$$

Carry out. Solving the system of equations we get (6,6).

Check. If there are six 30-sec and six 60-sec commercials, the total number of commercials is 12. The amount of time for six 30-sec commercials is 180 sec, or 3 min, and for six 60-sec commercials is 360 sec, or 6 min. The total commercial time is 9 min, and the amount of time for 30-sec commercials is 6 min less than this. The numbers check.

State. There were six 30-sec and six 60-sec commercials.

46.

47.

48. -1

49. The graph includes the points (-3,0), (0,-3), and (1,0). Substituting in $y = ax^2 + bx + c$, we have three equations.

$$0 = 9a - 3b + c,$$
$$-3 = c,$$
$$0 = a + b + c$$

The solution of this system of equations is $a = 1$, $b = 2$, $c = -3$.

50. a) 2

b) $\frac{11}{2}$

51. a) $kx^2 - 2x + k = 0$; one solution is -3

We first find k by substituting -3 for x.

$$k(-3)^2 - 2(-3) + k = 0$$
$$9k + 6 + k = 0$$
$$10k = -6$$
$$k = -\frac{6}{10}$$
$$k = -\frac{3}{5}$$

b) Now substitute $-\frac{3}{5}$ for k in the original equation.

$$-\frac{3}{5}x^2 - 2x + \left(-\frac{3}{5}\right) = 0$$
$$3x^2 + 10x + 3 = 0 \quad \text{Multiplying by -5}$$
$$(3x + 1)(x + 3) = 0$$
$$x = -\frac{1}{3} \quad \text{or} \quad x = -3$$

The other solution is $-\frac{1}{3}$.

52. a) 2

b) $1 - i$

53. a) $x^2 - (6 + 3i)x + k = 0$; one solution is 3.

We first find k by substituting 3 for x.

$$3^2 - (6 + 3i)3 + k = 0$$
$$9 - 18 - 9i + k = 0$$
$$-9 - 9i + k = 0$$
$$k = 9 + 9i$$

b) Now we substitute $9 + 9i$ for k in the original equation.

$$x^2 - (6 + 3i)x + (9 + 9i) = 0$$
$$x^2 - (6 + 3i)x + 3(3 + 3i) = 0$$
$$[x - (3 + 3i)][x - 3] = 0$$
$$x = 3 + 3i \quad \text{or} \quad x = 3$$

The other solution is $3 + 3i$.

54. $h = -36$, $k = 15$

55. From Exercise 47 we know that

$$-\frac{b}{a} = \sqrt{3} \quad \text{and} \quad \frac{c}{a} = 8.$$

Multiplying $ax^2 + bx + c = 0$ by $\frac{1}{a}$ we have:

$$x^2 + \frac{b}{a}x + \frac{c}{a} = 0$$
$$x^2 - \left(-\frac{b}{a}\right)x + \frac{c}{a} = 0$$
$$x^2 - \sqrt{3}x + 8 = 0 \quad \text{Substituting}$$

56. a) 4, -2

b) $1 \pm \sqrt{26}$

57. We substitute (-3,0), $\left(\frac{1}{2},0\right)$, and (0,-12) in $f(x) = ax^2 + bx + c$ and get three equations.

$$0 = 9a - 3b + c,$$
$$0 = \frac{1}{4}a + \frac{1}{2}b + c,$$
$$-12 = c$$

The solution of this system of equations is $a = 8$, $b = 20$, $c = -12$.

58. 2

59. $x^2 + kx + 8 = 0$ $\qquad$ $x^2 - kx + 8 = 0$

$$x = \frac{-k \pm \sqrt{k^2 - 4\cdot1\cdot8}}{2\cdot1} \qquad x = \frac{-(-k) \pm \sqrt{(-k)^2 - 4\cdot1\cdot8}}{2\cdot1}$$
$$x = \frac{-k \pm \sqrt{k^2 - 32}}{2} \qquad x = \frac{k \pm \sqrt{k^2 - 32}}{2}$$

$$\frac{-k + \sqrt{k^2 - 32}}{2} + 6 = \frac{k + \sqrt{k^2 - 32}}{2}$$
$$-k + \sqrt{k^2 - 32} + 12 = k + \sqrt{k^2 - 32}$$
$$-k + 12 = k$$
$$12 = 2k$$
$$6 = k$$

Exercise Set 8.5

1. $x^4 - 10x^2 + 25 = 0$

Let $u = x^2$ and think of x^4 as $(x^2)^2$.

$u^2 - 10u + 25 = 0$ Substituting u for x^2

$(u - 5)(u - 5) = 0$

$u - 5 = 0$ or $u - 5 = 0$

$u = 5$ or $u = 5$

Now we substitute x^2 for u and solve the equation.

$x^2 = 5$

$x = \pm\sqrt{5}$

Both $\sqrt{5}$ and $-\sqrt{5}$ check. They are the solutions.

2. $\pm\sqrt{2}$, ± 1

3. $x^4 - 12x^2 + 27 = 0$

Let $u = x^2$.

$u^2 - 12u + 27 = 0$ Substituting u for x^2

$(u - 9)(u - 3) = 0$

$u = 9$ or $u = 3$

Now substitute x^2 for u and solve these equations:

$x^2 = 9$ or $x^2 = 3$

$x = \pm 3$ or $x = \pm\sqrt{3}$

The numbers 3, -3, $\sqrt{3}$, and $-\sqrt{3}$ check. They are the solutions.

4. ± 2, $\pm\sqrt{5}$

5. $9x^4 - 14x^2 + 5 = 0$

Let $u = x^2$.

$9u^2 - 14u + 5 = 0$ Substituting u for x^2

$(9u - 5)(u - 1) = 0$

$u = \frac{5}{9}$ or $u = 1$

$x^2 = \frac{5}{9}$ or $x^2 = 1$ Substituting x^2 for u

$x = \pm\frac{\sqrt{5}}{3}$ or $x = \pm 1$

The numbers $\frac{\sqrt{5}}{3}$, $-\frac{\sqrt{5}}{3}$, 1, and -1 check. They are the solutions.

6. $\pm\frac{\sqrt{3}}{2}$, ± 2

7. $x - 10\sqrt{x} + 9 = 0$

Let $u = \sqrt{x}$ and think of x as $(\sqrt{x})^2$.

$u^2 - 10u + 9 = 0$ Substituting u for $\sqrt{x}$

$(u - 9)(u - 1) = 0$

$u - 9 = 0$ or $u - 1 = 0$

$u = 9$ or $u = 1$

Now we substitute $\sqrt{x}$ for u and solve these equations:

$\sqrt{x} = 9$ or $\sqrt{x} = 1$

$x = 81$ or $x = 1$

The numbers 81 and 1 both check. They are the solutions.

8. $\frac{1}{4}$, 16

9. $3x + 10\sqrt{x} - 8 = 0$

Let $u = \sqrt{x}$.

$3u^2 + 10u - 8 = 0$ Substituting u for $\sqrt{x}$

$(3u - 2)(u + 4) = 0$

$u = \frac{2}{3}$ or $u = -4$

$\sqrt{x} = \frac{2}{3}$ or $\sqrt{x} = -4$ Substituting $\sqrt{x}$ for u

Squaring the first equation, we get $x = \frac{4}{9}$. Note that $\sqrt{x} = -4$ has no real solution. The number $\frac{4}{9}$ checks and is the solution.

10. $\frac{4}{25}$

11. $(x^2 - 9)^2 + 3(x^2 - 9) + 2 = 0$

Let $u = x^2 - 9$.

$u^2 + 3u + 2 = 0$ Substituting u for $x^2 - 9$

$(u + 2)(u + 1) = 0$

$u = -2$ or $u = -1$

$x^2 - 9 = -2$ or $x^2 - 9 = -1$ Substituting $x^2 - 9$ for u

$x^2 = 7$ or $x^2 = 8$

$x = \pm\sqrt{7}$ or $x = \pm\sqrt{8}$

$x = \pm\sqrt{7}$ or $x = \pm 2\sqrt{2}$

The numbers $\sqrt{7}$, $-\sqrt{7}$, $2\sqrt{2}$, and $-2\sqrt{2}$ check. They are the solutions.

12. ± 1, $\pm\sqrt{2}$

13. $(x^2 - 6x)^2 - 2(x^2 - 6x) - 35 = 0$

Let $u = x^2 - 6x$.

$u^2 - 2u - 35 = 0$ Substituting u for $x^2 - 6x$

$(u - 7)(u + 5) = 0$

$u - 7 = 0$ or $u + 5 = 0$

$u = 7$ or $u = -5$

$x^2 - 6x = 7$ or $x^2 - 6x = -5$

Substituting $x^2 - 6x$ for u

$x^2 - 6x - 7 = 0$ or $x^2 - 6x + 5 = 0$

$(x - 7)(x + 1) = 0$ or $(x - 5)(x - 1) = 0$

$x = 7$ or $x = -1$ or $x = 5$ or $x = 1$

The numbers -1, 1, 5, and 7 check. They are the solutions.

14. $\frac{3 \pm \sqrt{33}}{2}$, 4, -1

15. $(3 + \sqrt{x})^2 - 3(3 + \sqrt{x}) - 10 = 0$

Let $u = 3 + \sqrt{x}$

$u^2 - 3u - 10 = 0$ Substituting u for $3 + \sqrt{x}$

$(u - 5)(u + 2) = 0$

$u = 5$ or $u = -2$

$3 + \sqrt{x} = 5$ or $3 + \sqrt{x} = -2$ Substituting $3 + \sqrt{x}$ for u

$\sqrt{x} = 2$ or $\sqrt{x} = -5$

$x = 4$ No real solution

The number 4 checks and is the solution.

16. 1

17. $x^{-2} - x^{-1} - 6 = 0$

Let $u = x^{-1}$ and think of x^{-2} as $(x^{-1})^2$.

$u^2 - u - 6 = 0$ Substituting u for x^{-1}

$(u - 3)(u + 2) = 0$

$u = 3$ or $u = -2$

Now we substitute x^{-1} for u and solve these equations:

$x^{-1} = 3$ or $x^{-1} = -2$

$\frac{1}{x} = 3$ or $\frac{1}{x} = -2$

$\frac{1}{3} = x$ or $-\frac{1}{2} = x$

Both $\frac{1}{3}$ and $-\frac{1}{2}$ check. They are the solutions.

18. $\frac{4}{5}$, -1

19. $2x^{-2} + x^{-1} - 1 = 0$

Let $u = x^{-1}$.

$2u^2 + u - 1 = 0$ Substituting u for x^{-1}

$(2u - 1)(u + 1) = 0$

$2u = 1$ or $u = -1$

$u = \frac{1}{2}$ or $u = -1$

$x^{-1} = \frac{1}{2}$ or $x^{-1} = -1$ Substituting x^{-1} for u

$\frac{1}{x} = \frac{1}{2}$ or $\frac{1}{x} = -1$

$x = 2$ or $x = -1$

Both 2 and -1 check. They are the solutions.

20. $-\frac{1}{10}$, 1

21. $t^{2/3} + t^{1/3} - 6 = 0$

Let $u = t^{1/3}$ and think of $t^{2/3}$ as $(t^{1/3})^2$.

$u^2 + u - 6 = 0$ Substituting u for $t^{1/3}$

$(u + 3)(u - 2) = 0$

$u = -3$ or $u = 2$

Now we substitute $t^{1/3}$ for u and solve these equations:

$t^{1/3} = -3$ or $t^{1/3} = 2$

$t = (-3)^3$ or $t = 2^3$ Raising to the third power

$t = -27$ or $t = 8$

Both -27 and 8 check. They are the solutions.

22. 64, -8

23. $z^{1/2} - z^{1/4} - 2 = 0$

Let $u = z^{1/4}$.

$u^2 - u - 2 = 0$ Substituting u for $z^{1/4}$

$(u - 2)(u + 1) = 0$

$u = 2$ or $u = -1$

$z^{1/4} = 2$ or $z^{1/4} = -1$ Substituting $z^{1/4}$ for u

$\sqrt[4]{z} = 2$ or $\sqrt[4]{z} = -1$

$z = 16$ This equation has no real solution since principal fourth roots are never negative.

The number 16 checks, so it is the solution.

24. 729

25. $x^{2/5} + x^{1/5} - 6 = 0$

Let $u = x^{1/5}$.

$u^2 + u - 6 = 0$ Substituting u for $x^{1/5}$

$(u + 3)(u - 2) = 0$

$u = -3$ or $u = 2$

$x^{1/5} = -3$ or $x^{1/5} = 2$ Substituting $x^{1/5}$ for u

$x = -243$ or $x = 32$ Raising to the fifth power

Both -243 and 32 check. They are the solutions.

26. 81

27. $t^{1/3} + 2t^{1/6} = 3$

$t^{1/3} + 2t^{1/6} - 3 = 0$

Let $u = t^{1/6}$.

$u^2 + 2u - 3 = 0$ Substituting u for $t^{1/6}$

$(u + 3)(u - 1) = 0$

$u = -3$ or $u = 1$

$t^{1/6} = -3$ or $t^{1/6} = 1$ Substituting $t^{1/6}$ for u

No real solution $t = 1$

The number 1 checks and is the solution.

28. 81, 16

29. $\left[\frac{x + 3}{x - 3}\right]^2 - \left[\frac{x + 3}{x - 3}\right] - 6 = 0$

Let $u = \frac{x + 3}{x - 3}$.

$u^2 - u - 6 = 0$ Substituting u for $\frac{x + 3}{x - 3}$

$(u - 3)(u + 2) = 0$

$u = 3$ or $u = -2$

$\frac{x + 3}{x - 3} = 3$ or $\frac{x + 3}{x - 3} = -2$ Substituting $\frac{x + 3}{x - 3}$ for u

$x + 3 = 3(x - 3)$ or $x + 3 = -2(x - 3)$ Multiplying by $(x - 3)$

$x + 3 = 3x - 9$ or $x + 3 = -2x + 6$

$-2x = -12$ or $3x = 3$

$x = 6$ or $x = 1$

Both 6 and 1 check. They are the solutions.

30. $-\frac{11}{6}, -\frac{1}{6}$

31. $9\left[\frac{x + 2}{x + 3}\right]^2 - 6\left[\frac{x + 2}{x + 3}\right] + 1 = 0$

Let $u = \frac{x + 2}{x + 3}$.

$9u^2 - 6u + 1 = 0$ Substituting u for $\frac{x + 2}{x + 3}$

$(3u - 1)(3u - 1) = 0$

$3u = 1$

$u = \frac{1}{3}$

$\frac{x + 2}{x + 3} = \frac{1}{3}$ Substituting $\frac{x + 2}{x + 3}$ for u

$3(x + 2) = x + 3$ Multiplying by $3(x + 3)$

$3x + 6 = x + 3$

$2x = -3$

$x = -\frac{3}{2}$

The number $-\frac{3}{2}$ checks. It is the solution.

32. $\frac{12}{5}$

33. $\left[\frac{y^2 - 1}{y}\right]^2 - 4\left[\frac{y^2 - 1}{y}\right] - 12 = 0$

Let $u = \frac{y^2 - 1}{y}$.

$u^2 - 4u - 12 = 0$ Substituting u for $\frac{y^2 - 1}{y}$

$(u - 6)(u + 2) = 0$

$u = 6$ or $u = -2$

$\frac{y^2 - 1}{y} = 6$ or $\frac{y^2 - 1}{y} = -2$ Substituting $\frac{y^2 - 1}{y}$ for u

$y^2 - 1 = 6y$ or $y^2 - 1 = -2y$ Multiplying by y

$y^2 - 6y - 1 = 0$ or $y^2 + 2y - 1 = 0$

$y = \frac{-(-6)\pm\sqrt{(-6)^2-4\cdot 1\cdot(-1)}}{2\cdot 1}$ or $y = \frac{-2\pm\sqrt{2^2-4\cdot 1\cdot(-1)}}{2\cdot 1}$

$y = \frac{6 \pm \sqrt{40}}{2}$ or $y = \frac{-2 \pm \sqrt{8}}{2}$

$y = \frac{6 \pm 2\sqrt{10}}{2}$ or $y = \frac{-2 \pm 2\sqrt{2}}{2}$

$y = 3 \pm \sqrt{10}$ or $y = -1 \pm \sqrt{2}$

All four numbers check. They are the solutions.

34. $\frac{9 \pm \sqrt{89}}{2}, -1 \pm \sqrt{3}$

35. $\sqrt{3x^2}\sqrt{3x^3} = \sqrt{3x^2\cdot 3x^3} = \sqrt{9x^5} = \sqrt{9x^4\cdot x} = 3x^2\sqrt{x}$

36. 4 L of A, 8 L of B

37. $\frac{x + 1}{x - 1} - \frac{x + 1}{x^2 + x + 1}$, LCD is $(x - 1)(x^2 + x + 1)$

$= \frac{x + 1}{x - 1}\cdot\frac{x^2 + x + 1}{x^2 + x + 1} - \frac{x + 1}{x^2 + x + 1}\cdot\frac{x - 1}{x - 1}$

$= \frac{(x^3 + 2x^2 + 2x + 1) - (x^2 - 1)}{(x - 1)(x^2 + x + 1)}$

$= \frac{x^3 + x^2 + 2x + 2}{x^3 - 1}$

38.

39. Solve $f(x) = 0$.

$x^4 - 8x^2 + 7 = 0$

Let $u = x^2$.

$u^2 - 8u + 7 = 0$

$(u - 7)(u - 1) = 0$

$u = 7$ or $u = 1$

$x^2 = 7$ or $x^2 = 1$

$x = \pm\sqrt{7}$ or $x = \pm 1$

All four numbers check. The x-intercepts are $(\sqrt{7},0)$, $(-\sqrt{7},0)$, $(1,0)$, and $(-1,0)$.

40. 28.45

41. $\pi x^4 - \pi^2 x^2 - \sqrt{99.3} = 0$

Let $u = x^2$.

$\pi u^2 - \pi^2 u - \sqrt{99.3} = 0$

$u = \frac{-(-\pi^2) \pm \sqrt{(-\pi^2)^2 - 4\cdot\pi(-\sqrt{99.3})}}{2\cdot\pi}$

$u \approx \frac{9.8696 \pm 14.9209}{6.2832}$

$u \approx 3.9455$ or $u \approx -0.8039$

$x^2 \approx 3.9455$ or $x^2 \approx -0.8039$

$x \approx \pm 1.99$ No real solutions

The numbers 1.99 and -1.99 check.

42. $\frac{100}{99}$

43. $\left[\sqrt{\frac{x}{x-3}}\right]^2 - 24 = 10\sqrt{\frac{x}{x-3}}$

Let $u = \sqrt{\frac{x}{x-3}}$.

$u^2 - 24 = 10u$ Substituting for $\sqrt{\frac{x}{x-3}}$

$u^2 - 10u - 24 = 0$ Standard form

$(u - 12)(u + 2) = 0$

$u = 12$ or $u = -2$

$\sqrt{\frac{x}{x-3}} = 12$ or $\sqrt{\frac{x}{x-3}} = -2$

$\frac{x}{x-3} = 144$ No real solutions

$x = 144x - 432$

$432 = 143x$

$\frac{432}{143} = x$

This number checks.

44. 259

45. $a^3 - 26a^{3/2} - 27 = 0$

Let $u = a^{3/2}$.

$u^2 - 26u - 27 = 0$

$(u - 27)(u + 1) = 0$

$u = 27$ or $u = -1$

$a^{3/2} = 27$ or $a^{3/2} = -1$

$a = 27^{2/3}$ No real solution

$a = (3^3)^{2/3}$

$a = 9$

The number 9 checks.

46. 3, 1

47. $x^6 + 7x^3 - 8 = 0$

Let $u = x^3$.

$u^2 + 7u - 8 = 0$

$(u + 8)(u - 1) = 0$

$u = -8$ or $u = 1$

$x^3 = -8$ or $x^3 = 1$

$x = -2$ or $x = 1$

Both numbers check.

48. 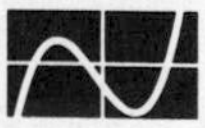

49. -3, -1, 1, 4

Exercise Set 8.6

1. $y = kx$

$24 = k\cdot 3$ Substituting

$8 = k$

The variation constant is 8.
The equation of variation is $y = 8x$.

2. $k = \frac{5}{12}$; $y = \frac{5}{12}x$

3. $y = kx$

$3.6 = k\cdot 1$ Substituting

$3.6 = k$

The variation constant is 3.6.
The equation of variation is $y = 3.6x$.

4. $k = \frac{2}{5}$; $y = \frac{2}{5}x$

5. $y = kx$

$30 = k\cdot 8$ Substituting

$\frac{30}{8} = k$

$\frac{15}{4} = k$

The variation constant is $\frac{15}{4}$.
The equation of variation is $y = \frac{15}{4}x$.

6. $k = 3$; $y = 3x$

7. $y = kx$

$0.8 = k(0.5)$ Substituting

$8 = k\cdot 5$ Clearing decimals

$\frac{8}{5} = k$

$1.6 = k$

The variation constant is 1.6.
The equation of variation is $y = 1.6x$.

8. $k = 1.5$; $y = 1.5x$

9. Familiarize. Because of the phrase "I. . .varies directly as. . .V," we express the current as a function of the voltage. Thus we have $I(V) = kV$. We know that $I(12) = 4$.

Translate. We find the variation constant and then find the equation of variation.

$I(V) = kV$

$I(12) = k\cdot 12$ Replacing V with 12

$4 = k\cdot 12$ Substituting

$\frac{4}{12} = k$

$\frac{1}{3} = k$

The equation of variation is $I(V) = \frac{1}{3}V$.

Carry out. We compute I(18).

$I(V) = \frac{1}{3}V$

$I(18) = \frac{1}{3} \cdot 18$ Replacing V with 18

$= 6$

Check. Reexamine the calculations. Note that the answer seems reasonable since 12/4 = 18/6.

State. The current is 6 amperes when 18 volts is applied.

10. $66\frac{2}{3}$ cm

11. Familiarize. Because A varies directly as G, we write A as a function of G: A(G) = kG. We know that A(9) = 9.66.

Translate.

$A(G) = kG$

$A(9) = k \cdot 9$ Replacing G with 9

$9.66 = k \cdot 9$ Substituting

$\frac{9.66}{9} = k$ Variation constant

$\frac{3.22}{3} = k$ Simplifying

$A(G) = \frac{3.22}{3}G$ Equation of variation

Carry out. Find A(4).

$A(G) = \frac{3.22}{3}G$

$A(4) = \frac{3.22}{3} \cdot 4$

≈ 4.29

Check. Reexamine the calculations.

State. The average weekly allowance of a 4th-grade student is about $4.29.

12. 204,000,000 cans

13. Familiarize. Because W varies directly as the total mass, we write W(m) = km. We know that W(96) = 64.

Translate.

$W(m) = km$

$W(96) = k \cdot 96$ Replacing m with 96

$64 = k \cdot 96$ Substituting

$\frac{2}{3} = k$ Variation constant

$W(m) = \frac{2}{3}m$ Equation of variation

Carry out. Find W(75).

$W(m) = \frac{2}{3}m$

$W(75) = \frac{2}{3} \cdot 75$

$= 50$

Check. Reexamine the calculations.

State. There are 50 kg of water in a 75 kg person.

14. 40 lb

15. Familiarize. Because the f-stop varies directly as F, we write f(F) = kF. We know that F(150) = 6.3.

Translate.

$f(F) = kF$

$f(150) = k \cdot 150$ Replacing F with 150

$6.3 = k \cdot 150$ Substituting

$0.042 = k$ Variation constant

$f(F) = 0.042F$ Equation of variation

Carry out. Find f(80).

$f(F) = 0.042F$

$f(80) = 0.042(80)$

$= 3.36$

Check. Reexamine the calculations.

State. An 80 mm focal length has an f-stop of 3.36.

16. 532,500 tons

17. $y = \frac{k}{x}$

$6 = \frac{k}{10}$ Substituting

$60 = k$

The variation constant is 60.

The equation of variation is $y = \frac{60}{x}$.

18. $k = 64$; $y = \frac{64}{x}$

19. $y = \frac{k}{x}$

$4 = \frac{k}{3}$ Substituting

$12 = k$

The variation constant is 12.

The equation of variation is $y = \frac{12}{x}$.

20. $k = 36$; $y = \frac{36}{x}$

21. $y = \frac{k}{x}$

$12 = \frac{k}{3}$ Substituting

$36 = k$

The variation constant is 36.

The equation of variation is $y = \frac{36}{x}$.

22. $k = 45$; $y = \frac{45}{x}$

23. $y = \frac{k}{x}$

$27 = \frac{k}{\frac{1}{3}}$ Substituting

$9 = k$

The variation constant is 9.

The equation of variation is $y = \frac{9}{x}$.

24. $k = 9$; $y = \frac{9}{x}$

25. Familiarize. Because I varies inversely as R, we express I as a function of R. Thus we write $I(R) = k/R$. We know that $I(240) = 1/2$.

Translate.

$I(R) = \frac{k}{R}$

$I(240) = \frac{k}{240}$ Replacing R with 240

$\frac{1}{2} = \frac{k}{240}$ Substituting

$120 = k$ Variation constant

$I(R) = \frac{120}{R}$ Equation of variation

Carry out. Find I(540).

$I(540) = \frac{120}{540}$

$= \frac{2}{9}$

Check. Reexamine the calculations. Note that, as expected, when the resistance increases the current decreases.

State. The current is $\frac{2}{9}$ ampere.

26. 27 min

27. Familiarize. Because V varies inversely as P, we write $V(P) = k/P$. We know that $V(32) = 200$.

Translate.

$V(P) = \frac{k}{P}$

$V(32) = \frac{k}{32}$ Replacing P with 32

$200 = \frac{k}{32}$ Substituting

$6400 = k$ Variation constant

$V(P) = \frac{6400}{P}$ Equation of variation

Carry out. Find V(40).

$V(40) = \frac{6400}{40}$

$= 160$

Check. Reexamine the calculations.

State. The volume will be 160 cm^3.

28. 3.5 hr

29. Familiarize. Because t varies inversely as r, we write $t(r) = k/r$. We know that $t(80) = 5$.

Translate.

$t(r) = \frac{k}{r}$

$t(80) = \frac{k}{80}$ Replacing r with 80

$5 = \frac{k}{80}$ Substituting

$400 = k$ Variation constant

$t(r) = \frac{400}{r}$ Equation of variation

Carry out. Find t(60).

$t(60) = \frac{400}{60}$

$= 6\frac{2}{3}$

Check. Reexamine the calculations.

State. It will take $6\frac{2}{3}$ hr.

30. 450 m

31. $y = kx^2$

$0.15 = k(0.1)^2$ Substituting

$0.15 = 0.01k$

$\frac{0.15}{0.01} = k$

$15 = k$ Variation constant

The equation of variation is $y = 15x^2$.

32. $y = \frac{2}{3}x^2$

33. $y = \frac{k}{x^2}$

$0.15 = \frac{k}{(0.1)^2}$ Substituting

$0.15 = \frac{k}{0.01}$

$0.15(0.01) = k$

$0.0015 = k$ Variation constant

The equation of variation is $y = \frac{0.0015}{x^2}$.

34. $y = \frac{54}{x^2}$

35. $y = kxz$

$56 = k\cdot 7\cdot 8$ Substituting 56 for y, 7 for x, and 8 for z

$56 = 56k$

$1 = k$ Variation constant

The equation of variation is $y = xz$.

36. $y = \frac{5x}{z}$

37. $y = kxz^2$

$105 = k \cdot 14 \cdot 5^2$ Substituting 105 for y, 14 for x, and 5 for z

$105 = 350k$

$\frac{105}{350} = k$

$0.3 = k$

The equation of variation is $y = 0.3xz^2$.

38. $y = \frac{xz}{w}$

39. $y = k \cdot \frac{wx^2}{z}$

$49 = k \cdot \frac{3 \cdot 7^2}{12}$ Substituting

$4 = k$ Variation constant

$y = \frac{4wx^2}{z}$

40. $y = \frac{6x}{wz^2}$

41. $y = k \cdot \frac{xz}{wp}$

$\frac{3}{28} = k \cdot \frac{3 \cdot 10}{7 \cdot 8}$ Substituting

$\frac{3}{28} = k \cdot \frac{30}{56}$

$\frac{3}{28} \cdot \frac{56}{30} = k$

$\frac{1}{5} = k$ Variation constant

The equation of variation is $y = \frac{xz}{5wp}$.

42. $y = \frac{5xz}{4w^2}$

43. Familiarize. Because d varies directly as the square of r, we write $d = kr^2$. We know that $d = 200$ when $r = 60$.

Translate. We first find k.

$d = kr^2$

$200 = k(60)^2$

$\frac{1}{18} = k$

$d = \frac{1}{18}r^2$ Equation of variation

Carry out. Substitute 72 for d and solve for r.

$72 = \frac{1}{18}r^2$

$1296 = r^2$

$36 = r$

Check. Recheck the calculations and perhaps make an estimate to see if the answer seems reasonable.

State. The car can go 36 mph.

44. 220 cm^3

45. Familiarize. I varies inversely as d^2, so we write $I = k/d^2$. We know that $I = 25$ when $d = 2$.

Translate. First we find k.

$I = \frac{k}{d^2}$

$25 = \frac{k}{2^2}$

$100 = k$

$I = \frac{100}{d^2}$ Equation of variation

Carry out. Substitute 2.56 for I and solve for d.

$2.56 = \frac{100}{d^2}$

$d^2 = \frac{100}{2.56}$

$d = 6.25$

Check. Recheck the calculations.

State. You are 6.25 km from the transmitter.

46. 5 sec

47. Familiarize. W varies inversely as d^2, so we write $W = k/d^2$. We know that $W = 100$ when $d = 6400$.

Translate. Find k.

$W = \frac{k}{d^2}$

$100 = \frac{k}{(6400)^2}$

$4{,}096{,}000{,}000 = k$

$W = \frac{4{,}096{,}000{,}000}{d^2}$ Equation of variation

Carry out. Substitute 64 for w and solve for d.

$64 = \frac{4{,}096{,}000{,}000}{d^2}$

$d^2 = \frac{4{,}096{,}000{,}000}{64}$

$d = 8000$

Note that a distance of 8000 km from the center of the earth is 8000 - 6400, or 1600 km, above the earth.

Check. Recheck the calculations.

State. The astronaut must be 1600 km above the earth in order to weigh 64 lb.

48. 2.5 m

49. Familiarize. R varies directly as ℓ and inversely as d^2, so we write $R = k\ell/d^2$. We know that $R = 0.1$ when $\ell = 50$ and $d = 1$.

Translate. Find k.

$R = \frac{k\ell}{d^2}$

$0.1 = \frac{k \cdot 50}{1^2}$

$0.002 = k$

$R = \frac{0.002\ell}{d^2}$ Equation of variation

Carry out. Substitute 1 for R and 2000 for ℓ and solve for d.

$$1 = \frac{0.002(2000)}{d^2}$$

$$d^2 = 4$$

$$d = 2$$

Check. Recheck the calculations.

State. The diameter is 2 mm.

50. 4.8 cm

51. Familiarize. Let v represent the boat's velocity. F varies jointly as A and v^2, so we write $F = kAv^2$. We know that F = 86 when A = 41.2 and v = 6.5.

Translate. Find k.

$$F = kAv^2$$

$$86 = k(41.2)(6.5)^2$$

$$0.0494 \approx k$$

$$F = 0.0494Av^2 \quad \text{Equation of variation}$$

Carry out. Substitute 28.5 for A and 94 for F and solve for v.

$$94 = 0.0494(28.5)v^2$$

$$66.7661 \approx v^2$$

$$8.17 \approx v$$

Check. Recheck the calculations.

State. The boat must go about 8.17 mph.

52. About 57.43 mph

53. Use the slope-intercept form, y = mx + b, where m is the slope and b is the y-intercept.

$$y = -\frac{2}{3}x - 5$$

54. $\frac{c - 2a}{3c + 4a}$

55. $f(x) = x^3 - 2x^2$

$f(3) = 3^3 - 2 \cdot 3^2 = 27 - 18 = 9$

56. $9x^2 - 12xy + 4y^2$

57.

58.

59. Let C represent the number of complaints, and let E represent the number of employees. Write C as a function of E.

$$C(E) = \frac{k}{E}$$

Now $C(5) = \frac{k}{5}$ and $C(10) = \frac{k}{10} = \frac{1}{2} \cdot \frac{k}{5}$, so $C(10) = \frac{1}{2} \cdot C(5)$, or expanding from 5 to 10 employees will result in half as many complaints. Also, $C(20) = \frac{k}{20}$ and $C(25) = \frac{k}{25} = \frac{4}{5} \cdot \frac{k}{20}$, so $C(25) = \frac{4}{5} \cdot C(20)$, or expanding from 20 to 25 employees will result in four-fifths as many complaints. Thus, the firm will reduce the number of complaints more by expanding from 5 to 10 employees. A graph of this function is shaped like the following:

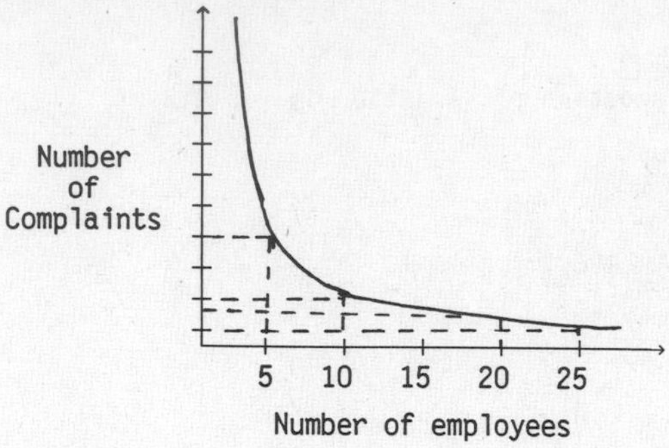

Projecting up and across, we see that the number of complaints decreases more as the number of employees changes from 5 to 10 than from 20 to 25. Thus, the graph also shows that the firm will reduce the number of complaints more by expanding from 5 to 10 employees.

60. $7.20

61. Write y as a function of x, and then substitute 0.5x for x.

$$y(x) = \frac{k}{x^3}$$

$$y(0.5x) = \frac{k}{(0.5x)^3} = \frac{k}{0.125x^3} = \frac{1}{0.125} \cdot \frac{k}{x^3}$$

$$= 8 \cdot y(x)$$

y is multiplied by 8.

62. a) $k \approx 0.001$; $N = \frac{0.001P_1P_2}{d^2}$

b) 1173 km

63. Familiarize. We write $T = km\ell^2f^2$. We know that T = 100 when m = 5, ℓ = 2, and f = 80.

Translate. Find k.

$$T = km\ell^2f^2$$

$$100 = k(5)(2)^2(80)^2$$

$$0.00078125 = k$$

$$T = 0.00078125m\ell^2f^2$$

Carry out. Substitute 72 for T, 5 for m, and 80 for f and solve for ℓ.

$$72 = 0.00078125(5)\ell^2(80)^2$$

$$2.88 = \ell^2$$

$$1.697 \approx \ell$$

Check. Recheck the calculations.

State. The string should be about 1.697 m long.

64. $d = \frac{28}{s}$; 70 yd

65. $Q = \frac{kp^2}{q^3}$

Q varies directly as the square of p and inversely as the cube of q.

66. W varies jointly as m_1 and M_1 and inversely as the square of d.

Exercise Set 8.7

1. $f(x) = x^2$

See Example 1 in the text.

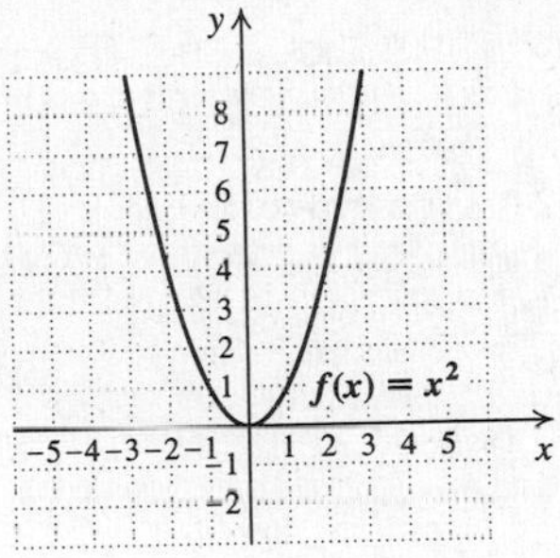

2.

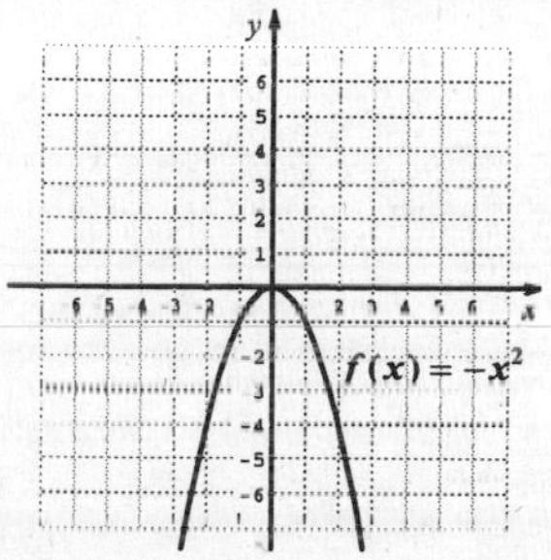

3. $f(x) = -4x^2$

We choose some numbers for x and compute f(x) for each one. Then we plot the ordered pairs (x,f(x)) and connect them with a smooth curve.

x	$f(x) = -4x^2$
0	0
1	-4
2	-16
-1	-4
-2	-16

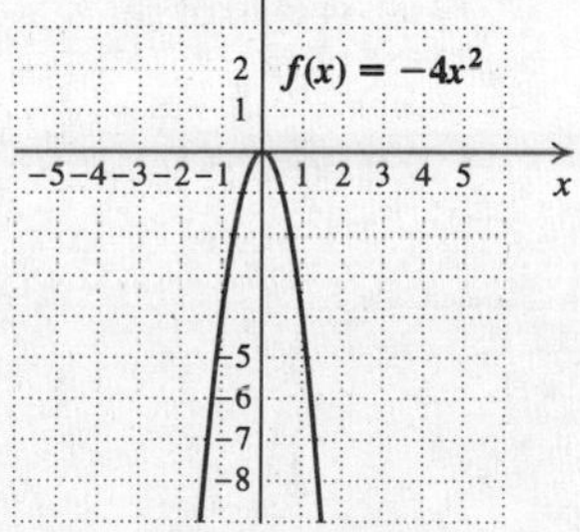

4.

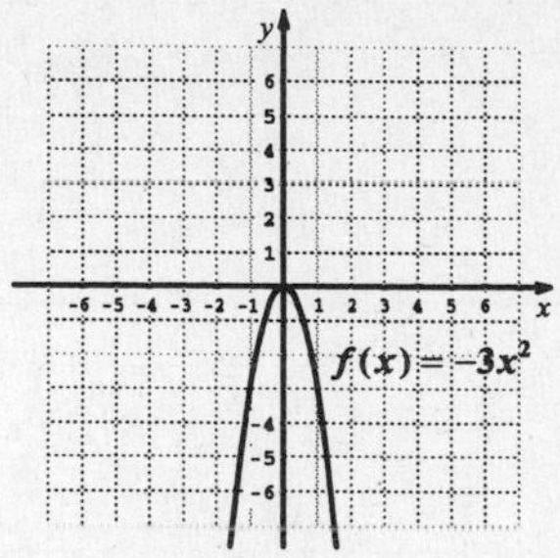

5. $g(x) = \frac{1}{4}x^2$

x	$g(x) = \frac{1}{4}x^2$
0	0
1	$\frac{1}{4}$
2	1
3	$\frac{9}{4}$
-1	$\frac{1}{4}$
-2	1
-3	$\frac{9}{4}$

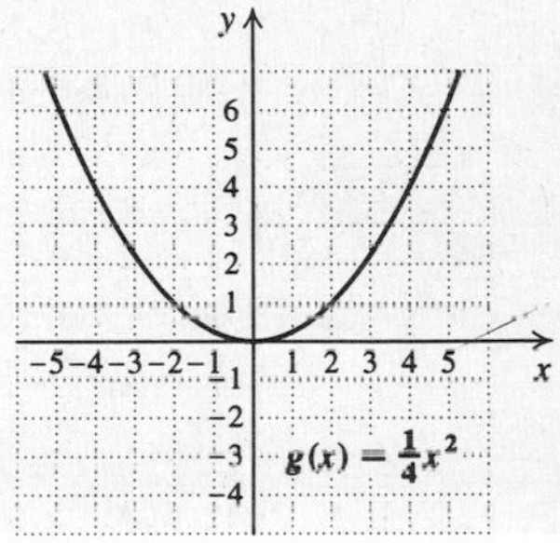

6.

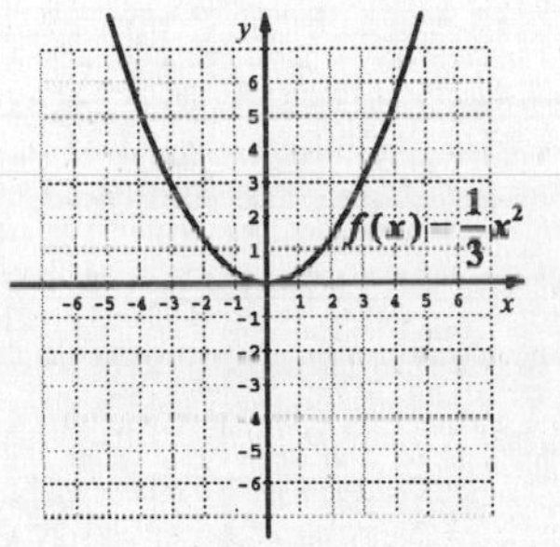

7. $h(x) = -\frac{1}{3}x^2$

x	$h(x) = -\frac{1}{3}x^2$
0	0
1	$-\frac{1}{3}$
2	$-\frac{4}{3}$
3	-3
-1	$-\frac{1}{3}$
-2	$-\frac{4}{3}$
-3	-3

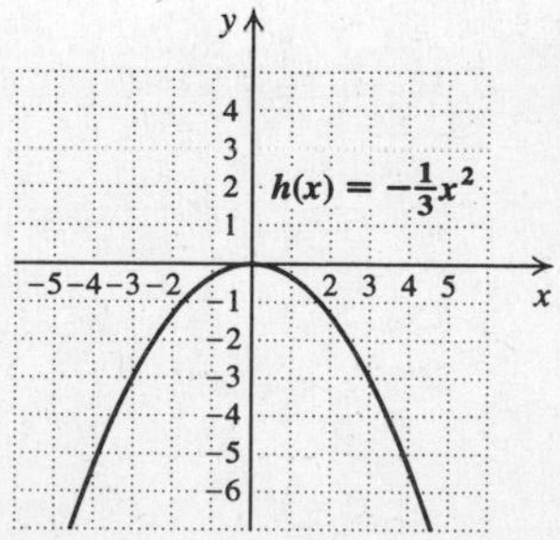

8.

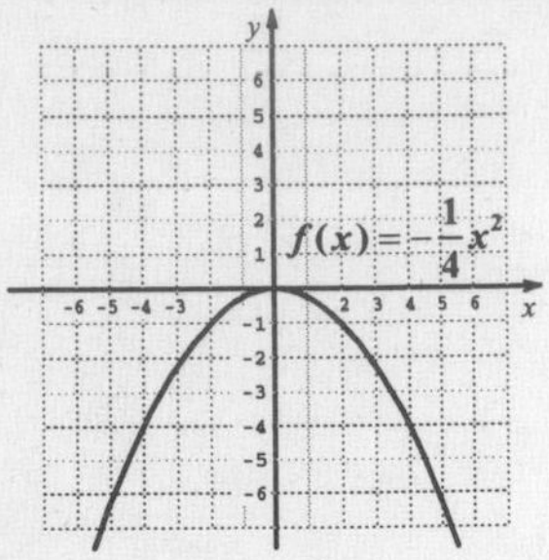

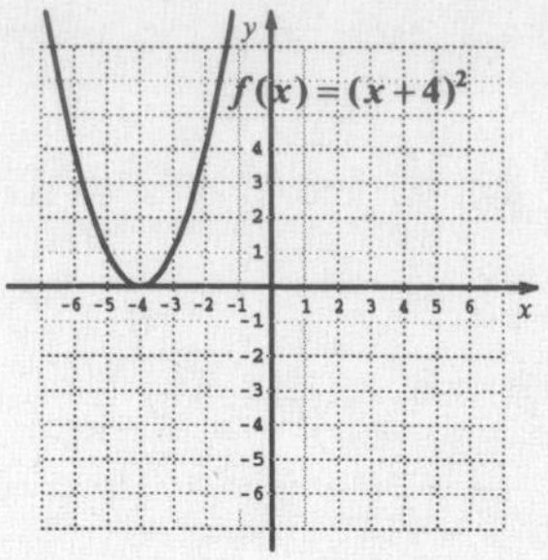

9. $f(x) = \frac{3}{2}x^2$

x	$f(x) = \frac{3}{2}x^2$
0	0
1	$\frac{3}{2}$
2	6
-1	$\frac{3}{2}$
-2	6

10.

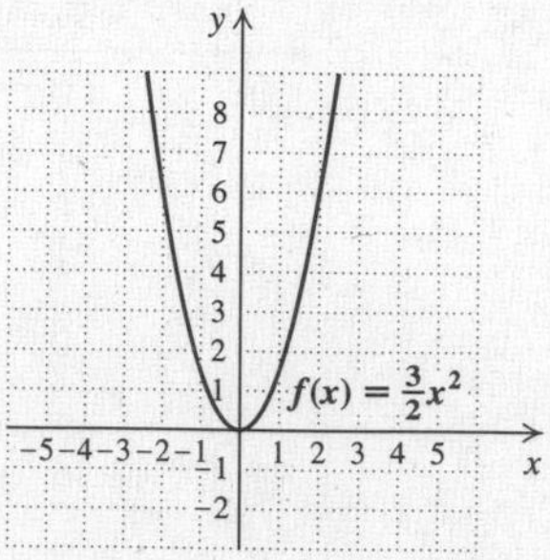

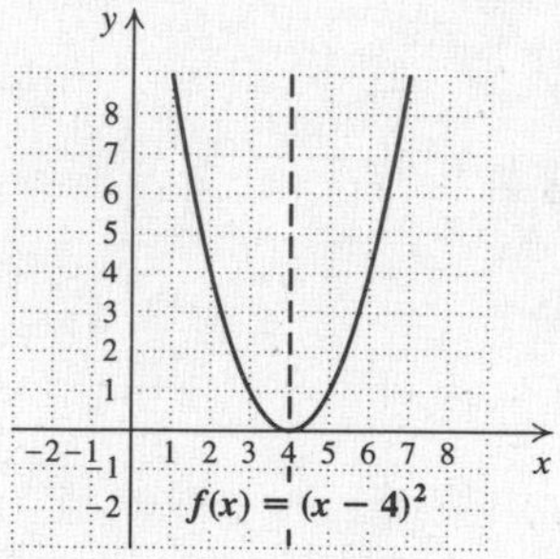

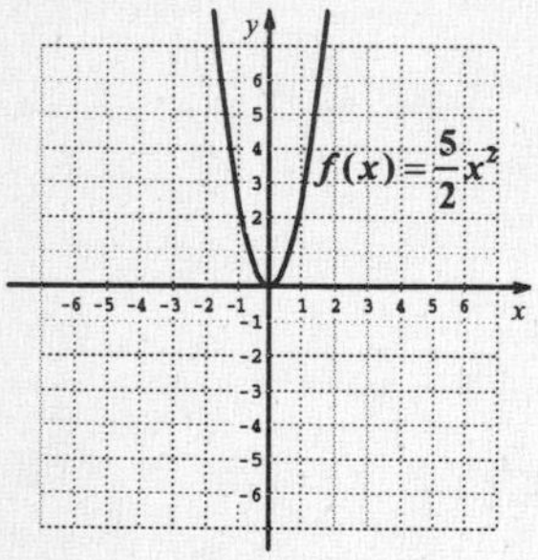

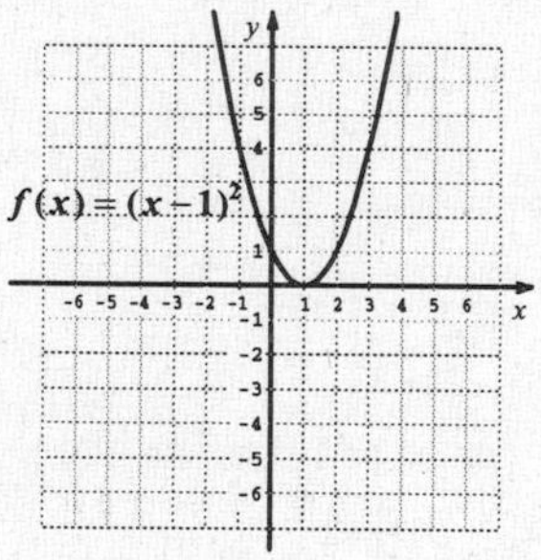

11. $g(x) = (x + 1)^2 = [x - (-1)]^2$

We know that the graph of $g(x) = (x + 1)^2$ looks like the graph of $f(x) = x^2$ (see Exercise 1) but moved to the left 1 unit.

Vertex: (-1,0), line of symmetry: x = -1

12. Vertex: (-4,0), line of symmetry: x = -4

13. $f(x) = (x - 4)^2$

The graph of $f(x) = (x - 4)^2$ looks like the graph of $f(x) = x^2$ (see Exercise 1) but moved to the right 4 units.

Vertex: (4,0), line of symmetry: x = 4

14. Vertex: (1,0), line of symmetry: x = 1

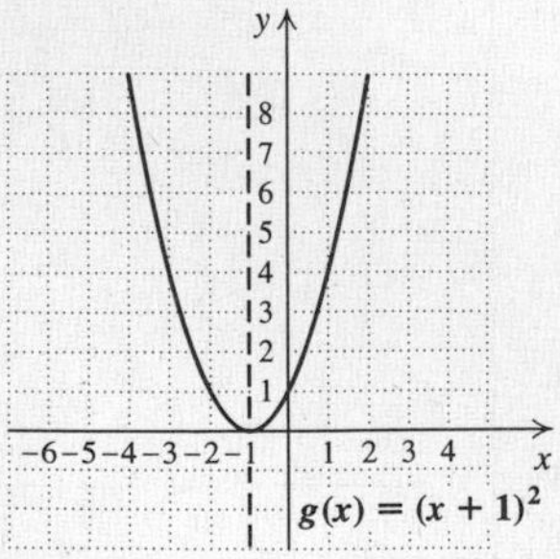

15. $h(x) = (x - 3)^2$

The graph of $h(x) = (x - 3)^2$ looks like the graph of $f(x) = x^2$ (see Exercise 1) but moved to the right 3 units.

Vertex: (3,0), line of symmetry: x = 3

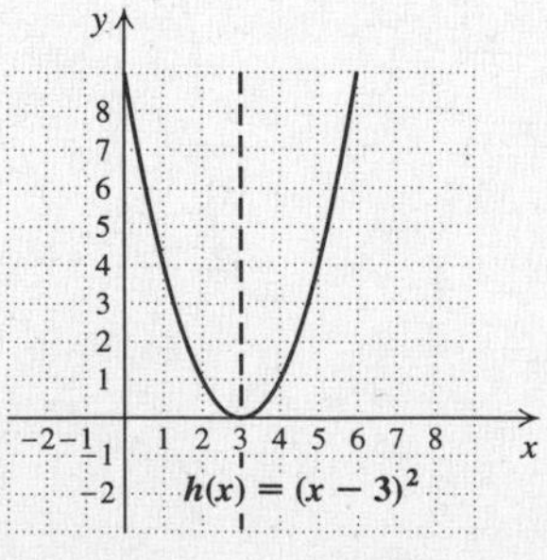

16. Vertex: (7,0), line of symmetry: x = 7

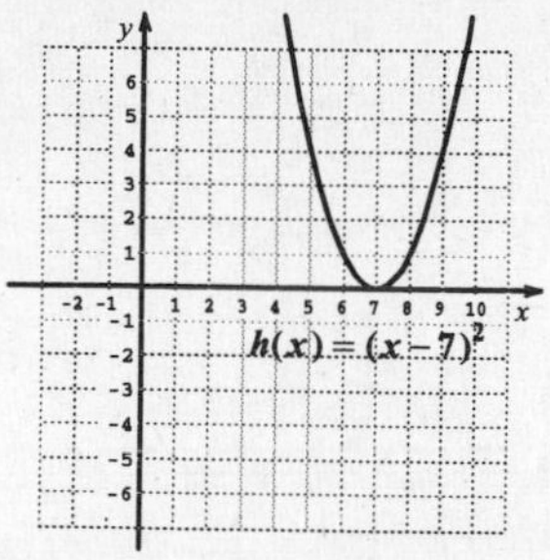

17. $f(x) = -(x + 4)^2 = -[x - (-4)]^2$

The graph of $f(x) = -(x + 4)^2$ looks like the graph of $f(x) = x^2$ (see Exercise 1) but moved to the left 4 units. It will also open downward because of the negative coefficient, -1.

Vertex: (-4,0), line of symmetry: x = -4

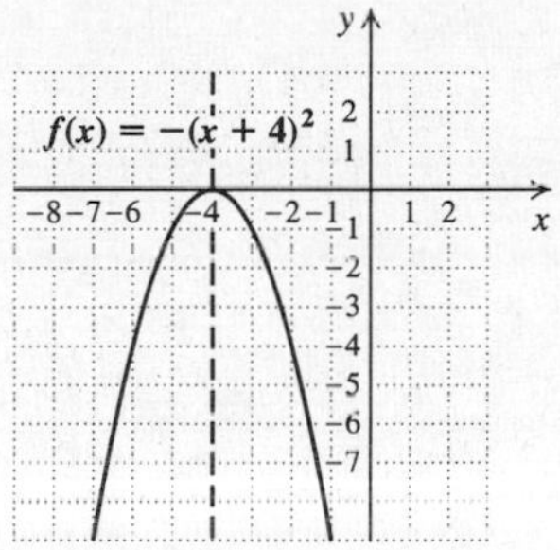

18. Vertex: (2,0), line of symmetry: x = 2

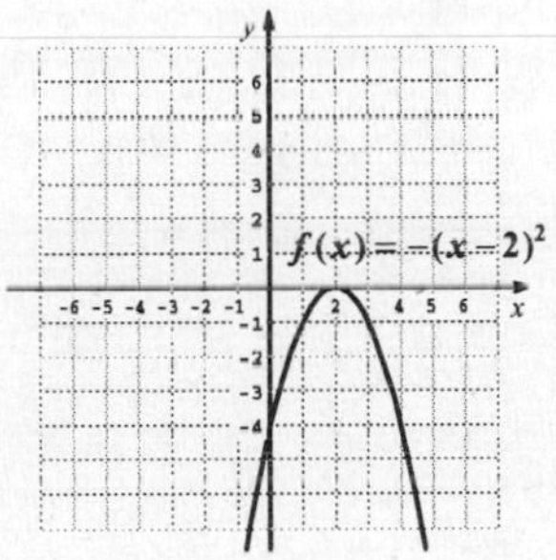

19. $g(x) = -(x - 1)^2$

The graph of $g(x) = -(x - 1)^2$ looks like the graph of $f(x) = x^2$ (see Exercise 1) but moved to the right 1 unit. It will also open downward because of the negative coefficient, -1.

Vertex: (1,0), line of symmetry: x = 1

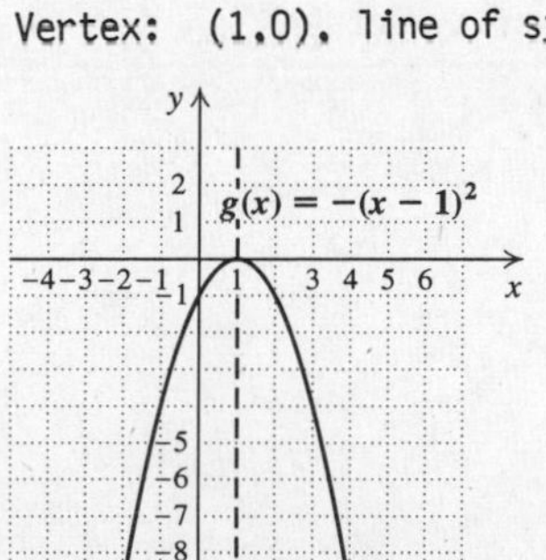

20. Vertex: (-5,0), line of symmetry: x = -5

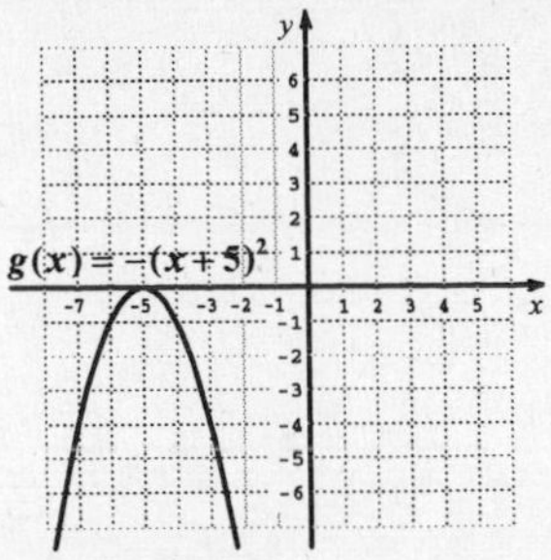

21. $f(x) = 2(x - 1)^2$

The graph of $f(x) = 2(x - 1)^2$ will look like the graph of $h(x) = 2x^2$ (see Example 1) but moved to the right 1 unit.

Vertex: (1,0), line of symmetry: x = 1

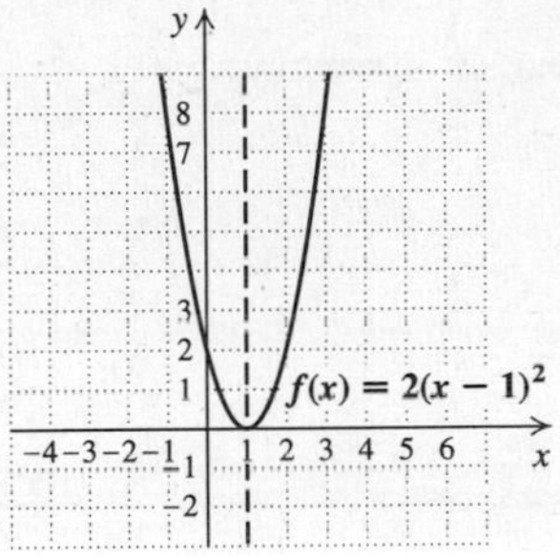

22. Vertex: (-4,0), line of symmetry: x = -4

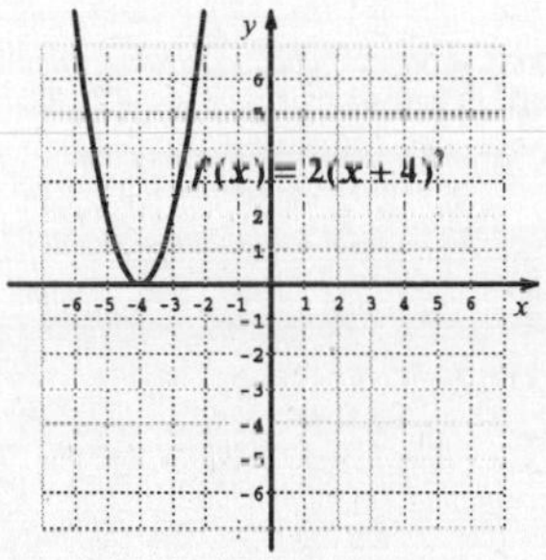

23. $h(x) = -\frac{1}{2}(x - 3)^2$

The graph of $h(x) = -\frac{1}{2}(x - 3)^2$ looks like the graph of $g(x) = \frac{1}{2}x^2$ (see Example 1) but moved to the right 3 units. It will also open downward because of the negative coefficient, $-\frac{1}{2}$.

Vertex: (3,0), line of symmetry: x = 3

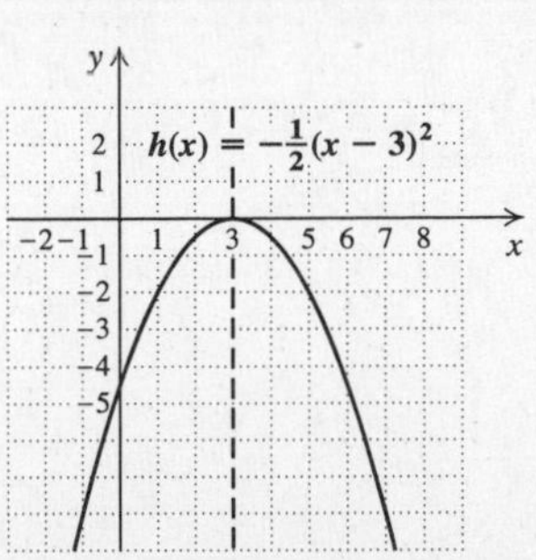

24. Vertex: (2,0), line of symmetry: $x = 2$

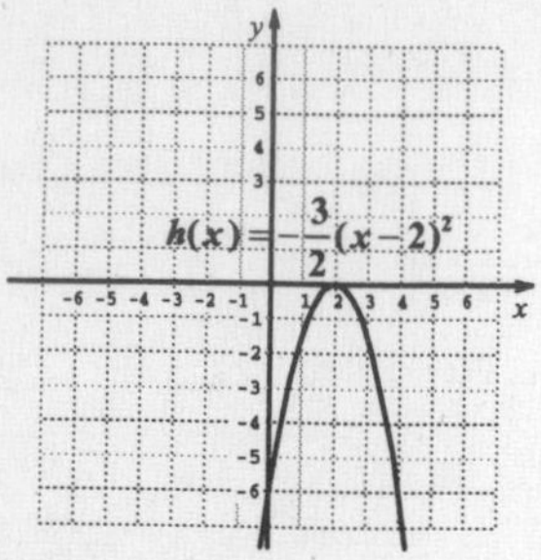

25. $f(x) = \frac{1}{2}(x + 1)^2 = \frac{1}{2}[x - (-1)]^2$

The graph of $f(x) = \frac{1}{2}(x + 1)^2$ looks like the graph of $g(x) = \frac{1}{2}x^2$ (see Example 1) but moved to the left 1 unit.

Vertex: (-1,0), line of symmetry: $x = -1$

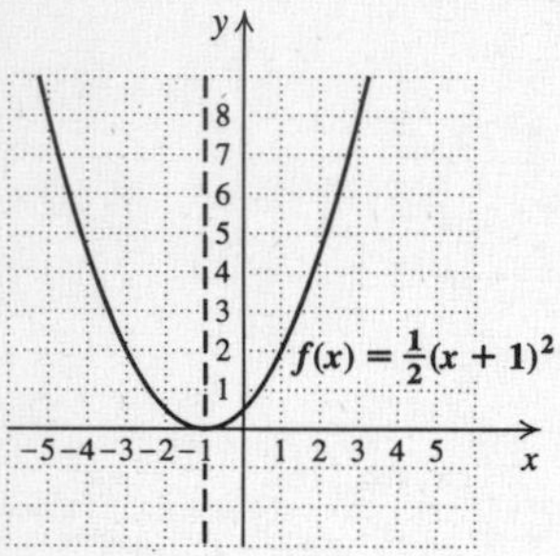

26. Vertex: (-2,0), line of symmetry: $x = -2$

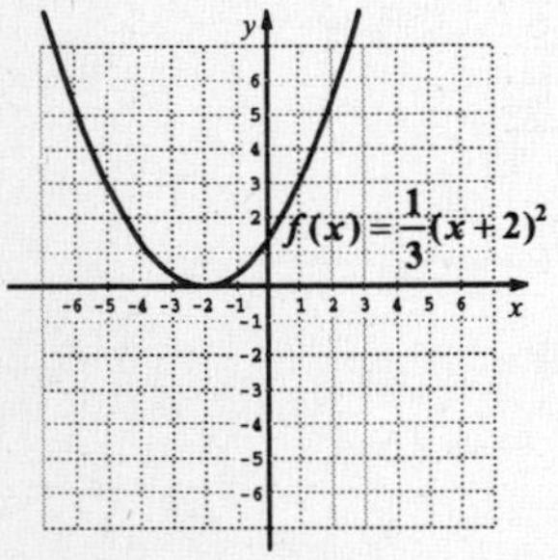

27. $g(x) = -3(x - 2)^2$

The graph of $g(x) = -3(x - 2)^2$ looks like the graph of $f(x) = -3x^2$ (see Exercise 4) but moved to the right 2 units.

Vertex: (2,0), line of symmetry: $x = 2$

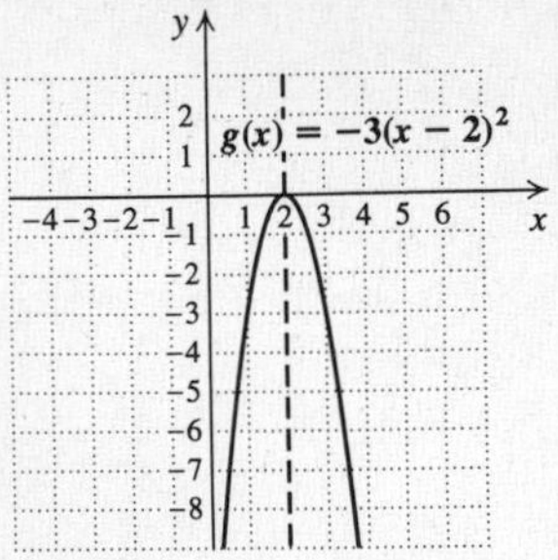

28. Vertex: (7,0), line of symmetry: $x = 7$

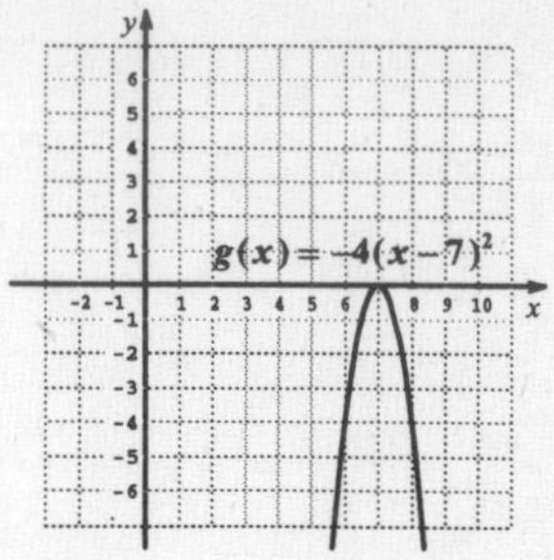

29. $f(x) = -2(x + 9)^2 = -2[x - (-9)]^2$

The graph of $f(x) = -2(x + 9)^2$ looks like the graph of $h(x) = 2x^2$ (see Example 1) but moved to the left 9 units. It will also open downward because of the negative coefficient, -2.

Vertex: (-9,0), line of symmetry: $x = -9$

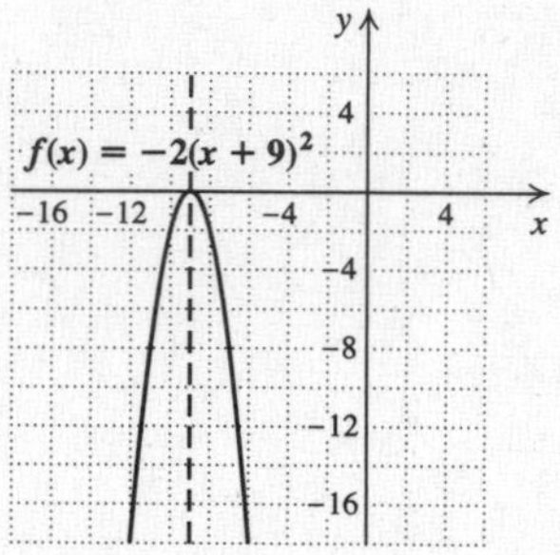

30. Vertex: (-7,0), line of symmetry: $x = -7$

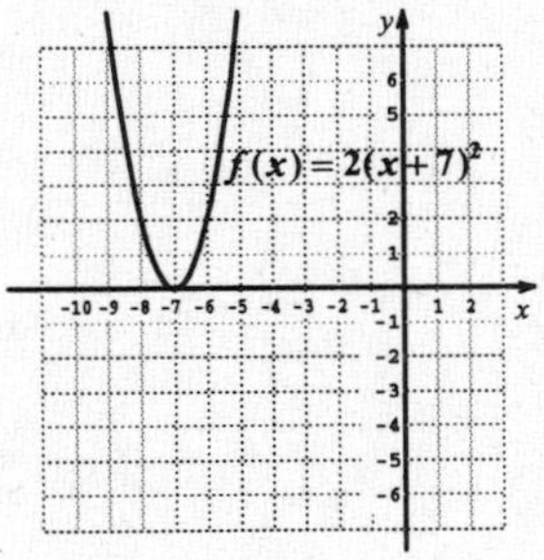

31. $h(x) = -3\left(x - \frac{1}{2}\right)^2$

The graph of $h(x) = -3\left(x - \frac{1}{2}\right)^2$ looks like the graph of $f(x) = -3x^2$ (see Exercise 4) but moved to the right $\frac{1}{2}$ unit.

Vertex: $\left(\frac{1}{2},0\right)$, line of symmetry: $x = \frac{1}{2}$

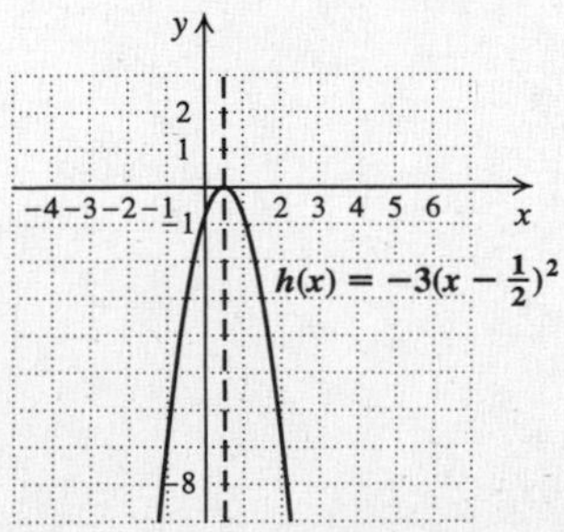

32. Vertex: $\left(-\frac{1}{2},0\right)$, line of symmetry: $x = -\frac{1}{2}$

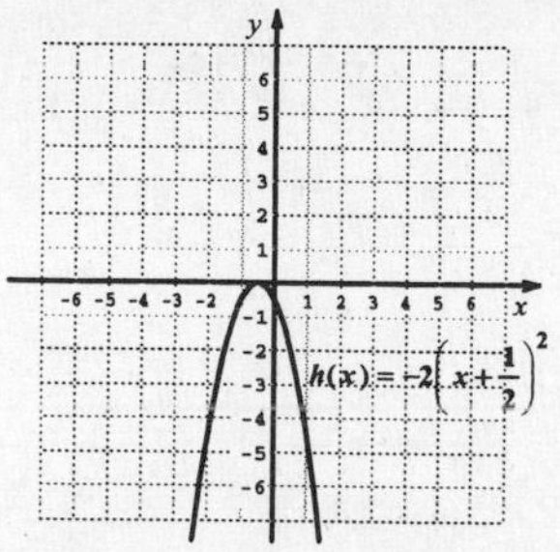

33. $f(x) = (x - 3)^2 + 1$

We know that the graph looks like the graph of $f(x) = x^2$ (see Example 1) but moved to the right 3 units and up 1 unit. The vertex is (3,1), and the line of symmetry is $x = 3$. Since the coefficient of $(x - 3)^2$ is positive $(1 > 0)$, there is a minimum function value, 1.

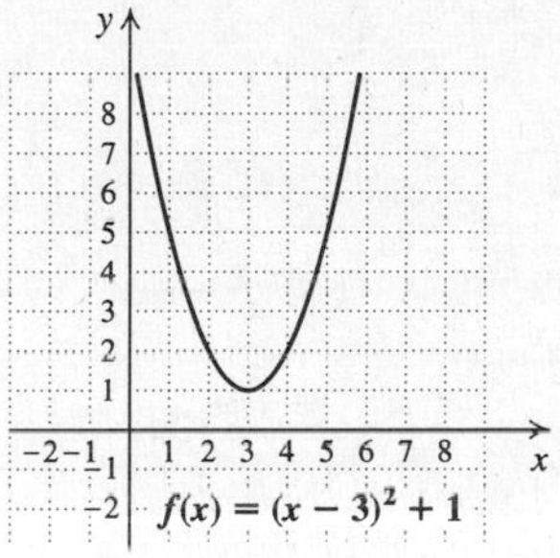

34. $f(x) = (x + 2)^2 - 3$

Vertex: (-2,-3), line of symmetry: $x = -2$

Minimum: -3

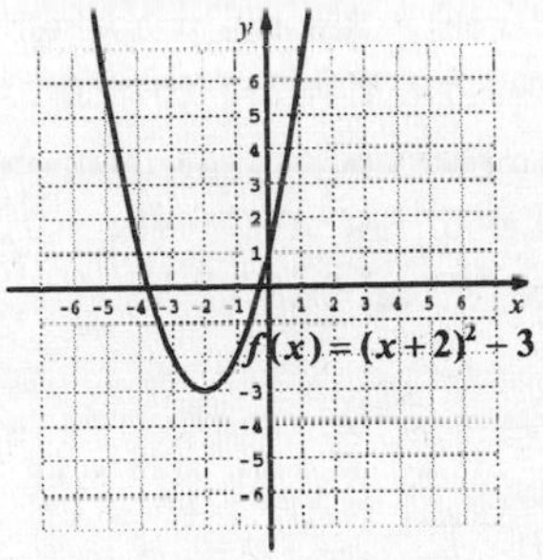

35. $f(x) = (x + 1)^2 - 2$

We know that the graph looks like the graph of $f(x) = x^2$ (see Example 1) but moved to the left 1 unit and down 2 units. The vertex is (-1,-2), and the line of symmetry is $x = -1$. Since the coefficient of $(x + 1)^2$ is positive $(1 > 0)$, there is a minimum function value, -2.

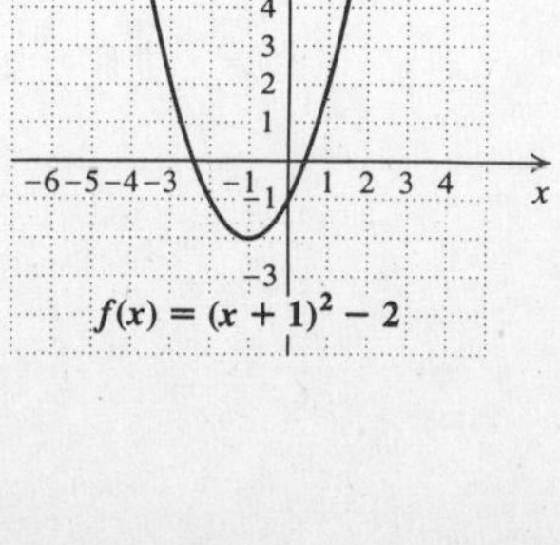

36. $f(x) = (x - 1)^2 + 2$

Vertex: (1,2), line of symmetry: $x = 1$

Minimum: 2

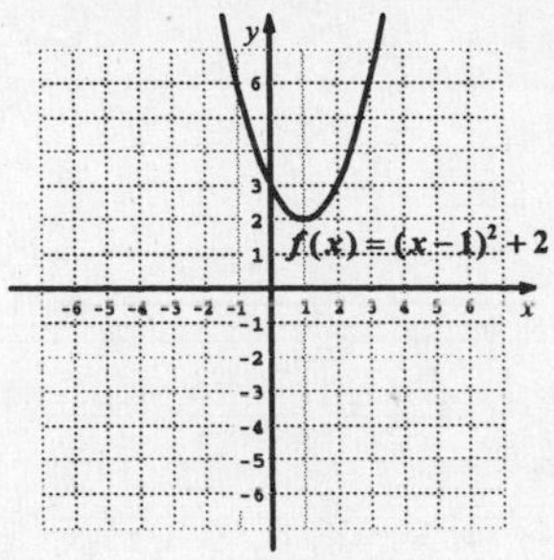

37. $g(x) = (x + 4)^2 + 1$

We know that the graph looks like the graph of $f(x) = x^2$ (see Example 1) but moved to the left 4 units and up 1 unit. The vertex is (-4,1), and the line of symmetry is $x = -4$. Since the coefficient of $(x + 4)^2$ is positive $(1 > 0)$, there is a minimum function value, 1.

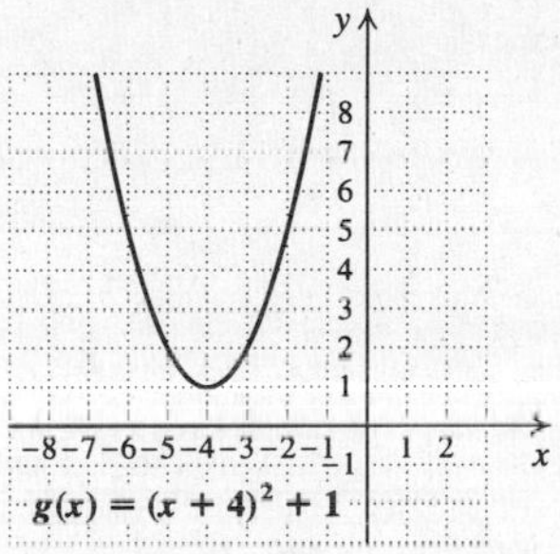

38. $g(x) = -(x - 2)^2 - 4$

Vertex: (2,-4), line of symmetry: $x = 2$

Maximum: -4

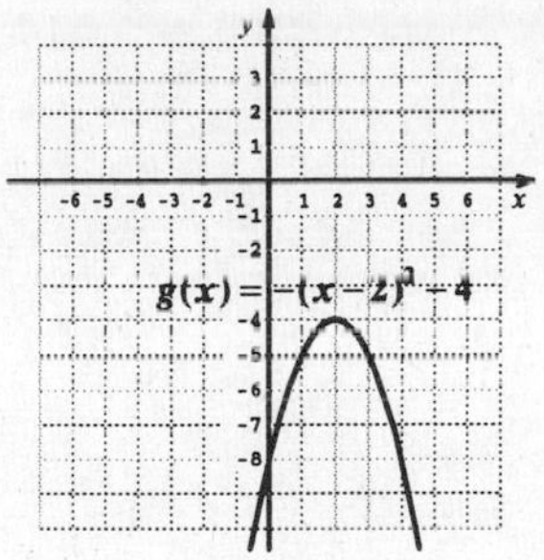

39. $f(x) = \frac{1}{2}(x - 5)^2 + 2$

We know that the graph looks like the graph of $g(x) = \frac{1}{2}x^2$ (see Example 1) but moved to the right 5 units and up 2 units. The vertex is (5,2), and the line of symmetry is $x = 5$. Since the coefficient of $(x - 5)^2$ is positive $\left(\frac{1}{2} > 0\right)$, there is a minimum function value, 2.

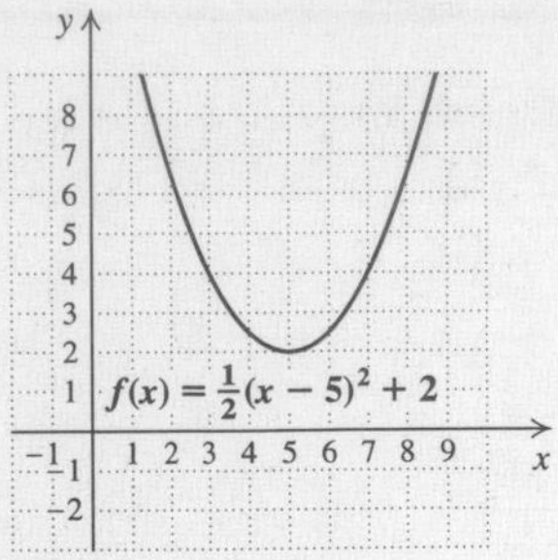

40. $f(x) = \frac{1}{2}(x + 1)^2 - 2$

Vertex: (-1,-2), line of symmetry: x = -1

Minimum: -2

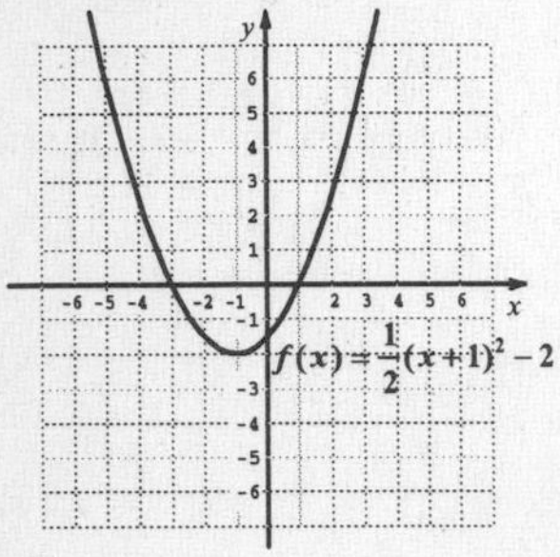

41. $h(x) = -2(x - 1)^2 - 3$

We know that the graph looks like the graph of $h(x) = 2x^2$ (see Example 1) but moved to the right 1 unit and down 3 units and turned upside down. The vertex is (1,-3), and the line of symmetry is x = 1. The maximum function value is -3.

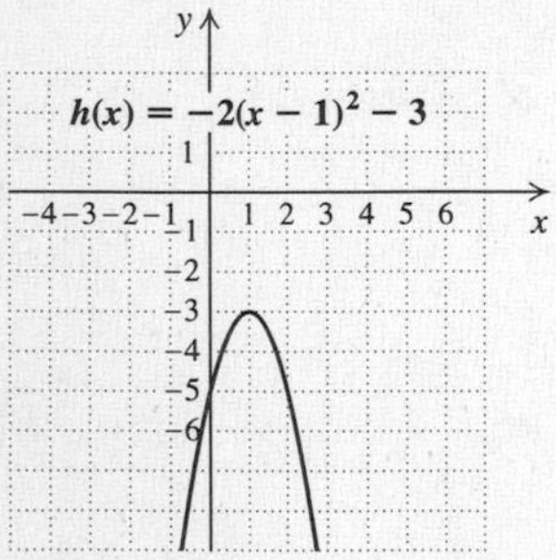

42. $h(x) = -2(x + 1)^2 + 4$

Vertex: (-1,4), line of symmetry: x = -1

Maximum: 4

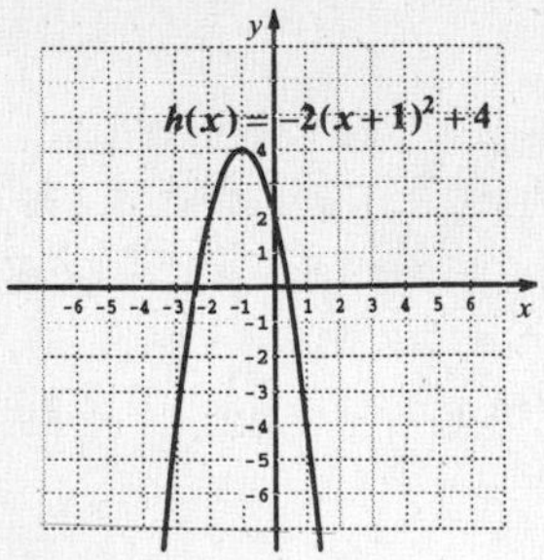

43. $f(x) = -3(x + 4)^2 + 1$

We know that the graph looks like the graph of $f(x) = -3x^2$ (see Exercise 4) but moved to the left 4 units and up 1 unit. The vertex is (-4,1), the line of symmetry is x = -4, and the maximum function value is 1.

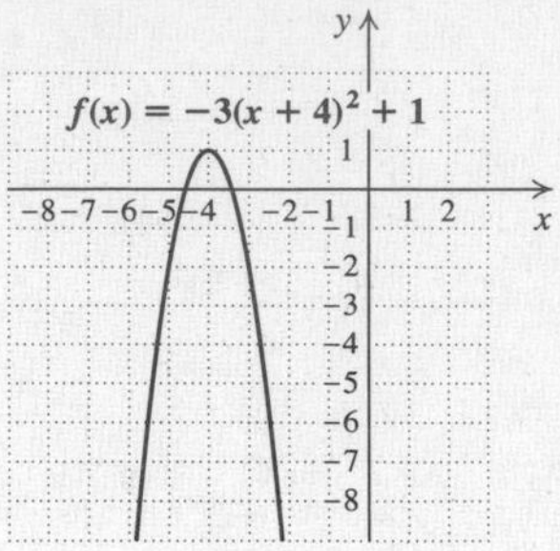

44. $f(x) = -2(x - 5)^2 - 3$

Vertex: (5,-3), line of symmetry: x = 5

Maximum: -3

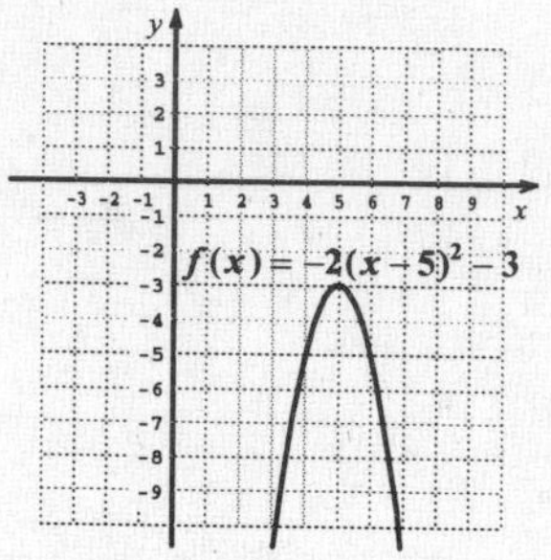

45. $g(x) = -\frac{3}{2}(x - 1)^2 + 2$

We know that the graph looks like the graph of $f(x) = \frac{3}{2}x^2$ (see Exercise 9) but moved to the right 1 unit and up 2 units and turned upside down. The vertex is (1,2), the line of symmetry is x = 1, and the maximum function value is 2.

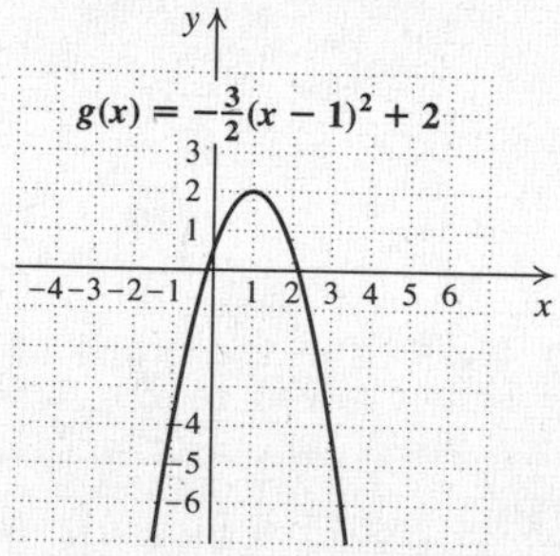

46. $g(x) = \frac{3}{2}(x + 2)^2 - 1$

Vertex: (-2,-1), line of symmetry: x = -2

Minimum: -1

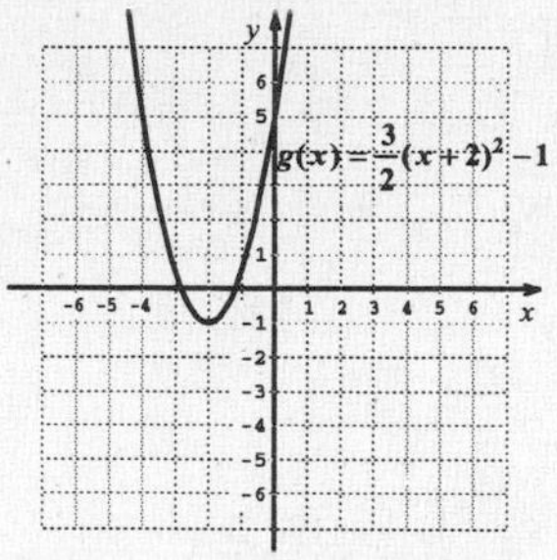

47. $f(x) = 8(x - 9)^2 + 5$

This function is of the form $f(x) = a(x - h)^2 + k$ with $a = 8$, $h = 9$, and $k = 5$. The vertex is (h,k), or $(9,5)$. The line of symmetry is $x = h$, or $x = 9$. Since $a > 0$, then k, or 5, is the minimum function value.

48. Vertex: $(-5,-8)$

Line of symmetry: $x = -5$

Minimum: -8

49. $h(x) = -\frac{2}{7}(x + 6)^2 + 11$

This function is of the form $f(x) = a(x - h)^2 + k$ with $a = -\frac{2}{7}$, $h = -6$, and $k = 11$. The vertex is (h,k), or $(-6,11)$. The line of symmetry is $x = h$, or $x = -6$. Since $a < 0$, then k, or 11, is the maximum function value.

50. Vertex: $(7,-9)$

Line of symmetry: $x = 7$

Maximum: -9

51. $f(x) = 5\left[x + \frac{1}{4}\right]^2 - 13$

This function is of the form $f(x) = a(x - h)^2 + k$ with $a = 5$, $h = -\frac{1}{4}$, and $k = -13$. The vertex is (h,k), or $\left[-\frac{1}{4},-13\right]$. The line of symmetry is $x = h$, or $x = -\frac{1}{4}$. Since $a > 0$, then k, or -13, is the minimum function value.

52. Vertex: $\left[\frac{1}{4},19\right]$

Line of symmetry: $x = \frac{1}{4}$

Minimum: 19

53. $f(x) = -7(x - 10)^2 - 20$

This function is of the form $f(x) = a(x - h)^2 + k$ with $a = -7$, $h = 10$, and $k = -20$. The vertex is (h,k), or $(10,-20)$. The line of symmetry is $x = h$, or $x = 10$. Since $a < 0$, then k, or -20, is the maximum function value.

54. Vertex: $(-12,23)$

Line of symmetry: $x = -12$

Maximum: 23

55. $f(x) = \sqrt{2}(x + 4.58)^2 + 65\pi$

This function is of the form $f(x) = a(x - h)^2 + k$ with $a = \sqrt{2}$, $h = -4.58$, and $k = 65\pi$. The vertex is (h,k), or $(-4.58,65\pi)$. The line of symmetry is $x = h$, or $x = -4.58$. Since $a > 0$, then k, or 65π, is the minimum function value.

56. Vertex: $(38.2,-\sqrt{34})$

Line of symmetry: $x = 38.2$

Minimum: $-\sqrt{34}$

57. $500 = 4a + 2b + c,$ (1)

$300 = a + b + c,$ (2)

$0 = c$ (3)

Substitute 0 for c in (1) and (2).

$500 = 4a + 2b$ (4)

$300 = a + b$ (5)

Multiply (5) by -2 and add the resulting equation to (4).

$$\begin{aligned} 500 &= 4a + 2b \\ -600 &= -2a - 2b \\ \hline -100 &= 2a \quad \text{Adding} \\ -50 &= a \end{aligned}$$

Substitute -50 for a in (5).

$300 = -50 + b$

$350 = b$

The solution is $(-50,350,0)$.

58. $\frac{1}{6}$, 2

59.

60.

61. Since there is a maximum at $(0,4)$, the parabola will have the same shape as $f(x) = -2x^2$. It will be of the form $f(x) = -2(x - h)^2 + k$ with $h = 0$ and $k = 4$: $f(x) = -2x^2 + 4$

62. $f(x) = 2(x - 2)^2$

63. Since there is a minimum at $(6,0)$, the parabola will have the same shape as $f(x) = 2x^2$. It will be of the form $f(x) = 2(x - h)^2 + k$ with $h = 6$ and $k = 0$: $f(x) = 2(x - 6)^2$

64. $f(x) = -2x^2 + 3$

65. Since there is a maximum at $(3,8)$, the parabola will have the same shape as $f(x) = -2x^2$. It will be of the form $f(x) = -2(x - h)^2 + k$ with $h = 3$ and $k = 8$: $f(x) = -2(x - 3)^2 + 8$

66. $f(x) = 2(x + 2)^2 + 3$

67. Since there is a minimum at (-3,6), the parabola will have the same shape as $f(x) = 2x^2$. It will be of the form $f(x) = 2(x - h)^2 + k$ with $h = -3$ and $k = 6$: $f(x) = 2[x - (-3)]^2 + 6$, or $f(x) = 2(x + 3)^2 + 6$

68. $f(x) = 6(x - 4)^2$

69. Since the parabola has the same shape as $f(x) = -\frac{1}{2}(x - 2)^2 + 4$, it is of the form $g(x) = -\frac{1}{2}(x - h)^2 + k$. Since it has a maximum value at the same point as $f(x) = -2(x - 1)^2 - 6$, its vertex is (1,-6). That is, $h = 1$ and $k = -6$. The equation is $g(x) = -\frac{1}{2}(x - 1)^2 - 6$.

70.

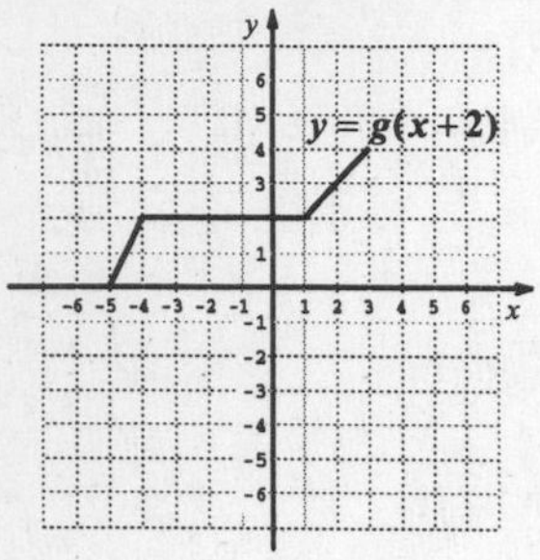

71. $y = g(x - 3)$

The graph looks just like the graph of $y = g(x)$ but moved 3 units to the right.

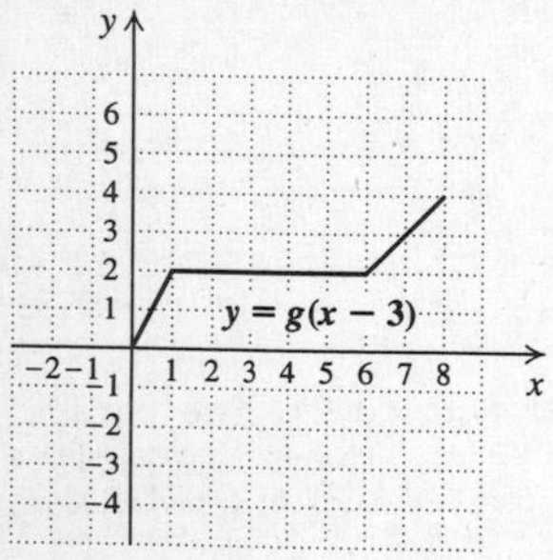

72.

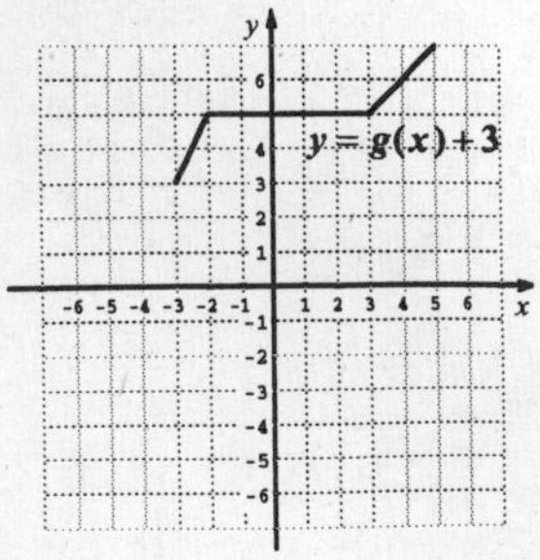

73. $y = g(x) + 4$

The graph looks just like the graph of $y = g(x)$ but moved up 4 units.

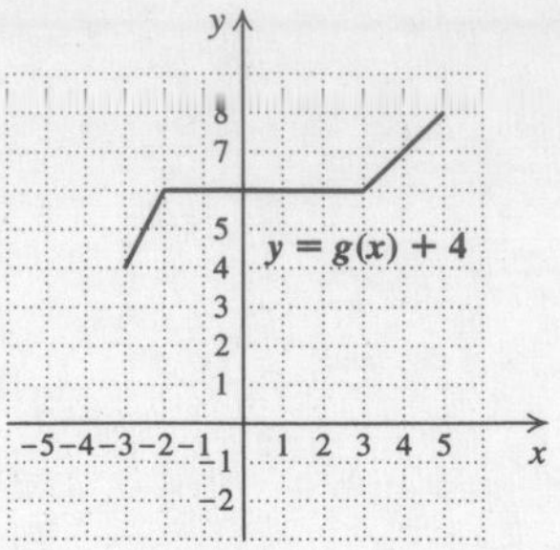

74.

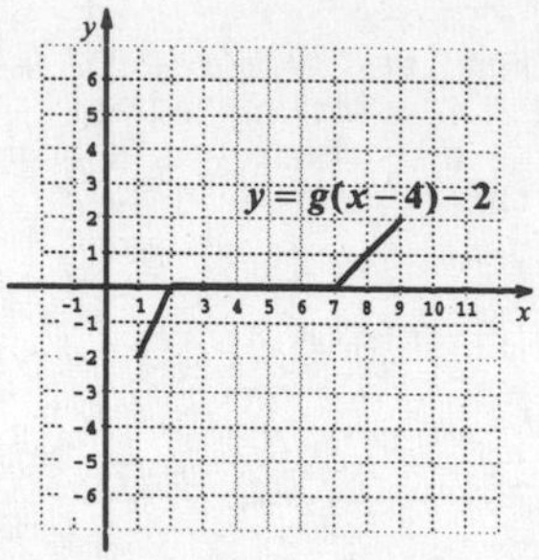

75. $y = g(x - 2) + 3$

The graph looks just like the graph of $y = g(x)$ but moved to the right 2 units and up 3 units.

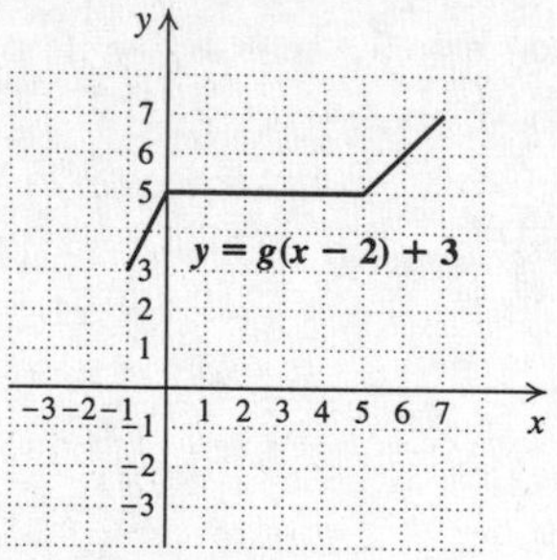

76. 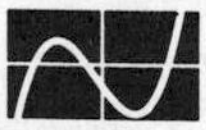

77. 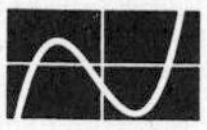

Exercise Set 8.8

1. $f(x) = x^2 - 2x - 3$

$= (x^2 - 2x + 1 - 1) - 3$ Adding 1 - 1

$= (x^2 - 2x + 1) - 1 - 3$ Regrouping

$= (x - 1)^2 - 4$

The vertex is (1,-4), the line of symmetry is $x = 1$, and the graph opens upward since the coefficient 1 is positive. We plot a few points as a check and draw the curve.

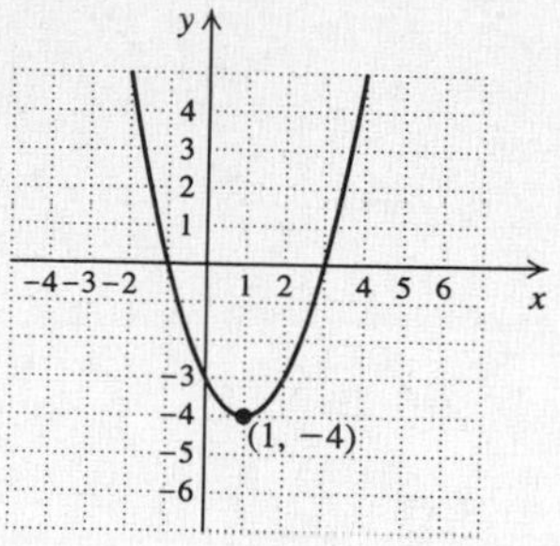

2. Vertex: (-1,-6), line of symmetry: x = -1

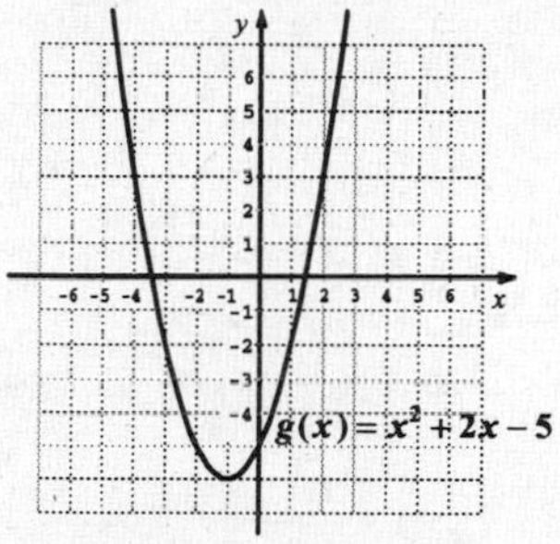

3. $g(x) = x^2 + 6x + 13$

$= (x^2 + 6x + 9 - 9) + 13$ Adding 9 - 9

$= (x^2 + 6x + 9) - 9 + 13$ Regrouping

$= (x + 3)^2 + 4$

The vertex is (-3,4), the line of symmetry is x = -3, and the graph opens upward since the coefficient 1 is positive. We plot a few points as a check and draw the curve.

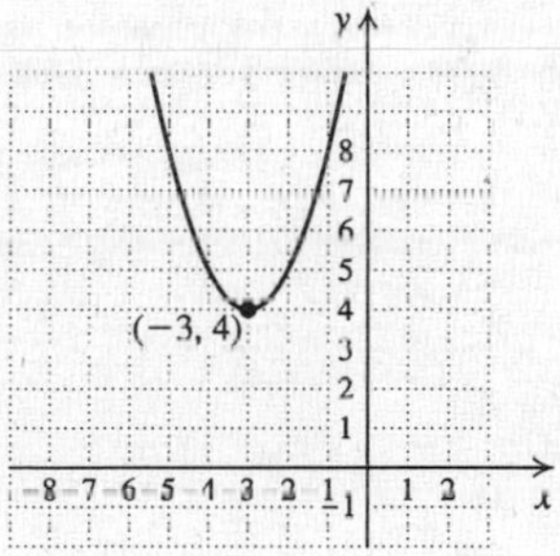

4. Vertex: (2,1), line of symmetry: x = 2

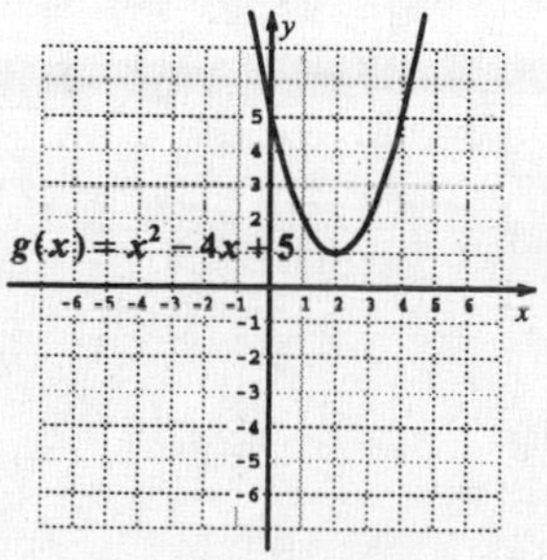

5. $f(x) = x^2 + 4x - 1$

$= (x^2 + 4x + 4 - 4) - 1$ Adding 4 - 4

$= (x^2 + 4x + 4) - 4 - 1$ Regrouping

$= (x + 2)^2 - 5$

The vertex is (-2,-5), the line of symmetry is x = -2, and the graph opens upward since the coefficient 1 is positive.

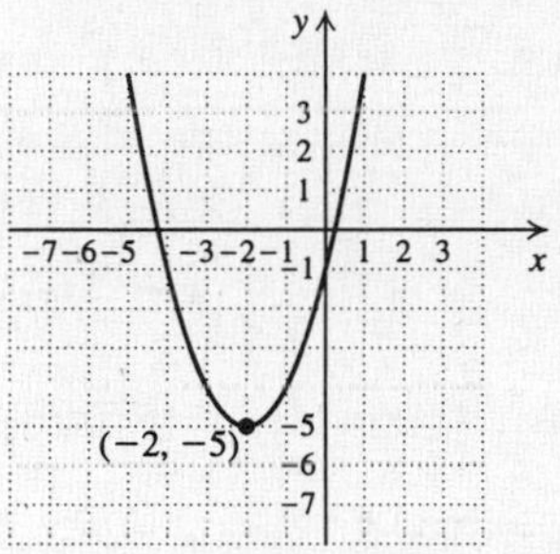

6. Vertex: (5,-4), line of symmetry: x = 5

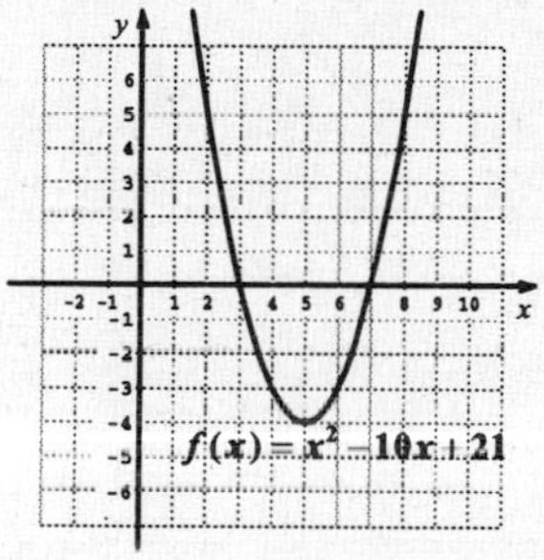

7. $h(x) = 2x^2 + 16x + 25$

$= 2(x^2 + 8x) + 25$ Factoring 2 from the first two terms

$= 2(x^2 + 8x + 16 - 16) + 25$ Adding 16 - 16 inside the parentheses

$= 2(x^2 + 8x + 16) + 2(-16) + 25$ Distributing to obtain a trinomial square

$= 2(x + 4)^2 - 7$

The vertex is (-4,-7), the line of symmetry is x = -4, and the graph opens upward since the coefficient 2 is positive.

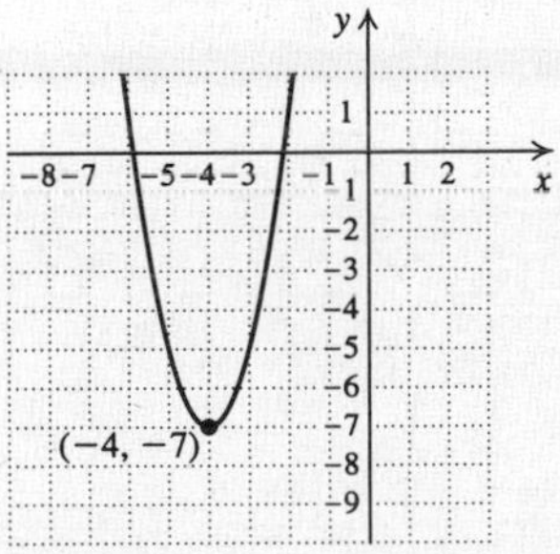

8. Vertex: (4,-9), line of symmetry: x = 4

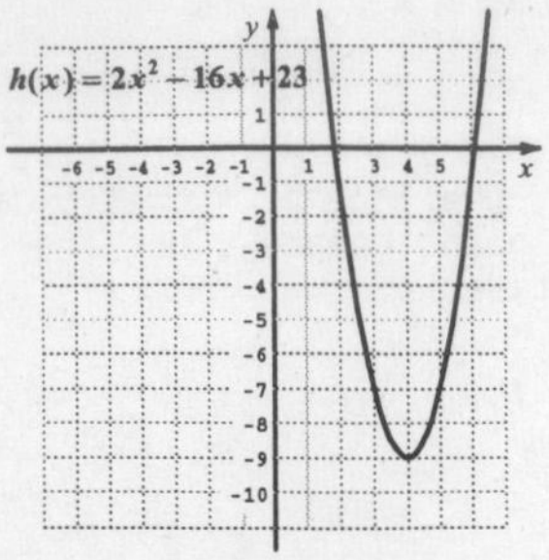

9. $f(x) = -x^2 + 4x + 6$

$= -(x^2 - 4x) + 6$ Factoring -1 from the first two terms

$= -(x^2 - 4x + 4 - 4) + 6$

Adding 4 - 4 inside the parentheses

$= -(x^2 - 4x + 4) - (-4) + 6$

$= -(x - 2)^2 + 10$

The vertex is (2,10), the line of symmetry is x = 2, and the graph opens downward since the coefficient -1 is negative.

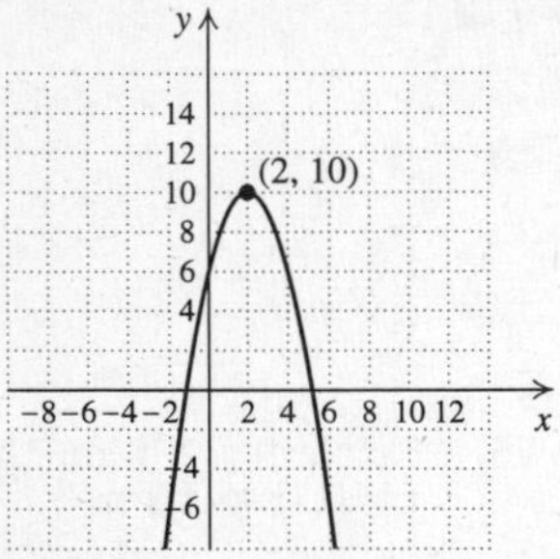

10. Vertex: (-2,7), line of symmetry: x = -2

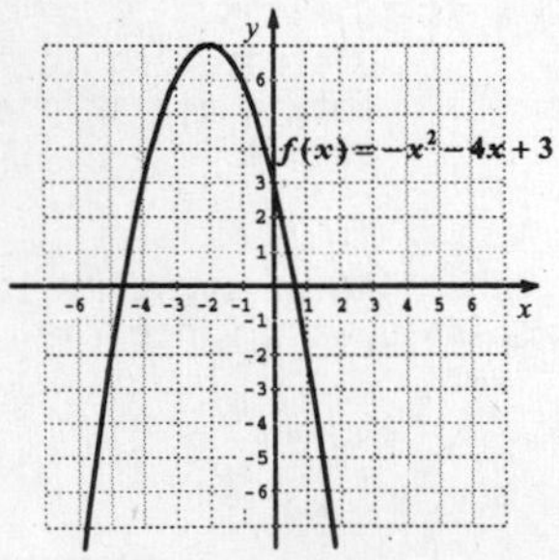

11. $g(x) = x^2 + 3x - 10$

$= \left[x^2 + 3x + \frac{9}{4} - \frac{9}{4}\right] - 10$

$= \left[x^2 + 3x + \frac{9}{4}\right] - \frac{9}{4} - 10$

$= \left[x + \frac{3}{2}\right]^2 - \frac{49}{4}$

The vertex is $\left[-\frac{3}{2}, -\frac{49}{4}\right]$, the line of symmetry is $x = -\frac{3}{2}$, and the graph opens upward since the coefficient 1 is positive.

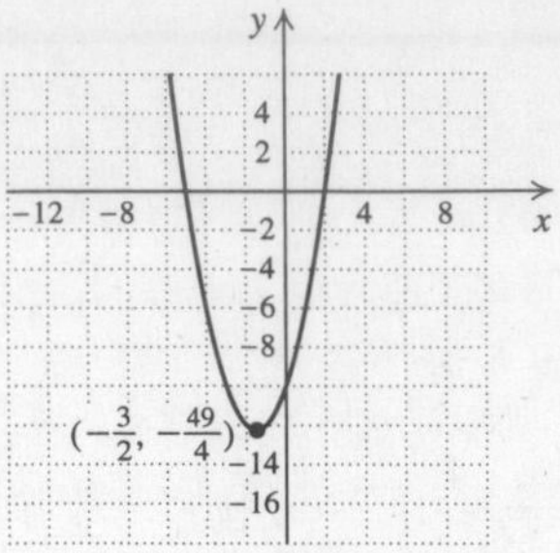

12. Vertex: $\left[-\frac{5}{2}, -\frac{9}{4}\right]$, line of symmetry: $x = -\frac{5}{2}$

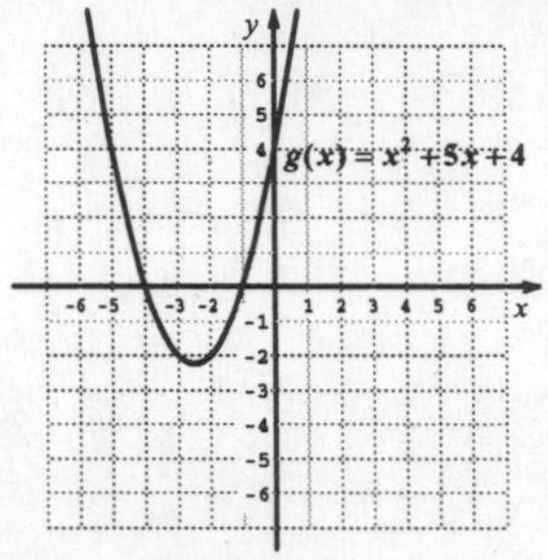

13. $f(x) = 3x^2 - 24x + 50$

$= 3(x^2 - 8x) + 50$ Factoring

$= 3(x^2 - 8x + 16 - 16) + 50$

Adding 16 - 16 inside the parentheses

$= 3(x^2 - 8x + 16) - 3 \cdot 16 + 50$

$= 3(x - 4)^2 + 2$

The vertex is (4,2), the line of symmetry is x = 4, and the graph opens upward since the coefficient 3 is positive.

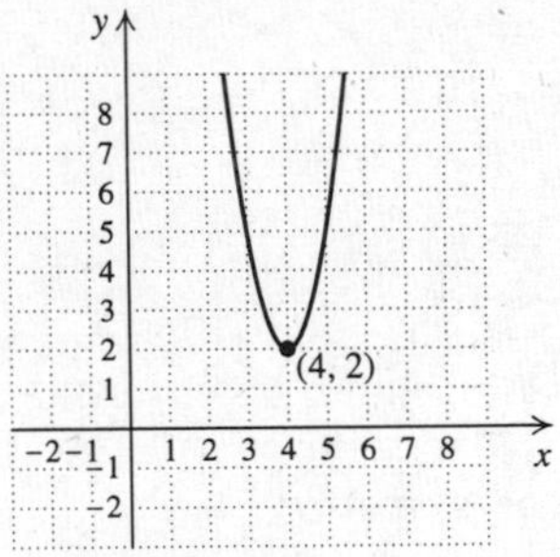

14. Vertex: (-1,-7), line of symmetry: x = -1

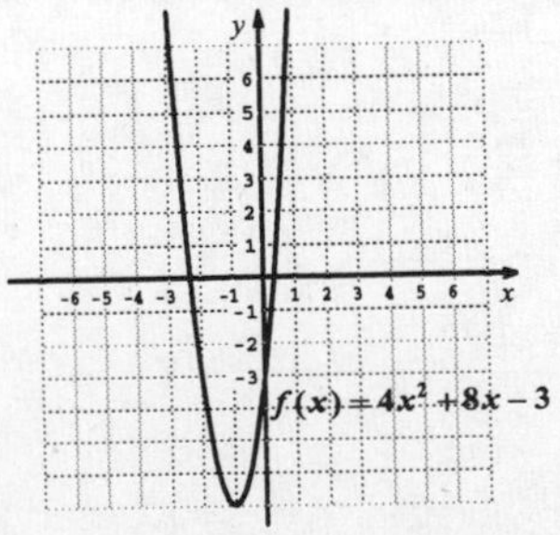

15. $h(x) = x^2 - 9x$

$= \left[x^2 - 9x + \frac{81}{4}\right] - \frac{81}{4}$

$= \left[x - \frac{9}{2}\right]^2 - \frac{81}{4}$

The vertex is $\left[\frac{9}{2}, -\frac{81}{4}\right]$, the line of symmetry is $x = \frac{9}{2}$, and the graph opens upward since the coefficient 1 is positive.

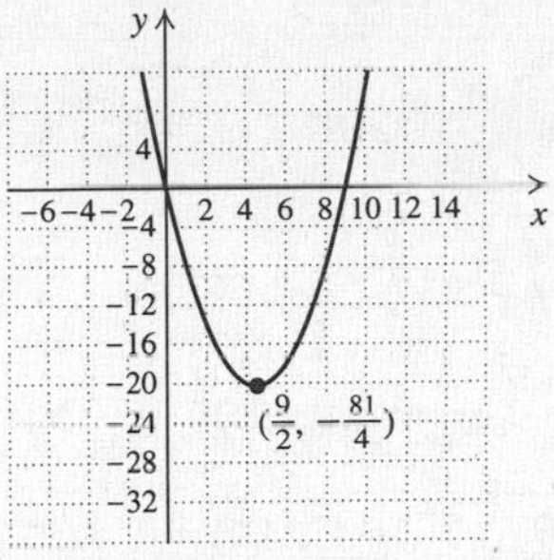

16. Vertex: $\left[-\frac{1}{2}, -\frac{1}{4}\right]$, line of symmetry: $x = -\frac{1}{2}$

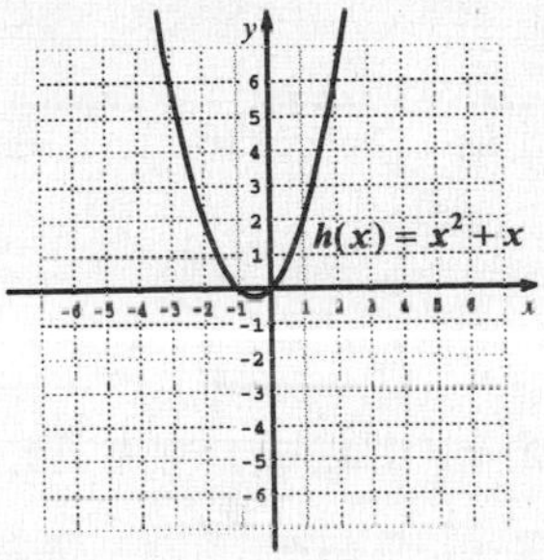

17. $f(x) = -2x^2 - 4x - 6$

$= -2(x^2 + 2x) - 6$ Factoring

$= -2(x^2 + 2x + 1 - 1) - 6$

Adding 1 - 1 inside the parentheses

$= -2(x^2 + 2x + 1) - 2(-1) - 6$

$= -2(x + 1)^2 - 4$

The vertex is (-1,-4), the line of symmetry is $x = -1$, and the graph opens downward since the coefficient -2 is negative.

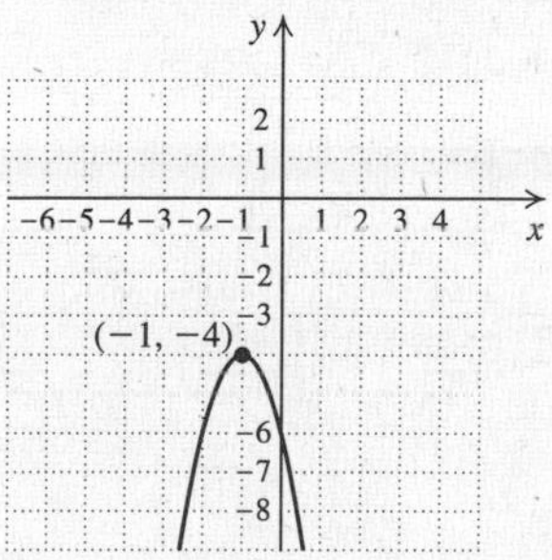

18. Vertex: (1,5), line of symmetry: $x = 1$

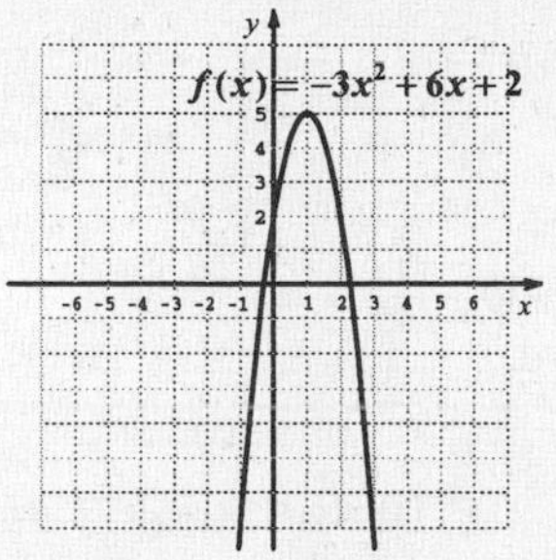

19. $g(x) = 2x^2 - 10x + 14$

$= 2(x^2 - 5x) + 14$ Factoring

$= 2\left[x^2 - 5x + \frac{25}{4} - \frac{25}{4}\right] + 14$

Adding $\frac{25}{4} - \frac{25}{4}$ inside the parentheses

$= 2\left[x^2 - 5x + \frac{25}{4}\right] + 2\left[-\frac{25}{4}\right] + 14$

$= 2\left[x - \frac{5}{2}\right]^2 + \frac{3}{2}$

The vertex is $\left[\frac{5}{2}, \frac{3}{2}\right]$, the line of symmetry is $x = \frac{5}{2}$, and the graph opens upward since the coefficient 2 is positive.

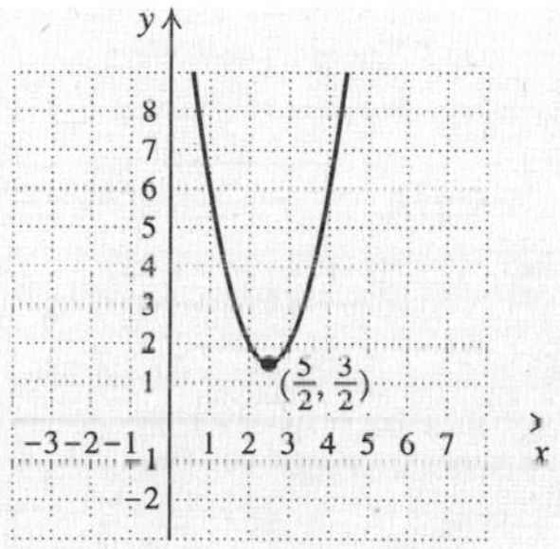

20. Vertex: $\left[-\frac{3}{2}, \frac{7}{2}\right]$, line of symmetry: $x = -\frac{3}{2}$

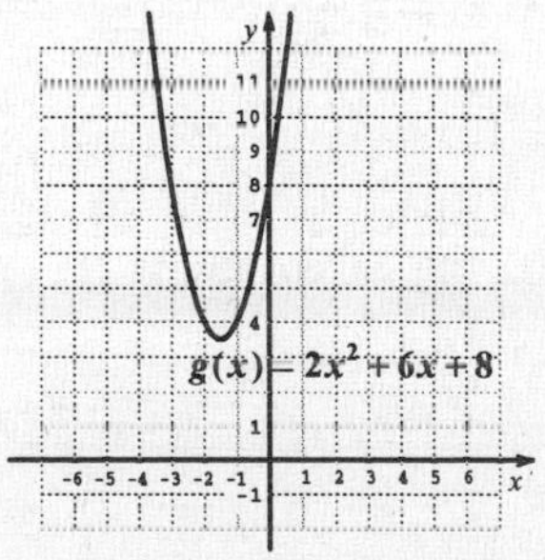

21. $f(x) = -3x^2 - 3x + 1$

$= -3(x^2 + x) + 1$ Factoring

$= -3\left(x^2 + x + \frac{1}{4} - \frac{1}{4}\right) + 1$

Adding $\frac{1}{4} - \frac{1}{4}$ inside the parentheses

$= -3\left(x^2 + x + \frac{1}{4}\right) - 3\left(-\frac{1}{4}\right) + 1$

$= -3\left(x + \frac{1}{2}\right)^2 + \frac{7}{4}$

The vertex is $\left(-\frac{1}{2},\frac{7}{4}\right)$, the line of symmetry is $x = -\frac{1}{2}$, and the graph opens downward since the coeffient -3 is negative.

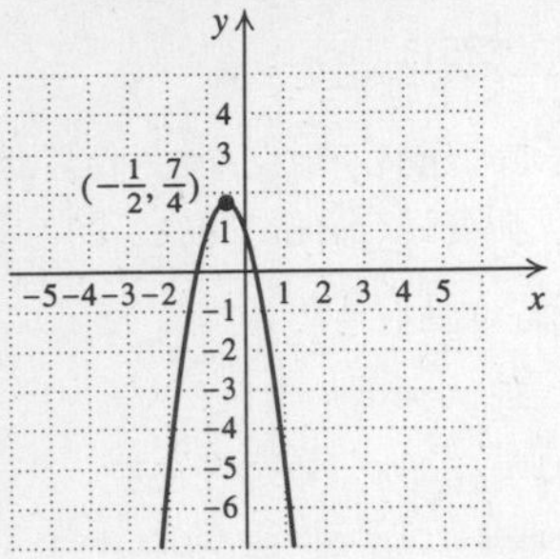

22. Vertex: $\left(\frac{1}{2},\frac{3}{2}\right)$, line of symmetry: $x = \frac{1}{2}$

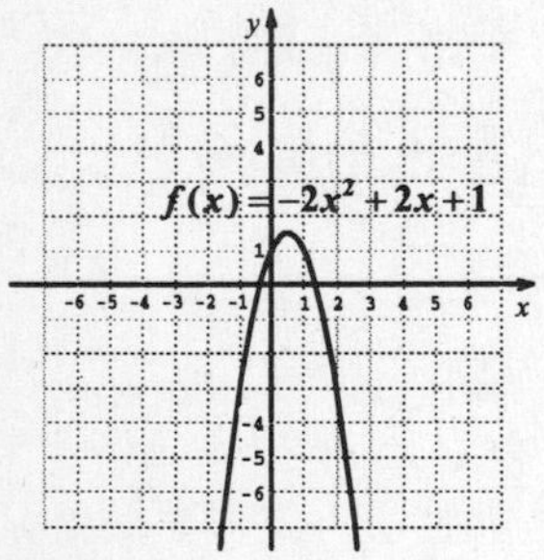

23. $h(x) = \frac{1}{2}x^2 + 4x + \frac{19}{3}$

$= \frac{1}{2}(x^2 + 8x) + \frac{19}{3}$ Factoring

$= \frac{1}{2}(x^2 + 8x + 16 - 16) + \frac{19}{3}$

Adding 16 - 16 inside the parentheses

$= \frac{1}{2}(x^2 + 8x + 16) + \frac{1}{2}(-16) + \frac{19}{3}$

$= \frac{1}{2}(x + 4)^2 - \frac{5}{3}$

The vertex is $\left(-4,-\frac{5}{3}\right)$, the line of symmetry is $x = -4$, and the graph opens upward since the coefficient $\frac{1}{2}$ is positive.

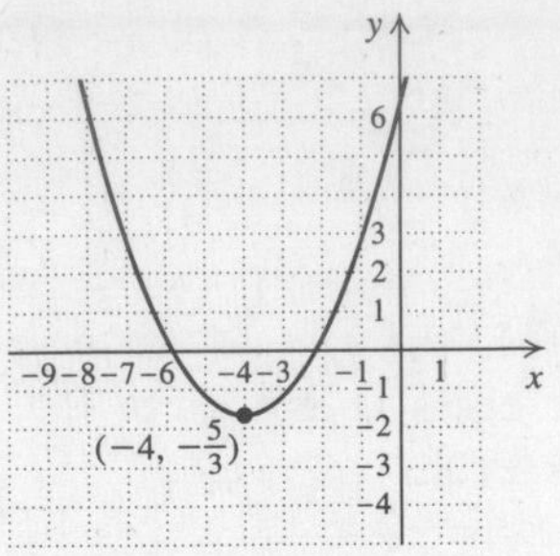

24. Vertex: $\left(3,-\frac{5}{2}\right)$, line of symmetry: $x = 3$

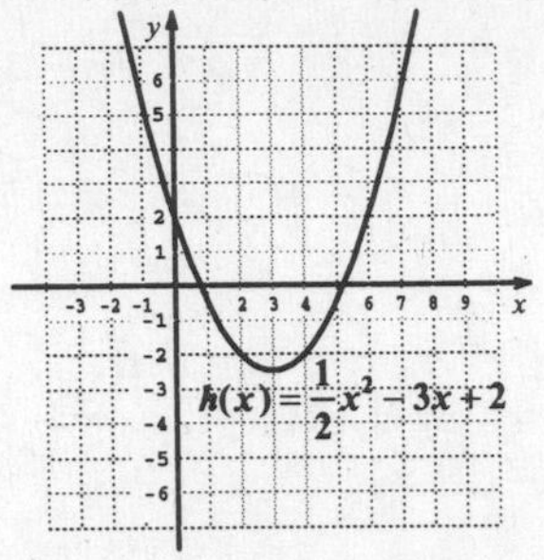

25. Solve $0 = x^2 - 4x + 1$. Use the quadratic formula.

$x = \frac{-(-4) \pm \sqrt{(-4)^2 - 4\cdot1\cdot1}}{2\cdot1}$

$x = \frac{4 \pm \sqrt{12}}{2} = \frac{4 \pm 2\sqrt{3}}{2} = 2 \pm \sqrt{3}$

The x-intercepts are $(2 + \sqrt{3},0)$ and $(2 - \sqrt{3},0)$.

26. None

27. Solve: $0 = -x^2 + 2x + 3$

$0 = x^2 - 2x - 3$ Multiplying by -1

$0 = (x - 3)(x + 1)$

$x = 3$ or $x = -1$

The x-intercepts are (3,0) and (-1,0).

28. $(1 + \sqrt{6},0)$, $(1 - \sqrt{6},0)$

29. Solve: $0 = x^2 - 3x - 4$

$0 = (x - 4)(x + 1)$

$x = 4$ or $x = -1$

The x-intercepts are (4,0) and (-1,0).

30. $(4 + \sqrt{11},0)$, $(4 - \sqrt{11},0)$

31. Solve: $0 = -x^2 + 3x - 2$

$0 = x^2 - 3x + 2$

$0 = (x - 2)(x - 1)$

$x = 2$ or $x = 1$

The x-intercepts are (2,0) and (1,0).

32. None

33. Solve $0 = 2x^2 + 4x - 1$. Using the quadratic formula we get $x = \frac{-2 \pm \sqrt{6}}{2}$. The x-intercepts are $\left(\frac{-2 + \sqrt{6}}{2}, 0\right)$ and $\left(\frac{-2 - \sqrt{6}}{2}, 0\right)$.

34. None

35. Solve: $0 = x^2 - x + 1$

$$x = \frac{-(-1) \pm \sqrt{(-1)^2 - 4\cdot1\cdot1}}{2\cdot1}$$

$$x = \frac{1 \pm \sqrt{-3}}{2} = \frac{1 \pm i\sqrt{3}}{2}$$

Since the equation has no real solutions, no x-intercepts exist.

36. $\left(-\frac{3}{2}, 0\right)$

37.

$$\sqrt{4x - 4} = \sqrt{x + 4} + 1$$

$$4x - 4 = x + 4 + 2\sqrt{x + 4} + 1$$ Squaring both sides

$$3x - 9 = 2\sqrt{x + 4}$$

$$9x^2 - 54x + 81 = 4(x + 4)$$ Squaring both sides again

$$9x^2 - 54x + 81 = 4x + 16$$

$$9x^2 - 58x + 65 = 0$$

$$(9x - 13)(x - 5) = 0$$

$$x = \frac{13}{9} \quad \text{or} \quad x = 5$$

Check: For $x = \frac{13}{9}$:

$\sqrt{4x - 4}$	=	$\sqrt{x + 4} + 1$	
$\sqrt{4\left(\frac{13}{9}\right) - 4}$	?	$\sqrt{\frac{13}{9} + 4} + 1$	
$\sqrt{\frac{16}{9}}$		$\sqrt{\frac{49}{9}} + 1$	
$\frac{4}{3}$		$\frac{7}{3} + 1$	
		$\frac{10}{3}$	FALSE

For $x = 5$:

$\sqrt{4x - 4}$	=	$\sqrt{x + 4} + 1$	
$\sqrt{4\cdot5 - 4}$	?	$\sqrt{5 + 4} + 1$	
$\sqrt{16}$		$\sqrt{9} + 1$	
4		3 + 1	
		4	TRUE

5 checks, but $\frac{13}{9}$ does not. The solution is 5.

38. 4

39. ◈

40. ◈

41. a) $f(x) = 2.31x^2 - 3.135x - 5.89$

$= 2.31(x^2 - 1.357142857x) - 5.89$

$= 2.31(x^2 - 1.357142857x + 0.460459183 - 0.460459183) - 5.89$

$= 2.31(x^2 - 1.357142857x + 0.460459183) + 2.31(-0.460459183) - 5.89$

$= 2.31(x - 0.678571428)^2 - 6.953660714$

Since the coefficient 2.31 is positive, the function has a minimum value. It is -6.953660714.

b) $2.31x^2 - 3.135x - 5.89 = 0$

$$x = \frac{-(-3.135) \pm \sqrt{(-3.135)^2 - 4(2.31)(-5.89)}}{2(2.31)}$$

$$x \approx \frac{3.135 \pm 8.015723611}{4.62}$$

$x \approx -1.056433682$ or $x \approx 2.413576539$

The x-intercepts are (-1.056433682, 0) and (2.413576539, 0).

42. a) Maximum: 7.01412766

b) (-0.400174191, 0), (0.821450786, 0)

43. $f(x) = x^2 - x - 6$

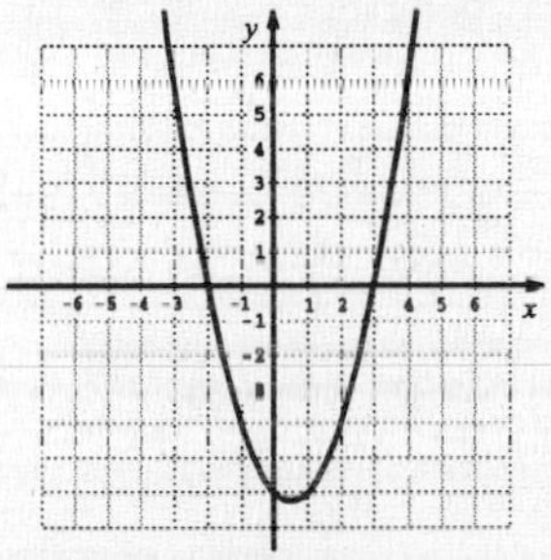

a) $x^2 - x - 6 = 2$

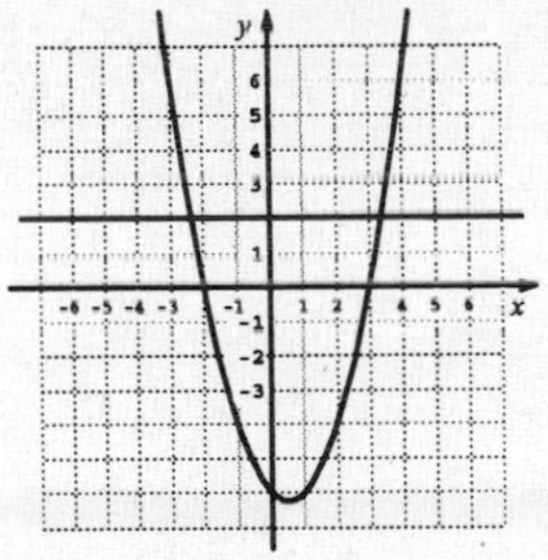

The solutions are approximately -2.4 and 3.4.

b) $x^2 - x - 6 = -3$

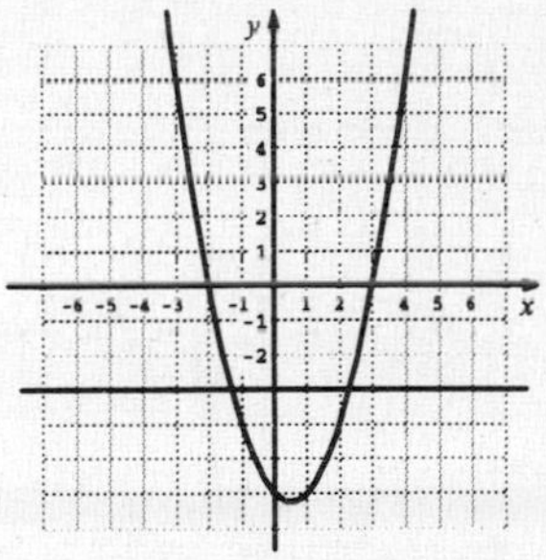

The solutions are approximately -1.3 and 2.3.

44. a) -3, 1

b) -4.5, 2.5

c) -5.5, 3.5

45. $f(x) = mx^2 - nx + p$

$$= m\left[x^2 - \frac{n}{m}x\right] + p$$

$$= m\left[x^2 - \frac{n}{m}x + \frac{n^2}{4m^2} - \frac{n^2}{4m^2}\right] + p$$

$$= m\left[x - \frac{n}{2m}\right]^2 - \frac{n^2}{4m} + p$$

$$= m\left[x - \frac{n}{2m}\right]^2 + \frac{-n^2 + 4mp}{4m}, \text{ or}$$

$$m\left[x - \frac{n}{2m}\right]^2 + \frac{4mp - n^2}{4m}$$

46. $f(x) = 3\left[x - \left(-\frac{m}{6}\right)\right]^2 + \frac{11m^2}{12}$

47. $f(x) = |x^2 - 1|$

We plot some points and draw the curve. Note that it will lie entirely on or above the x-axis since absolute value is never negative. For positive values of $x^2 - 1$ it will look like the graph of $f(x) = x^2$ but moved down 1 unit.

x	$f(x) = \lvert x^2 - 1\rvert$
-2	3
-1	0
0	1
1	0
2	3

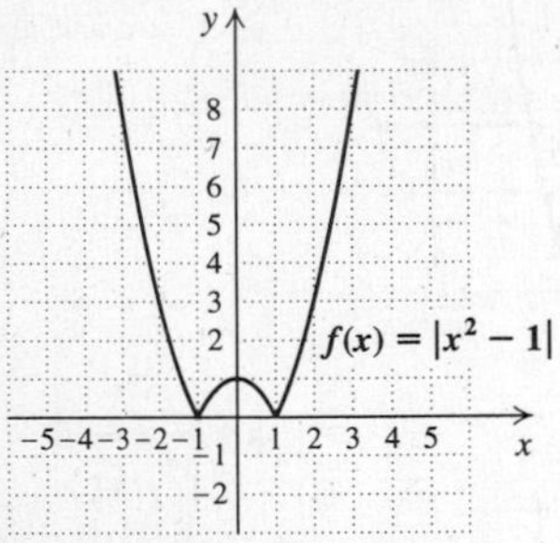

48. $f(x) = |3 - 2x - x^2|$

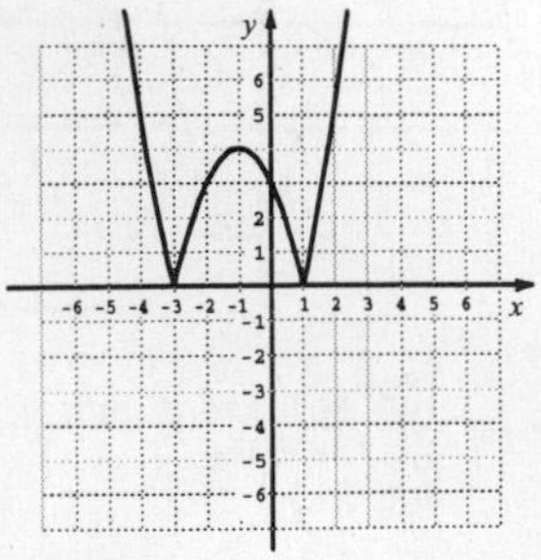

49. 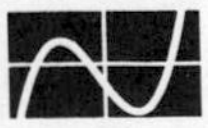

Exercise Set 8.9

1. Familiarize. We make a drawing and label it.

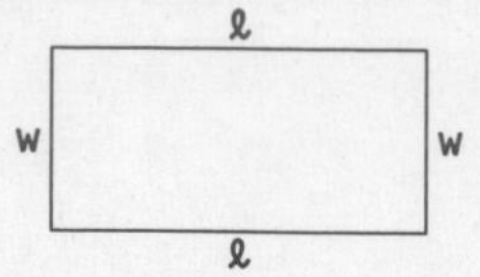

Perimeter: $2\ell + 2w = 76$ ft

Area: $A = \ell \cdot w$

Translate. We have a system of equations.

$2\ell + 2w = 76,$

$A = \ell w$

Carry out. Solving the first equation for ℓ, we get $\ell = 38 - w$. Substituting for ℓ in the second equation we get a quadratic function A:

$A = (38 - w)w$

$A = -w^2 + 38w$

Completing the square, we get

$A = -(w - 19)^2 + 361.$

The maximum function value is 361. It occurs when w is 19. When w = 19, ℓ = 38 - 19, or 19.

Check. We check a function value for w less than 19 and for w greater than 19.

$A(18) = -(18)^2 + 38 \cdot 18 = 360$

$A(20) = -(20)^2 + 38 \cdot 20 = 360$

Since 361 is greater than these numbers, it looks as though we have a maximum.

State. The maximum area of 361 ft^2 occurs when the dimensions are 19 ft by 19 ft.

2. Dimensions: 17 ft by 17 ft; Area: 289 ft^2

3. Familiarize. We let x and y represent the numbers, and we let P represent their product.

Translate. We have two equations.

$x + y = 16,$

$P = xy$

Carry out. Solving the first equation for y, we get $y = 16 - x$. Substituting for y in the second equation we get a quadratic function P:

$P = x(16 - x)$

$P = -x^2 + 16x$

Completing the square, we get

$P = -(x - 8)^2 + 64.$

The maximum function value is 64. It occurs when x = 8. When x = 8, y = 16 - 8, or 8.

Check. We can check a function value for x less than 8 and for x greater than 8.

$P(7) = -(7)^2 + 16 \cdot 7 = 63$

$P(9) = -(9)^2 + 16 \cdot 9 = 63$

Since 64 is greater than these numbers, it looks as though we have a maximum.

State. The maximum product of 64 occurs for the numbers 8 and 8.

4. Maximum product: 196; Numbers: 14 and 14

5. Familiarize. Let x and y represent the two numbers, and let P represent their product.

   Translate. We have two equations.

   $x + y = 22,$

   $P = xy$

   Carry out. Solve the first equation for y.

   $y = 22 - x$

   Substitute for y in the second equation.

   $P = x(22 - x)$

   $P = -x^2 + 22x$

   Completing the square, we get

   $P = -(x - 11)^2 + 121$

   The maximum function value is 121. It occurs when x = 11. When x = 11, y = 22 - 11, or 11.

   Check. Check a function value for x less than 11 and for x greater than 11.

   $P(10) = -(10)^2 + 22\cdot 10 = 120$

   $P(12) = -(12)^2 + 22\cdot 12 = 120$

   Since 121 is greater than these numbers, it looks as though we have a maximum.

   State. The maximum product of 121 occurs for the numbers 11 and 11.

6. Maximum product: $\frac{2025}{4}$; Numbers: $\frac{45}{2}$ and $\frac{45}{2}$

7. Familiarize. Let x and y represent the two numbers, and let P represent their product.

   Translate. We have two equations.

   $x - y = 4,$

   $P = xy$

   Carry out. Solve the first equation for x.

   $x = 4 + y$

   Substitute for x in the second equation.

   $P = (4 + y)y$

   $P = y^2 + 4y$

   Completing the square, we get

   $P = (y + 2)^2 - 4.$

   The minimum function value is -4. It occurs when y = -2. When y = -2, x = 4 + (-2), or 2.

   Check. Check a function value for y less than -2 and for y greater than -2.

   $P(-3) = (-3)^2 + 4(-3) = -3$

   $P(-1) = (-1)^2 + 4(-1) = -3$

   Since -4 is less than these numbers, it looks as though we have a minimum.

   State. The minimum product of -4 occurs for the numbers 2 and -2.

8. Minimum product: -25; Numbers: 5 and -5

9. Familiarize. Let x and y represent the two numbers, and let P represent their product.

   Translate. We have two equations.

   $x - y = 9,$

   $P = xy$

   Carry out. Solve the first equation for x.

   $x = y + 9$

   Substitute for x in the second equation.

   $P = (y + 9)y$

   $P = y^2 + 9y$

   Completing the square, we get

   $P = \left(y + \frac{9}{2}\right)^2 - \frac{81}{4}.$

   The minimum function value is $-\frac{81}{4}$. It occurs when $y = -\frac{9}{2}$. When $y = -\frac{9}{2}$, $x = -\frac{9}{2} + 9 = \frac{9}{2}$.

   Check. Check a function value for y less than $-\frac{9}{2}$ and for y greater than $-\frac{9}{2}$.

   $P(-5) = (-5)^2 + 9(-5) = -20$

   $P(-4) = (-4)^2 + 9(-4) = -20$

   Since $-\frac{81}{4}$ is less than these numbers, it looks as though we have a minimum.

   State. The minimum product of $-\frac{81}{4}$ occurs for the numbers $\frac{9}{2}$ and $-\frac{9}{2}$.

10. Minimum product: $-\frac{49}{4}$; Numbers: $\frac{7}{2}$ and $-\frac{7}{2}$

11. Familiarize. Let x and y represent the two numbers, and let P represent their product.

    Translate. We have two equations.

    $x + y = -7,$

    $P = xy$

    Carry out. Solve the first equation for y.

    $y = -x - 7$

    Substitute for y in the second equation.

    $P = x(-x - 7)$

    $P = -x^2 - 7x$

    Completing the square, we get

    $P = -\left(x + \frac{7}{2}\right)^2 + \frac{49}{4}$

    The maximum function value is $\frac{49}{4}$. It occurs when $x = -\frac{7}{2}$. When $x = -\frac{7}{2}$, $y = -\left(-\frac{7}{2}\right) - 7 = -\frac{7}{2}$.

    Check. Check a function value for x less than $-\frac{7}{2}$ and for x greater than $-\frac{7}{2}$.

    $P(-4) = -(-4)^2 - 7(-4) = 12$

    $P(-3) = -(-3)^2 - 7(-3) = 12$

    Since $\frac{49}{4}$ is greater than these numbers, it looks as though we have a maximum.

    State. The maximum product of $\frac{49}{4}$ occurs for the numbers $-\frac{7}{2}$ and $-\frac{7}{2}$.

12. Maximum product: $\frac{81}{4}$; Numbers: $-\frac{9}{2}$ and $-\frac{9}{2}$

13. Familiarize. We make a drawing and label it.

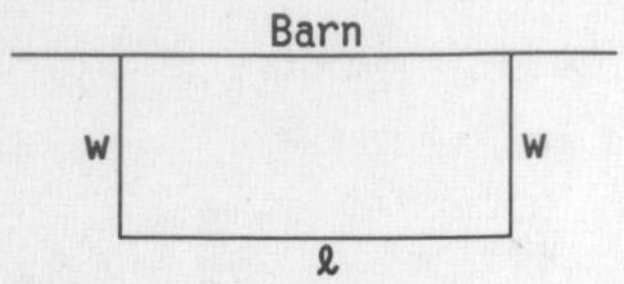

Translate. We have two equations.

$\ell + 2w = 40,$

$A = \ell w$

Carry out. Solve the first equation for ℓ.

$\ell = 40 - 2w$

Substitute for ℓ in the second equation.

$A = (40 - 2w)w$

$A = -2w^2 + 40w$

Completing the square, we get

$A = -2(w - 10)^2 + 200.$

The maximum function value of 200 occurs when $w = 10$. When $w = 10$, $\ell = 40 - 2\cdot10 = 20$.

Check. Check a function value for w less than 10 and for w greater than 10.

$A(9) = -2\cdot9^2 + 40\cdot9 = 198$

$A(11) = -2\cdot11^2 + 40\cdot11 = 198$

Since 200 is greater than these numbers, it looks as though we have a maximum.

State. The maximum area of 200 ft^2 will occur when the dimensions are 10 ft by 20 ft.

14. Maximum area: 450 ft^2;

Dimensions: 15 ft by 30 ft

15. Familiarize. We make an overhead drawing and label it.

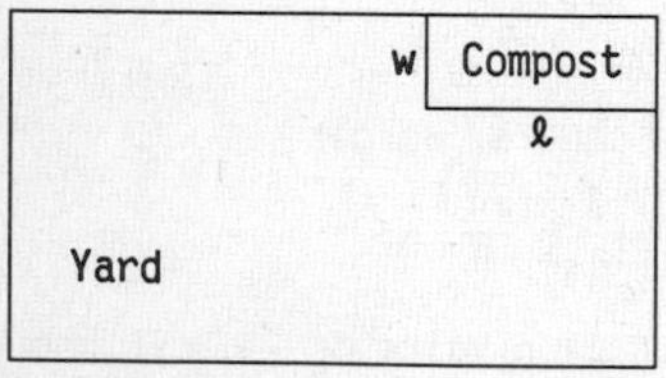

Translate. We have two equations.

$\ell + w = 8,$

$V = \ell\cdot w\cdot 3$, or $V = 3\ell w$

Carry out. Solve the first equation for ℓ.

$\ell = 8 - w$

Substitute for ℓ in the second equation.

$V = 3(8 - w)w$

$V = -3w^2 + 24w$

Completing the square, we get

$V = -3(w - 4)^2 + 48.$

The maximum function value of 48 occurs when $w = 4$. When $w = 4$, $\ell = 8 - 4 = 4$.

Check. Check a function value for w less than 4 and for w greater than 4.

$V(3) = -3\cdot3^2 + 24\cdot3 = 45$

$V(5) = -3\cdot5^2 + 24\cdot5 = 45$

Since 48 is greater than these numbers, it looks as though we have a maximum.

State. The dimensions of the base that will maximize the container's volume are 4 ft by 4 ft.

16. 3.5 in.

17. We look for a function of the form $f(x) = ax^2 + bx + c$. Substituting the data points, we get

$4 = a(1)^2 + b(1) + c,$

$-2 = a(-1)^2 + b(-1) + c,$

$13 = a(2)^2 + b(2) + c,$

or

$4 = a + b + c,$

$-2 = a - b + c,$

$13 = 4a + 2b + c.$

Solving this system, we get

$a = 2,\ b = 3,$ and $c = -1.$

Therefore the function we are looking for is

$f(x) = 2x^2 + 3x - 1.$

18. $f(x) = 3x^2 - x + 2$

19. We look for a function of the form $f(x) = ax^2 + bx + c$. Substituting the data points, we get

$0 = a(2)^2 + b(2) + c,$

$3 = a(4)^2 + b(4) + c,$

$-5 = a(12)^2 + b(12) + c,$

or

$0 = 4a + 2b + c,$

$3 = 16a + 4b + c,$

$-5 = 144a + 12b + c.$

Solving this system, we get

$a = -\frac{1}{4},\ b = 3,\ c = -5.$

Therefore the function we are looking for is

$f(x) = -\frac{1}{4}x^2 + 3x - 5.$

20. $f(x) = -\frac{1}{3}x^2 + 5x - 12$

21. a) Familiarize. We look for a function of the form $f(x) = ax^2 + bx + c$, where $f(x)$ represents the earnings for week x.

Translate. We substitute the given values of x and $f(x)$.

$38 = a(1)^2 + b(1) + c,$

$66 = a(2)^2 + b(2) + c,$

$86 = a(3)^2 + b(3) + c,$

or

$38 = a + b + c,$

$66 = 4a + 2b + c,$

$86 = 9a + 3b + c.$

Carry out. Solving the system of equations, we get

$a = -4, \quad b = 40, \quad c = 2.$

Check. Recheck the calculations.

State. The function $f(x) = -4x^2 + 40x + 2$ fits the data.

b) Find f(4).

$f(4) = -4(4)^2 + 40(4) + 2 = 98$

The predicted earnings for the fourth week are $98.

22. a) $f(x) = 2500x^2 - 6500x + 5000$

b) $19,000

23. a) Familiarize. We look for a function of the form $A(s) = as^2 + bs + c$, where A(s) represents the number of daytime accidents (for every 200 million km) and s represents the travel speed (in km/h).

Translate. We substitute the given values of s and A(s).

$100 = a(60)^2 + b(60) + c,$

$130 = a(80)^2 + b(80) + c,$

$200 = a(100)^2 + b(100) + c,$

or

$100 = 3600a + 60b + c,$

$130 = 6400a + 80b + c,$

$200 = 10{,}000a + 100b + c.$

Carry out. Solving the system of equations, we get

$a = 0.05, \quad b = -5.5, \quad c = 250.$

Check. Recheck the calculations.

State. The function $A(s) = 0.05s^2 - 5.5s + 250$ fits the data.

b) Find A(50).

$A(50) = 0.05(50)^2 - 5.5(50) + 250 = 100$

100 accidents occur at 50 km/h.

24. a) $A(s) = \frac{3}{16}s^2 - \frac{135}{4}s + 1750$

b) 531.25

25. a) Familiarize. We look for a function of the form $P(d) = ad^2 + bd + c$, where P(d) represents the price for a pizza with diameter d.

Translate. Substitute the given values of d and P(d).

$6 = a(8)^2 + b(8) + c,$

$8.50 = a(12)^2 + b(12) + c,$

$11.50 = a(16)^2 + b(16) + c,$

or

$6 = 64a + 8b + c,$

$8.50 = 144a + 12b + c,$

$11.50 = 256a + 16b + c.$

Carry out. Solving the system of equations, we get

$a = \frac{1}{64}, \quad b = \frac{5}{16}, \quad c = \frac{5}{2}$

Check. Recheck the calculations.

State. The function $P(d) = \frac{1}{64}d^2 + \frac{5}{16}d + \frac{5}{2}$ fits the data.

b) Find P(14).

$P(14) = \frac{1}{64}(14)^2 + \frac{5}{16}(14) + \frac{5}{2}$

$= 9.9375$

The price of a 14-in. pizza is approximately $9.94.

26. $P(x) = -x^2 + 980x - 3000$; $237,100 at $x = 490$

27. Find the total profit:

$P(x) = R(x) - C(x)$

$P(x) = (200x - x^2) - (5000 + 8x)$

$P(x) = -x^2 + 192x - 5000$

To find the maximum value of the total profit and the value of x at which it occurs we complete the square:

$P(x) = -(x^2 - 192x) - 5000$

$= -(x^2 - 192x + 9216 - 9216) - 5000$

$= -(x^2 - 192x + 9216) - (-9216) - 5000$

$= -(x - 96)^2 + 4216$

The maximum profit of $4216 occurs at $x = 96$.

28. $P(x) = -x^2 + 220x - 50$; $12,050 at $x = 110$

29. $A = 6s^2$

$\frac{A}{6} = s^2$

$\sqrt{\frac{A}{6}} = s$

30. $r = \frac{1}{2}\sqrt{\frac{A}{\pi}}$

31. $F = \frac{Gm_1m_2}{r^2}$

$Fr^2 = Gm_1m_2$

$r^2 = \frac{Gm_1m_2}{F}$

$r = \sqrt{\frac{Gm_1m_2}{F}}$

32. $s = \sqrt{\frac{kQ_1Q_2}{N}}$

33. $E = mc^2$

$\frac{E}{m} = c^2$

$\sqrt{\frac{E}{m}} = c$

34. $r = \sqrt{\frac{A}{\pi}}$

35. $a^2 + b^2 = c^2$

$b^2 = c^2 - a^2$

$b = \sqrt{c^2 - a^2}$

36. $c = \sqrt{d^2 - a^2 - b^2}$

37. $N = \frac{k^2 - 3k}{2}$

$2N = k^2 - 3k$

$0 = k^2 - 3k - 2N$

$a = 1, \quad b = -3, \quad c = -2N$

$k = \frac{-(-3) \pm \sqrt{(-3)^2 - 4\cdot1\cdot(-2N)}}{2\cdot1}$

$= \frac{3 \pm \sqrt{9 + 8N}}{2}$

Since taking the negative square root would result in a negative answer, we take the positive one.

$k = \frac{3 + \sqrt{9 + 8N}}{2}$

38. $t = \frac{-v_0 + \sqrt{v_0^2 + 2gs}}{g}$

39. $A = 2\pi r^2 + 2\pi rh$

$0 = 2\pi r^2 + 2\pi rh - A$

$a = 2\pi, \quad b = 2\pi h, \quad c = -A$

$r = \frac{-2\pi h \pm \sqrt{(2\pi h)^2 - 4\cdot2\pi\cdot(-A)}}{2\cdot2\pi}$

$= \frac{-2\pi h \pm \sqrt{4\pi^2h^2 + 8\pi A}}{4\pi}$

$= \frac{-2\pi h \pm 2\sqrt{\pi^2h^2 + 2\pi A}}{4\pi}$

$= \frac{-\pi h \pm \sqrt{\pi^2h^2 + 2\pi A}}{2\pi}$

Since taking the negative square root would result in a negative answer, we take the positive one.

$r = \frac{-\pi h + \sqrt{\pi^2h^2 + 2\pi A}}{2\pi}$

40. $r = \frac{-\pi s + \sqrt{\pi^2s^2 + 4\pi A}}{2\pi}$

41. $N = \frac{1}{2}(n^2 - n)$

$N = \frac{1}{2}n^2 - \frac{1}{2}n$

$0 = \frac{1}{2}n^2 - \frac{1}{2}n - N$

$a = \frac{1}{2}, \quad b = -\frac{1}{2}, \quad c = -N$

$n = \frac{-\left(-\frac{1}{2}\right) \pm \sqrt{\left(-\frac{1}{2}\right)^2 - 4\cdot\frac{1}{2}\cdot(-N)}}{2\left(\frac{1}{2}\right)}$

$= \frac{1}{2} \pm \sqrt{\frac{1}{4} + 2N}$

$= \frac{1}{2} \pm \sqrt{\frac{1 + 8N}{4}}$

$= \frac{1}{2} \pm \frac{1}{2}\sqrt{1 + 8N}$

Since taking the negative square root would result in a negative answer, we take the positive one.

$n = \frac{1}{2} + \frac{1}{2}\sqrt{1 + 8N}$, or $\frac{1 + \sqrt{1 + 8N}}{2}$

42. $r = 1 - \sqrt{\frac{A}{A_0}}$

43. $A = 2w^2 + 4\ell w$

$0 = 2w^2 + 4\ell w - A$

$a = 2, \quad b = 4\ell, \quad c = -A$

$w = \frac{-4\ell \pm \sqrt{(4\ell)^2 - 4\cdot2\cdot(-A)}}{2\cdot2}$

$= \frac{-4\ell \pm \sqrt{16\ell^2 + 8A}}{4}$

$= \frac{-4\ell \pm 2\sqrt{4\ell^2 + 2A}}{4}$

$= \frac{-2\ell \pm \sqrt{4\ell^2 + 2A}}{2}$

Since taking the negative square root would result in a negative answer, we take the positive one.

$w = \frac{-2\ell + \sqrt{4\ell^2 + 2A}}{2}$

44. $r = \frac{-\pi h + \sqrt{\pi^2h^2 + 4\pi A}}{4\pi}$

45. $T = 2\pi\sqrt{\frac{\ell}{g}}$

$\frac{T}{2\pi} = \sqrt{\frac{\ell}{g}}$ Multiplying by $\frac{1}{2\pi}$

$\frac{T^2}{4\pi^2} = \frac{\ell}{g}$ Squaring

$gT^2 = 4\pi^2\ell$ Multiplying by $4\pi^2 g$

$g = \frac{4\pi^2\ell}{T^2}$ Multiplying by $\frac{1}{T^2}$

46. $L = \frac{1}{W^2C}$

47. $A = P_1(1 + r)^2 + P_2(1 + r)$

$0 = P_1(1 + r)^2 + P_2(1 + r) - A$

Let $u = 1 + r$.

$0 = P_1u^2 + P_2u - A$ Substituting

$u = \frac{-P_2 \pm \sqrt{P_2^2 - 4(P_1)(-A)}}{2P_1}$

$u = \frac{-P_2 + \sqrt{P_2^2 + 4AP_1}}{2P_1}$ Simplifying and taking the positive square root

$$1 + r = \frac{-P_2 + \sqrt{P_2{}^2 + 4AP_1}}{2P_1}$$ Substituting $1 + r$ for u

$$r = -1 + \frac{-P_2 + \sqrt{P_2{}^2 + 4AP_1}}{2P_1}$$

48. $r = -2 + \dfrac{-P_2 + \sqrt{P_2{}^2 + 4AP_1}}{P_1}$

49.
$$m = \frac{m_0}{\sqrt{1 - \frac{v^2}{c^2}}}$$
$$m^2 = \frac{m_0{}^2}{1 - \frac{v^2}{c^2}}$$
$$m^2\left(1 - \frac{v^2}{c^2}\right) = m_0{}^2$$
$$m^2 - \frac{m^2v^2}{c^2} = m_0{}^2$$
$$m^2 - m_0{}^2 = \frac{m^2v^2}{c^2}$$
$$c^2(m^2 - m_0{}^2) = m^2v^2$$
$$\frac{c^2(m^2 - m_0{}^2)}{m^2} = v^2$$
$$\sqrt{\frac{c^2(m^2 - m_0{}^2)}{m^2}} = v$$
$$\frac{c}{m}\sqrt{m^2 - m_0{}^2} = v$$

50. $c = \dfrac{mv}{\sqrt{m^2 - m_0{}^2}}$

51. a) Familiarize and Translate. From Example 6, we know

$$t = \frac{-v_0 + \sqrt{v_0{}^2 + 19.6s}}{9.8}.$$

Carry out. Substituting 75 for s and 0 for v_0, we have

$$t = \frac{0 + \sqrt{0^2 + 19.6(75)}}{9.8}$$
$$t \approx 3.9$$

Check. Substitute 3.9 for t and 0 for v_0 in the original formula. (See Example 6.)

$$s = 4.9t^2 + v_0t = 4.9(3.9)^2 + 0\cdot(3.9)^2$$
$$\approx 75$$

The answer checks.

State. It takes about 3.9 sec to reach the ground.

b) Familiarize and Translate. From Example 6, we know

$$t = \frac{-v_0 + \sqrt{v_0{}^2 + 19.6s}}{9.8}.$$

Carry out. Substitute 75 for s and 30 for v_0.

$$t = \frac{-30 + \sqrt{30^2 + 19.6(75)}}{9.8}$$
$$t \approx 1.9$$

Check. Substitute 30 for v_0 and 1.9 for t in the original formula. (See Example 6.)

$$s = 4.9t^2 + v_0t = 4.9(1.9)^2 + (30)(1.9)$$
$$\approx 75$$

The answer checks.

State. It takes about 1.9 sec to reach the ground.

c) Familiarize and Translate. We will use the formula $s = 4.9t^2 + v_0t$.

Carry out. Substitute 2 for t and 30 for v_0.

$$s = 4.9(2)^2 + 30(2) = 79.6$$

Check. We can substitute 30 for v_0 and 79.6 for s in the form of the formula we used in part b).

$$t = \frac{-v_0 + \sqrt{v_0{}^2 + 19.6s}}{9.8} =$$
$$\frac{-30 + \sqrt{(30)^2 + 19.6(79.6)}}{9.8} = 2$$

The answer checks.

State. The object will fall 79.6 m.

52. a) 10.1 sec

b) 7.49 sec

c) 272.5 m

53. Familiarize and Translate. From Example 5, we know

$$T = \frac{\sqrt{3V}}{12}.$$

Carry out. Substituting 38 for V, we have

$$T = \frac{\sqrt{3\cdot 38}}{12} = \frac{\sqrt{114}}{12}$$
$$T \approx 0.89$$

Check. Substitute 0.89 in the original formula. (See Example 5.)

$$48T^2 = V$$
$$48(0.89)^2 = V$$
$$38 \approx V$$

The answer checks.

State. The hang time is about 0.89 sec.

54. 1 m

55. Familiarize and Translate. We will use the formula in Exercise 41. (If we did not know that this is the correct formula, we could derive it ourselves.)

$$N = \frac{1}{2}(n^2 - n)$$

Carry out. Substitute 91 for N and solve for n.

$$91 = \frac{1}{2}(n^2 - n)$$
$$182 = n^2 - n$$
$$0 = n^2 - n - 182$$
$$0 = (n - 14)(n + 13)$$
$$n = 14 \quad \text{or} \quad n = -13$$

Check. Since a negative solution has no meaning in this problem, we check 14. Substitute 14 for n in the original formula.

$$N = \frac{1}{2}(14^2 - 14) = \frac{1}{2}(182) = 91$$

The solution checks.

State. There are 14 teams in the league.

56. 12

57. Familiarize and Translate. From Example 6, we know

$$t = \frac{-v_0 + \sqrt{v_0{}^2 + 19.6s}}{9.8}.$$

Carry out. Substituting 40 for s and 0 for v_0 we have

$$t = \frac{0 + \sqrt{0^2 + 19.6(40)}}{9.8}$$
$$t \approx 2.9$$

Check. Substitute 2.9 for t and 0 for v_0 in the original formula. (See Example 6.)

$$s = 4.9t^2 + v_0t = 4.9(2.9)^2 + 0(2.9) \approx 41 \approx 40.$$

The answer checks.

State. He will be falling for about 2.9 sec.

58. 30.625 m

59. Familiarize and Translate. We will use the formula in Example 6, $s = 4.9t^2 + v_0t$.

Carry out. Solve the formula for v_0.

$$s - 4.9t^2 = v_0t$$
$$\frac{s - 4.9t^2}{t} = v_0$$

Now substitute 51.6 for s and 3 for t.

$$\frac{51.6 - 4.9(3)^2}{3} = v_0$$
$$2.5 = v_0$$

Check. Substitute 3 for t and 2.5 for v_0 in the original formula.

$$s = 4.9(3)^2 + 2.5(3) = 51.6$$

The solution checks.

State. The initial velocity is 2.5 m/sec.

60. 3.2 m/sec

61. Familiarize and Translate. From Exercise 47 we know that

$$r = -1 + \frac{-P_2 + \sqrt{P_2^2 + 4P_1A}}{2P_1},$$

where A is the total amount in the account after two years, P_1 is the amount of the original deposit, P_2 is deposited at the beginning of the second year, and r is the annual interest rate.

Carry out. Substitute 3000 for P_1, 1700 for P_2, and 5253.70 for A

$$r = -1 + \frac{-1700 + \sqrt{(1700)^2 + 4(3000)(5253.70)}}{2(3000)}$$

Using a calculator, we have r = 0.07.

Check. Substitute in the original formula in Exercise 47.

$$P_1(1 + r)^2 + P_2(1 + r) = A$$
$$3000(1.07)^2 + 1700(1.07) = A$$
$$5253.70 = A$$

The answer checks.

State. The annual interest rate is 0.07, or 7%.

62. 8.5%

63. $\sqrt{-20} = \sqrt{-4\cdot 5} = 2i\sqrt{5}$, or $2\sqrt{5}i$

64. $\dfrac{x - 9}{(x + 9)(x + 7)}$

65. Familiarize. Recall that the formula for the area of a triangle is $A = \frac{1}{2}bh$.

Translate. We have two equations.

$$b + h = 38,$$
$$A = \frac{1}{2}bh$$

Carry out. Solve the first equation for h.

$$h = 38 - b$$

Substitute for h in the second equation.

$$A = \frac{1}{2}b(38 - b)$$
$$A = -\frac{1}{2}b^2 + 19b$$

Completing the square, we get

$$A = -\frac{1}{2}(b - 19)^2 + \frac{361}{2}$$

The maximum function value of $\frac{361}{2}$, or 180.5, occurs when b = 19. When b = 19, h = 38 - 19, or 19.

Check. Check a function value for b less than 19 and for b greater than 19.

$$A(18) = -\frac{1}{2}(18)^2 + 19(18) = 180$$
$$A(20) = -\frac{1}{2}(20)^2 + 19(20) = 180$$

Since 180.5 is greater than these numbers, it looks as though we have a maximum.

State. The maximum area of 180.5 cm^2 occurs when the base and height are both 19 cm.

66. $11\sqrt{2}$ ft

67. $mn^4 - r^2pm^3 - r^2n^2 + p = 0$

Let $u = n^2$. Substitute and rearrange.

$mu^2 - r^2u - r^2pm^3 + p = 0$

$a = m,\ b = -r^2,\ c = -r^2pm^3 + p$

$$u = \frac{-(-r^2) \pm \sqrt{(-r^2)^2 - 4\cdot m(-r^2pm^3 + p)}}{2\cdot m}$$

$$u = \frac{r^2 \pm \sqrt{r^4 + 4m^4r^2p - 4mp}}{2m}$$

$$n^2 = \frac{r^2 \pm \sqrt{r^4 + 4m^4r^2p - 4mp}}{2m}$$

$$n = \pm\sqrt{\frac{r^2 \pm \sqrt{r^4 + 4m^4r^2p - 4mp}}{2m}}$$

68. $t = \dfrac{r + s \pm \sqrt{r^2 + 2rs + s^2 - 4r^2s^2 + 4s^3r}}{2r - 2s}$

69. Familiarize. Let x represent the number of 10¢ increases in the admission price. Then $2.00 + 0.1x$ represents the admission price and $100 - x$ represents the corresponding average attendance.

Translate. Since total revenue is admission price times the attendance, we have the following function for the revenue.

$R(x) = (2.00 + 0.1x)(100 - x)$

$R(x) = -0.1x^2 + 8x + 200$

Carry out. Completing the square, we get

$R(x) = -0.1(x - 40)^2 + 360$

The maximum function value of 360 occurs when $x = 40$. When $x = 40$ the admission price is $2.00 + 0.1(40)$, or \$6.

Check. We check a function value for x less than 40 and for x greater than 40.

$R(39) = -0.1(39)^2 + 8(39) + 200 = 359.9$

$R(41) = -0.1(41)^2 + 8(41) + 200 = 359.9$

Since 360 is greater than these numbers, it looks as though we have a maximum.

State. In order to maximize revenue the theater owner should charge \$6.00 for admission.

70. 30

71. Familiarize. We want to find the maximum value of a function of the form $h(t) = at^2 + bt + c$ that fits the following data.

Time (sec)	Height (ft)
0	64
3	64
3 + 2, or 5	0

Translate. Substitute the given values for t and h(t).

$64 = a(0)^2 + b(0) + c,$

$64 = a(3)^2 + b(3) + c,$

$0 = a(5)^2 + b(5) + c,$

or

$64 = c,$

$64 = 9a + 3b + c,$

$0 = 25a + 5b + c.$

Carry out. Solving the system of equations, we get $a = -6.4$, $b = 19.2$, $c = 64$. The function $h(t) = -6.4t^2 + 19.2t + 64$ fits the data.

Completing the square, we get

$h(t) = -6.4(t - 1.5)^2 + 78.4.$

The maximum function value of 78.4 occurs at $t = 1.5$.

Check. Recheck the calculations. Also check a function value for t less than 1.5 and for t greater than 1.5.

$h(1) = -6.4(1)^2 + 19.2(1) + 64 = 76.8$

$h(2) = -6.4(2)^2 + 19.2(2) + 64 = 76.8$

Since 78.4 is greater than these numbers, it looks as though we have a maximum.

State. The maximum height is 78.4 ft.

72. 158 ft

73. $A = 2\pi r^2 + 2\pi rh$ (See Exercise 39.)

Substitute $\frac{d}{2}$ for r.

$A = 2\pi\left(\frac{d}{2}\right)^2 + 2\pi\left(\frac{d}{2}\right)h$

$A = \frac{\pi}{2}d^2 + \pi dh$

$0 = \pi d^2 + 2\pi dh - 2A$

$a = \pi,\ b = 2\pi h,\ c = -2A$

$$d = \frac{2\pi h \pm \sqrt{(2\pi h)^2 - 4\pi(-2A)}}{2\pi}$$

$$d = \frac{-2\pi h \pm \sqrt{4\pi^2h^2 + 8\pi A}}{2\pi}$$

We take the positive square root and simplify.

$$d = \frac{-\pi h + \sqrt{\pi^2h^2 + 2\pi A}}{\pi}$$

(Since $d = 2r$, we could also have multiplied the result of Exercise 39 by 2 to find this formula.)

74. $L = \sqrt{3V^{2/3}}$, or $\sqrt{3}\,\sqrt[3]{V}$.

75. Let s represent the length of a side of the cube, let A represent the surface area of the cube, and let L represent the length of the cube's three-dimensional diagonal. From the Pythagorean formula in three dimensions (See Exercise 36.) we know that $L^2 = s^2 + s^2 + s^2$, or $L^2 = 3s^2$. From the formula for the surface area of a cube (See Exercise 29.) we know that $A = 6s^2$, so $\frac{A}{2} = 3s^2$ and $L^2 = \frac{A}{2}$. Then $L = \sqrt{\frac{A}{2}}$.

Exercise Set 8.10

1. $(x - 5)(x + 3) > 0$

The solutions of $(x - 5)(x + 3) = 0$ are 5 and -3. They are not solutions of the inequality, but they divide the real-number line in a natural way. The product $(x - 5)(x + 3)$ is positive or negative, for values other than 5 and -3, depending on the signs of the factors $x - 5$ and $x + 3$.

Sign of $x - 5$:	- - - -	- - - - -	+ + + +
Sign of $x + 3$:	- - - -	+ + + + +	+ + + +
Sign of product:	+ + + +	- - - - -	+ + + +
	-3		5

$x - 5 > 0$ when $x > 5$
$x - 5 < 0$ when $x < 5$

$x + 3 > 0$ when $x > -3$
$x + 3 < 0$ when $x < -3$

For the product $(x - 5)(x + 3)$ to be positive, both factors must be positive or both factors must be negative. We see from the diagram that numbers satisfying $x < -3$ or $x > 5$ are solutions. The solution set of the inequality is $\{x | x < -3 \text{ or } x > 5\}$, or $(-\infty,-3) \cup (5,\infty)$.

2. $\{x | x < -1 \text{ or } x > 4\}$, or $(-\infty,-1) \cup (4,\infty)$

3. $(x + 1)(x - 2) \leqslant 0$

The solutions of $(x + 1)(x - 2) = 0$ are -1 and 2. They divide the number line into three intervals as shown:

A B C

-1 2

We try test numbers in each interval.

A: Test -2, $f(-2) = (-2 + 1)(-2 - 2) = 4$;

B: Test 0, $f(0) = (0 + 1)(0 - 2) = -2$;

C: Test 3, $f(3) = (3 + 1)(3 - 2) = 4$

Since $f(0)$ is negative, the function value will be negative for all numbers in the interval containing 0. The inequality symbol is $\leqslant$, so we need to include the intercepts.

The solution set is $\{x | -1 \leqslant x \leqslant 2\}$, or $[-1,2]$.

4. $\{x | -3 \leqslant x \leqslant 5\}$, or $[-3,5]$

5. $x^2 - x - 2 < 0$

$(x + 1)(x - 2) < 0$ Factoring

See the diagram and test numbers in Exercise 3. The solution set is $\{x | -1 < x < 2\}$, or $(-1,2)$.

6. $\{x | -2 < x < 1\}$, or $(-2,1)$

7. $9 - x^2 \leqslant 0$

$(3 - x)(3 + x) \leqslant 0$

The solutions of $(3 - x)(3 + x) = 0$ are 3 and -3. They divide the real-number line in a natural way. The product $(3 - x)(3 + x)$ is positive or negative, for values other than 3 and -3, depending on the signs of the factors $3 - x$ and $3 + x$.

Sign of $3 - x$:	+ + + +	+ + + + +	- - - -
Sign of $3 + x$:	- - - -	+ + + + +	+ + + +
Sign of product:	- - - -	+ + + + +	- - - -
	-3		3

$3 - x > 0$ when $x < 3$
$3 - x < 0$ when $x > 3$

$3 + x > 0$ when $x > -3$
$3 + x < 0$ when $x < -3$

For the product $(3 - x)(3 + x)$ to be negative, one factor must be positive and the other negative. We see from the diagram that numbers satisfying $x < -3$ or $x > 3$ are solutions. The intercepts are also solutions. The solution set of the inequality is $\{x | x \leqslant -3 \text{ or } x \geqslant 3\}$, or $(-\infty,-3] \cup [3,\infty)$.

8. $\{x | -2 \leqslant x \leqslant 2\}$, or $[-2,2]$

9. $x^2 - 2x + 1 \geqslant 0$

$(x - 1)^2 \geqslant 0$

The solution of $(x - 1)^2 = 0$ is 1. For all real-number values of x except 1, $(x - 1)^2$ will be positive. Thus the solution set is $\{x | x \text{ is a real number}\}$, or $(-\infty,\infty)$.

10. $\emptyset$

11. $x^2 + 8 < 6x$

$x^2 - 6x + 8 < 0$

$(x - 4)(x - 2) < 0$

The solutions of $(x - 4)(x - 2) = 0$ are 4 and 2. They are not solutions of the inequality, but they divide the real-number line in a natural way. The product $(x - 4)(x - 2)$ is positive or negative, for values other than 4 and 2, depending on the signs of the factors $x - 4$ and $x - 2$.

Sign of $x - 4$:	- - - -	- - - - -	+ + + +
Sign of $x - 2$:	- - - -	+ + + + +	+ + + +
Sign of product:	+ + + +	- - - - -	+ + + +
	2		4

$x - 4 > 0$ when $x > 4$
$x - 4 < 0$ when $x < 4$

$x - 2 > 0$ when $x > 2$
$x - 2 < 0$ when $x < 2$

For the product $(x - 4)(x - 2)$ to be negative one factor must be positive and the other negative. The only situation in the table for which this happens is when $2 < x < 4$. The solution set of the inequality is $\{x | 2 < x < 4\}$, or $(2,4)$.

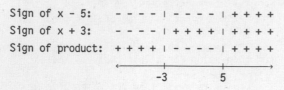

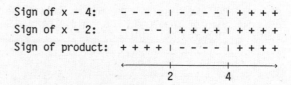

12. $\{x \mid x < -2 \text{ or } x > 6\}$, or $(-\infty,-2) \cup (6,\infty)$

13. $3x(x + 2)(x - 2) < 0$

The solutions of $3x(x + 2)(x - 2) = 0$ are 0, -2, and 2. They divide the number line into four intervals as shown.

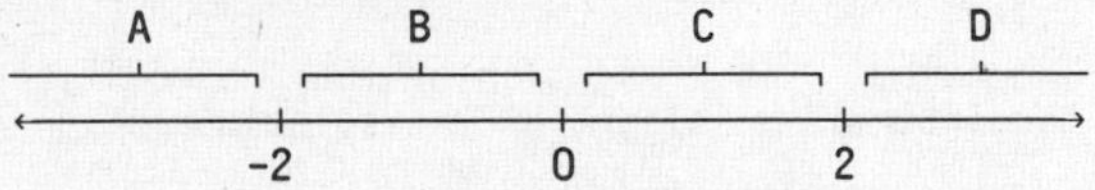

We try test numbers in each interval.

A: Test -3, $f(-3) = 3(-3)(-3 + 2)(-3 - 2) = -45$

B: Test -1, $f(-1) = 3(-1)(-1 + 2)(-1 - 2) = 9$

C: Test 1, $f(1) = 3(1)(1 + 2)(1 - 2) = -9$

D: Test 3, $f(3) = 3(3)(3 + 2)(3 - 2) = 45$

Since $f(-3)$ and $f(1)$ are negative, the function value will be negative for all numbers in the intervals containing -3 and 1. The solution set is $\{x \mid x < -2 \text{ or } 0 < x < 2\}$, or $(-\infty,-2) \cup (0,2)$.

14. $\{x \mid -1 < x < 0 \text{ or } x > 1\}$, or $(-1,0) \cup (1,\infty)$

15. $(x + 3)(x - 2)(x + 1) > 0$

The solutions of $(x + 3)(x - 2)(x + 1) = 0$ are -3, 2, and -1. They are not solutions of the inequality, but they divide the real number line in a natural way. The product $(x + 3)(x - 2)(x + 1)$ is positive or negative, for values other than -3, 2, and -1, depending on the signs of the factors $x + 3$, $x - 2$, and $x + 1$.

Sign of $x + 3$: - - - - | + + + + | + + + + | + + + +

Sign of $x - 2$: - - - - | - - - - | - - - - | + + + +

Sign of $x + 1$: - - - - | - - - - | + + + + | + + + +

Sign of product: - - - - | + + + + | - - - - | + + + +

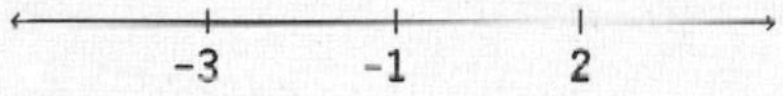

$x + 3 > 0$ when $x > -3$
$x + 3 < 0$ when $x < -3$

$x - 2 > 0$ when $x > 2$
$x - 2 < 0$ when $x < 2$

$x + 1 > 0$ when $x > -1$
$x + 1 < 0$ when $x < -1$

The product of three numbers is positive when all three are positive or two are negative and one is positive. We see from the diagram that numbers satisfying $-3 < x < -1$ or $x > 2$ are solutions. The solution set of the inequality is $\{x \mid -3 < x < -1 \text{ or } x > 2\}$, or $(-3,-1) \cup (2,\infty)$.

16. $\{x \mid x < -2 \text{ or } 1 < x < 4\}$, or $(-\infty,-2) \cup (1,4)$

17. $(x + 3)(x + 2)(x - 1) < 0$

The solutions of $(x + 3)(x + 2)(x - 1) = 0$ are -3, -2, and 1. They divide the number line into four intervals as shown:

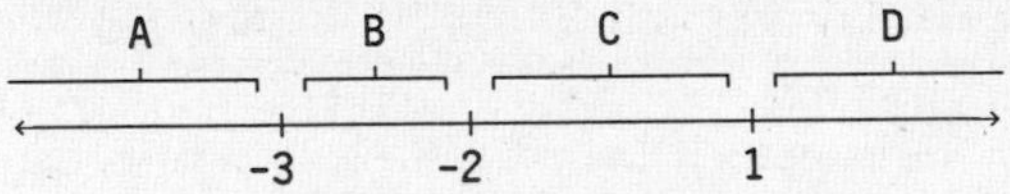

We try test numbers in each interval.

A: Test -4, $f(-4) = (-4 + 3)(-4 + 2)(-4 - 1) = -10$

B: Test $-\frac{5}{2}$, $f\left(-\frac{5}{2}\right) =$

$$\left(-\frac{5}{2} + 3\right)\left(-\frac{5}{2} + 2\right)\left(-\frac{5}{2} - 1\right) = \frac{7}{8}$$

C: Test 0, $f(0) = (0 + 3)(0 + 2)(0 - 1) = -6$

D: Test 2, $f(2) = (2 + 3)(2 + 2)(2 - 1) = 20$

The function value will be negative for all numbers in intervals A and C. The solution set is $\{x \mid x < -3 \text{ or } -2 < x < 1\}$, or $(-\infty,-3) \cup (-2,1)$.

18. $\{x \mid x < -1 \text{ or } 2 < x < 3\}$, or $(-\infty,-1) \cup (2,3)$

19. $\frac{1}{x - 4} < 0$

We write the related equation by changing the $<$ symbol to $=$:

$$\frac{1}{x - 4} = 0$$

We solve the related equation.

$$(x - 4) \cdot \frac{1}{x - 4} = (x - 4) \cdot 0$$

$$1 = 0$$

The related equation has no solution.

Next we find the values that make the denominator 0 by setting the denominator equal to 0 and solving:

$$x - 4 = 0$$
$$x = 4$$

We use 4 to divide the number line into two intervals as shown:

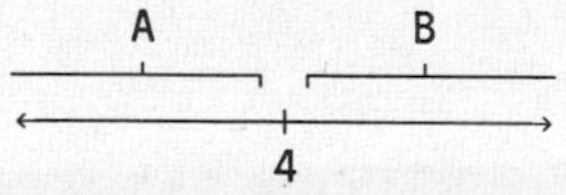

We try test numbers in each interval.

A: Test 0, $\frac{1}{0 - 4} = \frac{1}{-4} = -\frac{1}{4} < 0$

The number 0 is a solution of the inequality, so the interval A is part of the solution set.

B: Test 5, $\frac{1}{5 - 4} = \frac{1}{1} = 1 \not< 0$

The number 5 is not a solution of the inequality, so the interval B is not part of the solution set. The solution set is $\{x \mid x < 4\}$, or $(-\infty,4)$.

20. $\{x \mid x > -5\}$, or $(-5,\infty)$

21. $\frac{x + 1}{x - 3} > 0$

Solve the related equation.

$\frac{x + 1}{x - 3} = 0$

$x + 1 = 0$

$x = -1$

Find the values that make the denominator 0.

$x - 3 = 0$

$x = 3$

Use the numbers -1 and 3 to divide the number line into intervals as shown:

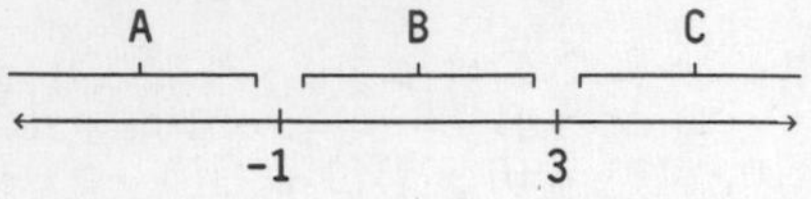

Try test numbers in each interval.

A: Test -2, $\frac{-2 + 1}{-2 - 3} = \frac{-1}{-5} = \frac{1}{5} > 0$

The number -2 is a solution of the inequality so the interval A is part of the solution set.

B: Test 0, $\frac{0 + 1}{0 - 3} = \frac{1}{-3} = -\frac{1}{3} \not> 0$

The number 0 is not a solution of the inequality, so the interval B is not part of the solution set.

C: Test 4, $\frac{4 + 1}{4 - 3} = \frac{5}{1} = 5 > 0$

The number 4 is a solution of the inequality, so the interval C is part of the solution set. The solution set is $\{x | x < -1 \text{ or } x > 3\}$, or $(-\infty,-1) \cup (3,\infty)$.

22. $\{x | -5 < x < 2\}$, or $(-5,2)$

23. $\frac{3x + 2}{x - 3} \leq 0$

Solve the related equation.

$\frac{3x + 2}{x - 3} = 0$

$3x + 2 = 0$

$3x = -2$

$x = -\frac{2}{3}$

Find the values that make the denominator 0.

$x - 3 = 0$

$x = 3$

Use the numbers $-\frac{2}{3}$ and 3 to divide the number line into intervals as shown.

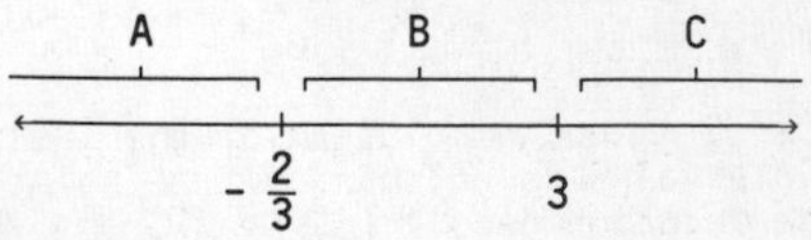

Try test numbers in each interval.

A: Test -1, $\frac{3(-1) + 2}{-1 - 3} = \frac{-1}{-4} = \frac{1}{4} \not\leq 0$

The number -1 is not a solution of the inequality, so the interval A is not part of the solution set.

B: Test 0, $\frac{3 \cdot 0 + 2}{0 - 3} = \frac{2}{-3} = -\frac{2}{3} \leq 0$

The number 0 is a solution of the inequality, so the interval B is part of the solution set.

C: Test 4, $\frac{3 \cdot 4 + 2}{4 - 3} = \frac{14}{1} = 14 \not\leq 0$

The number 4 is not a solution of the inequality, so the interval C is not part of the solution set. The solution set includes the interval B. The number $-\frac{2}{3}$ is also included since the inequality symbol is $\leq$ and $-\frac{2}{3}$ is the solution of the related equation. The number 3 is not included since $\frac{3x + 2}{x - 3}$ is undefined for x = 3. The solution set is $\left\{x \middle| -\frac{2}{3} \leq x < 3\right\}$, or $\left[-\frac{2}{3},3\right)$.

24. $\left\{x \middle| x < -\frac{3}{4} \text{ or } x \geq \frac{5}{2}\right\}$, or $\left(-\infty,-\frac{3}{4}\right) \cup \left[\frac{5}{2},\infty\right)$

25. $\frac{x - 1}{x - 2} > 3$

Solve the related equation.

$\frac{x - 1}{x - 2} = 3$

$x - 1 = 3(x - 2)$

$x - 1 = 3x - 6$

$5 = 2x$

$\frac{5}{2} = x$

Find the values that make the denominator 0.

$x - 2 = 0$

$x = 2$

Use the numbers $\frac{5}{2}$ and 2 to divide the number line into intervals as shown:

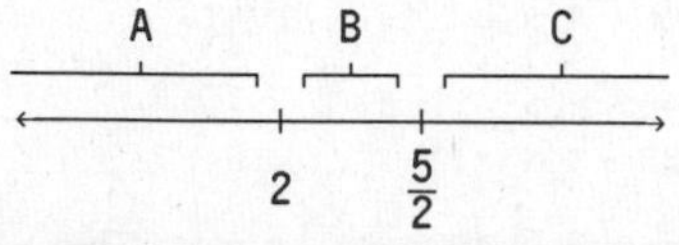

Try test numbers in each interval.

A: Test 0, $\frac{0 - 1}{0 - 2} = \frac{-1}{-2} = \frac{1}{2} \not> 3$

The number 0 is not a solution of the inequality, so the interval A is not part of the solution set.

B: Test $\frac{9}{4}$, $\frac{\frac{9}{4} - 1}{\frac{9}{4} - 2} = \frac{\frac{5}{4}}{\frac{1}{4}} = 5 > 3$

The number $\frac{9}{4}$ is a solution of the inequality, so the interval B is part of the solution set.

C: Test 3, $\frac{3 - 1}{3 - 2} = \frac{2}{1} = 2 \not> 3$

The number 3 is not a solution of the inequality, so the interval C is not part of the solution set. The solution set is $\left\{x \middle| 2 < x < \frac{5}{2}\right\}$, or $\left(2,\frac{5}{2}\right]$.

26. $\left\{x \middle| x < \frac{3}{2} \text{ or } x > 4\right\}$, or $\left(-\infty,\frac{3}{2}\right] \cup (4,\infty)$

27. $\frac{(x - 2)(x + 1)}{x - 5} < 0$

Solve the related equation.

$\frac{(x - 2)(x + 1)}{x - 5} = 0$

$(x - 2)(x + 1) = 0$

$x = 2$ or $x = -1$

Find the values that make the denominator 0.

$x - 5 = 0$

$x = 5$

Use the numbers 2, -1, and 5 to divide the number line into intervals as shown:

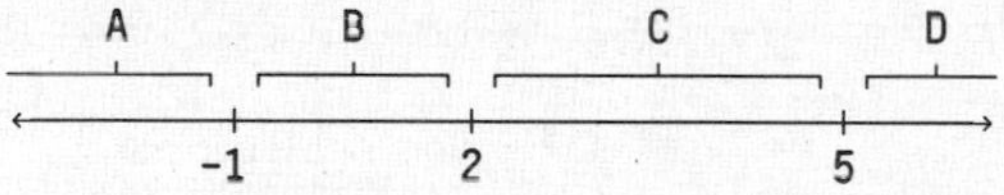

Try test numbers in each interval.

A: Test -2, $\frac{(-2 - 2)(-2 + 1)}{-2 - 5} = \frac{-4(-1)}{-7} =$

$-\frac{4}{7} < 0$

Interval A is part of the solution set.

B: Test 0, $\frac{(0 - 2)(0 + 1)}{0 - 5} = \frac{-2\cdot 1}{-5} = \frac{2}{5} \not< 0$

Interval B is not part of the solution set.

C: Test 3, $\frac{(3 - 2)(3 + 1)}{3 - 5} = \frac{1\cdot 4}{-2} = -2 < 0$

Interval C is part of the solution set.

D: Test 6, $\frac{(6 - 2)(6 + 1)}{6 - 5} = \frac{4\cdot 7}{1} = 28 \not< 0$

Interval D is not part of the solution set.

The solution set is $\{x | x < -1 \text{ or } 2 < x < 5\}$, or $(-\infty,-1) \cup (2,5)$.

28. $\{x | -4 < x < -3 \text{ or } x > 1\}$, or $(-4,-3) \cup (1,\infty)$

29. $\frac{x}{x - 2} \geq 0$

Solve the related equation.

$\frac{x}{x - 2} = 0$

$x = 0$

Find the values that make the denominator 0.

$x - 2 = 0$

$x = 2$

Use the numbers 0 and 2 to divide the number line into intervals as shown.

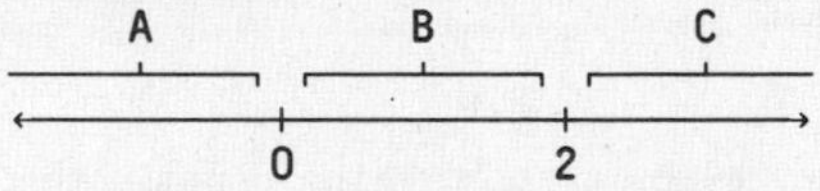

Try test numbers in each interval.

A: Test -1, $\frac{-1}{-1 - 2} = \frac{-1}{-3} = \frac{1}{3} \geq 0$

Interval A is part of the solution set.

B: Test 1, $\frac{1}{1 - 2} = \frac{1}{-1} = -1 \ngeq 0$

Interval B is not part of the solution set.

C: Test 3, $\frac{3}{3 - 2} = \frac{3}{1} = 3 \geq 0$

Interval C is part of the solution set.

The solution set includes intervals A and C. The number 0 is also included since the inequality symbol is $\geq$ and 0 is the solution of the related equation. The number 2 is not included since $\frac{x}{x - 2}$ is undefined for $x = 2$. The solution set is $\{x | x \leq 0 \text{ or } x > 2\}$, or $(-\infty,0] \cup (2,\infty)$.

30. $\{x | -3 \leq x < 0\}$, or $[-3,0)$

31. $\frac{x - 5}{x} < 1$

Solve the related equation.

$\frac{x - 5}{x} = 1$

$x - 5 = x$

$-5 = 0$

The related equation has no solution.

Find the values that make the denominator 0.

$x = 0$

Use the number 0 to divide the number line into two intervals as shown.

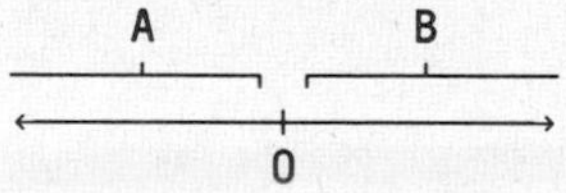

Try test numbers in each interval.

A: Test -1, $\frac{-1 - 5}{-1} = \frac{-6}{-1} = 6 \not< 1$

Interval A is not part of the solution set.

B: Test 1, $\frac{1 - 5}{1} = \frac{-4}{1} = -4 < 1$

Interval B is part of the solution set.

The solution set is $\{x | x > 0\}$, or $(0,\infty)$.

32. $\{x | 1 < x < 2\}$, or $(1,2)$

33. $\dfrac{x - 1}{(x - 3)(x + 4)} < 0$

Solve the related equation.

$$\frac{x - 1}{(x - 3)(x + 4)} = 0$$
$$x - 1 = 0$$
$$x = 1$$

Find the values that make the denominator 0.

$$(x - 3)(x + 4) = 0$$
$$x = 3 \text{ or } x = -4$$

Use the numbers 1, 3, and -4 to divide the number line into intervals as shown:

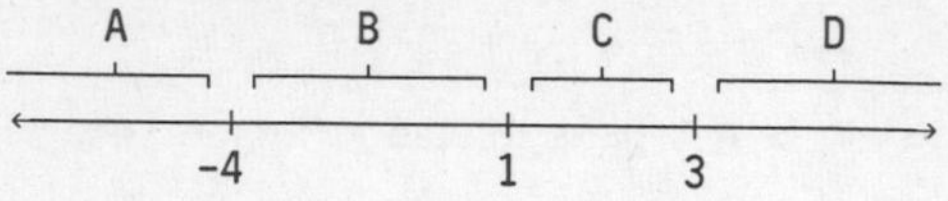

Try test numbers in each interval.

A: Test -5, $\dfrac{-5 - 1}{(-5 - 3)(-5 + 4)} = \dfrac{-6}{-8(-1)} = -\dfrac{3}{4} < 0$

Interval A is part of the solution set.

B: Test 0, $\dfrac{0 - 1}{(0 - 3)(0 + 4)} = \dfrac{-1}{-3 \cdot 4} = \dfrac{1}{12} \not< 0$

Interval B is not part of the solution set.

C: Test 2, $\dfrac{2 - 1}{(2 - 3)(2 + 4)} = \dfrac{1}{-1 \cdot 6} = -\dfrac{1}{6} < 0$

Interval C is part of the solution set.

D: Test 4, $\dfrac{4 - 1}{(4 - 3)(4 + 4)} = \dfrac{3}{1 \cdot 8} = \dfrac{3}{8} \not< 0$

Interval D is not part of the solution set.

The solution set is $\{x | x < -4 \text{ or } 1 < x < 3\}$, or $(-\infty,-4) \cup (1,3)$.

34. $\{x | -7 < x < -2 \text{ or } x > 2\}$, or $(-7,-2) \cup (2,\infty)$

35. $2 < \dfrac{1}{x}$

Solve the related equation.

$$2 = \frac{1}{x}$$
$$x = \frac{1}{2}$$

Find the values that make the denominator 0.

$$x = 0$$

Use the numbers $\frac{1}{2}$ and 0 to divide the number line into intervals as shown.

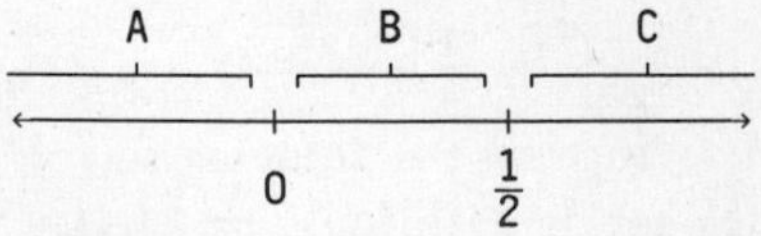

Try test numbers in each interval.

A: Test -1, $\dfrac{1}{-1} = -1 \not> 2$

Interval A is not part of the solution set.

B: Test $\frac{1}{4}$, $\dfrac{1}{\frac{1}{4}} = 4 > 2$

Interval B is part of the solution set.

C: Test 1, $\dfrac{1}{1} = 1 \not> 2$

Interval C is not part of the solution set.

The solution set is $\left\{x \middle| 0 < x < \frac{1}{2}\right\}$, or $\left(0,\frac{1}{2}\right)$.

36. $\left\{x \middle| x < 0 \text{ or } x \geq \frac{1}{3}\right\}$, or $(-\infty,0) \cup \left[\frac{1}{3},\infty\right)$

37. $\sqrt[5]{a^2b}\,\sqrt[3]{ab^2} = (a^2b)^{1/5}(ab^2)^{1/3}$ Converting to exponential notation

$= a^{2/5}b^{1/5}a^{1/3}b^{2/3}$

$= a^{2/5+1/3}b^{1/5+2/3}$

$= a^{11/15}b^{13/15}$

$= (a^{11}b^{13})^{1/15}$

$= \sqrt[15]{a^{11}b^{13}}$ Converting back to radical notation

38. 6

39. 

40. $\{x | x < -1 - \sqrt{5} \text{ or } x > -1 + \sqrt{5}\}$, or $(-\infty, -1 - \sqrt{5}) \cup (-1 + \sqrt{5}, \infty)$

41. $x^4 + 2x^2 \geq 0$

$x^2(x^2 + 2) \geq 0$

$x^2 \geq 0$ for all values of x, and $x^2 + 2 > 0$ for all values of x. Then $x^2(x^2 + 2) \geq 0$ for all values of x. The solution set is the set of all real numbers.

42. {0}

43. $\left|\dfrac{x + 2}{x - 1}\right| < 3$

$$-3 < \frac{x + 2}{x - 1} < 3$$

First we solve $-3 < \dfrac{x + 2}{x - 1}$, or $\dfrac{x + 2}{x - 1} > -3$.

Solve the related equation.

$$\frac{x + 2}{x - 1} = -3$$
$$x + 2 = -3x + 3$$
$$4x = 1$$
$$x = \frac{1}{4}$$

Find the values that make the denominator 0.

$x - 1 = 0$

$x = 1$

Use the numbers $\frac{1}{4}$ and 1 to divide the number line into intervals as shown:

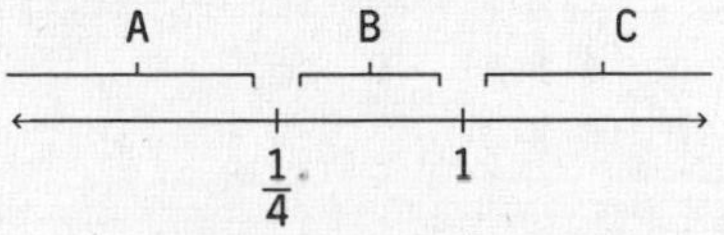

Try test numbers in each interval.

A: Test 0, $\frac{0 + 2}{0 - 1} = \frac{2}{-1} = -2 > -3$

Interval A is in the solution set.

B: Test $\frac{1}{2}$, $\frac{\frac{1}{2} + 2}{\frac{1}{2} - 1} = \frac{\frac{5}{2}}{-\frac{1}{2}} = -5 \not> -3$

Interval B is not in the solution set.

C: Test 2: $\frac{2 + 2}{2 - 1} = \frac{4}{1} = 4 > -3$

Interval C is in the solution set.

The solution set for this portion of the compound inequality is $\left\{x \middle| x < \frac{1}{4} \text{ or } x > 1\right\}$.

Now we solve $\frac{x + 2}{x - 1} < 3$.

Solve the related equation.

$\frac{x + 2}{x - 1} = 3$

$x + 2 = 3x - 3$

$5 = 2x$

$\frac{5}{2} = x$

As above, the denominator is 0 for $x = 1$.

Use the numbers 1 and $\frac{5}{2}$ to divide the number line into intervals as shown:

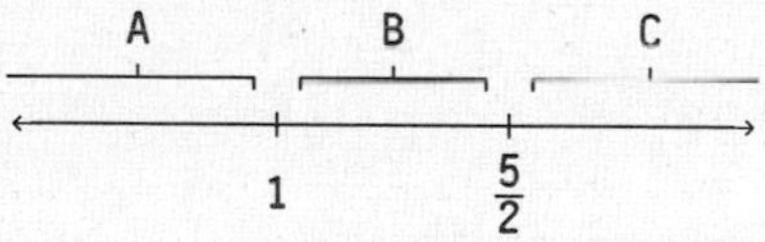

Try test numbers in each interval.

A: Test 0, $\frac{0 + 2}{0 - 1} = \frac{2}{-1} = -2 < 3$

Interval A is in the solution set.

B: Test 2, $\frac{2 + 2}{2 - 1} = \frac{4}{1} = 4 \not< 3$

Interval B is not in the solution set.

C: Test 3, $\frac{3 + 2}{3 - 1} = \frac{5}{2} < 3$

Interval C is in the solution set.

The solution set for this portion of the compound inequality is $\left\{x \middle| x < 1 \text{ or } x > \frac{5}{2}\right\}$.

The solution set for the original inequality is $\left\{x \middle| x < \frac{1}{4} \text{ or } x > 1\right\} \cap \left\{x \middle| x < 1 \text{ or } x > \frac{5}{2}\right\}$, or $\left\{x \middle| x < \frac{1}{4} \text{ or } x > \frac{5}{2}\right\}$.

44. a) The company makes a profit for values of x such that $10 < x < 200$, or for values of x in the interval (10,200).

b) The company loses money for values of x such that $0 \leqslant x < 10$ or $x > 200$, or for values of x in the interval $[0,10) \cup (200,\infty)$.

45. a) $-16t^2 + 32t + 1920 > 1920$

$-16t^2 + 32t > 0$

$t^2 - 2t < 0$

$t(t - 2) < 0$

The solutions of $f(t) = t(t - 2) = 0$ are 0 and 2. They divide the number line as shown:

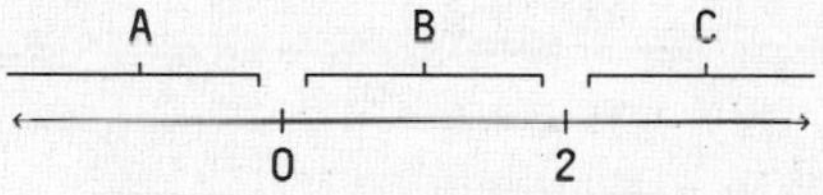

A: Test -1, $f(-1) = (-1)^2 - 2(-1) = 3$

B: Test 1, $f(1) = 1^2 - 2\cdot 1 = -1$

C: Test 3, $f(3) = 3^2 - 2\cdot 3 = 3$

The height is greater than 1920 ft for $\{t|0 < t < 2\}$, or for values of t in the interval (0,2).

b) $-16t^2 + 32t + 1920 < 640$

$-16t^2 + 32t + 1280 < 0$

$t^2 - 2t - 80 > 0$

$(t - 10)(t + 8) > 0$

The solutions of $f(t) = (t - 10)(t + 8) = 0$ are 10 and -8. They divide the number line as shown:

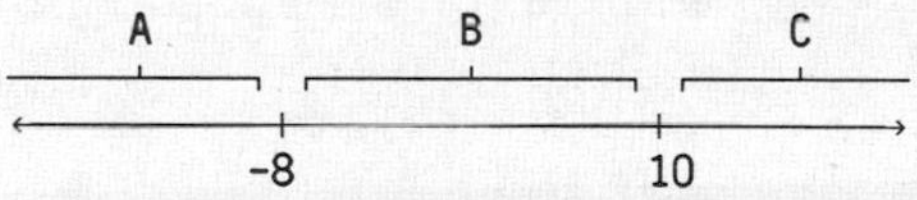

A: Test -10, $f(-10) = (-10)^2 - 2(-10) - 80 = 40$

B: Test 0, $f(0) = 0^2 - 2\cdot 0 - 80 = -80$

C: Test 20, $f(20) = 20^2 - 2\cdot 20 - 80 = 280$

Since negative values of t have no meaning in this problem, the solution set is $\{t|t > 10\}$, or $(10,\infty)$.

46. $\{n|12 \leqslant n \leqslant 25\}$

47. Find the values of n for which $D \geqslant 27$ and $D \leqslant 230$.

For $D \geqslant 27$:

$$\frac{n(n-3)}{2} \geqslant 27$$

$$n(n-3) \geqslant 54$$

$$n^2 - 3n - 54 \geqslant 0$$

$$(n-9)(n+6) \geqslant 0$$

The solutions of $f(n) = (n-9)(n+6) = 0$ are 9 and -6. Considering only positive values of n, we use the intervals below:

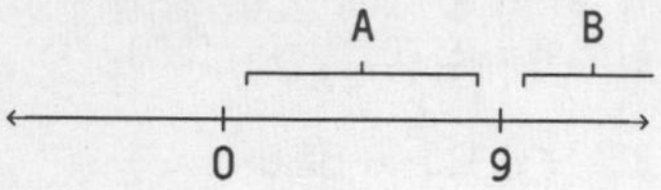

A: Test 1, $f(1) = 1^2 - 3\cdot 1 - 54 = -56$

B: Test 10, $f(10) = 10^2 - 3\cdot 10 - 54 = 16$

$D(n) \geqslant 27$ for $\{n \mid n \geqslant 9\}$

For $D \leqslant 230$:

$$\frac{n(n-3)}{2} \leqslant 230$$

$$n(n-3) \leqslant 460$$

$$n^2 - 3n - 460 \leqslant 0$$

$$(n-23)(n+20) \leqslant 0$$

The solutions of $f(n) = (n-23)(n+20) = 0$ are 23 and -20. Again considering only positive values of n, we use the intervals below:

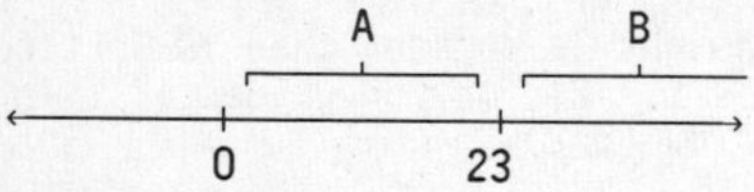

A: Test 1, $f(1) = 1^2 - 3\cdot 1 - 460 = -462$

B: Test 25, $f(25) = 25^2 - 3\cdot 25 - 460 = 90$

$D(n) \leqslant 230$ (and $n > 0$) for $\{n \mid 0 < n \leqslant 23\}$.

Then $27 \leqslant D \leqslant 230$ for $\{n \mid 9 \leqslant n \leqslant 23\}$, or for values of n in the interval $[9,23]$.

48. The solutions of $f(x) = 0$ are -2, 1, and 3.

The solution of $f(x) < 0$ is $\{x \mid x < -2 \text{ or } 1 < x < 3\}$, or $(-\infty,-2) \cup (1,3)$.

The solution of $f(x) > 0$ is $\{x \mid -2 < x < 1 \text{ or } x > 3\}$, or $(-2,1) \cup (3,\infty)$.

49. $f(x) = 0$ for $x = -2$ and $x = 1$;

$f(x) < 0$ for $\{x \mid x < -2\}$, or $(-\infty,-2)$;

$f(x) > 0$ for $\{x \mid -2 < x < 1 \text{ or } x > 1\}$, or $(-2,1) \cup (1,\infty)$

50. $f(x)$ has no zeros.

The solutions $f(x) < 0$ are $\{x \mid x < 0\}$, or $(-\infty,0)$.

The solutions of $f(x) > 0$ are $\{x \mid x > 0\}$, or $(0,\infty)$.

51. $f(x) = 0$ for $x = 0$ and $x = 1$;

$f(x) < 0$ for $\{x \mid 0 < x < 1\}$, or $(0,1)$;

$f(x) > 0$ for $\{x \mid x > 1\}$, or $(1,\infty)$

52. The solutions of $f(x) = 0$ are -2, 1, 2, and 3.

The solutions of $f(x) < 0$ are $\{x \mid -2 < x < 1 \text{ or } 2 < x < 3\}$, or $(-2,1) \cup (2,3)$.

The solutions of $f(x) > 0$ are $\{x \mid x < -2 \text{ or } 1 < x < 2 \text{ or } x > 3\}$, or $(-\infty,-2) \cup (1,2) \cup (3,\infty)$.

53. $f(x) = 0$ for $x = -2$, $x = 0$, and $x = 1$;

$f(x) < 0$ for $\{x \mid x < -3 \text{ or } -2 < x < 0 \text{ or } 1 < x < 2\}$, or $(-\infty,-3) \cup (-2,0) \cup (1,2)$;

$f(x) > 0$ for $\{x \mid -3 < x < -2 \text{ or } 0 < x < 1 \text{ or } x > 2\}$, or $(-3,-2) \cup (0,1) \cup (2,\infty)$

54. 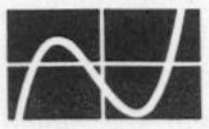

Exercise Set 9.1

1. Graph: $y = 2^x$

We compute some function values, thinking of y as f(x), and keep the results in a table.

$f(0) = 2^0 = 1$

$f(1) = 2^1 = 2$

$f(2) = 2^2 = 4$

$f(-1) = 2^{-1} = \frac{1}{2^1} = \frac{1}{2}$

$f(-2) = 2^{-2} = \frac{1}{2^2} = \frac{1}{4}$

x	y, or f(x)
0	1
1	2
2	4
-1	$\frac{1}{2}$
-2	$\frac{1}{4}$

Next we plot these points and connect them with a smooth curve.

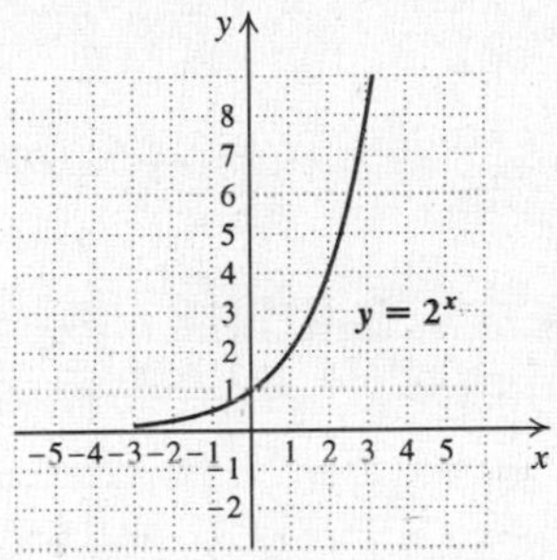

2.

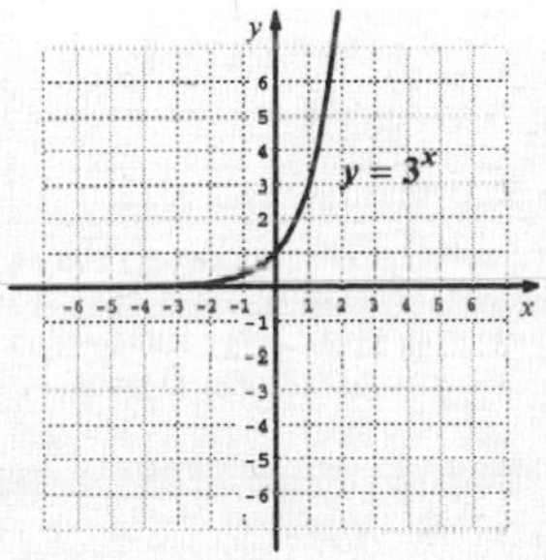

3. Graph: $y = 5^x$

We compute some function values, thinking of y as f(x), and keep the results in a table.

$f(0) = 5^0 = 1$

$f(1) = 5^1 = 5$

$f(2) = 5^2 = 25$

$f(-1) = 5^{-1} = \frac{1}{5^1} = \frac{1}{5}$

$f(-2) = 5^{-2} = \frac{1}{5^2} = \frac{1}{25}$

x	y, or f(x)
0	1
1	5
2	25
-1	$\frac{1}{5}$
-2	$\frac{1}{25}$

Next we plot these points and connect them with a smooth curve.

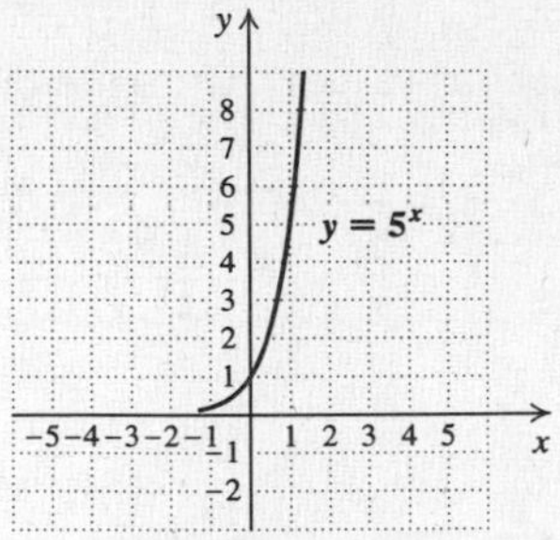

4.

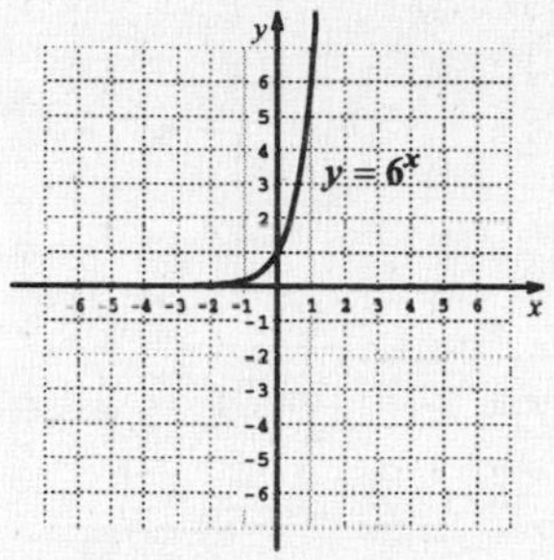

5. Graph: $y = 2^{x+1}$

We compute some function values, thinking of y as f(x), and keep the results in a table.

$f(0) = 2^{0+1} = 2^1 = 2$

$f(-1) = 2^{-1+1} = 2^0 = 1$

$f(-2) = 2^{-2+1} = 2^{-1} = \frac{1}{2^1} = \frac{1}{2}$

$f(-3) = 2^{-3+1} = 2^{-2} = \frac{1}{2^2} = \frac{1}{4}$

$f(1) = 2^{1+1} = 2^2 = 4$

$f(2) = 2^{2+1} = 2^3 = 8$

x	y, or f(x)
0	2
-1	1
-2	$\frac{1}{2}$
-3	$\frac{1}{4}$
1	4
2	8

Next we plot these points and connect them with a smooth curve.

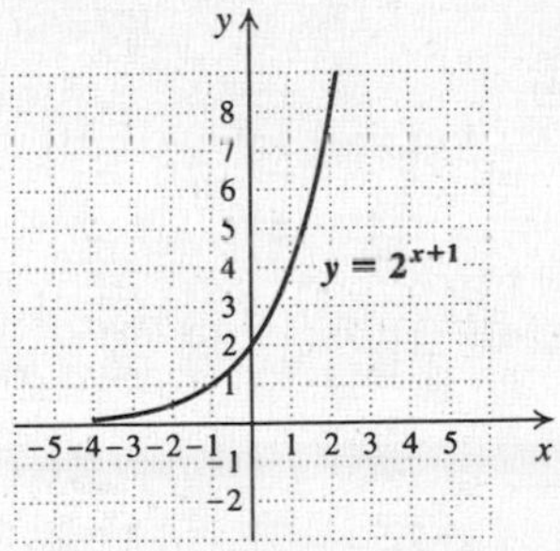

6.

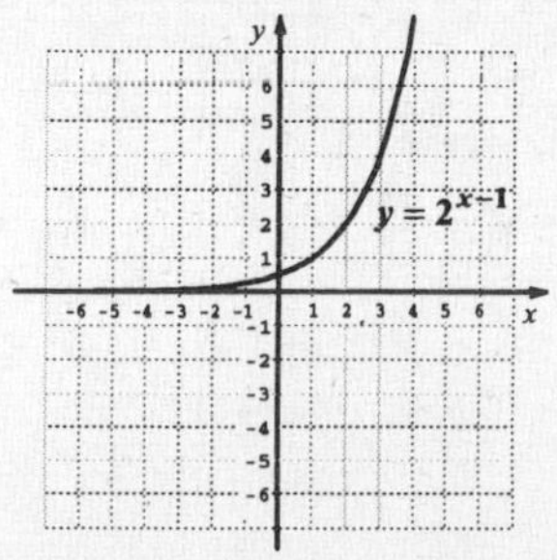

7. Graph: $y = 3^{x-2}$

We compute some function values, thinking of y as f(x), and keep the results in a table.

$f(0) = 3^{0-2} = 3^{-2} = \frac{1}{3^2} = \frac{1}{9}$

$f(1) = 3^{1-2} = 3^{-1} = \frac{1}{3^1} = \frac{1}{3}$

$f(2) = 3^{2-2} = 3^0 = 1$

$f(3) = 3^{3-2} = 3^1 = 3$

$f(4) = 3^{4-2} = 3^2 = 9$

$f(-1) = 3^{-1-2} = 3^{-3} = \frac{1}{3^3} = \frac{1}{27}$

$f(-2) = 3^{-2-2} = 3^{-4} = \frac{1}{3^4} = \frac{1}{81}$

x	y, or f(x)
0	$\frac{1}{9}$
1	$\frac{1}{3}$
2	1
3	3
4	9
-1	$\frac{1}{27}$
-2	$\frac{1}{81}$

Next we plot these points and connect them with a smooth curve.

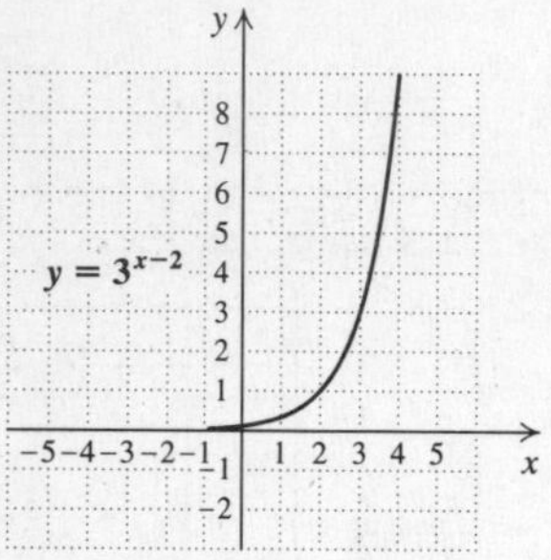

8.

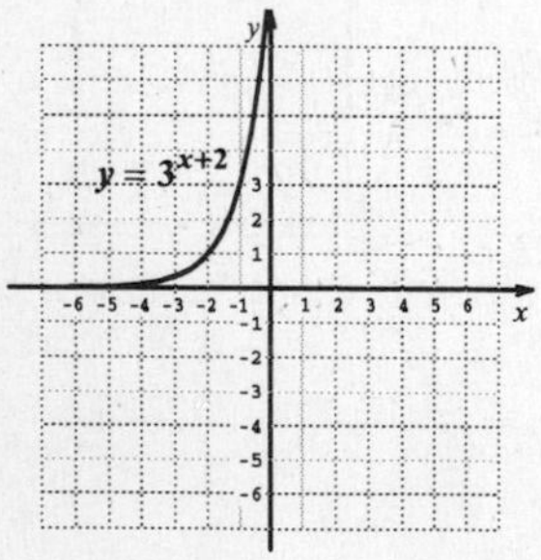

9. Graph: $y = 2^x - 3$

We construct a table of values, thinking of y as f(x). Then we plot the points and connect them with a smooth curve.

$f(0) = 2^0 - 3 = 1 - 3 = -2$

$f(1) = 2^1 - 3 = 2 - 3 = -1$

$f(2) = 2^2 - 3 = 4 - 3 = 1$

$f(3) = 2^3 - 3 = 8 - 3 = 5$

$f(-1) = 2^{-1} - 3 = \frac{1}{2} - 3 = -\frac{5}{2}$

$f(-2) = 2^{-2} - 3 = \frac{1}{4} - 3 = -\frac{11}{4}$

x	y, or f(x)
0	-2
1	-1
2	1
3	5
-1	$-\frac{5}{2}$
-2	$-\frac{11}{4}$

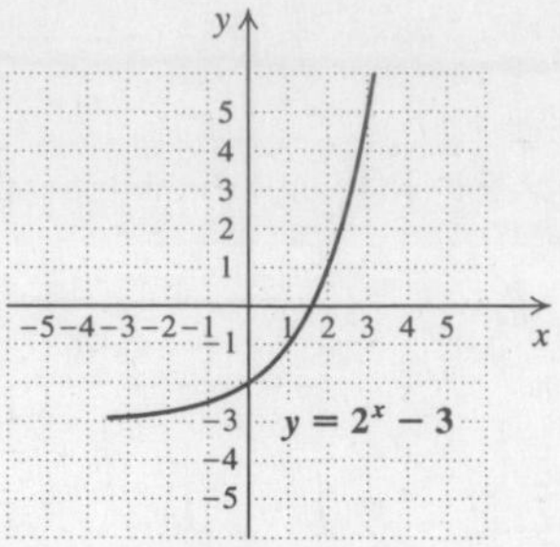

10.

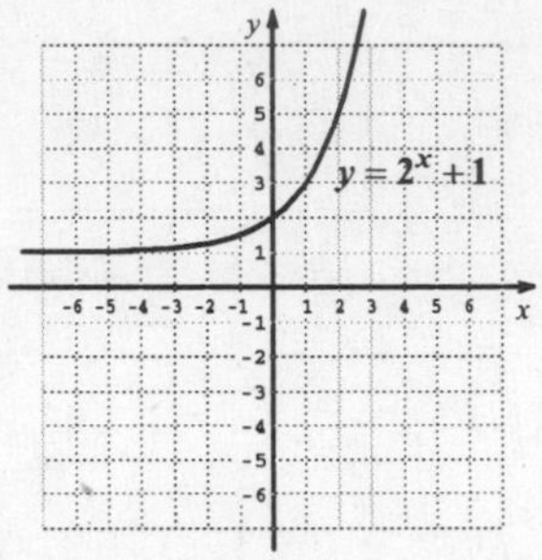

11. Graph: $y = 5^{x+3}$

We construct a table of values, thinking of y as f(x). Then we plot the points and connect them with a smooth curve.

$f(0) = 5^{0+3} = 5^3 = 125$

$f(-1) = 5^{-1+3} = 5^2 = 25$

$f(-2) = 5^{-2+3} = 5^1 = 5$

$f(-3) = 5^{-3+3} = 5^0 = 1$

$f(-4) = 5^{-4+3} = 5^{-1} = \frac{1}{5}$

$f(-5) = 5^{-5+3} = 5^{-2} = \frac{1}{25}$

x	y, or f(x)
0	125
-1	25
-2	5
-3	1
-4	$\frac{1}{5}$
-5	$\frac{1}{25}$

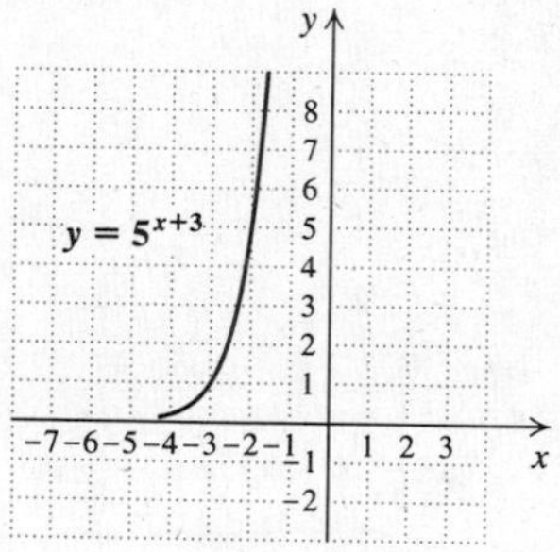

12.

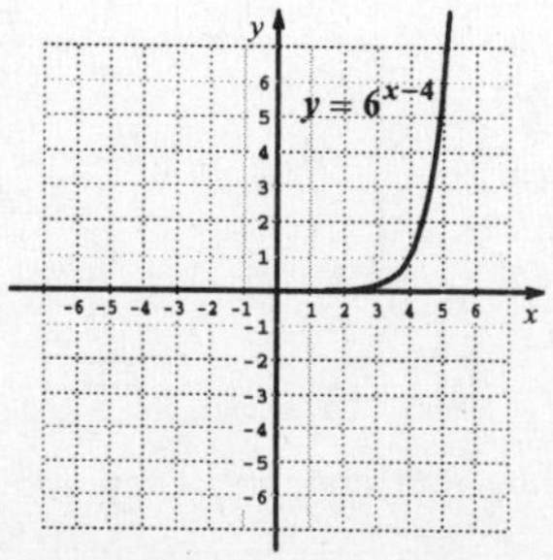

13. Graph: $y = \left(\frac{1}{2}\right)^x$

We construct a table of values, thinking of y as f(x). Then we plot the points and connect them with a smooth curve.

$f(0) = \left(\frac{1}{2}\right)^0 = 1$

$f(1) = \left(\frac{1}{2}\right)^1 = \frac{1}{2}$

$f(2) = \left(\frac{1}{2}\right)^2 = \frac{1}{4}$

$f(3) = \left(\frac{1}{2}\right)^3 = \frac{1}{8}$

$f(-1) = \left(\frac{1}{2}\right)^{-1} = \frac{1}{\left(\frac{1}{2}\right)^1} = \frac{1}{\frac{1}{2}} = 2$

$f(-2) = \left(\frac{1}{2}\right)^{-2} = \frac{1}{\left(\frac{1}{2}\right)^2} = \frac{1}{\frac{1}{4}} = 4$

$f(-3) = \left(\frac{1}{2}\right)^{-3} = \frac{1}{\left(\frac{1}{2}\right)^3} = \frac{1}{\frac{1}{8}} = 8$

x	y, or f(x)
0	1
1	$\frac{1}{2}$
2	$\frac{1}{4}$
3	$\frac{1}{8}$
-1	2
-2	4
-3	8

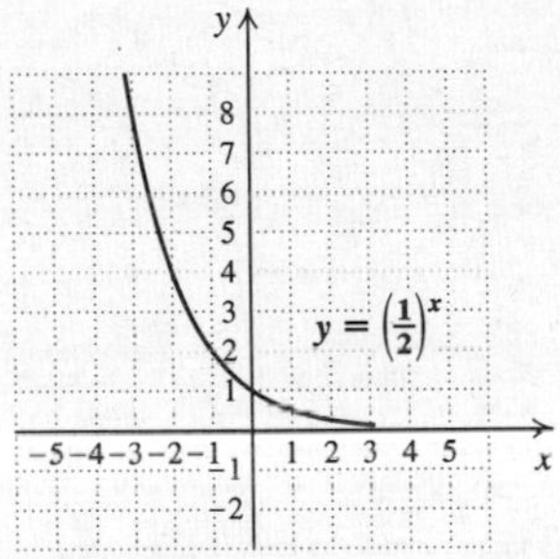

14.

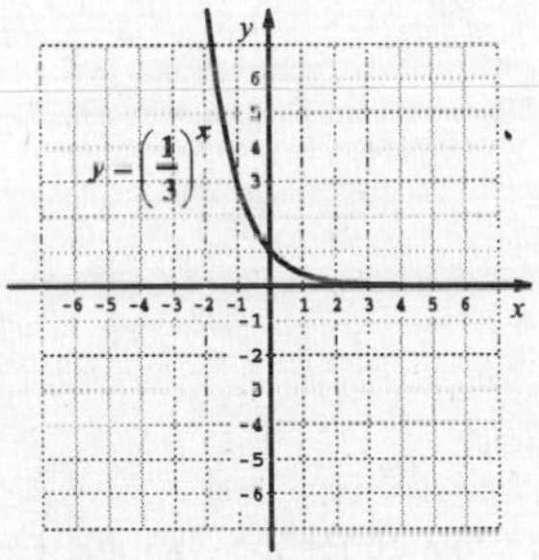

15. Graph: $y = \left(\frac{1}{5}\right)^x$

We construct a table of values, thinking of y as f(x). Then we plot the points and connect them with a smooth curve.

$f(0) = \left(\frac{1}{5}\right)^0 = 1$

$f(1) = \left(\frac{1}{5}\right)^1 = \frac{1}{5}$

$f(2) = \left(\frac{1}{5}\right)^2 = \frac{1}{25}$

$f(-1) = \left(\frac{1}{5}\right)^{-1} = \frac{1}{\frac{1}{5}} = 5$

$f(-2) = \left(\frac{1}{5}\right)^{-2} = \frac{1}{\frac{1}{25}} = 25$

x	y, or f(x)
0	1
1	$\frac{1}{5}$
2	$\frac{1}{25}$
-1	5
-2	25

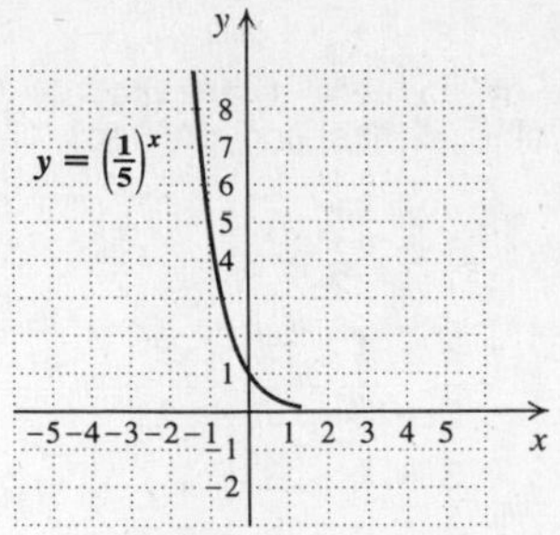

16.

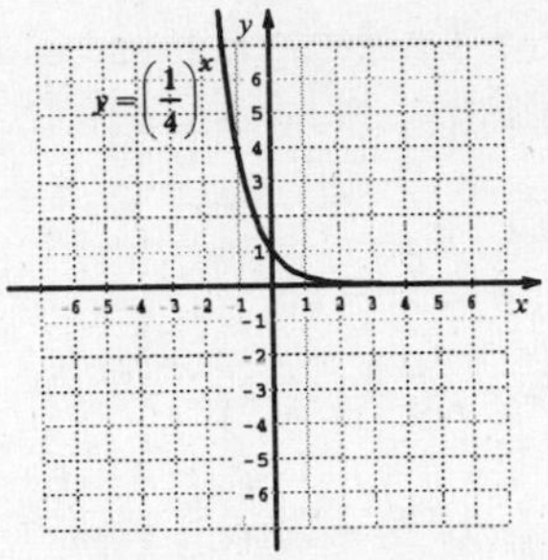

17. Graph: $y = 2^{2x-1}$

We construct a table of values, thinking of y as f(x). Then we plot the points and connect them with a smooth curve.

$f(0) = 2^{2\cdot 0-1} = 2^{-1} = \frac{1}{2}$

$f(1) = 2^{2\cdot 1-1} = 2^1 = 2$

$f(2) = 2^{2\cdot 2-1} = 2^3 = 8$

$f(-1) = 2^{2(-1)-1} = 2^{-3} = \frac{1}{8}$

$f(-2) = 2^{2(-2)-1} = 2^{-5} = \frac{1}{32}$

x	y, or f(x)
0	$\frac{1}{2}$
1	2
2	8
-1	$\frac{1}{8}$
-2	$\frac{1}{32}$

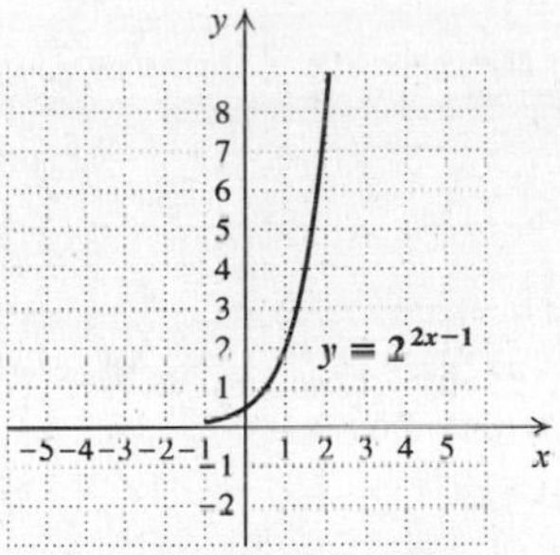

18.

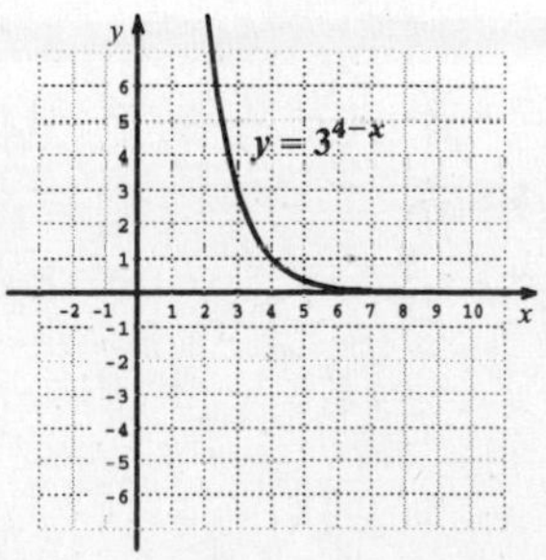

19. Graph: $y = 2^{x-1} - 3$

We construct a table of values, thinking of y as f(x). Then we plot the points and connect them with a smooth curve.

$f(0) = 2^{0-1} - 3 = 2^{-1} - 3 = \frac{1}{2} - 3 = -\frac{5}{2}$

$f(1) = 2^{1-1} - 3 = 2^{0} - 3 = 1 - 3 = -2$

$f(2) = 2^{2-1} - 3 = 2^{1} - 3 = 2 - 3 = -1$

$f(3) = 2^{3-1} - 3 = 2^{2} - 3 = 4 - 3 = 1$

$f(4) = 2^{4-1} - 3 = 2^{3} - 3 = 8 - 3 = 5$

$f(-1) = 2^{-1-1} - 3 = 2^{-2} - 3 = \frac{1}{4} - 3 = -\frac{11}{4}$

$f(-2) = 2^{-2-1} - 3 = 2^{-3} - 3 = \frac{1}{8} - 3 = -\frac{23}{8}$

x	y, or f(x)
0	$-\frac{5}{2}$
1	-2
2	-1
3	1
4	5
-1	$-\frac{11}{4}$
-2	$-\frac{23}{8}$

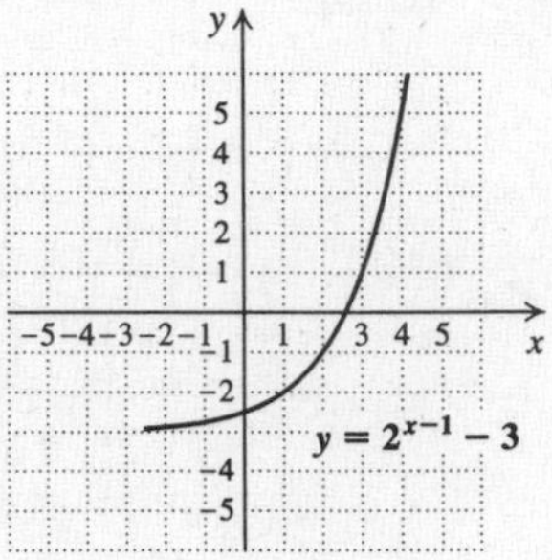

20.

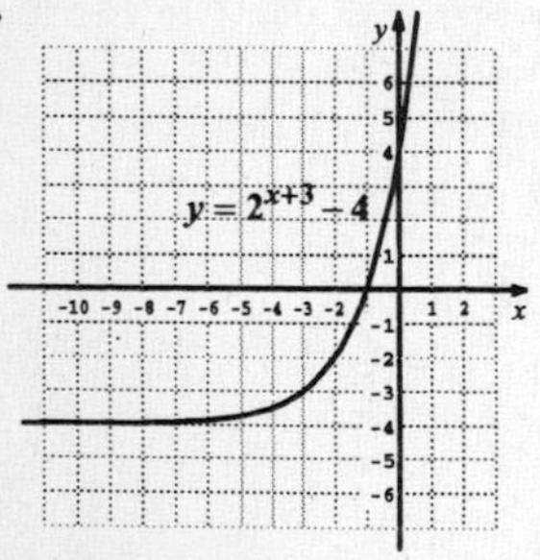

21. Graph: $x = 3^y$

We can find ordered pairs by choosing values for y and then computing values for x.

For y = 0, $x = 3^0 = 1$.

For y = 1, $x = 3^1 = 3$.

For y = 2, $x = 3^2 = 9$.

For y = 3, $x = 3^3 = 27$.

For y = -1, $x = 3^{-1} = \frac{1}{3^1} = \frac{1}{3}$.

For y = -2, $x = 3^{-2} = \frac{1}{3^2} = \frac{1}{9}$.

For y = -3, $x = 3^{-3} = \frac{1}{3^3} = \frac{1}{27}$.

x	y
1	0
3	1
9	2
27	3
$\frac{1}{3}$	-1
$\frac{1}{9}$	-2
$\frac{1}{27}$	-3

(1) Choose values for y.

(2) Compute values for x.

We plot these points and connect them with a smooth curve.

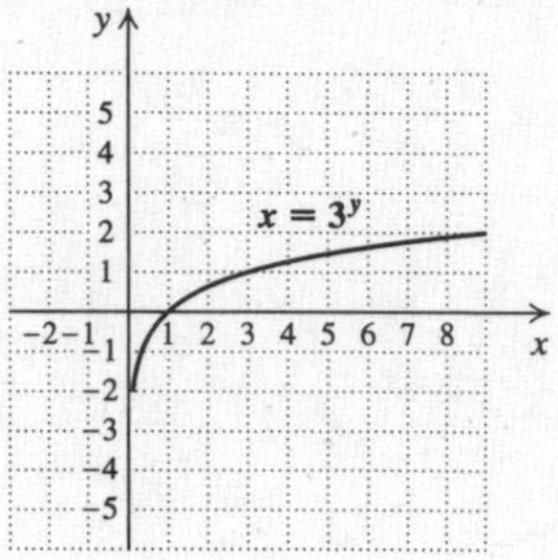

22.

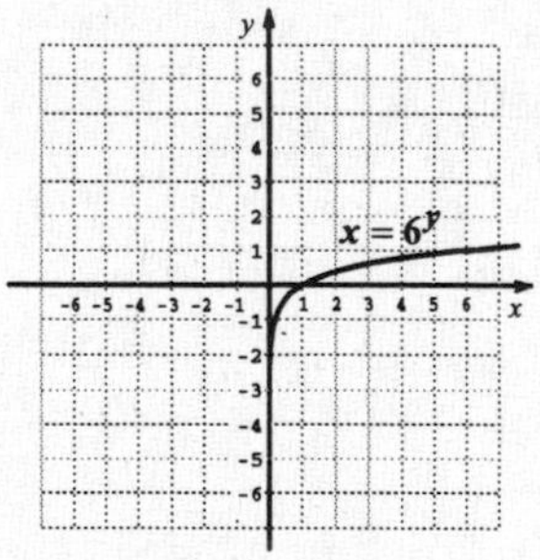

23. Graph: $x = \left(\frac{1}{2}\right)^y$

We can find ordered pairs by choosing values for y and then computing values for x. Then we plot these points and connect them with a smooth curve.

For y = 0, $x = \left(\frac{1}{2}\right)^0 = 1$.

For y = 1, $x = \left(\frac{1}{2}\right)^1 = \frac{1}{2}$.

For y = 2, $x = \left(\frac{1}{2}\right)^2 = \frac{1}{4}$.

For y = 3, $x = \left(\frac{1}{2}\right)^3 = \frac{1}{8}$.

For y = -1, $x = \left(\frac{1}{2}\right)^{-1} = \frac{1}{\frac{1}{2}} = 2$.

For y = -2, $x = \left(\frac{1}{2}\right)^{-2} = \frac{1}{\frac{1}{4}} = 4$.

For y = -3, $x = \left(\frac{1}{2}\right)^{-3} = \frac{1}{\frac{1}{8}} = 8$.

x	y
1	0
$\frac{1}{2}$	1
$\frac{1}{4}$	2
$\frac{1}{8}$	3
2	-1
4	-2
8	-3

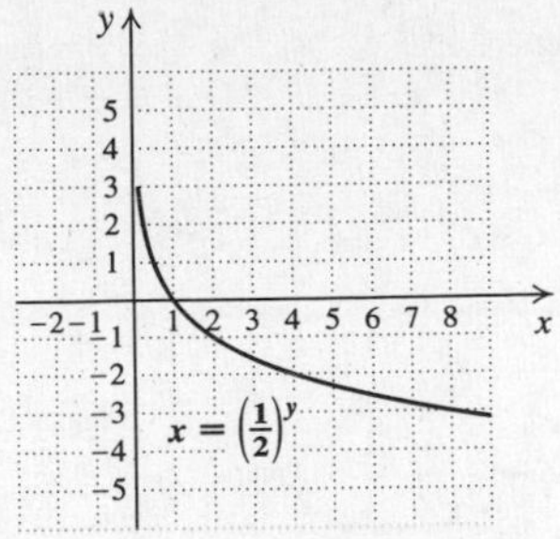

24.

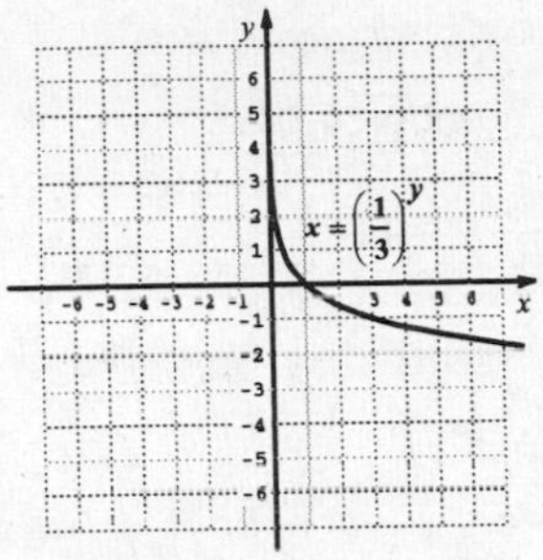

25. Graph: $x = 5^y$

We can find ordered pairs by choosing values for y and then computing values for x. Then we plot these points and connect them with a smooth curve.

For $y = 0$, $x = 5^0 = 1$.

For $y = 1$, $x = 5^1 = 5$.

For $y = 2$, $x = 5^2 = 25$.

For $y = -1$, $x = 5^{-1} = \frac{1}{5}$.

For $y = -2$, $x = 5^{-2} = \frac{1}{25}$.

x	y
1	0
5	1
25	2
$\frac{1}{5}$	-1
$\frac{1}{25}$	-2

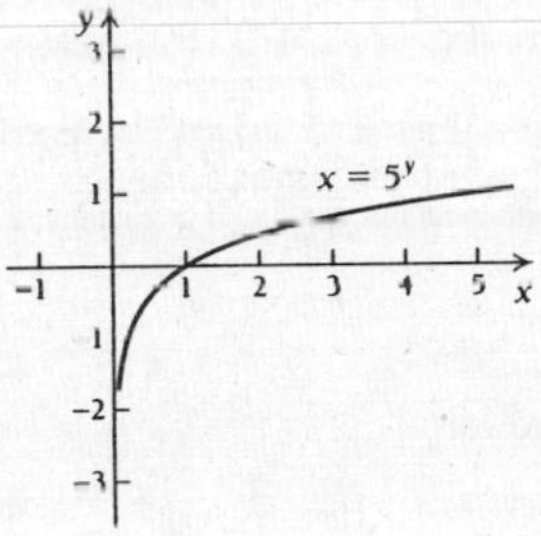

26.

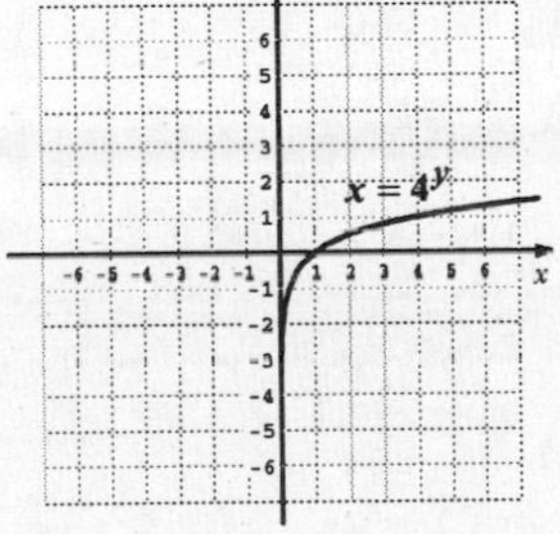

27. Graph: $x = \left(\frac{2}{3}\right)^y$

We can find ordered pairs by choosing values for y and then computing values for x. Then we plot these points and connect them with a smooth curve.

For $y = 0$, $x = \left(\frac{2}{3}\right)^0 = 1$.

For $y = 1$, $x = \left(\frac{2}{3}\right)^1 = \frac{2}{3}$.

For $y = 2$, $x = \left(\frac{2}{3}\right)^2 = \frac{4}{9}$.

For $y = 3$, $x = \left(\frac{2}{3}\right)^3 = \frac{8}{27}$.

For $y = -1$, $x = \left(\frac{2}{3}\right)^{-1} = \frac{1}{\frac{2}{3}} = \frac{3}{2}$.

For $y = -2$, $x = \left(\frac{2}{3}\right)^{-2} = \frac{1}{\frac{4}{9}} = \frac{9}{4}$.

For $y = -3$, $x = \left(\frac{2}{3}\right)^{-3} = \frac{1}{\frac{8}{27}} = \frac{27}{8}$.

x	y
1	0
$\frac{2}{3}$	1
$\frac{4}{9}$	2
$\frac{8}{27}$	3
$\frac{3}{2}$	-1
$\frac{9}{4}$	-2
$\frac{27}{8}$	-3

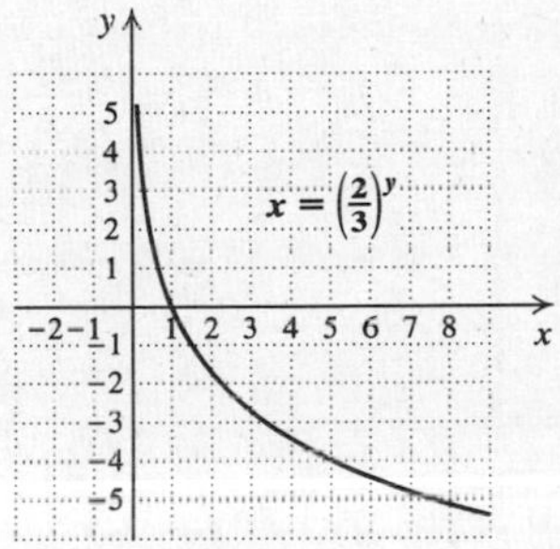

28.

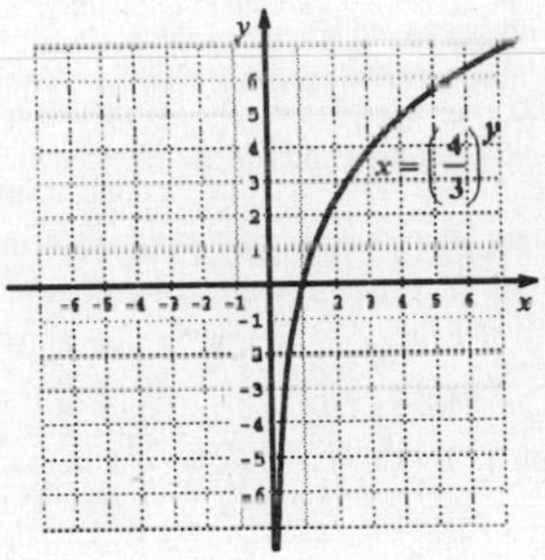

29. Graph $y = 2^x$ (see Exercise 1) and $x = 2^y$ (see Example 4) using the same set of axes.

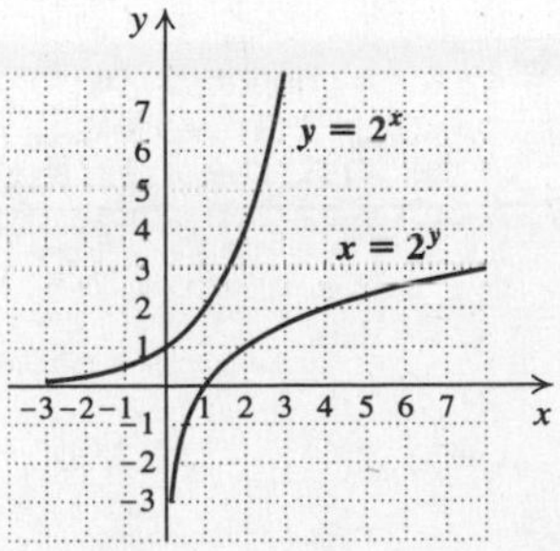

30. 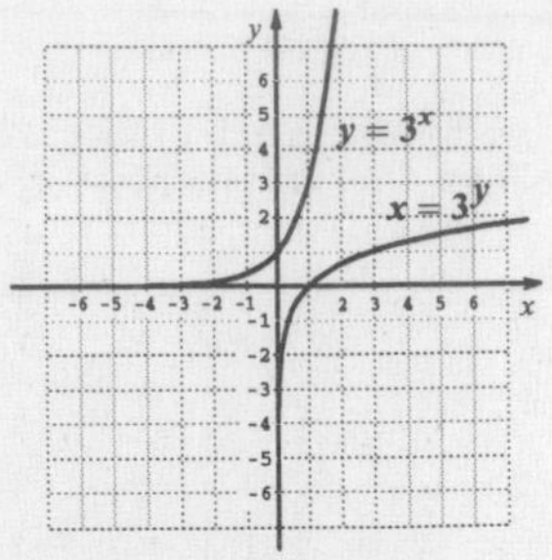

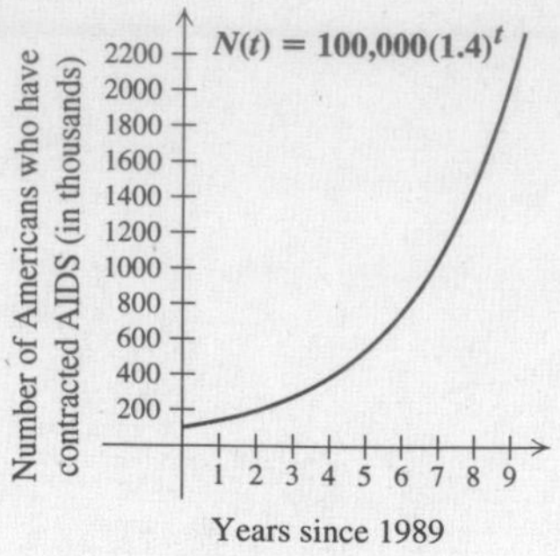

31. Graph $y = \left(\frac{1}{2}\right)^x$ (see Exercise 13) and $x = \left(\frac{1}{2}\right)^y$ (see Exercise 23) using the same set of axes.

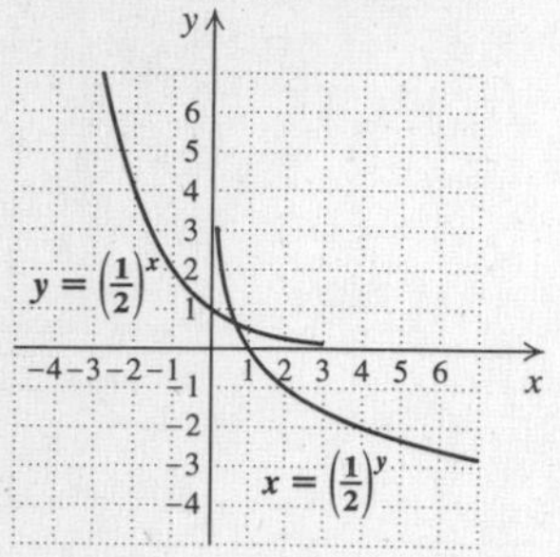

32. 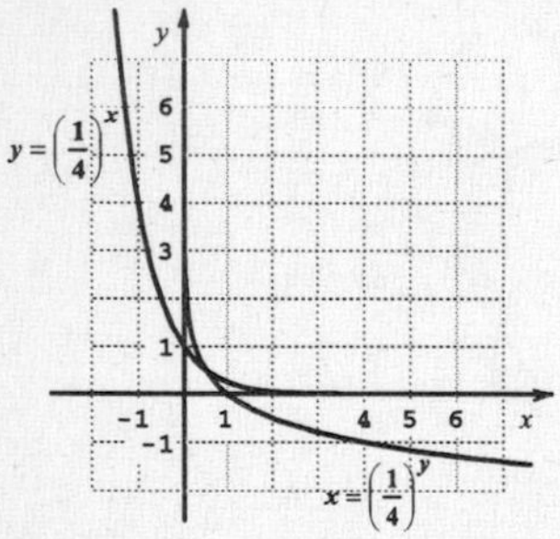

33. Keep in mind that t represents the number of years after 1989.

a) For 1993, t = 1993 - 1989, or 4:

$N(4) = 100{,}000(1.4)^4$

$= 384{,}160$

b) For 1998, t = 1998 - 1989, or 9:

$N(9) = 100{,}000(1.4)^9$

$= 2{,}066{,}105$

c) We use the function values computed in parts (a) and (b), and others if we wish, to draw the graph. Note that the axes are scaled differently because of the large numbers.

34. a) 4243; 6000; 8485; 12,000; 24,000

b) 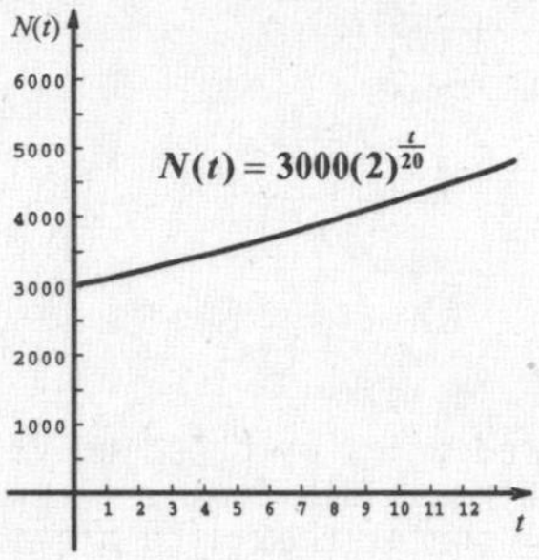

35. a) Substitute for t.

$N(0) = 250{,}000\left(\frac{2}{3}\right)^0 = 250{,}000(1) = 250{,}000;$

$N(1) = 250{,}000\left(\frac{2}{3}\right)^1 = 250{,}000\left(\frac{2}{3}\right) \approx 166{,}667;$

$N(4) = 250{,}000\left(\frac{2}{3}\right)^4 = 250{,}000\left(\frac{16}{81}\right) \approx 49{,}383;$

$N(10) = 250{,}000\left(\frac{2}{3}\right)^{10} = 250{,}000\left(\frac{1024}{59{,}049}\right) \approx 4335$

b) Use the function values computed in part (a) to draw the graph of the function.

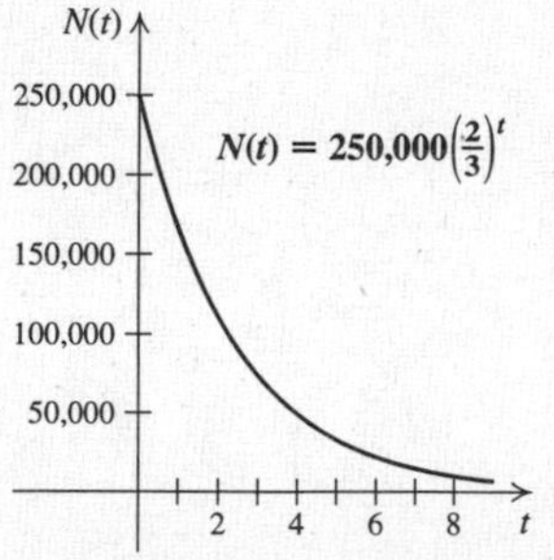

36. a) \$5200; \$4160; \$3328; \$1703.94; \$558.35

b) 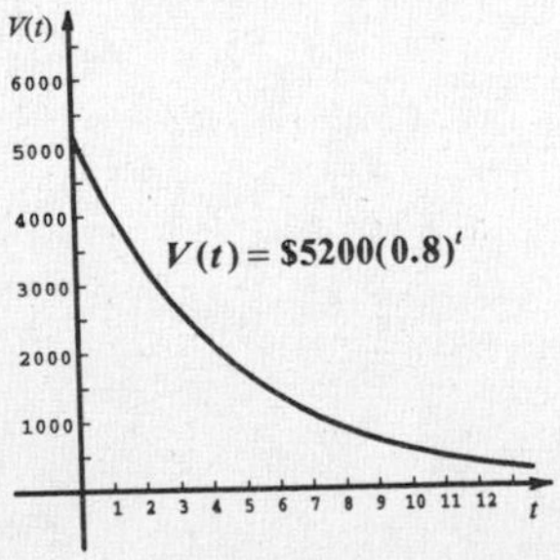

37\. a) Keep in mind that t represents the number of years after 1985 and that N is given in millions.

For 1985, t = 0:

$N(0) = 7.5(6)^{0.5(0)} = 7.5(6)^0 = 7.5(1) = 7.5$ million;

For 1986, t = 1:

$N(1) = 7.5(6)^{0.5(1)} = 7.5(6)^{0.5} \approx 7.5(2.449489743) \approx 18.4$ million;

For 1988, t = 1988 - 1985, or 3:

$N(3) = 7.5(6)^{0.5(3)} = 7.5(6)^{1.5} \approx 7.5(14.69693846) \approx 110.2$ million;

For 1990, t = 1990 - 1985, or 5:

$N(5) = 7.5(6)^{0.5(5)} = 7.5(6)^{2.5} \approx 7.5(88.18163074) \approx 661.4$ million;

For 1995, t = 1995 - 1985, or 10:

$N(10) = 7.5(6)^{0.5(10)} = 7.5(6)^5 = 7.5(7776) = 58{,}320$ million;

For 2000, t = 2000 - 1985, or 15:

$N(15) = 7.5(6)^{0.5(15)} = 7.5(6)^{7.5} \approx 7.5(685{,}700.3606) \approx 5{,}142{,}752.7$ million

b) Use the function values computed in part (a) to draw the graph of the function.

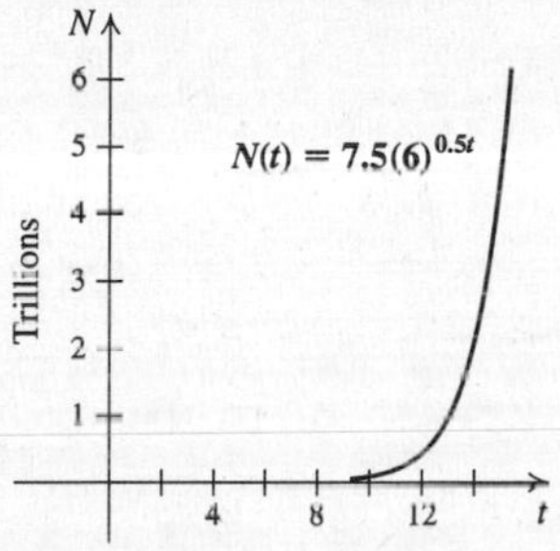

38\. a) 2.6 lb; 3.7 lb; 10.9 lb; 14.6 lb

b) 22.6 lb

c)

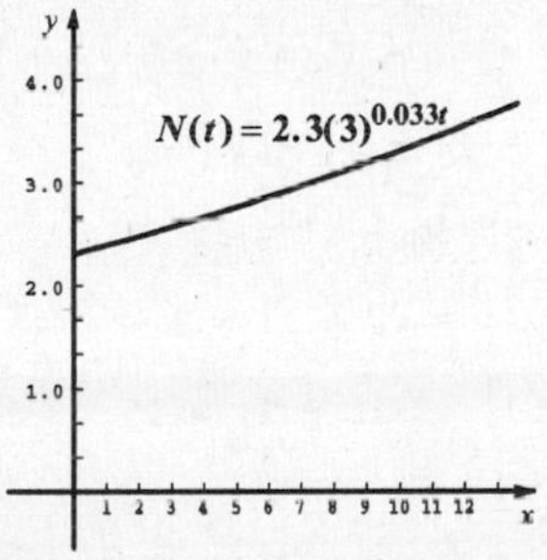

39\. $x^{-5} \cdot x^3 = x^{-5+3} = x^{-2}$, or $\frac{1}{x^2}$

40\. x^{-12}, or $\frac{1}{x^{12}}$

41\. $\frac{x^{-3}}{x^4} = x^{-3-4} = x^{-7}$, or $\frac{1}{x^7}$

42\. 1

43\. 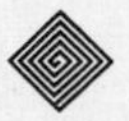

44\. 

45\. Since the bases are the same, the one with the larger exponent is the larger number. Thus $\pi^{2.4}$ is larger.

46\. $8^{\sqrt{3}}$

47\. Graph: $f(x) = (2.3)^x$

Use a calculator with a power key to construct a table of values. (We will round values of f(x) to the nearest hundredth.) Then plot these points and connect them with a smooth curve.

x	f(x)
0	1
1	2.3
2	5.29
3	12.17
-1	0.43
-2	0.19

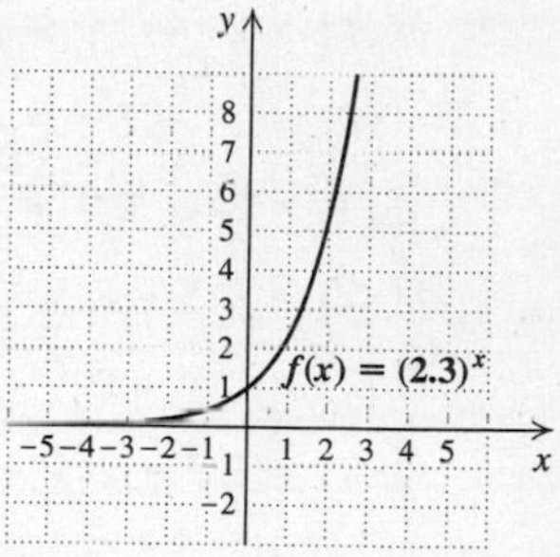

48\. 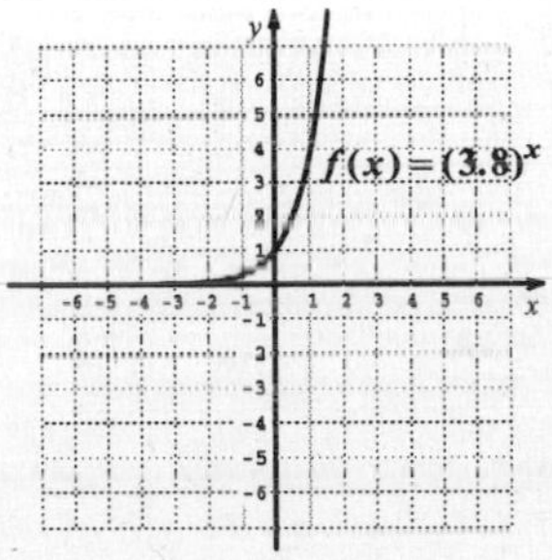

49\. Graph: $g(x) = (0.125)^x$

Use the procedure described in Exercise 47, rounding values of g(x) to the nearest thousandth.

x	g(x)
0	1
1	0.125
2	0.016
3	0.002
-1	8
-2	64

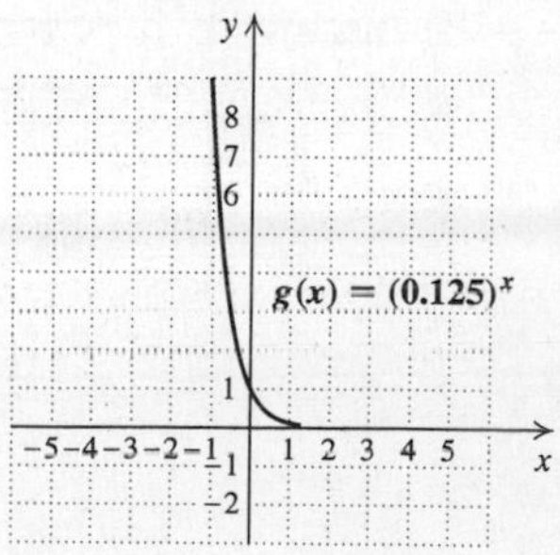

50.

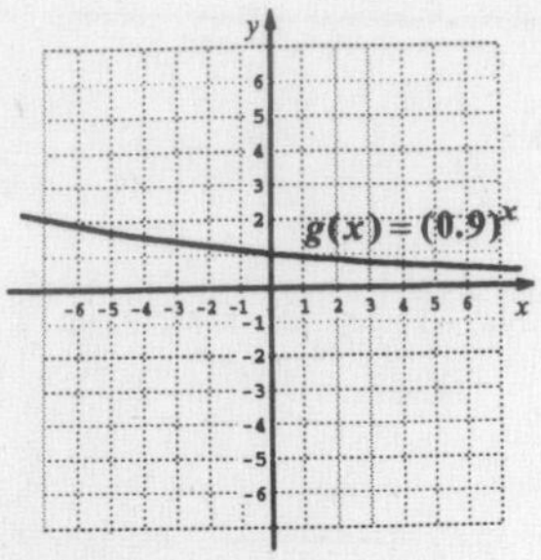

51. Graph: $y = 2^x + 2^{-x}$

Construct a table of values, thinking of y as f(x). Then plot these points and connect them with a curve.

$f(0) = 2^0 + 2^{-0} = 1 + 1 = 2$

$f(1) = 2^1 + 2^{-1} = 2 + \frac{1}{2} = 2\frac{1}{2}$

$f(2) = 2^2 + 2^{-2} = 4 + \frac{1}{4} = 4\frac{1}{4}$

$f(3) = 2^3 + 2^{-3} = 8 + \frac{1}{8} = 8\frac{1}{8}$

$f(-1) = 2^{-1} + 2^{-(-1)} = \frac{1}{2} + 2 = 2\frac{1}{2}$

$f(-2) = 2^{-2} + 2^{-(-2)} = \frac{1}{4} + 4 = 4\frac{1}{4}$

$f(-3) = 2^{-3} + 2^{-(-3)} = \frac{1}{8} + 8 = 8\frac{1}{8}$

x	y, or f(x)
0	2
1	$2\frac{1}{2}$
2	$4\frac{1}{4}$
3	$8\frac{1}{8}$
-1	$2\frac{1}{2}$
-2	$4\frac{1}{4}$
-3	$8\frac{1}{8}$

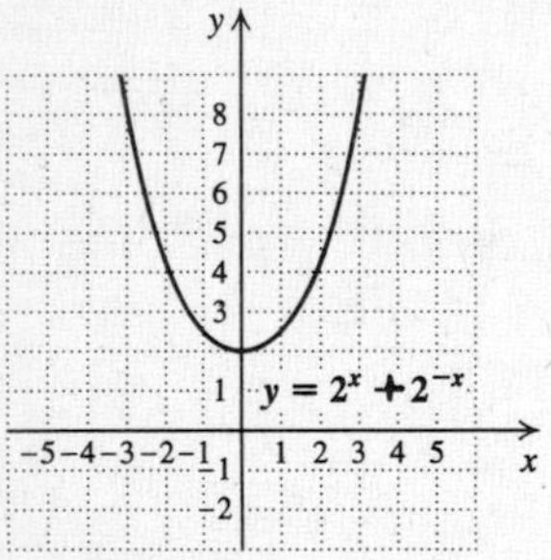

52.

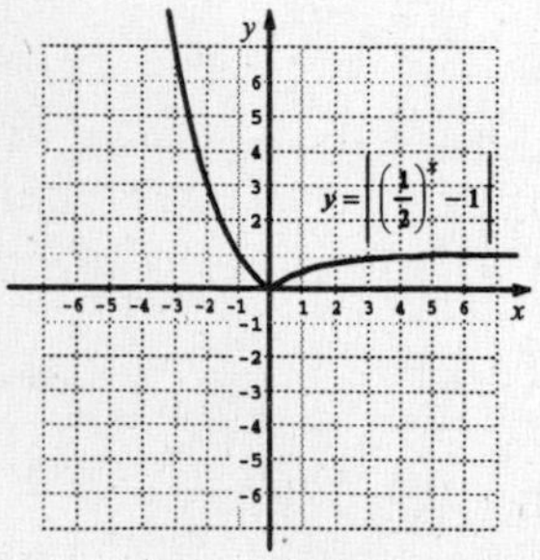

53. Graph: $y = 3^x + 3^{-x}$

We construct a table of values, thinking of y as f(x). Then plot these points and connect them with a curve.

$f(0) = 3^0 + 3^{-0} = 1 + 1 = 2$

$f(1) = 3^1 + 3^{-1} = 3 + \frac{1}{3} = 3\frac{1}{3}$

$f(2) = 3^2 + 3^{-2} = 9 + \frac{1}{9} = 9\frac{1}{9}$

$f(-1) = 3^{-1} + 3^{-(-1)} = \frac{1}{3} + 3 = 3\frac{1}{3}$

$f(-2) = 3^{-2} + 3^{-(-2)} = \frac{1}{9} + 9 = 9\frac{1}{9}$

x	y, or f(x)
0	2
1	$3\frac{1}{3}$
2	$9\frac{1}{9}$
-1	$3\frac{1}{3}$
-2	$9\frac{1}{9}$

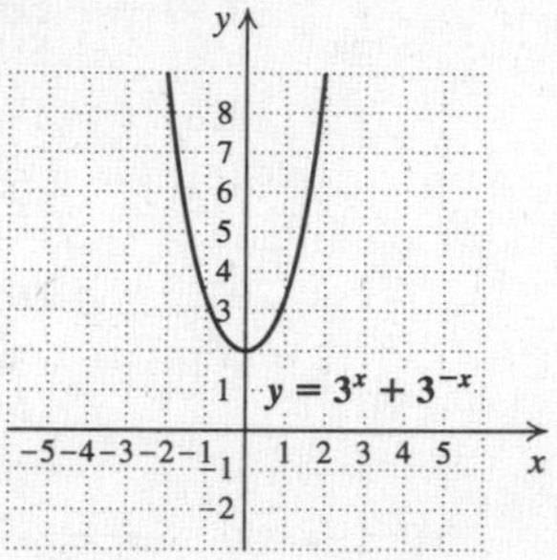

54.

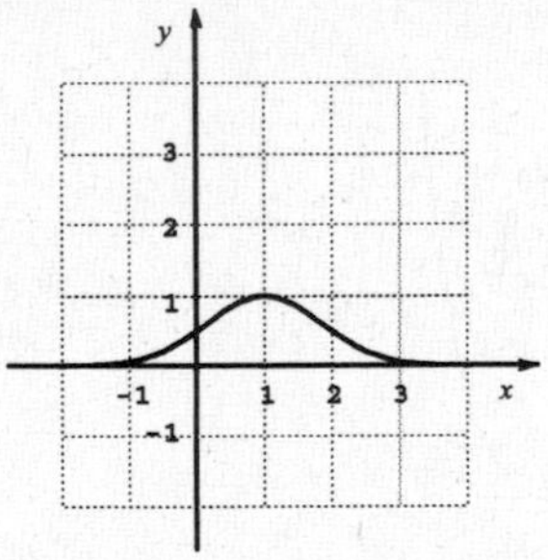

55. Graph: $y = |2x^2 - 1|$

We construct a table of values, thinking of y as f(x). Then we plot these points and connect them with a curve.

$f(0) = |2 0^2 - 1| = |1 - 1| = 0$

$f(1) = |2 1^2 - 1| = |2 - 1| = 1$

$f(2) = |2 2^2 - 1| = |16 - 1| = 15$

$f(-1) = |2(-1)^2 - 1| = |2 - 1| = 1$

$f(-2) = |2(-2)^2 - 1| = |16 - 1| = 15$

x	y, or f(x)
0	0
1	1
2	15
-1	1
-2	15

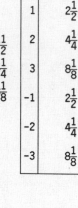

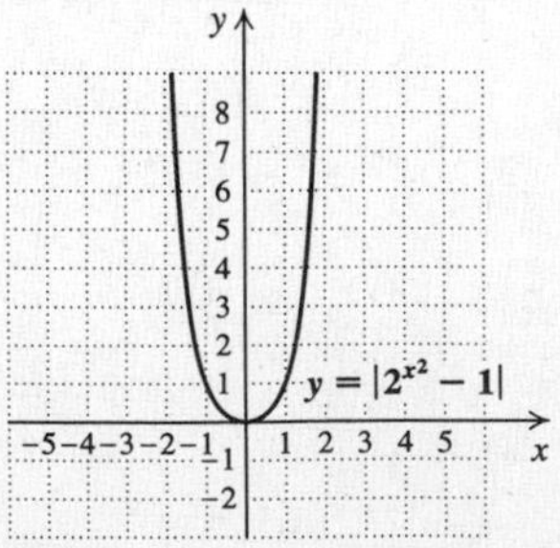

56.

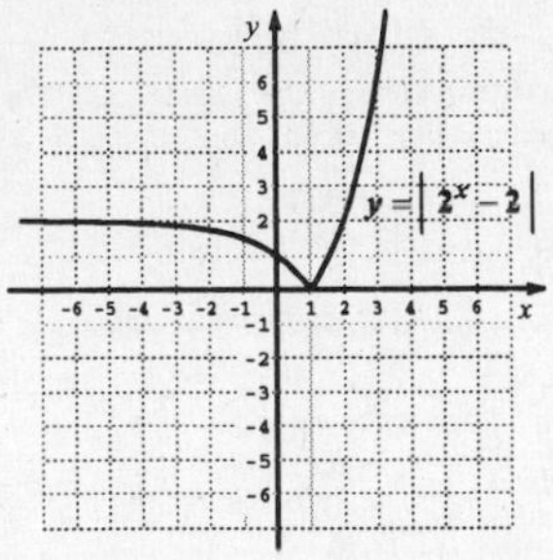

57. $y = 3^{-(x-1)}$ $x = 3^{-(y-1)}$

x	y
0	3
1	1
2	$\frac{1}{3}$
3	$\frac{1}{9}$
-1	9

x	y
3	0
1	1
$\frac{1}{3}$	2
$\frac{1}{9}$	3
9	-1

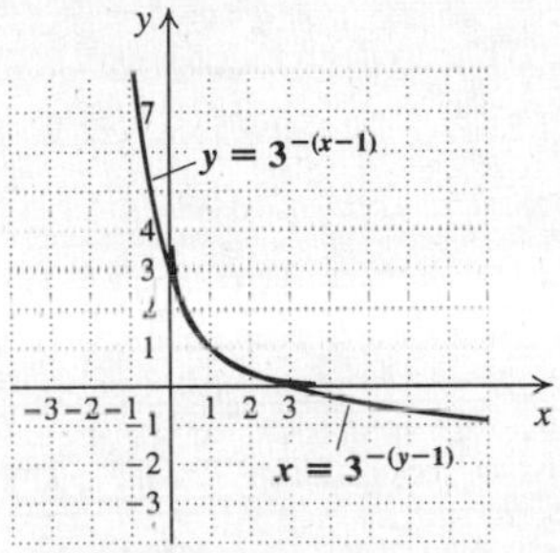

58.

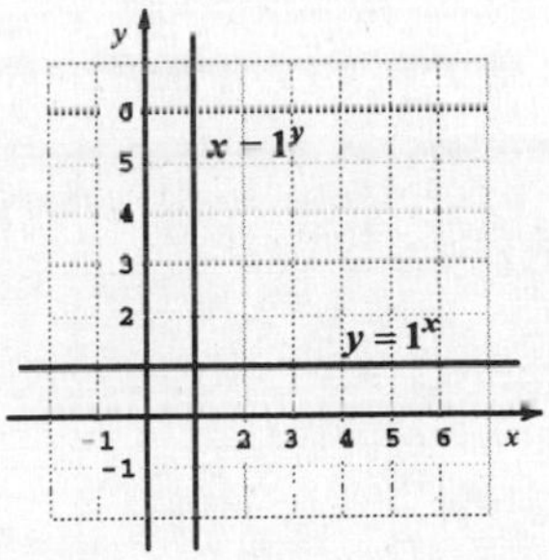

59. a) $S(10) = 200[1 - (0.99)^{10}] \approx 19$ words per minute;

$S(20) = 200[1 - (0.99)^{20}] \approx 36$ words per minute;

$S(40) = 200[1 - (0.99)^{40}] \approx 66$ words per minute;

$S(85) = 200[1 - (0.99)^{85}] \approx 115$ words per minute

b) Use the function values calculated in part (a) to draw the graph.

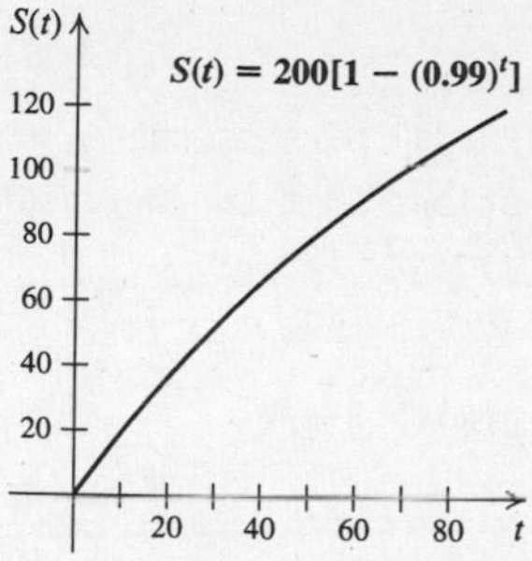

Exercise Set 9.2

1. $f \circ g(x) = f(g(x)) = f(2x - 1) = 3(2x - 1)^2 + 2 =$
$3(4x^2 - 4x + 1) + 2 =$
$12x^2 - 12x + 3 + 2 = 12x^2 - 12x + 5$

$g \circ f(x) = g(f(x)) = g(3x^2 + 2) = 2(3x^2 + 2) - 1 =$
$6x^2 + 4 - 1 = 6x^2 + 3$

2. $f \circ g(x) = 8x^2 - 17$; $g \circ f(x) = 32x^2 + 48x + 13$

3. $f \circ g(x) = f(g(x)) = f\left(\frac{2}{x}\right) = 4\left(\frac{2}{x}\right)^2 - 1 =$
$4\left(\frac{4}{x^2}\right) - 1 = \frac{16}{x^2} - 1$

$g \circ f(x) = g(f(x)) = g(4x^2 - 1) = \frac{2}{4x^2 - 1}$

4. $f \circ g(x) = \frac{3}{2x^2 + 3}$; $g \circ f(x) = \frac{18}{x^2} + 3$

5. $f \circ g(x) = f(g(x)) = f(x^2 - 1) = (x^2 - 1)^2 + 1 =$
$x^4 - 2x^2 + 1 + 1 = x^4 - 2x^2 + 2$

$g \circ f(x) = g(f(x)) = g(x^2 + 1) = (x^2 + 1)^2 - 1 =$
$x^4 + 2x^2 + 1 - 1 = x^4 + 2x^2$

6. $f \circ g(x) = \frac{1}{x^2 + 4x + 4}$; $g \circ f(x) = \frac{1}{x^2} + 2$

7. $h(x) = (5 - 3x)^2$
This is $5 - 3x$ to the 2nd power, so the two most obvious functions are $f(x) = x^2$ and $g(x) = 5 - 3x$.

8. $f(x) = 4x^2 + 9$, $g(x) = 3x - 1$

9. $h(x) = (3x^2 - 7)^5$
This is $3x^2 - 7$ to the 5th power, so the two most obvious functions are $f(x) = x^5$ and $g(x) = 3x^2 - 7$.

10. $f(x) = \sqrt{x}$, $g(x) = 5x + 2$

11. $h(x) = \frac{1}{x - 1}$
This is the reciprocal of $x - 1$, so the two most obvious functions are $f(x) = \frac{1}{x}$ and $g(x) = x - 1$.

12. $f(x) = x + 4$, $g(x) = \frac{3}{x}$

13. $h(x) = \dfrac{1}{\sqrt{7x + 2}}$

This is the reciprocal of the square root of $7x + 2$. Two functions that can be used are $f(x) = \dfrac{1}{\sqrt{x}}$ and $g(x) = 7x + 2$.

14. $f(x) = \sqrt{x} - 3$, $g(x) = x - 7$

15. $h(x) = \dfrac{x^3 + 1}{x^3 - 1}$

Two functions that can be used are $f(x) = \dfrac{x + 1}{x - 1}$ and $g(x) = x^3$.

16. $f(x) = x^4$, $g(x) = \sqrt{x} + 5$

17. The graph of $f(x) = 3x - 4$ is shown below.

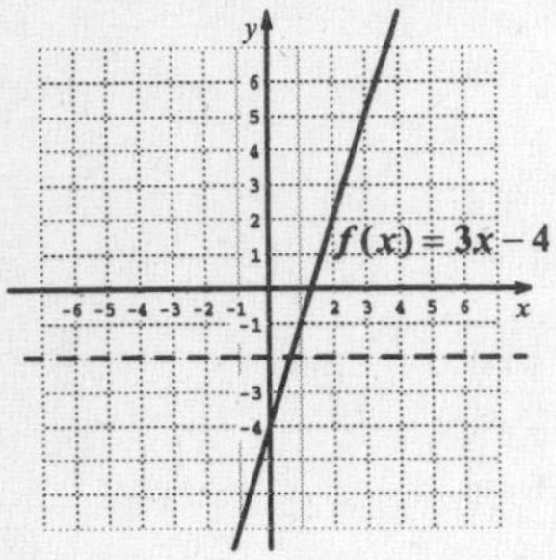

Since there is no horizontal line that crosses the graph more than once, the function is one-to-one.

18. Yes

19. The graph of $f(x) = x^2 - 3$ is shown below.

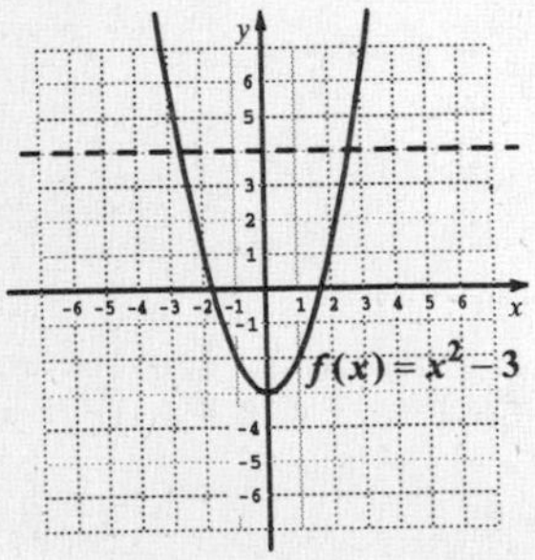

There are many horizontal lines that cross the graph more than once. In particular, the line $y = 4$ crosses the graph more than once. The function is not one-to-one.

20. No

21. The graph of $g(x) = 3^x$ is shown below.

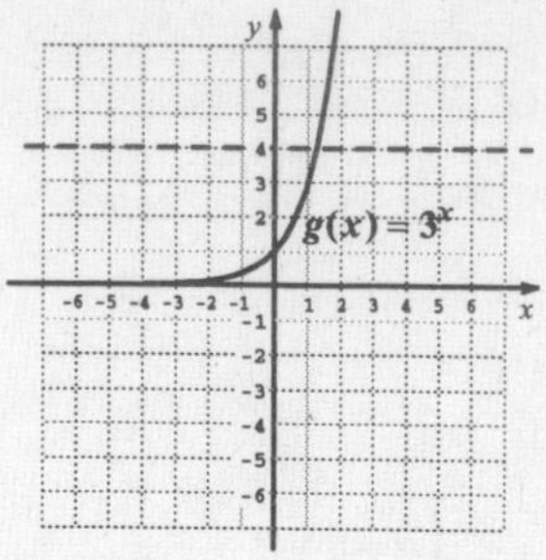

Since no horizontal line crosses the graph more than once, the function is one-to-one.

22. Yes

23. The graph of $g(x) = |x|$ is shown below.

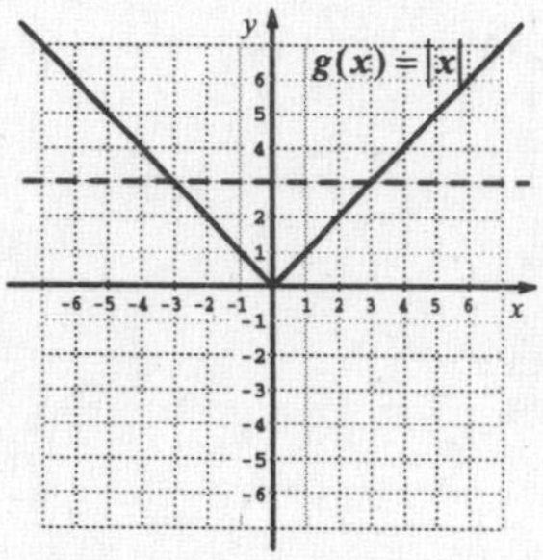

There are many horizontal lines that cross the graph more than once. In particular, the line $y = 3$ crosses the graph more than once. The function is not one-to-one.

24. No

25. a) The function $f(x) = x + 4$ is a linear function that is not constant, so it passes the horizontal line test. Thus, f is one-to-one.

b) Replace $f(x)$ by y: $y = x + 4$

Interchange x and y: $x = y + 4$

Solve for y: $x - 4 = y$

Replace y by $f^{-1}(x)$: $f^{-1}(x) = x - 4$

26. a) Yes

b) $f^{-1}(x) = x - 7$

27. a) The function $f(x) = 5 - x$ is a linear function that is not constant, so it passes the horizontal line test. Thus, f is one-to-one.

b) Replace $f(x)$ by y: $y = 5 - x$

Interchange x and y: $x = 5 - y$

Solve for y: $y = 5 - x$

Replace y by $f^{-1}(x)$: $f^{-1}(x) = 5 - x$

28. a) Yes

b) $f^{-1}(x) = 9 - x$

29. a) The function $g(x) = x - 5$ is a linear function that is not constant, so it passes the horizontal line test. Thus, g is one-to-one.

b) Replace g(x) by y: $y = x - 5$

Interchange x and y: $x = y - 5$

Solve for y: $x + 5 = y$

Replace y by $g^{-1}(x)$: $g^{-1}(x) = x + 5$

30. a) Yes

b) $g^{-1}(x) = x + 8$

31. a) The function $f(x) = 3x$ is a linear function that is not constant, so it passes the horizontal line test. Thus, f is one-to-one.

b) Replace f(x) by y: $y = 3x$

Interchange x and y: $x = 3y$

Solve for y: $\frac{x}{3} = y$

Replace y by $f^{-1}(x)$: $f^{-1}(x) = \frac{x}{3}$

32. a) Yes

b) $f^{-1}(x) = \frac{x}{4}$

33. a) The function $g(x) = 3x + 2$ is a linear function that is not constant, so it passes the horizontal line test. Thus, g is one-to-one.

b) Replace g(x) by y: $y = 3x + 2$

Interchange variables: $x = 3y + 2$

Solve for y: $x - 2 = 3y$

$\frac{x - 2}{3} = y$

Replace y by $g^{-1}(x)$: $g^{-1}(x) = \frac{x - 2}{3}$

34. a) Yes

b) $g^{-1}(x) = \frac{x - 7}{4}$

35. a) The graph of $h(x) = 7$ is shown below. The horizontal line $y = 7$ crosses the graph more than once, so the function is not one-to-one.

36. a) No

37. a) The graph of $f(x) = \frac{1}{x}$ is shown below. It passes the horizontal line test, so the function is one-to-one.

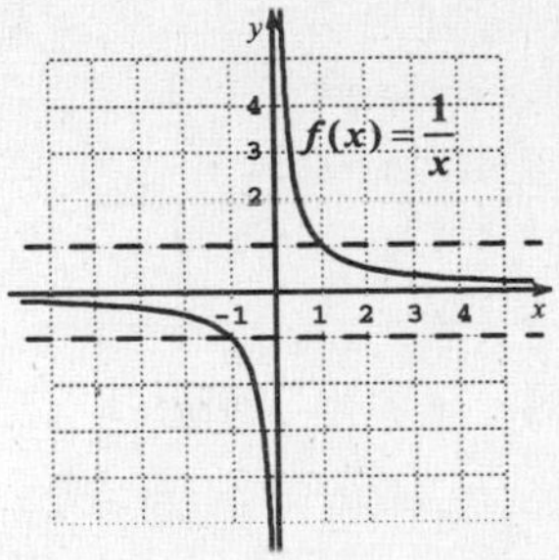

b) Replace f(x) by y: $y = \frac{1}{x}$

Interchange x and y: $x = \frac{1}{y}$

Solve for y: $xy = 1$

$y = \frac{1}{x}$

Replace y by $f^{-1}(x)$: $f^{-1}(x) = \frac{1}{x}$

38. a) Yes

b) $f^{-1}(x) = \frac{3}{x}$

39. a) The function $f(x) = \frac{2x + 1}{3} = \frac{2}{3}x + \frac{1}{3}$ is a linear function that is not constant, so it passes the horizontal line test. Thus, f is one-to-one.

b) Replace f(x) by y: $y = \frac{2x + 1}{3}$

Interchange x and y: $x = \frac{2y + 1}{3}$

Solve for y: $3x = 2y + 1$

$3x - 1 = 2y$

$\frac{3x - 1}{2} = y$

Replace y by $f^{-1}(x)$: $f^{-1}(x) = \frac{3x - 1}{2}$

40. a) Yes

b) $f^{-1}(x) = \frac{5x - 2}{3}$

41. a) The graph of $f(x) = x^3 - 1$ is shown below. It passes the horizontal line test, so the function is one-to-one.

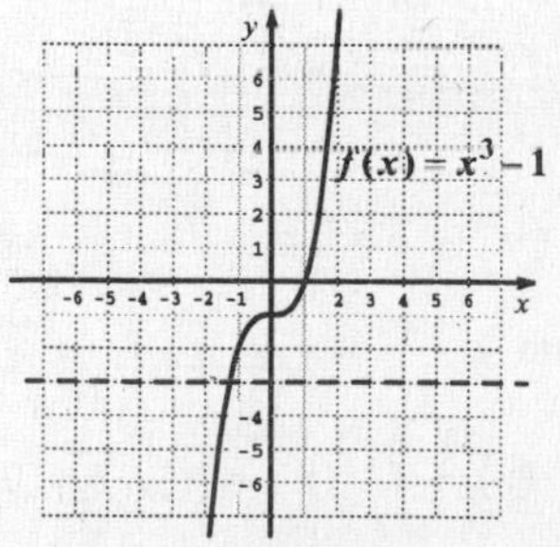

b) Replace f(x) by y: $y = x^3 - 1$

Interchange x and y: $x = y^3 - 1$

Solve for y: $x = y^3 - 1$

$x + 1 = y^3$

$\sqrt[3]{x + 1} = y$

Replace y by $f^{-1}(x)$: $f^{-1}(x) = \sqrt[3]{x + 1}$

42. a) Yes

b) $f^{-1}(x) = \sqrt[3]{x - 5}$

43. a) The graph of $g(x) = (x - 2)^3$ is shown below. It passes the horizontal line test, so the function is one-to-one.

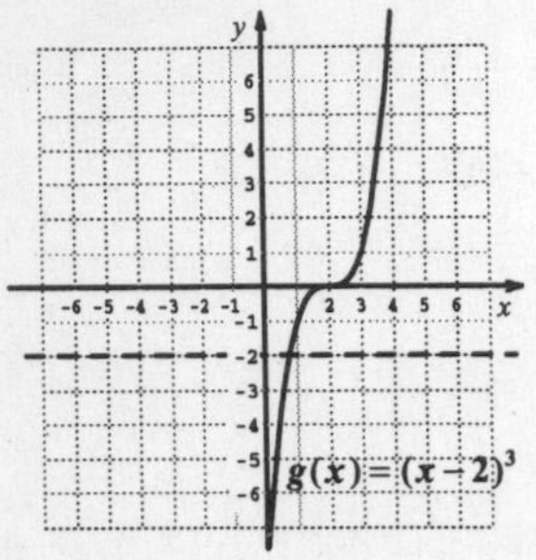

b) Replace g(x) by y: $y = (x - 2)^3$

Interchange x and y: $x = (y - 2)^3$

Solve for y: $\sqrt[3]{x} = y - 2$

$\sqrt[3]{x} + 2 = y$

Replace y by $g^{-1}(x)$: $g^{-1}(x) = \sqrt[3]{x} + 2$

44. a) Yes

b) $g^{-1}(x) = \sqrt[3]{x} - 7$

45. a) The graph of $f(x) = \sqrt{x}$ is shown below. It passes the horizontal line test, so the function is one-to-one.

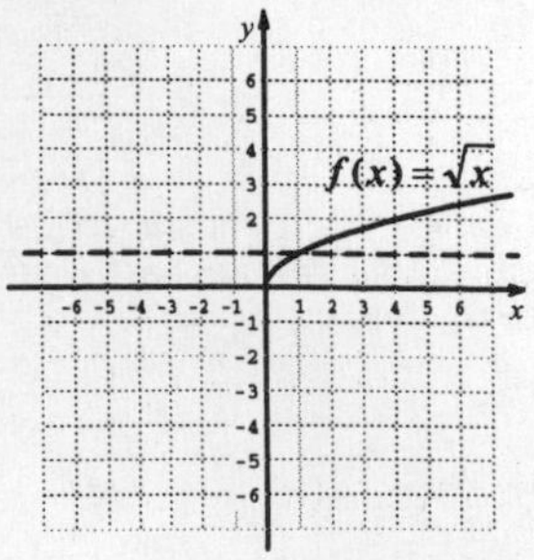

b) Replace f(x) by y: $y = \sqrt{x}$ (Note that $f(x) \geqslant 0$.)

Interchange x and y: $x = \sqrt{y}$

Solve for y: $x^2 = y$

Replace y by $f^{-1}(x)$: $f^{-1}(x) = x^2,\ x \geqslant 0$

46. a) Yes

b) $f^{-1}(x) = x^2 + 1,\ x \geqslant 0$

47. a) The graph of $f(x) = 2x^2 + 3,\ x \geqslant 0$, is shown below. It passes the horizontal line test, so the function is one-to-one.

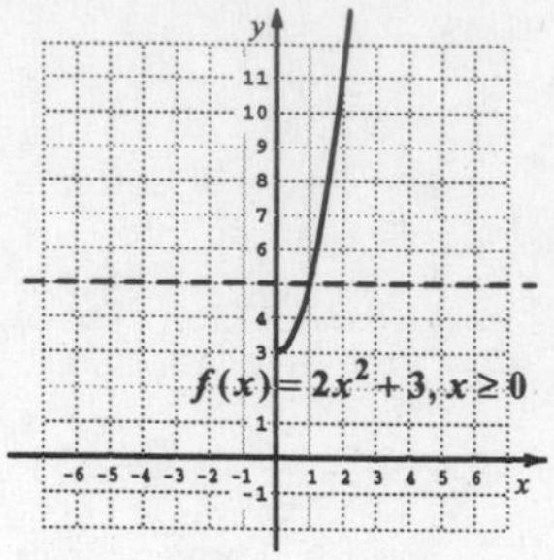

b) Replace f(x) by y: $y = 2x^2 + 3$

Interchange x and y: $x = 2y^2 + 3$

Solve for y: $x - 3 = 2y^2$

$\frac{x - 3}{2} = y^2$

$\sqrt{\frac{x - 3}{2}} = y$

(We take the principal square root since $y \geqslant 0$.)

Replace y by $f^{-1}(x)$: $f^{-1}(x) = \sqrt{\frac{x - 3}{2}}$

48. a) Yes

b) $f^{-1}(x) = \sqrt{\frac{x + 2}{3}}$

49. First graph $f(x) = \frac{1}{2}x - 3$. Then graph the inverse function by reflecting the graph of $f(x) = \frac{1}{2}x - 3$ across the line $y = x$. The graph of the inverse function can also be found by first finding a formula for the inverse, substituting to find function values, and then plotting points.

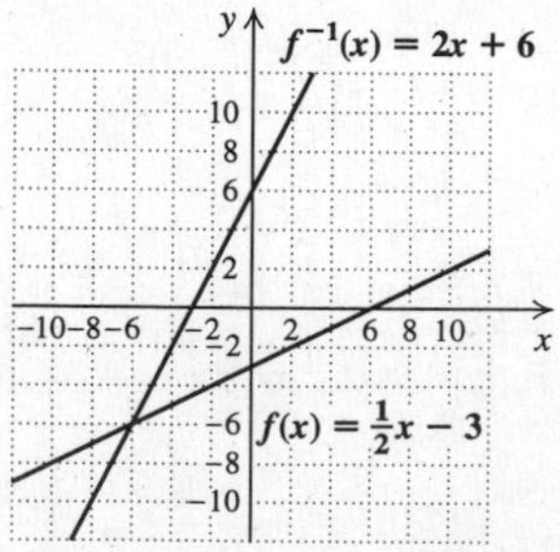

50.

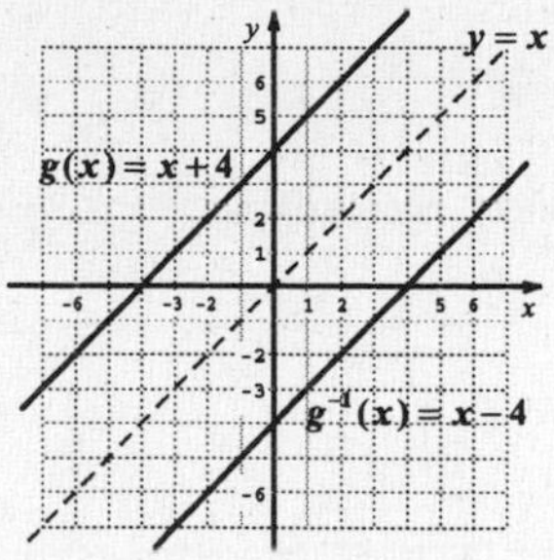

51. Follow the procedure described in Exercise 49 to graph the function and its inverse.

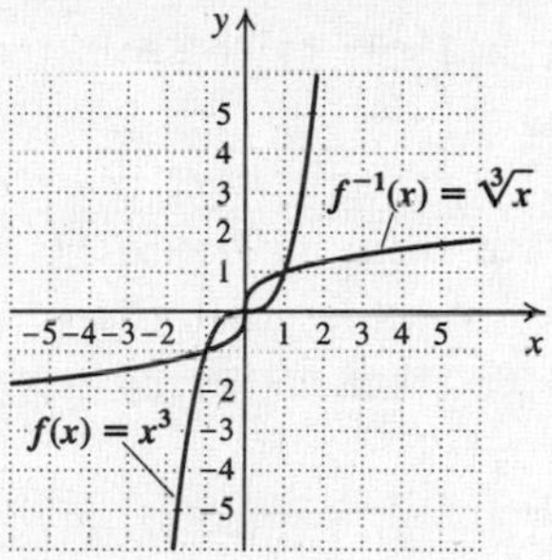

52.

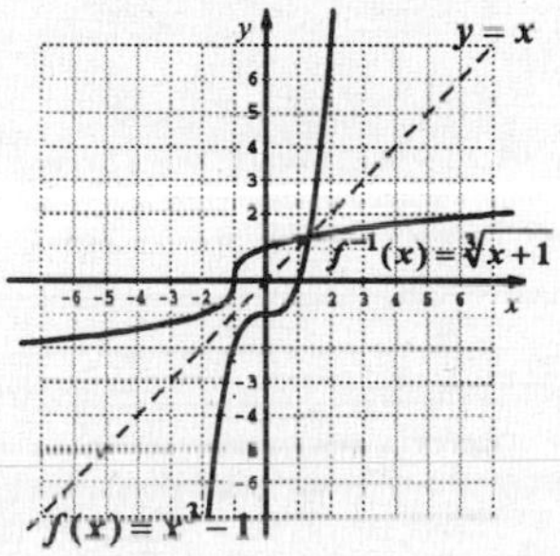

53. Use the procedure described in Exercise 49 to graph the function and its inverse.

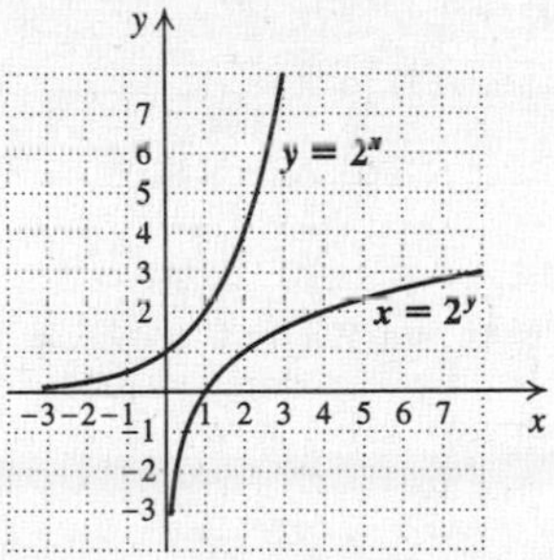

54.

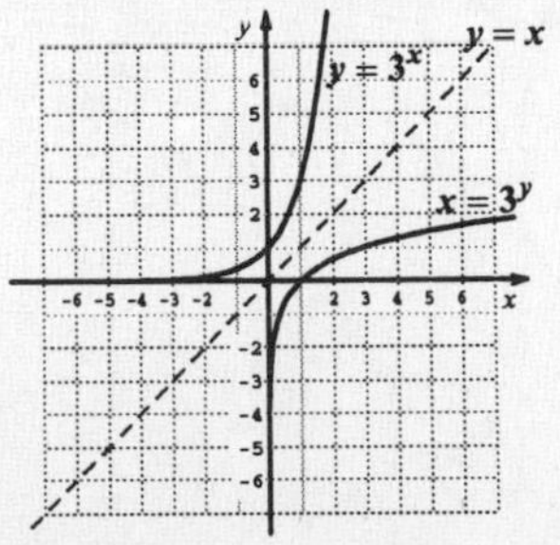

55. Use the procedure described in Exercise 49 to graph the function and its inverse.

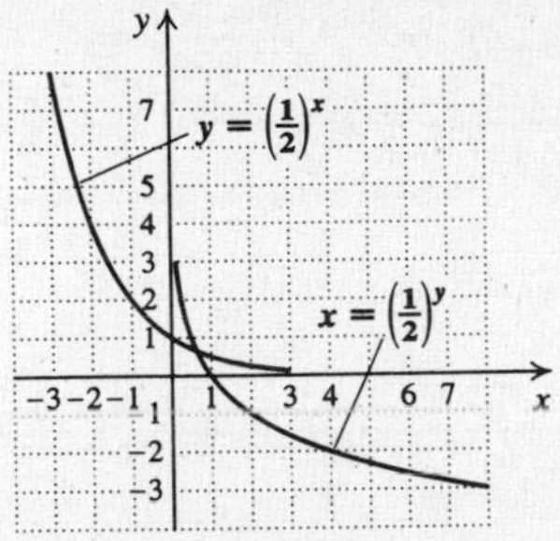

56.

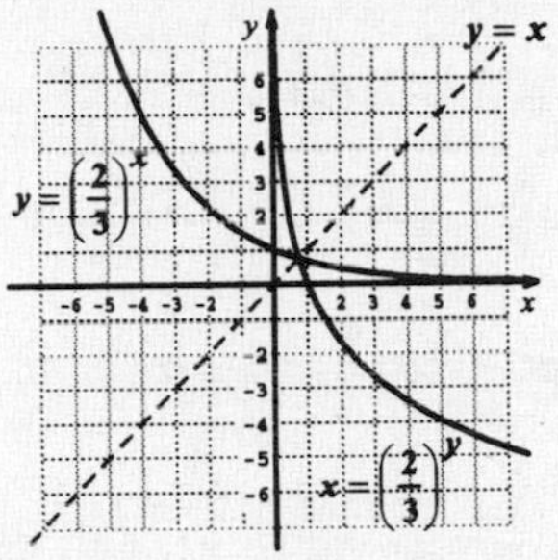

57. Use the procedure described in Exercise 49 to graph the function and its inverse.

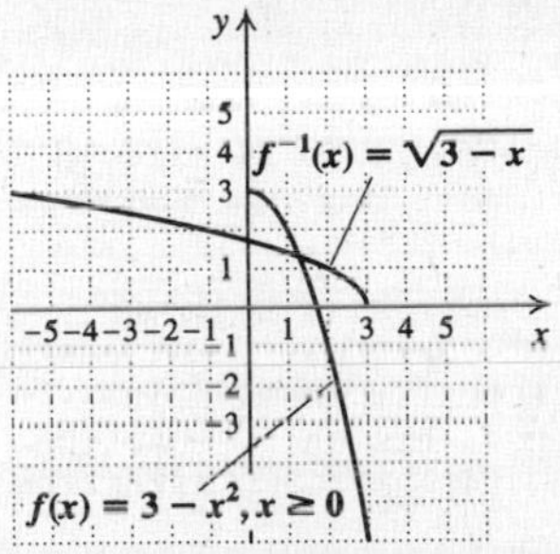

58.

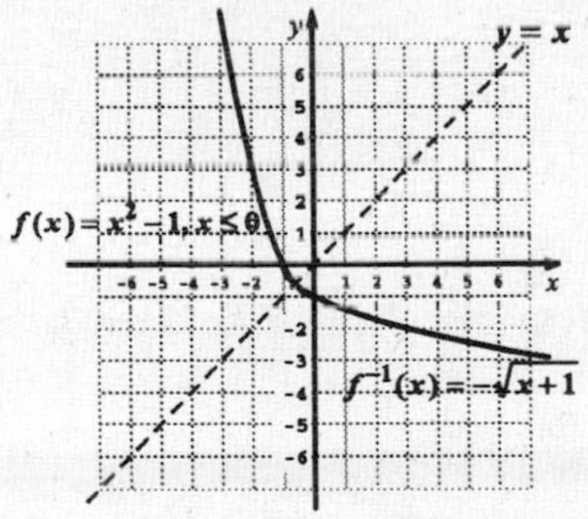

59. We check to see that $f^{-1} \circ f(x) = x$ and $f \circ f^{-1}(x) = x$.

a) $f^{-1} \circ f(x) = f^{-1}(f(x)) = f^{-1}\left[\frac{4}{5}x\right] = \frac{5}{4} \cdot \frac{4}{5}x = x$

b) $f \circ f^{-1}(x) = f(f^{-1}(x)) = f\left[\frac{5}{4}x\right] = \frac{4}{5} \cdot \frac{5}{4}x = x$

60. a) $f^{-1} \circ f(x) = 3\left[\frac{x + 7}{3}\right] - 7 = x + 7 - 7 = x$

b) $f \circ f^{-1}(x) = \frac{(3x - 7) + 7}{3} = \frac{3x}{3} = x$

61. We check to see that $f^{-1} \circ f(x) = x$ and $f \circ f^{-1}(x) = x$.

a) $f^{-1} \circ f(x) = f^{-1}(f(x)) = f^{-1}\left(\frac{1-x}{x}\right) =$

$$\frac{1}{\frac{1-x}{x}+1} = \frac{1}{\frac{1-x}{x}+1} \cdot \frac{x}{x} = \frac{x}{1-x+x} = \frac{x}{1} = x$$

b) $f \circ f^{-1}(x) = f(f^{-1}(x)) = f\left(\frac{1}{x+1}\right) =$

$$\frac{1 - \frac{1}{x+1}}{\frac{1}{x+1}} = \frac{1 - \frac{1}{x+1}}{\frac{1}{x+1}} \cdot \frac{x+1}{x+1} = \frac{x+1-1}{1} =$$

$$\frac{x}{1} = x$$

62. a) $f^{-1} \circ f(x) = \sqrt[3]{x^3 - 5 + 5} = \sqrt[3]{x^3} = x$

b) $f \circ f^{-1}(x) = (\sqrt[3]{x+5})^3 - 5 = x + 5 - 5 = x$

63. a) $f(8) = 8 + 32 = 40$

Size 40 in France corresponds to size 8 in the U.S.

$f(10) = 10 + 32 = 42$

Size 42 in France corresponds to size 10 in the U.S.

$f(14) = 14 + 32 = 46$

Size 46 in France corresponds to size 14 in the U.S.

$f(18) = 18 + 32 = 50$

Size 50 in France corresponds to size 18 in the U.S.

b) The function $f(x) = x + 32$ is a linear function that is not constant, so it passes the horizontal line test. Thus, f is one-to-one and, hence, has an inverse that is a function. We now find a formula for the inverse.

Replace $f(x)$ by y: $y = x + 32$

Interchange x and y: $x = y + 32$

Solve for y: $x - 32 = y$

Replace y by $f^{-1}(x)$: $f^{-1}(x) = x - 32$

c) $f^{-1}(40) = 40 - 32 = 8$

Size 8 in the U.S. corresponds to size 40 in France.

$f^{-1}(42) = 42 - 32 = 10$

Size 10 in the U.S. corresponds to size 42 in France.

$f^{-1}(46) = 46 - 32 = 14$

Size 14 in the U.S. corresponds to size 46 in France.

$f^{-1}(50) = 50 - 32 = 18$

Size 18 in the U.S. corresponds to size 50 in France.

64. a) 40, 44, 52, 60

b) $f^{-1}(x) = \frac{x - 24}{2}$, or $\frac{x}{2} - 12$

c) 8, 10, 14, 18

65.

$y = kx$		
$7.2 = k(0.8)$	Substituting	
$9 = k$	Variation constant	
$y = 9x$	Equation of variation	

66. $y = \frac{21.35}{x}$

67.

68.

69. From Exercise 64(b), we know that a function that converts dress sizes in Italy to those in the United States is $g(x) = \frac{x - 24}{2}$. From Exercise 63, we know that a function that converts dress sizes in the United States to those in France is $f(x) = x + 32$. Then a function that converts dress sizes in Italy to those in France is

$h(x) = f \circ g(x)$

$h(x) = f\left(\frac{x - 24}{2}\right)$

$h(x) = \frac{x - 24}{2} + 32$

$h(x) = \frac{x}{2} - 12 + 32$

$h(x) = \frac{x}{2} + 20.$

70. No

71. Yes

72. Yes

73. No

Exercise Set 9.3

1. Graph: $y = \log_2 x$

The equation $y = \log_2 x$ is equivalent to $2^y = x$. We can find ordered pairs by choosing values for y and computing the corresponding x-values.

For $y = 0$, $x = 2^0 = 1$.

For $y = 1$, $x = 2^1 = 2$.

For $y = 2$, $x = 2^2 = 4$.

For $y = 3$, $x = 2^3 = 8$.

For $y = -1$, $x = 2^{-1} = \frac{1}{2}$.

For $y = -2$, $x = 2^{-2} = \frac{1}{4}$.

x, or 2^y	y
1	0
2	1
4	2
8	3
$\frac{1}{2}$	-1
$\frac{1}{4}$	-2

↑ (1) Select y.
↑ (2) Compute x.

We plot the set of ordered pairs and connect the points with a smooth curve.

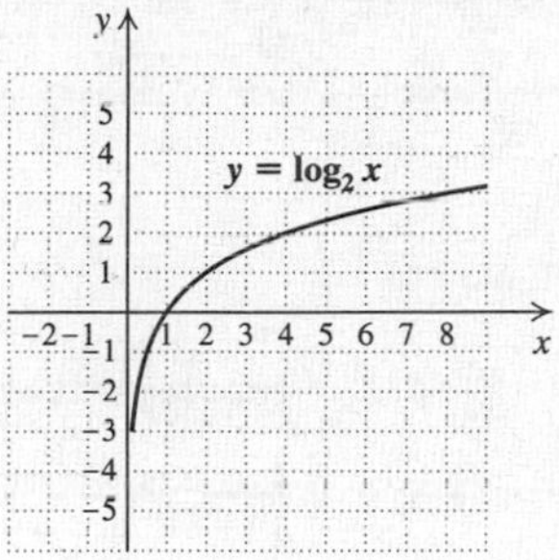

2.

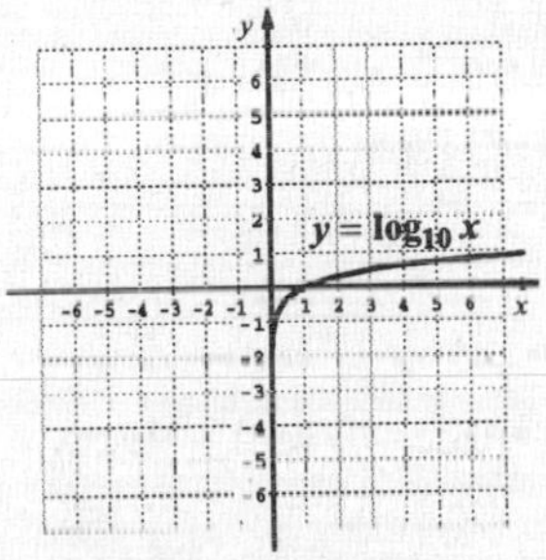

3. Graph: $y = \log_6 x$

The equation $y = \log_6 x$ is equivalent to $6^y = x$. We can find ordered pairs by choosing values for y and computing the corresponding x-values.

For $y = 0$, $x = 6^0 = 1$.

For $y = 1$, $x = 6^1 = 6$.

For $y = 2$, $x = 6^2 = 36$.

For $y = -1$, $x = 6^{-1} = \frac{1}{6}$.

For $y = -2$, $x = 6^{-2} = \frac{1}{36}$.

x, or 6^y	y
1	0
6	1
36	2
$\frac{1}{6}$	-1
$\frac{1}{36}$	-2

We plot the set of ordered pairs and connect the points with a smooth curve.

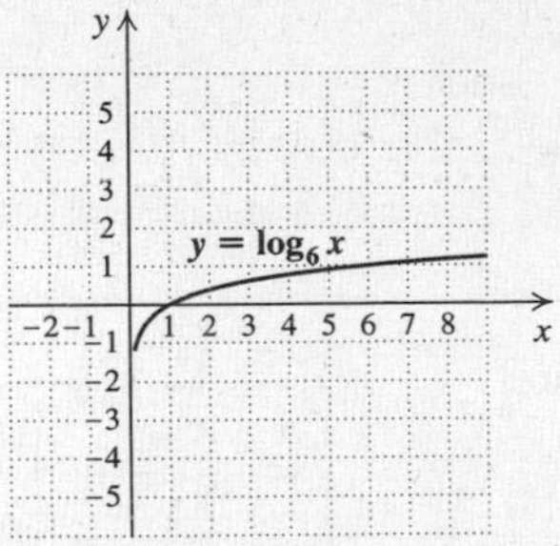

4.

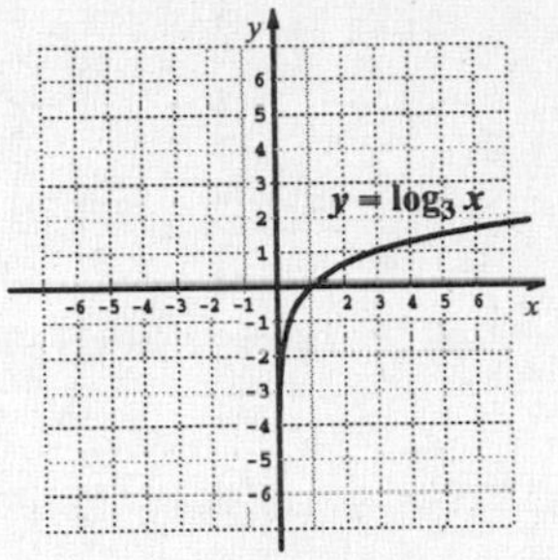

5. Graph: $f(x) = \log_4 x$

Think of f(x) as y. Then $y = \log_4 x$ is equivalent to $4^y = x$. We find ordered pairs by choosing values for y and computing the corresponding x-values. Then we plot the points and connect them with a smooth curve.

For $y = 0$, $x = 4^0 = 1$.

For $y = 1$, $x = 4^1 = 4$.

For $y = 2$, $x = 4^2 = 16$.

For $y = -1$, $x = 4^{-1} = \frac{1}{4}$.

For $y = -2$, $x = 4^{-2} = \frac{1}{16}$.

x, or 4^y	y
1	0
4	1
16	2
$\frac{1}{4}$	-1
$\frac{1}{16}$	-2

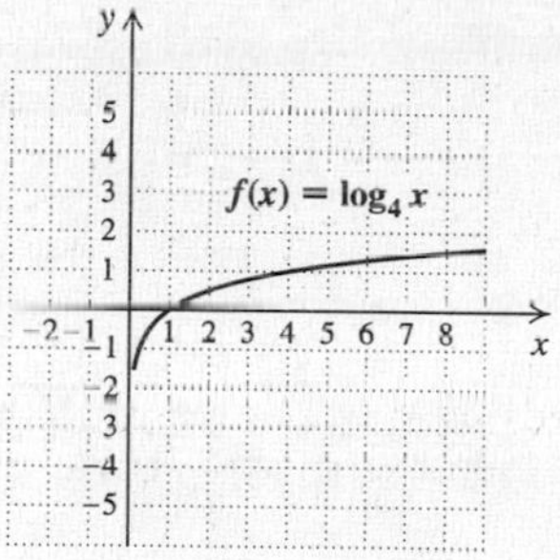

6.

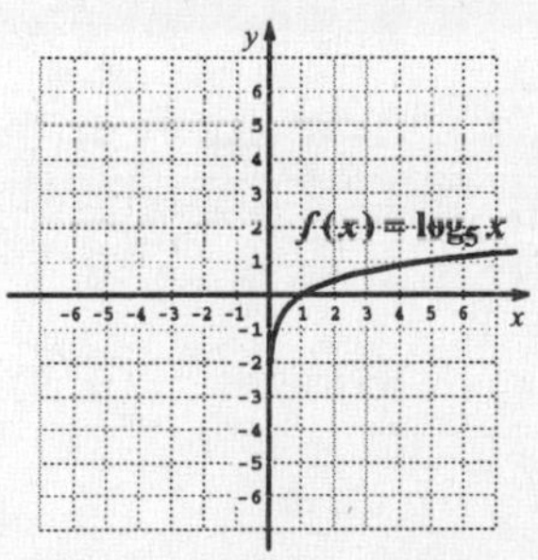

7. Graph: $f(x) = \log_{1/2} x$
Think of $f(x)$ as y. Then $y = \log_{1/2} x$ is equivalent to $\left(\frac{1}{2}\right)^y = x$. We construct a table of values, plot these points, and connect them with a smooth curve.

For $y = 0$, $x = \left(\frac{1}{2}\right)^0 = 1$.

For $y = 1$, $x = \left(\frac{1}{2}\right)^1 = \frac{1}{2}$.

For $y = 2$, $x = \left(\frac{1}{2}\right)^2 = \frac{1}{4}$.

For $y = -1$, $x = \left(\frac{1}{2}\right)^{-1} = 2$.

For $y = -2$, $x = \left(\frac{1}{2}\right)^{-2} = 4$.

For $y = -3$, $x = \left(\frac{1}{2}\right)^{-3} = 8$.

x, or $\left(\frac{1}{2}\right)^y$	y
1	0
$\frac{1}{2}$	1
$\frac{1}{4}$	2
2	-1
4	-2
8	-3

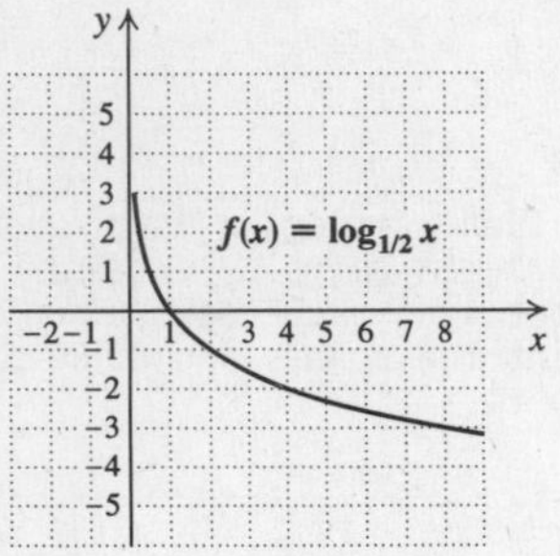

8.

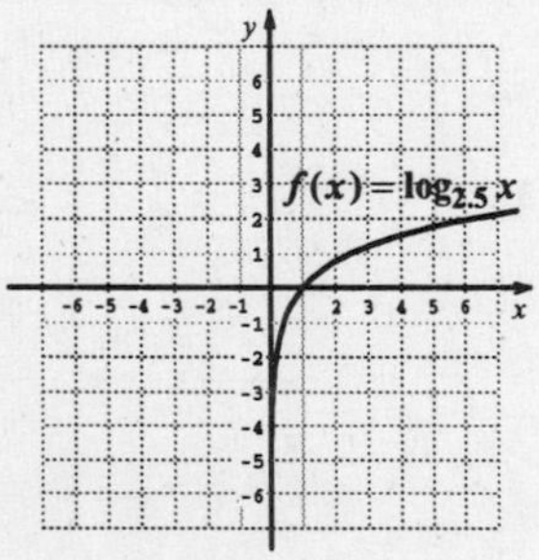

9. Graph $f(x) = 3^x$ (see Exercise Set 9.1, Exercise 2) and $f^{-1}(x) = \log_3 x$ (see Exercise 4 above) on the same set of axes.

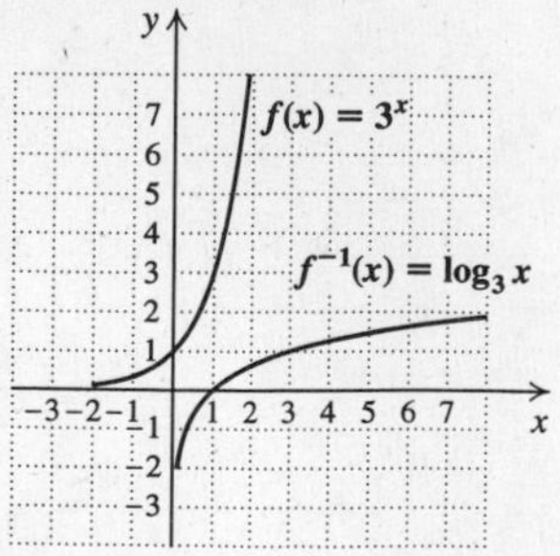

10.

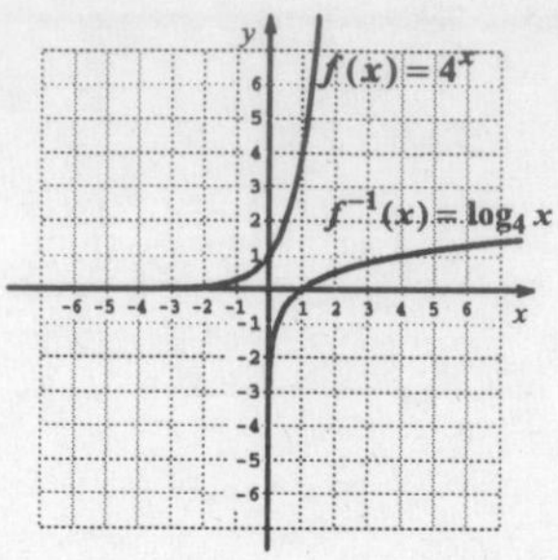

11. $10^3 = 1000$ is equivalent to

$3 = \log_{10} 1000$

The exponent is the logarithm.

The base remains the same.

12. $2 = \log_{10} 100$

13. $5^{-3} = \frac{1}{125}$ is equivalent to

$-3 = \log_5 \frac{1}{125}$

The exponent is the logarithm.

The base remains the same.

14. $-5 = \log_4 \frac{1}{1024}$

15. $8^{1/3} = 2$ is equivalent to $\frac{1}{3} = \log_8 2$

16. $\frac{1}{4} = \log_{16} 2$

17. $10^{0.3010} = 2$ is equivalent to $0.3010 = \log_{10} 2$

18. $0.4771 = \log_{10} 3$

19. $e^2 = t$ is equivalent to $2 = \log_e t$

20. $k = \log_p 3$

21. $Q^t = x$ is equivalent to $t = \log_Q x$

22. $m = \log_p V$

23. $e^2 = 7.3891$ is equivalent to $2 = \log_e 7.3891$

24. $3 = \log_e 20.0855$

25. $e^{-2} = 0.1353$ is equivalent to $-2 = \log_e 0.1353$

26. $-4 = \log_e 0.0183$

27. $t = \log_3 8$ is equivalent to

$3^t = 8$

The base remains the same.

The logarithm is the exponent.

28. $7^h = 10$

29. $\log_5 25 = 2$ is equivalent to

$5^2 = 25$

The logarithm is the exponent.

The base remains the same.

30. $6^1 = 6$

31. $\log_{10} 0.1 = -1$ is equivalent to $10^{-1} = 0.1$

32. $10^{-2} = 0.01$

33. $\log_{10} 7 = 0.845$ is equivalent to $10^{0.845} = 7$

34. $10^{0.4771} = 3$

35. $\log_e 20 = 2.9957$ is equivalent to $e^{2.9957} = 20$

36. $e^{2.3026} = 10$

37. $\log_t Q = k$ is equivalent to $t^k = Q$

38. $m^a = P$

39. $\log_e 0.25 = -1.3863$ is equivalent to $e^{-1.3863} = 0.25$

40. $e^{-0.0111} = 0.989$

41. $\log_r T = -x$ is equivalent to $r^{-x} = T$

42. $c^{-w} = M$

43. $\log_3 x = 2$

$3^2 = x$	Converting to an exponential equation
$9 = x$	Computing 3^2

44. 64

45. $\log_x 125 = 3$

$x^3 = 125$	Converting to an exponential equation
$x = 5$	Taking cube roots

46. 4

47. $\log_2 16 = x$

$2^x = 16$	Converting to an exponential equation
$2^x = 2^4$	
$x = 4$	The exponents must be the same.

48. 2

49. $\log_3 27 = x$

$3^x = 27$	Converting to an exponential equation
$3^x = 3^3$	
$x = 3$	The exponents must be the same.

50. 2

51. $\log_x 13 = 1$

$x^1 = 13$	Converting to an exponential equation
$x = 13$	Simplifying x^1

52. 23

53. $\log_6 x = 0$

$6^0 = x$	Converting to an exponential equation
$1 = x$	Computing 6^0

54. 9

55. $\log_2 x = -1$

$2^{-1} = x$	Converting to an exponential equation
$\frac{1}{2} = x$	Simplifying

56. $\frac{1}{9}$

57. $\log_8 x = \frac{1}{3}$

$8^{1/3} = x$

$2 = x$

58. 2

59. Let $\log_{10} 100 = x$.

Then $10^x = 100$

$10^x = 10^2$

$x = 2.$

Thus, $\log_{10} 100 = 2.$

60. 5

61. Let $\log_{10} 0.1 = x$.

Then $10^x = 0.1 = \frac{1}{10}$

$10^x = 10^{-1}$

$x = -1.$

Thus, $\log_{10} 0.1 = -1.$

62. -3

63. Let $\log_{10} 1 = x$.
Then $10^x = 1$
$10^x = 10^0 \quad (10^0 = 1)$
$x = 0$.
Thus, $\log_{10} 1 = 0$.

64. 1

65. Let $\log_5 625 = x$.
Then $5^x = 625$
$5^x = 5^4$
$x = 4$.
Thus, $\log_5 625 = 4$.

66. 6

67. Let $\log_5 \frac{1}{25} = x$.
Then $5^x = \frac{1}{25}$
$5^x = 5^{-2}$
$x = -2$.
Thus, $\log_5 \frac{1}{25} = -2$.

68. -4

69. Let $\log_3 1 = x$.
Then $3^x = 1$
$3^x = 3^0 \quad (3^0 = 1)$
$x = 0$.
Thus, $\log_3 1 = 0$.

70. 1

71. Let $\log_e e = x$.
Then $e^x = e$
$e^x = e^1$
$x = 1$.
Thus, $\log_e e = 1$.

72. 0

73. Let $\log_{27} 9 = x$.
Then $27^x = 9$
$(3^3)^x = 3^2$
$3^{3x} = 3^2$
$3x = 2$
$x = \frac{2}{3}$.
Thus, $\log_{27} 9 = \frac{2}{3}$.

74. $\frac{1}{3}$

75. Let $\log_e e^3 = x$.
Then $e^x = e^3$
$x = 3$.
Thus, $\log_e e^3 = 3$.

76. -4

77. Let $\log_{10} 10^t = x$.
Then $10^x = 10^t$
$x = t$.
Thus, $\log_{10} 10^t = t$.

78. p

79. $\dfrac{\frac{3}{x} - \frac{2}{xy}}{\frac{2}{x^2} + \frac{1}{xy}}$

The LCD of all the denominators is x^2y. We multiply numerator and denominator by the LCD.

$$\frac{\frac{3}{x} - \frac{2}{xy}}{\frac{2}{x^2} + \frac{1}{xy}} \cdot \frac{x^2y}{x^2y} = \frac{\left(\frac{3}{x} - \frac{2}{xy}\right)x^2y}{\left(\frac{2}{x^2} + \frac{1}{xy}\right)x^2y}$$

$$= \frac{\frac{3}{x} \cdot x^2y - \frac{2}{xy} \cdot x^2y}{\frac{2}{x^2} \cdot x^2y + \frac{1}{xy} \cdot x^2y}$$

$$= \frac{3xy - 2x}{2y + x}, \text{ or}$$

$$= \frac{x(3y - 2)}{2y + x}$$

80. $\frac{x + 2}{x + 1}$

81. $8^{-4} = \frac{1}{8^4}$, or $\frac{1}{4096}$

82. $\sqrt[5]{x^4}$

83. $t^{-2/3} = \frac{1}{t^{2/3}} = \frac{1}{\sqrt[3]{t^2}}$

84. 5

85.

86.

87. Graph: $y = \left(\frac{3}{2}\right)^x$

x	y, or $\left(\frac{3}{2}\right)^x$
0	1
1	$\frac{3}{2}$
2	$\frac{9}{4}$
3	$\frac{27}{8}$
-1	$\frac{2}{3}$
-2	$\frac{4}{9}$

Graph: $y = \log_{3/2} x$, or $x = \left(\frac{3}{2}\right)^y$

x, or $\left(\frac{3}{2}\right)^y$	y
1	0
$\frac{3}{2}$	1
$\frac{9}{4}$	2
$\frac{27}{8}$	3
$\frac{2}{3}$	-1
$\frac{4}{9}$	-2

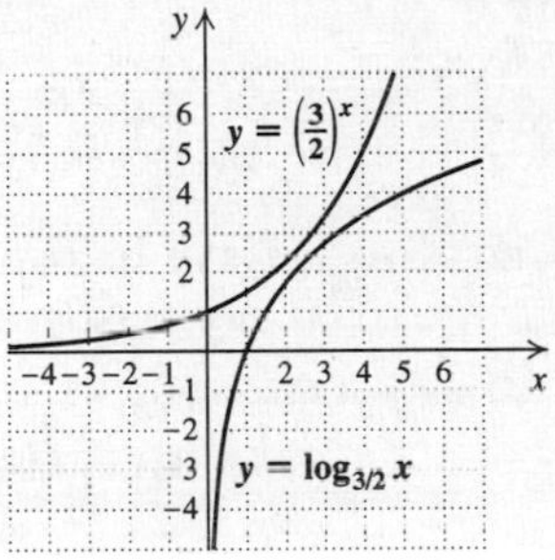

88.

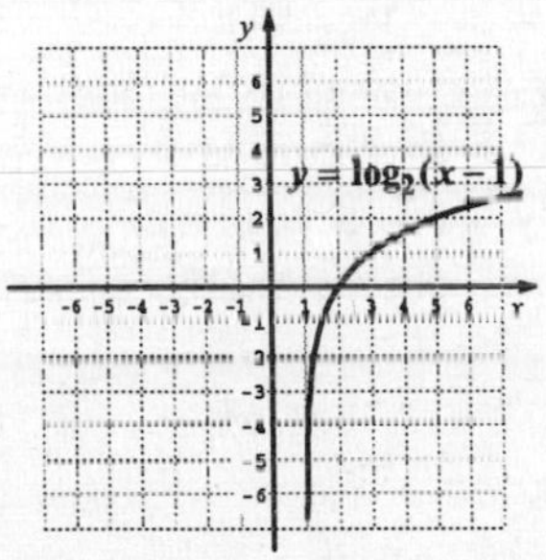

89. Graph: $y = \log_3 |x + 1|$

x	y
0	0
2	1
8	2
-2	0
-4	1
-9	2

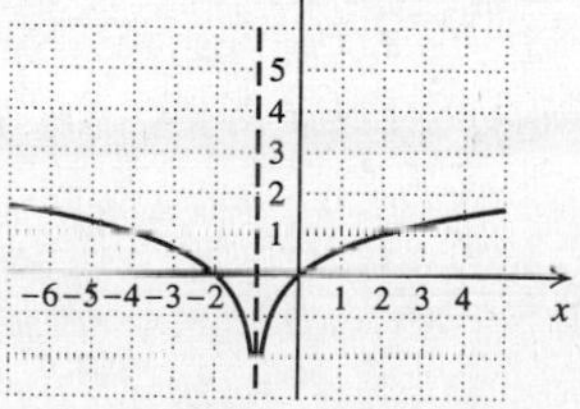

$y = \log_3 |x + 1|$

90. $27, \frac{1}{27}$

91. $\log_{125} x = \frac{2}{3}$

$125^{2/3} = x$

$(5^3)^{2/3} = x$

$5^2 = x$

$25 = x$

92. 6

93. $\log_8 (2x + 1) = -1$

$8^{-1} = 2x + 1$

$\frac{1}{8} = 2x + 1$

$1 = 16x + 8$

$-7 = 16x$

$-\frac{7}{16} = x$

94. -25, 4

95. Let $\log_{1/4} \frac{1}{64} = x$.

Then $\left(\frac{1}{4}\right)^x = \frac{1}{64}$

$\left(\frac{1}{4}\right)^x = \left(\frac{1}{4}\right)^3$

$x = 3$.

Thus, $\log_{1/4} \frac{1}{64} = 3$.

96. 1

97. $\log_{10} (\log_4 (\log_3 81))$

$= \log_{10} (\log_4 4)$ $(\log_3 81 = 4)$

$= \log_{10} 1$ $(\log_4 4 = 1)$

$= 0$

98. 1

99. Let $\log_{1/5} 25 = x$.

Then $\left(\frac{1}{5}\right)^x = 25$

$(5^{-1})^x = 25$

$5^{-x} = 5^2$

$-x = 2$

$x = -2$

Exercise Set 9.4

1. $\log_2 (32 \cdot 8) = \log_2 32 + \log_2 8$ Using the product rule

2. $\log_3 27 + \log_3 81$

3. $\log_4 (64 \cdot 16) = \log_4 64 + \log_4 16$ Using the product rule

4. $\log_5 25 + \log_5 125$

5. $\log_c Bx = \log_c B + \log_c x$ Using the product rule

6. $\log_t 5 + \log_t Y$

7. $\log_a 6 + \log_a 70 = \log_a (6\cdot 70)$ Using the product rule

8. $\log_b (65\cdot 2)$

9. $\log_c K + \log_c y = \log_c(K\cdot y)$ Using the product rule

10. $\log_t HM$

11. $\log_a x^3 = 3 \log_a x$ Using the power rule

12. $5 \log_b t$

13. $\log_c y^6 = 6 \log_c y$ Using the power rule

14. $7 \log_{10} y$

15. $\log_b C^{-3} = -3 \log_b C$ Using the power rule

16. $-5 \log_c M$

17. $\log_a \frac{67}{5} = \log_a 67 - \log_a 5$ Using the quotient rule

18. $\log_t T - \log_t 7$

19. $\log_b \frac{3}{4} = \log_b 3 - \log_b 4$ Using the quotient rule

20. $\log_a y - \log_a x$

21. $\log_a 15 - \log_a 7 = \log_a \frac{15}{7}$ Using the quotient rule

22. $\log_b \frac{42}{7}$, or $\log_b 6$

23. $\log_a x^2y^3z$
$= \log_a x^2 + \log_a y^3 + \log_a z$ Using the product rule
$= 2 \log_a x + 3 \log_a y + \log_a z$ Using the power rule

24. $\log_a x + 4 \log_a y + 3 \log_a z$

25. $\log_b \frac{xy^2}{z^3}$
$= \log_b xy^2 - \log_b z^3$ Using the quotient rule
$= \log_b x + \log_b y^2 - \log_b z^3$ Using the product rule
$= \log_b x + 2 \log_b y - 3 \log_b z$ Using the power rule

26. $2 \log_b x + 5 \log_b y - 4 \log_b w - 7 \log_b z$

27. $\log_c \sqrt[3]{\frac{x^4}{y^3z^2}}$
$= \log_c \left(\frac{x^4}{y^3z^2}\right)^{1/3}$
$= \frac{1}{3} \log_c \frac{x^4}{y^3z^2}$ Using the power rule
$= \frac{1}{3} (\log_c x^4 - \log_c y^3z^2)$ Using the quotient rule
$= \frac{1}{3}[\log_c x^4 - (\log_c y^3 + \log_c z^2)]$ Using the product rule
$= \frac{1}{3}(\log_c x^4 - \log_c y^3 - \log_c z^2)$ Removing parentheses
$= \frac{1}{3}(4 \log_c x - 3 \log_c y - 2 \log_c z)$ Using the power rule

28. $\frac{1}{2}(6 \log_a x - 5 \log_a y - 8 \log_a z)$

29. $\log_a \sqrt[4]{\frac{x^8y^{12}}{a^3z^5}}$
$= \frac{1}{4}\left(\log_a \frac{x^8y^{12}}{a^3z^5}\right)$ Using the power rule
$= \frac{1}{4}(\log_a x^8y^{12} - \log_a a^3z^5)$ Using the quotient rule
$= \frac{1}{4}[\log_a x^8 + \log_a y^{12} - (\log_a a^3 + \log_a z^5)]$
Using the product rule
$= \frac{1}{4}(\log_a x^8 + \log_a y^{12} - \log_a a^3 - \log_a z^5)$
Removing parentheses
$= \frac{1}{4}(\log_a x^8 + \log_a y^{12} - 3 - \log_a z^5)$
3 is the power to which you raise a to get a^3.
$= \frac{1}{4}(8 \log_a x + 12 \log_a y - 3 - 5 \log_a z)$
Using the power rule

30. $\frac{1}{3}(6 \log_a x + 3 \log_a y - 2 - 7 \log_a z)$

31. $\frac{2}{3} \log_a x - \frac{1}{2} \log_a y$
$= \log_a x^{2/3} - \log_a y^{1/2}$ Using the power rule
$= \log_a \sqrt[3]{x^2} - \log_a \sqrt{y}$
$= \log_a \frac{\sqrt[3]{x^2}}{\sqrt{y}}$ Using the quotient rule
$= \log_a \frac{\sqrt[3]{x^2}\sqrt{y}}{y}$ Multiplying by $\frac{\sqrt{y}}{\sqrt{y}}$

32. $\log_a \frac{\sqrt{x}\, y^3}{x^2}$

33. $\log_a 2x + 3(\log_a x - \log_a y)$
$= \log_a 2x + 3 \log_a x - 3 \log_a y$
$= \log_a 2x + \log_a x^3 - \log_a y^3$ Using the power rule
$= \log_a 2x^4 - \log_a y^3$ Using the product rule
$= \log_a \frac{2x^4}{y^3}$ Using the quotient rule

34. $\log_a x$

35. $\log_a \frac{a}{\sqrt{x}} - \log_a \sqrt{ax}$

$= \log_a ax^{-1/2} - \log_a a^{1/2}x^{1/2}$

$= \log_a \frac{ax^{-1/2}}{a^{1/2}x^{1/2}}$ Using the quotient rule

$= \log_a \frac{a^{1/2}}{x}$

$= \log_a \frac{\sqrt{a}}{x}$

36. $\log_a (x + 2)$

37. $\log_b 15 = \log_b (3\cdot5)$

$= \log_b 3 + \log_b 5$ Using the product rule

$= 1.099 + 1.609$

$= 2.708$

38. -0.51

39. $\log_b \frac{5}{3} = \log_b 5 - \log_b 3$ Using the quotient rule

$= 1.609 - 1.099$

$= 0.51$

40. -1.099

41. $\log_b \frac{1}{5} = \log_b 1 - \log_b 5$ Using the quotient rule

$= 0 - 1.609$ ($\log_b 1 = 0$)

$= -1.609$

42. $\frac{1}{2}$

43. $\log_b \sqrt{b^3} = \log_b b^{3/2} = \frac{3}{2}$ 3/2 is the power to which you raise b to get $b^{3/2}$.

44. 2.099

45. $\log_b 5b = \log_b 5 + \log_b b$ Using the product rule

$= 1.609 + 1$ ($\log_b b = 1$)

$= 2.609$

46. 2.198

47. $\log_b 25 = \log_b 5^2$

$= 2 \log_b 5$ Using the power rule

$= 2(1.609)$

$= 3.218$

48. 4.317

49. $\log_t t^9 = 9$ 9 is the power to which you raise t to get t^9.

50. 4

51. $\log_e e^m = m$

52. -2

53. $\log_3 3^4 = x$

$4 = x$

54. 7

55. $\log_e e^x = -7$

$x = -7$

56. 2.7

57. $i^{29} = i^{28}\cdot i = (i^4)^7\cdot i = 1^7\cdot i = 1\cdot i = i$

58. 5

59. $\frac{2+i}{2-i} = \frac{2+i}{2-i}\cdot\frac{2+i}{2+i} = \frac{4+4i+i^2}{4-i^2} = \frac{4+4i-1}{4-(-1)} =$

$\frac{3+4i}{5} = \frac{3}{5} + \frac{4}{5}i$

60. 23 - 18i

61.

62.

63. $\log_a (x^8 - y^8) - \log_a (x^2 + y^2)$

$= \log_a \frac{x^8 - y^8}{x^2 + y^2}$ Using the quotient rule

$= \log_a \frac{(x^4 + y^4)(x^2 + y^2)(x + y)(x - y)}{x^2 + y^2}$ Factoring

$= \log_a [(x^4 + y^4)(x^2 - y^2)]$ Simplifying

$= \log_a (x^6 - x^4y^2 + x^2y^4 - y^6)$ Multiplying

64. $\log_a (x^3 + y^3)$

65. $\log_a \sqrt{1 - s^2}$

$= \log_a (1 - s^2)^{1/2}$

$= \frac{1}{2} \log_a (1 - s^2)$

$= \frac{1}{2} \log_a [(1 - s)(1 + s)]$

$= \frac{1}{2} \log_a (1 - s) + \frac{1}{2} \log_a (1 + s)$

66. $\frac{1}{2} \log_a (c - d) - \frac{1}{2} \log_a (c + d)$

67. $\log_a \dfrac{\sqrt[3]{x^2z}}{\sqrt[3]{y^2z^{-2}}}$

$= \log_a \left(\dfrac{x^2z^3}{y^2}\right)^{1/3}$

$= \dfrac{1}{3}(\log_a x^2z^3 - \log_a y^2)$

$= \dfrac{1}{3}(2 \log_a x + 3 \log_a z - 2 \log_a y)$

$= \dfrac{1}{3}[2\cdot 2 + 3\cdot 4 - 2\cdot 3]$ Substituting

$= \dfrac{1}{3}(10)$

$= \dfrac{10}{3}$

68. -2

69. $\log_a x = 2$ Given

$a^2 = x$ Definition

Let $\log_{1/a} x = n$ and solve for n.

$\log_{1/a} a^2 = n$ Substituting a^2 for x

$\left(\dfrac{1}{a}\right)^n = a^2$

$(a^{-1})^n = a^2$

$a^{-n} = a^2$

$-n = 2$

$n = -2$

Thus, $\log_{1/a} x = -2$ when $\log_a x = 2$.

70. False. For a counterexample, let a = 10, P = 100, and Q = 10.

71. The statement is false. For example, let a = 10, P = 100, and Q = 10. Then

$\dfrac{\log_a P}{\log_a Q} = \dfrac{\log_{10} 100}{\log_{10} 10} = \dfrac{2}{1} = 2$, but

$\log_a P - \log_a Q = \log_{10} 100 - \log_{10} 10 = 2 - 1 = 1$.

72. True

73. The statement is false. For example, let a = 3 and x = 9. Then

$\log_a 3x = \log_3 3\cdot 9 = \log_3 27 = 3$, but

$3 \log_a x = 3 \log_3 9 = 3\cdot 2 = 6$.

74. False. For a counterexample, let a = 2, P = 1, and Q = 1.

75. The statement is true, by the power rule.

76. $\log_a \dfrac{x + \sqrt{x^2 - 3}}{3} =$

$\log_a \left[\dfrac{x + \sqrt{x^2 - 3}}{3} \cdot \dfrac{x - \sqrt{x^2 - 3}}{x - \sqrt{x^2 - 3}}\right] =$

$\log_a \left[\dfrac{3}{3(x - \sqrt{x^2 - 3})}\right] = \log_a \left[\dfrac{1}{x - \sqrt{x^2 - 3}}\right] =$

$\log_a 1 - \log_a (x - \sqrt{x^2 - 3}) =$

$0 - \log_a (x - \sqrt{x^2 - 3}) = -\log_a (x - \sqrt{x^2 - 3})$

Exercise Set 9.5

1. 0.3010

2. 0.6990

3. 0.8021

4. 0.7007

5. 1.6532

6. 1.8692

7. 2.6405

8. 2.4698

9. 4.1271

10. 4.9689

11. -1.2840

12. -0.4123

13. 1000

14. 100,000

15. 501.1872

16. 6.3096×10^{14}

17. 3.0001

18. 1.1623

19. 0.2841

20. 0.4567

21. 0.0011

22. 79,104.2833

23. 0.6931

24. 1.0986

25. 4.1271

26. 3.4012

27. 8.3814

28. 6.8037

29. -5.0832

30. -7.2225

31. 36.7890

32. 138.5457

33. 0.0023

34. 0.1002

35. 1.0057

36. 1.0112

37. 5.8346×10^{14}

38. 2.0917×10^{24}

39. 7.6331

40. 0.2520

41. We will use common logarithms for the conversion. Let a = 10, b = 6, and M = 100 and substitute into the change-of-base formula.

$$\log_b M = \frac{\log_a M}{\log_a b}$$

$$\log_6 100 = \frac{\log_{10} 100}{\log_{10} 6} \approx \frac{2}{0.7782} \approx 2.5702$$

42. 2.6309

43. We will use common logarithms for the conversion. Let a = 10, b = 2, and M = 10 and substitute in the change-of-base formula.

$$\log_2 10 = \frac{\log_{10} 10}{\log_{10} 2} \approx \frac{1}{0.3010} \approx 3.3219$$

44. 2.0104

45. We will use natural logarithms for the conversion. Let a = e, b = 200, and M = 30 and substitute in the change-of-base formula.

$$\log_{200} 30 = \frac{\ln 30}{\ln 200} \approx \frac{3.4012}{5.2983} \approx 0.6419$$

46. 0.7386

47. We will use natural logarithms for the conversion. Let a = e, b = 0.5, and M = 5 and substitute in the change-of-base formula.

$$\log_{0.5} 5 = \frac{\ln 5}{\ln 0.5} \approx \frac{1.6094}{-0.6931} \approx -2.3219$$

48. -0.4771

49. We will use common logarithms for the conversion. Let a = 10, b = 2, and M = 0.2 and substitute in the change-of-base formula.

$$\log_2 0.2 = \frac{\log_{10} 0.2}{\log_{10} 2} \approx \frac{-0.6990}{0.3010} \approx -2.3219$$

50. -3.6439

51. We will use natural logarithms for the conversion. Let a = e, b = π, and M = 58 and substitute in the change-of-base formula.

$$\log_\pi 58 = \frac{\ln 58}{\ln \pi} \approx \frac{4.0604}{1.1447} \approx 3.5471$$

52. 4.6284

53. Graph: $f(x) = e^x$

We find some function values with a calculator. We use these values to plot points and draw the graph.

x	e^x
0	1
1	2.7
2	7.4
3	20.1
-1	0.4
-2	0.1

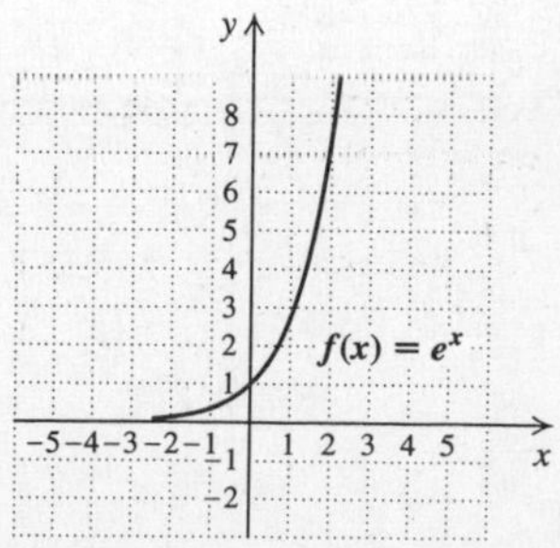

54.

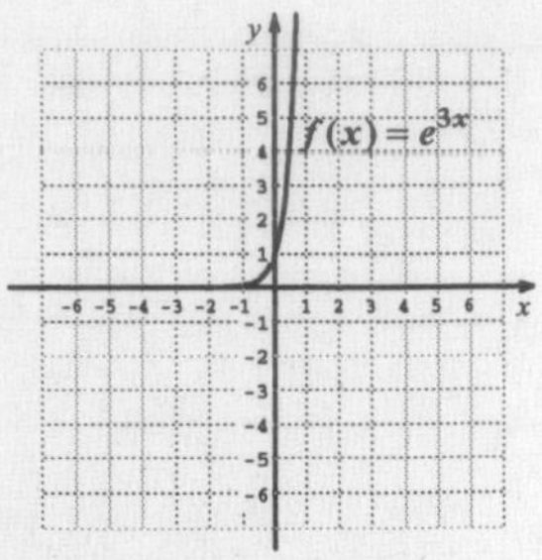

55. Graph: $f(x) = e^{-3x}$

We find some function values, plot points, and draw the graph.

x	e^{-3x}
0	1
1	0.05
2	0.002
-1	20.1
-2	403.4

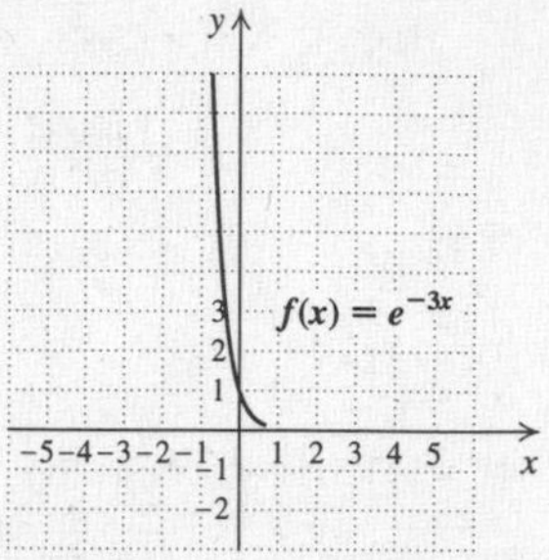

56.

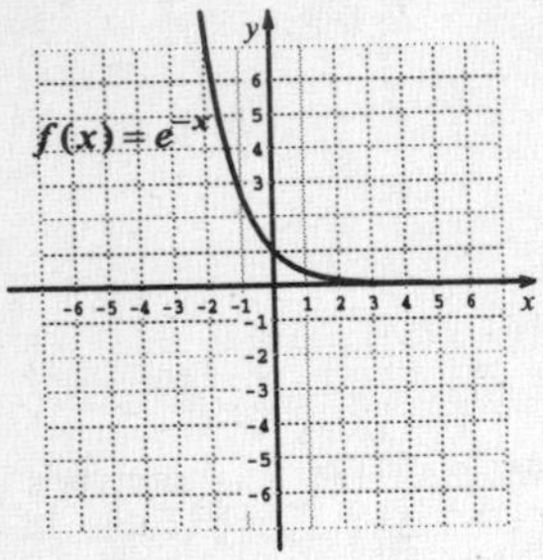

57. Graph: $f(x) = e^{x-1}$

We find some function values, plot points, and draw the graph.

x	e^{x-1}
0	0.4
1	1
2	2.7
3	7.4
4	20.1
-1	0.1
-2	0.05

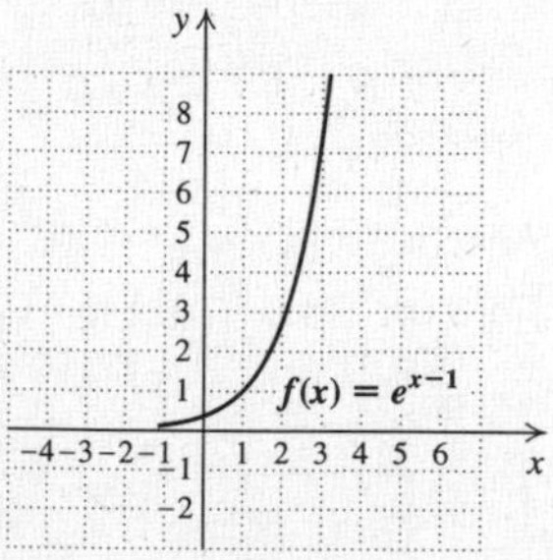

58.

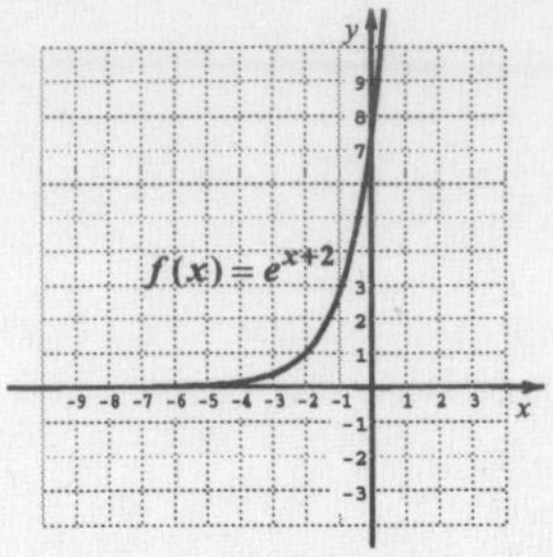

59. Graph: $f(x) = e^{-x} - 3$

We find some function values, plot points, and draw the graph.

x	$e^{-x} - 3$
0	-2
1	-2.6
2	-2.9
3	-3.0
-1	-0.3
-2	4.4
-3	17.1

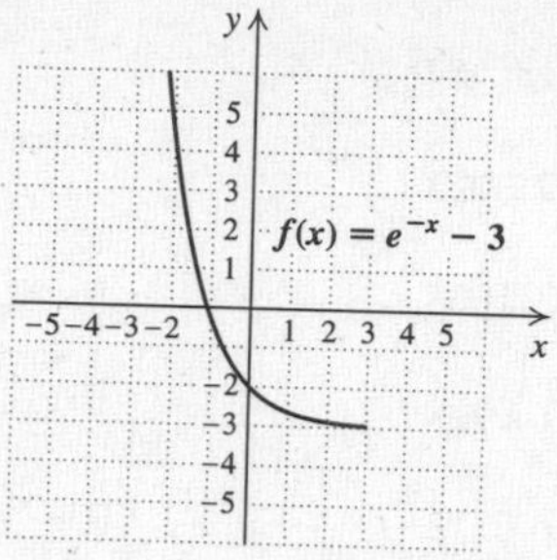

60.

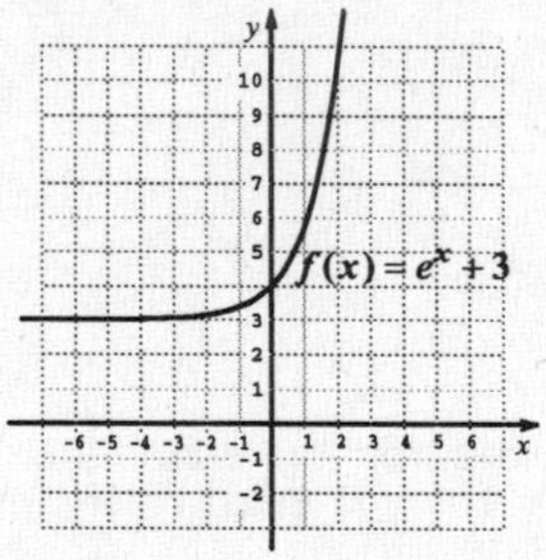

61. Graph: $f(x) = 5e^{0.2x}$

We find some function values, plot points, and draw the graph.

x	$5e^{0.2x}$
0	5
1	6.1
2	7.5
3	9.1
4	11.1
-1	4.1
-2	3.3
-3	2.7
-4	2.2

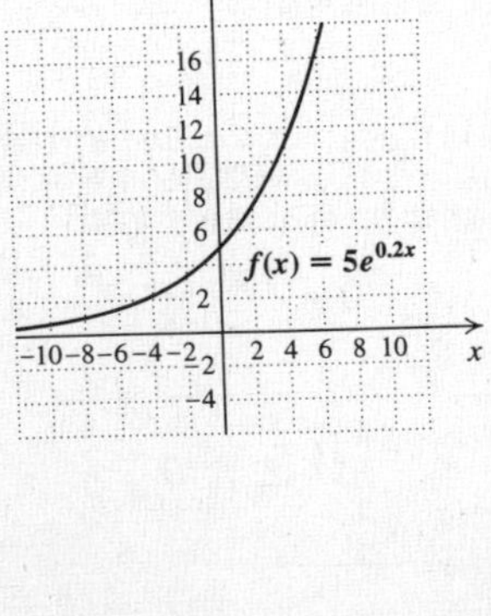

62.

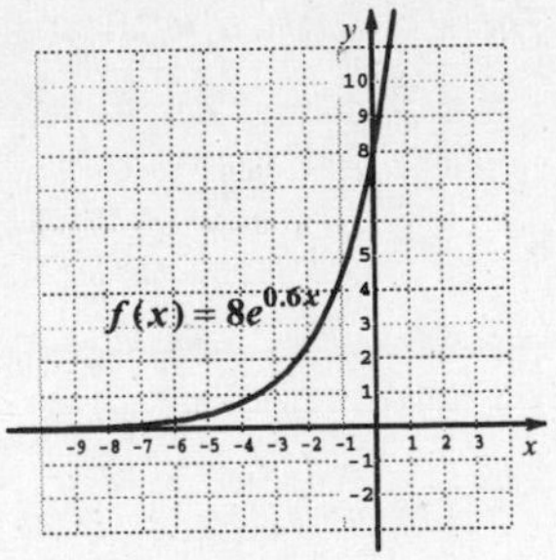

63. Graph: $f(x) = 20e^{-0.5x}$

We find some function values, plot points, and draw the graph.

x	$20e^{-0.5x}$
0	20
1	12.1
2	7.4
3	4.5
4	2.7
-1	33.0
-2	54.4
-3	89.6
-4	147.8

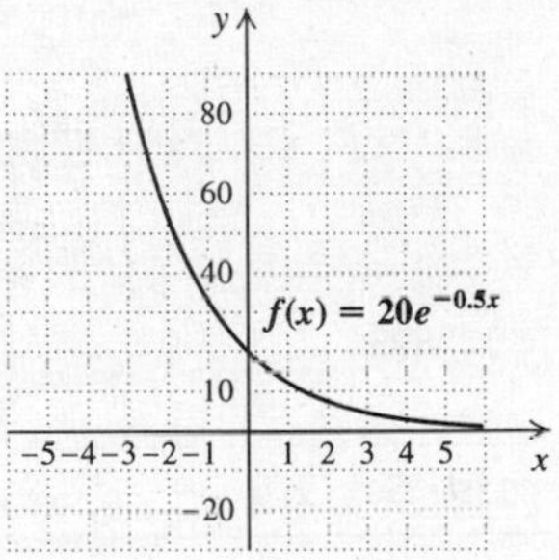

64.

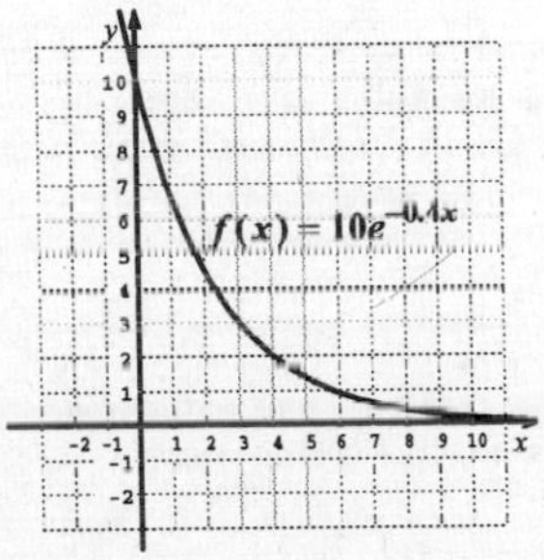

65. Graph: $f(x) = \ln (x + 4)$

x	$\ln(x + 4)$
0	1.4
1	1.6
2	1.8
3	1.9
4	2.1
-1	1.1
-2	0.7
-3	0
-4	Undefined

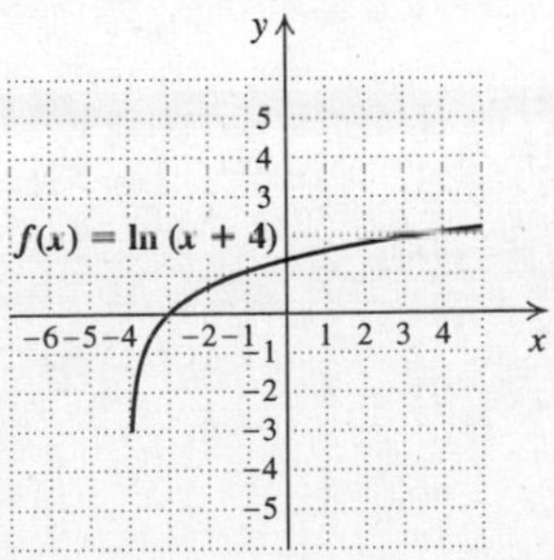

66.

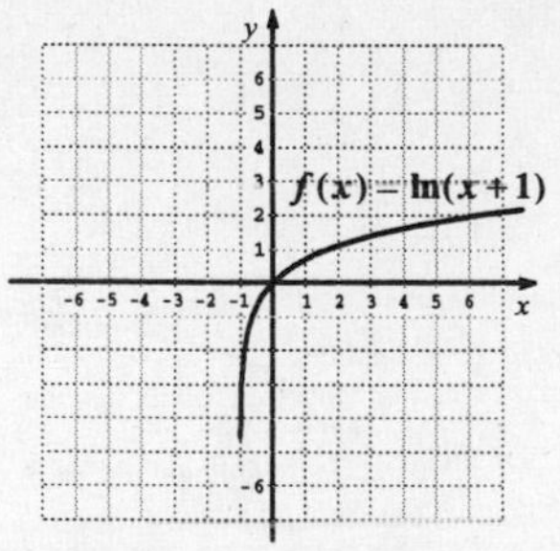

67. Graph: $f(x) = 3 \ln x$

x	$3 \ln x$
0.5	-2.1
1	0
2	2.1
3	3.3
4	4.2
5	4.8
6	5.4

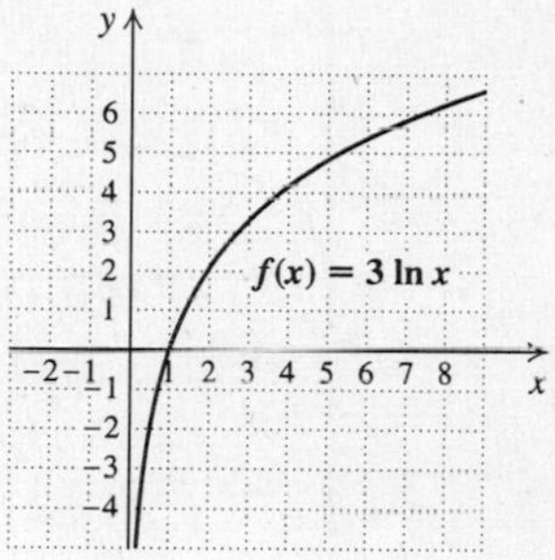

68.

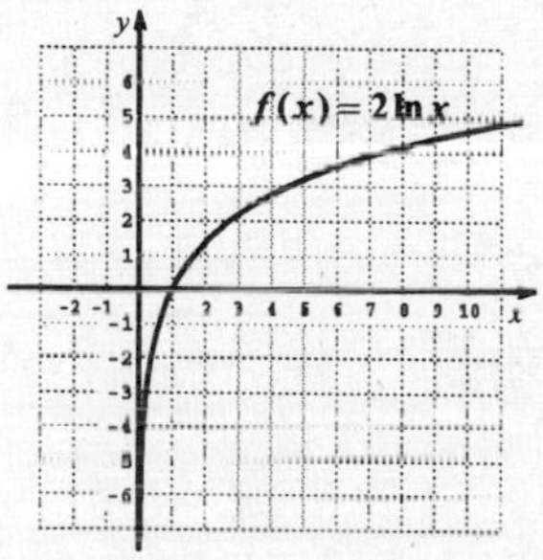

69. Graph: $f(x) = \ln (x - 2)$

x	$\ln (x - 2)$
2.5	-0.7
3	0
4	0.7
5	1.1
7	1.6
9	1.9
10	2.1

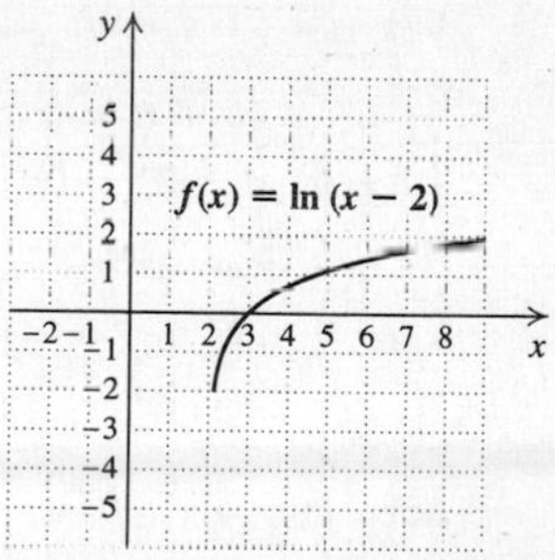

70.

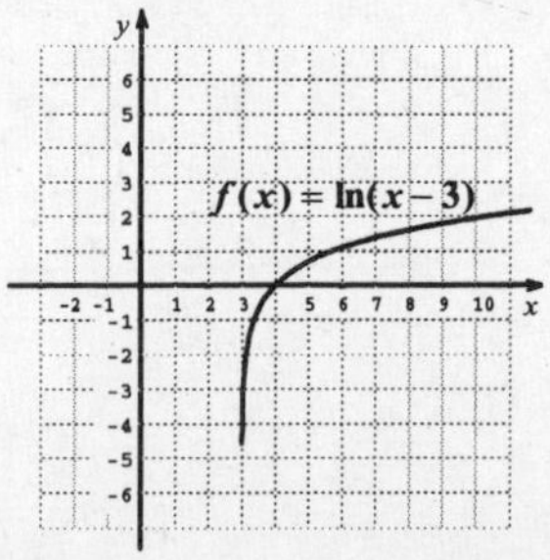

71. Graph: $f(x) = \ln x - 3$

x	$\ln x - 3$
0.5	-3.7
1	-3
2	-2.3
3	-1.9
4	-1.6
5	-1.4
6	-1.2

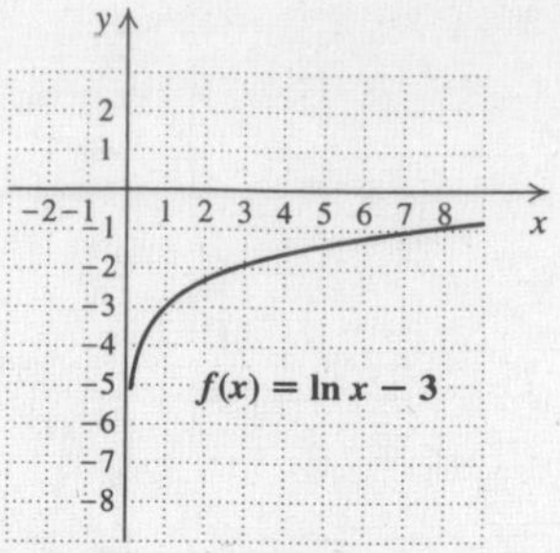

72.

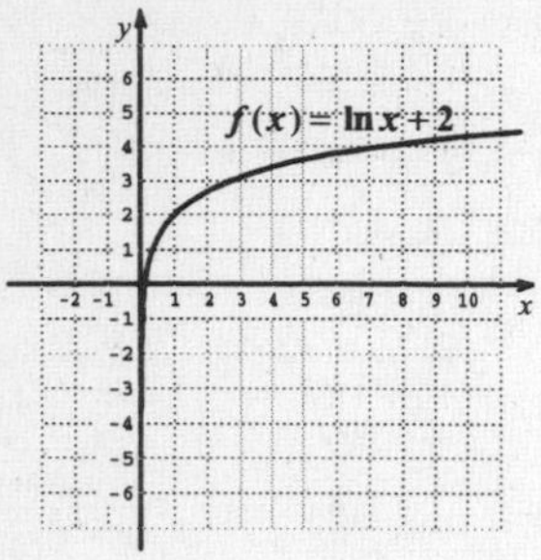

73. $4x^2 - 25 = 0$

$(2x + 5)(2x - 5) = 0$

$2x + 5 = 0$ or $2x - 5 = 0$

$2x = -5$ or $2x = 5$

$x = -\frac{5}{2}$ or $x = \frac{5}{2}$

The solutions are $-\frac{5}{2}$ and $\frac{5}{2}$.

74. $0, \frac{7}{5}$

75. $x^{1/2} - 6x^{1/4} + 8 = 0$

Let $u = x^{1/4}$.

$u^2 - 6u + 8 = 0$ Substituting

$(u - 4)(u - 2) = 0$

$u = 4$ or $u = 2$

$x^{1/4} = 4$ or $x^{1/4} = 2$

$x = 256$ or $x = 16$ Raising both sides to the fourth power

Both numbers check. The solutions are 256 and 16.

76. $\frac{1}{4}, 9$

77.

78.

79. Use the change-of-base formula with $a = 10$ and $b = e$. We obtain

$$\ln M = \frac{\log M}{\log e}.$$

80. $\log M = \frac{\ln M}{\ln 10}$

81. $\log 374x = 4.2931$

$374x = 10^{4.2931}$

$374x \approx 19{,}638.12409$

$x \approx 52.5084$

82. ± 3.3112

83. $\log 692 + \log x = \log 3450$

$\log x = \log 3450 - \log 692$

$\log x = \log \frac{3450}{692}$ Using the quotient rule

$x = \frac{3450}{692}$

$x \approx 4.9855$

84. 1.5893

Exercise Set 9.6

1. $2^x = 8$

$2^x = 2^3$

$x = 3$ The exponents must be the same.

2. 4

3. $4^x = 256$

$4^x = 4^4$

$x = 4$ The exponents must be the same.

4. 3

5. $2^{2x} = 32$

$2^{2x} = 2^5$

$2x = 5$

$x = \frac{5}{2}$

6. 1

7. $3^{5x} = 27$

$3^{5x} = 3^3$

$5x = 3$

$x = \frac{3}{5}$

8. $\frac{4}{7}$

9. $2^x = 9$

$\log 2^x = \log 9$ Taking the common logarithm on both sides

$x \log 2 = \log 9$ Using the power rule

$x = \dfrac{\log 9}{\log 2}$ Solving for x

$x \approx 3.170$ Using a calculator

10. 4.907

11. $2^x = 10$

$\log 2^x = \log 10$ Taking the common logarithm on both sides

$x \log 2 = \log 10$ Using the power rule

$x = \dfrac{\log 10}{\log 2}$ Solving for x

$x \approx 3.322$ Using a calculator

12. 5.044

13. $5^{4x-7} = 125$

$5^{4x-7} = 5^3$

$4x - 7 = 3$ The exponents must be the same.

$4x = 10$

$x = \dfrac{10}{4}$, or $\dfrac{5}{2}$

14. -1

15. $3^{x^2} \cdot 3^{4x} = \dfrac{1}{27}$

$3^{x^2+4x} = 3^{-3}$

$x^2 + 4x = -3$

$x^2 + 4x + 3 = 0$

$(x + 3)(x + 1) = 0$

$x = -3$ or $x = -1$

16. $\frac{1}{2}$, -3

17. $4^x = 7$

$\log 4^x = \log 7$

$x \log 4 = \log 7$

$x = \dfrac{\log 7}{\log 4}$

$x \approx 1.404$

18. 1.107

19. $e^t = 100$

$\ln e^t = \ln 100$ Taking $\ln$ on both sides

$t = \ln 100$ Finding the logarithm of the base to a power

$t \approx 4.605$ Using a calculator

20. 6.908

21. $e^{-t} = 0.1$

$\ln e^{-t} = \ln 0.1$ Taking $\ln$ on both sides

$-t = \ln 0.1$ Finding the logarithm of the base to a power

$-t \approx -2.303$

$t \approx 2.303$

22. 4.605

23. $e^{-0.02t} = 0.06$

$\ln e^{-0.02t} = \ln 0.06$ Taking $\ln$ on both sides

$-0.02t = \ln 0.06$ Finding the logarithm of the base to a power

$t = \dfrac{\ln 0.06}{-0.02}$

$t \approx \dfrac{-2.8134}{-0.02}$

$t \approx 140.671$

24. 9.902

25. $2^x = 3^{x-1}$

$\log 2^x = \log 3^{x-1}$

$x \log 2 = (x - 1) \log 3$

$x \log 2 = x \log 3 - \log 3$

$\log 3 = x \log 3 - x \log 2$

$\log 3 = x(\log 3 - \log 2)$

$\dfrac{\log 3}{\log 3 - \log 2} = x$

$\dfrac{0.4771}{0.4771 - 0.3010} \approx x$

$2.710 \approx x$

26. 7.162

27. $(2.8)^x = 41$

$\log (2.8)^x = \log 41$

$x \log 2.8 = \log 41$

$x = \dfrac{\log 41}{\log 2.8}$

$x \approx \dfrac{1.6128}{0.4472}$

$x \approx 3.607$

28. 3.581

29. $20 - (1.7)^x = 0$

$20 = (1.7)^x$

$\log 20 = \log (1.7)^x$

$\log 20 = x \log 1.7$

$\dfrac{\log 20}{\log 1.7} = x$

$\dfrac{1.3010}{0.2304} \approx x$

$5.646 \approx x$

30. 3.210

31. $\log_3 x = 3$

$x = 3^3$ Writing an equivalent exponential equation

$x = 27$

32. 625

33. $\log_2 x = -3$

$x = 2^{-3}$ Writing an equivalent exponential equation

$x = \frac{1}{8}$

34. 2

35. $\log x = 1$ The base is 10.

$x = 10^1$

$x = 10$

36. 1000

37. $\log x = -2$ The base is 10.

$x = 10^{-2}$

$x = \frac{1}{100}$

38. $\frac{1}{1000}$

39. $\ell n\ x = 2$

$x = e^2 \approx 7.389$

40. 2.718

41. $\ell n\ x = -1$

$x = e^{-1}$

$x = \frac{1}{e} \approx 0.368$

42. 0.050

43. $\log_5 (2x - 7) = 3$

$2x - 7 = 5^3$

$2x - 7 = 125$

$2x = 132$

$x = 66$

The answer checks. The solution is 66.

44. $-\frac{25}{6}$

45. $\log x + \log (x - 9) = 1$ The base is 10.

$\log_{10} [x(x - 9)] = 1$ Using the product rule

$x(x - 9) = 10^1$

$x^2 - 9x = 10$

$x^2 - 9x - 10 = 0$

$(x - 10)(x + 1) = 0$

$x = 10$ or $x = -1$

Check: For 10:

$\log x + \log (x - 9) = 1$

$\log 10 + \log (10 - 9) \ ?\ 1$

$\log 10 + \log 1$

$1 + 0$

1 TRUE

For -1:

$\log x + \log (x - 9) = 1$

$\log (-1) + \log (-1 - 9) \ ?\ 1$

The number -1 does not check, because negative numbers do not have logarithms. The solution is 10.

46. 1

47. $\log x - \log (x + 3) = -1$ The base is 10.

$\log_{10} \frac{x}{x + 3} = -1$ Using the quotient rule

$\frac{x}{x + 3} = 10^{-1}$

$\frac{x}{x + 3} = \frac{1}{10}$

$10x = x + 3$

$9x = 3$

$x = \frac{1}{3}$

The answer checks. The solution is $\frac{1}{3}$.

48. 1

49. $\log_2 (x + 1) + \log_2 (x - 1) = 3$

$\log_2 [(x + 1)(x - 1)] = 3$ Using the product rule

$(x + 1)(x - 1) = 2^3$

$x^2 - 1 = 8$

$x^2 = 9$

$x = \pm 3$

The number 3 checks, but -3 does not. The solution is 3.

50. $\frac{83}{15}$

51. $\log_4 (x + 6) - \log_4 x = 2$

$\log_4 \frac{x + 6}{x} = 2$ Using the quotient rule

$\frac{x + 6}{x} = 4^2$

$\frac{x + 6}{x} = 16$

$x + 6 = 16x$

$6 = 15x$

$\frac{2}{5} = x$

The answer checks. The solution is $\frac{2}{5}$.

52. 4

53. $\log_4 (x + 3) + \log_4 (x - 3) = 2$

$\log_4 [(x + 3)(x - 3)] = 2$ Using the product rule

$(x + 3)(x - 3) = 4^2$

$x^2 - 9 = 16$

$x^2 = 25$

$x = \pm 5$

The number 5 checks, but -5 does not. The solution is 5.

54. $\sqrt{41}$

55. $(125x^7y^{-2}z^6)^{-2/3} =$

$(5^3)^{-2/3}(x^7)^{-2/3}(y^{-2})^{-2/3}(z^6)^{-2/3} =$

$5^{-2}x^{-14/3}y^{4/3}z^{-4} = \frac{1}{25}x^{-14/3}y^{4/3}z^{-4}$, or

$\frac{y^{4/3}}{25x^{14/3}z^4}$

56. $-i$

57. $E = mc^2$

$\frac{E}{m} = c^2$

$\sqrt{\frac{E}{m}} = c$ Taking the principal square root

58. $\pm 10, \pm 2$

59.

60.

61. $8^x = 16^{3x+9}$

$(2^3)^x = (2^4)^{3x+9}$

$2^{3x} = 2^{12x+36}$

$3x = 12x + 36$

$-36 = 9x$

$-4 = x$

62. $\frac{12}{5}$

63. $\log_6 (\log_2 x) = 0$

$\log_2 x = 6^0$

$\log_2 x = 1$

$x = 2^1$

$x = 2$

64. $\sqrt[3]{3}$

65. $\log_5 \sqrt{x^2 - 9} = 1$

$\sqrt{x^2 - 9} = 5^1$

$x^2 - 9 = 25$ Squaring both sides

$x^2 = 34$

$x = \pm\sqrt{34}$

Both numbers check. The solutions are $\pm\sqrt{34}$.

66. -1

67. $\log (\log x) = 5$ The base is 10.

$\log x = 10^5$

$\log x = 100{,}000$

$x = 10^{100{,}000}$

The number checks. The solution is $10^{100{,}000}$.

68. -3, -1

69. $\log x^2 = (\log x)^2$

$2 \log x = (\log x)^2$

$0 = (\log x)^2 - 2 \log x$

Let $u = \log x$.

$0 = u^2 - 2u$

$0 = u(u - 2)$

$u = 0$ or $u = 2$

$\log x = 0$ or $\log x = 2$

$x = 10^0$ or $x = 10^2$

$x = 1$ or $x = 100$

Both numbers check. The solutions are 1 and 100.

70. 625, -625

71. $\log x^{\log x} = 25$

$\log x (\log x) = 25$ Using the power rule

$(\log x)^2 = 25$

$\log x = \pm 5$

$x = 10^5$ or $x = 10^{-5}$

$x = 100{,}000$ or $x = \frac{1}{100{,}000}$

Both numbers check. The solutions are 100,000 and $\frac{1}{100{,}000}$.

72. $\frac{1}{2}$, 5000

73. $$(81^{x-2})(27^{x+1}) = 9^{2x-3}$$
$$[(3^4)^{x-2}][(3^3)^{x+1}] = (3^2)^{2x-3}$$
$$(3^{4x-8})(3^{3x+3}) = 3^{4x-6}$$
$$3^{7x-5} = 3^{4x-6}$$
$$7x - 5 = 4x - 6$$
$$3x = -1$$
$$x = -\frac{1}{3}$$

74. 1.465, 1

75. $$3^{2x} - 3^{2x-1} = 18$$
$$3^{2x}(1 - 3^{-1}) = 18 \quad \text{Factoring}$$
$$3^{2x}\left(1 - \frac{1}{3}\right) = 18$$
$$3^{2x}\left(\frac{2}{3}\right) = 18$$
$$3^{2x} = 27 \quad \text{Multiplying by } \frac{3}{2}$$
$$3^{2x} = 3^3$$
$$2x = 3$$
$$x = \frac{3}{2}$$

76. 38

77. $\log_5 125 = 3$ and $\log_{125} 5 = \frac{1}{3}$, so $x = (\log_{125} 5)^{\log_5 125}$ is equivalent to $x = \left(\frac{1}{3}\right)^3 = \frac{1}{27}$. Then $\log_3 x = \log_3 \frac{1}{27} = -3$.

78. 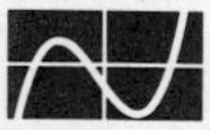

Exercise Set 9.7

1. a) One billion is 1000 million, so we set $N(t) = 1000$ and solve for t:
$$1000 = 7.5(6)^{0.5t}$$
$$\frac{1000}{7.5} = (6)^{0.5t}$$
$$\log \frac{1000}{7.5} = \log(6)^{0.5t}$$
$$\log 1000 - \log 7.5 = 0.5t \log 6$$
$$t = \frac{\log 1000 - \log 7.5}{0.5 \log 6}$$
$$t \approx \frac{3 - 0.87506}{0.5(0.77815)} \approx 5.5$$
After about 5.5 years, one billion compact discs will be sold in a year.

b) When $t = 0$, $N(t) = 7.5(6)^{0.5(0)} = 7.5(6)^0 = 7.5(1) = 7.5$. Twice this initial number is 15, so we set $N(t) = 15$ and solve for t:
$$15 = 7.5(6)^{0.5t}$$
$$2 = (6)^{0.5t}$$
$$\log 2 = \log(6)^{0.5t}$$
$$\log 2 = 0.5t \log 6$$
$$t = \frac{\log 2}{0.5 \log 6} \approx \frac{0.30103}{0.5(0.77815)} \approx 0.8$$
The doubling time is about 0.8 year.

2. a) 3.6 days

b) 0.6 days

3. a) We set $A(t) = \$450{,}000$ and solve for t:
$$450{,}000 = 50{,}000(1.06)^t$$
$$\frac{450{,}000}{50{,}000} = (1.06)^t$$
$$9 = (1.06)^t$$
$$\log 9 = \log (1.06)^t \quad \text{Taking the common logarithm on both sides}$$
$$\log 9 = t \log 1.06 \quad \text{Using the power rule}$$
$$t = \frac{\log 9}{\log 1.06} \approx \frac{0.95424}{0.02531} \approx 37.7$$
It will take about 37.7 years for the \$50,000 to grow to \$450,000.

b) We set $A(t) = \$100{,}000$ and solve for t:
$$100{,}000 = 50{,}000(1.06)^t$$
$$2 = (1.06)^t$$
$$\log 2 = \log (1.06)^t$$
Taking the common logarithm on both sides
$$\log 2 = t \log 1.06 \quad \text{Using the power rule}$$
$$t = \frac{\log 2}{\log 1.06} \approx \frac{0.30103}{0.02531} \approx 11.9$$
The doubling time is about 11.9 years.

4. a) 59.7 years

b) 19.1 years

5. a) We set $N(t) = 60{,}000$ and solve for t:
$$60{,}000 = 250{,}000\left(\frac{2}{3}\right)^t$$
$$\frac{60{,}000}{250{,}000} = \left(\frac{2}{3}\right)^t$$
$$0.24 = \left(\frac{2}{3}\right)^t$$
$$\log 0.24 = \log \left(\frac{2}{3}\right)^t$$
$$\log 0.24 = t \log \frac{2}{3}$$
$$t = \frac{\log 0.24}{\log \frac{2}{3}} \approx \frac{-0.61979}{-0.17609} \approx 3.5$$

After about 3.5 years 60,000 cans will still be in use.

b) We set N(t) = 1000 and solve for t.

$$1000 = 250{,}000\left(\frac{2}{3}\right)^t$$

$$\frac{1000}{250{,}000} = \left(\frac{2}{3}\right)^t$$

$$0.004 = \left(\frac{2}{3}\right)^t$$

$$\log 0.004 = \log \left(\frac{2}{3}\right)^t$$

$$\log 0.004 = t \log \frac{2}{3}$$

$$t = \frac{\log 0.004}{\log \frac{2}{3}} \approx \frac{-2.39794}{-0.17609} \approx 13.6$$

After about 13.6 years only 1000 cans will still be in use.

6. a) 6.6 years

b) 3.1 years

7. a) Substitute 0.09 for k:

$$P(t) = P_0\, e^{0.09t}$$

b) To find the balance after one year, set $P_0 = 1000$ and $t = 1$. We find P(1):

$P(1) = 1000\, e^{0.09(1)} = 1000\, e^{0.09} \approx$
$1000(1.094174284) \approx \$1094.17$

To find the balance after 2 years, set $P_0 = 1000$ and $t = 2$. We find P(2):

$P(2) = 1000\, e^{0.09(2)} = 1000\, e^{0.18} \approx$
$1000(1.197217363) \approx \$1197.22$

c) To find the doubling time, we set $P_0 = 1000$ and $P(t) = 2000$ and solve for t.

$2000 = 1000\, e^{0.09t}$

$2 = e^{0.09t}$

$\ln 2 = \ln e^{0.09t}$ Taking the natural logarithm on both sides

$\ln 2 = 0.09t$ Finding the logarithm of the base to a power

$\frac{\ln 2}{0.09} = t$

$7.7 \approx t$

The investment will double in about 7.7 years.

8. a) $P(t) = P_0\, e^{0.1t}$

b) \$22,103.42, \$24,428.06

c) 6.9 years

9. We start with the exponential growth equation

$P(t) = P_0\, e^{kt}$.

Substitute $2\,P_0$ for P(t) and 0.01 for k and solve for t.

$2\,P_0 = P_0\, e^{0.01t}$

$2 = e^{0.01t}$ Dividing by P_0 on both sides

$\ln 2 = \ln e^{0.01t}$ Taking the natural logarithm on both sides

$\ln 2 = 0.01t$ Finding the logarithm of the base to a power

$\frac{\ln 2}{0.01} = t$

$69.3 \approx t$

The doubling time is about 69.3 years.

10. About 19.8 years

11. a) The equation $P(t) = P_0\, e^{kt}$ can be used to model population growth. At t = 0 (1990), the population was 5.2 billion. We substitute 5.2 for P_0 and 1.6%, or 0.016, for k to obtain the exponential growth function:

$P(t) = 5.2\, e^{0.016t}$, where t is the number of years after 1990.

b) In 2000, t = 2000 - 1990, or 10. To estimate the population in 2000, we find P(10).

$P(10) = 5.2\, e^{0.016(10)}$

$= 5.2\, e^{0.16}$

$\approx 5.2(1.17351)$ Finding $e^{0.16}$ using a calculator

≈ 6.1

The population of the world in 2000 will be about 6.1 billion.

c) We set P(t) = 8 and solve for t.

$8 = 5.2\, e^{0.016t}$

$\frac{8}{5.2} = e^{0.016t}$

$\ln \left(\frac{8}{5.2}\right) = \ln e^{0.016t}$

$\ln \left(\frac{8}{5.2}\right) = 0.016t$

$\frac{\ln \left(\frac{8}{5.2}\right)}{0.016} = t$

$26.9 \approx t$

The population will be 8.0 billion about 26.9 years after 1990, or in 2017.

12. a) 86.4 minutes

b) 300.5 minutes

c) 20 minutes

13. a) $S(0) = 68 - 20 \log (0 + 1) = 68 - 20 \log 1 =$
$68 - 20(0) = 68\%$

b) $S(4) = 68 - 20 \log (4 + 1) = 68 - 20 \log 5 \approx$
$68 - 20(0.69897) \approx 54\%$

$S(24) = 68 - 20 \log (24 + 1) =$
$68 - 20 \log 25 \approx 68 - 20\,(1.39794) \approx$
40%

c) Using the values we computed in parts (a) and (b) and any others we wish to calculate, we sketch the graph:

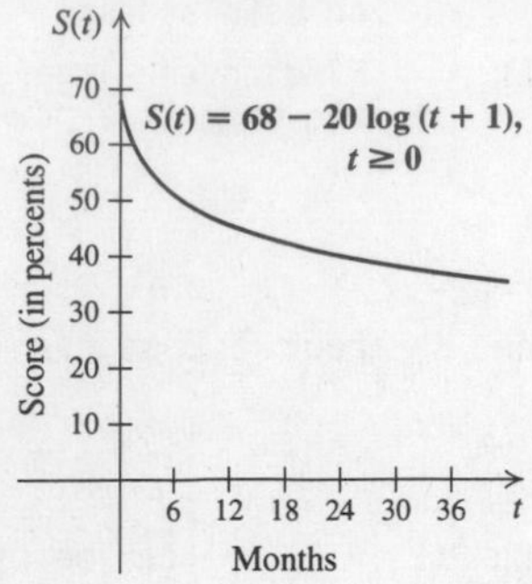

d) We set S(t) = 50 and solve for t:

$50 = 68 - 20 \log (t + 1)$

$-18 = -20 \log (t + 1)$

$0.9 = \log (t + 1)$

$10^{0.9} = t + 1$ Using the definition of logarithms or taking the antilogarithm

$7.9 \approx t + 1$

$6.9 \approx t$

After about 6.9 months, the average score was 50.

14. a) 2000

b) 2452

c)

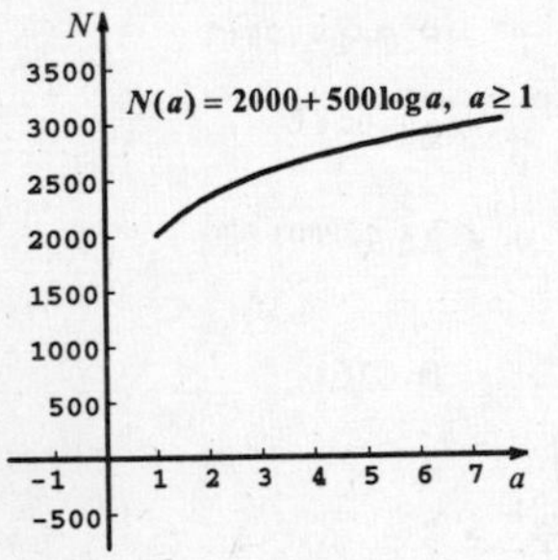

d) \$1,000,000 thousand, or \$1,000,000,000

15. We substitute 6.3×10^{-7} for H^+ in the formula for pH.

$pH = -\log [H^+] = -\log [6.3 \times 10^{-7}] =$

$-[\log 6.3 + \log 10^{-7}] \approx -[0.7993 - 7] =$

$-[-6.2007] = 6.2007 \approx 6.2$

The pH of a common brand of mouthwash is about 6.2.

16. 7.4

17. We substitute 1.6×10^{-8} for H^+ in the formula for pH.

$pH = -\log [H^+] = -\log [1.6 \times 10^{-8}] =$

$-[\log 1.6 + \log 10^{-8}] \approx -[0.2041 - 8] =$

$-[-7.7959] = 7.7959 \approx 7.8$

The pH of eggs is about 7.8.

18. 4.2

19. We substitute 7 for pH in the formula and solve for H^+.

$7 = -\log [H^+]$

$-7 = \log [H^+]$

$10^{-7} = H^+$ Using the definition of logarithm or taking the antilogarithm

The hydrogen ion concentration of tap water is 10^{-7} moles per liter.

20. 4.0×10^{-6} moles per liter

21. We substitute 3.2 for pH in the formula and solve for H^+.

$3.2 = -\log [H^+]$

$-3.2 = \log [H^+]$

$10^{-3.2} = H^+$ Using the definition of logarithm or taking the antilogarithm

$0.00063 \approx H^+$

$6.3 \times 10^{-4} \approx H^+$ Writing in scientific notation

The hydrogen ion concentration of orange juice is about 6.3×10^{-4} moles per liter.

22. 1.6×10^{-5} moles per liter

23. We substitute into the formula.

$R = \log \frac{10^{8.25} I_0}{I_0} = \log 10^{8.25} = 8.25$

24. 5

25. We start with the exponential growth equation

$D(t) = D_0 e^{kt}$, where t is the number of years after 1993.

Substitute $2 D_0$ for D(t) and 0.1 for k and solve for t.

$2 D_0 = D_0 e^{0.1t}$

$2 = e^{0.1t}$

$\ln 2 = \ln e^{0.1t}$

$\ln 2 = 0.1t$

$\frac{\ln 2}{0.1} = t$

$6.9 \approx t$

The demand will be double that of 1993 about 6.9 years after 1993, or in 2000.

26. 2011

27. a) We start with the exponential growth equation

$N(t) = N_0 e^{kt}$, where t is the number of years after 1967.

Substituting 1 for N_0, we have

$N(t) = e^{kt}$.

To find the exponential growth rate k, observe that there were 1418 heart transplants in 1987, or 20 years after 1967. We substiute and solve for k.

$$N(20) = e^{k \cdot 20}$$
$$1418 = e^{20k}$$
$$\ln 1418 = \ln e^{20k}$$
$$\ln 1418 = 20k$$
$$\frac{\ln 1418}{20} = k$$
$$0.363 \approx k$$

Thus the exponential growth function is

$N(t) = e^{0.363t}$, where t is the number of years after 1967.

b) In 1998, t = 1998 - 1967, or 31. We find N(31).

$$N(31) = e^{0.363(31)} \approx 77{,}111$$

The number of heart transplants in 1998 will be about 77,111.

28. a) 28.2¢

b) 40¢

c) 2025

29. We will use the function derived in Example 7:

$P(t) = P_0 e^{-0.00012t}$

If the tusk has lost 20% of its carbon-14 from an initial amount P_0, then 80% (P_0) is the amount present. To find the age of the tusk t, we substitute 80% (P_0), or $0.8P_0$, for P(t) in the function above and solve for t.

$$0.8P_0 = P_0 e^{-0.00012t}$$
$$0.8 = e^{-0.00012t}$$
$$\ln 0.8 = \ln e^{-0.00012t}$$
$$-0.2231 \approx -0.00012t$$
$$t \approx \frac{-0.2231}{-0.00012} \approx 1860$$

The tusk is about 1860 years old.

30. 878 years

31. The function $P(t) = P_0 e^{-kt}$, $k > 0$, can be used to model decay. For iodine-131, k = 9.6%, or 0.096. To find the half-life we substitute 0.096 for k and $\frac{1}{2} P_0$ for P(t), and solve for t.

$$\frac{1}{2} P_0 = P_0 e^{-0.096t}, \text{ or } \frac{1}{2} = e^{-0.096t}$$
$$\ln \frac{1}{2} = \ln e^{-0.096t} = -0.096t$$
$$t = \frac{\ln 0.5}{-0.096} \approx \frac{-0.6931}{-0.096} \approx 7.2 \text{ days}$$

32. 11 years

33. We use the function $P(t) = P_0 e^{-kt}$, $k > 0$. When $t = 3$, $P(t) = \frac{1}{2} P_0$. We substitute and solve for k.

$$\frac{1}{2} P_0 = P_0 e^{-k(3)}, \text{ or } \frac{1}{2} = e^{-3k}$$
$$\ln \frac{1}{2} = \ln e^{-3k} = -3k$$
$$k = \frac{\ln 0.5}{-3} \approx \frac{-0.6931}{-3} \approx 0.23$$

The decay rate is 0.23, or 23%, per minute.

34. 3.2% per year

35. a) We start with the exponential growth equation

$V(t) = V_0 e^{kt}$, where t is the number of years after 1986.

Substituting 110,000 for V_0, we have

$V(t) = 110{,}000\, e^{kt}$.

To find the exponential growth rate k, observe that the card sold for $451,000 in 1991, or 5 years after 1986. We substitute and solve for k.

$$V(5) = 110{,}000\, e^{k \cdot 5}$$
$$451{,}000 = 110{,}000\, e^{5k}$$
$$4.1 = e^{5k}$$
$$\ln 4.1 = \ln e^{5k}$$
$$\ln 4.1 = 5k$$
$$\frac{\ln 4.1}{5} = k$$
$$0.2822 \approx k$$

Thus the exponential growth function is $V(t) = 110{,}000\, e^{0.2822t}$, where t is the number of years after 1986.

b) In 1998, t = 1998 - 1986, or 12.

$$V(12) = 110{,}000\, e^{0.2822(12)} \approx 3{,}251{,}528$$

The card's value in 1998 will be about $3,251,528.

c) Substitute $220,000 for V(t) and solve for t.

$$220{,}000 = 110{,}000\, e^{0.2822t}$$
$$2 = e^{0.2822t}$$
$$\ln 2 = \ln e^{0.2822t}$$
$$\ln 2 = 0.2822t$$
$$\frac{\ln 2}{0.2822} = t$$
$$2.5 \approx t$$

The doubling time is about 2.5 years.

d) Substitute \$9,000,000 for V(t) and solve for t.

$$9{,}000{,}000 = 110{,}000\, e^{0.2822t}$$

$$\frac{900}{11} = e^{0.2822t}$$

$$\ln \frac{900}{11} = \ln e^{0.2822t}$$

$$\ln \frac{900}{11} = 0.2822t$$

$$\frac{\ln \frac{900}{11}}{0.2822} = t$$

$$15.6 \approx t$$

The value of the card will be \$9,000,000 about 15.6 years after 1986, or in 2002.

36. a) $k \approx 0.16$; $V(t) = 84\, e^{0.16t}$, where V is in thousands of dollars and t is the number of years after 1947

b) \$250,400 thousand, or \$250,400,000

c) 4.3 years

d) About 58.7 years

37. $$\frac{\frac{x-5}{x+3}}{\frac{x}{x-3} + \frac{2}{x+3}}$$

$$= \frac{\frac{x-5}{x+3}}{\frac{x}{x-3}\cdot\frac{x+3}{x+3} + \frac{2}{x+3}\cdot\frac{x-3}{x-3}}$$ Finding the LCD and adding in the denominator

$$= \frac{\frac{x-5}{x+3}}{\frac{x^2+3x+2x-6}{(x-3)(x+3)}}$$

$$= \frac{\frac{x-5}{x+3}}{\frac{x^2+5x-6}{(x-3)(x+3)}}$$

$$= \frac{x-5}{x+3}\cdot\frac{(x-3)(x+3)}{x^2+5x-6}$$ Multiplying by the reciprocal of the denominator

$$= \frac{(x-5)(x-3)(x+3)}{(x+3)(x+6)(x-1)}$$ Factoring $x^2 + 5x - 6$ and multiplying

$$= \frac{(x-5)(x-3)\cancel{(x+3)}}{\cancel{(x+3)}(x+6)(x-1)}$$

$$= \frac{(x-5)(x-3)}{(x+6)(x-1)}, \text{ or } \frac{(x-5)(x-3)}{x^2+5x-6}$$

38. $\dfrac{3ab^2 + 5a^2b}{2b^2 - 4a^2}$

39.

40.

41.

42. (6,\$403)

Exercise Set 10.1

1. $y = x^2$

   a) This is equivalent to $y = (x - 0)^2 + 0$. The vertex is (0,0).

   b) We choose some x-values on both sides of the vertex and compute the corresponding values of y. The graph opens upward, because the coefficient of x^2, 1, is positive.

x	y
0	0
1	1
2	4
-1	1
-2	4

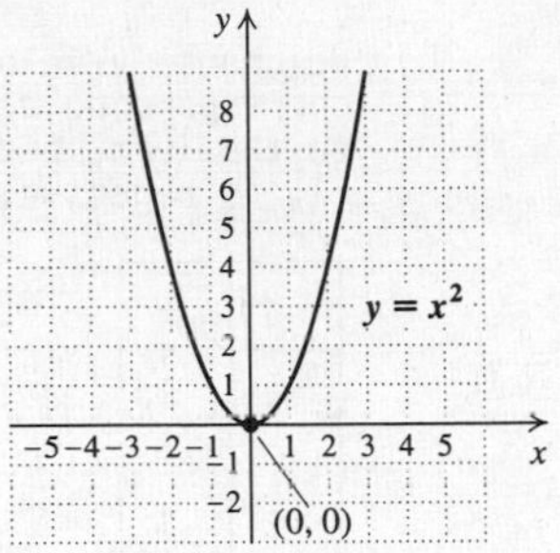

2. $x = y^2$

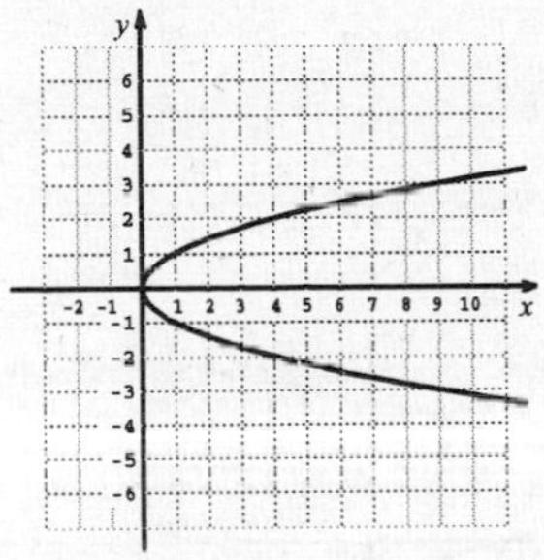

3. $x = y^2 + 4y + 1$

   a) We find the vertex by completing the square.

$$x = (y^2 + 4y + 4) + 1 - 4$$
$$x = (y + 2)^2 - 3$$

   The vertex is (-3,-2).

   b) To find ordered pairs, we choose values for y and compute the corresponding values for x. The graph opens to the right, because the coefficient of y^2, 1, is positive.

x	y
-3	-2
-2	-3
-2	-1
1	-4
1	0

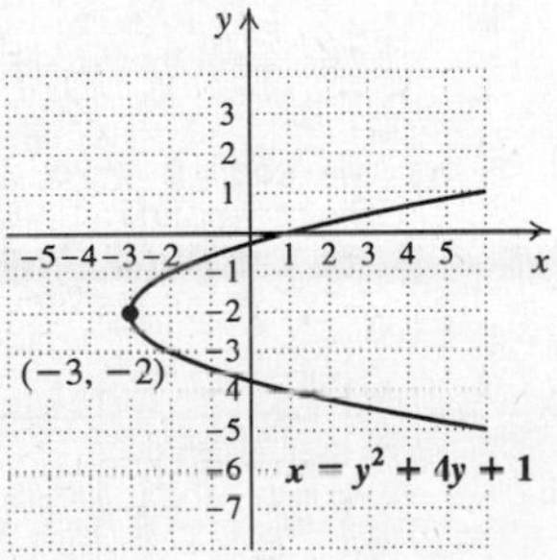

4. $y = x^2 - 2x + 3$

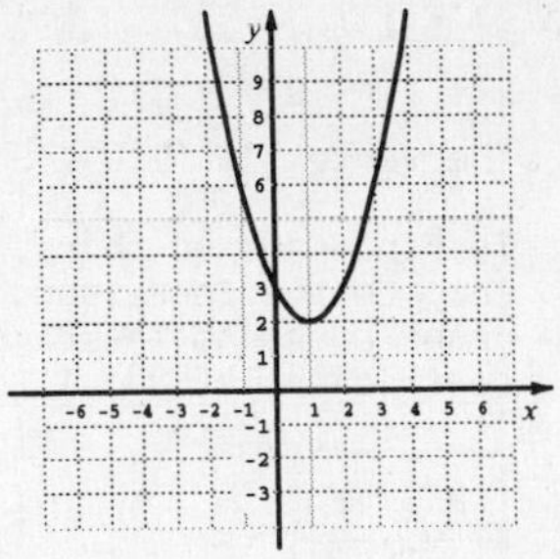

5. $y = -x^2 + 4x - 5$

   a) We can find the vertex by computing the first coordinate, $x = -b/2a$, and then substituting to find the second coordinate:

$$x = -\frac{b}{2a} = -\frac{4}{2(-1)} = 2$$

$$y = -x^2 + 4x - 5 = -(2)^2 + 4(2) - 5 = -1$$

   The vertex is (2,-1).

   b) We choose some x-values and compute the corresponding values for y. The graph opens downward because the coefficient of x^2, -1, is negative.

x	y
2	-1
3	-2
4	-5
1	-2
0	-5

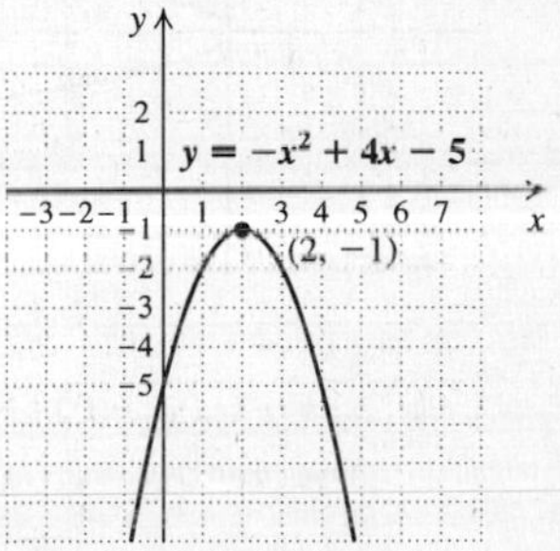

6. $x = 4 - 3y - y^2$

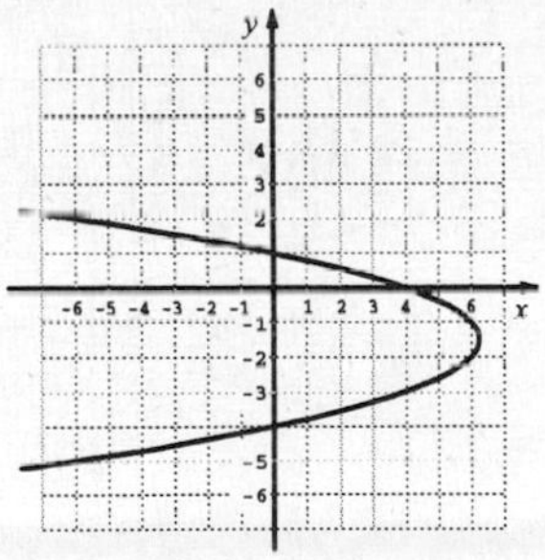

7. $x = y^2 + 1$

a) $x = (y - 0)^2 + 1$

The vertex is (1,0).

b) To find ordered pairs, we choose y-values and compute the corresponding values for x. The graph opens to the right, because the coefficient of y^2, 1, is positive.

x	y
1	0
2	1
5	2
2	-1
5	-2

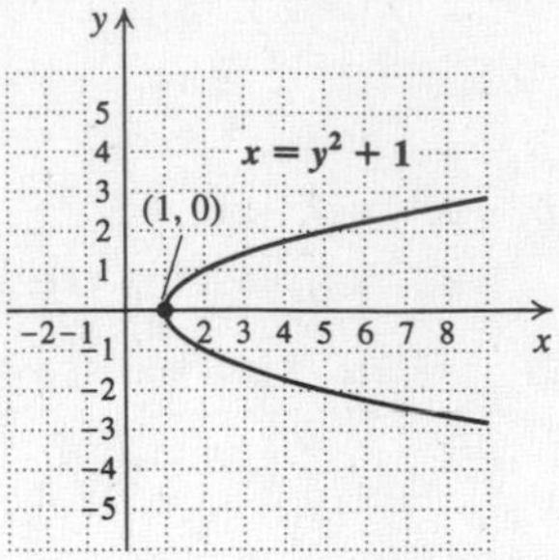

8. $x = 2y^2$

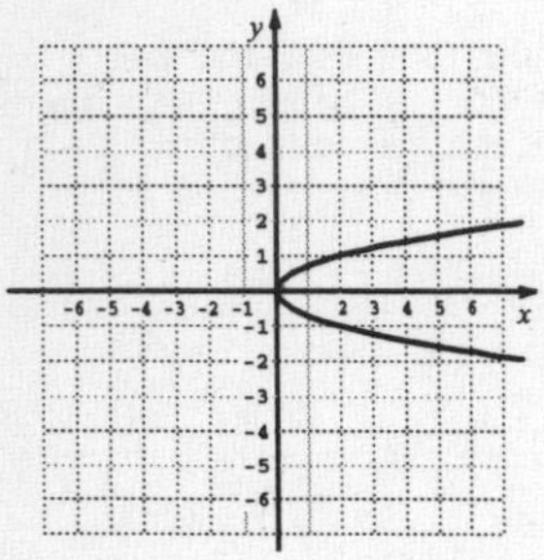

9. $x = -1 \cdot y^2$

a) $x = -1 \cdot (y - 0)^2 + 0$

The vertex is (0,0).

b) We choose y-values and compute the corresponding values for x. The graph opens to the left, because the coefficient of y^2, -1, is negative.

x	y
0	0
-1	1
-4	2
-1	-1
-4	-2

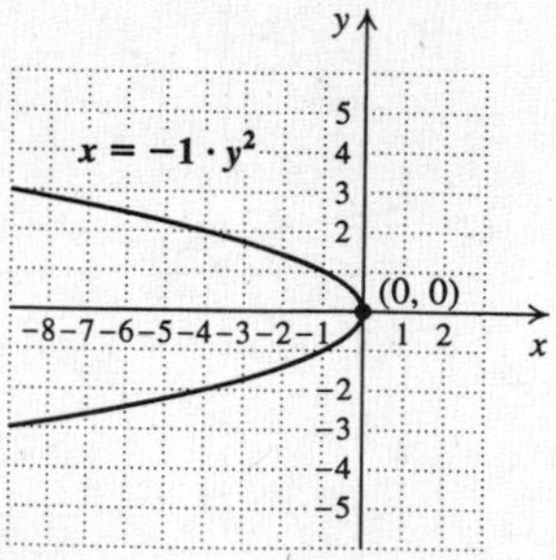

10. $x = y^2 - 1$

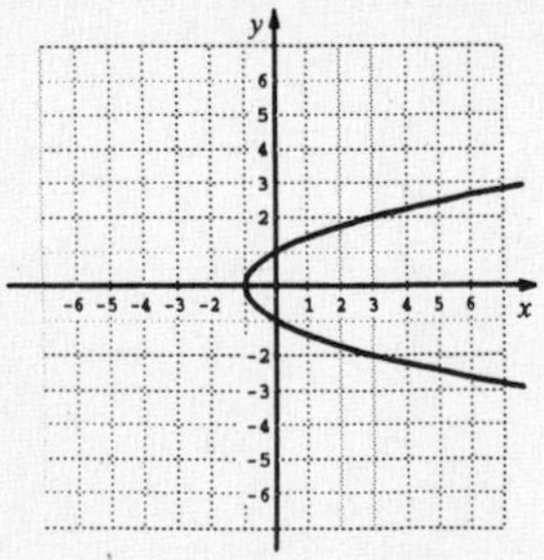

11. $x = -y^2 + 2y$

a) We find the vertex by computing the second coordinate, $y = -b/2a$ and then substituting to find the first coordinate:

$$y = -\frac{b}{2a} = -\frac{2}{2(-1)} = 1$$

$$x = -y^2 + 2y = -(1)^2 + 2(1) = 1$$

The vertex is (1,1).

b) We choose y-values and compute the corresponding values for x. The graph opens to the left, because the coefficient of y^2, -1, is negative.

x	y
1	1
0	0
-3	-1
0	2
-3	3

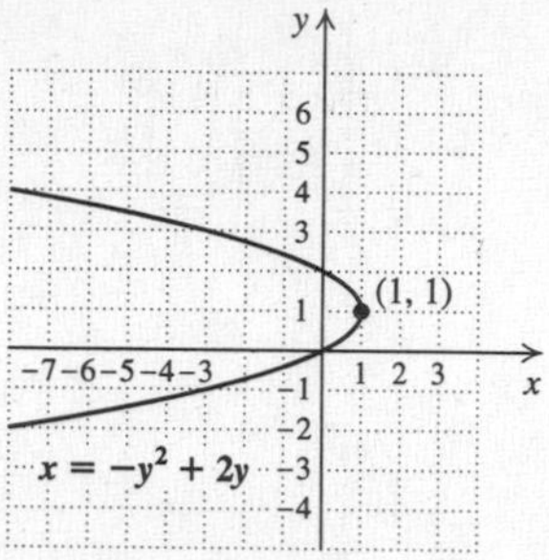

12. $x = y^2 + y - 6$

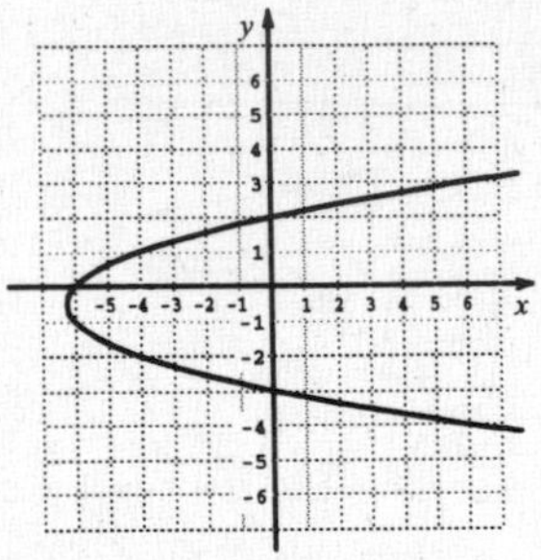

13. $x = 8 - y - y^2$

a) We find the vertex by completing the square.

$$x = -(y^2 + y) + 8$$

$$x = -\left(y^2 + y + \frac{1}{4}\right) + 8 + \frac{1}{4}$$

$$x = -\left(y + \frac{1}{2}\right)^2 + \frac{33}{4}$$

The vertex is $\left(\frac{33}{4}, -\frac{1}{2}\right)$.

b) We choose y-values and compute the corresponding values for x. The graph opens to the left, because the coefficient of y^2, -1, is negative.

x	y
$\frac{33}{4}$	$-\frac{1}{2}$
8	0
6	1
2	2
8	-1
6	-2
2	-3

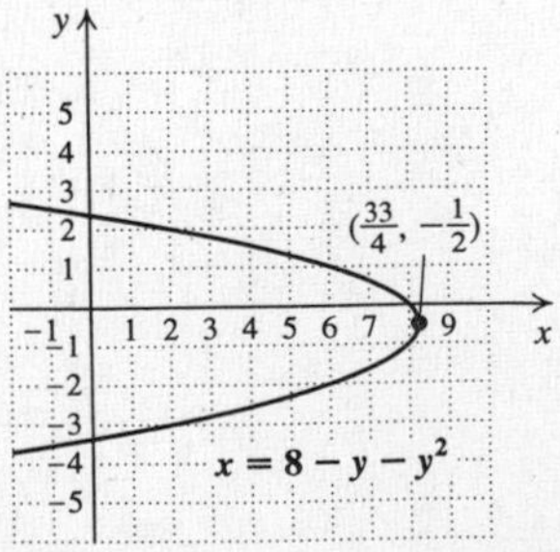

14. $y = x^2 + 2x + 1$

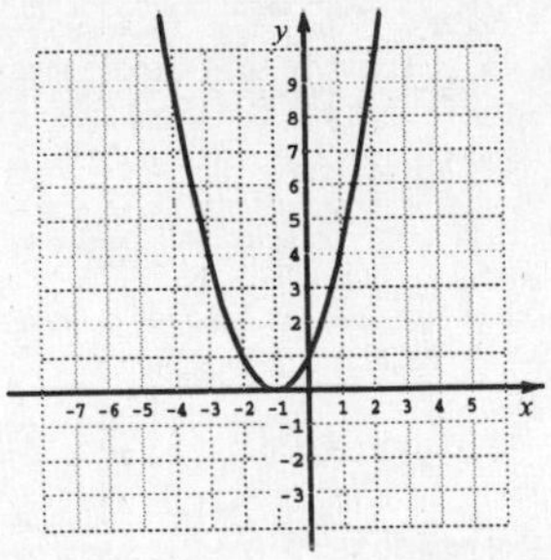

15. $y = x^2 - 2x + 1$

a) $y = (x - 1)^2 + 0$

The vertex is (1,0).

b) We choose x-values and compute the corresponding values for y. The graph opens upward, because the coefficient of x^2, 1, is positive.

x	y
1	0
0	1
-1	4
2	1
3	4

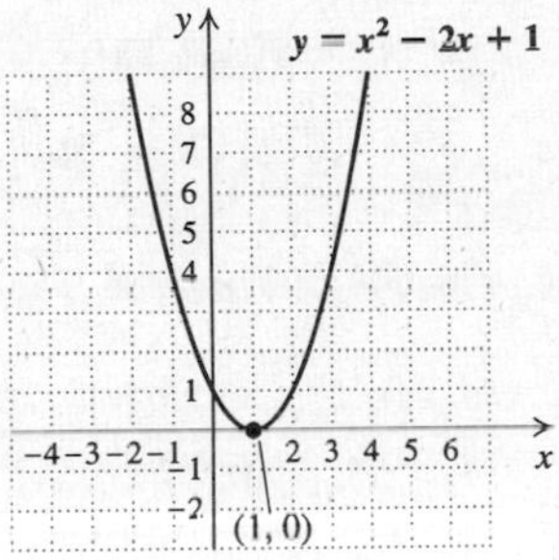

16. $y = -\frac{1}{2}x^2$

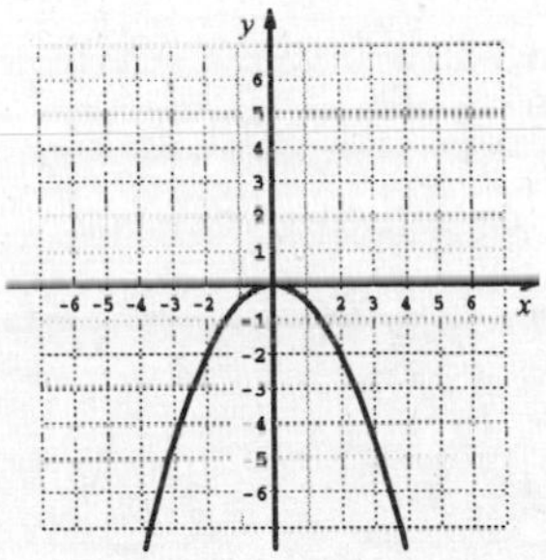

17. $x = -y^2 + 2y + 3$

a) We find the vertex by computing the second coordinate, $y = -b/2a$, and then substituting to find the first coordinate.

$$y = -\frac{b}{2a} = -\frac{2}{2(-1)} = 1$$

$$x = -y^2 + 2y + 3 = -(1)^2 + 2(1) + 3 = 4$$

The vertex is (4,1).

b) We choose y-values and compute the corresponding values for x. The graph opens to the left, because the coefficient of y^2, -1, is negative.

x	y
4	1
3	0
0	-1
3	2
0	3

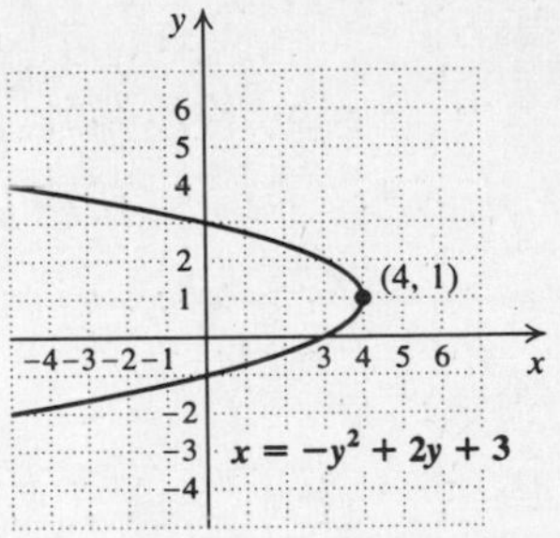

18. $x = -y^2 - 2y + 3$

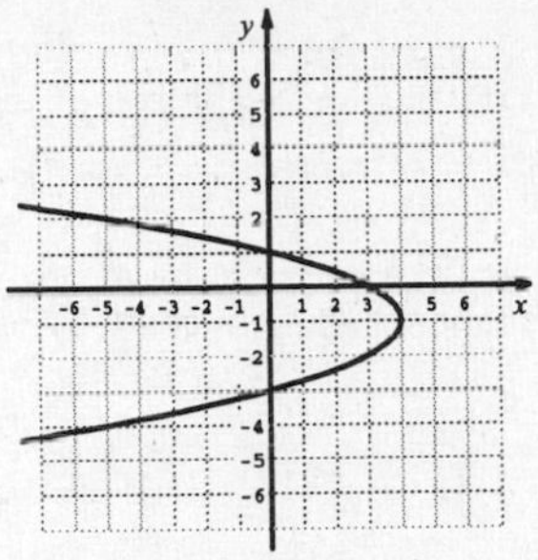

19. $x = -2y^2 - 4y + 1$

a) We find the vertex by completing the square.

$$x = -2(y^2 + 2y) + 1$$
$$x = -2(y^2 + 2y + 1) + 1 + 2$$
$$x = -2(y + 1)^2 + 3$$

The vertex is (3,-1).

b) We choose y-values and compute the corresponding values for x. The graph opens to the left, because the coefficient of y^2, -2, is negative.

x	y
3	-1
1	-2
-5	-3
1	0
-5	1

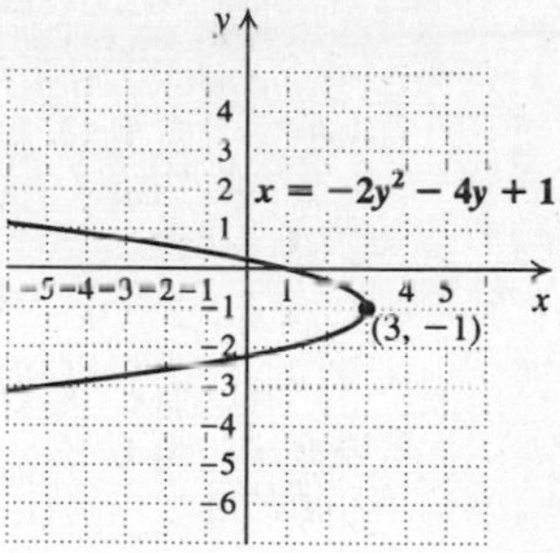

20. $x = 2y^2 + 4y - 1$

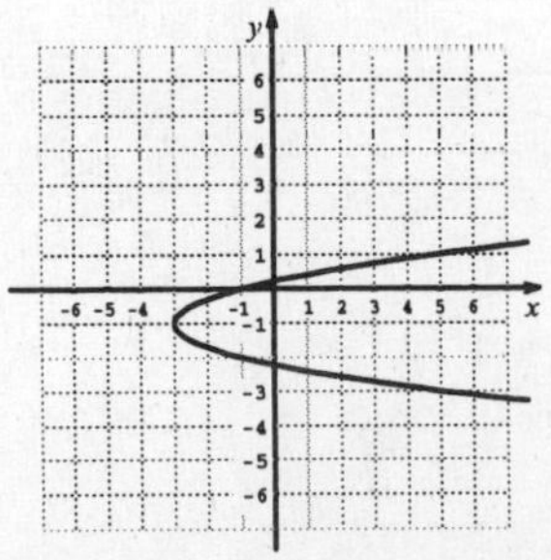

21. $d = \sqrt{(x_2 - x_1)^2 + (y_2 - y_1)^2}$ Distance formula

$= \sqrt{(6 - 9)^2 + (1 - 5)^2}$ Substituting

$= \sqrt{(-3)^2 + (-4)^2}$

$= \sqrt{25} = 5$

22. 10

23. $d = \sqrt{(x_2 - x_1)^2 + (y_2 - y_1)^2}$ Distance formula

$= \sqrt{(3 - 0)^2 + [-4 - (-7)]^2}$ Substituting

$= \sqrt{3^2 + 3^2}$

$= \sqrt{18} \approx 4.243$ Simplifying and approximating

24. 10

(Since these points are on a vertical line, we could have found the distance between them by computing $|-8 - 2|$.)

25. $d = \sqrt{(x_2 - x_1)^2 + (y_2 - y_1)^2}$

$= \sqrt{(-2 - 2)^2 + (-2 - 2)^2}$

$= \sqrt{32} \approx 5.657$

26. $\sqrt{464} \approx 21.541$

27. $d = \sqrt{(x_2 - x_1)^2 + (y_2 - y_1)^2}$

$= \sqrt{(-9.2 - 8.6)^2 + [-3.4 - (-3.4)]^2}$

$= \sqrt{(-17.8)^2 + 0^2}$

$= \sqrt{316.84} = 17.8$

(Since these points are on a horizontal line, we could have found the distance between them by subtracting their first coordinates and taking the absolute value: $d = |-9.2 - 8.6| = |-17.8| = 17.8$.)

28. $\sqrt{98.93} \approx 9.946$

29. $d = \sqrt{(x_2 - x_1)^2 + (y_2 - y_1)^2}$

$= \sqrt{\left(\frac{1}{7} - \frac{5}{7}\right)^2 + \left(\frac{11}{14} - \frac{1}{14}\right)^2}$

$= \sqrt{\frac{41}{49}}$

$= \frac{\sqrt{41}}{7} \approx 0.915$

30. $\sqrt{13} \approx 3.606$

31. $d = \sqrt{(x_2 - x_1)^2 + (y_2 - y_1)^2}$

$= \sqrt{[56 - (-23)]^2 + (-17 - 10)^2}$

$= \sqrt{6970} \approx 83.487$

32. $\sqrt{6800} \approx 82.462$

33. $d = \sqrt{(x_2 - x_1)^2 + (y_2 - y_1)^2}$

$= \sqrt{(0 - a)^2 + (0 - b)^2}$

$= \sqrt{a^2 + b^2}$

34. $\sqrt{p^2 + q^2}$

35. $d = \sqrt{(x_2 - x_1)^2 + (y_2 - y_1)^2}$

$= \sqrt{(-\sqrt{7} - \sqrt{2})^2 + [\sqrt{5} - (-\sqrt{3})]^2}$

$= \sqrt{7 + 2\sqrt{14} + 2 + 5 + 2\sqrt{15} + 3}$

$= \sqrt{17 + 2\sqrt{14} + 2\sqrt{15}} \approx 5.677$

36. $\sqrt{22 + 2\sqrt{40} + 2\sqrt{18}} \approx 6.568$

37. $d = \sqrt{(x_2 - x_1)^2 + (y_2 - y_1)^2}$

$= \sqrt{(-2000 - 1000)^2 + [580 - (-240)]^2}$

$= \sqrt{9,672,400} \approx 3110.048$

38. $\sqrt{8,044,900} \approx 2836.353$

39. We use the midpoint formula:

$\left(\frac{x_1 + x_2}{2}, \frac{y_1 + y_2}{2}\right) = \left(\frac{-3 + 2}{2}, \frac{6 + (-8)}{2}\right)$, or

$\left(\frac{-1}{2}, \frac{-2}{2}\right)$, or $\left(-\frac{1}{2}, -1\right)$

40. $\left(\frac{13}{2}, -1\right)$

41. We use the midpoint formula:

$\left(\frac{x_1 + x_2}{2}, \frac{y_1 + y_2}{2}\right) = \left(\frac{8 + (-1)}{2}, \frac{5 + 2}{2}\right)$, or $\left(\frac{7}{2}, \frac{7}{2}\right)$

42. $\left(0, -\frac{1}{2}\right)$

43. We use the midpoint formula:

$\left(\frac{x_1 + x_2}{2}, \frac{y_1 + y_2}{2}\right) = \left(\frac{-8 + 6}{2}, \frac{-5 + (-1)}{2}\right)$, or

$\left(\frac{-2}{2}, \frac{-6}{2}\right)$, or $(-1, -3)$

44. $\left(\frac{5}{2}, 1\right)$

45. $\left(\frac{x_1 + x_2}{2}, \frac{y_1 + y_2}{2}\right) = \left(\frac{-3.4 + 2.9}{2}, \frac{8.1 - 8.7}{2}\right)$, or

$\left(\frac{-0.5}{2}, \frac{-0.6}{2}\right)$, or $(-0.25, -0.3)$

46. $(4.65, 0)$

47. $\left(\frac{x_1 + x_2}{2}, \frac{y_1 + y_2}{2}\right) = \left(\frac{\frac{1}{6} + \left(-\frac{1}{3}\right)}{2}, \frac{-\frac{3}{4} + \frac{5}{6}}{2}\right)$, or

$\left(\frac{-\frac{1}{6}}{2}, \frac{\frac{1}{12}}{2}\right)$, or $\left(-\frac{1}{12}, \frac{1}{24}\right)$

48. $\left(-\frac{27}{80}, \frac{1}{24}\right)$

49. $\left(\frac{x_1 + x_2}{2}, \frac{y_1 + y_2}{2}\right) = \left(\frac{\sqrt{2} + \sqrt{3}}{2}, \frac{-1 + 4}{2}\right)$, or

$\left(\frac{\sqrt{2} + \sqrt{3}}{2}, \frac{3}{2}\right)$

50. $\left(\frac{5}{2}, \frac{7\sqrt{3}}{2}\right)$

51. $(x - h)^2 + (y - k)^2 = r^2$ Standard form

$(x - 0)^2 + (y - 0)^2 = 7^2$ Substituting

$x^2 + y^2 = 49$ Simplifying

52. $x^2 + y^2 = 16$

53. $(x - h)^2 + (y - k)^2 = r^2$ Standard form

$[x - (-2)]^2 + (y - 7)^2 = (\sqrt{5})^2$ Substituting

$(x + 2)^2 + (y - 7)^2 = 5$

54. $(x - 5)^2 + (y - 6)^2 = 12$

55. $(x - h)^2 + (y - k)^2 = r^2$ Standard form

$[x - (-4)]^2 + (y - 3)^2 = (4\sqrt{3})^2$ Substituting

$(x + 4)^2 + (y - 3)^2 = 48$

$[(4\sqrt{3})^2 = 16 \cdot 3 = 48]$

56. $(x + 2)^2 + (y - 7)^2 = 20$

57. $(x - h)^2 + (y - k)^2 = r^2$

$[x - (-7)]^2 + [y - (-2)]^2 = (5\sqrt{2})^2$

$(x + 7)^2 + (y + 2)^2 = 50$

58. $(x + 5)^2 + (y + 8)^2 = 18$

59. Since the center is (0, 0), we have

$(x - 0)^2 + (y - 0)^2 = r^2$ or $x^2 + y^2 = r^2$

The circle passes through (-3, 4). We find r^2 by substituting -3 for x and 4 for y.

$(-3)^2 + 4^2 = r^2$

$9 + 16 = r^2$

$25 = r^2$

Then $x^2 + y^2 = 25$ is an equation of the circle.

60. $(x - 3)^2 + (y + 2)^2 = 64$

61. Since the center is (-4, 1), we have

$[x - (-4)]^2 + (y - 1)^2 = r^2$, or

$(x + 4)^2 + (y - 1)^2 = r^2$.

The circle passes through (-2, 5). We find r^2 by substituting -2 for x and 5 for y.

$(-2 + 4)^2 + (5 - 1)^2 = r^2$

$4 + 16 = r^2$

$20 = r^2$

Then $(x + 4)^2 + (y - 1)^2 = 20$ is an equation of the circle.

62. $(x + 3)^2 + (y + 3)^2 = 54.4$

63. We write standard form.

$(x - 0)^2 + (y - 0)^2 = 6^2$

The center is (0, 0), and the radius is 6.

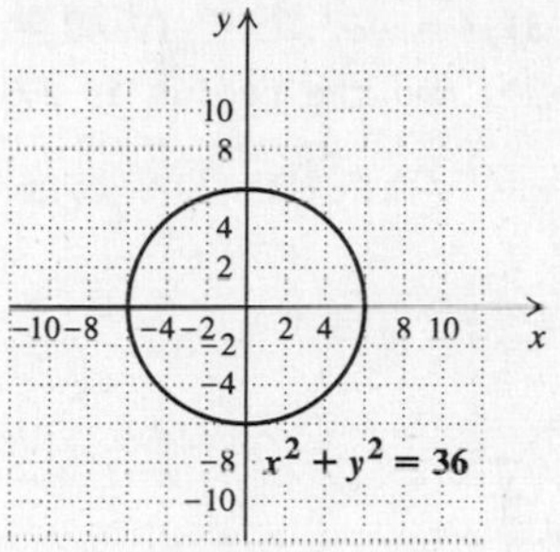

64. Center: (0,0); radius: 5

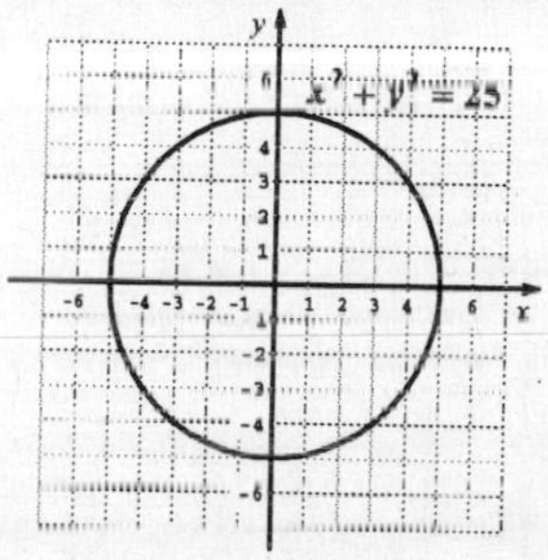

65. $(x + 1)^2 + (y + 3)^2 = 4$

$[x - (-1)]^2 + [y - (-3)]^2 = 2^2$ Standard form

The center is (-1,-3), and the radius is 2.

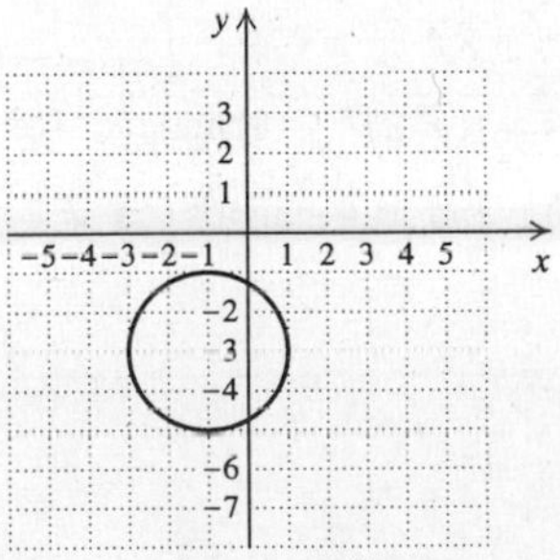

$(x + 1)^2 + (y + 3)^2 = 4$

66. Center: (2,-3)
Radius: 1

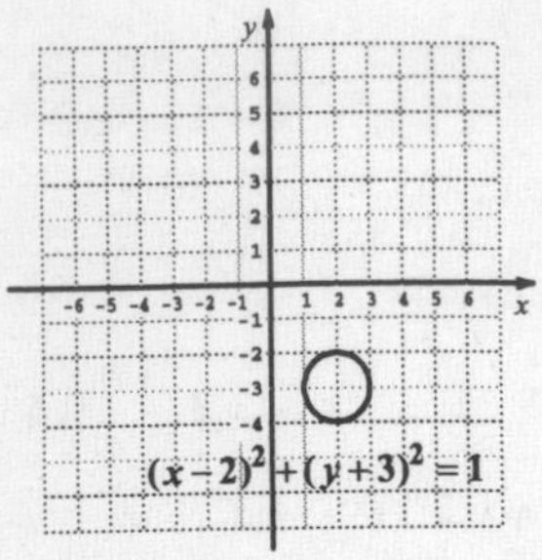

67. $(x - 8)^2 + (y + 3)^2 = 40$

$(x - 8)^2 + [y - (-3)]^2 = (2\sqrt{10})^2 \quad (\sqrt{40} = 2\sqrt{10})$

The center is (8,-3), and the radius is $2\sqrt{10}$.

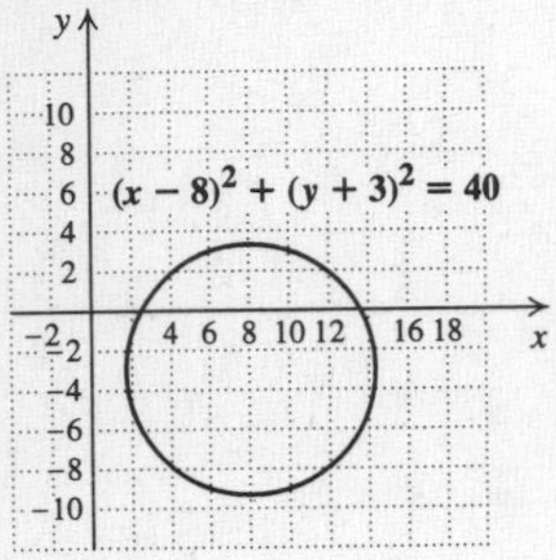

68. Center: (-5,1)
Radius: $5\sqrt{3}$

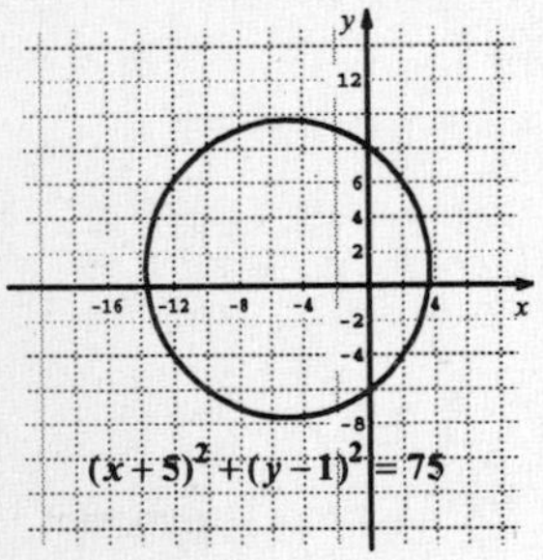

69. $x^2 + y^2 = 2$

$(x - 0)^2 + (y - 0)^2 = (\sqrt{2})^2$ Standard form

The center is (0,0), and the radius is $\sqrt{2}$.

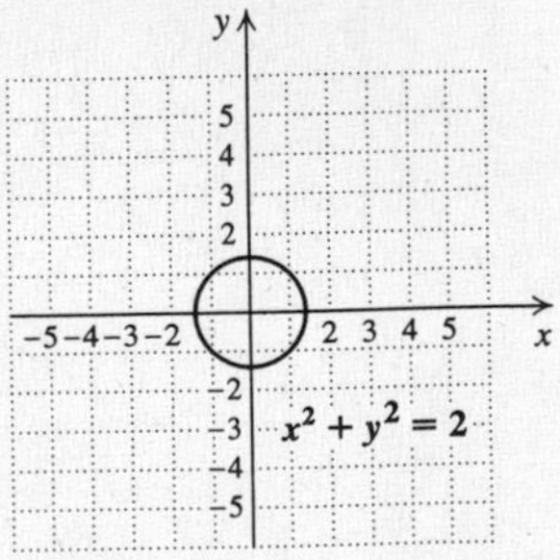

70. Center: (0,0)
Radius: $\sqrt{3}$

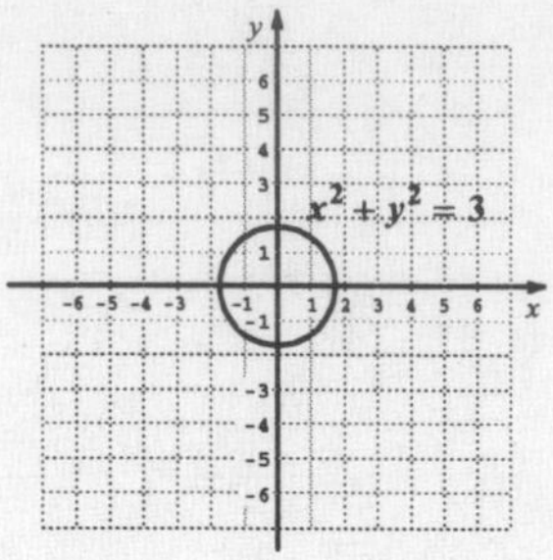

71. $(x - 5)^2 + y^2 = \frac{1}{4}$

$(x - 5)^2 + (y - 0)^2 = \left(\frac{1}{2}\right)^2$ Standard form

The center is (5,0), and the radius is $\frac{1}{2}$.

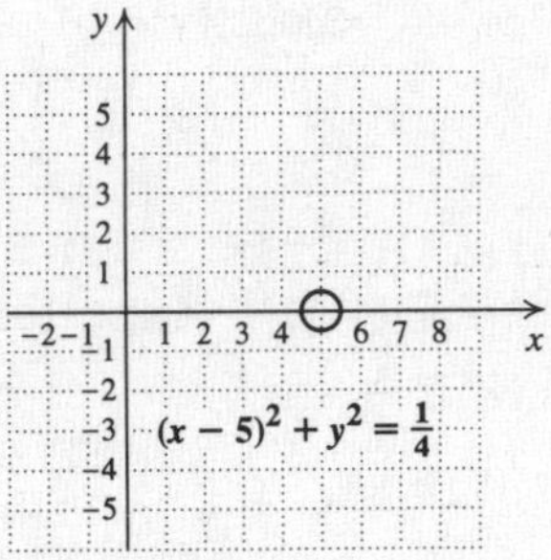

72. Center: (0,1)
Radius: $\frac{1}{5}$

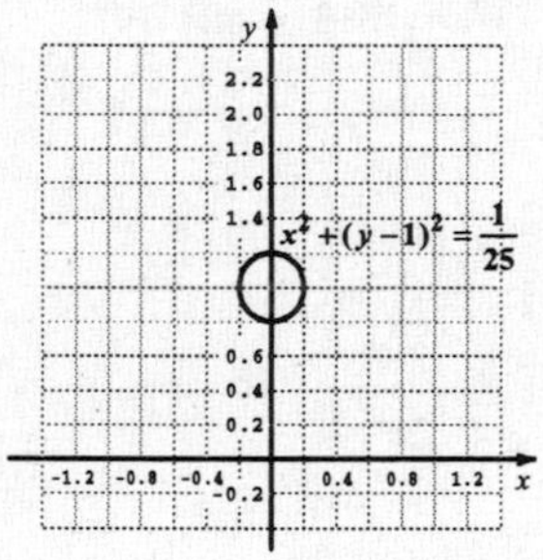

73. $x^2 + y^2 + 8x - 6y - 15 = 0$

$x^2 + 8x + y^2 - 6y = 15$

$(x^2 + 8x + 16) + (y^2 - 6y + 9) = 15 + 16 + 9$

Completing the square twice

$(x + 4)^2 + (y - 3)^2 = 40$

$[x - (-4)]^2 + (y - 3)^2 = (2\sqrt{10})^2$

Standard form

The center is (-4,3), and the radius is $2\sqrt{10}$.

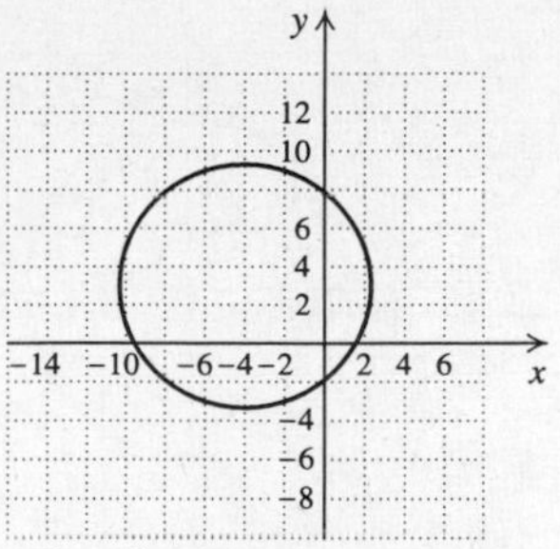

$x^2 + y^2 + 8x - 6y - 15 = 0$

74. Center: (-3,2)

Radius: $2\sqrt{7}$

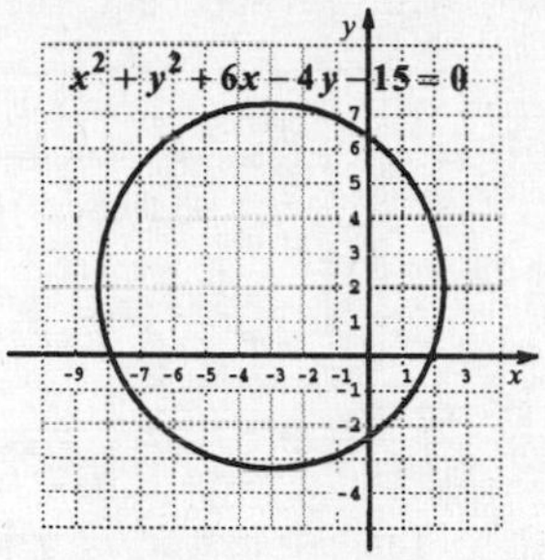

75.

$$x^2 + y^2 - 8x + 2y + 13 = 0$$
$$x^2 - 8x + y^2 + 2y = -13$$
$$(x^2 - 8x + 16) + (y^2 + 2y + 1) = -13 + 16 + 1$$

Completing the square twice

$$(x - 4)^2 + (y + 1)^2 = 4$$
$$(x - 4)^2 + [y - (-1)]^2 = 2^2 \quad \text{Standard form}$$

The center is (4,-1), and the radius is 2.

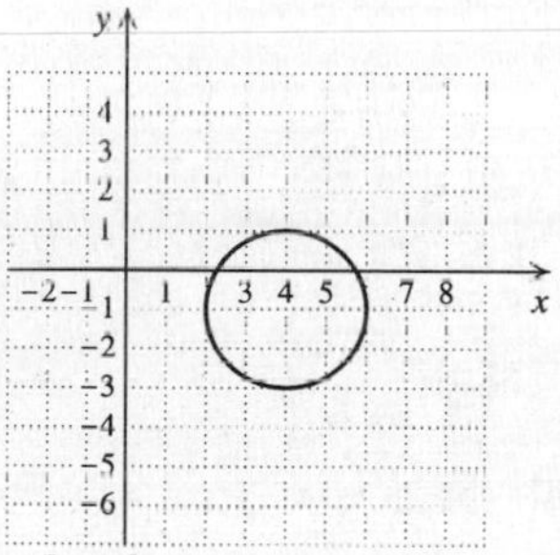

$x^2 + y^2 - 8x + 2y + 13 = 0$

76. Center: (-3,-2)

Radius: 1

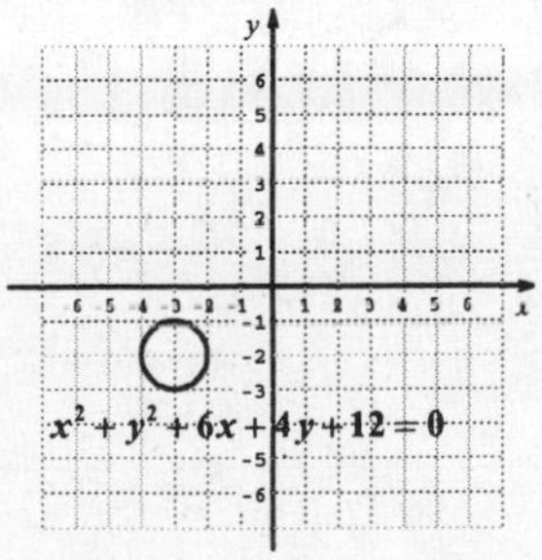

77.

$$x^2 + y^2 - 4x = 0$$
$$x^2 - 4x + y^2 = 0$$
$$(x^2 - 4x + 4) + y^2 = 4 \quad \text{Completing the square}$$
$$(x - 2)^2 + (y - 0)^2 = 2^2 \quad \text{Standard form}$$

The center is (2,0), and the radius is 2.

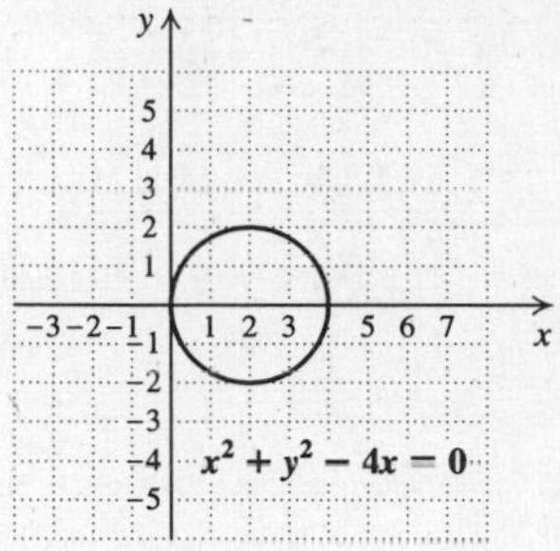

78. Center: (-3,0)

Radius: 3

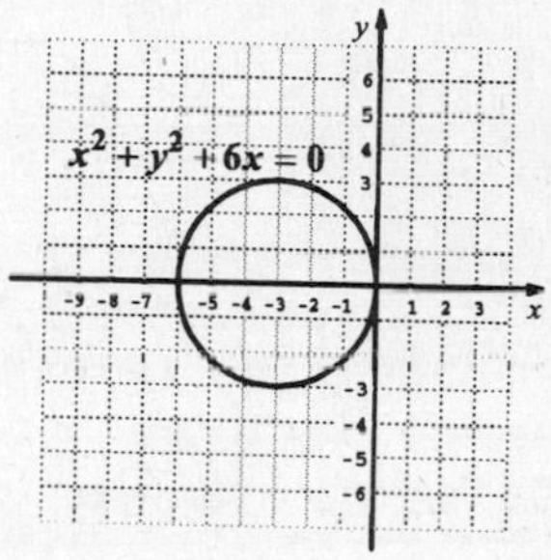

79.

$$x^2 + y^2 + 10y - 75 = 0$$
$$x^2 + y^2 + 10y = 75$$
$$x^2 + (y^2 + 10y + 25) = 75 + 25$$
$$(x - 0)^2 + (y + 5)^2 = 100$$
$$(x - 0)^2 + [y - (-5)]^2 = 10^2$$

The center is (0,-5), and the radius is 10.

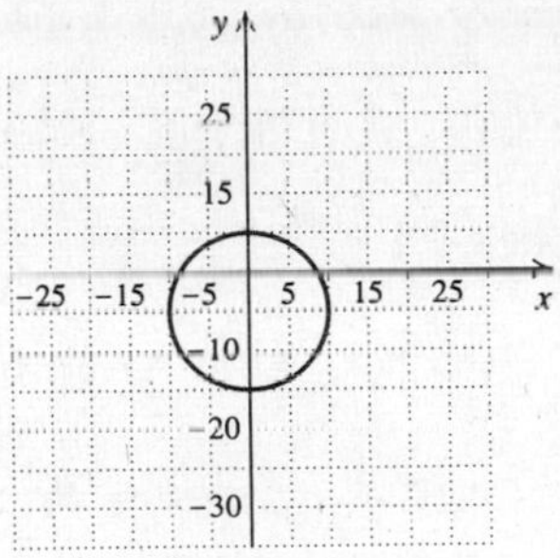

$x^2 + y^2 + 10y - 75 = 0$

80. Center: (4,0)

Radius: 10

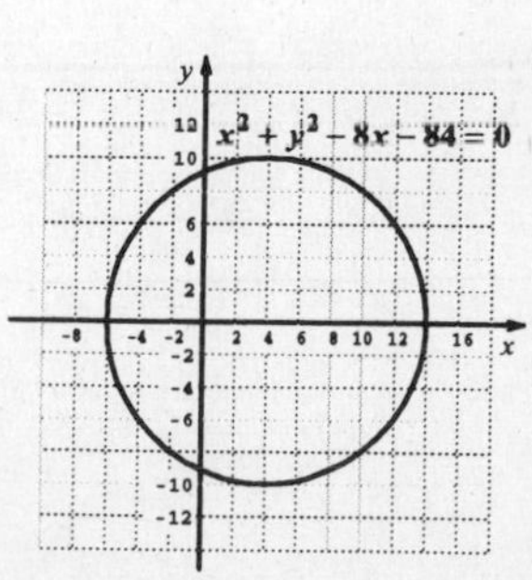

81. $x^2 + y^2 + 7x - 3y - 10 = 0$

$$x^2 + 7x + y^2 - 3y = 10$$

$$\left(x^2 + 7x + \frac{49}{4}\right) + \left(y^2 - 3y + \frac{9}{4}\right) = 10 + \frac{49}{4} + \frac{9}{4}$$

$$\left(x + \frac{7}{2}\right)^2 + \left(y - \frac{3}{2}\right)^2 = \frac{98}{4}$$

$$\left[x - \left(-\frac{7}{2}\right)\right]^2 + \left(y - \frac{3}{2}\right)^2 = \left(\sqrt{\frac{98}{4}}\right)^2$$

The center is $\left(-\frac{7}{2}, \frac{3}{2}\right)$, and the radius is $\sqrt{\frac{98}{4}}$, or $\frac{\sqrt{98}}{2}$, or $\frac{7\sqrt{2}}{2}$.

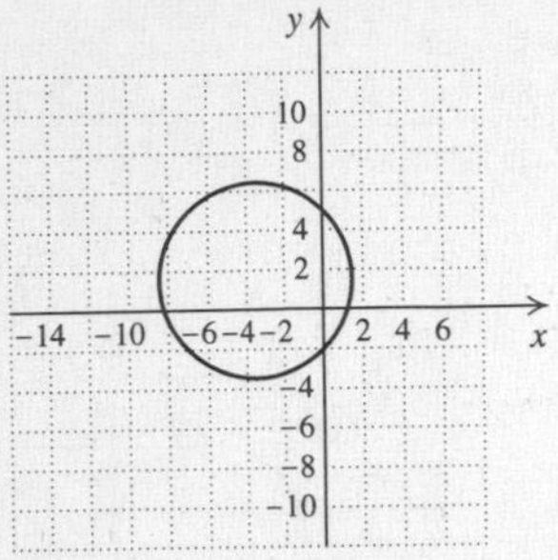

$x^2 + y^2 + 7x - 3y - 10 = 0$

82. Center: $\left(\frac{21}{2}, \frac{33}{2}\right)$

Radius: $\frac{\sqrt{1462}}{2}$

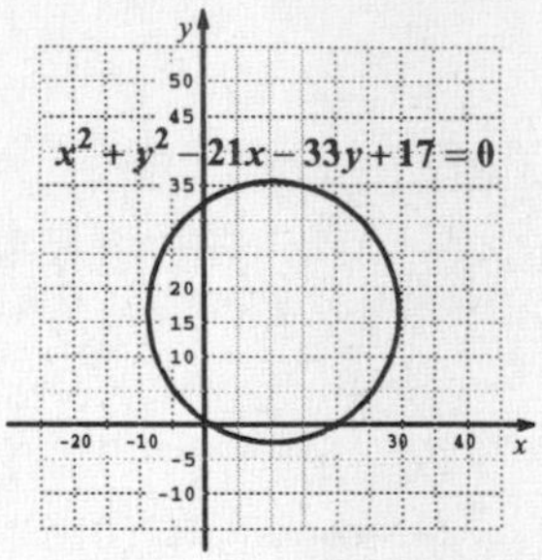

83. $4x^2 + 4y^2 = 1$

$x^2 + y^2 = \frac{1}{4}$ Multiplying by $\frac{1}{4}$ on both sides

$$(x - 0)^2 + (y - 0)^2 = \left(\frac{1}{2}\right)^2$$

The center is (0,0), and the radius is $\frac{1}{2}$.

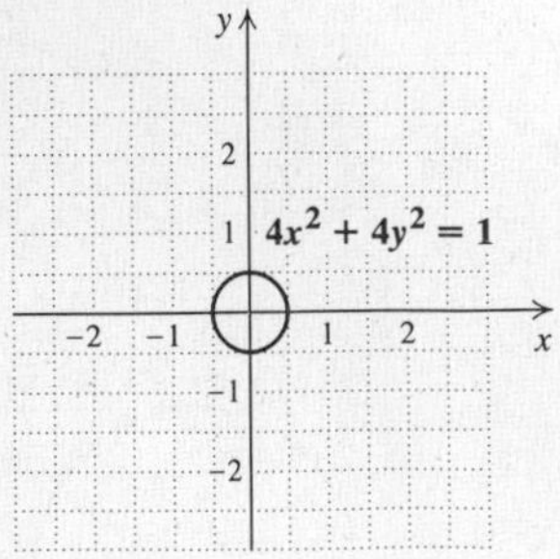

84. Center: (0,0)

Radius: $\frac{1}{5}$

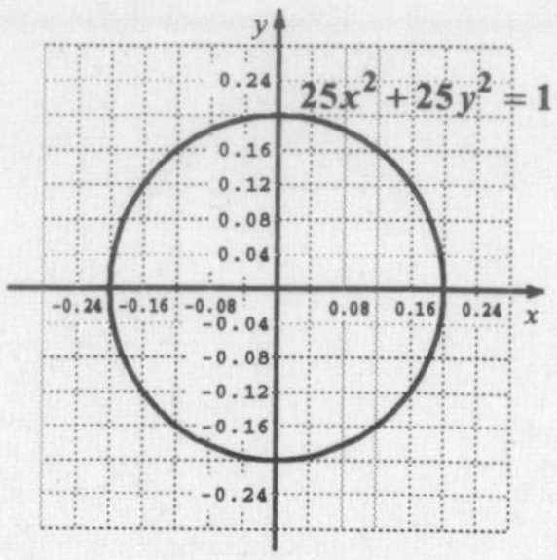

85. Familiarize. We make a drawing and label it. Let x represent the width of the border.

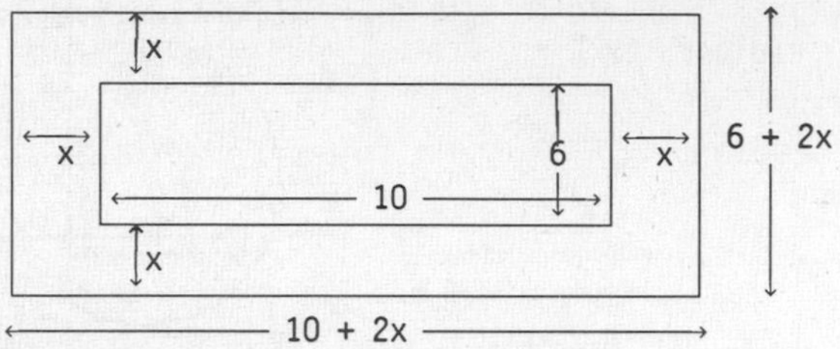

The perimeter of the larger rectangle is

$2(10 + 2x) + 2(6 + 2x)$, or $8x + 32$.

The perimeter of the smaller rectangle is

$2(10) + 2(6)$, or 32.

Translate. The perimeter of the larger rectangle is twice the perimeter of the smaller rectangle.

$$8x + 32 = 2 \cdot 32$$

Carry out. We solve the equation.

$$8x + 32 = 64$$
$$8x = 32$$
$$x = 4$$

Check. If the width of the border is 4 in., then the length and width of the larger rectangle are 18 in. and 14 in. Thus its perimeter is 2(18) + 2(14), or 64 in. The perimeter of the smaller rectangle is 32 in. The perimeter of the larger rectangle is twice the perimeter of the smaller rectangle.

State. The width of the border is 4 in.

86. 2640 mi

87.

88.

89. $x = y^2 - y - 6$

$$x = \left(y^2 - y + \frac{1}{4}\right) - 6 - \frac{1}{4}$$

$$x = \left(y - \frac{1}{2}\right)^2 - \frac{25}{4}$$

The vertex is $\left(-\frac{25}{4}, \frac{1}{2}\right)$.

x	y
$-\frac{25}{4}$	$\frac{1}{2}$
-6	1
-4	2
0	3
-6	0
-4	-1
0	-2

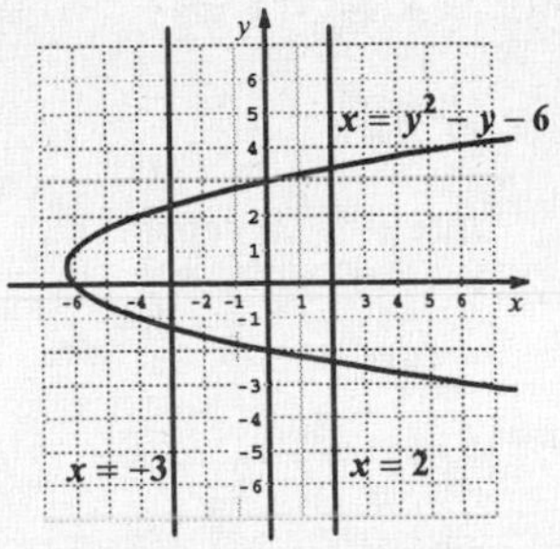

a) Graph x = 2 on the same set of axes as $x = y^2 - y - 6$ and approximate the y-coordinates of the points of intersection. (See the graph above.) The solutions are approximately 3.4 and -2.4.

b) Graph x = -3 on the same set of axes as $x = y^2 - y - 6$ and approximate the y-coordinates of the points of intersection. (See the graph above.) The solutions are approximately 2.3 and -1.3.

90.

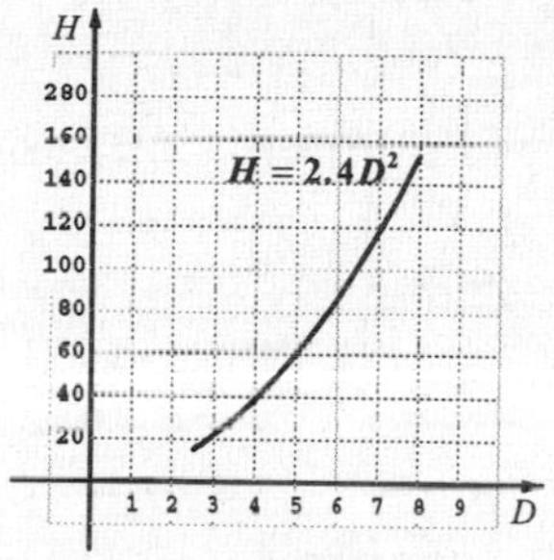

91.

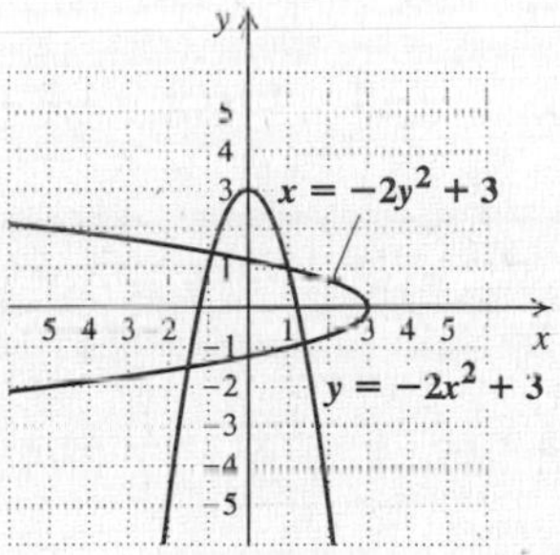

Reflect one graph across the line y = x to obtain the other.

92.

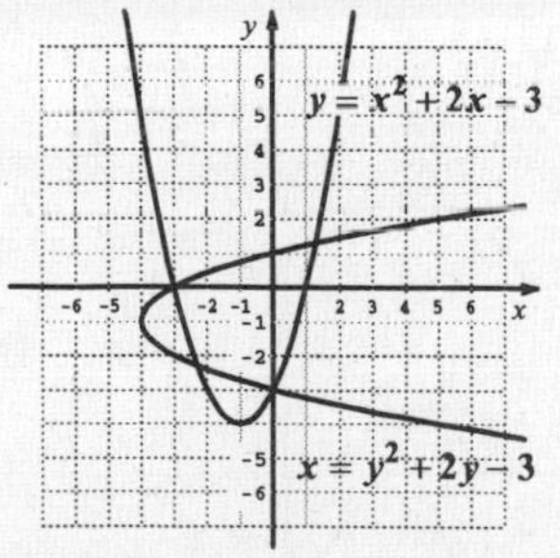

Reflect one graph across the line y = x to obtain the other.

93. Let (0,y) be the point on the y-axis that is equidistant from (2,10) and (6,2). Then the distance between (2,10) and (0,y) is the same as the distance between (6,2) and (0,y).

$$\sqrt{(0-2)^2 + (y-10)^2} = \sqrt{(0-6)^2 + (y-2)^2}$$

$$(-2)^2 + (y-10)^2 = (-6)^2 + (y-2)^2 \quad \text{Squaring both sides}$$

$$4 + y^2 - 20y + 100 = 36 + y^2 - 4y + 4$$

$$64 = 16y$$

$$4 = y$$

This number checks. The point is (0,4).

94. (-5,0)

95. a) When the circle is positioned on a coordinate system as shown in the text, the center lies on the y-axis. To find the center, we will find the point on the y-axis that is equidistant from (-4,0) and (0,2). Let (0,y) be this point.

$$\sqrt{[0-(-4)]^2 + (y-0)^2} = \sqrt{(0-0)^2 + (y-2)^2}$$

$$4^2 + y^2 = 0^2 + (y-2)^2 \quad \text{Squaring both sides}$$

$$16 + y^2 = y^2 - 4y + 4$$

$$12 = -4y$$

$$-3 = y$$

The center of the circle is (0,-3).

b) We find the radius of the circle.

$$(x-0)^2 + [y-(-3)]^2 = r^2 \quad \text{Standard form}$$

$$x^2 + (y+3)^2 = r^2$$

$$(-4)^2 + (0+3)^2 = r^2 \quad \text{Substituting (-4,0) for (x,y)}$$

$$16 + 9 = r^2$$

$$25 = r^2$$

$$5 = r$$

The radius is 5 ft.

96. $(x-3)^2 + (y+5)^2 = 9$

97. We make a drawing of a circle with center (-7,-4) and tangent to the x-axis.

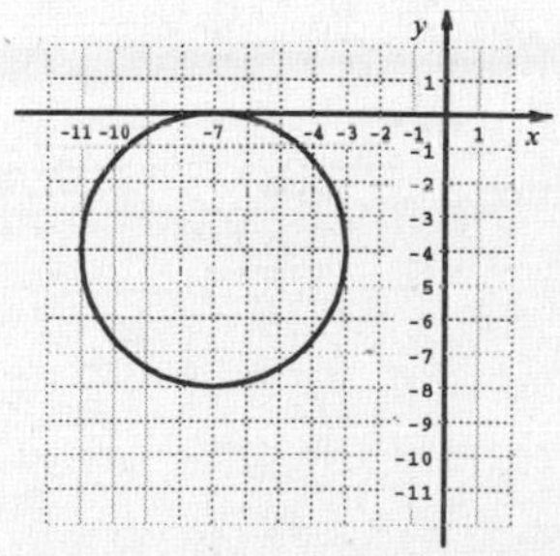

We see that the circle touches the x-axis at (-7,0). Hence, the radius is the distance between (-7,0) and (-7,-4), or |-4 - 0|, or 4. Now we write the equation of the circle.

$$(x - h)^2 + (y - k)^2 = r^2$$
$$[x - (-7)]^2 + [y - (-4)]^2 = 4^2$$
$$(x + 7)^2 + (y + 4)^2 = 16$$

98. $(x - 3)^2 + y^2 = 25$

99. We use the formula $C = 2\pi r$ to find the radius.

$$2\pi r = 8\pi$$
$$r = 4$$

Write the equation of a circle with center (-3,5) and radius 4.

$$[x - (-3)]^2 + (y - 5)^2 = 4^2$$
$$(x + 3)^2 + (y - 5)^2 = 16$$

100. $x^2 + (y - 30.6)^2 = 590.49$

101. See the answer section in the text.

Exercise Set 10.2

1. $\frac{x^2}{4} + \frac{y^2}{1} = 1$

$$\frac{x^2}{2^2} + \frac{y^2}{1^2} = 1$$

The x-intercepts are (2,0) and (-2,0), and the y-intercepts are (0,1) and (0,-1). We plot these points and connect them with an oval-shaped curve.

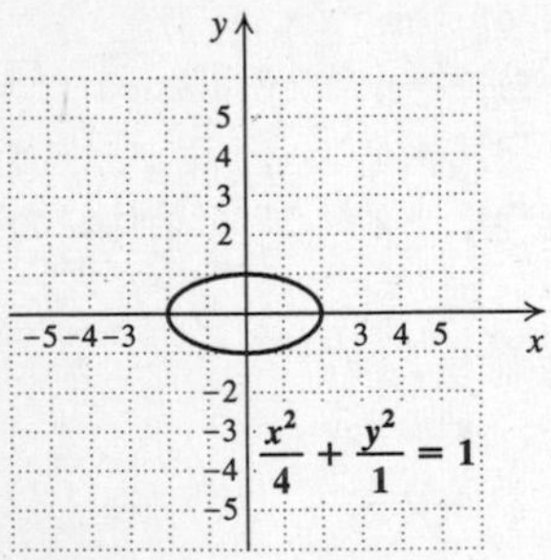

2. $\frac{x^2}{1} + \frac{y^2}{4} = 1$

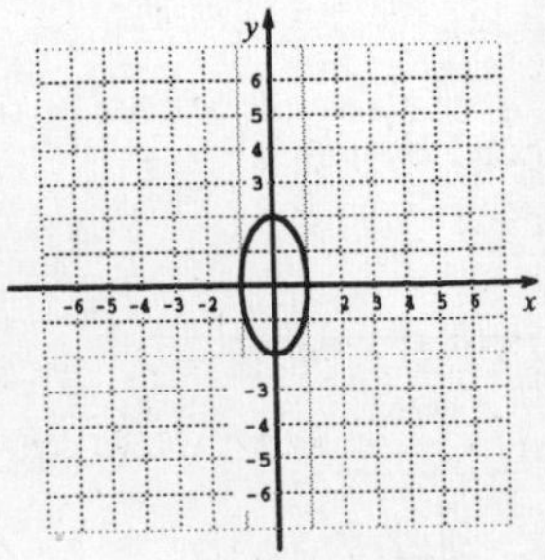

3. $\frac{x^2}{16} + \frac{y^2}{25} = 1$

$$\frac{x^2}{4^2} + \frac{y^2}{5^2} = 1$$

The x-intercepts are (4,0) and (-4,0), and the y-intercepts are (0,5) and (0,-5). We plot these points and connect them with an oval-shaped curve.

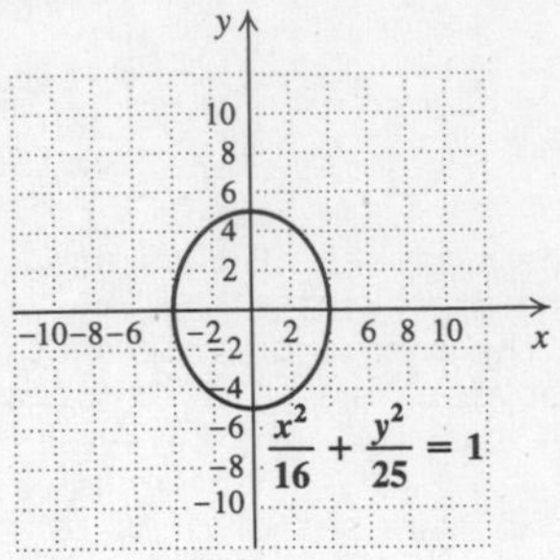

4. $\frac{x^2}{9} + \frac{y^2}{25} = 1$

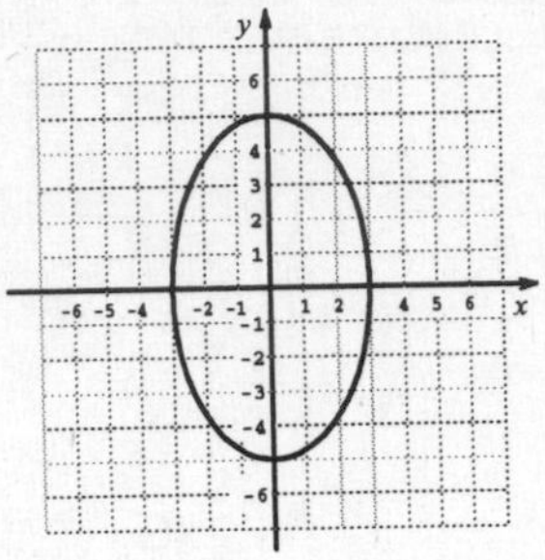

5. $4x^2 + 9y^2 = 36$

$$\frac{1}{36}(4x^2 + 9y^2) = \frac{1}{36}(36) \quad \text{Multiplying by } \frac{1}{36}$$
$$\frac{x^2}{9} + \frac{y^2}{4} = 1$$
$$\frac{x^2}{3^2} + \frac{y^2}{2^2} = 1$$

The x-intercepts are (3,0) and (-3,0), and the y-intercepts are (0,2) and (0,-2). We plot these points and connect them with an oval-shaped curve.

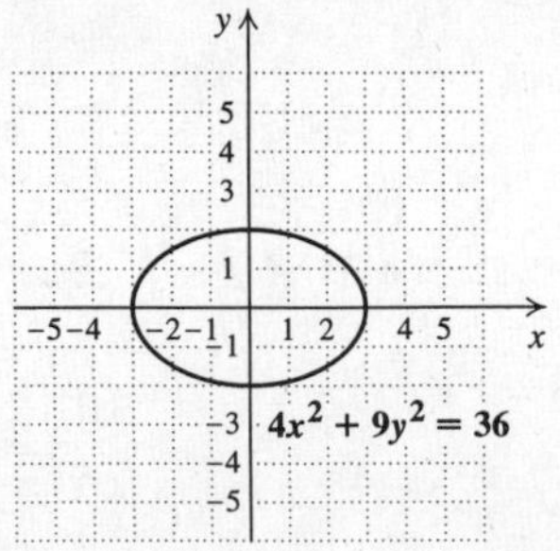

6. $9x^2 + 4y^2 = 36$

$$\frac{x^2}{4} + \frac{y^2}{9} = 1$$

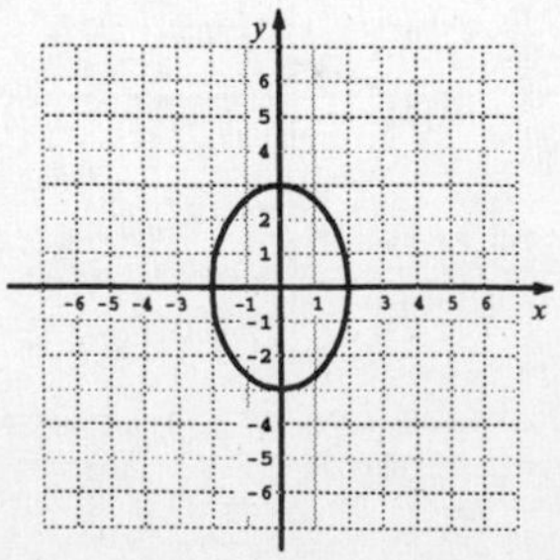

7. $16x^2 + 9y^2 = 144$

$$\frac{x^2}{9} + \frac{y^2}{16} = 1 \quad \text{Multiplying by } \frac{1}{144}$$

$$\frac{x^2}{3^2} + \frac{y^2}{4^2} = 1$$

The x-intercepts are (3,0) and (-3,0), and the y-intercepts are (0,4) and (0,-4). We plot these points and connect them with an oval-shaped curve.

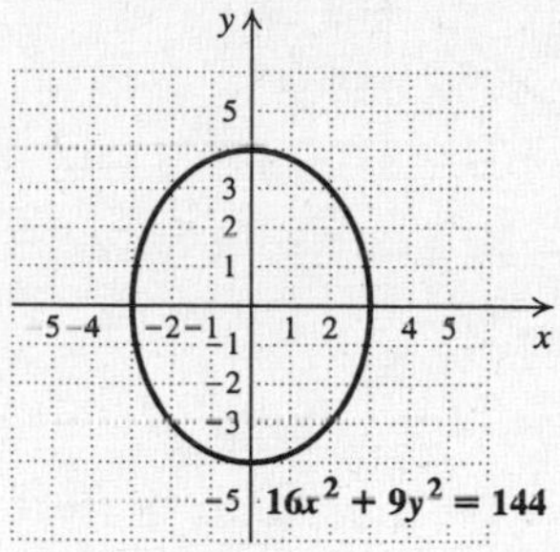

8. $9x^2 + 16y^2 = 144$

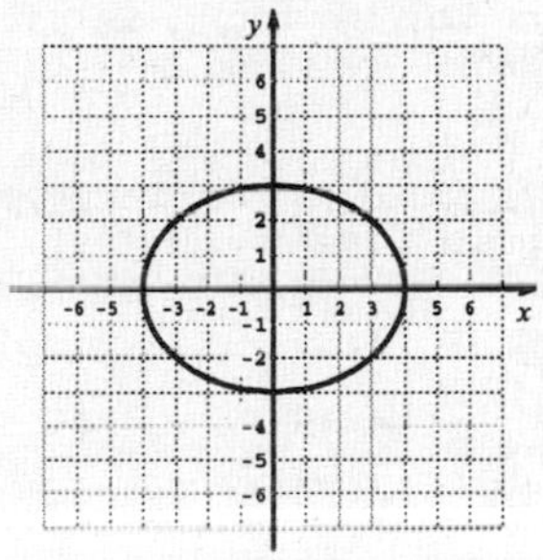

9. $2x^2 + 3y^2 = 6$

$$\frac{x^2}{3} + \frac{y^2}{2} = 1 \quad \text{Multiplying by } \frac{1}{6}$$

$$\frac{x^2}{(\sqrt{3})^2} + \frac{y^2}{(\sqrt{2})^2} = 1$$

The x-intercepts are$(\sqrt{3},0)$ and $(-\sqrt{3},0)$, and the y-intercepts are $(0,\sqrt{2})$ and $(0,-\sqrt{2})$. We plot these points and connect them with an oval-shaped curve.

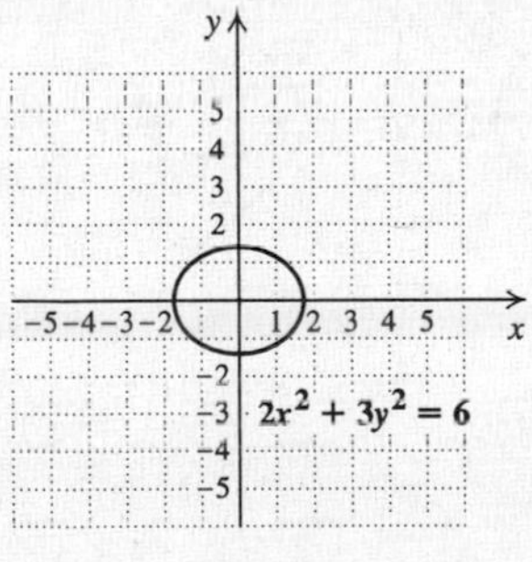

10. $5x^2 + 7y^2 = 35$

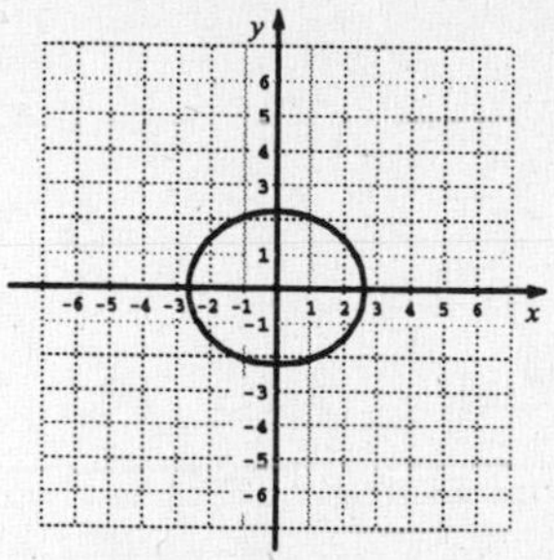

11. $4x^2 + 9y^2 = 1$

$$\frac{x^2}{\frac{1}{4}} + \frac{y^2}{\frac{1}{9}} = 1 \qquad 4x^2 = \frac{x^2}{\frac{1}{4}},\quad 9y^2 = \frac{y^2}{\frac{1}{9}}$$

$$\frac{x^2}{\left(\frac{1}{2}\right)^2} + \frac{y^2}{\left(\frac{1}{3}\right)^2} = 1$$

The x-intercepts are $\left(\frac{1}{2},0\right)$ and $\left(-\frac{1}{2},0\right)$, and the y-intercepts are $\left(0,\frac{1}{3}\right)$ and $\left(0,-\frac{1}{3}\right)$. We plot these points and connect them with an oval-shaped curve.

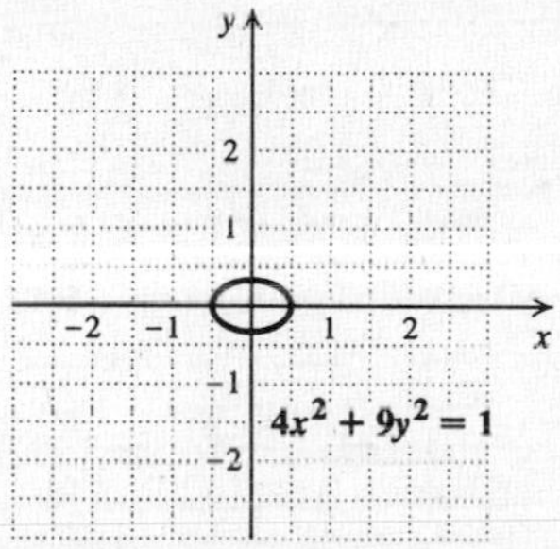

12. $25x^2 + 16y^2 = 1$

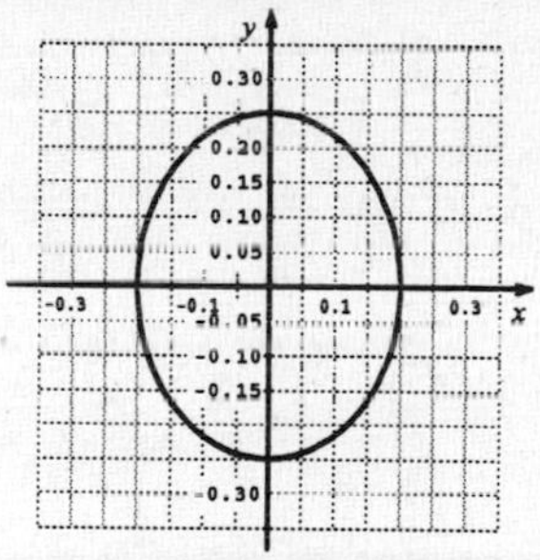

13. $5x^2 + 12y^2 = 60$

$$\frac{x^2}{12} + \frac{y^2}{5} = 1 \quad \text{Multiplying by } \frac{1}{60}$$

$$\frac{x^2}{(\sqrt{12})^2} + \frac{y^2}{(\sqrt{5})^2} = 1$$

The x-intercepts are $(\sqrt{12},0)$ and $(-\sqrt{12},0)$, or $(2\sqrt{3},0)$ and $(-2\sqrt{3},0)$, and the y-intercepts are $(0,\sqrt{5})$ and $(0,-\sqrt{5})$. We plot these points and connect them with an oval-shaped curve.

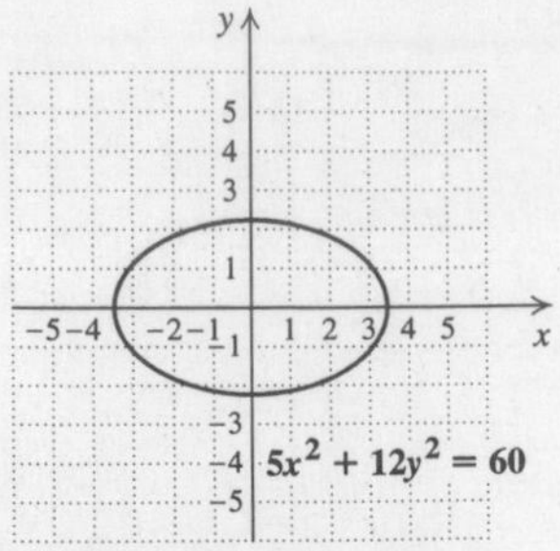

14. $8x^2 + 3y^2 = 24$

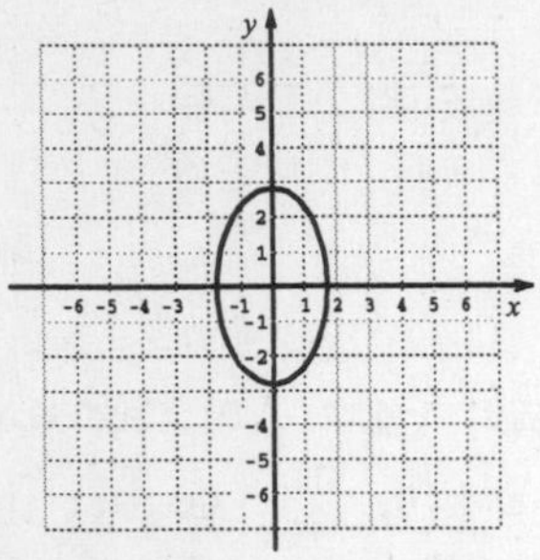

15. $\frac{x^2}{16} - \frac{y^2}{16} = 1$

$\frac{x^2}{4^2} - \frac{y^2}{4^2} = 1$

a) $a = 4$ and $b = 4$, so the asymptotes are $y = \frac{4}{4}x$ and $y = -\frac{4}{4}x$, or $y = x$ and $y = -x$. We sketch them.

b) Replacing y with 0 and solving for x, we get $x = \pm 4$, so the intercepts are (4,0) and (-4,0).

c) We plot the intercepts and draw smooth curves through them that approach the asymptotes.

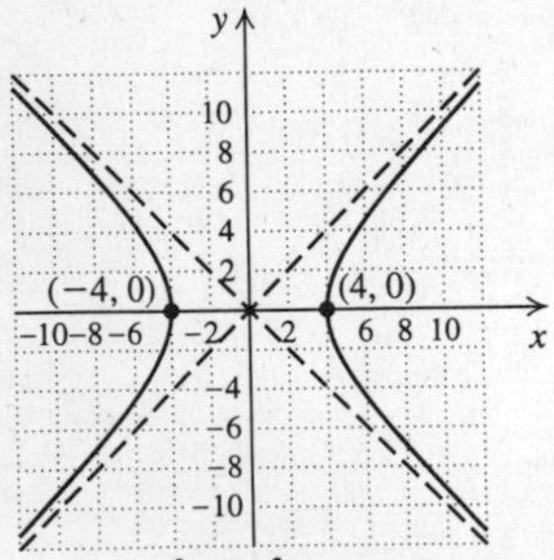

$\frac{x^2}{16} - \frac{y^2}{16} = 1$

16. $\frac{y^2}{9} - \frac{x^2}{9} = 1$

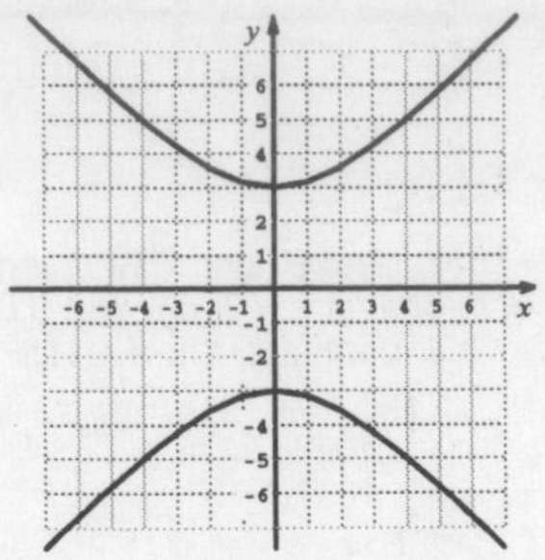

17. $\frac{y^2}{16} - \frac{x^2}{9} = 1$

$\frac{y^2}{4^2} - \frac{x^2}{3^2} = 1$

a) $a = 3$ and $b = 4$, so the asymptotes are $y = \frac{4}{3}x$ and $y = -\frac{4}{3}x$. We sketch them.

b) Replacing x with 0 and solving for y, we get $y = \pm 4$, so the intercepts are (0,4) and (0,-4).

c) We plot the intercepts and draw smooth curves through them that approach the asymptotes.

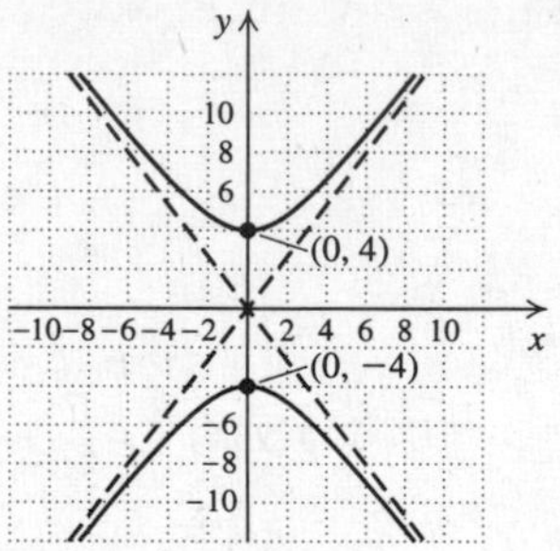

$\frac{y^2}{16} - \frac{x^2}{9} = 1$

18. $\frac{x^2}{9} - \frac{y^2}{4} = 1$

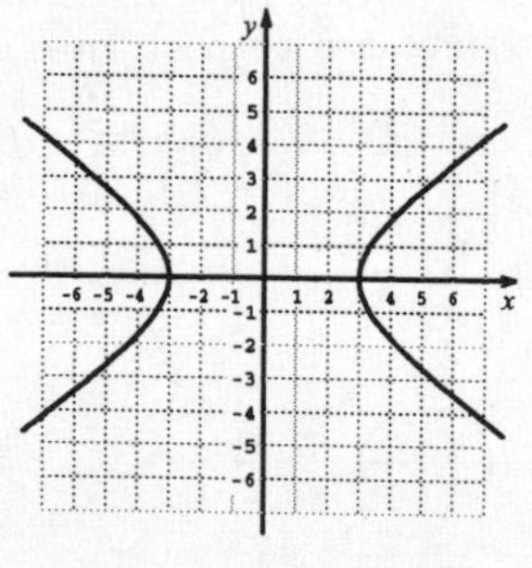

19. $\frac{x^2}{25} - \frac{y^2}{36} = 1$

$\frac{x^2}{5^2} - \frac{y^2}{6^2} = 1$

a) $a = 5$ and $b = 6$, so the asymptotes are $y = \frac{6}{5}x$ and $y = -\frac{6}{5}x$. We sketch them.

b) Replacing y with 0 and solving for x, we get $x = \pm 5$, so the intercepts are (5,0) and (-5,0).

c) We plot the intercepts and draw smooth curves through them that approach the asymptotes.

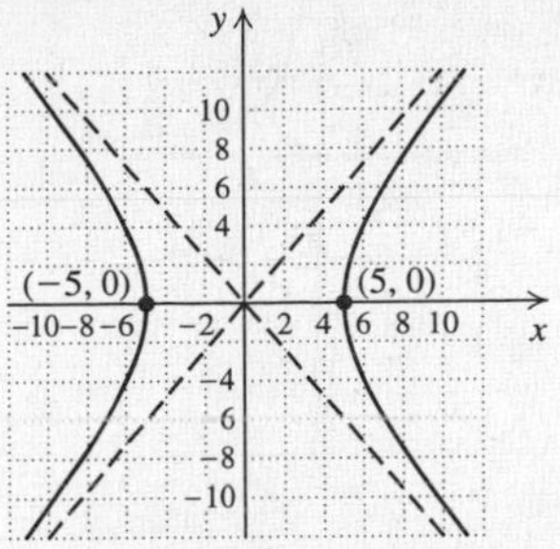

$\frac{x^2}{25} - \frac{y^2}{36} = 1$

20. $\frac{y^2}{9} - \frac{x^2}{25} = 1$

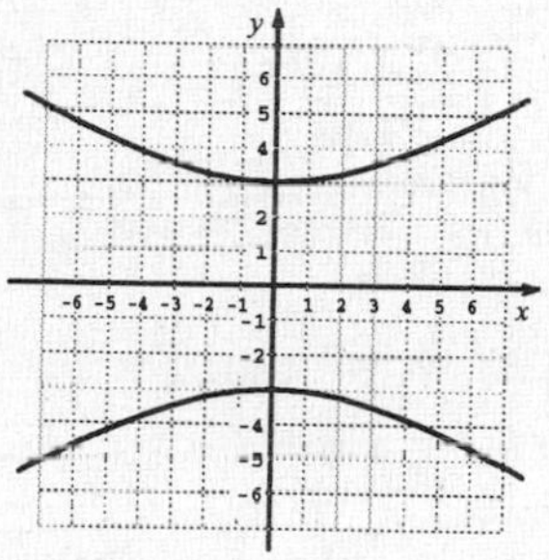

21. $x^2 - y^2 = 4$

$\frac{x^2}{4} - \frac{y^2}{4} = 1$ Multiplying by $\frac{1}{4}$

$\frac{x^2}{2^2} - \frac{y^2}{2^2} = 1$

a) a = 2 and b = 2, so the asymptotes are $y = \frac{2}{2}x$ and $y = -\frac{2}{2}x$, or y = x and y = -x. We sketch them.

b) Replacing y with 0 and solving for x, we get x = ±2, so the intercepts are (2,0) and (-2,0).

c) We plot the intercepts and draw smooth curves through them that approach the asymptotes.

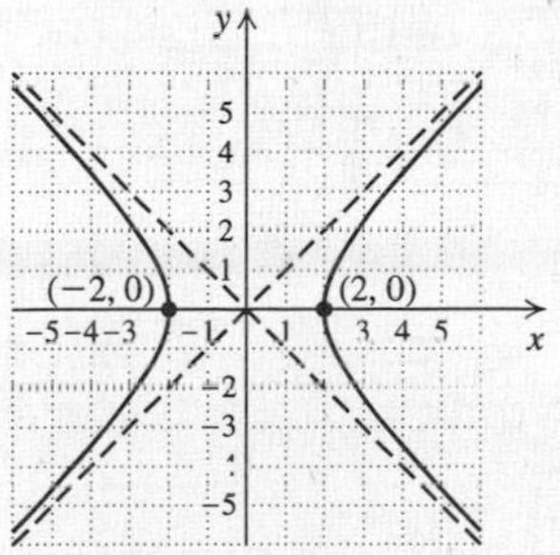

$x^2 - y^2 = 4$

22. $y^2 - x^2 = 25$

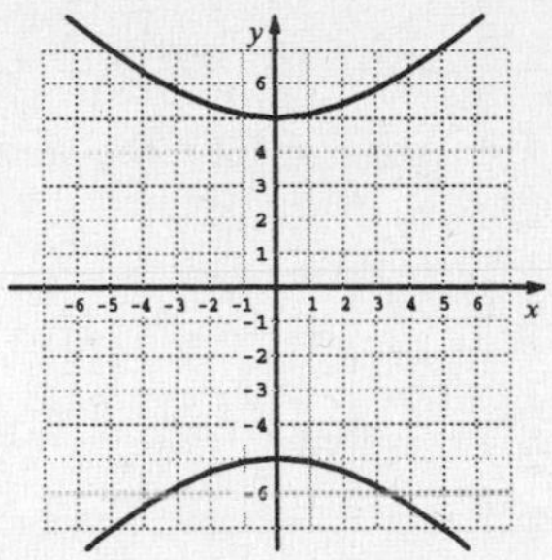

23. $4y^2 - 9x^2 = 36$

$\frac{y^2}{9} - \frac{x^2}{4} = 1$ Multiplying by $\frac{1}{36}$

$\frac{y^2}{3^2} - \frac{x^2}{2^2} = 1$

a) a = 2 and b = 3, so the asymptotes are $y = \frac{3}{2}x$ and $y = -\frac{3}{2}x$. We sketch them.

b) Replacing x with 0 and solving for y, we get y = ±3, so the intercepts are (0,3) and (0,-3).

c) We plot the intercepts and draw smooth curves through them that approach the asymptotes.

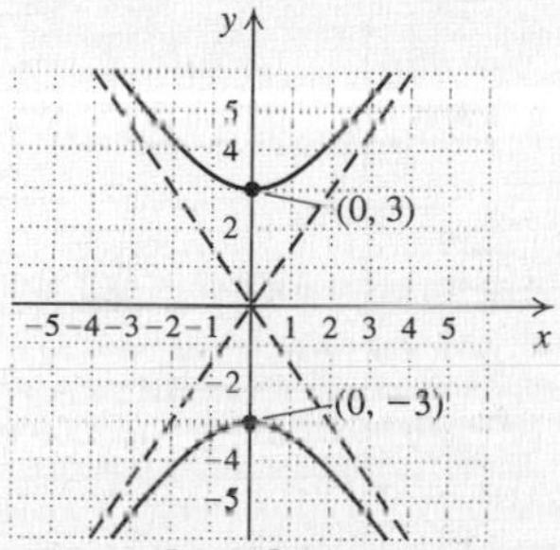

$4y^2 - 9x^2 = 36$

24. $25x^2 - 16y^2 = 400$

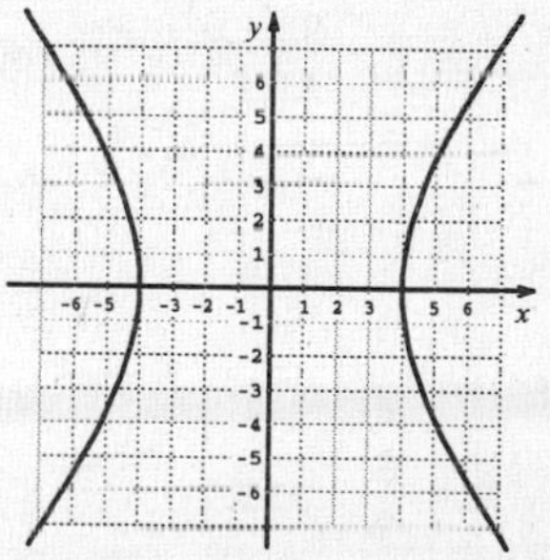

25. xy = 6

$y = \frac{6}{x}$ Solving for y

We find some solutions, keeping the results in a table.

x	y
1	6
2	3
3	2
6	1
$\frac{1}{2}$	12
$\frac{1}{3}$	18
-1	-6
-2	-3
-3	-2
-6	-1
$-\frac{1}{2}$	-12
$-\frac{1}{3}$	-18

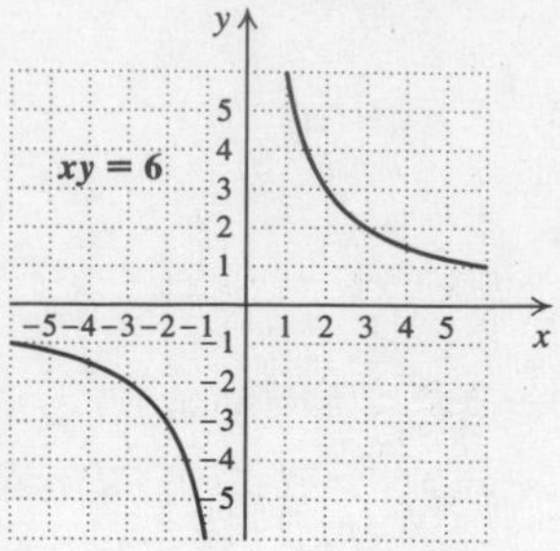

Note that we cannot use 0 for x. The x-axis and the y-axis are the asymptotes.

26. xy = -4

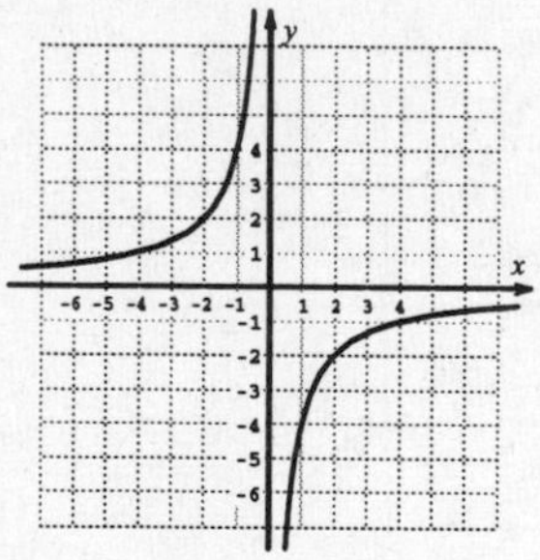

27. xy = -9

$y = -\frac{9}{x}$ Solving for y

x	y
1	-9
3	-3
9	-1
$\frac{1}{2}$	-18
-1	9
-3	3
-9	1
$-\frac{1}{2}$	18

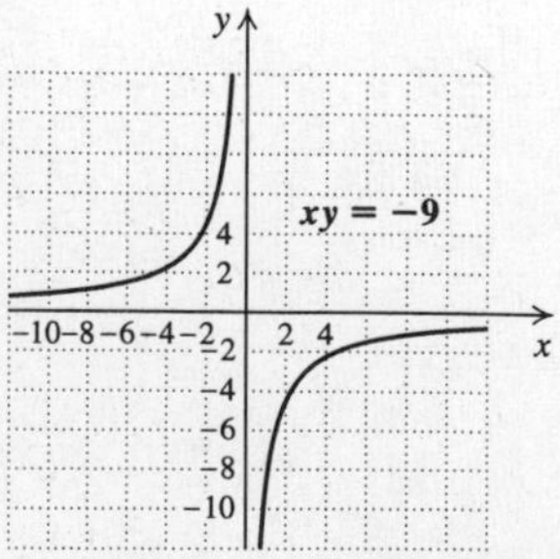

Note that we cannot use 0 for x. The x-axis and the y-axis are the asymptotes.

28. xy = 3

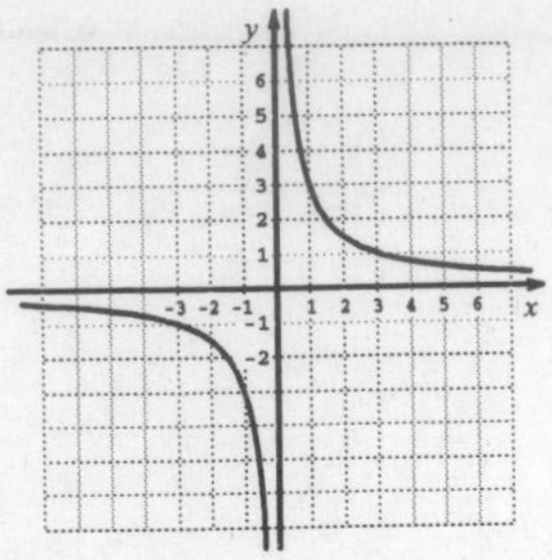

29. xy = -1

$y = -\frac{1}{x}$ Solving for y

x	y
1	-1
2	$-\frac{1}{2}$
4	$-\frac{1}{4}$
$\frac{1}{2}$	-2
$\frac{1}{4}$	-4
-1	1
-2	$\frac{1}{2}$
-4	$\frac{1}{4}$
$-\frac{1}{2}$	2
$-\frac{1}{4}$	4

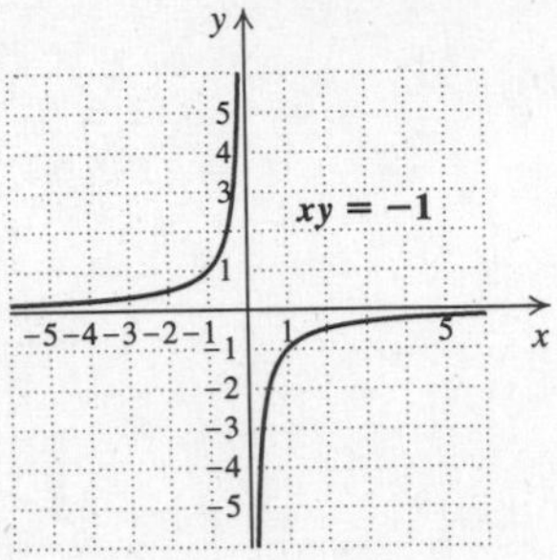

Note that we cannot use 0 for x. The x-axis and the y-axis are the asymptotes.

30. xy = -2

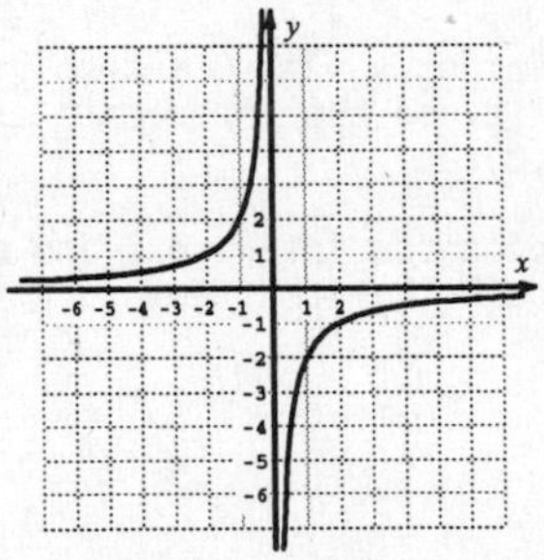

31. $xy = 2$

$y = \frac{2}{x}$ Solving for x

x	y
1	2
2	1
4	$\frac{1}{2}$
$\frac{1}{2}$	4
-1	-2
-2	-1
-4	$-\frac{1}{2}$
$-\frac{1}{2}$	-4

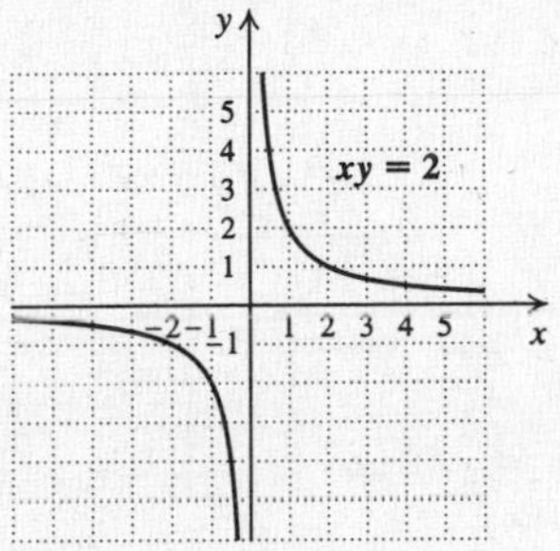

Note that we cannot use 0 for x. The x-axis and the y-axis are the asymptotes.

32. $xy = 1$

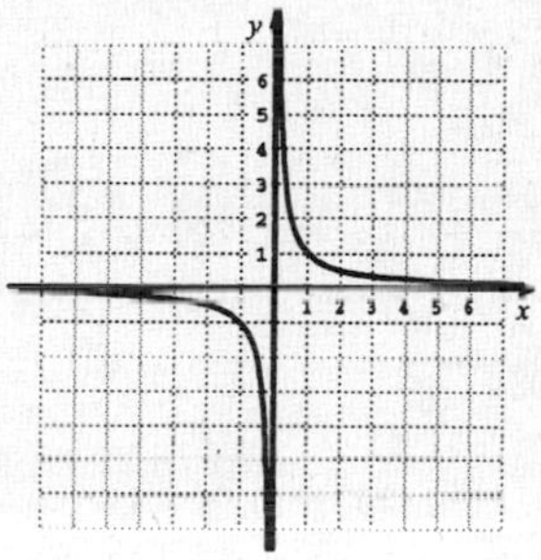

33. $x^2 + y^2 - 10x + 8y - 40 = 0$

Completing the square twice, we obtain an equivalent equation:

$$(x^2 - 10x) + (y^2 + 8y) = 40$$
$$(x^2 - 10x + 25) + (y^2 + 8y + 16) = 40 + 25 + 16$$
$$(x - 5)^2 + (y + 4)^2 = 81$$

The graph is a circle.

34. Parabola

35. $9x^2 - 4y^2 - 36 = 0$

$$9x^2 - 4y^2 = 36$$
$$\frac{x^2}{4} - \frac{y^2}{9} = 1$$

The graph is a hyperbola.

36. Parabola

37. $4x^2 + 25y^2 - 100 = 0$

$$4x^2 + 25y^2 = 100$$
$$\frac{x^2}{25} + \frac{y^2}{4} = 1$$

The graph is an ellipse.

38. Circle

39.
$$x^2 + y^2 = 2x + 4y + 4$$
$$x^2 - 2x + y^2 - 4y = 4$$
$$(x^2 - 2x + 1) + (y^2 - 4y + 4) = 4 + 1 + 4$$
$$(x - 1)^2 + (y - 2)^2 = 9$$

The graph is a circle.

40. Circle

41.
$$4x^2 = 64 - y^2$$
$$4x^2 + y^2 = 64$$
$$\frac{x^2}{16} + \frac{y^2}{64} = 1$$

The graph is an ellipse.

42. Hyperbola

43. $x - \frac{3}{y} = 0$

$$x = \frac{3}{y}$$
$$xy = 3$$

The graph is a hyperbola.

44. Parabola

45. $y + 6x = x^2 + 6$

$$y = x^2 - 6x + 6$$

The graph is a parabola.

46. Hyperbola

47.
$$9y^2 = 36 + 4x^2$$
$$9y^2 - 4x^2 = 36$$
$$\frac{y^2}{4} - \frac{x^2}{9} = 1$$

The graph is a hyperbola.

48. Circle

49. $\sqrt[3]{125t^{15}} = \sqrt[3]{5^3 \cdot (t^5)^3} = 5t^5$

50. $\pm\, i\sqrt{5}$

51.
$$\frac{4\sqrt{2} - 5\sqrt{3}}{6\sqrt{3} - 8\sqrt{2}} = \frac{4\sqrt{2} - 5\sqrt{3}}{6\sqrt{3} - 8\sqrt{2}} \cdot \frac{6\sqrt{3} + 8\sqrt{2}}{6\sqrt{3} + 8\sqrt{2}}$$
$$= \frac{24\sqrt{6} + 32\cdot 2 - 30\cdot 3 - 40\sqrt{6}}{36\cdot 3 - 64\cdot 2}$$
$$= \frac{-26 - 16\sqrt{6}}{-20}$$
$$= \frac{-2(13 + 8\sqrt{6})}{-2\cdot 10}$$
$$= \frac{13 + 8\sqrt{6}}{10}$$

52. Smaller plane: 400 mph, larger plane: 720 mph

53. 

54. 2.134×10^8 mi

55. a), b) See the answer section in the text.

56.

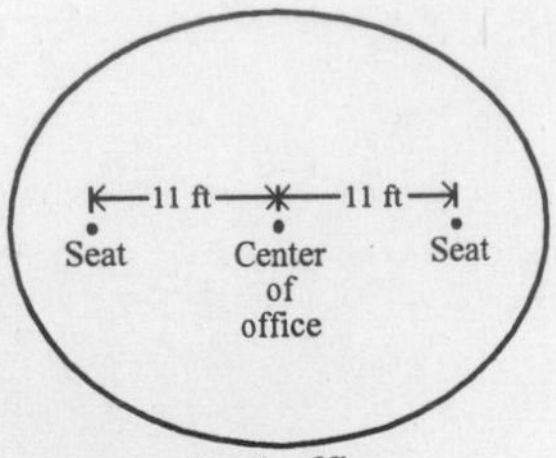

Oval office

57. For the given ellipse, a = 6/2, or 3, and b = 2/2, or 1. The patient's mouth should be at a distance 2c from the light source, where the coordinates of the foci of the ellipse are (-c,0) and (c,0). From Exercise 55(b), we know $b^2 = a^2 - c^2$. We use this to find c.

$b^2 = a^2 - c^2$

$1^2 = 3^2 - c^2$ Substituting

$c^2 = 8$

$c = \sqrt{8}$

Then $2c = 2\sqrt{8} \approx 5.66$. The patient's mouth should be about 5.66 ft from the light source.

58. $\dfrac{x^2}{81} + \dfrac{y^2}{121} = 1$

59.

$$16x^2 + y^2 + 96x - 8y + 144 = 0$$
$$16x^2 + 96x + y^2 - 8y = -144$$
$$16(x^2 + 6x) + (y^2 - 8y) = -144$$
$$16(x^2 + 6x + 9) + (y^2 - 8y + 16) = -144 + 144 + 16$$
$$16(x + 3)^2 + (y - 4)^2 = 16$$
$$\frac{(x + 3)^2}{1} + \frac{(y - 4)^2}{16} = 1 \quad \text{Standard form}$$

h = -3, k = 4, a = 1, b = 4

Center: (-3,4)

Vertices: (-2,4), (-4,4), (-3,8), (-3,0)

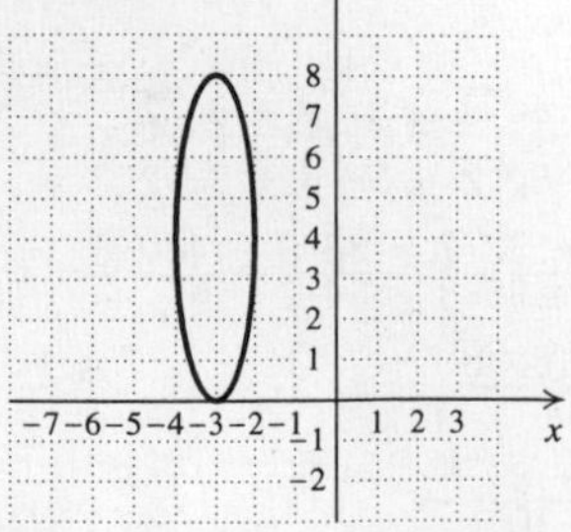

$16x^2 + y^2 + 96x - 8y + 144 = 0$

60. Center: (1,-1)

Vertices: (6,-1), (-4,-1), (1,1), (1,-3)

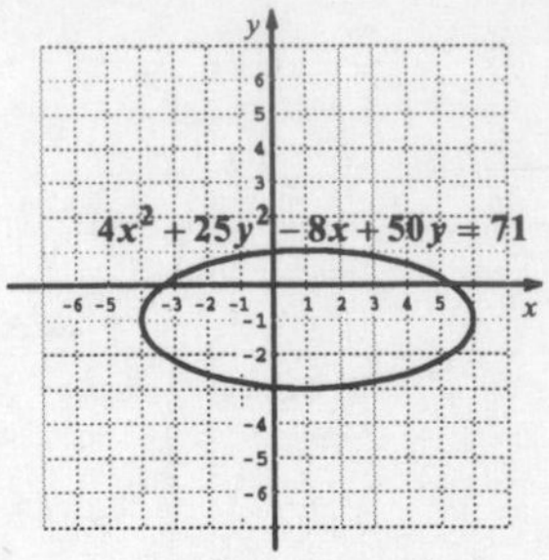

61. Since the intercepts are (0,8) and (0,-8), we know that the hyperbola is of the form $\dfrac{y^2}{b^2} - \dfrac{x^2}{a^2} = 1$ and that b = 8. The equations of the asymptotes tell us that b/a = 4, so

$\dfrac{8}{a} = 4$

a = 2.

The equation is $\dfrac{y^2}{8^2} - \dfrac{x^2}{2^2} = 1$, or $\dfrac{y^2}{64} - \dfrac{x^2}{4} = 1$.

62. $\dfrac{x^2}{64} - \dfrac{y^2}{1024} = 1$

63. Center: (2,-1)

Vertices: (-1,-1), (5,-1)

Asymptotes: $y + 1 = \frac{4}{3}(x - 2)$, $y + 1 = -\frac{4}{3}(x - 2)$

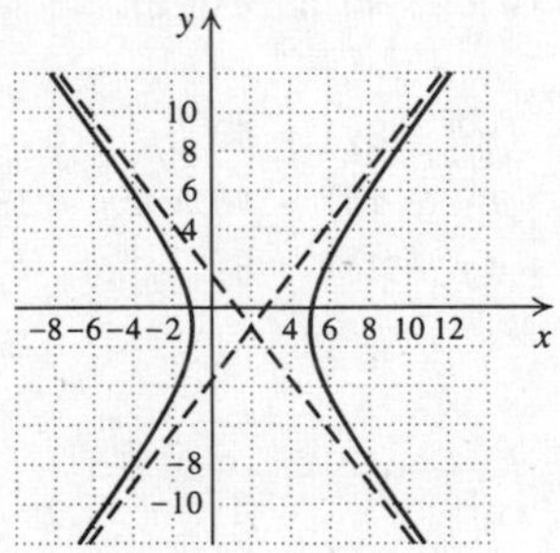

64. $\dfrac{(x + 3)^2}{1} - \dfrac{(y - 2)^2}{4} = 1$

Center: (-3,2)

Vertices: (-4,2), (-2,2)

Asymptotes: y - 2 = 2(x + 3), y - 2 = -2(x + 3)

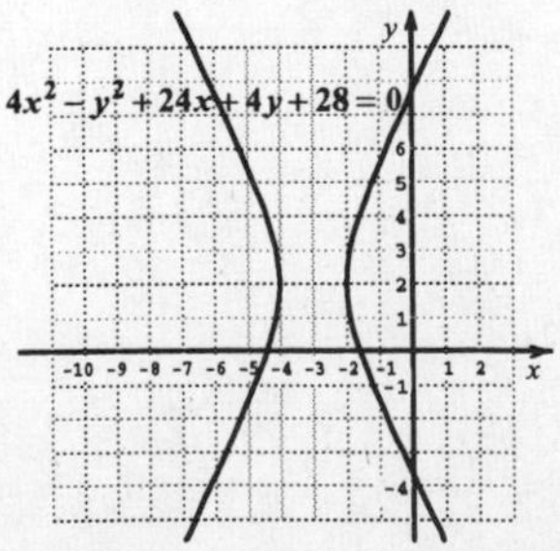

65. $4y^2 - 25x^2 - 8y - 100x - 196 = 0$

$$4(y^2 - 2y) - 25(x^2 + 4x) = 196$$

$$4(y^2 - 2y + 1) - 25(x^2 + 4x + 4) = 196 + 4 - 100$$

$$4(y - 1)^2 - 25(x + 2)^2 = 100$$

$$\frac{(y - 1)^2}{25} - \frac{(x + 2)^2}{4} = 1 \quad \text{Standard form}$$

Center: $(-2,1)$

Vertices: $(-2,6)$, $(-2,-4)$

Asymptotes: $y - 1 = \frac{5}{2}(x + 2)$, $y - 1 = -\frac{5}{2}(x + 2)$

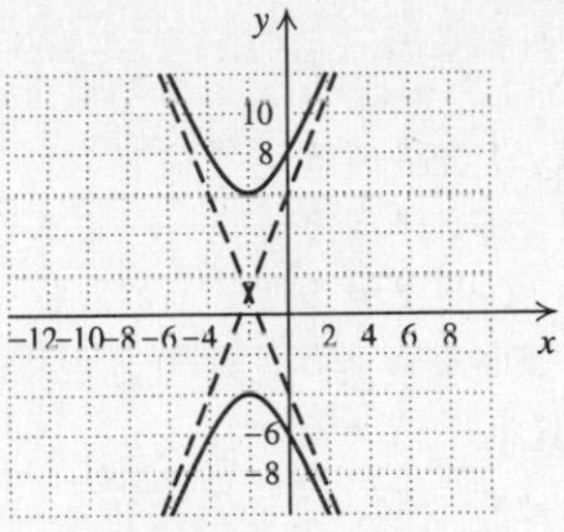

66. $\frac{(x - 2)^2}{9} - \frac{(y + 1)^2}{9} = 1$

Center: $(2,-1)$

Vertices: $(5,-1)$, $(-1,-1)$

Asymptotes: $y + 1 = x - 2$, $y + 1 = -(x - 2)$

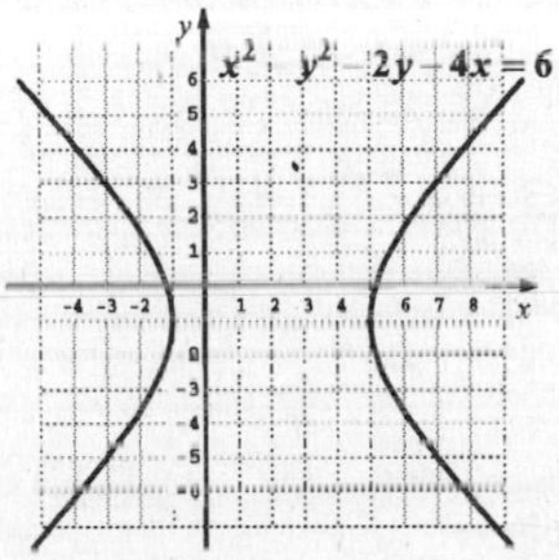

Exercise Set 10.3

1. $x^2 + y^2 = 25$, (1)

$y - x = 1$ (2)

First solve Eq. (2) for y.

$y = x + 1$ (3)

Then substitute $x + 1$ for y in Eq. (1) and solve for x.

$$x^2 + y^2 = 25$$

$$x^2 + (x + 1)^2 = 25$$

$$x^2 + x^2 + 2x + 1 = 25$$

$$2x^2 + 2x - 24 = 0$$

$$x^2 + x - 12 = 0 \quad \text{Multiplying by } \frac{1}{2}$$

$$(x + 4)(x - 3) = 0 \quad \text{Factoring}$$

$x + 4 = 0$ or $x - 3 = 0$ Principle of zero products

$x = -4$ or $x = 3$

Now substitute these numbers into Eq. (3) and solve for y.

$y = -4 + 1 = -3$

$y = 3 + 1 = 4$

The pairs $(-4,-3)$ and $(3,4)$ check, so they are the solutions.

2. $(-8,-6)$, $(6,8)$

3. $4x^2 + 9y^2 = 36$, (1)

$3y + 2x = 6$ (2)

First solve Eq. (2) for y.

$3y = -2x + 6$

$y = -\frac{2}{3}x + 2$ (3)

Then substitute $-\frac{2}{3}x + 2$ for y in Eq. (1) and solve for x.

$$4x^2 + 9y^2 = 36$$

$$4x^2 + 9\left[-\frac{2}{3}x + 2\right]^2 = 36$$

$$4x^2 + 9\left[\frac{4}{9}x^2 - \frac{8}{3}x + 4\right] = 36$$

$$4x^2 + 4x^2 - 24x + 36 = 36$$

$$8x^2 - 24x = 0$$

$$x^2 - 3x = 0$$

$$x(x - 3) = 0$$

$x = 0$ or $x = 3$

Now substitute these numbers in Eq. (3) and solve for y.

$y = -\frac{2}{3} \cdot 0 + 2 = 2$

$y = -\frac{2}{3} \cdot 3 + 2 = 0$

The pairs $(0,2)$ and $(3,0)$ check, so they are the solutions.

4. $(2,0)$, $(0,3)$

5. $y^2 = x + 3$, (1)

$2y = x + 4$ (2)

First solve Eq. (2) for x.

$2y - 4 = x$ (3)

Then substitute $2y - 4$ for x in Eq. (1) and solve for y.

$$y^2 = x + 3$$

$$y^2 = (2y - 4) + 3$$

$$y^2 = 2y - 1$$

$$y^2 - 2y + 1 = 0$$

$$(y - 1)(y - 1) = 0$$

$y = 1$ or $y = 1$

Now substitute 1 for y in Eq. (3) and solve for x.

$2 \cdot 1 - 4 = x$

$-2 = x$

The pair $(-2,1)$ checks. It is the solution.

6. (2,4), (1,1)

7. $x^2 - xy + 3y^2 = 27,$ (1)

$x - y = 2$ (2)

First solve Eq. (2) for y.

$x - 2 = y$ (3)

Then substitute x - 2 for y in Eq. (1) and solve for x.

$$x^2 - xy + 3y^2 = 27$$
$$x^2 - x(x - 2) + 3(x - 2)^2 = 27$$
$$x^2 - x^2 + 2x + 3x^2 - 12x + 12 = 27$$
$$3x^2 - 10x - 15 = 0$$

$$x = \frac{-(-10) \pm \sqrt{(-10)^2 - 4(3)(-15)}}{2\cdot 3}$$
$$= \frac{10 \pm \sqrt{100 + 180}}{6}$$
$$= \frac{10 \pm \sqrt{280}}{6}$$
$$= \frac{10 \pm 2\sqrt{70}}{6}$$
$$= \frac{5 \pm \sqrt{70}}{3}$$

Now substitute these numbers in Eq. (3) and solve for y.

$$y = \frac{5 + \sqrt{70}}{3} - 2 = \frac{-1 + \sqrt{70}}{3}$$
$$y = \frac{5 - \sqrt{70}}{3} - 2 = \frac{-1 - \sqrt{70}}{3}$$

The pairs $\left(\frac{5 + \sqrt{70}}{3}, \frac{-1 + \sqrt{70}}{3}\right)$ and $\left(\frac{5 - \sqrt{70}}{3}, \frac{-1 - \sqrt{70}}{3}\right)$ check, so they are the solutions.

8. $\left(\frac{11}{4}, -\frac{9}{8}\right)$, (1,-2)

9. $x^2 + 4y^2 = 25,$ (1)

$x + 2y = 7$ (2)

First solve Eq. (2) for x.

$x = -2y + 7$ (3)

Then substitute -2y + 7 for x in Eq. (1) and solve for y.

$$x^2 + 4y^2 = 25$$
$$(-2y + 7)^2 + 4y^2 = 25$$
$$4y^2 - 28y + 49 + 4y^2 = 25$$
$$8y^2 - 28y + 24 = 0$$
$$2y^2 - 7y + 6 = 0$$
$$(2y - 3)(y - 2) = 0$$

$y = \frac{3}{2}$ or $y = 2$

Now substitute these numbers in Eq. (3) and solve for x.

$$x = -2\cdot\frac{3}{2} + 7 = 4$$
$$x = -2\cdot 2 + 7 = 3$$

The pairs $\left(4, \frac{3}{2}\right)$ and (3,2) check, so they are the solutions.

10. $\left(-\frac{5}{3}, -\frac{13}{3}\right)$, (3,5)

11. $x^2 - xy + 3y^2 = 5,$ (1)

$x - y = 2$ (2)

First solve Eq. (2) for y.

$x - 2 = y$ (3)

Then substitute x - 2 for y in Eq. (1) and solve for x.

$$x^2 - xy + 3y^2 = 5$$
$$x^2 - x(x - 2) + 3(x - 2)^2 = 5$$
$$x^2 - x^2 + 2x + 3x^2 - 12x + 12 = 5$$
$$3x^2 - 10x + 7 = 0$$
$$(3x - 7)(x - 1) = 0$$

$x = \frac{7}{3}$ or $x = 1$

Now substitute these numbers in Eq. (3) and solve for y.

$$y = \frac{7}{3} - 2 = \frac{1}{3}$$
$$y = 1 - 2 = -1$$

The pairs $\left(\frac{7}{3}, \frac{1}{3}\right)$ and (1,-1) check, so they are the solutions.

12. $\left(\frac{3 + \sqrt{7}}{2}, \frac{-1 + \sqrt{7}}{2}\right)$, $\left(\frac{3 - \sqrt{7}}{2}, \frac{-1 - \sqrt{7}}{2}\right)$

13. $3x + y = 7,$ (1)

$4x^2 + 5y = 24$ (2)

First solve Eq. (1) for y.

$y = 7 - 3x$ (3)

Then substitute 7 - 3x for y in Eq. (2) and solve for x.

$$4x^2 + 5y = 24$$
$$4x^2 + 5(7 - 3x) = 24$$
$$4x^2 + 35 - 15x = 24$$
$$4x^2 - 15x + 11 = 0$$
$$(4x - 11)(x - 1) = 0$$

$x = \frac{11}{4}$ or $x = 1$

Now substitute these numbers in Eq. (3) and solve for y.

$$y = 7 - 3\cdot\frac{11}{4} = -\frac{5}{4}$$
$$y = 7 - 3\cdot 1 = 4$$

The pairs $\left(\frac{11}{4}, -\frac{5}{4}\right)$ and (1,4) check, so they are the solutions.

14. $\left(-3, \frac{5}{2}\right)$, (3,1)

15. $a + b = 7$, (1)
$ab = 4$ (2)

First solve Eq. (1) for a.

$a = -b + 7$ (3)

Then substitute $-b + 7$ for a in Eq. (2) and solve for b.

$(-b + 7)b = 4$

$-b^2 + 7b = 4$

$0 = b^2 - 7b + 4$

$$b = \frac{-(-7) \pm \sqrt{(-7)^2 - 4\cdot1\cdot4}}{2\cdot1}$$

$$b = \frac{7 \pm \sqrt{33}}{2}$$

Now substitute these numbers in Eq. (3) and solve for a.

$$a = -\left(\frac{7 + \sqrt{33}}{2}\right) + 7 = \frac{7 - \sqrt{33}}{2}$$

$$a = -\left(\frac{7 - \sqrt{33}}{2}\right) + 7 = \frac{7 + \sqrt{33}}{2}$$

The pairs $\left(\frac{7 - \sqrt{33}}{2}, \frac{7 + \sqrt{33}}{2}\right)$ and $\left(\frac{7 + \sqrt{33}}{2}, \frac{7 - \sqrt{33}}{2}\right)$ check, so they are the solutions.

16. (1,-7), (-7,1)

17. $2a + b = 1$, (1)
$b = 4 - a^2$ (2)

Eq. (2) is already solved for b. Substitute $4 - a^2$ for b in Eq. (1) and solve for a.

$2a + 4 - a^2 = 1$

$0 = a^2 - 2a - 3$

$0 = (a - 3)(a + 1)$

$a = 3$ or $a = -1$

Substitute these numbers in Eq. (2) and solve for b.

$b = 4 - 3^2 = -5$

$b = 4 - (-1)^2 = 3$

The pairs (3,-5) and (-1,3) check.

18. (3,0), $\left(-\frac{9}{5}, \frac{8}{5}\right)$

19. $a^2 + b^2 = 89$, (1)
$a - b = 3$ (2)

First solve Eq. (2) for a.

$a = b + 3$ (3)

Then substitute $b + 3$ for a in Eq. (1) and solve for b.

$(b + 3)^2 + b^2 = 89$

$b^2 + 6b + 9 + b^2 = 89$

$2b^2 + 6b - 80 = 0$

$b^2 + 3b - 40 = 0$

$(b + 8)(b - 5) = 0$

$b = -8$ or $b = 5$

Substitute these numbers in Eq. (3) and solve for a.

$a = -8 + 3 = -5$

$a = 5 + 3 = 8$

The pairs (-5,-8) and (8,5) check.

20. (1,4), (4,1)

21. $x^2 + y^2 = 25$, (1)
$y^2 = x + 5$ (2)

We substitute $x + 5$ for y^2 in Eq. (1) and solve for x.

$x^2 + y^2 = 25$

$x^2 + (x + 5) = 25$

$x^2 + x - 20 = 0$

$(x + 5)(x - 4) = 0$

$x + 5 = 0$ or $x - 4 = 0$

$x = -5$ or $x = 4$

Next we substitute these numbers for x in either Eq. (1) or Eq. (2) and solve for y. Here we use Eq. (2).

$y^2 = -5 + 5 = 0$ and $y = 0$.

$y^2 = 4 + 5 = 9$ and $y = \pm3$.

The pairs (-5,0), (4,3), and (4,-3) check. They are the solutions.

22. (0,0), (1,1), $\left(-\frac{1}{2} + \frac{\sqrt{3}}{2}i, -\frac{1}{2} - \frac{\sqrt{3}}{2}i\right)$,
$\left(-\frac{1}{2} - \frac{\sqrt{3}}{2}i, -\frac{1}{2} + \frac{\sqrt{3}}{2}i\right)$

23. $x^2 + y^2 = 9$, (1)
$x^2 - y^2 = 9$ (2)

Here we use the addition method.

$x^2 + y^2 = 9$

$x^2 - y^2 = 9$

$2x^2 = 18$ Adding

$x^2 = 9$

$x = \pm3$

If $x = 3$, $x^2 = 9$, and if $x = -3$, $x^2 = 9$, so substituting 3 or -3 in Eq. (1) give us

$x^2 + y^2 = 9$

$9 + y^2 = 9$

$y^2 = 0$

$y = 0$.

The pairs (3,0) and (-3,0) check. They are the solutions.

24. (0,2), (0,-2)

25. $x^2 + y^2 = 25,$ (1)
$xy = 12$ (2)

First we solve Eq. (2) for y.

$xy = 12$

$y = \frac{12}{x}$

Then we substitute $\frac{12}{x}$ for y in Eq. (1) and solve for x.

$$x^2 + y^2 = 25$$
$$x^2 + \left(\frac{12}{x}\right)^2 = 25$$
$$x^2 + \frac{144}{x^2} = 25$$
$$x^4 + 144 = 25x^2 \quad \text{Multiplying by } x^2$$
$$x^4 - 25x^2 + 144 = 0$$
$$u^2 - 25u + 144 = 0 \quad \text{Letting } u = x^2$$
$$(u - 9)(u - 16) = 0$$
$$u = 9 \text{ or } u = 16$$

We now substitute x^2 for u and solve for x.

$$x^2 = 9 \text{ or } x^2 = 16$$
$$x = \pm 3 \text{ or } x = \pm 4$$

Since $y = 12/x$, if $x = 3$, $y = 4$; if $x = -3$, $y = -4$; if $x = 4$, $y = 3$; and if $x = -4$, $y = -3$. The pairs (3,4), (-3,-4), (4,3), (-4,-3) check. They are the solutions.

26. (-5,3), (-5,-3), (4,0)

27. $x^2 + y^2 = 4,$ (1)
$16x^2 + 9y^2 = 144$ (2)

$$-9x^2 - 9y^2 = -36 \quad \text{Multiplying (1) by -9}$$
$$\underline{16x^2 + 9y^2 = 144}$$
$$7x^2 = 108 \quad \text{Adding}$$
$$x^2 = \frac{108}{7}$$
$$x = \pm\sqrt{\frac{108}{7}} = \pm 6\sqrt{\frac{3}{7}}$$
$$x = \pm\frac{6\sqrt{21}}{7} \quad \text{Rationalizing the denominator}$$

Substituting $\frac{6\sqrt{21}}{7}$ or $-\frac{6\sqrt{21}}{7}$ for x in Eq. (1) gives us

$$\frac{36 \cdot 21}{49} + y^2 = 4$$
$$y^2 = 4 - \frac{108}{7}$$
$$y^2 = -\frac{80}{7}$$
$$y = \pm\sqrt{-\frac{80}{7}} = \pm 4i\sqrt{\frac{5}{7}}$$
$$y = \pm\frac{4i\sqrt{35}}{7}. \quad \text{Rationalizing the denominator}$$

The pairs $\left(\frac{6\sqrt{21}}{7}, \frac{4i\sqrt{35}}{7}\right)$, $\left(\frac{6\sqrt{21}}{7}, -\frac{4i\sqrt{35}}{7}\right)$, $\left(-\frac{6\sqrt{21}}{7}, \frac{4i\sqrt{35}}{7}\right)$, and $\left(-\frac{6\sqrt{21}}{7}, -\frac{4i\sqrt{35}}{7}\right)$ check. They are the solutions.

28. (0,5), (0,-5)

29. $x^2 + y^2 = 16,$ or $x^2 + y^2 = 16,$ (1)
$y^2 - 2x^2 = 10$ $\quad -2x^2 + y^2 = 10$ (2)

Here we use the addition method.

$$2x^2 + 2y^2 = 32 \quad \text{Multiplying (1) by 2}$$
$$\underline{-2x^2 + y^2 = 10}$$
$$3y^2 = 42 \quad \text{Adding}$$
$$y^2 = 14$$
$$y = \pm\sqrt{14}$$

Substituting $\sqrt{14}$ or $-\sqrt{14}$ for y in Eq. (1) gives us

$$x^2 + 14 = 16$$
$$x^2 = 2$$
$$x = \pm\sqrt{2}$$

The pairs $(-\sqrt{2},-\sqrt{14})$, $(-\sqrt{2},\sqrt{14})$, $(\sqrt{2},-\sqrt{14})$, and $(\sqrt{2},\sqrt{14})$ check. They are the solutions.

30. $(-3,-\sqrt{5})$, $(-3,\sqrt{5})$, $(3,-\sqrt{5})$, $(3,\sqrt{5})$

31. $x^2 + y^2 = 5,$ (1)
$xy = 2$ (2)

First we solve Eq. (2) for y.

$xy = 2$

$y = \frac{2}{x}$

Then we substitute $\frac{2}{x}$ for y in Eq. (1) and solve for x.

$$x^2 + y^2 = 5$$
$$x^2 + \left(\frac{2}{x}\right)^2 = 5$$
$$x^2 + \frac{4}{x^2} = 5$$
$$x^4 + 4 = 5x^2 \quad \text{Multiplying by } x^2$$
$$x^4 - 5x^2 + 4 = 0$$
$$u^2 - 5u + 4 = 0 \quad \text{Letting } u = x^2$$
$$(u - 4)(u - 1) = 0$$
$$u = 4 \text{ or } u = 1$$

We now substitute x^2 for u and solve for x.

$$x^2 = 4 \text{ or } x^2 = 1$$
$$x = \pm 2 \qquad x = \pm 1$$

Since $y = 2/x$, if $x = 2$, $y = 1$; if $x = -2$, $y = -1$; if $x = 1$, $y = 2$; and if $x = -1$, $y = -2$. The pairs (2,1), (-2,-1), (1,2), and (-1,-2) check. They are the solutions.

32. (4,2), (-4,-2), (2,4), (-2,-4)

33. $x^2 + y^2 = 13,$ (1)
$xy = 6$ (2)

First we solve Eq. (2) for y.

$xy = 6$

$y = \frac{6}{x}$

Then we substitute $\frac{6}{x}$ for y in Eq. (1) and solve for x.

$$x^2 + y^2 = 13$$
$$x^2 + \left(\frac{6}{x}\right)^2 = 13$$
$$x^2 + \frac{36}{x^2} = 13$$
$$x^4 + 36 = 13x^2 \quad \text{Multiplying by } x^2$$
$$x^4 - 13x^2 + 36 = 0$$
$$u^2 - 13u + 36 = 0 \quad \text{Letting } u = x^2$$
$$(u - 9)(u - 4) = 0$$
$$u = 9 \text{ or } u = 4$$

We now substitute x^2 for u and solve for x.

$$x^2 = 9 \text{ or } x^2 = 4$$
$$x = \pm 3 \qquad x = \pm 2$$

Since $y = 6/x$, if $x = 3$, $y = 2$; if $x = -3$, $y = -2$; if $x = 2$, $y = 3$; and if $x = -2$, $y = -3$. The pairs (3,2), (-3,-2), (2,3), (-2,-3) check. They are the solutions.

34. (4,1), (-4,-1), (2,2), (-2,-2)

35. $3xy + x^2 = 34,$ (1)
$2xy - 3x^2 = 8$ (2)

$$6xy + 2x^2 = 68 \quad \text{Multiplying (1) by 2}$$
$$\underline{-6xy + 9x^2 = -24} \quad \text{Multiplying (2) by -3}$$
$$11x^2 = 44 \quad \text{Adding}$$
$$x^2 = 4$$
$$x = \pm 2$$

Substitute for x in Eq. (1) and solve for y.

When $x = 2$: $3 \cdot 2 \cdot y + 2^2 = 34$
$$6y + 4 = 34$$
$$6y = 30$$
$$y = 5$$

When $x = -2$: $3(-2)(y) + (-2)^2 = 34$
$$-6y + 4 = 34$$
$$-6y = 30$$
$$y = -5$$

The pairs (2,5) and (-2,-5) check. They are the solutions.

36. (2,1), (-2,-1)

37. $xy - y^2 = 2,$ (1)
$2xy - 3y^2 = 0$ (2)

$$-2xy + 2y^2 = -4 \quad \text{Multiplying (1) by -2}$$
$$\underline{2xy - 3y^2 = 0}$$
$$-y^2 = -4$$
$$y^2 = 4$$
$$y = \pm 2$$

We substitute for y in Eq. (1) and solve for x.

When $y = 2$: $x \cdot 2 - 2^2 = 2$
$$2x - 4 = 2$$
$$2x = 6$$
$$x = 3$$

When $y = -2$: $x(-2) - (-2)^2 = 2$
$$-2x - 4 = 2$$
$$-2x = 6$$
$$x = -3$$

The pairs (3,2) and (-3,-2) check. They are the solutions.

38. $\left(2,-\frac{4}{5}\right)$, $\left(-2,-\frac{4}{5}\right)$, (5,2), (-5,2)

39. $x^2 - y = 5,$ (1)
$x^2 + y^2 = 25$ (2)

We solve Eq. (1) for y.

$x^2 - 5 = y$ (3)

Substitute $x^2 - 5$ for y in Eq. (2) and solve for x.

$$x^2 + (x^2 - 5)^2 = 25$$
$$x^2 + x^4 - 10x^2 + 25 = 25$$
$$x^4 - 9x^2 = 0$$
$$u^2 - 9u = 0 \quad \text{Letting } u = x^2$$
$$u(u - 9) = 0$$
$$u = 0 \text{ or } u = 9$$
$$x^2 = 0 \text{ or } x^2 = 9$$
$$x = 0 \text{ or } x = \pm 3$$

Substitute in Eq. (3) and solve for y.

When $x = 0$: $y = 0^2 - 5 = -5$

When $x = 3$ or -3: $y = 9 - 5 = 4$

The pairs (0,-5), (3,4), and (-3,4) check. They are the solutions.

(This exercise could also be solved using the addition method.)

40. $(-\sqrt{2},\sqrt{2})$, $(\sqrt{2},-\sqrt{2})$

41. Familiarize. We first make a drawing. We let ℓ and w represent the length and width, respectively.

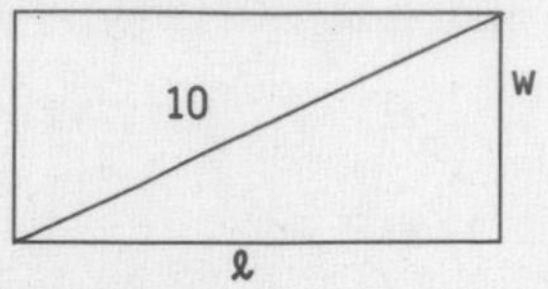

Translate. The perimeter is 28 cm.

$2\ell + 2w = 28$, or $\ell + w = 14$

Using the Pythagorean theorem we have another equation.

$\ell^2 + w^2 = 10^2$, or $\ell^2 + w^2 = 100$

Carry out. We solve the system:

$\ell + w = 14,$ (1)

$\ell^2 + w^2 = 100$ (2)

First solve Eq. (1) for w.

$w = 14 - \ell$ (3)

Then substitute $14 - \ell$ for w in Eq. (2) and solve for ℓ.

$$\ell^2 + w^2 = 100$$
$$\ell^2 + (14 - \ell)^2 = 100$$
$$\ell^2 + 196 - 28\ell + \ell^2 = 100$$
$$2\ell^2 - 28\ell + 96 = 0$$
$$\ell^2 - 14\ell + 48 = 0$$
$$(\ell - 8)(\ell - 6) = 0$$

$\ell = 8$ or $\ell = 6$

If $\ell = 8$, then $w = 14 - 8$, or 6. If $\ell = 6$, then $w = 14 - 6$, or 8. Since the length is usually considered to be longer than the width, we have the solution $\ell = 8$ and $w = 6$, or (8,6).

Check. If $\ell = 8$ and $w = 6$, then the perimeter is $2\cdot 8 + 2\cdot 6$, or 28. The length of a diagonal is $\sqrt{8^2 + 6^2}$, or $\sqrt{100}$, or 10. The numbers check.

State. The length is 8 cm, and the width is 6 cm.

42. 2 in. by 1 in.

43. Familiarize. We first make a drawing. Let ℓ = the length and w = the width of the rectangle.

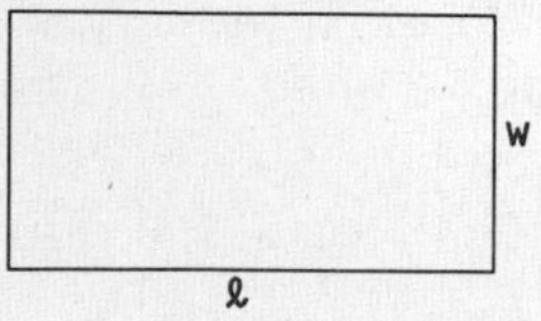

Translate.

Area: $\ell w = 20$

Perimeter: $2\ell + 2w = 18$, or $\ell + w = 9$

Carry out. We solve the system:

Solve the second equation for ℓ: $\ell = 9 - w$

Substitute $9 - w$ for ℓ in the first equation and solve for w.

$$(9 - w)w = 20$$
$$9w - w^2 = 20$$
$$0 = w^2 - 9w + 20$$
$$0 = (w - 5)(w - 4)$$

$w = 5$ or $w = 4$

If $w = 5$, then $\ell = 9 - w$, or 4. If $w = 4$, then $\ell = 9 - 4$, or 5. Since length is usually considered to be longer than width, we have the solution $\ell = 5$ and $w = 4$, or (5,4).

Check. If $\ell = 5$ and $w = 4$, the area is $5\cdot 4$, or 20. The perimeter is $2\cdot 5 + 2\cdot 4$, or 18. The numbers check.

State. The length is 5 in. and the width is 4 in.

44. 2 yd by 1 yd

45. Familiarize. We make a drawing of the field. Let ℓ = the length and w = the width.

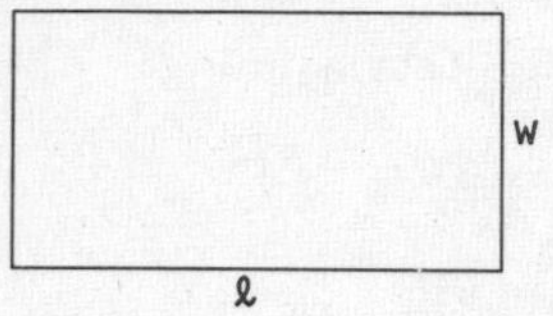

Since it takes 210 yd of fencing to enclose the field, we know that the perimeter is 210 yd.

Translate.

Perimeter: $2\ell + 2w = 210$, or $\ell + w = 105$

Area: $\ell w = 2250$

Carry out. We solve the system:

Solve the first equation for ℓ: $\ell = 105 - w$

Substitute $105 - w$ for ℓ in the second equation and solve for w.

$$(105 - w)w = 2250$$
$$105w - w^2 = 2250$$
$$0 = w^2 - 105w + 2250$$
$$0 = (w - 30)(w - 75)$$

$w = 30$ or $w = 75$

If $w = 30$, then $\ell = 105 - 30$, or 75. If $w = 75$, then $\ell = 105 - 75$, or 30. Since length is usually considered to be longer than width, we have the solution $\ell = 75$ and $w = 30$, or (75,30).

Check. If $\ell = 75$ and $w = 30$, the perimeter is $2\cdot 75 + 2\cdot 30$, or 210. The area is 75(30), or 2250. The numbers check.

State. The length is 75 yd and the width is 30 yd.

46. 12 ft by 5 ft

47. Familiarize. We make a drawing and label it. Let x and y represent the lengths of the legs of the triangle.

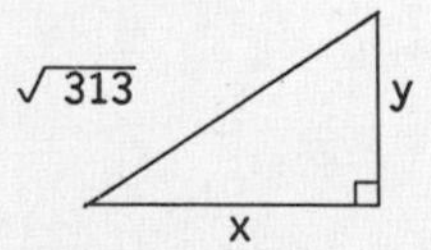

Translate. The product of the lengths of the legs is 156, so we have:

$xy = 156$

We use the Pythagorean theorem to get a second equation:

$$x^2 + y^2 = (\sqrt{313})^2, \text{ or } x^2 + y^2 = 313$$

Carry out. We solve the system of equations

$$xy = 156, \quad (1)$$
$$x^2 + y^2 = 313. \quad (2)$$

First solve Eq. (1) for y.

$$xy = 156$$
$$y = \frac{156}{x}$$

Then we substitute $\frac{156}{x}$ for y in Eq. (2) and solve for x.

$$x^2 + y^2 = 313 \quad (2)$$
$$x^2 + \left(\frac{156}{x}\right)^2 = 313$$
$$x^2 + \frac{24,336}{x^2} = 313$$
$$x^4 + 24,336 = 313x^2$$
$$x^4 - 313x^2 + 24,336 = 0$$
$$u^2 - 313u + 24,336 = 0 \quad \text{Letting } u = x^2$$
$$(u - 169)(u - 144) = 0$$
$$u = 169 \text{ or } u = 144$$

We now substitute x^2 for u and solve for x.

$$x^2 = 169 \text{ or } x^2 = 144$$
$$x = \pm 13 \text{ or } x = \pm 12$$

Since $y = 156/x$, if $x = 13$, $y = 12$; if $x = -13$, $y = -12$; if $x = 12$, $y = 13$; and if $x = -12$, $y = -13$. The possible solutions are (13,12), (-13,-12), (12,13), and (-12,-13).

Check. Since measurements cannot be negative, we consider only (13,12) and (12,13). Since both possible solutions give the same pair of legs, we only need to check (13,12). If $x = 13$ and $y = 12$, their product is 156. Also, $\sqrt{13^2 + 12^2} = \sqrt{313}$. The numbers check.

State. The lengths of the legs are 13 and 12.

48. 6 and 10 or -6 and -10

49. Familiarize. We let x = the length of a side of one peanut bed and y = the length of a side of the other peanut bed. Make a drawing.

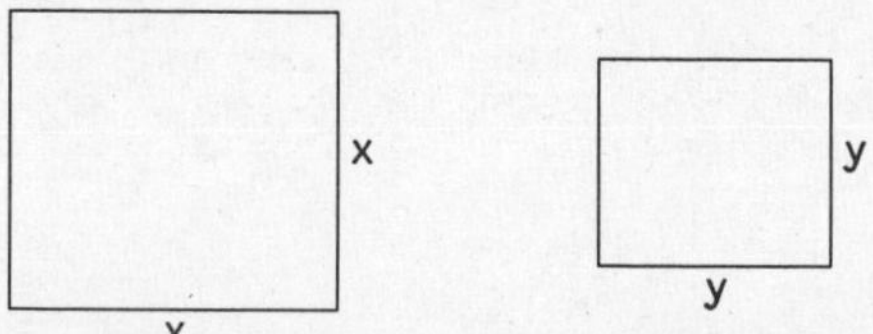

Area: x^2 Area: y^2

Translate.

The sum of the areas is 832 ft².

$$x^2 + y^2 = 832$$

The difference of the areas is 320 ft².

$$x^2 - y^2 = 320$$

Carry out. We solve the system of equations.

$$x^2 + y^2 = 832$$
$$x^2 - y^2 = 320$$
$$2x^2 = 1152 \quad \text{Adding}$$
$$x^2 = 576$$
$$x = \pm 24$$

Since measurements cannot be negative, we consider only $x = 24$. Substitute 24 for x in the first equation and solve for y.

$$24^2 + y^2 = 832$$
$$576 + y^2 = 832$$
$$y^2 = 256$$
$$y = \pm 16$$

Again, we consider only the positive value, 16. The possible solution is (24,16).

Check. The areas of the peanut beds are 24^2, or 576, and 16^2, or 256. The sum of the areas is 576 + 256, or 832. The difference of the areas is 576 - 256, or 320. The values check.

State. The lengths of the beds are 24 ft and 16 ft.

50. \$125, 6%

51. Familiarize. We first make a drawing. Let ℓ = the length and w = the width.

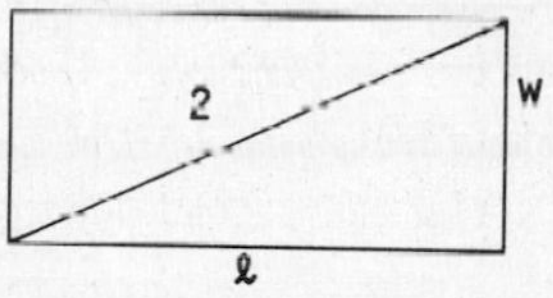

Translate.

Area: $\ell w = \sqrt{3}$ (1)

From the Pythagorean theorem: $\ell^2 + w^2 = 2^2$ (2)

Carry out. We solve the system of equations.

We first solve Eq. (1) for w.

$$\ell w = \sqrt{3}$$
$$w = \frac{\sqrt{3}}{\ell}$$

Then we substitute $\frac{\sqrt{3}}{\ell}$ for w in Eq. (2) and solve for ℓ.

$$\ell^2 + \left(\frac{\sqrt{3}}{\ell}\right)^2 = 4$$
$$\ell^2 + \frac{3}{\ell^2} = 4$$
$$\ell^4 + 3 = 4\ell^2$$
$$\ell^4 - 4\ell^2 + 3 = 0$$
$$u^2 - 4u + 3 = 0 \quad \text{Letting } u = \ell^2$$
$$(u - 3)(u - 1) = 0$$

$u = 3$ or $u = 1$

We now substitute ℓ^2 for u and solve for ℓ.

$\ell^2 = 3$ or $\ell^2 = 1$

$\ell = \pm\sqrt{3}$ or $\ell = \pm 1$

Measurements cannot be negative, so we only need to consider $\ell = \sqrt{3}$ and $\ell = 1$. Since $w = \sqrt{3}/\ell$, if $\ell = \sqrt{3}$, $w = 1$ and if $\ell = 1$, $w = \sqrt{3}$. Length is usually considered to be longer than width, so we have the solution $\ell = \sqrt{3}$ and $w = 1$, or $(\sqrt{3},1)$.

Check. If $\ell = \sqrt{3}$ and $w = 1$, the area is $\sqrt{3}\cdot 1 = \sqrt{3}$. Also $(\sqrt{3})^2 + 1^2 = 3 + 1 = 4 = 2^2$. The numbers check.

State. The length is $\sqrt{3}$ m, and the width is 1 m.

52. $\sqrt{2}$ m by 1 m

53. $\sqrt{48} = \sqrt{16\cdot 3} = \sqrt{16}\,\sqrt{3} = 4\sqrt{3}$

54. $2a^6d^2\sqrt[4]{2d}$

55. Familiarize. Let r represent the speed of the boat in still water and t represent the time of the trip upstream. Organize the information in a table.

	Speed	Time	Distance
Upstream	r - 2	t	4
Downstream	r + 2	3 - t	4

Recall that $rt = d$, or $t = d/r$.

Translate. From the first line of the table we obtain $t = \frac{4}{r - 2}$. From the second line we obtain $3 - t = \frac{4}{r + 2}$.

Carry out. Substitute $\frac{4}{r - 2}$ for t in the second equation and solve for r.

$$3 - \frac{4}{r - 2} = \frac{4}{r + 2}$$

LCD is $(r - 2)(r + 2)$

$$(r - 2)(r + 2)\left[3 - \frac{4}{r - 2}\right] = (r - 2)(r + 2)\cdot\frac{4}{r + 2}$$

$$3(r - 2)(r + 2) - 4(r + 2) = 4(r - 2)$$

$$3(r^2 - 4) - 4r - 8 = 4r - 8$$

$$3r^2 - 12 - 4r - 8 = 4r - 8$$

$$3r^2 - 8r - 12 = 0$$

$$r = \frac{-(-8) \pm \sqrt{(-8)^2 - 4(3)(-12)}}{2\cdot 3}$$

$$r = \frac{8 \pm \sqrt{208}}{6} = \frac{8 \pm 4\sqrt{13}}{6}$$

$$r = \frac{4 \pm 2\sqrt{13}}{3}$$

Since negative speed has no meaning in this problem, we consider only the positive square root.

$$r = \frac{4 + 2\sqrt{13}}{3} \approx 3.7$$

Check. Left to the student.

State. The speed of the boat in still water is approximately 3.7 mph.

56. $\dfrac{x - 2\sqrt{xh} + h}{x - h}$

57. 

58. One piece should be 61.52 cm long, and then the other will be 38.48 cm long.

59. Let (h,k) represent the point on the line $5x + 8y = -2$ which is the center of a circle that passes through the points (-2,3) and (-4,1). The distance between (h,k) and (-2,3) is the same as the distance between (h,k) and (-4,1). This gives us one equation:

$$\sqrt{[h - (-2)]^2 + (k - 3)^2} = \sqrt{[h - (-4)]^2 + (k - 1)^2}$$

$$(h + 2)^2 + (k - 3)^2 = (h + 4)^2 + (k - 1)^2$$

$$h^2 + 4h + 4 + k^2 - 6k + 9 = h^2 + 8h + 16 + k^2 - 2k + 1$$

$$4h - 6k + 13 = 8h - 2k + 17$$

$$-4h - 4k = 4$$

$$h + k = -1$$

We get a second equation by substituting (h,k) in $5x + 8y = -2$.

$$5h + 8k = -2$$

We now solve the following system:

$h + k = -1,$

$5h + 8k = -2$

The solution, which is the center of the circle, is (-2,1).

Next we find the length of the radius. We can find the distance between either (-2,3) or (-4,1) and the center (-2,1). We use (-2,3).

$$r = \sqrt{[-2 - (-2)]^2 + (1 - 3)^2}$$

$$= \sqrt{0^2 + (-2)^2}$$

$$= \sqrt{4}$$

$$= 2$$

We can write the equation of the circle with center (-2,1) and radius 2.

$$(x - h)^2 + (y - k)^2 = r^2$$

$$[x - (-2)]^2 + (y - 1)^2 = 2^2$$

$$(x + 2)^2 + (y - 1)^2 = 4$$

60. $4x^2 + 3y^2 = 43$

61. Familiarize. We let x and y represent the length and width of the base of the box, respectively. Then the dimensions of the metal sheet are x + 10 and y + 10. Make a drawing.

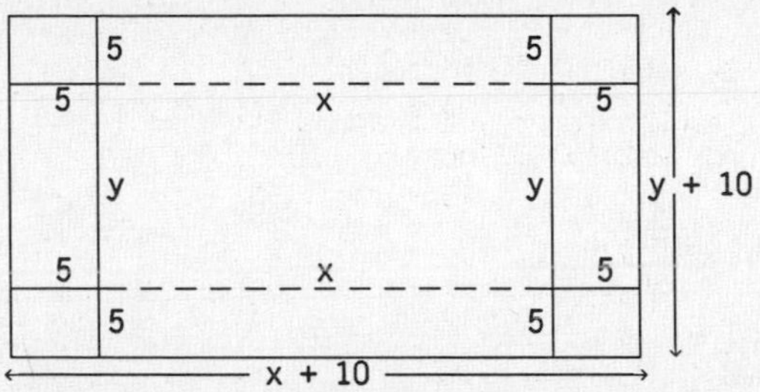

Translate. The area of the metal sheet is 340 in^2, so we have:

$$(x + 10)(y + 10) = 340$$

The volume of the box is 350 in^3, so we have another equation:

$$x \cdot y \cdot 5 = 350$$

Carry out. Solve the system

$$(x + 10)(y + 10) = 340,$$
$$x \cdot y \cdot 5 = 350.$$

The solutions are (10,7) and (7,10).

Check. Since length is usually considered to be longer than width, we check $\ell = 10$ and $w = 7$, or (10,7). The numbers check.

State. The dimensions of the box are 10 in. by 7 in. by 5 in.

62. 30 and 50

63. $p^2 + q^2 = 13,$ (1)

$\frac{1}{pq} = -\frac{1}{6}$ (2)

Solve Eq. (2) for p.

$$\frac{1}{q} = -\frac{p}{6}$$

$$-\frac{6}{q} = p$$

Substitute $-6/q$ for p in Eq. (1) and solve for q.

$$\left(-\frac{6}{q}\right)^2 + q^2 = 13$$

$$\frac{36}{q^2} + q^2 = 13$$

$$36 + q^4 = 13q^2$$

$$q^4 - 13q^2 + 36 = 0$$

$$u^2 - 13u + 36 = 0 \quad \text{Letting } u = q^2$$

$$(u - 9)(u - 4) = 0$$

$$u = 9 \quad \text{or} \quad u = 4$$

$$x^2 = 9 \quad \text{or} \quad x^2 = 4$$

$$x = \pm 3 \quad \text{or} \quad x = \pm 2$$

Since $p = -6/q$, if $q = 3$, $p = -2$; if $q = -3$, $p = 2$; if $q = 2$, $p = -3$, and if $q = -2$, $p = 3$. The pairs (-2,3), (2,-3), (-3,2), and (3,-2) check. They are the solutions.

64. $\left(\frac{1}{3},\frac{1}{2}\right)$, $\left(\frac{1}{2},\frac{1}{3}\right)$

Exercise Set 11.1

1. $a_n = 3n + 1$
   $a_1 = 3\cdot1 + 1 = 4,\quad a_4 = 3\cdot4 + 1 = 13;$
   $a_2 = 3\cdot2 + 1 = 7,\quad a_{10} = 3\cdot10 + 1 = 31;$
   $a_3 = 3\cdot3 + 1 = 10,\quad a_{15} = 3\cdot15 + 1 = 46$

2. 2, 5, 8, 11; 29; 44

3. $a_n = \frac{n}{n+1}$
   $a_1 = \frac{1}{1+1} = \frac{1}{2},\quad a_4 = \frac{4}{4+1} = \frac{4}{5};$
   $a_2 = \frac{2}{2+1} = \frac{2}{3},\quad a_{10} = \frac{10}{10+1} = \frac{10}{11};$
   $a_3 = \frac{3}{3+1} = \frac{3}{4},\quad a_{15} = \frac{15}{15+1} = \frac{15}{16}$

4. 2, 5, 10, 17; 101; 226

5. $a_n = n^2 - 2n$
   $a_1 = 1^2 - 2\cdot1 = -1,\quad a_4 = 4^2 - 2\cdot4 = 8;$
   $a_2 = 2^2 - 2\cdot2 = 0,\quad a_{10} = 10^2 - 2\cdot10 = 80;$
   $a_3 = 3^2 - 2\cdot3 = 3,\quad a_{15} = 15^2 - 2\cdot15 = 195$

6. $0, \frac{3}{5}, \frac{4}{5}, \frac{15}{17}; \frac{99}{101}; \frac{112}{113}$

7. $a_n = n + \frac{1}{n}$
   $a_1 = 1 + \frac{1}{1} = 2,\quad a_4 = 4 + \frac{1}{4} = 4\frac{1}{4};$
   $a_2 = 2 + \frac{1}{2} = 2\frac{1}{2},\quad a_{10} = 10 + \frac{1}{10} = 10\frac{1}{10};$
   $a_3 = 3 + \frac{1}{3} = 3\frac{1}{3},\quad a_{15} = 15 + \frac{1}{15} = 15\frac{1}{15}$

8. $1, -\frac{1}{2}, \frac{1}{4}, -\frac{1}{8}; -\frac{1}{512}; \frac{1}{16,384}$

9. $a_n = (-1)^n n^2$
   $a_1 = (-1)^1 1^2 = -1,\quad a_4 = (-1)^4 4^2 = 16;$
   $a_2 = (-1)^2 2^2 = 4,\quad a_{10} = (-1)^{10} 10^2 = 100;$
   $a_3 = (-1)^3 3^2 = -9,\quad a_{15} = (-1)^{15} 15^2 = -225$

10. -4, 5, -6, 7; 13; -18

11. $a_n = (-1)^{n+1}(3n - 5)$
    $a_1 = (-1)^{1+1}(3\cdot1 - 5) = -2,$
    $a_2 = (-1)^{2+1}(3\cdot2 - 5) = -1,$
    $a_3 = (-1)^{3+1}(3\cdot3 - 5) = 4,$
    $a_4 = (-1)^{4+1}(3\cdot4 - 5) = -7;$
    $a_{10} = (-1)^{10+1}(3\cdot10 - 5) = -25;$
    $a_{15} = (-1)^{15+1}(3\cdot15 - 5) = 40$

12. 0, 7, -26, 63; 999; -3374

13. $a_n = 4n - 7$
    $a_8 = 4\cdot8 - 7 = 32 - 7 = 25$

14. 56

15. $a_n = (3n + 4)(2n - 5)$
    $a_7 = (3\cdot7 + 4)(2\cdot7 - 5) = 25\cdot9 = 225$

16. 400

17. $a_n = (-1)^{n-1}(3.4n - 17.3)$
    $a_{12} = (-1)^{12-1}[3.4(12) - 17.3] = -23.5$

18. -37,916,508.16

19. $a_n = 5n^2(4n - 100)$
    $a_{11} = 5(11)^2(4\cdot11 - 100) = 5(121)(-56) = -33,880$

20. 528,528

21. $a_n = \left(1 + \frac{1}{n}\right)^2$
    $a_{20} = \left(1 + \frac{1}{20}\right)^2 = \left(\frac{21}{20}\right)^2 = \frac{441}{400}$

22. $\frac{2744}{3375}$

23. $a_n = \log 10^n$
    $a_{43} = \log 10^{43} = 43$

24. 67

25. 1, 3, 5, 7, 9, . . .
    These are odd integers, so the general term may be $2n - 1$.

26. 3^n

27. -2, 6, -18, 54, . . .
    We can see a pattern if we write the sequence as
    $-1\cdot2\cdot1,\ 1\cdot2\cdot3,\ -1\cdot2\cdot9,\ 1\cdot2\cdot27, \ldots$
    The general term may be $(-1)^n 2(3)^{n-1}$.

28. $5n - 7$

29. $\frac{2}{3}, \frac{3}{4}, \frac{4}{5}, \frac{5}{6}, \frac{6}{7}, \ldots$
    These are fractions in which the denominator is 1 greater than the numerator. Also, each numerator is 1 greater than the preceding numerator. The general term may be $\frac{n+1}{n+2}$.

30. $\sqrt{2n}$

31. $\sqrt{3}, 3, 3\sqrt{3}, 9, 9\sqrt{3}, \ldots$
    These are powers of $\sqrt{3}$. The general term may be $(\sqrt{3})^n$, or $3^{n/2}$.

32. $n(n + 1)$

33. $-1, -4, -7, -10, -13, \ldots$

Each term is 3 less than the preceding term. The general term may be $-1 - 3(n - 1)$. After removing parentheses and simplifying, we can express the general term as $-3n + 2$, or $-(3n - 2)$.

34. $\log 10^{n-1}$, or $n - 1$

35. $1, 2, 3, 4, 5, 6, 7, \ldots$

$S_7 = 1 + 2 + 3 + 4 + 5 + 6 + 7 = 28$

36. -8

37. $2, 4, 6, 8, \ldots$

$S_5 = 2 + 4 + 6 + 8 + 10 = 30$

38. $\frac{5269}{3600}$

39. $\sum_{k=1}^{5} \frac{1}{2k} = \frac{1}{2\cdot1} + \frac{1}{2\cdot2} + \frac{1}{2\cdot3} + \frac{1}{2\cdot4} + \frac{1}{2\cdot5}$

$= \frac{1}{2} + \frac{1}{4} + \frac{1}{6} + \frac{1}{8} + \frac{1}{10}$

$= \frac{60}{120} + \frac{30}{120} + \frac{20}{120} + \frac{15}{120} + \frac{12}{120}$

$= \frac{137}{120}$

40. $\frac{1}{3} + \frac{1}{5} + \frac{1}{7} + \frac{1}{9} + \frac{1}{11} + \frac{1}{13} = \frac{43,024}{45,045}$

41. $\sum_{k=0}^{5} 2^k = 2^0 + 2^1 + 2^2 + 2^3 + 2^4 + 2^5$

$= 1 + 2 + 4 + 8 + 16 + 32$

$= 63$

42. $\sqrt{7} + \sqrt{9} + \sqrt{11} + \sqrt{13} \approx 12.5679$

43. $\sum_{k=7}^{10} \log k = \log 7 + \log 8 + \log 9 + \log 10 \approx 3.7024$

44. $0 + \pi + 2\pi + 3\pi + 4\pi \approx 31.4159$

45. $\sum_{k=1}^{8} \frac{k}{k+1} = \frac{1}{1+1} + \frac{2}{2+1} + \frac{3}{3+1} + \frac{4}{4+1} +$

$\frac{5}{5+1} + \frac{6}{6+1} + \frac{7}{7+1} + \frac{8}{8+1}$

$= \frac{1}{2} + \frac{2}{3} + \frac{3}{4} + \frac{4}{5} + \frac{5}{6} + \frac{6}{7} + \frac{7}{8} + \frac{8}{9}$

$= \frac{15,551}{2520}$

46. $-\frac{1}{4} + 0 + \frac{1}{6} + \frac{2}{7} = \frac{17}{84}$

47. $\sum_{k=1}^{5} (-1)^k = (-1)^1 + (-1)^2 + (-1)^3 + (-1)^4 + (-1)^5$

$= -1 + 1 - 1 + 1 - 1$

$= -1$

48. $1 - 1 + 1 - 1 + 1 = 1$

49. $\sum_{k=1}^{8} (-1)^{k+1}3^k = (-1)^2 3^1 + (-1)^3 3^2 + (-1)^4 3^3 +$

$(-1)^5 3^4 + (-1)^6 3^5 + (-1)^7 3^6 +$

$(-1)^8 3^7 + (-1)^9 3^8$

$= 3 - 9 + 27 - 81 + 243 - 729 +$

$2187 - 6561$

$= -4920$

50. $-4^2 + 4^3 - 4^4 + 4^5 - 4^6 + 4^7 - 4^8 = -52,432$

51. $\sum_{k=0}^{5} (k^2 - 2k + 3) = (0^2 - 2\cdot0 + 3) +$

$(1^2 - 2\cdot1 + 3) + (2^2 - 2\cdot2 + 3) +$

$(3^2 - 2\cdot3 + 3) + (4^2 - 2\cdot4 + 3) +$

$(5^2 - 2\cdot5 + 3)$

$= 3 + 2 + 3 + 6 + 11 + 18$

$= 43$

52. $4 + 2 + 2 + 4 + 8 + 14 = 34$

53. $\sum_{k=1}^{10} \frac{1}{k(k+1)} = \frac{1}{1(1+1)} + \frac{1}{2(2+1)} + \frac{1}{3(3+1)} +$

$\frac{1}{4(4+1)} + \frac{1}{5(5+1)} + \frac{1}{6(6+1)} +$

$\frac{1}{7(7+1)} + \frac{1}{8(8+1)} + \frac{1}{9(9+1)} +$

$\frac{1}{10(10+1)}$

$= \frac{1}{1\cdot2} + \frac{1}{2\cdot3} + \frac{1}{3\cdot4} + \frac{1}{4\cdot5} + \frac{1}{5\cdot6} +$

$\frac{1}{6\cdot7} + \frac{1}{7\cdot8} + \frac{1}{8\cdot9} + \frac{1}{9\cdot10} + \frac{1}{10\cdot11}$

$= \frac{1}{2} + \frac{1}{6} + \frac{1}{12} + \frac{1}{20} + \frac{1}{30} + \frac{1}{42} + \frac{1}{56} +$

$\frac{1}{72} + \frac{1}{90} + \frac{1}{110}$

$= \frac{10}{11}$

54. $\frac{2}{3} + \frac{4}{5} + \frac{8}{9} + \frac{16}{17} + \frac{32}{33} + \frac{64}{65} + \frac{128}{129} + \frac{256}{257} + \frac{512}{513} +$

$\frac{1024}{1025}$

55. $\frac{1}{2} + \frac{2}{3} + \frac{3}{4} + \frac{4}{5} + \frac{5}{6} + \frac{6}{7}$

This is a sum of fractions in which the denominator is one greater than the numerator. Also, each numerator is 1 greater than the preceding numerator. Sigma notation is

$\sum_{k=1}^{6} \frac{k}{k+1}$.

56. $\sum_{k=1}^{5} 3k$

57. $-2 + 4 - 8 + 16 - 32 + 64$

This is a sum of powers of 2 with alternating signs. Sigma notation is

$$\sum_{k=1}^{6} (-1)^k 2^k, \text{ or } \sum_{k=1}^{6} (-2)^k.$$

58. $\sum_{k=1}^{5} \frac{1}{k^2}$

59. $4 - 9 + 16 - 25 + \ldots + (-1)^n n^2$

This is a sum of terms of the form $(-1)^k k^2$, beginning with $k = 2$ and continuing through $k = n$. Sigma notation is

$$\sum_{k=2}^{n} (-1)^k k^2.$$

60. $\sum_{k=3}^{n} (-1)^{k+1} k^2$

61. $5 + 10 + 15 + 20 + 25 + \ldots$

This is a sum of multiples of 5, and it is an infinite series. Sigma notation is

$$\sum_{k=1}^{\infty} 5k.$$

62. $\sum_{k=1}^{\infty} 7k$

63. $\frac{1}{1\cdot2} + \frac{1}{2\cdot3} + \frac{1}{3\cdot4} + \frac{1}{4\cdot5} + \ldots$

This is a sum of fractions in which the numerator is 1 and the denominator is a product of two consecutive integers. The larger integer in each product is the smaller integer in the succeeding product. It is an infinite series. Sigma notation is

$$\sum_{k=1}^{\infty} \frac{1}{k(k+1)}.$$

64. $\sum_{k=1}^{\infty} \frac{1}{k(k+1)^2}$

65. $\log_3 3 = 1$

1 is the power to which you raise 3 to get 3.

66. 0

67. $\log_3 3^7 = 7$

7 is the power to which you raise 3 to get 3^7.

68. 1

69.

70.

71. $a_n = \frac{1}{2^n} \log 1000^n$

$a_1 = \frac{1}{2^1} \log 1000^1 = \frac{1}{2} \log 10^3 = \frac{1}{2} \cdot 3 = \frac{3}{2}$

$a_2 = \frac{1}{2^2} \log 1000^2 = \frac{1}{4} \log (10^3)^2 = \frac{1}{4} \log 10^6 =$

$\frac{1}{4} \cdot 6 = \frac{3}{2}$

$a_3 = \frac{1}{2^3} \log 1000^3 = \frac{1}{8} \log (10^3)^3 = \frac{1}{8} \log 10^9 =$

$\frac{1}{8} \cdot 9 = \frac{9}{8}$

$a_4 = \frac{1}{2^4} \log 1000^4 = \frac{1}{16} \log (10^3)^4 =$

$\frac{1}{16} \log 10^{12} = \frac{1}{16} \cdot 12 = \frac{3}{4}$

$a_5 = \frac{1}{2^5} \log 1000^5 = \frac{1}{32} \log (10^3)^5 = \frac{1}{32} \log 10^{15} =$

$\frac{1}{32} \cdot 15 = \frac{15}{32}$

$S_5 = \frac{3}{2} + \frac{3}{2} + \frac{9}{8} + \frac{3}{4} + \frac{15}{32} = \frac{171}{32}$

72. $i, -1, -i, 1, i; i$

73. $a_n = \ln(1\cdot2\cdot3\cdot \cdot \cdot n)$

$a_1 = \ln 1 = 0$

$a_2 = \ln(1\cdot2) = \ln 2$

$a_3 = \ln(1\cdot2\cdot3) = \ln 6$

$a_4 = \ln(1\cdot2\cdot3\cdot4) = \ln 24$

$a_5 = \ln(1\cdot2\cdot3\cdot4\cdot5) = \ln 120$

$S_5 = 0 + \ln 2 + \ln 6 + \ln 24 + \ln 120$

$= \ln(2\cdot6\cdot24\cdot120) = \ln 34{,}560$

74. 0.414214, 0.317837, 0.267949, 0.236068, 0.213422, 0.196262

75. $a_n = \left(1 + \frac{1}{n}\right)^n$

$a_1 = \left(1 + \frac{1}{1}\right)^1 = 2$

$a_2 = \left(1 + \frac{1}{2}\right)^2 = (1.5)^2 = 2.25$

$a_3 = \left(1 + \frac{1}{3}\right)^3 = 2.370370$

$a_4 = \left(1 + \frac{1}{4}\right)^4 = 2.441406$

$a_5 = \left(1 + \frac{1}{5}\right)^5 = 2.488320$

$a_6 = \left(1 + \frac{1}{6}\right)^6 = 2.521626$

76. 1, 1, 1, 1, 1, 1

77. $a_1 = 0$, $a_{n+1} = a_n^2 + 4$
$a_1 = 0$
$a_2 = a_1^2 + 4 = 0^2 + 4 = 4$
$a_3 = a_2^2 + 4 = 4^2 + 4 = 20$
$a_4 = a_3^2 + 4 = 20^2 + 4 = 404$
$a_5 = a_4^2 + 4 = 404^2 + 4 = 163,220$
$a_6 = a_5^2 + 4 = 163,220^2 + 4 = 26,640,768,404$

78. 1, 2, 4, 8, 16, 32, 64, 128, 256, 512, 1024, 2048, 4096, 8192, 16,384, 32,768, 65,536

79. Find each term by multiplying the preceding term by 0.75:
\$5200, \$3900, \$2925, \$2193.75, \$1645.31, \$1233.98, \$925.49, \$694.12, \$520.59, \$390.44

80. \$6.20, \$6.60, \$7.00, \$7.40, \$7.80, \$8.20, \$8.60, \$9.00, \$9.40, \$9.80

Exercise Set 11.2

1. 2, 7, 12, 17, . . .
$a_1 = 2$
$d = 5$ (7 - 2 = 5, 12 - 7 = 5, 17 - 12 = 5)

2. $a_1 = 1.06$, $d = 0.06$

3. 7, 3, -1, -5, . . .
$a_1 = 7$
$d = -4$ (3 - 7 = -4, -1 - 3 = -4, -5 - (-1) = -4)

4. $a_1 = -9$, $d = 3$

5. $\frac{3}{2}, \frac{9}{4}, 3, \frac{15}{4}, \ldots$
$a_1 = \frac{3}{2}$
$d = \frac{3}{4}$ $\left[\frac{9}{4} - \frac{3}{2} = \frac{3}{4}, 3 - \frac{9}{4} = \frac{3}{4}\right]$

6. $a_1 = \frac{3}{5}$, $d = -\frac{1}{2}$

7. \$2.12, \$2.24, \$2.36, \$2.48, . . .
$a_1 = \$2.12$
$d = \$0.12$ (\$2.24 - \$2.12 = \$0.12, \$2.36 - \$2.24 = \$0.12, \$2.48 - \$2.36 = \$0.12)

8. $a_1 = \$214$, $d = -\$3$

9. 2, 6, 10, . . .
$a_1 = 2$, $d = 4$, and $n = 12$
$a_n = a_1 + (n - 1)d$
$a_{12} = 2 + (12 - 1)4 = 2 + 11\cdot4 = 2 + 44 = 46$

10. 0.57

11. 7, 4, 1, . . .
$a_1 = 7$, $d = -3$, and $n = 17$
$a_n = a_1 + (n - 1)d$
$a_{17} = 7 + (17 - 1)(-3) = 7 + 16(-3) = 7 - 48 = -41$

12. $-\frac{17}{3}$

13. \$1200, \$964.32, \$728.64, . . .
$a_1 = \$1200$, $d = \$964.32 - \$1200 = -\$235.68$, and $n = 13$
$a_n = a_1 + (n - 1)d$
$a_{13} = \$1200 + (13 - 1)(-\$235.68) = \$1200 + 12(-\$235.68) = \$1200 - \$2828.16 = -\$1628.16$

14. \$7941.62

15. $a_1 = 2$, $d = 4$
$a_n = a_1 + (n - 1)d$
Let $a_n = 106$, and solve for n.
$106 = 2 + (n - 1)(4)$
$106 = 2 + 4n - 4$
$108 = 4n$
$27 = n$
The 27th term is 106.

16. 33rd

17. $a_1 = 7$, $d = -3$
$a_n = a_1 + (n - 1)d$
$-296 = 7 + (n - 1)(-3)$
$-296 = 7 - 3n + 3$
$-306 = -3n$
$102 = n$
The 102nd term is -296.

18. 46th

19. $a_n = a_1 + (n - 1)d$
$a_{17} = 5 + (17 - 1)6$ Substituting 17 for n, 5 for a_1, and 6 for d
$= 5 + 16\cdot6$
$= 5 + 96$
$= 101$

20. -43

21. $a_n = a_1 + (n - 1)d$

$33 = a_1 + (8 - 1)4$ Substituting 33 for a_8, 8 for n, and 4 for d

$33 = a_1 + 28$

$5 = a_1$

(Note that this procedure is equivalent to subtracting d from a_8 seven times to get a_1: $33 - 7(4) = 33 - 28 = 5$)

22. -54

23. $a_n = a_1 + (n - 1)d$

$-76 = 5 + (n - 1)(-3)$ Substituting -76 for a_n, 5 for a_1 and -3 for d

$-76 = 5 - 3n + 3$

$-76 = 8 - 3n$

$-84 = -3n$

$28 = n$

24. 39

25. We know that $a_{17} = -40$ and $a_{28} = -73$. We would have to add d eleven times to get from a_{17} to a_{28}. That is,

$-40 + 11d = -73$

$11d = -33$

$d = -3.$

Since $a_{17} = -40$, we subtract d sixteen times to get to a_1.

$a_1 = -40 - 16(-3) = -40 + 48 = 8$

We write the first five terms of the sequence:

8, 5, 2, -1, -4

26. $\frac{1}{3}, \frac{5}{6}, \frac{4}{3}, \frac{11}{6}, \frac{7}{3}$

27. $5 + 8 + 11 + 14 + \ldots$

Note that $a_1 = 5$, $d = 3$, and $n = 20$. Before using Formula 2, we find a_{20}:

$a_{20} = 5 + (20 - 1)3$ Substituting into Formula 1

$= 5 + 19 \cdot 3 = 62$

Then

$S_{20} = \frac{20}{2}(5 + 62)$ Using Formula 2

$= 10(67) = 670.$

28. -210

29. The sum is $1 + 2 + 3 + \ldots + 299 + 300$. This is the sum of the arithmetic sequence for which $a_1 = 1$, $a_n = 300$, and $n = 300$. We use Formula 2.

$S_n = \frac{n}{2}(a_1 + a_n)$

$S_{300} = \frac{300}{2}(1 + 300) = 150(301) = 45,150$

30. 80,200

31. The sum is $2 + 4 + 6 + \ldots + 98 + 100$. This is the sum of the arithmetic sequence for which $a_1 = 2$, $a_n = 100$, and $n = 50$. We use Formula 2.

$S_n = \frac{n}{2}(a_1 + a_n)$

$S_{50} = \frac{50}{2}(2 + 100) = 25(102) = 2550$

32. 2500

33. The sum is $7 + 14 + 21 + \ldots + 91 + 98$. This is the sum of the arithmetic sequence for which $a_1 = 7$, $a_n = 98$, and $n = 14$. We use Formula 2.

$S_n = \frac{n}{2}(a_1 + a_n)$

$S_{14} = \frac{14}{2}(7 + 98) = 7(105) = 735$

34. 34,036

35. Before using Formula 2, we find a_{20}:

$a_{20} = 2 + (20 - 1)5$ Substituting into Formula 1

$= 2 + 19 \cdot 5 = 97$

Then

$S_{20} = \frac{20}{2}(2 + 97)$ Using Formula 2

$= 10(99) = 990.$

36. -1264

37. We first find how many plants will be in the last row.

Familiarize. The sequence is 35, 31, 27, ... It is an arithmetic sequence with $a_1 = 35$ and $d = -4$. Since each row must contain a positive number of plants, we must determine how many times we can add -4 to 35 and still have a positive result.

Translate. We find the largest integer x for which $35 + x(-4) > 0$. Then we evaluate the expression $35 - 4x$ for that value of x.

Carry out. We solve the inequality.

$35 - 4x > 0$

$35 > 4x$

$\frac{35}{4} > x$

$8\frac{3}{4} > x$

The integer we are looking for is 8. Thus $35 - 4x = 35 - 4(8) = 3$.

Check. If we add -4 to 35 eight times we get 3, a positive number, but if we add -4 to 35 more than eight times we get a negative number.

State. There will be 3 plants in the last row.

Next we find how many plants there are altogether.

Familiarize. We want to find the sum 35 + 31 + 27 + . . . + 3. We know $a_1 = 35$, $a_n = 3$, and, since we add -4 to 35 eight times, n = 9. (There are 8 terms after a_1, for a total of 9 terms.) We will use the formula $S_n = \frac{n}{2}(a_1 + a_n)$.

Translate. We want to find the sum of the first 9 terms of an arithmetic sequence in which $a_1 = 35$ and $a_9 = 3$.

Carry out. Substituting into Formula 2 we have,

$$S_9 = \frac{9}{2}(35 + 3)$$
$$= \frac{9}{2} \cdot 38 = 171$$

Check. We can check the calculations by doing them again. We could also do the entire addition:

35 + 31 + 27 + . . . + 3.

State. There are 171 plants altogether.

38. 62; 950

39. Familiarize. We go from 50 poles in a row, down to six poles in the top row, so there must be 45 rows. We want the sum 50 + 49 + 48 + . . . + 6. Thus we want the sum of an arithmetic sequence. We will use the formula $S_n = \frac{n}{2}(a_1 + a_n)$.

Translate. We want to find the sum of the first 45 terms of an arithmetic sequence with $a_1 = 50$ and $a_{45} = 6$.

Carry out. Substituting into Formula 2, we have

$$S_{45} = \frac{45}{2}(50 + 6)$$
$$= \frac{45}{2} \cdot 56 = 1260$$

Check. We can do the calculation again, or we can do the entire addition:
50 + 49 + 48 + . . . + 6.

State. There will be 1260 poles in the pile.

40. \$49.60

41. Familiarize. We want to find the sum of an arithmetic sequence with $a_1 = \$600$, $d = \$100$, and n = 20. We will use Formula 1 to find a_{20}, and then we will use Formula 2 to find S_{20}.

Translate. Substituting into Formula 1, we have

$$a_{20} = 600 + (20 - 1)(100).$$

Carry out. We first find a_{20}.

$$a_{20} = 600 + 19 \cdot 100 = 600 + 1900 = 2500$$

Then we use Formula 2 to find S_{20}.

$$S_{20} = \frac{20}{2}(600 + 2500) = 10(3100) = 31{,}000$$

Check. We can do the calculation again.

State. They save \$31,000 (disregarding interest).

42. \$10,230

43. Familiarize. We want to find the sum of an arithmetic sequence with $a_1 = 28$, d = 4, and n = 50. We will use Formula 1 to find a_{50}, and then we will use Formula 2 to find S_{50}.

Translate. Substituting into Formula 1, we have

$$a_{50} = 28 + (50 - 1)(4).$$

Carry out. We find find a_{50}.

$$a_{50} = 28 + 49 \cdot 4 = 28 + 196 = 224$$

Then we use Formula 2 to find S_{50}.

$$S_{50} = \frac{50}{2}(28 + 224) = 25 \cdot 252 = 6300$$

Check. We can do the calculation again.

State. There are 6300 seats.

44. \$462,500

45. $\log_a P = k$, $a^k = P$

The logarithm is the exponent.

The base does not change.

46. $e^a = t$

47. Standard form for the equation of a circle with center (h,k) and radius r is

$$(x - h)^2 + (y - k)^2 = r^2.$$

We substitute 0 for h, 0 for k, and 9 for r:

$$(x - 0)^2 + (y - 0)^2 = 9^2$$
$$x^2 + y^2 = 81$$

48. $(x + 2)^2 + (y - 5)^2 = 18$

49.

50.

51. Familiarize. Let x represent the first number in the sequence, and let d represent the common difference. Then the three numbers in the sequence are x, x + d, and x + 2d.

Translate.

The sum of the first and third numbers is 10.

$$x + x + 2d = 10$$

The product of the first and second numbers is 15.

$$x(x + d) = 15$$

Carry out. Solving the system of equations we get x = 3 and d = 2. Thus the numbers are 3, 5, and 7.

Check. The numbers are in an arithmetic sequence. Also 3 + 7 = 10 and 3·5 = 15. The numbers check.

State. The numbers are 3, 5, and 7.

52. $S_n = n^2$

53. $a_1 = \$8760$
$a_2 = \$8760 + (-\$798.23) = \$7961.67$
$a_3 = \$8760 + 2(-\$798.23) = \$7163.54$
$a_4 = \$8760 + 3(-\$798.23) = \$6365.31$
$a_5 = \$8760 + 4(-\$798.23) = \$5567.08$
$a_6 = \$8760 + 5(-\$798.23) = \$4768.85$
$a_7 = \$8760 + 6(-\$798.23) = \$3970.62$
$a_8 = \$8760 + 7(-\$798.23) = \$3172.39$
$a_9 = \$8760 + 8(-\$798.23) = \$2374.16$
$a_{10} = \$8760 + 9(-\$798.23) = \$1575.93$

54. \$51,679.65

55. See the answer section in the text.

56. a) $a_t = \$5200 - \$512.50t$
b) \$5200, \$4687.50, \$4175, \$3662.50, \$3150, \$1612.50, \$1100

Exercise Set 11.3

1. 2, 4, 8, 16, . . .
$\frac{4}{2} = 2$, $\frac{8}{4} = 2$, $\frac{16}{8} = 2$
$r = 2$

2. $-\frac{1}{3}$

3. 1, -1, 1, -1, . . .
$\frac{-1}{1} = -1$, $\frac{1}{-1} = -1$, $\frac{-1}{1} = -1$
$r = -1$

4. 0.1

5. $\frac{1}{2}, -\frac{1}{4}, \frac{1}{8}, -\frac{1}{16}, \ldots$

$$\frac{-\frac{1}{4}}{\frac{1}{2}} = -\frac{1}{4} \cdot \frac{2}{1} = -\frac{2}{4} = -\frac{1}{2}$$

$$\frac{\frac{1}{8}}{-\frac{1}{4}} = \frac{1}{8} \cdot \left(-\frac{4}{1}\right) = -\frac{4}{8} = -\frac{1}{2}$$

$r = -\frac{1}{2}$

6. -2

7. $75, 15, 3, \frac{3}{5}, \ldots$

$$\frac{15}{75} = \frac{1}{5},\quad \frac{3}{15} = \frac{1}{5},\quad \frac{\frac{3}{5}}{3} = \frac{3}{5} \cdot \frac{1}{3} = \frac{1}{5}$$

$r = \frac{1}{5}$

8. 0.1

9. $\frac{1}{x}, \frac{1}{x^2}, \frac{1}{x^3}, \ldots$

$$\frac{\frac{1}{x^2}}{\frac{1}{x}} = \frac{1}{x^2} \cdot \frac{x}{1} = \frac{x}{x^2} = \frac{1}{x}$$

$$\frac{\frac{1}{x^3}}{\frac{1}{x^2}} = \frac{1}{x^3} \cdot \frac{x^2}{1} = \frac{x^2}{x^3} = \frac{1}{x}$$

$r = \frac{1}{x}$

10. $\frac{m}{2}$

11. \$780, \$858, \$943.80, \$1038.18, . . .
$\frac{\$858}{\$780} = 1.1$, $\frac{\$943.80}{\$858} = 1.1$,
$\frac{\$1038.18}{\$943.80} = 1.1$
$r = 1.1$

12. 0.95

13. 2, 4, 8, 16, . . .
$a_1 = 2$, $n = 6$, and $r = \frac{4}{2}$, or 2.
We use the formula $a_n = a_1r^{n-1}$.
$a_6 = 2(2)^{6-1} = 2 \cdot 2^5 = 2 \cdot 32 = 64$

14. 781,250

15. $2, 2\sqrt{3}, 6, \ldots$
$a_1 = 2$, $n = 9$, and $r = \frac{2\sqrt{3}}{2}$, or $\sqrt{3}$
$a_n = a_1r^{n-1}$
$a_9 = 2(\sqrt{3})^{9-1} = 2(\sqrt{3})^8 = 2 \cdot 81 = 162$

16. 1

17. $\frac{8}{243}, \frac{8}{81}, \frac{8}{27}, \ldots$

$$a_1 = \frac{8}{243},\ n = 10, \text{ and } r = \frac{\frac{8}{81}}{\frac{8}{243}} = \frac{8}{81} \cdot \frac{243}{8} = 3$$

$a_n = a_1r^{n-1}$

$$a_{10} = \frac{8}{243}(3)^{10-1} = \frac{8}{243}(3)^9 = \frac{8}{243} \cdot 19{,}683 = 648$$

18. 2,734,375

19. \$1000, \$1080, \$1166.40, . . .

$a_1 = \$1000$, $n = 12$, and $r = \frac{\$1080}{\$1000} = 1.08$

$a_n = a_1r^{n-1}$

$a_{12} = \$1000(1.08)^{12-1} \approx \$1000(2.331638997) \approx$ \$2331.64

20. \$1967.15

21. 1, 3, 9, . . .

$a_1 = 1$ and $r = \frac{3}{1}$, or 3

$a_n = a_1r^{n-1}$

$a_n = 1(3)^{n-1} = 3^{n-1}$

22. 5^{3-n}

23. 1, -1, 1, -1, . . .

$a_1 = 1$ and $r = \frac{-1}{1} = -1$

$a_n = a_1r^{n-1}$

$a_n = 1(-1)^{n-1} = (-1)^{n-1}$

24. 2^n

25. $\frac{1}{x}, \frac{1}{x^2}, \frac{1}{x^3}, \ldots$

$a_1 = \frac{1}{x}$ and $r = \frac{1}{x}$ (see Exercise 9)

$a_n = a_1r^{n-1}$

$a_n = \frac{1}{x}\left(\frac{1}{x}\right)^{n-1} = \frac{1}{x} \cdot \frac{1}{x^{n-1}} = \frac{1}{x^{1+n-1}} = \frac{1}{x^n}$

26. $5\left(\frac{m}{2}\right)^{n-1}$

27. 6 + 12 + 24 + . . .

$a_1 = 6$, $n = 7$, and $r = \frac{12}{6}$, or 2

$S_n = \frac{a_1(1 - r^n)}{1 - r}$

$S_7 = \frac{6(1 - 2^7)}{1 - 2} = \frac{6(1 - 128)}{-1} = \frac{6(-127)}{-1} = 762$

28. $\frac{21}{2}$, or 10.5

29. $\frac{1}{18} - \frac{1}{6} + \frac{1}{2} - \ldots$

$a_1 = \frac{1}{18}$, $n = 7$, and $r = \frac{-\frac{1}{6}}{\frac{1}{18}} = -\frac{1}{6} \cdot \frac{18}{1} = -3$

$S_n = \frac{a_1(1 - r^n)}{1 - r}$

$S_7 = \frac{\frac{1}{18}\left[1 - (-3)^7\right]}{1 - (-3)} = \frac{\frac{1}{18}(2 + 2187)}{4} = \frac{\frac{1}{18}(2188)}{4} =$

$\frac{1}{18}(2188)\left(\frac{1}{4}\right) = \frac{547}{18}$

30. 6.6666

31. $1 + x + x^2 + x^3 + \ldots$

$a_1 = 1$, $n = 8$, and $r = \frac{x}{1}$, or x

$S_n = \frac{a_1(1 - r^n)}{1 - r}$

$S_8 = \frac{1(x - x^8)}{1 - x} = \frac{(1 + x^4)(1 - x^4)}{1 - x} =$

$\frac{(1 + x^4)(1 + x^2)(1 - x^2)}{1 - x} =$

$\frac{(1 + x^4)(1 + x^2)(1 + x)(1 - x)}{1 - x} =$

$(1 + x^4)(1 + x^2)(1 + x)$

32. $\frac{1 - x^{20}}{1 - x^2}$

33. \$200, \$200(1.06), \$200(1.06)², . . .

$a_1 = \$200$, $n = 16$, and $r = \frac{\$200(1.06)}{\$200} = 1.06$

$S_n = \frac{a_1(1 - r^n)}{1 - r}$

$S_{16} = \frac{\$200[1 - (1.06)^{16}]}{1 - 1.06} \approx \frac{\$200(1 - 2.540351685)}{-0.06} \approx$

\$5134.51

34. \$60,893.30

35. 4 + 2 + 1 + . . .

$|r| = \left|\frac{2}{4}\right| = \left|\frac{1}{2}\right| = \frac{1}{2}$, and since $|r| < 1$, the series does have a sum.

$S_\infty = \frac{a_1}{1 - r} = \frac{4}{1 - \frac{1}{2}} = \frac{4}{\frac{1}{2}} = 4 \cdot \frac{2}{1} = 8$

36. $\frac{49}{4}$

37. 25 + 20 + 16 + . . .

$|r| = \left|\frac{20}{25}\right| = \left|\frac{4}{5}\right| = \frac{4}{5}$, and since $|r| < 1$, the series does have a sum.

$S_\infty = \frac{a_1}{1 - r} = \frac{25}{1 - \frac{4}{5}} = \frac{25}{\frac{1}{5}} = 25 \cdot \frac{5}{1} = 125$

38. 48

39. $100 - 10 + 1 - \frac{1}{10} + \ldots$

$|r| = \left|\frac{-10}{100}\right| = \left|-\frac{1}{10}\right| = \frac{1}{10}$, and since $|r| < 1$, the series does have a sum.

$S_\infty = \frac{a_1}{1 - r} = \frac{100}{1 - \left(-\frac{1}{10}\right)} = \frac{100}{\frac{11}{10}} = 100 \cdot \frac{10}{11} = \frac{1000}{11}$

40. $|r| = \left|\frac{18}{-6}\right| = |-3| = 3 \nless 1$, so the series does not have a sum.

41. $8 + 40 + 200 + \ldots$

$|r| = \left|\frac{40}{8}\right| = |5| = 5$, and since $|r| \nless 1$ the series does not have a sum.

42. No

43. $0.3 + 0.03 + 0.003 + \ldots$

$|r| = \left|\frac{0.03}{0.3}\right| = |0.1| = 0.1$, and since $|r| < 1$, the series does have a sum.

$$S_\infty = \frac{a_1}{1 - r} = \frac{0.3}{1 - 0.1} = \frac{0.3}{0.9} = \frac{3}{9} = \frac{1}{3}$$

44. $\frac{37}{99}$

45. $\$500(1.02)^{-1} + \$500(1.02)^{-2} + \$500(1.02)^{-3} + \ldots$

$|r| = \left|\frac{\$500(1.02)^{-2}}{\$500(1.02)^{-1}}\right| = |(1.02)^{-1}| = (1.02)^{-1}$, or $\frac{1}{1.02}$, and since $|r| < 1$, the series does have a sum.

$$S_\infty = \frac{a_1}{1 - r} = \frac{\$500(1.02)^{-1}}{1 - \left(\frac{1}{1.02}\right)} = \frac{\frac{\$500}{1.02}}{\frac{0.02}{1.02}} =$$

$$\frac{\$500}{1.02} \cdot \frac{1.02}{0.02} = \$25{,}000$$

46. $12,500

47. $0.4444\ldots = 0.4 + 0.04 + 0.004 + 0.0004 + \ldots$

This is an infinite geometric series with $a_1 = 0.4$.

$|r| = \left|\frac{0.04}{0.4}\right| = |0.1| = 0.1 < 1$, so the series has a sum.

$$S_\infty = \frac{a_1}{1 - r} = \frac{0.4}{1 - 0.1} = \frac{0.4}{0.9} = \frac{4}{9}$$

Fractional notation for $0.4444\ldots$ is $\frac{4}{9}$.

48. 10

49. $0.55555 = 0.5 + 0.05 + 0.005 + 0.0005 + \ldots$

This is an infinite geometric series with $a_1 = 0.5$.

$|r| = \left|\frac{0.05}{0.5}\right| = |0.1| = 0.1 < 1$, so the series has a sum.

$$S_\infty = \frac{a_1}{1 - r} = \frac{0.5}{1 - 0.1} = \frac{0.5}{0.9} = \frac{5}{9}$$

50. $\frac{2}{3}$

51. $0.15151515\ldots = 0.15 + 0.0015 + 0.000015 + \ldots$

This is an infinite geometric series with $a_1 = 0.15$.

$|r| = \left|\frac{0.0015}{0.15}\right| = |0.01| = 0.01 < 1$, so the series has a sum.

$$S_\infty = \frac{a_1}{1 - r} = \frac{0.15}{1 - 0.01} = \frac{0.15}{0.99} = \frac{15}{99} = \frac{5}{33}$$

52. $\frac{4}{33}$

53. Familiarize. The rebound distances form a geometric sequence:

$$\frac{1}{4} \times 16,\quad \left(\frac{1}{4}\right)^2 \times 16,\quad \left(\frac{1}{4}\right)^3 \times 16,\ \ldots,$$

$$\text{or } 4,\quad \frac{1}{4} \times 4,\quad \left(\frac{1}{4}\right)^2 \times 4,\ \ldots$$

The height of the 6th rebound is the 6th term of the sequence.

Translate. We will use the formula $a_n = a_1r^{n-1}$, with $a_1 = 4$, $r = \frac{1}{4}$, and $n = 6$:

$$a_6 = 4\left(\frac{1}{4}\right)^{6-1}$$

Carry out. We calculate to obtain $a_6 = \frac{1}{256}$.

Check. We can do the calculation again.

State. It rebounds $\frac{1}{256}$ ft the 6th time.

54. $5\frac{1}{3}$ ft

55. Familiarize. In one year, the population will be $100{,}000 + 0.03(100{,}000)$, or $(1.03)100{,}000$. In two years, the population will be $(1.03)100{,}000 + 0.03(1.03)100{,}000$, or $(1.03)^2100{,}000$. Thus the populations form a geometric sequence:

$$100{,}000,\quad (1.03)100{,}000,\quad (1.03)^2100{,}000,\ \ldots$$

The population in 15 years will be the 16th term of the sequence.

Translate. We will use the formula $a_n = a_1r^{n-1}$ with $a_1 = 100{,}000$, $r = 1.03$, and $n = 16$:

$$a_{16} = 100{,}000(1.03)^{16-1}$$

Carry out. We calculate to obtain $a_{16} \approx 155{,}797$.

Check. We can do the calculation again.

State. In 15 years the population will be about 155,797.

56. About 24 years

57. Familiarize. The amounts owed at the beginning of successive years form a geometric sequence:

$\$1200$, $(1.12)\$1200$, $(1.12)^2\$1200$, $(1.12)^3\$1200, \ldots$

The amount to be repaid at the end of 13 years is the amount owed at the beginning of the 14th year.

Translate. We use the formula $a_n = a_1r^{n-1}$ with $a_1 = 1200$, $r = 1.12$, and $n = 14$:

$$a_{14} = 1200(1.12)^{14-1}$$

Carry out. We calculate to obtain $a_{14} \approx 5236.19$.

Check. We can do the calculation again.

State. At the end of 13 years, $5236.19 will be repaid.

58. 10,485.76 in.

59. Familiarize. The lengths of the falls form a geometric sequence:

$556, \left(\frac{3}{4}\right)556, \left(\frac{3}{4}\right)^2 556, \left(\frac{3}{4}\right)^3 556, \ldots$

The total length of the first 6 falls is the sum of the first six terms of this sequence. The heights of the rebounds also form a geometric sequence:

$\left(\frac{3}{4}\right)556, \left(\frac{3}{4}\right)^2 556, \left(\frac{3}{4}\right)^3 556, \ldots,$ or

$417, \left(\frac{3}{4}\right)417, \left(\frac{3}{4}\right)^2 417, \ldots$

When the ball hits the ground for the 6th time, it will have rebounded 5 times. Thus the total length of the rebounds is the sum of the first five terms of this sequence.

Translate. We use the formula $S_n = \frac{a_1(1 - r^n)}{1 - r}$ twice, once with $a_1 = 556$, $r = \frac{3}{4}$, and $n = 6$ and a second time with $a_1 = 417$, $r = \frac{3}{4}$, and $n = 5$.

D = Length of falls + length of rebounds

$$= \frac{556\left[1 - \left(\frac{3}{4}\right)^6\right]}{1 - \frac{3}{4}} + \frac{417\left[1 - \left(\frac{3}{4}\right)^5\right]}{1 - \frac{3}{4}}.$$

Carry out. We use a calculator to obtain $D \approx 3100.35$.

Check. We can do the calculations again.

State. The ball will have traveled about 3100.35 ft.

60. 3892 ft

61. Familiarize. The amounts form a geometric series:

$\$0.01, \$0.01(2), \$0.01(2)^2, \$0.01(2^3) + \ldots + (\$0.01)(2)^{27}$

Translate. We use the formula $S_n = \frac{a_1(1 - r^n)}{1 - r}$ to find the sum of the geometric series with $a_1 = 0.01$, $r = 2$, and $n = 28$:

$$S_{28} = \frac{0.01(1 - 2^{28})}{1 - 2}$$

Carry out. We use a calculator to obtain $S_{28} = \$2,684,354.55$.

Check. We can do the calculation again.

State. You would earn \$2,684,354.55.

62. \$645,826.93

63. $5x - 2y = -3,$ (1)

$2x + 5y = -24$ (2)

Multiply Eq. (1) by 5 and Eq. (2) by 2 and add.

$$\begin{aligned} 25x - 10y &= -15 \\ 4x + 10y &= -48 \\ \hline 29x \quad &= -63 \\ x &= -\frac{63}{29} \end{aligned}$$

Substitute $-\frac{63}{29}$ for x in the second equation and solve for y.

$$\begin{aligned} 2\left(-\frac{63}{29}\right) + 5y &= -24 \\ -\frac{126}{29} + 5y &= -24 \\ 5y &= -\frac{570}{29} \\ y &= -\frac{114}{29} \end{aligned}$$

The solution is $\left(-\frac{63}{29}, -\frac{114}{29}\right)$.

64. (-1,2,3)

65.

66.

67. $1 + x + x^2 + \ldots$

This is a geometric series with $a_1 = 1$ and $r = x$.

$$S_n = \frac{a_1(1 - r^n)}{1 - r} = \frac{1(1 - x^n)}{1 - x} = \frac{1 - x^n}{1 - x}$$

68. $\frac{x^2[1 - (-x)^n]}{1 + x}$

69. Familiarize. The length of a side of the first square is 16 cm. The length of a side of the next square is the length of the hypotenuse of a right triangle with legs 8 cm and 8 cm, or $8\sqrt{2}$ cm. The length of a side of the next square is the length of the hypotenuse of a right triangle with legs $4\sqrt{2}$ cm and $4\sqrt{2}$ cm, or 8 cm. The areas of the squares form a sequence:

$(16)^2, (8\sqrt{2})^2, (8)^2, \ldots,$ or

$256, 128, 64, \ldots.$

This is a geometric sequence with $a_1 = 256$ and $r = \frac{1}{2}$.

Translate. We find the sum of the infinite geometric series $256 + 128 + 64 + \ldots.$

$$S_\infty = \frac{a_1}{1 - r}$$

$$S_\infty = \frac{256}{1 - \frac{1}{2}}$$

Carry out. We calculate to obtain $S_\infty = 512$.

Check. We can do the calculation again.

State. The sum of the areas is 512 cm^2.

Exercise Set 11.4

1. $9! = 9 \cdot 8 \cdot 7 \cdot 6 \cdot 5 \cdot 4 \cdot 3 \cdot 2 \cdot 1 = 362,880$

2. 3,628,800

3. $11! = 11 \cdot 10 \cdot 9 \cdot 8 \cdot 7 \cdot 6 \cdot 5 \cdot 4 \cdot 3 \cdot 2 \cdot 1 = 39,916,800$

4. 479,001,600

5. $\frac{7!}{4!} = \frac{7\cdot6\cdot5\cdot4!}{4!}$

$= 7\cdot6\cdot5$ Removing a factor of 1: $\frac{4!}{4!}$

$= 210$

6. 56

7. $\frac{9!}{5!} = \frac{9\cdot8\cdot7\cdot6\cdot5!}{5!} = 9\cdot8\cdot7\cdot6 = 3024$

8. 720

9. $\binom{8}{2} = \frac{8!}{(8-2)!2!} = \frac{8!}{6!2!} = \frac{8\cdot7\cdot6!}{6!\cdot2\cdot1} = \frac{8\cdot7}{2} = 4\cdot7 = 28$

10. 35

11. $\binom{9}{6} = \frac{9!}{(9-6)!6!} = \frac{9!}{3!6!} = \frac{9\cdot8\cdot7\cdot6!}{3\cdot2\cdot1\cdot6!} = \frac{9\cdot8\cdot7}{3\cdot2} =$
$3\cdot4\cdot7 = 84$

12. 120

13. $\binom{20}{18} = \frac{20!}{(20-18)!18!} = \frac{20!}{2!18!} = \frac{20\cdot19\cdot18!}{2\cdot1\cdot18!} =$
$\frac{20\cdot19}{2} = 10\cdot19 = 190$

14. 4060

15. $\binom{35}{2} = \frac{35!}{(35-2)!2!} = \frac{35!}{33!2!} = \frac{35\cdot34\cdot33!}{33!\cdot2\cdot1} =$
$\frac{35\cdot34}{2} = 35\cdot17 = 595$

16. 780

17. Expand $(m + n)^5$.

Form 1: The expansion of $(m + n)^5$ has 5 + 1, or 6, terms. The sum of the exponents in each term is 5. The exponents of m start with 5 and decrease to 0. The last term has no factor of m. The first term has no factor of n. The exponents of n start in the second term with 1 and increase to 5. We get the coefficients from the 6th row of Pascal's Triangle.

1
1 1
1 2 1
1 3 3 1
1 4 6 4 1
1 5 10 10 5 1

$(m + n)^5 = 1\cdot m^5 + 5\cdot m^4n^1 + 10\cdot m^3\cdot n^2 + 10\cdot m^2\cdot n^3 + 5\cdot m\cdot n^4 + 1\cdot n^5$
$= m^5 + 5m^4n + 10m^3n^2 + 10m^2n^3 + 5mn^4 + n^5$

Form 2: We have a = m, b = n, and n = 5.

$(m + n)^5 = \binom{5}{0}m^5 + \binom{5}{1}m^4n + \binom{5}{2}m^3n^2 + \binom{5}{3}m^2n^3 + \binom{5}{4}mn^4 + \binom{5}{5}n^5$
$= \frac{5!}{5!0!}m^5 + \frac{5!}{4!1!}m^4n + \frac{5!}{3!2!}m^3n^2 + \frac{5!}{2!3!}m^2n^3 + \frac{5!}{1!4!}mn^4 + \frac{5!}{0!5!}m^5$
$= m^5 + 5m^4n + 10m^3n^2 + 10m^2n^3 + 5mn^4 + n^5$

18. $a^4 - 4a^3b + 6a^2b^2 - 4ab^3 + b^4$

19. Expand $(x - y)^6$.

Form 1: The expansion of $(x - y)^6$ has 6 + 1, or 7 terms. The sum of the exponents in each term is 6. The exponents of x start with 6 and decrease to 0. The last term has no factor of x. The first term has no factor of -y. The exponents of -y start in the second term with 1 and increase to 6. We get the coefficients from the 7th row of Pascal's Triangle.

1
1 1
1 2 1
1 3 3 1
1 4 6 4 1
1 5 10 10 5 1
1 6 15 20 15 6 1

$(x - y)^6 = 1\cdot x^6 + 6\cdot x^5\cdot(-y) + 15\cdot x^4\cdot(-y)^2 + 20\cdot x^3\cdot(-y)^3 + 15\cdot x^2\cdot(-y)^4 + 6\cdot x\cdot(-y)^5 + 1\cdot(-y)^6$
$= x^6 - 6x^5y + 15x^4y^2 - 20x^3y^3 + 15x^2y^4 - 6xy^5 + y^6$

Form 2: We have a = x, b = -y, and n = 6.

$(x - y)^6 = \binom{6}{0}x^6 + \binom{6}{1}x^5(-y) + \binom{6}{2}x^4(-y)^2 + \binom{6}{3}x^3(-y)^3 + \binom{6}{4}x^2(-y)^4 + \binom{6}{5}x(-y)^5 + \binom{6}{6}(-y)^6$
$= \frac{6!}{6!0!}x^6 + \frac{6!}{5!1!}x^5(-y) + \frac{6!}{4!2!}x^4y^2 + \frac{6!}{3!3!}x^3(-y^3) + \frac{6!}{2!4!}x^2y^4 + \frac{6!}{1!5!}x(-y^5) + \frac{6!}{0!6!}y^6$
$= x^6 - 6x^5y + 15x^4y^2 - 20x^3y^3 + 15x^2y^4 - 6xy^5 + y^6$

20. $p^7 + 7p^6q + 21p^5q^2 + 35p^4q^3 + 35p^3q^4 + 21p^2q^5 + 7pq^6 + q^7$

21. Expand $(x^2 - 3y)^5$.

We have $a = x^2$, $b = -3y$, and $n = 5$.

Form 1: We get the coefficients from the 6th row of Pascal's Triangle.

$$\begin{array}{ccccccccccc} & & & & & 1 & & & & & \\ & & & & 1 & & 1 & & & & \\ & & & 1 & & 2 & & 1 & & & \\ & & 1 & & 3 & & 3 & & 1 & & \\ & 1 & & 4 & & 6 & & 4 & & 1 & \\ 1 & & 5 & & 10 & & 10 & & 5 & & 1 \end{array}$$

$$(x^2 - 3y)^5 = 1\cdot(x^2)^5 + 5\cdot(x^2)^4\cdot(-3y) + 10\cdot(x^2)^3\cdot(-3y)^2 + 10\cdot(x^2)^2\cdot(-3y)^3 + 5\cdot(x^2)\cdot(-3y)^4 + 1\cdot(-3y)^5$$

$$= x^{10} - 15x^8y + 90x^6y^2 - 270x^4y^3 + 405x^2y^4 - 243y^5$$

Form 2:

$$(x^2 - 3y)^5 = \binom{5}{0}(x^2)^5 + \binom{5}{1}(x^2)^4(-3y) + \binom{5}{2}(x^2)^3(-3y)^2 + \binom{5}{3}(x^2)^2(-3y)^3 + \binom{5}{4}x^2(-3y)^4 + \binom{5}{5}(-3y)^5$$

$$= \frac{5!}{5!0!}x^{10} + \frac{5!}{4!1!}x^8(-3y) + \frac{5!}{3!2!}x^6(9y^2) + \frac{5!}{2!3!}x^4(-27y^3) + \frac{5!}{1!4!}x^2(81y^4) + \frac{5!}{0!5!}(-243y^5)$$

$$= x^{10} - 15x^8y + 90x^6y^2 - 270x^4y^3 + 405x^2y^4 - 243y^5$$

22. $2187c^7 - 5103c^6d + 5103c^5d^2 - 2835c^4d^3 + 945c^3d^4 - 189c^2d^5 + 21cd^6 - d^7$

23. Expand $(3c - d)^6$.

We have $a = 3c$, $b = -d$, and $n = 6$.

Form 1: We get the coefficients from the 7th row of Pascal's Triangle.

$$\begin{array}{ccccccccccccc} & & & & & & 1 & & & & & & \\ & & & & & 1 & & 1 & & & & & \\ & & & & 1 & & 2 & & 1 & & & & \\ & & & 1 & & 3 & & 3 & & 1 & & & \\ & & 1 & & 4 & & 6 & & 4 & & 1 & & \\ & 1 & & 5 & & 10 & & 10 & & 5 & & 1 & \\ 1 & & 6 & & 15 & & 20 & & 15 & & 6 & & 1 \end{array}$$

$$(3c - d)^6 = 1\cdot(3c)^6 + 6\cdot(3c)^5\cdot(-d) + 15\cdot(3c)^4\cdot(-d)^2 + 20\cdot(3c)^3\cdot(-d)^3 + 15\cdot(3c)^2\cdot(-d)^4 + 6\cdot(3c)\cdot(-d)^5 + 1\cdot(-d)^6$$

$$= 3^6c^6 - 6\cdot3^5c^5d + 15\cdot3^4c^4d^2 - 20\cdot3^3c^3d^3 + 15\cdot3^2c^2d^4 - 6\cdot3cd^5 + d^6$$

$$= 729c^6 - 6\cdot243c^5d + 15\cdot81c^4d^2 - 20\cdot27c^3d^3 + 15\cdot9c^2d^4 - 6\cdot3cd^5 + d^6$$

$$= 729c^6 - 1458c^5d + 1215c^4d^2 - 540c^3d^3 + 135c^2d^4 - 18cd^5 + d^6$$

Form 2:

$$(3c - d)^6 = \binom{6}{0}(3c)^6 + \binom{6}{1}(3c)^5(-d) + \binom{6}{2}(3c)^4(-d)^2 + \binom{6}{3}(3c)^3(-d)^3 + \binom{6}{4}(3c)^2(-d)^4 + \binom{6}{5}(3c)(-d)^5 + \binom{6}{6}(-d)^6$$

$$= \frac{6!}{6!0!}(729c^6) + \frac{6!}{5!1!}(243c^5)(-d) + \frac{6!}{4!2!}(81c^4)(d^2) + \frac{6!}{3!3!}(27c^3)(-d^3) + \frac{6!}{2!4!}(9c^2)(d^4) + \frac{6!}{1!5!}(3c)(-d^5) + \frac{6!}{0!6!}d^6$$

$$= 729c^6 - 1458c^5d + 1215c^4d^2 - 540c^3d^3 + 135c^2d^4 - 18cd^5 + d^6$$

24. $t^{-12} + 12t^{-10} + 60t^{-8} + 160t^{-6} + 240t^{-4} + 192t^{-2} + 64$

25. Expand $(x - y)^3$.

We have $a = x$, $b = -y$, and $n = 3$.

Form 1: We get the coefficients from the 4th row of Pascal's Triangle.

$$\begin{array}{ccccccc} & & & 1 & & & \\ & & 1 & & 1 & & \\ & 1 & & 2 & & 1 & \\ 1 & & 3 & & 3 & & 1 \end{array}$$

$$(x - y)^3 = 1\cdot x^3 + 3\cdot x^2\cdot(-y) + 3\cdot x\cdot(-y)^2 + 1\cdot(-y)^3$$
$$= x^3 - 3x^2y + 3xy^2 - y^3$$

Form 2:

$$(x - y)^3 = \binom{3}{0}x^3 + \binom{3}{1}x^2(-y) + \binom{3}{2}x(-y)^2 + \binom{3}{3}(-y)^3$$

$$= \frac{3!}{3!0!}x^3 + \frac{3!}{2!1!}x^2(-y) + \frac{3!}{1!2!}xy^2 + \frac{3!}{0!3!}(-y^3)$$

$$= x^3 - 3x^2y + 3xy^2 - y^3$$

26. $x^5 - 5x^4y + 10x^3y^2 - 10x^2y^3 + 5xy^4 - y^5$

27. Expand $\left(\frac{1}{x} + y\right)^7$.

We have $a = \frac{1}{x}$, $b = y$, and $n = 7$.

Form 1: We get the coefficients from the 8th row of Pascal's Triangle.

1
1 1
1 2 1
1 3 3 1
1 4 6 4 1
1 5 10 10 5 1
1 6 15 20 15 6 1
1 7 21 35 35 21 7 1

$$\left(\frac{1}{x} + y\right)^7 = 1\cdot\left(\frac{1}{x}\right)^7 + 7\left(\frac{1}{x}\right)^6\cdot y + 21\left(\frac{1}{x}\right)^5\cdot y^2 + 35\left(\frac{1}{x}\right)^4\cdot y^3 + 35\left(\frac{1}{x}\right)^3\cdot y^4 + 21\left(\frac{1}{x}\right)^2\cdot y^5 + 7\left(\frac{1}{x}\right)\cdot y^6 + 1\cdot y^7$$

$$= x^{-7} + 7x^{-6}y + 21x^{-5}y^2 + 35x^{-4}y^3 + 35x^{-3}y^4 + 21x^{-2}y^5 + 7x^{-1}y^6 + y^7$$

Form 2:

$$\left(\frac{1}{x} + y\right)^7 = \binom{7}{0}\left(\frac{1}{x}\right)^7 + \binom{7}{1}\left(\frac{1}{x}\right)^6 y + \binom{7}{2}\left(\frac{1}{x}\right)^5 y^2 + \binom{7}{3}\left(\frac{1}{x}\right)^4 y^3 + \binom{7}{4}\left(\frac{1}{x}\right)^3 y^4 + \binom{7}{5}\left(\frac{1}{x}\right)^2 y^5 + \binom{7}{6}\left(\frac{1}{x}\right) y^6 + \binom{7}{7} y^7$$

$$= \frac{7!}{7!0!}\left(\frac{1}{x}\right)^7 + \frac{7!}{6!1!}\left(\frac{1}{x}\right)^6 y + \frac{7!}{5!2!}\left(\frac{1}{x}\right)^5 y^2 + \frac{7!}{4!3!}\left(\frac{1}{x}\right)^4 y^3 + \frac{7!}{3!4!}\left(\frac{1}{x}\right)^3 y^4 + \frac{7!}{2!5!}\left(\frac{1}{x}\right)^2 y^5 + \frac{7!}{1!6!}\left(\frac{1}{x}\right) y^6 + \frac{7!}{0!7!} y^7$$

$$= x^{-7} + 7x^{-6}y + 21x^{-5}y^2 + 35x^{-4}y^3 + 35x^{-3}y^4 + 21x^{-2}y^5 + 7x^{-1}y^6 + y^7$$

28. $8s^3 - 36s^2t^2 + 54st^4 - 27t^6$

29. Expand $\left(a - \frac{2}{a}\right)^9$.

We have $a = a$, $b = -\frac{2}{a}$, and $n = 9$.

Form 1: We get the coefficients from the 10th row of Pascal's Triangle.

1
1 1
1 2 1
1 3 3 1
1 4 6 4 1
1 5 10 10 5 1
1 6 15 20 15 6 1
1 7 21 35 35 21 7 1
1 8 28 56 70 56 28 8 1
1 9 36 84 126 126 84 36 9 1

$$\left(a - \frac{2}{a}\right)^9 = 1\cdot a^9 + 9a^8\left(-\frac{2}{a}\right) + 36a^7\left(-\frac{2}{a}\right)^2 + 84a^6\left(-\frac{2}{a}\right)^3 + 126a^5\left(-\frac{2}{a}\right)^4 + 126a^4\left(-\frac{2}{a}\right)^5 + 84a^3\left(-\frac{2}{a}\right)^6 + 36a^2\left(-\frac{2}{a}\right)^7 + 9a\left(-\frac{2}{a}\right)^8 + 1\cdot\left(-\frac{2}{a}\right)^9$$

$$= a^9 - 18a^7 + 144a^5 - 672a^3 + 2016a - 4032a^{-1} + 5376a^{-3} - 4608a^{-5} + 2304a^{-7} - 512a^{-9}$$

Form 2:

$$\left(a - \frac{2}{a}\right)^9 = \binom{9}{0}a^9 + \binom{9}{1}a^8\left(-\frac{2}{a}\right) + \binom{9}{2}a^7\left(-\frac{2}{a}\right)^2 + \binom{9}{3}a^6\left(-\frac{2}{a}\right)^3 + \binom{9}{4}a^5\left(-\frac{2}{a}\right)^4 + \binom{9}{5}a^4\left(-\frac{2}{a}\right)^5 + \binom{9}{6}a^3\left(-\frac{2}{a}\right)^6 + \binom{9}{7}a^2\left(-\frac{2}{a}\right)^7 + \binom{9}{8}a\left(-\frac{2}{a}\right)^8 + \binom{9}{9}\left(-\frac{2}{a}\right)^9$$

$$= \frac{9!}{9!0!}a^9 + \frac{9!}{8!1!}a^8\left(-\frac{2}{a}\right) + \frac{9!}{7!2!}a^7\left(\frac{4}{a^2}\right) + \frac{9!}{6!3!}a^6\left(-\frac{8}{a^3}\right) + \frac{9!}{5!4!}a^5\left(\frac{16}{a^4}\right) + \frac{9!}{4!5!}a^4\left(-\frac{32}{a^5}\right) + \frac{9!}{3!6!}a^3\left(\frac{64}{a^6}\right) + \frac{9!}{2!7!}a^2\left(-\frac{128}{a^7}\right) + \frac{9!}{1!8!}a\left(\frac{256}{a^8}\right) + \frac{9!}{0!9!}\left(-\frac{512}{a^9}\right)$$

$$= a^9 - 9(2a^7) + 36(4a^5) - 84(8a^3) + 126(16a) - 126(32a^{-1}) + 84(64a^{-3}) - 36(128a^{-5}) + 9(256a^{-7}) - 512a^{-9}$$

$$= a^9 - 18a^7 + 144a^5 - 672a^3 + 2016a - 4032a^{-1} + 5376a^{-3} - 4608a^{-5} + 2304a^{-7} - 512a^{-9}$$

30. $512x^9 + 2304x^7 + 4608x^5 + 5376x^3 + 4032x + 2016x^{-1} + 672x^{-3} + 144x^{-5} + 18x^{-7} + x^{-9}$

31. Expand $(a^2 + b^3)^5$.

We have $a = a^2$, $b = b^3$, and $n = 5$.

Form 1: We get the coefficients from the sixth row of Pascal's Triangle. From Exercise 17 we know that the coefficients are

1 5 10 10 5 1.

$$(a^2 + b^3)^5 = 1\cdot(a^2)^5 + 5(a^2)^4(b^3) + 10(a^2)^3(b^3)^2 + 10(a^2)^2(b^3)^3 + 5a^2(b^3)^4 + 1\cdot(b^3)^5$$
$$= a^{10} + 5a^8b^3 + 10a^6b^6 + 10a^4b^9 + 5a^2b^{12} + b^{15}$$

Form 2:

$$(a^2 + b^3)^5 = \binom{5}{0}(a^2)^5 + \binom{5}{1}(a^2)^4(b^3) + \binom{5}{2}(a^2)^3(b^3)^2 + \binom{5}{3}(a^2)^2(b^3)^3 + \binom{5}{4}(a^2)(b^3)^4 + \binom{5}{5}(b^3)^5$$
$$= \frac{5!}{5!0!}a^{10} + \frac{5!}{4!1!}a^8b^3 + \frac{5!}{3!2!}a^6b^6 + \frac{5!}{2!3!}a^4b^9 + \frac{5!}{1!4!}a^2b^{12} + \frac{5!}{0!5!}b^{15}$$
$$= a^{10} + 5a^8b^3 + 10a^6b^6 + 10a^4b^9 + 5a^2b^{12} + b^{15}$$

32. $x^{18} + 12x^{15} + 60x^{12} + 160x^9 + 240x^6 + 192x^3 + 64$

33. Expand $(\sqrt{3} - t)^4$.

We have $a = \sqrt{3}$, $b = -t$, and $n = 4$.

Form 1: We get the coefficients from the fifth row of Pascal's Triangle. From Exercise 17 we know that the coefficients are

1 4 6 4 1.

$$(\sqrt{3} - t)^4 = 1\cdot(\sqrt{3})^4 + 4(\sqrt{3})^3(-t) + 6(\sqrt{3})^2(-t)^2 + 4\sqrt{3}(-t)^3 + 1\cdot(-t)^4$$
$$= 9 - 12\sqrt{3}\,t + 18t^2 - 4\sqrt{3}\,t^3 + t^4$$

Form 2:

$$(\sqrt{3} - t)^4 = \binom{4}{0}(\sqrt{3})^4 + \binom{4}{1}(\sqrt{3})^3(-t) + \binom{4}{2}(\sqrt{3})^2(-t)^2 + \binom{4}{3}(\sqrt{3})(-t)^3 + \binom{4}{4}(-t)^4$$
$$= \frac{4!}{4!0!}\cdot 9 + \frac{4!}{3!1!}\cdot(3\sqrt{3})(-t) + \frac{4!}{2!2!}\cdot 3t^2 + \frac{4!}{1!3!}\cdot\sqrt{3}(-t^3) + \frac{4!}{0!4!}\cdot t^4$$
$$= 9 - 12\sqrt{3}\,t + 18t^2 - 4\sqrt{3}\,t^3 + t^4$$

34. $125 + 150\sqrt{5}\,t + 375t^2 + 100\sqrt{5}\,t^3 + 75t^4 + 6\sqrt{5}\,t^5 + t^6$

35. Expand $(x^{-2} + x^2)^4$.

We have $a = x^{-2}$, $b = x^2$, and $n = 4$.

Form 1: We get the coefficients from the fifth row of Pascal's Triangle. From Exercise 17 we know that the coefficients are

1 4 6 4 1.

$$(x^{-2} + x^2)^4 = 1\cdot(x^{-2})^4 + 4(x^{-2})^3(x^2) + 6(x^{-2})^2(x^2)^2 + 4(x^{-2})(x^2)^3 + 1\cdot(x^2)^4$$
$$= x^{-8} + 4x^{-4} + 6 + 4x^4 + x^8$$

Form 2:

$$(x^{-2} + x^2)^4 = \binom{4}{0}(x^{-2})^4 + \binom{4}{1}(x^{-2})^3(x^2) + \binom{4}{2}(x^{-2})^2(x^2)^2 + \binom{4}{3}(x^{-2})(x^2)^3 + \binom{4}{4}(x^2)^4$$
$$= \frac{4!}{4!0!}(x^{-8}) + \frac{4!}{3!1!}(x^{-6})(x^2) + \frac{4!}{2!2!}(x^{-4})(x^4) + \frac{4!}{1!3!}(x^{-2})(x^6) + \frac{4!}{0!4!}(x^8)$$
$$= x^{-8} + 4x^{-4} + 6 + 4x^4 + x^8$$

36. $x^{-3} - 6x^{-2} + 15x^{-1} - 20 + 15x - 6x^2 + x^3$

37. Find the 3rd term of $(a + b)^6$.

First we note that $3 = 2 + 1$, $a = a$, $b = b$, and $n = 6$. Then the 3rd term of the expansion of $(a + b)^6$ is

$\binom{6}{2}a^{6-2}b^2$, or $\frac{6!}{4!2!}a^4b^2$, or $15a^4b^2$.

38. $21x^2y^5$

39. Find the 12th term of $(a - 2)^{14}$.

First, we note that $12 = 11 + 1$, $a = a$, $b = -2$, and $n = 14$. Then the 12th term of the expansion of $(a - 2)^{14}$ is

$$\binom{14}{11}a^{14-11}\cdot(-2)^{11} = \frac{14!}{3!11!}a^3(-2048)$$
$$= 364a^3(-2048)$$
$$= -745{,}472a^3$$

40. $3{,}897{,}234x^2$

41. Find the 5th term of $(2x^3 - \sqrt{y})^8$.

First, we note that $5 = 4 + 1$, $a = 2x^3$, $b = -\sqrt{y}$, and $n = 8$. Then the 5th term of the expansion of $(2x^3 - \sqrt{y})^8$ is

$$\binom{8}{4}(2x^3)^{8-4}(-\sqrt{y})^4$$
$$= \frac{8!}{4!4!}(2x^3)^4(-\sqrt{y})^4$$
$$= 70(16x^{12})(y^2)$$
$$= 1120x^{12}y^2$$

42. $\frac{35}{27}b^{-5}$

43. The middle term of the expansion of $(2u - 3v^2)^{10}$ is the 6th term. Note that $6 = 5 + 1$, $a = 2u$, $b = -3v^2$, and $n = 10$. Then the 6th term of the expansion of $(2u - 3v^2)^{10}$ is

$$\binom{10}{5}(2u)^{10-5}(-3v^2)^5$$

$$= \frac{10!}{5!5!}(2u)^5(-3v^2)^5$$

$$= 252(32u^5)(-243v^{10})$$

$$= -1{,}959{,}552u^5v^{10}$$

44. $30x\sqrt{x}$, $30x\sqrt{3}$

45.
$$\log_2 x + \log_2(x - 2) = 3$$
$$\log_2 x(x - 2) = 3$$
$$x(x - 2) = 2^3$$
$$x^2 - 2x - 8 = 0$$
$$(x - 4)(x + 2) = 0$$

$x = 4$ or $x = -2$

Only 4 checks.

46. $\frac{5}{2}$

47.
$$e^t = 280$$
$$\ln e^t = \ln 280$$
$$t = \ln 280$$
$$t \approx 5.6348$$

48. 15

49. Find the third term of $(0.313 + 0.687)^5$:

$$\binom{5}{2}(0.313)^{5-2}(0.687)^2 = \frac{5!}{3!2!}(0.313)^3(0.687)^2 \approx$$

0.145

50. 0.084

51. Find and add the 3rd through 6th terms of $(0.313 + 0.687)^5$:

$$\binom{5}{2}(0.313)^3(0.687)^2 + \binom{5}{3}(0.313)^2(0.687)^3 +$$

$$\binom{5}{4}(0.313)(0.687)^4 + \binom{5}{5}(0.687)^5 \approx 0.964$$

52. $\binom{8}{6}(0.15)^2(0.85)^6 + \binom{8}{7}(0.15)(0.85)^7 +$

$\binom{8}{8}(0.85)^8 \approx 0.89$

53. See the answer section in the text.

54. $\frac{55}{144}$

55. The expansion of (... 4th term is the midd...

$\binom{6}{3}(x^2)^3(-6y^{3/2})^3 = \frac{6!}{3!3!}$ ...

$-4320x^6y^{9/2}$

56. $-\frac{\sqrt[3]{q}}{2p}$

57. The $(r + 1)$st term of $\left(\sqrt[3]{x} - \frac{1}{\sqrt{x}}\right)^7$ is $\binom{7}{r}(\sqrt[3]{x})^{7-r}\left(-\frac{1}{\sqrt{x}}\right)^r$. The term containing $\frac{1}{x^{1/6}}$ is the term in which the sum of the exponents is $-1/6$. That is,

$$\left(\frac{1}{3}\right)(7 - r) + \left(-\frac{1}{2}\right)(r) = -\frac{1}{6}$$

$$\frac{7}{3} - \frac{r}{3} - \frac{r}{2} = -\frac{1}{6}$$

$$-\frac{5r}{6} = -\frac{15}{6}$$

$$r = 3$$

Find the $(3 + 1)$st, or 4th term.

$$\binom{7}{3}(\sqrt[3]{x})^4\left(-\frac{1}{\sqrt{x}}\right)^3 = \frac{7!}{4!3!}(x^{4/3})(-x^{-3/2}) = -35x^{-1/6},$$

or $-\frac{35}{x^{1/6}}$.

58. 8